## ALGEBRA

### Arithmetic Operations

$$a(b + c) = ab + ac$$

$$\frac{a}{b} + \frac{c}{d} = \frac{ad + bc}{bd}$$

$$\frac{a + c}{b} = \frac{a}{b} + \frac{c}{b}$$

$$\frac{\dfrac{a}{b}}{\dfrac{c}{d}} = \frac{a}{b} \times \frac{d}{c} = \frac{ad}{bc}$$

### Exponents and Radicals

$$x^m x^n = x^{m+n}$$

$$\frac{x^m}{x^n} = x^{m-n}$$

$$(x^m)^n = x^{mn}$$

$$x^{-n} = \frac{1}{x^n}$$

$$(xy)^n = x^n y^n$$

$$\left(\frac{x}{y}\right)^n = \frac{x^n}{y^n}$$

$$x^{1/n} = \sqrt[n]{x}$$

$$x^{m/n} = \sqrt[n]{x^m} = \left(\sqrt[n]{x}\right)^m$$

$$\sqrt[n]{xy} = \sqrt[n]{x}\,\sqrt[n]{y}$$

$$\sqrt[n]{\frac{x}{y}} = \frac{\sqrt[n]{x}}{\sqrt[n]{y}}$$

### Factoring Special Polynomials

$$x^2 - y^2 = (x + y)(x - y)$$

$$x^3 + y^3 = (x + y)(x^2 - xy + y^2)$$

$$x^3 - y^3 = (x - y)(x^2 + xy + y^2)$$

### Binomial Theorem

$$(x + y)^2 = x^2 + 2xy + y^2 \qquad (x - y)^2 = x^2 - 2xy + y^2$$

$$(x + y)^3 = x^3 + 3x^2 y + 3xy^2 + y^3$$

$$(x - y)^3 = x^3 - 3x^2 y + 3xy^2 - y^3$$

$$(x + y)^n = x^n + nx^{n-1}y + \frac{n(n - 1)}{2} x^{n-2}y^2$$

$$+ \cdots + \binom{n}{k} x^{n-k}y^k + \cdots + nxy^{n-1} + y^n$$

$$\text{where } \binom{n}{k} = \frac{n(n - 1) \cdots (n - k + 1)}{1 \cdot 2 \cdot 3 \cdots \cdot k}$$

### Quadratic Formula

If $ax^2 + bx + c = 0$, then $x = \dfrac{-b \pm \sqrt{b^2 - 4ac}}{2a}$.

### Inequalities and Absolute Value

If $a < b$ and $b < c$, then $a < c$.

If $a < b$, then $a + c < b + c$.

If $a < b$ and $c > 0$, then $ca < cb$.

If $a < b$ and $c < 0$, then $ca > cb$.

If $a > 0$, then

$$|x| = a \quad \text{means} \quad x = a \quad \text{or} \quad x = -a$$

$$|x| < a \quad \text{means} \quad -a < x < a$$

$$|x| > a \quad \text{means} \quad x > a \quad \text{or} \quad x < -a$$

## GEOMETRY

### Geometric Formulas

Formulas for area $A$, circumference $C$

**Triangle**
$$A = \tfrac{1}{2}bh$$
$$= \tfrac{1}{2}ab \sin \theta$$

**Circle**
$$A = \pi r^2$$
$$C = 2\pi r$$

$$A = \tfrac{1}{2}r^2\theta$$
$$s = r\theta \quad (\theta \text{ in radians})$$

**Sphere**
$$V = \tfrac{4}{3}\pi r^3$$
$$A = 4\pi r^2$$

**Cylinder**
$$V = \pi r^2 h$$

**Cone**
$$V = \tfrac{1}{3}\pi r^2 h$$
$$A = \pi r \sqrt{r^2 + h^2}$$

### Distance and Midpoint Formulas

Distance between $P_1(x_1, y_1)$ and $P_2(x_2, y_2)$:

$$d = \sqrt{(x_2 - x_1)^2 + (y_2 - y_1)^2}$$

Midpoint of $\overline{P_1 P_2}$: $\left(\dfrac{x_1 + x_2}{2}, \dfrac{y_1 + y_2}{2}\right)$

### Lines

Slope of line through $P_1(x_1, y_1)$ and $P_2(x_2, y_2)$:

$$m = \frac{y_2 - y_1}{x_2 - x_1}$$

Point-slope equation of line through $P_1(x_1, y_1)$ with slope $m$:

$$y - y_1 = m(x - x_1)$$

Slope-intercept equation of line with slope $m$ and $y$-intercept $b$:

$$y = mx + b$$

### Circles

Equation of the circle with center $(h, k)$ and radius $r$:

$$(x - h)^2 + (y - k)^2 = r^2$$

# TRIGONOMETRY

## Angle Measurement

$\pi$ radians $= 180°$

$1° = \dfrac{\pi}{180}$ rad $\qquad$ 1 rad $= \dfrac{180°}{\pi}$

$s = r\theta$

($\theta$ in radians)

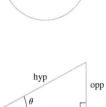

## Right Angle Trigonometry

$\sin \theta = \dfrac{\text{opp}}{\text{hyp}} \qquad \csc \theta = \dfrac{\text{hyp}}{\text{opp}}$

$\cos \theta = \dfrac{\text{adj}}{\text{hyp}} \qquad \sec \theta = \dfrac{\text{hyp}}{\text{adj}}$

$\tan \theta = \dfrac{\text{opp}}{\text{adj}} \qquad \cot \theta = \dfrac{\text{adj}}{\text{opp}}$

## Trigonometric Functions

$\sin \theta = \dfrac{y}{r} \qquad \csc \theta = \dfrac{r}{y}$

$\cos \theta = \dfrac{x}{r} \qquad \sec \theta = \dfrac{r}{x}$

$\tan \theta = \dfrac{y}{x} \qquad \cot \theta = \dfrac{x}{y}$

## Graphs of Trigonometric Functions

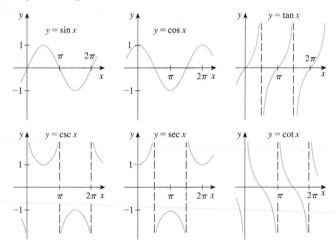

## Trigonometric Functions of Important Angles

| $\theta$ | radians | $\sin \theta$ | $\cos \theta$ | $\tan \theta$ |
|---|---|---|---|---|
| 0° | 0 | 0 | 1 | 0 |
| 30° | $\pi/6$ | $1/2$ | $\sqrt{3}/2$ | $\sqrt{3}/3$ |
| 45° | $\pi/4$ | $\sqrt{2}/2$ | $\sqrt{2}/2$ | 1 |
| 60° | $\pi/3$ | $\sqrt{3}/2$ | $1/2$ | $\sqrt{3}$ |
| 90° | $\pi/2$ | 1 | 0 | — |

## Fundamental Identities

$\csc \theta = \dfrac{1}{\sin \theta} \qquad\qquad \sec \theta = \dfrac{1}{\cos \theta}$

$\tan \theta = \dfrac{\sin \theta}{\cos \theta} \qquad\qquad \cot \theta = \dfrac{\cos \theta}{\sin \theta}$

$\cot \theta = \dfrac{1}{\tan \theta} \qquad\qquad \sin^2\theta + \cos^2\theta = 1$

$1 + \tan^2\theta = \sec^2\theta \qquad 1 + \cot^2\theta = \csc^2\theta$

$\sin(-\theta) = -\sin \theta \qquad\quad \cos(-\theta) = \cos \theta$

$\tan(-\theta) = -\tan \theta \qquad\quad \sin\left(\dfrac{\pi}{2} - \theta\right) = \cos \theta$

$\cos\left(\dfrac{\pi}{2} - \theta\right) = \sin \theta \qquad \tan\left(\dfrac{\pi}{2} - \theta\right) = \cot \theta$

## The Law of Sines

$\dfrac{\sin A}{a} = \dfrac{\sin B}{b} = \dfrac{\sin C}{c}$

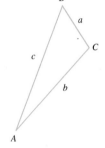

## The Law of Cosines

$a^2 = b^2 + c^2 - 2bc \cos A$

$b^2 = a^2 + c^2 - 2ac \cos B$

$c^2 = a^2 + b^2 - 2ab \cos C$

## Addition and Subtraction Formulas

$\sin(x + y) = \sin x \cos y + \cos x \sin y$

$\sin(x - y) = \sin x \cos y - \cos x \sin y$

$\cos(x + y) = \cos x \cos y - \sin x \sin y$

$\cos(x - y) = \cos x \cos y + \sin x \sin y$

$\tan(x + y) = \dfrac{\tan x + \tan y}{1 - \tan x \tan y}$

$\tan(x - y) = \dfrac{\tan x - \tan y}{1 + \tan x \tan y}$

## Double-Angle Formulas

$\sin 2x = 2 \sin x \cos x$

$\cos 2x = \cos^2 x - \sin^2 x = 2 \cos^2 x - 1 = 1 - 2 \sin^2 x$

$\tan 2x = \dfrac{2 \tan x}{1 - \tan^2 x}$

## Half-Angle Formulas

$\sin^2 x = \dfrac{1 - \cos 2x}{2} \qquad \cos^2 x = \dfrac{1 + \cos 2x}{2}$

# CALCULUS

Special Version for MTH 132 Calculus

7th Edition

James Stewart

CENGAGE
Learning

Australia • Brazil • Japan • Korea • Mexico • Singapore • Spain • United Kingdom • United States

**CALCULUS: Special Version for MTH 132**
**Calculus, 7th Edition**

Calculus, 7th Edition
James Stewart

© 2012, 2008 Cengage Learning. All Rights Reserved.

Senior Manager, Student Engagement:

Linda deStefano

Janey Moeller

Manager, Student Engagement:

Julie Dierig

Marketing Manager:

Rachael Kloos

Manager, Production Editorial:

Kim Fry

Manager, Intellectual Property Project Manager:

Brian Methe

Senior Manager, Production and Manufacturing:

Donna M. Brown

Manager, Production:

Terri Daley

For product information and technology assistance, contact us at
**Cengage Learning Customer & Sales Support, 1-800-354-9706**

For permission to use material from this text or product,
submit all requests online at **cengage.com/permissions**
Further permissions questions can be emailed to
**permissionrequest@cengage.com**

This book contains select works from existing Cengage Learning resources and was produced by Cengage Learning Custom Solutions for collegiate use. As such, those adopting and/or contributing to this work are responsible for editorial content accuracy, continuity and completeness.

**Compilation © 2014 Cengage Learning**

ISBN-13: 978-1-305-00738-3
ISBN-10: 1-305-00738-7

**WCN: 01-100-101**

**Cengage Learning**
5191 Natorp Boulevard
Mason, Ohio 45040
USA

Cengage Learning is a leading provider of customized learning solutions with office locations around the globe, including Singapore, the United Kingdom, Australia, Mexico, Brazil, and Japan. Locate your local office at:
**international.cengage.com/region.**

Cengage Learning products are represented in Canada by Nelson Education, Ltd.
For your lifelong learning solutions, visit **www.cengage.com/custom.**
Visit our corporate website at **www.cengage.com.**

Printed in the United States of America

# Contents

## 3  Applications of Differentiation    197

## 4  Integrals    283

## Appendixes    A1

## Index    A135

## MTH 132 Final Exam Study Guide

# Preface

The art of teaching, Mark Van Doren said, is the art of assisting discovery. I have tried to write a book that assists students in discovering calculus—both for its practical power and its surprising beauty. In this edition, as in the first six editions, I aim to convey to the student a sense of the utility of calculus and develop technical competence, but I also strive to give some appreciation for the intrinsic beauty of the subject. Newton undoubtedly experienced a sense of triumph when he made his great discoveries. I want students to share some of that excitement.

The emphasis is on understanding concepts. I think that nearly everybody agrees that this should be the primary goal of calculus instruction. In fact, the impetus for the current calculus reform movement came from the Tulane Conference in 1986, which formulated as their first recommendation:

*Focus on conceptual understanding.*

I have tried to implement this goal through the *Rule of Three:* "Topics should be presented geometrically, numerically, and algebraically." Visualization, numerical and graphical experimentation, and other approaches have changed how we teach conceptual reasoning in fundamental ways. The Rule of Three has been expanded to become the *Rule of Four* by emphasizing the verbal, or descriptive, point of view as well.

In writing the seventh edition my premise has been that it is possible to achieve conceptual understanding and still retain the best traditions of traditional calculus. The book contains elements of reform, but within the context of a traditional curriculum.

## Alternative Versions

I have written several other calculus textbooks that might be preferable for some instructors. Most of them also come in single variable and multivariable versions.

- *Calculus,* Seventh Edition, Hybrid Version, is similar to the present textbook in content and coverage except that all end-of-section exercises are available only in Enhanced WebAssign. The printed text includes all end-of-chapter review material.

- *Calculus: Early Transcendentals,* Seventh Edition, is similar to the present textbook except that the exponential, logarithmic, and inverse trigonometric functions are covered in the first semester.

- *Calculus: Early Transcendentals*, Seventh Edition, Hybrid Version, is similar to *Calculus: Early Transcendentals*, Seventh Edition, in content and coverage except that all end-of-section exercises are available only in Enhanced WebAssign. The printed text includes all end-of-chapter review material.

- *Essential Calculus* is a much briefer book (800 pages), though it contains almost all of the topics in *Calculus,* Seventh Edition. The relative brevity is achieved through briefer exposition of some topics and putting some features on the website.

- *Essential Calculus: Early Transcendentals* resembles *Essential Calculus,* but the exponential, logarithmic, and inverse trigonometric functions are covered in Chapter 3.

- *Calculus: Concepts and Contexts*, Fourth Edition, emphasizes conceptual understanding even more strongly than this book. The coverage of topics is not encyclopedic and the material on transcendental functions and on parametric equations is woven throughout the book instead of being treated in separate chapters.

- *Calculus: Early Vectors* introduces vectors and vector functions in the first semester and integrates them throughout the book. It is suitable for students taking Engineering and Physics courses concurrently with calculus.

- *Brief Applied Calculus* is intended for students in business, the social sciences, and the life sciences.

## What's New in the Seventh Edition?

The changes have resulted from talking with my colleagues and students at the University of Toronto and from reading journals, as well as suggestions from users and reviewers. Here are some of the many improvements that I've incorporated into this edition:

- Some material has been rewritten for greater clarity or for better motivation. See, for instance, the introduction to maximum and minimum values on page 198, the introduction to series on page 727, and the motivation for the cross product on page 832.

- New examples have been added (see Example 4 on page 1045 for instance). And the solutions to some of the existing examples have been amplified. A case in point: I added details to the solution of Example 1.6.11 because when I taught Section 1.6 from the sixth edition I realized that students need more guidance when setting up inequalities for the Squeeze Theorem.

- Chapter 1, *Functions and Limits*, consists of most of the material from Chapters 1 and 2 of the sixth edition. The section on Graphing Calculators and Computers is now Appendix G.

- The art program has been revamped: New figures have been incorporated and a substantial percentage of the existing figures have been redrawn.

- The data in examples and exercises have been updated to be more timely.

- Three new projects have been added: *The Gini Index* (page 351) explores how to measure income distribution among inhabitants of a given country and is a nice application of areas between curves. (I thank Klaus Volpert for suggesting this project.) *Families of Implicit Curves* (page 163) investigates the changing shapes of implicitly defined curves as parameters in a family are varied. *Families of Polar Curves* (page 688) exhibits the fascinating shapes of polar curves and how they evolve within a family.

- The section on the surface area of the graph of a function of two variables has been restored as Section 15.6 for the convenience of instructors who like to teach it after double integrals, though the full treatment of surface area remains in Chapter 16.

- I continue to seek out examples of how calculus applies to so many aspects of the real world. On page 933 you will see beautiful images of the earth's magnetic field strength and its second vertical derivative as calculated from Laplace's equation. I thank Roger Watson for bringing to my attention how this is used in geophysics and mineral exploration.

- More than 25% of the exercises are new. Here are some of my favorites: 2.2.13–14, 2.4.56, 2.5.67, 2.6.53–56, 2.7.22, 3.3.70, 3.4.43, 4.2.51–53, 5.4.30, 6.3.58, 11.2.49–50, 11.10.71–72, 12.1.44, 12.4.43–44, and Problems 4, 5, and 8 on pages 861–62.

## Technology Enhancements

- The media and technology to support the text have been enhanced to give professors greater control over their course, to provide extra help to deal with the varying levels of student preparedness for the calculus course, and to improve support for conceptual understanding. New Enhanced WebAssign features including a customizable Cengage YouBook, *Just in Time* review, *Show Your Work,* Answer Evaluator, Personalized Study Plan, Master Its, solution videos, lecture video clips (with associated questions), and *Visualizing Calculus* (TEC animations with associated questions) have been developed to facilitate improved student learning and flexible classroom teaching.

- *Tools for Enriching Calculus* (TEC) has been completely redesigned and is accessible in Enhanced WebAssign, CourseMate, and PowerLecture. Selected Visuals and Modules are available at www.stewartcalculus.com.

## Features

CONCEPTUAL EXERCISES
The most important way to foster conceptual understanding is through the problems that we assign. To that end I have devised various types of problems. Some exercise sets begin with requests to explain the meanings of the basic concepts of the section. (See, for instance, the first few exercises in Sections 1.5, 1.8, 11.2, 14.2, and 14.3.) Similarly, all the review sections begin with a Concept Check and a True-False Quiz. Other exercises test conceptual understanding through graphs or tables (see Exercises 2.1.17, 2.2.33–38, 2.2.41–44, 9.1.11–13, 10.1.24–27, 11.10.2, 13.2.1–2, 13.3.33–39, 14.1.1–2, 14.1.32–42, 14.3.3–10, 14.6.1–2, 14.7.3–4, 15.1.5–10, 16.1.11–18, 16.2.17–18, and 16.3.1–2).

Another type of exercise uses verbal description to test conceptual understanding (see Exercises 1.8.10, 2.2.56, 3.3.51–52, and 7.8.67). I particularly value problems that combine and compare graphical, numerical, and algebraic approaches (see Exercises 3.4.31–32, 2.7.25, and 9.4.2).

GRADED EXERCISE SETS
Each exercise set is carefully graded, progressing from basic conceptual exercises and skill-development problems to more challenging problems involving applications and proofs.

REAL-WORLD DATA
My assistants and I spent a great deal of time looking in libraries, contacting companies and government agencies, and searching the Internet for interesting real-world data to introduce, motivate, and illustrate the concepts of calculus. As a result, many of the examples and exercises deal with functions defined by such numerical data or graphs. See, for instance, Figure 1 in Section 1.1 (seismograms from the Northridge earthquake), Exercise

2.2.34 (percentage of the population under age 18), Exercise 4.1.16 (velocity of the space shuttle *Endeavour*), and Figure 4 in Section 4.4 (San Francisco power consumption). Functions of two variables are illustrated by a table of values of the wind-chill index as a function of air temperature and wind speed (Example 2 in Section 14.1). Partial derivatives are introduced in Section 14.3 by examining a column in a table of values of the heat index (perceived air temperature) as a function of the actual temperature and the relative humidity. This example is pursued further in connection with linear approximations (Example 3 in Section 14.4). Directional derivatives are introduced in Section 14.6 by using a temperature contour map to estimate the rate of change of temperature at Reno in the direction of Las Vegas. Double integrals are used to estimate the average snowfall in Colorado on December 20–21, 2006 (Example 4 in Section 15.1). Vector fields are introduced in Section 16.1 by depictions of actual velocity vector fields showing San Francisco Bay wind patterns.

PROJECTS One way of involving students and making them active learners is to have them work (perhaps in groups) on extended projects that give a feeling of substantial accomplishment when completed. I have included four kinds of projects: *Applied Projects* involve applications that are designed to appeal to the imagination of students. The project after Section 9.3 asks whether a ball thrown upward takes longer to reach its maximum height or to fall back to its original height. (The answer might surprise you.) The project after Section 14.8 uses Lagrange multipliers to determine the masses of the three stages of a rocket so as to minimize the total mass while enabling the rocket to reach a desired velocity. *Laboratory Projects* involve technology; the one following Section 10.2 shows how to use Bézier curves to design shapes that represent letters for a laser printer. *Writing Projects* ask students to compare present-day methods with those of the founders of calculus—Fermat's method for finding tangents, for instance. Suggested references are supplied. *Discovery Projects* anticipate results to be discussed later or encourage discovery through pattern recognition (see the one following Section 7.6). Others explore aspects of geometry: tetrahedra (after Section 12.4), hyperspheres (after Section 15.7), and intersections of three cylinders (after Section 15.8). Additional projects can be found in the *Instructor's Guide* (see, for instance, Group Exercise 4.1: Position from Samples).

PROBLEM SOLVING Students usually have difficulties with problems for which there is no single well-defined procedure for obtaining the answer. I think nobody has improved very much on George Polya's four-stage problem-solving strategy and, accordingly, I have included a version of his problem-solving principles following Chapter 1. They are applied, both explicitly and implicitly, throughout the book. After the other chapters I have placed sections called *Problems Plus*, which feature examples of how to tackle challenging calculus problems. In selecting the varied problems for these sections I kept in mind the following advice from David Hilbert: "A mathematical problem should be difficult in order to entice us, yet not inaccessible lest it mock our efforts." When I put these challenging problems on assignments and tests I grade them in a different way. Here I reward a student significantly for ideas toward a solution and for recognizing which problem-solving principles are relevant.

DUAL TREATMENT OF EXPONENTIAL AND LOGARITHMIC FUNCTIONS There are two possible ways of treating the exponential and logarithmic functions and each method has its passionate advocates. Because one often finds advocates of both approaches teaching the same course, I include full treatments of both methods. In Sections 6.2, 6.3, and 6.4 the exponential function is defined first, followed by the logarithmic function as its inverse. (Students have seen these functions introduced this way since high school.) In the alternative approach, presented in Sections 6.2*, 6.3*, and 6.4*, the logarithm is defined as an integral and the exponential function is its inverse. This latter method is, of course, less intuitive but more elegant. You can use whichever treatment you prefer.

If the first approach is taken, then much of Chapter 6 can be covered before Chapters 4 and 5, if desired. To accommodate this choice of presentation there are specially identified problems involving integrals of exponential and logarithmic functions at the end of the appropriate sections of Chapters 4 and 5. This order of presentation allows a faster-paced course to teach the transcendental functions and the definite integral in the first semester of the course.

For instructors who would like to go even further in this direction I have prepared an alternate edition of this book, called *Calculus, Early Transcendentals,* Seventh Edition, in which the exponential and logarithmic functions are introduced in the first chapter. Their limits and derivatives are found in the second and third chapters at the same time as polynomials and the other elementary functions.

**TOOLS FOR ENRICHING™ CALCULUS**

TEC is a companion to the text and is intended to enrich and complement its contents. (It is now accessible in Enhanced WebAssign, CourseMate, and PowerLecture. Selected Visuals and Modules are available at www.stewartcalculus.com.) Developed by Harvey Keynes, Dan Clegg, Hubert Hohn, and myself, TEC uses a discovery and exploratory approach. In sections of the book where technology is particularly appropriate, marginal icons direct students to TEC modules that provide a laboratory environment in which they can explore the topic in different ways and at different levels. **Visuals are animations of figures in text; Modules are more elaborate activities and include exercises**. Instructors can choose to become involved at several different levels, ranging from simply encouraging students to use the Visuals and Modules for independent exploration, to assigning specific exercises from those included with each Module, or to creating additional exercises, labs, and projects that make use of the Visuals and Modules.

**HOMEWORK HINTS**

Homework Hints presented in the form of questions try to imitate an effective teaching assistant by functioning as a silent tutor. Hints for representative exercises (usually odd-numbered) are included in every section of the text, indicated by printing the exercise number in red. They are constructed so as not to reveal any more of the actual solution than is minimally necessary to make further progress, and are available to students at stewartcalculus.com and in CourseMate and Enhanced WebAssign.

**ENHANCED WEBASSIGN**

Technology is having an impact on the way homework is assigned to students, particularly in large classes. The use of online homework is growing and its appeal depends on ease of use, grading precision, and reliability. With the seventh edition we have been working with the calculus community and WebAssign to develop a more robust online homework system. Up to 70% of the exercises in each section are assignable as online homework, including free response, multiple choice, and multi-part formats.

The system also includes Active Examples, in which students are guided in step-by-step tutorials through text examples, with links to the textbook and to video solutions. New enhancements to the system include a customizable eBook, a *Show Your Work* feature, *Just in Time* review of precalculus prerequisites, an improved Assignment Editor, and an Answer Evaluator that accepts more mathematically equivalent answers and allows for homework grading in much the same way that an instructor grades.

**www.stewartcalculus.com**

This site includes the following.

- Homework Hints
- Algebra Review
- Lies My Calculator and Computer Told Me
- History of Mathematics, with links to the better historical websites
- Additional Topics (complete with exercise sets): Fourier Series, Formulas for the Remainder Term in Taylor Series, Rotation of Axes
- Archived Problems (Drill exercises that appeared in previous editions, together with their solutions)
- Challenge Problems (some from the Problems Plus sections from prior editions)

- Links, for particular topics, to outside web resources
- Selected Tools for Enriching Calculus (TEC) Modules and Visuals

## Content

**Diagnostic Tests**  The book begins with four diagnostic tests, in Basic Algebra, Analytic Geometry, Functions, and Trigonometry.

**A Preview of Calculus**  This is an overview of the subject and includes a list of questions to motivate the study of calculus.

**1  Functions and Limits**  From the beginning, multiple representations of functions are stressed: verbal, numerical, visual, and algebraic. A discussion of mathematical models leads to a review of the standard functions from these four points of view. The material on limits is motivated by a prior discussion of the tangent and velocity problems. Limits are treated from descriptive, graphical, numerical, and algebraic points of view. Section 1.7, on the precise epsilon-delta definition of a limit, is an optional section.

**2  Derivatives**  The material on derivatives is covered in two sections in order to give students more time to get used to the idea of a derivative as a function. The examples and exercises explore the meanings of derivatives in various contexts. Higher derivatives are introduced in Section 2.2.

**3  Applications of Differentiation**  The basic facts concerning extreme values and shapes of curves are deduced from the Mean Value Theorem. Graphing with technology emphasizes the interaction between calculus and calculators and the analysis of families of curves. Some substantial optimization problems are provided, including an explanation of why you need to raise your head 42° to see the top of a rainbow.

**4  Integrals**  The area problem and the distance problem serve to motivate the definite integral, with sigma notation introduced as needed. (Full coverage of sigma notation is provided in Appendix E.) Emphasis is placed on explaining the meanings of integrals in various contexts and on estimating their values from graphs and tables.

**5  Applications of Integration**  Here I present the applications of integration—area, volume, work, average value—that can reasonably be done without specialized techniques of integration. General methods are emphasized. The goal is for students to be able to divide a quantity into small pieces, estimate with Riemann sums, and recognize the limit as an integral.

**6  Inverse Functions: Exponential, Logarithmic, and Inverse Trigonometric Functions**  As discussed more fully on page xiv, only one of the two treatments of these functions need be covered. Exponential growth and decay are covered in this chapter.

**7  Techniques of Integration**  All the standard methods are covered but, of course, the real challenge is to be able to recognize which technique is best used in a given situation. Accordingly, in Section 7.5, I present a strategy for integration. The use of computer algebra systems is discussed in Section 7.6.

**8  Further Applications of Integration**  Here are the applications of integration—arc length and surface area—for which it is useful to have available all the techniques of integration, as well as applications to biology, economics, and physics (hydrostatic force and centers of mass). I have also included a section on probability. There are more applications here than can realistically be covered in a given course. Instructors should select applications suitable for their students and for which they themselves have enthusiasm.

| 9  Differential Equations | Modeling is the theme that unifies this introductory treatment of differential equations. Direction fields and Euler's method are studied before separable and linear equations are solved explicitly, so that qualitative, numerical, and analytic approaches are given equal consideration. These methods are applied to the exponential, logistic, and other models for population growth. The first four or five sections of this chapter serve as a good introduction to first-order differential equations. An optional final section uses predator-prey models to illustrate systems of differential equations. |
|---|---|

10  **Parametric Equations and Polar Coordinates**

This chapter introduces parametric and polar curves and applies the methods of calculus to them. Parametric curves are well suited to laboratory projects; the three presented here involve families of curves and Bézier curves. A brief treatment of conic sections in polar coordinates prepares the way for Kepler's Laws in Chapter 13.

11  **Infinite Sequences and Series**

The convergence tests have intuitive justifications (see page 738) as well as formal proofs. Numerical estimates of sums of series are based on which test was used to prove convergence. The emphasis is on Taylor series and polynomials and their applications to physics. Error estimates include those from graphing devices.

12  **Vectors and The Geometry of Space**

The material on three-dimensional analytic geometry and vectors is divided into two chapters. Chapter 12 deals with vectors, the dot and cross products, lines, planes, and surfaces.

13  **Vector Functions**

This chapter covers vector-valued functions, their derivatives and integrals, the length and curvature of space curves, and velocity and acceleration along space curves, culminating in Kepler's laws.

14  **Partial Derivatives**

Functions of two or more variables are studied from verbal, numerical, visual, and algebraic points of view. In particular, I introduce partial derivatives by looking at a specific column in a table of values of the heat index (perceived air temperature) as a function of the actual temperature and the relative humidity.

15  **Multiple Integrals**

Contour maps and the Midpoint Rule are used to estimate the average snowfall and average temperature in given regions. Double and triple integrals are used to compute probabilities, surface areas, and (in projects) volumes of hyperspheres and volumes of intersections of three cylinders. Cylindrical and spherical coordinates are introduced in the context of evaluating triple integrals.

16  **Vector Calculus**

Vector fields are introduced through pictures of velocity fields showing San Francisco Bay wind patterns. The similarities among the Fundamental Theorem for line integrals, Green's Theorem, Stokes' Theorem, and the Divergence Theorem are emphasized.

17  **Second-Order Differential Equations**

Since first-order differential equations are covered in Chapter 9, this final chapter deals with second-order linear differential equations, their application to vibrating springs and electric circuits, and series solutions.$\pi$

## Ancillaries

*Calculus,* Seventh Edition, is supported by a complete set of ancillaries developed under my direction. Each piece has been designed to enhance student understanding and to facilitate creative instruction. With this edition, new media and technologies have been developed that help students to visualize calculus and instructors to customize content to better align with the way they teach their course. The tables on pages xxi–xxii describe each of these ancillaries.

## Acknowledgments

The preparation of this and previous editions has involved much time spent reading the reasoned (but sometimes contradictory) advice from a large number of astute reviewers. I greatly appreciate the time they spent to understand my motivation for the approach taken. I have learned something from each of them.

**SEVENTH EDITION REVIEWERS**

Amy Austin, *Texas A&M University*
Anthony J. Bevelacqua, *University of North Dakota*
Zhen-Qing Chen, *University of Washington—Seattle*
Jenna Carpenter, *Louisiana Tech University*
Le Baron O. Ferguson, *University of California—Riverside*
Shari Harris, *John Wood Community College*
Amer Iqbal, *University of Washington—Seattle*
Akhtar Khan, *Rochester Institute of Technology*
Marianne Korten, *Kansas State University*
Joyce Longman, *Villanova University*
Richard Millspaugh, *University of North Dakota*
Lon H. Mitchell, *Virginia Commonwealth University*
Ho Kuen Ng, *San Jose State University*
Norma Ortiz-Robinson, *Virginia Commonwealth University*
Qin Sheng, *Baylor University*
Magdalena Toda, *Texas Tech University*
Ruth Trygstad, *Salt Lake Community College*
Klaus Volpert, *Villanova University*
Peiyong Wang, *Wayne State University*

**TECHNOLOGY REVIEWERS**

Maria Andersen, *Muskegon Community College*
Eric Aurand, *Eastfield College*
Joy Becker, *University of Wisconsin–Stout*
Przemyslaw Bogacki, *Old Dominion University*
Amy Elizabeth Bowman, *University of Alabama in Huntsville*
Monica Brown, *University of Missouri–St. Louis*
Roxanne Byrne, *University of Colorado at Denver and Health Sciences Center*
Teri Christiansen, *University of Missouri–Columbia*
Bobby Dale Daniel, *Lamar University*
Jennifer Daniel, *Lamar University*
Andras Domokos, *California State University, Sacramento*
Timothy Flaherty, *Carnegie Mellon University*
Lee Gibson, *University of Louisville*
Jane Golden, *Hillsborough Community College*
Semion Gutman, *University of Oklahoma*
Diane Hoffoss, *University of San Diego*
Lorraine Hughes, *Mississippi State University*
Jay Jahangiri, *Kent State University*
John Jernigan, *Community College of Philadelphia*

Brian Karasek, *South Mountain Community College*
Jason Kozinski, *University of Florida*
Carole Krueger, *The University of Texas at Arlington*
Ken Kubota, *University of Kentucky*
John Mitchell, *Clark College*
Donald Paul, *Tulsa Community College*
Chad Pierson, *University of Minnesota, Duluth*
Lanita Presson, *University of Alabama in Huntsville*
Karin Reinhold, *State University of New York at Albany*
Thomas Riedel, *University of Louisville*
Christopher Schroeder, *Morehead State University*
Angela Sharp, *University of Minnesota, Duluth*
Patricia Shaw, *Mississippi State University*
Carl Spitznagel, *John Carroll University*
Mohammad Tabanjeh, *Virginia State University*
Capt. Koichi Takagi, *United States Naval Academy*
Lorna TenEyck, *Chemeketa Community College*
Roger Werbylo, *Pima Community College*
David Williams, *Clayton State University*
Zhuan Ye, *Northern Illinois University*

**PREVIOUS EDITION REVIEWERS**

B. D. Aggarwala, *University of Calgary*
John Alberghini, *Manchester Community College*
Michael Albert, *Carnegie-Mellon University*
Daniel Anderson, *University of Iowa*
Donna J. Bailey, *Northeast Missouri State University*
Wayne Barber, *Chemeketa Community College*
Marilyn Belkin, *Villanova University*
Neil Berger, *University of Illinois, Chicago*
David Berman, *University of New Orleans*
Richard Biggs, *University of Western Ontario*
Robert Blumenthal, *Oglethorpe University*
Martina Bode, *Northwestern University*
Barbara Bohannon, *Hofstra University*
Philip L. Bowers, *Florida State University*
Amy Elizabeth Bowman, *University of Alabama in Huntsville*
Jay Bourland, *Colorado State University*
Stephen W. Brady, *Wichita State University*
Michael Breen, *Tennessee Technological University*
Robert N. Bryan, *University of Western Ontario*
David Buchthal, *University of Akron*
Jorge Cassio, *Miami-Dade Community College*
Jack Ceder, *University of California, Santa Barbara*
Scott Chapman, *Trinity University*
James Choike, *Oklahoma State University*
Barbara Cortzen, *DePaul University*
Carl Cowen, *Purdue University*
Philip S. Crooke, *Vanderbilt University*
Charles N. Curtis, *Missouri Southern State College*
Daniel Cyphert, *Armstrong State College*
Robert Dahlin
M. Hilary Davies, *University of Alaska Anchorage*
Gregory J. Davis, *University of Wisconsin–Green Bay*
Elias Deeba, *University of Houston–Downtown*
Daniel DiMaria, *Suffolk Community College*
Seymour Ditor, *University of Western Ontario*
Greg Dresden, *Washington and Lee University*
Daniel Drucker, *Wayne State University*
Kenn Dunn, *Dalhousie University*
Dennis Dunninger, *Michigan State University*
Bruce Edwards, *University of Florida*
David Ellis, *San Francisco State University*
John Ellison, *Grove City College*
Martin Erickson, *Truman State University*
Garret Etgen, *University of Houston*
Theodore G. Faticoni, *Fordham University*
Laurene V. Fausett, *Georgia Southern University*
Norman Feldman, *Sonoma State University*
Newman Fisher, *San Francisco State University*
José D. Flores, *The University of South Dakota*
William Francis, *Michigan Technological University*
James T. Franklin, *Valencia Community College, East*
Stanley Friedlander, *Bronx Community College*
Patrick Gallagher, *Columbia University–New York*
Paul Garrett, *University of Minnesota–Minneapolis*
Frederick Gass, *Miami University of Ohio*

Bruce Gilligan, *University of Regina*
Matthias K. Gobbert, *University of Maryland, Baltimore County*
Gerald Goff, *Oklahoma State University*
Stuart Goldenberg, *California Polytechnic State University*
John A. Graham, *Buckingham Browne & Nichols School*
Richard Grassl, *University of New Mexico*
Michael Gregory, *University of North Dakota*
Charles Groetsch, *University of Cincinnati*
Paul Triantafilos Hadavas, *Armstrong Atlantic State University*
Salim M. Haïdar, *Grand Valley State University*
D. W. Hall, *Michigan State University*
Robert L. Hall, *University of Wisconsin–Milwaukee*
Howard B. Hamilton, *California State University, Sacramento*
Darel Hardy, *Colorado State University*
Gary W. Harrison, *College of Charleston*
Melvin Hausner, *New York University/Courant Institute*
Curtis Herink, *Mercer University*
Russell Herman, *University of North Carolina at Wilmington*
Allen Hesse, *Rochester Community College*
Randall R. Holmes, *Auburn University*
James F. Hurley, *University of Connecticut*
Matthew A. Isom, *Arizona State University*
Gerald Janusz, *University of Illinois at Urbana-Champaign*
John H. Jenkins, *Embry-Riddle Aeronautical University, Prescott Campus*
Clement Jeske, *University of Wisconsin, Platteville*
Carl Jockusch, *University of Illinois at Urbana-Champaign*
Jan E. H. Johansson, *University of Vermont*
Jerry Johnson, *Oklahoma State University*
Zsuzsanna M. Kadas, *St. Michael's College*
Nets Katz, *Indiana University Bloomington*
Matt Kaufman
Matthias Kawski, *Arizona State University*
Frederick W. Keene, *Pasadena City College*
Robert L. Kelley, *University of Miami*
Virgil Kowalik, *Texas A&I University*
Kevin Kreider, *University of Akron*
Leonard Krop, *DePaul University*
Mark Krusemeyer, *Carleton College*
John C. Lawlor, *University of Vermont*
Christopher C. Leary, *State University of New York at Geneseo*
David Leeming, *University of Victoria*
Sam Lesseig, *Northeast Missouri State University*
Phil Locke, *University of Maine*
Joan McCarter, *Arizona State University*
Phil McCartney, *Northern Kentucky University*
James McKinney, *California State Polytechnic University, Pomona*
Igor Malyshev, *San Jose State University*
Larry Mansfield, *Queens College*
Mary Martin, *Colgate University*
Nathaniel F. G. Martin, *University of Virginia*
Gerald Y. Matsumoto, *American River College*
Tom Metzger, *University of Pittsburgh*

Michael Montaño, *Riverside Community College*
Teri Jo Murphy, *University of Oklahoma*
Martin Nakashima, *California State Polytechnic University, Pomona*
Richard Nowakowski, *Dalhousie University*
Hussain S. Nur, *California State University, Fresno*
Wayne N. Palmer, *Utica College*
Vincent Panico, *University of the Pacific*
F. J. Papp, *University of Michigan–Dearborn*
Mike Penna, *Indiana University–Purdue University Indianapolis*
Mark Pinsky, *Northwestern University*
Lothar Redlin, *The Pennsylvania State University*
Joel W. Robbin, *University of Wisconsin–Madison*
Lila Roberts, *Georgia College and State University*
E. Arthur Robinson, Jr., *The George Washington University*
Richard Rockwell, *Pacific Union College*
Rob Root, *Lafayette College*
Richard Ruedemann, *Arizona State University*
David Ryeburn, *Simon Fraser University*
Richard St. Andre, *Central Michigan University*
Ricardo Salinas, *San Antonio College*
Robert Schmidt, *South Dakota State University*
Eric Schreiner, *Western Michigan University*
Mihr J. Shah, *Kent State University–Trumbull*
Theodore Shifrin, *University of Georgia*

Wayne Skrapek, *University of Saskatchewan*
Larry Small, *Los Angeles Pierce College*
Teresa Morgan Smith, *Blinn College*
William Smith, *University of North Carolina*
Donald W. Solomon, *University of Wisconsin–Milwaukee*
Edward Spitznagel, *Washington University*
Joseph Stampfli, *Indiana University*
Kristin Stoley, *Blinn College*
M. B. Tavakoli, *Chaffey College*
Paul Xavier Uhlig, *St. Mary's University, San Antonio*
Stan Ver Nooy, *University of Oregon*
Andrei Verona, *California State University–Los Angeles*
Russell C. Walker, *Carnegie Mellon University*
William L. Walton, *McCallie School*
Jack Weiner, *University of Guelph*
Alan Weinstein, *University of California, Berkeley*
Theodore W. Wilcox, *Rochester Institute of Technology*
Steven Willard, *University of Alberta*
Robert Wilson, *University of Wisconsin–Madison*
Jerome Wolbert, *University of Michigan–Ann Arbor*
Dennis H. Wortman, *University of Massachusetts, Boston*
Mary Wright, *Southern Illinois University–Carbondale*
Paul M. Wright, *Austin Community College*
Xian Wu, *University of South Carolina*

In addition, I would like to thank Jordan Bell, George Bergman, Leon Gerber, Mary Pugh, and Simon Smith for their suggestions; Al Shenk and Dennis Zill for permission to use exercises from their calculus texts; COMAP for permission to use project material; George Bergman, David Bleecker, Dan Clegg, Victor Kaftal, Anthony Lam, Jamie Lawson, Ira Rosenholtz, Paul Sally, Lowell Smylie, and Larry Wallen for ideas for exercises; Dan Drucker for the roller derby project; Thomas Banchoff, Tom Farmer, Fred Gass, John Ramsay, Larry Riddle, Philip Straffin, and Klaus Volpert for ideas for projects; Dan Anderson, Dan Clegg, Jeff Cole, Dan Drucker, and Barbara Frank for solving the new exercises and suggesting ways to improve them; Marv Riedesel and Mary Johnson for accuracy in proofreading; and Jeff Cole and Dan Clegg for their careful preparation and proofreading of the answer manuscript.

In addition, I thank those who have contributed to past editions: Ed Barbeau, Fred Brauer, Andy Bulman-Fleming, Bob Burton, David Cusick, Tom DiCiccio, Garret Etgen, Chris Fisher, Stuart Goldenberg, Arnold Good, Gene Hecht, Harvey Keynes, E.L. Koh, Zdislav Kovarik, Kevin Kreider, Emile LeBlanc, David Leep, Gerald Leibowitz, Larry Peterson, Lothar Redlin, Carl Riehm, John Ringland, Peter Rosenthal, Doug Shaw, Dan Silver, Norton Starr, Saleem Watson, Alan Weinstein, and Gail Wolkowicz.

I also thank Kathi Townes, Stephanie Kuhns, and Rebekah Million of TECHarts for their production services and the following Brooks/Cole staff: Cheryll Linthicum, content project manager; Liza Neustaetter, assistant editor; Maureen Ross, media editor; Sam Subity, managing media editor; Jennifer Jones, marketing manager; and Vernon Boes, art director. They have all done an outstanding job.

I have been very fortunate to have worked with some of the best mathematics editors in the business over the past three decades: Ron Munro, Harry Campbell, Craig Barth, Jeremy Hayhurst, Gary Ostedt, Bob Pirtle, Richard Stratton, and now Liz Covello. All of them have contributed greatly to the success of this book.

JAMES STEWART

## Ancillaries for Instructors

### PowerLecture
ISBN 0-8400-5414-9

*This comprehensive DVD contains all art from the text in both jpeg and PowerPoint formats, key equations and tables from the text, complete pre-built PowerPoint lectures, an electronic version of the Instructor's Guide, Solution Builder, ExamView testing software, Tools for Enriching Calculus, video instruction, and JoinIn on TurningPoint clicker content.*

### Instructor's Guide
by Douglas Shaw
ISBN 0-8400-5407-6

*Each section of the text is discussed from several viewpoints. The Instructor's Guide contains suggested time to allot, points to stress, text discussion topics, core materials for lecture, workshop/discussion suggestions, group work exercises in a form suitable for handout, and suggested homework assignments. An electronic version of the Instructor's Guide is available on the PowerLecture DVD.*

### Complete Solutions Manual

#### Single Variable
By Daniel Anderson, Jeffery A. Cole, and Daniel Drucker
ISBN 0-8400-5302-9

#### Multivariable
By Dan Clegg and Barbara Frank
ISBN 0-8400-4947-1

*Includes worked-out solutions to all exercises in the text.*

### Solution Builder
www.cengage.com/solutionbuilder

*This online instructor database offers complete worked out solutions to all exercises in the text. Solution Builder allows you to create customized, secure solutions printouts (in PDF format) matched exactly to the problems you assign in class.*

### Printed Test Bank
By William Steven Harmon
ISBN 0-8400-5408-4

*Contains text-specific multiple-choice and free response test items.*

### ExamView Testing

*Create, deliver, and customize tests in print and online formats with ExamView, an easy-to-use assessment and tutorial software. ExamView contains hundreds of multiple-choice and free response test items. ExamView testing is available on the PowerLecture DVD.*

## Ancillaries for Instructors and Students

### Stewart Website
www.stewartcalculus.com

Contents: *Homework Hints* ▪ *Algebra Review* ▪ *Additional Topics* ▪ *Drill exercises* ▪ *Challenge Problems* ▪ *Web Links* ▪ *History of Mathematics* ▪ *Tools for Enriching Calculus (TEC)*

### TEC Tools for Enriching™ Calculus
By James Stewart, Harvey Keynes, Dan Clegg, and developer Hu Hohn

*Tools for Enriching Calculus (TEC) functions as both a powerful tool for instructors, as well as a tutorial environment in which students can explore and review selected topics. The Flash simulation modules in TEC include instructions, written and audio explanations of the concepts, and exercises. TEC is accessible in CourseMate, WebAssign, and PowerLecture. Selected Visuals and Modules are available at www.stewartcalculus.com.*

### WebAssign Enhanced WebAssign
www.webassign.net

*WebAssign's homework delivery system lets instructors deliver, collect, grade, and record assignments via the web. Enhanced WebAssign for Stewart's* Calculus *now includes opportunities for students to review prerequisite skills and content both at the start of the course and at the beginning of each section. In addition, for selected problems, students can get extra help in the form of "enhanced feedback" (rejoinders) and video solutions.* **Other key features include:** *thousands of problems from Stewart's* Calculus, *a customizable Cengage YouBook, Personal Study Plans, Show Your Work, Just in Time Review, Answer Evaluator, Visualizing Calculus animations and modules, quizzes, lecture videos (with associated questions), and more!*

### Cengage Customizable YouBook

*YouBook is a Flash-based eBook that is interactive and customizable! Containing all the content from Stewart's* Calculus, *YouBook features a text edit tool that allows instructors to modify the textbook narrative as needed. With YouBook, instructors can quickly re-order entire sections and chapters or hide any content they don't teach to create an eBook that perfectly matches their syllabus. Instructors can further customize the text by adding instructor-created or YouTube video links. Additional media assets include: animated figures, video clips, highlighting, notes, and more! YouBook is available in Enhanced WebAssign.*

■ Electronic items   ■ Printed items

*(Table continues on page xxii.)*

CourseMate CourseMate

www.cengagebrain.com

*CourseMate is a perfect self-study tool for students, and requires no set up from instructors. CourseMate brings course concepts to life with interactive learning, study, and exam preparation tools that support the printed textbook. CourseMate for Stewart's* Calculus *includes: an interactive eBook, Tools for Enriching Calculus, videos, quizzes, flashcards, and more! For instructors, CourseMate includes Engagement Tracker, a first-of-its-kind tool that monitors student engagement.*

### Maple CD-ROM

*Maple provides an advanced, high performance mathematical computation engine with fully integrated numerics & symbolics, all accessible from a WYSIWYG technical document environment.*

### CengageBrain.com

*To access additional course materials and companion resources, please visit www.cengagebrain.com. At the CengageBrain.com home page, search for the ISBN of your title (from the back cover of your book) using the search box at the top of the page. This will take you to the product page where free companion resources can be found.*

## Ancillaries for Students

### Student Solutions Manual

#### Single Variable

By Daniel Anderson, Jeffery A. Cole, and Daniel Drucker
ISBN 0-8400-4949-8

#### Multivariable

By Dan Clegg and Barbara Frank
ISBN 0-8400-4945-5

*Provides completely worked-out solutions to all odd-numbered exercises in the text, giving students a chance to check their answers and ensure they took the correct steps to arrive at an answer.*

### Study Guide

#### Single Variable

By Richard St. Andre
ISBN 0-8400-5409-2

#### Multivariable

By Richard St. Andre
ISBN 0-8400-5410-6

*For each section of the text, the Study Guide provides students with a brief introduction, a short list of concepts to master, as well as summary and focus questions with explained answers. The Study Guide also contains "Technology Plus" questions, and multiple-choice "On Your Own" exam-style questions.*

### CalcLabs with Maple

#### Single Variable By Philip B. Yasskin and Robert Lopez
ISBN 0-8400-5811-X

#### Multivariable By Philip B. Yasskin and Robert Lopez
ISBN 0-8400-5812-8

### CalcLabs with Mathematica

#### Single Variable By Selwyn Hollis
ISBN 0-8400-5814-4

#### Multivariable By Selwyn Hollis
ISBN 0-8400-5813-6

*Each of these comprehensive lab manuals will help students learn to use the technology tools available to them. CalcLabs contain clearly explained exercises and a variety of labs and projects to accompany the text.*

### A Companion to Calculus

By Dennis Ebersole, Doris Schattschneider, Alicia Sevilla, and Kay Somers
ISBN 0-495-01124-X

*Written to improve algebra and problem-solving skills of students taking a Calculus course, every chapter in this companion is keyed to a calculus topic, providing conceptual background and specific algebra techniques needed to understand and solve calculus problems related to that topic. It is designed for calculus courses that integrate the review of precalculus concepts or for individual use.*

### Linear Algebra for Calculus

by Konrad J. Heuvers, William P. Francis, John H. Kuisti, Deborah F. Lockhart, Daniel S. Moak, and Gene M. Ortner
ISBN 0-534-25248-6

*This comprehensive book, designed to supplement the calculus course, provides an introduction to and review of the basic ideas of linear algebra.*

# To the Student

Reading a calculus textbook is different from reading a newspaper or a novel, or even a physics book. Don't be discouraged if you have to read a passage more than once in order to understand it. You should have pencil and paper and calculator at hand to sketch a diagram or make a calculation.

Some students start by trying their homework problems and read the text only if they get stuck on an exercise. I suggest that a far better plan is to read and understand a section of the text before attempting the exercises. In particular, you should look at the definitions to see the exact meanings of the terms. And before you read each example, I suggest that you cover up the solution and try solving the problem yourself. You'll get a lot more from looking at the solution if you do so.

Part of the aim of this course is to train you to think logically. Learn to write the solutions of the exercises in a connected, step-by-step fashion with explanatory sentences—not just a string of disconnected equations or formulas.

The answers to the odd-numbered exercises appear at the back of the book, in Appendix I. Some exercises ask for a verbal explanation or interpretation or description. In such cases there is no single correct way of expressing the answer, so don't worry that you haven't found the definitive answer. In addition, there are often several different forms in which to express a numerical or algebraic answer, so if your answer differs from mine, don't immediately assume you're wrong. For example, if the answer given in the back of the book is $\sqrt{2} - 1$ and you obtain $1/(1 + \sqrt{2})$, then you're right and rationalizing the denominator will show that the answers are equivalent.

The icon ⌂ indicates an exercise that definitely requires the use of either a graphing calculator or a computer with graphing software. (Appendix G discusses the use of these graphing devices and some of the pitfalls that you may encounter.) But that doesn't mean that graphing devices can't be used to check your work on the other exercises as well. The symbol CAS is reserved for problems in which the full resources of a computer algebra system (like Derive, Maple, Mathematica, or the TI-89/92) are required.

You will also encounter the symbol ⊘, which warns you against committing an error. I have placed this symbol in the margin in situations where I have observed that a large proportion of my students tend to make the same mistake.

*Tools for Enriching Calculus,* which is a companion to this text, is referred to by means of the symbol TEC and can be accessed in Enhanced WebAssign and CourseMate (selected Visuals and Modules are available at www.stewartcalculus.com). It directs you to modules in which you can explore aspects of calculus for which the computer is particularly useful.

*Homework Hints* for representative exercises are indicated by printing the exercise number in red: 5. These hints can be found on stewartcalculus.com as well as Enhanced WebAssign and CourseMate. The homework hints ask you questions that allow you to make progress toward a solution without actually giving you the answer. You need to pursue each hint in an active manner with pencil and paper to work out the details. If a particular hint doesn't enable you to solve the problem, you can click to reveal the next hint.

I recommend that you keep this book for reference purposes after you finish the course. Because you will likely forget some of the specific details of calculus, the book will serve as a useful reminder when you need to use calculus in subsequent courses. And, because this book contains more material than can be covered in any one course, it can also serve as a valuable resource for a working scientist or engineer.

Calculus is an exciting subject, justly considered to be one of the greatest achievements of the human intellect. I hope you will discover that it is not only useful but also intrinsically beautiful.

JAMES STEWART

# Diagnostic Tests

Success in calculus depends to a large extent on knowledge of the mathematics that precedes calculus: algebra, analytic geometry, functions, and trigonometry. The following tests are intended to diagnose weaknesses that you might have in these areas. After taking each test you can check your answers against the given answers and, if necessary, refresh your skills by referring to the review materials that are provided.

## A    Diagnostic Test: Algebra

1. Evaluate each expression without using a calculator.

   (a) $(-3)^4$ (b) $-3^4$ (c) $3^{-4}$

   (d) $\dfrac{5^{23}}{5^{21}}$ (e) $\left(\dfrac{2}{3}\right)^{-2}$ (f) $16^{-3/4}$

2. Simplify each expression. Write your answer without negative exponents.

   (a) $\sqrt{200} - \sqrt{32}$

   (b) $(3a^3b^3)(4ab^2)^2$

   (c) $\left(\dfrac{3x^{3/2}y^3}{x^2y^{-1/2}}\right)^{-2}$

3. Expand and simplify.

   (a) $3(x + 6) + 4(2x - 5)$ (b) $(x + 3)(4x - 5)$

   (c) $(\sqrt{a} + \sqrt{b})(\sqrt{a} - \sqrt{b})$ (d) $(2x + 3)^2$

   (e) $(x + 2)^3$

4. Factor each expression.

   (a) $4x^2 - 25$ (b) $2x^2 + 5x - 12$

   (c) $x^3 - 3x^2 - 4x + 12$ (d) $x^4 + 27x$

   (e) $3x^{3/2} - 9x^{1/2} + 6x^{-1/2}$ (f) $x^3y - 4xy$

5. Simplify the rational expression.

   (a) $\dfrac{x^2 + 3x + 2}{x^2 - x - 2}$ (b) $\dfrac{2x^2 - x - 1}{x^2 - 9} \cdot \dfrac{x + 3}{2x + 1}$

   (c) $\dfrac{x^2}{x^2 - 4} - \dfrac{x + 1}{x + 2}$ (d) $\dfrac{\dfrac{y}{x} - \dfrac{x}{y}}{\dfrac{1}{y} - \dfrac{1}{x}}$

**6.** Rationalize the expression and simplify.

(a) $\dfrac{\sqrt{10}}{\sqrt{5} - 2}$

(b) $\dfrac{\sqrt{4 + h} - 2}{h}$

**7.** Rewrite by completing the square.

(a) $x^2 + x + 1$

(b) $2x^2 - 12x + 11$

**8.** Solve the equation. (Find only the real solutions.)

(a) $x + 5 = 14 - \frac{1}{2}x$

(b) $\dfrac{2x}{x + 1} = \dfrac{2x - 1}{x}$

(c) $x^2 - x - 12 = 0$

(d) $2x^2 + 4x + 1 = 0$

(e) $x^4 - 3x^2 + 2 = 0$

(f) $3|x - 4| = 10$

(g) $2x(4 - x)^{-1/2} - 3\sqrt{4 - x} = 0$

**9.** Solve each inequality. Write your answer using interval notation.

(a) $-4 < 5 - 3x \le 17$

(b) $x^2 < 2x + 8$

(c) $x(x - 1)(x + 2) > 0$

(d) $|x - 4| < 3$

(e) $\dfrac{2x - 3}{x + 1} \le 1$

**10.** State whether each equation is true or false.

(a) $(p + q)^2 = p^2 + q^2$

(b) $\sqrt{ab} = \sqrt{a}\,\sqrt{b}$

(c) $\sqrt{a^2 + b^2} = a + b$

(d) $\dfrac{1 + TC}{C} = 1 + T$

(e) $\dfrac{1}{x - y} = \dfrac{1}{x} - \dfrac{1}{y}$

(f) $\dfrac{1/x}{a/x - b/x} = \dfrac{1}{a - b}$

## Answers to Diagnostic Test A: Algebra

**1.** (a) 81  (b) $-81$  (c) $\frac{1}{81}$
(d) 25  (e) $\frac{9}{4}$  (f) $\frac{1}{8}$

**2.** (a) $6\sqrt{2}$  (b) $48a^5b^7$  (c) $\dfrac{x}{9y^7}$

**3.** (a) $11x - 2$  (b) $4x^2 + 7x - 15$
(c) $a - b$  (d) $4x^2 + 12x + 9$
(e) $x^3 + 6x^2 + 12x + 8$

**4.** (a) $(2x - 5)(2x + 5)$  (b) $(2x - 3)(x + 4)$
(c) $(x - 3)(x - 2)(x + 2)$  (d) $x(x + 3)(x^2 - 3x + 9)$
(e) $3x^{-1/2}(x - 1)(x - 2)$  (f) $xy(x - 2)(x + 2)$

**5.** (a) $\dfrac{x + 2}{x - 2}$  (b) $\dfrac{x - 1}{x - 3}$

(c) $\dfrac{1}{x - 2}$  (d) $-(x + y)$

**6.** (a) $5\sqrt{2} + 2\sqrt{10}$  (b) $\dfrac{1}{\sqrt{4 + h} + 2}$

**7.** (a) $\left(x + \frac{1}{2}\right)^2 + \frac{3}{4}$  (b) $2(x - 3)^2 - 7$

**8.** (a) 6  (b) 1  (c) $-3, 4$
(d) $-1 \pm \frac{1}{2}\sqrt{2}$  (e) $\pm 1, \pm\sqrt{2}$  (f) $\frac{2}{3}, \frac{22}{3}$
(g) $\frac{12}{5}$

**9.** (a) $[-4, 3)$  (b) $(-2, 4)$
(c) $(-2, 0) \cup (1, \infty)$  (d) $(1, 7)$
(e) $(-1, 4]$

**10.** (a) False  (b) True  (c) False
(d) False  (e) False  (f) True

If you have had difficulty with these problems, you may wish to consult the Review of Algebra on the website www.stewartcalculus.com

## B    Diagnostic Test: Analytic Geometry

1. Find an equation for the line that passes through the point $(2, -5)$ and
   (a) has slope $-3$
   (b) is parallel to the $x$-axis
   (c) is parallel to the $y$-axis
   (d) is parallel to the line $2x - 4y = 3$

2. Find an equation for the circle that has center $(-1, 4)$ and passes through the point $(3, -2)$.

3. Find the center and radius of the circle with equation $x^2 + y^2 - 6x + 10y + 9 = 0$.

4. Let $A(-7, 4)$ and $B(5, -12)$ be points in the plane.
   (a) Find the slope of the line that contains $A$ and $B$.
   (b) Find an equation of the line that passes through $A$ and $B$. What are the intercepts?
   (c) Find the midpoint of the segment $AB$.
   (d) Find the length of the segment $AB$.
   (e) Find an equation of the perpendicular bisector of $AB$.
   (f) Find an equation of the circle for which $AB$ is a diameter.

5. Sketch the region in the $xy$-plane defined by the equation or inequalities.
   (a) $-1 \leqslant y \leqslant 3$          (b) $|x| < 4$ and $|y| < 2$
   (c) $y < 1 - \frac{1}{2}x$          (d) $y \geqslant x^2 - 1$
   (e) $x^2 + y^2 < 4$          (f) $9x^2 + 16y^2 = 144$

## Answers to Diagnostic Test B: Analytic Geometry

1. (a) $y = -3x + 1$          (b) $y = -5$
   (c) $x = 2$          (d) $y = \frac{1}{2}x - 6$

2. $(x + 1)^2 + (y - 4)^2 = 52$

3. Center $(3, -5)$, radius 5

4. (a) $-\frac{4}{3}$
   (b) $4x + 3y + 16 = 0$; $x$-intercept $-4$, $y$-intercept $-\frac{16}{3}$
   (c) $(-1, -4)$
   (d) 20
   (e) $3x - 4y = 13$
   (f) $(x + 1)^2 + (y + 4)^2 = 100$

5.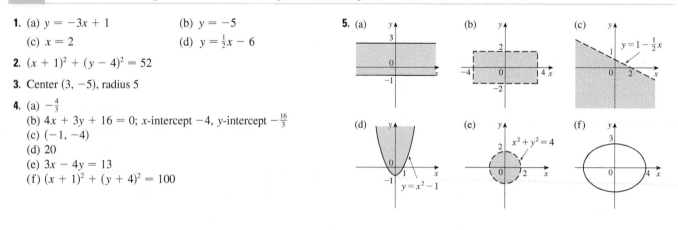

If you have had difficulty with these problems, you may wish to consult the review of analytic geometry in Appendixes B and C.

## C    Diagnostic Test: Functions

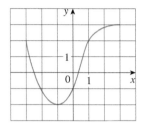

y

1

0    1    x

**FIGURE FOR PROBLEM 1**

**1.** The graph of a function $f$ is given at the left.
(a)  State the value of $f(-1)$.
(b)  Estimate the value of $f(2)$.
(c)  For what values of $x$ is $f(x) = 2$?
(d)  Estimate the values of $x$ such that $f(x) = 0$.
(e)  State the domain and range of $f$.

**2.** If $f(x) = x^3$, evaluate the difference quotient $\dfrac{f(2 + h) - f(2)}{h}$ and simplify your answer.

**3.** Find the domain of the function.

(a)  $f(x) = \dfrac{2x + 1}{x^2 + x - 2}$          (b)  $g(x) = \dfrac{\sqrt[3]{x}}{x^2 + 1}$          (c)  $h(x) = \sqrt{4 - x} + \sqrt{x^2 - 1}$

**4.** How are graphs of the functions obtained from the graph of $f$?

(a)  $y = -f(x)$          (b)  $y = 2f(x) - 1$          (c)  $y = f(x - 3) + 2$

**5.** Without using a calculator, make a rough sketch of the graph.

(a)  $y = x^3$          (b)  $y = (x + 1)^3$          (c)  $y = (x - 2)^3 + 3$
(d)  $y = 4 - x^2$          (e)  $y = \sqrt{x}$          (f)  $y = 2\sqrt{x}$
(g)  $y = -2^x$          (h)  $y = 1 + x^{-1}$

**6.** Let $f(x) = \begin{cases} 1 - x^2 & \text{if } x \leq 0 \\ 2x + 1 & \text{if } x > 0 \end{cases}$

(a)  Evaluate $f(-2)$ and $f(1)$.          (b)  Sketch the graph of $f$.

**7.** If $f(x) = x^2 + 2x - 1$ and $g(x) = 2x - 3$, find each of the following functions.
(a)  $f \circ g$          (b)  $g \circ f$          (c)  $g \circ g \circ g$

## Answers to Diagnostic Test C: Functions

**1.** (a)  $-2$          (b)  2.8
(c)  $-3, 1$          (d)  $-2.5, 0.3$
(e)  $[-3, 3], [-2, 3]$

**2.** $12 + 6h + h^2$

**3.** (a)  $(-\infty, -2) \cup (-2, 1) \cup (1, \infty)$
(b)  $(-\infty, \infty)$
(c)  $(-\infty, -1] \cup [1, 4]$

**4.** (a)  Reflect about the $x$-axis
(b)  Stretch vertically by a factor of 2, then shift 1 unit downward
(c)  Shift 3 units to the right and 2 units upward

**5.** (a)          (b)          (c)

**6.** (a)  $-3, 3$
(b)

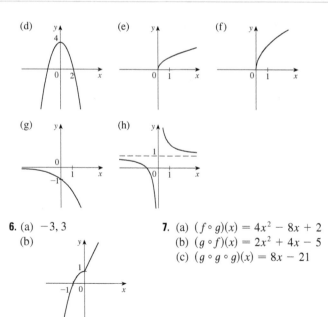

**7.** (a)  $(f \circ g)(x) = 4x^2 - 8x + 2$
(b)  $(g \circ f)(x) = 2x^2 + 4x - 5$
(c)  $(g \circ g \circ g)(x) = 8x - 21$

If you have had difficulty with these problems, you should look at Sections 1.1–1.3 of this book.

## D   Diagnostic Test: Trigonometry

1. Convert from degrees to radians.
   (a) 300°      (b) −18°

2. Convert from radians to degrees.
   (a) $5\pi/6$      (b) 2

3. Find the length of an arc of a circle with radius 12 cm if the arc subtends a central angle of 30°.

4. Find the exact values.
   (a) $\tan(\pi/3)$     (b) $\sin(7\pi/6)$     (c) $\sec(5\pi/3)$

5. Express the lengths $a$ and $b$ in the figure in terms of $\theta$.

6. If $\sin x = \frac{1}{3}$ and $\sec y = \frac{5}{4}$, where $x$ and $y$ lie between 0 and $\pi/2$, evaluate $\sin(x + y)$.

7. Prove the identities.
   (a) $\tan\theta \sin\theta + \cos\theta = \sec\theta$

   (b) $\dfrac{2\tan x}{1 + \tan^2 x} = \sin 2x$

8. Find all values of $x$ such that $\sin 2x = \sin x$ and $0 \leqslant x \leqslant 2\pi$.

9. Sketch the graph of the function $y = 1 + \sin 2x$ without using a calculator.

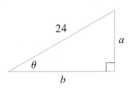

24

$a$

$\theta$

$b$

**FIGURE FOR PROBLEM 5**

## Answers to Diagnostic Test D: Trigonometry

1. (a) $5\pi/3$        (b) $-\pi/10$

2. (a) 150°        (b) $360°/\pi \approx 114.6°$

3. $2\pi$ cm

4. (a) $\sqrt{3}$        (b) $-\frac{1}{2}$        (c) 2

5. (a) $24\sin\theta$        (b) $24\cos\theta$

6. $\frac{1}{15}\left(4 + 6\sqrt{2}\right)$

8. $0,\ \pi/3,\ \pi,\ 5\pi/3,\ 2\pi$

9.

If you have had difficulty with these problems, you should look at Appendix D of this book.

# A Preview of Calculus

© Ziga Camernik / Shutterstock

© Pichugin Dmitry / Shutterstock

By the time you finish this course, you will be able to estimate the number of laborers needed to build a pyramid, explain the formation and location of rainbows, design a roller coaster for a smooth ride, and calculate the force on a dam.

© Brett Mulcahy / Shutterstock

© iofoto / Shutterstock

Calculus is fundamentally different from the mathematics that you have studied previously: calculus is less static and more dynamic. It is concerned with change and motion; it deals with quantities that approach other quantities. For that reason it may be useful to have an overview of the subject before beginning its intensive study. Here we give a glimpse of some of the main ideas of calculus by showing how the concept of a limit arises when we attempt to solve a variety of problems.

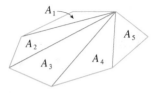

$$A = A_1 + A_2 + A_3 + A_4 + A_5$$

**FIGURE 1**

## The Area Problem

The origins of calculus go back at least 2500 years to the ancient Greeks, who found areas using the "method of exhaustion." They knew how to find the area $A$ of any polygon by dividing it into triangles as in Figure 1 and adding the areas of these triangles.

It is a much more difficult problem to find the area of a curved figure. The Greek method of exhaustion was to inscribe polygons in the figure and circumscribe polygons about the figure and then let the number of sides of the polygons increase. Figure 2 illustrates this process for the special case of a circle with inscribed regular polygons.

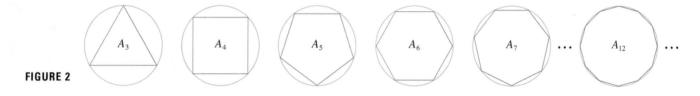

**FIGURE 2**

Let $A_n$ be the area of the inscribed polygon with $n$ sides. As $n$ increases, it appears that $A_n$ becomes closer and closer to the area of the circle. We say that the area of the circle is the *limit* of the areas of the inscribed polygons, and we write

**TEC** In the Preview Visual, you can see how areas of inscribed and circumscribed polygons approximate the area of a circle.

$$A = \lim_{n \to \infty} A_n$$

The Greeks themselves did not use limits explicitly. However, by indirect reasoning, Eudoxus (fifth century BC) used exhaustion to prove the familiar formula for the area of a circle: $A = \pi r^2$.

We will use a similar idea in Chapter 4 to find areas of regions of the type shown in Figure 3. We will approximate the desired area $A$ by areas of rectangles (as in Figure 4), let the width of the rectangles decrease, and then calculate $A$ as the limit of these sums of areas of rectangles.

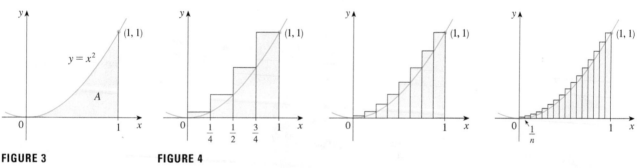

**FIGURE 3**          **FIGURE 4**

The area problem is the central problem in the branch of calculus called *integral calculus*. The techniques that we will develop in Chapter 4 for finding areas will also enable us to compute the volume of a solid, the length of a curve, the force of water against a dam, the mass and center of gravity of a rod, and the work done in pumping water out of a tank.

## The Tangent Problem

Consider the problem of trying to find an equation of the tangent line $t$ to a curve with equation $y = f(x)$ at a given point $P$. (We will give a precise definition of a tangent line in

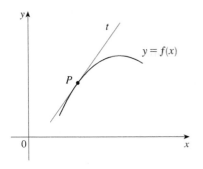

**FIGURE 5**

The tangent line at $P$

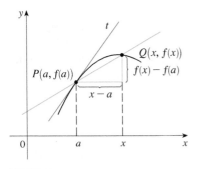

**FIGURE 6**

The secant line $PQ$

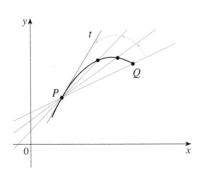

**FIGURE 7**

Secant lines approaching the
tangent line

Chapter 1. For now you can think of it as a line that touches the curve at $P$ as in Figure 5.) Since we know that the point $P$ lies on the tangent line, we can find the equation of $t$ if we know its slope $m$. The problem is that we need two points to compute the slope and we know only one point, $P$, on $t$. To get around the problem we first find an approximation to $m$ by taking a nearby point $Q$ on the curve and computing the slope $m_{PQ}$ of the secant line $PQ$. From Figure 6 we see that

$$\boxed{1} \qquad m_{PQ} = \frac{f(x) - f(a)}{x - a}$$

Now imagine that $Q$ moves along the curve toward $P$ as in Figure 7. You can see that the secant line rotates and approaches the tangent line as its limiting position. This means that the slope $m_{PQ}$ of the secant line becomes closer and closer to the slope $m$ of the tangent line. We write

$$m = \lim_{Q \to P} m_{PQ}$$

and we say that $m$ is the limit of $m_{PQ}$ as $Q$ approaches $P$ along the curve. Since $x$ approaches $a$ as $Q$ approaches $P$, we could also use Equation 1 to write

$$\boxed{2} \qquad m = \lim_{x \to a} \frac{f(x) - f(a)}{x - a}$$

Specific examples of this procedure will be given in Chapter 1.

The tangent problem has given rise to the branch of calculus called *differential calculus*, which was not invented until more than 2000 years after integral calculus. The main ideas behind differential calculus are due to the French mathematician Pierre Fermat (1601–1665) and were developed by the English mathematicians John Wallis (1616–1703), Isaac Barrow (1630–1677), and Isaac Newton (1642–1727) and the German mathematician Gottfried Leibniz (1646–1716).

The two branches of calculus and their chief problems, the area problem and the tangent problem, appear to be very different, but it turns out that there is a very close connection between them. The tangent problem and the area problem are inverse problems in a sense that will be described in Chapter 4.

## Velocity

When we look at the speedometer of a car and read that the car is traveling at 48 mi/h, what does that information indicate to us? We know that if the velocity remains constant, then after an hour we will have traveled 48 mi. But if the velocity of the car varies, what does it mean to say that the velocity at a given instant is 48 mi/h?

In order to analyze this question, let's examine the motion of a car that travels along a straight road and assume that we can measure the distance traveled by the car (in feet) at 1-second intervals as in the following chart:

| $t =$ Time elapsed (s) | 0 | 1 | 2 | 3 | 4 | 5 |
|---|---|---|---|---|---|---|
| $d =$ Distance (ft) | 0 | 2 | 9 | 24 | 42 | 71 |

As a first step toward finding the velocity after 2 seconds have elapsed, we find the average velocity during the time interval $2 \leqslant t \leqslant 4$:

$$\text{average velocity} = \frac{\text{change in position}}{\text{time elapsed}}$$

$$= \frac{42 - 9}{4 - 2}$$

$$= 16.5 \text{ ft/s}$$

Similarly, the average velocity in the time interval $2 \leqslant t \leqslant 3$ is

$$\text{average velocity} = \frac{24 - 9}{3 - 2} = 15 \text{ ft/s}$$

We have the feeling that the velocity at the instant $t = 2$ can't be much different from the average velocity during a short time interval starting at $t = 2$. So let's imagine that the distance traveled has been measured at 0.l-second time intervals as in the following chart:

| $t$ | 2.0 | 2.1 | 2.2 | 2.3 | 2.4 | 2.5 |
|-----|------|-------|-------|-------|-------|-------|
| $d$ | 9.00 | 10.02 | 11.16 | 12.45 | 13.96 | 15.80 |

Then we can compute, for instance, the average velocity over the time interval [2, 2.5]:

$$\text{average velocity} = \frac{15.80 - 9.00}{2.5 - 2} = 13.6 \text{ ft/s}$$

The results of such calculations are shown in the following chart:

| Time interval | [2, 3] | [2, 2.5] | [2, 2.4] | [2, 2.3] | [2, 2.2] | [2, 2.1] |
|---------------|--------|----------|----------|----------|----------|----------|
| Average velocity (ft/s) | 15.0 | 13.6 | 12.4 | 11.5 | 10.8 | 10.2 |

The average velocities over successively smaller intervals appear to be getting closer to a number near 10, and so we expect that the velocity at exactly $t = 2$ is about 10 ft/s. In Chapter 1 we will define the instantaneous velocity of a moving object as the limiting value of the average velocities over smaller and smaller time intervals.

In Figure 8 we show a graphical representation of the motion of the car by plotting the distance traveled as a function of time. If we write $d = f(t)$, then $f(t)$ is the number of feet traveled after $t$ seconds. The average velocity in the time interval $[2, t]$ is

$$\text{average velocity} = \frac{\text{change in position}}{\text{time elapsed}} = \frac{f(t) - f(2)}{t - 2}$$

which is the same as the slope of the secant line $PQ$ in Figure 8. The velocity $v$ when $t = 2$ is the limiting value of this average velocity as $t$ approaches 2; that is,

$$v = \lim_{t \to 2} \frac{f(t) - f(2)}{t - 2}$$

and we recognize from Equation 2 that this is the same as the slope of the tangent line to the curve at $P$.

**FIGURE 8**

Thus, when we solve the tangent problem in differential calculus, we are also solving problems concerning velocities. The same techniques also enable us to solve problems involving rates of change in all of the natural and social sciences.

## The Limit of a Sequence

In the fifth century BC the Greek philosopher Zeno of Elea posed four problems, now known as *Zeno's paradoxes,* that were intended to challenge some of the ideas concerning space and time that were held in his day. Zeno's second paradox concerns a race between the Greek hero Achilles and a tortoise that has been given a head start. Zeno argued, as follows, that Achilles could never pass the tortoise: Suppose that Achilles starts at position $a_1$ and the tortoise starts at position $t_1$. (See Figure 9.) When Achilles reaches the point $a_2 = t_1$, the tortoise is farther ahead at position $t_2$. When Achilles reaches $a_3 = t_2$, the tortoise is at $t_3$. This process continues indefinitely and so it appears that the tortoise will always be ahead! But this defies common sense.

**FIGURE 9**

One way of explaining this paradox is with the idea of a *sequence.* The successive positions of Achilles $(a_1, a_2, a_3, \ldots)$ or the successive positions of the tortoise $(t_1, t_2, t_3, \ldots)$ form what is known as a sequence.

In general, a sequence $\{a_n\}$ is a set of numbers written in a definite order. For instance, the sequence

$$\left\{1, \tfrac{1}{2}, \tfrac{1}{3}, \tfrac{1}{4}, \tfrac{1}{5}, \ldots\right\}$$

can be described by giving the following formula for the $n$th term:

$$a_n = \frac{1}{n}$$

We can visualize this sequence by plotting its terms on a number line as in Figure 10(a) or by drawing its graph as in Figure 10(b). Observe from either picture that the terms of the sequence $a_n = 1/n$ are becoming closer and closer to 0 as $n$ increases. In fact, we can find terms as small as we please by making $n$ large enough. We say that the limit of the sequence is 0, and we indicate this by writing

$$\lim_{n \to \infty} \frac{1}{n} = 0$$

In general, the notation

$$\lim_{n \to \infty} a_n = L$$

is used if the terms $a_n$ approach the number $L$ as $n$ becomes large. This means that the numbers $a_n$ can be made as close as we like to the number $L$ by taking $n$ sufficiently large.

**FIGURE 10**

The concept of the limit of a sequence occurs whenever we use the decimal representation of a real number. For instance, if

$$a_1 = 3.1$$

$$a_2 = 3.14$$

$$a_3 = 3.141$$

$$a_4 = 3.1415$$

$$a_5 = 3.14159$$

$$a_6 = 3.141592$$

$$a_7 = 3.1415926$$

$$\vdots$$

then

$$\lim_{n \to \infty} a_n = \pi$$

The terms in this sequence are rational approximations to $\pi$.

Let's return to Zeno's paradox. The successive positions of Achilles and the tortoise form sequences $\{a_n\}$ and $\{t_n\}$, where $a_n < t_n$ for all $n$. It can be shown that both sequences have the same limit:

$$\lim_{n \to \infty} a_n = p = \lim_{n \to \infty} t_n$$

It is precisely at this point $p$ that Achilles overtakes the tortoise.

## The Sum of a Series

Another of Zeno's paradoxes, as passed on to us by Aristotle, is the following: "A man standing in a room cannot walk to the wall. In order to do so, he would first have to go half the distance, then half the remaining distance, and then again half of what still remains. This process can always be continued and can never be ended." (See Figure 11.)

**FIGURE 11**

Of course, we know that the man can actually reach the wall, so this suggests that perhaps the total distance can be expressed as the sum of infinitely many smaller distances as follows:


$$\boxed{3} \qquad 1 = \frac{1}{2} + \frac{1}{4} + \frac{1}{8} + \frac{1}{16} + \cdots + \frac{1}{2^n} + \cdots$$

Zeno was arguing that it doesn't make sense to add infinitely many numbers together. But there are other situations in which we implicitly use infinite sums. For instance, in decimal notation, the symbol $0.\overline{3} = 0.3333\ldots$ means

$$\frac{3}{10} + \frac{3}{100} + \frac{3}{1000} + \frac{3}{10,000} + \cdots$$

and so, in some sense, it must be true that

$$\frac{3}{10} + \frac{3}{100} + \frac{3}{1000} + \frac{3}{10,000} + \cdots = \frac{1}{3}$$

More generally, if $d_n$ denotes the $n$th digit in the decimal representation of a number, then

$$0.d_1 d_2 d_3 d_4 \ldots = \frac{d_1}{10} + \frac{d_2}{10^2} + \frac{d_3}{10^3} + \cdots + \frac{d_n}{10^n} + \cdots$$

Therefore some infinite sums, or infinite series as they are called, have a meaning. But we must define carefully what the sum of an infinite series is.

Returning to the series in Equation 3, we denote by $s_n$ the sum of the first $n$ terms of the series. Thus

$$s_1 = \tfrac{1}{2} = 0.5$$

$$s_2 = \tfrac{1}{2} + \tfrac{1}{4} = 0.75$$

$$s_3 = \tfrac{1}{2} + \tfrac{1}{4} + \tfrac{1}{8} = 0.875$$

$$s_4 = \tfrac{1}{2} + \tfrac{1}{4} + \tfrac{1}{8} + \tfrac{1}{16} = 0.9375$$

$$s_5 = \tfrac{1}{2} + \tfrac{1}{4} + \tfrac{1}{8} + \tfrac{1}{16} + \tfrac{1}{32} = 0.96875$$

$$s_6 = \tfrac{1}{2} + \tfrac{1}{4} + \tfrac{1}{8} + \tfrac{1}{16} + \tfrac{1}{32} + \tfrac{1}{64} = 0.984375$$

$$s_7 = \tfrac{1}{2} + \tfrac{1}{4} + \tfrac{1}{8} + \tfrac{1}{16} + \tfrac{1}{32} + \tfrac{1}{64} + \tfrac{1}{128} = 0.9921875$$

$$\vdots$$

$$s_{10} = \tfrac{1}{2} + \tfrac{1}{4} + \cdots + \tfrac{1}{1024} \approx 0.99902344$$

$$\vdots$$

$$s_{16} = \frac{1}{2} + \frac{1}{4} + \cdots + \frac{1}{2^{16}} \approx 0.99998474$$

Observe that as we add more and more terms, the partial sums become closer and closer to 1. In fact, it can be shown that by taking $n$ large enough (that is, by adding sufficiently many terms of the series), we can make the partial sum $s_n$ as close as we please to the number 1. It therefore seems reasonable to say that the sum of the infinite series is 1 and to write

$$\frac{1}{2} + \frac{1}{4} + \frac{1}{8} + \cdots + \frac{1}{2^n} + \cdots = 1$$

In other words, the reason the sum of the series is 1 is that

$$\lim_{n \to \infty} s_n = 1$$

In Chapter 11 we will discuss these ideas further. We will then use Newton's idea of combining infinite series with differential and integral calculus.

## Summary

We have seen that the concept of a limit arises in trying to find the area of a region, the slope of a tangent to a curve, the velocity of a car, or the sum of an infinite series. In each case the common theme is the calculation of a quantity as the limit of other, easily calculated quantities. It is this basic idea of a limit that sets calculus apart from other areas of mathematics. In fact, we could define calculus as the part of mathematics that deals with limits.

After Sir Isaac Newton invented his version of calculus, he used it to explain the motion of the planets around the sun. Today calculus is used in calculating the orbits of satellites and spacecraft, in predicting population sizes, in estimating how fast oil prices rise or fall, in forecasting weather, in measuring the cardiac output of the heart, in calculating life insurance premiums, and in a great variety of other areas. We will explore some of these uses of calculus in this book.

In order to convey a sense of the power of the subject, we end this preview with a list of some of the questions that you will be able to answer using calculus:

1. How can we explain the fact, illustrated in Figure 12, that the angle of elevation from an observer up to the highest point in a rainbow is 42°? (See page 206.)

2. How can we explain the shapes of cans on supermarket shelves? (See page 262.)

3. Where is the best place to sit in a movie theater? (See page 461.)

4. How can we design a roller coaster for a smooth ride? (See page 140.)

5. How far away from an airport should a pilot start descent? (See page 156.)

6. How can we fit curves together to design shapes to represent letters on a laser printer? (See page 677.)

7. How can we estimate the number of workers that were needed to build the Great Pyramid of Khufu in ancient Egypt? (See page 373.)

8. Where should an infielder position himself to catch a baseball thrown by an outfielder and relay it to home plate? (See page 658.)

9. Does a ball thrown upward take longer to reach its maximum height or to fall back to its original height? (See page 628.)

10. How can we explain the fact that planets and satellites move in elliptical orbits? (See page 892.)

11. How can we distribute water flow among turbines at a hydroelectric station so as to maximize the total energy production? (See page 990.)

12. If a marble, a squash ball, a steel bar, and a lead pipe roll down a slope, which of them reaches the bottom first? (See page 1063.)

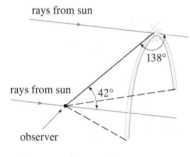

**FIGURE 12**

# 1 Functions and Limits

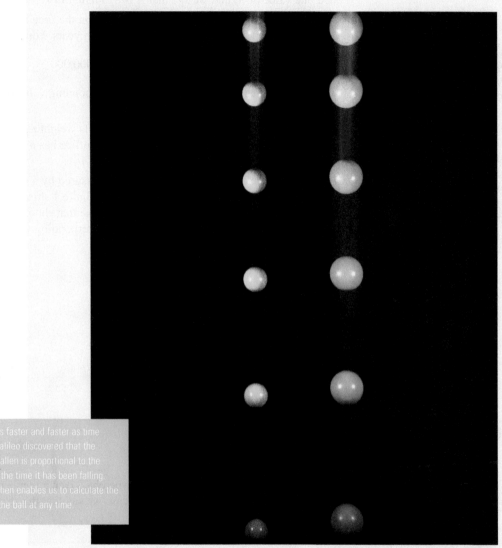

A ball falls faster and faster as time passes. Galileo discovered that the distance fallen is proportional to the square of the time it has been falling. Calculus then enables us to calculate the speed of the ball at any time.

© 1986 Peticolas / Megna, Fundamental Photographs, NYC

The fundamental objects that we deal with in calculus are functions. We stress that a function can be represented in different ways: by an equation, in a table, by a graph, or in words. We look at the main types of functions that occur in calculus and describe the process of using these functions as mathematical models of real-world phenomena.

In *A Preview of Calculus* (page 1) we saw how the idea of a limit underlies the various branches of calculus. It is therefore appropriate to begin our study of calculus by investigating limits of functions and their properties.

## 1.1 Four Ways to Represent a Function

Functions arise whenever one quantity depends on another. Consider the following four situations.

**A.** The area $A$ of a circle depends on the radius $r$ of the circle. The rule that connects $r$ and $A$ is given by the equation $A = \pi r^2$. With each positive number $r$ there is associated one value of $A$, and we say that $A$ is a *function* of $r$.

**B.** The human population of the world $P$ depends on the time $t$. The table gives estimates of the world population $P(t)$ at time $t$, for certain years. For instance,

$$P(1950) \approx 2{,}560{,}000{,}000$$

But for each value of the time $t$ there is a corresponding value of $P$, and we say that $P$ is a function of $t$.

**C.** The cost $C$ of mailing an envelope depends on its weight $w$. Although there is no simple formula that connects $w$ and $C$, the post office has a rule for determining $C$ when $w$ is known.

**D.** The vertical acceleration $a$ of the ground as measured by a seismograph during an earthquake is a function of the elapsed time $t$. Figure 1 shows a graph generated by seismic activity during the Northridge earthquake that shook Los Angeles in 1994. For a given value of $t$, the graph provides a corresponding value of $a$.

| Year | Population (millions) |
|------|------------------------|
| 1900 | 1650 |
| 1910 | 1750 |
| 1920 | 1860 |
| 1930 | 2070 |
| 1940 | 2300 |
| 1950 | 2560 |
| 1960 | 3040 |
| 1970 | 3710 |
| 1980 | 4450 |
| 1990 | 5280 |
| 2000 | 6080 |
| 2010 | 6870 |

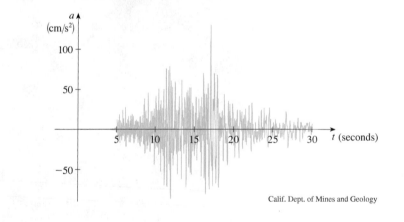

**FIGURE 1**
Vertical ground acceleration during the Northridge earthquake

Calif. Dept. of Mines and Geology

Each of these examples describes a rule whereby, given a number ($r$, $t$, $w$, or $t$), another number ($A$, $P$, $C$, or $a$) is assigned. In each case we say that the second number is a function of the first number.

> A **function** $f$ is a rule that assigns to each element $x$ in a set $D$ exactly one element, called $f(x)$, in a set $E$.

We usually consider functions for which the sets $D$ and $E$ are sets of real numbers. The set $D$ is called the **domain** of the function. The number $f(x)$ is the **value of $f$ at $x$** and is read "$f$ of $x$." The **range** of $f$ is the set of all possible values of $f(x)$ as $x$ varies throughout the domain. A symbol that represents an arbitrary number in the *domain* of a function $f$ is called an **independent variable**. A symbol that represents a number in the *range* of $f$ is called a **dependent variable**. In Example A, for instance, $r$ is the independent variable and $A$ is the dependent variable.

**FIGURE 2**

Machine diagram for a function $f$

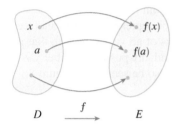

**FIGURE 3**

Arrow diagram for $f$

It's helpful to think of a function as a **machine** (see Figure 2). If $x$ is in the domain of the function $f$, then when $x$ enters the machine, it's accepted as an input and the machine produces an output $f(x)$ according to the rule of the function. Thus we can think of the domain as the set of all possible inputs and the range as the set of all possible outputs.

The preprogrammed functions in a calculator are good examples of a function as a machine. For example, the square root key on your calculator computes such a function. You press the key labeled $\sqrt{\phantom{x}}$ (or $\sqrt{x}$) and enter the input $x$. If $x < 0$, then $x$ is not in the domain of this function; that is, $x$ is not an acceptable input, and the calculator will indicate an error. If $x \geq 0$, then an *approximation* to $\sqrt{x}$ will appear in the display. Thus the $\sqrt{x}$ key on your calculator is not quite the same as the exact mathematical function $f$ defined by $f(x) = \sqrt{x}$.

Another way to picture a function is by an **arrow diagram** as in Figure 3. Each arrow connects an element of $D$ to an element of $E$. The arrow indicates that $f(x)$ is associated with $x$, $f(a)$ is associated with $a$, and so on.

The most common method for visualizing a function is its graph. If $f$ is a function with domain $D$, then its **graph** is the set of ordered pairs

$$\{(x, f(x)) \mid x \in D\}$$

(Notice that these are input-output pairs.) In other words, the graph of $f$ consists of all points $(x, y)$ in the coordinate plane such that $y = f(x)$ and $x$ is in the domain of $f$.

The graph of a function $f$ gives us a useful picture of the behavior or "life history" of a function. Since the $y$-coordinate of any point $(x, y)$ on the graph is $y = f(x)$, we can read the value of $f(x)$ from the graph as being the height of the graph above the point $x$ (see Figure 4). The graph of $f$ also allows us to picture the domain of $f$ on the $x$-axis and its range on the $y$-axis as in Figure 5.

**FIGURE 4**          **FIGURE 5**

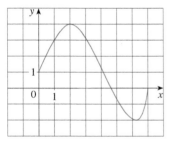

**FIGURE 6**

The notation for intervals is given in Appendix A.

**EXAMPLE 1** The graph of a function $f$ is shown in Figure 6.
(a) Find the values of $f(1)$ and $f(5)$.
(b) What are the domain and range of $f$?

SOLUTION
(a) We see from Figure 6 that the point $(1, 3)$ lies on the graph of $f$, so the value of $f$ at 1 is $f(1) = 3$. (In other words, the point on the graph that lies above $x = 1$ is 3 units above the $x$-axis.)

When $x = 5$, the graph lies about 0.7 unit below the $x$-axis, so we estimate that $f(5) \approx -0.7$.

(b) We see that $f(x)$ is defined when $0 \leq x \leq 7$, so the domain of $f$ is the closed interval $[0, 7]$. Notice that $f$ takes on all values from $-2$ to 4, so the range of $f$ is

$$\{y \mid -2 \leq y \leq 4\} = [-2, 4]$$

**FIGURE 7**

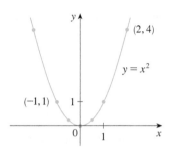

**FIGURE 8**

**EXAMPLE 2** Sketch the graph and find the domain and range of each function.

(a) $f(x) = 2x - 1$           (b) $g(x) = x^2$

SOLUTION

(a) The equation of the graph is $y = 2x - 1$, and we recognize this as being the equation of a line with slope 2 and $y$-intercept $-1$. (Recall the slope-intercept form of the equation of a line: $y = mx + b$. See Appendix B.) This enables us to sketch a portion of the graph of $f$ in Figure 7. The expression $2x - 1$ is defined for all real numbers, so the domain of $f$ is the set of all real numbers, which we denote by $\mathbb{R}$. The graph shows that the range is also $\mathbb{R}$.

(b) Since $g(2) = 2^2 = 4$ and $g(-1) = (-1)^2 = 1$, we could plot the points $(2, 4)$ and $(-1, 1)$, together with a few other points on the graph, and join them to produce the graph (Figure 8). The equation of the graph is $y = x^2$, which represents a parabola (see Appendix C). The domain of $g$ is $\mathbb{R}$. The range of $g$ consists of all values of $g(x)$, that is, all numbers of the form $x^2$. But $x^2 \geq 0$ for all numbers $x$ and any positive number $y$ is a square. So the range of $g$ is $\{y \mid y \geq 0\} = [0, \infty)$. This can also be seen from Figure 8. ▪

**EXAMPLE 3** If $f(x) = 2x^2 - 5x + 1$ and $h \neq 0$, evaluate $\dfrac{f(a + h) - f(a)}{h}$.

SOLUTION We first evaluate $f(a + h)$ by replacing $x$ by $a + h$ in the expression for $f(x)$:

$$f(a + h) = 2(a + h)^2 - 5(a + h) + 1$$
$$= 2(a^2 + 2ah + h^2) - 5(a + h) + 1$$
$$= 2a^2 + 4ah + 2h^2 - 5a - 5h + 1$$

Then we substitute into the given expression and simplify:

$$\frac{f(a + h) - f(a)}{h} = \frac{(2a^2 + 4ah + 2h^2 - 5a - 5h + 1) - (2a^2 - 5a + 1)}{h}$$
$$= \frac{2a^2 + 4ah + 2h^2 - 5a - 5h + 1 - 2a^2 + 5a - 1}{h}$$
$$= \frac{4ah + 2h^2 - 5h}{h} = 4a + 2h - 5$$ ▪

The expression

$$\frac{f(a + h) - f(a)}{h}$$

in Example 3 is called a **difference quotient** and occurs frequently in calculus. As we will see in Chapter 2, it represents the average rate of change of $f(x)$ between $x = a$ and $x = a + h$.

## Representations of Functions

There are four possible ways to represent a function:

- verbally          (by a description in words)
- numerically      (by a table of values)
- visually           (by a graph)
- algebraically     (by an explicit formula)

If a single function can be represented in all four ways, it's often useful to go from one representation to another to gain additional insight into the function. (In Example 2, for instance, we started with algebraic formulas and then obtained the graphs.) But certain functions are described more naturally by one method than by another. With this in mind, let's reexamine the four situations that we considered at the beginning of this section.

**A.** The most useful representation of the area of a circle as a function of its radius is probably the algebraic formula $A(r) = \pi r^2$, though it is possible to compile a table of values or to sketch a graph (half a parabola). Because a circle has to have a positive radius, the domain is $\{r \mid r > 0\} = (0, \infty)$, and the range is also $(0, \infty)$.

**B.** We are given a description of the function in words: $P(t)$ is the human population of the world at time $t$. Let's measure $t$ so that $t = 0$ corresponds to the year 1900. The table of values of world population provides a convenient representation of this function. If we plot these values, we get the graph (called a *scatter plot*) in Figure 9. It too is a useful representation; the graph allows us to absorb all the data at once. What about a formula? Of course, it's impossible to devise an explicit formula that gives the exact human population $P(t)$ at any time $t$. But it is possible to find an expression for a function that *approximates* $P(t)$. In fact, using methods explained in Section 1.2, we obtain the approximation

$$P(t) \approx f(t) = (1.43653 \times 10^9) \cdot (1.01395)^t$$

Figure 10 shows that it is a reasonably good "fit." The function $f$ is called a *mathematical model* for population growth. In other words, it is a function with an explicit formula that approximates the behavior of our given function. We will see, however, that the ideas of calculus can be applied to a table of values; an explicit formula is not necessary.

| $t$ | Population (millions) |
|-----|-----------------------|
| 0   | 1650 |
| 10  | 1750 |
| 20  | 1860 |
| 30  | 2070 |
| 40  | 2300 |
| 50  | 2560 |
| 60  | 3040 |
| 70  | 3710 |
| 80  | 4450 |
| 90  | 5280 |
| 100 | 6080 |
| 110 | 6870 |

**FIGURE 9**

**FIGURE 10**

A function defined by a table of values is called a *tabular* function.

The function $P$ is typical of the functions that arise whenever we attempt to apply calculus to the real world. We start with a verbal description of a function. Then we may be able to construct a table of values of the function, perhaps from instrument readings in a scientific experiment. Even though we don't have complete knowledge of the values of the function, we will see throughout the book that it is still possible to perform the operations of calculus on such a function.

**C.** Again the function is described in words: Let $C(w)$ be the cost of mailing a large envelope with weight $w$. The rule that the US Postal Service used as of 2010 is as follows: The cost is 88 cents for up to 1 oz, plus 17 cents for each additional ounce (or less) up to 13 oz. The table of values shown in the margin is the most convenient representation for this function, though it is possible to sketch a graph (see Example 10).

**D.** The graph shown in Figure 1 is the most natural representation of the vertical acceleration function $a(t)$. It's true that a table of values could be compiled, and it is even possible to devise an approximate formula. But everything a geologist needs to know—amplitudes and patterns—can be seen easily from the graph. (The same is true for the patterns seen in electrocardiograms of heart patients and polygraphs for lie-detection.)

| $w$ (ounces) | $C(w)$ (dollars) |
|--------------|------------------|
| $0 < w \leqslant 1$ | 0.88 |
| $1 < w \leqslant 2$ | 1.05 |
| $2 < w \leqslant 3$ | 1.22 |
| $3 < w \leqslant 4$ | 1.39 |
| $4 < w \leqslant 5$ | 1.56 |
| ⋮ | ⋮ |

**FIGURE 11**

In the next example we sketch the graph of a function that is defined verbally.

**EXAMPLE 4** When you turn on a hot-water faucet, the temperature $T$ of the water depends on how long the water has been running. Draw a rough graph of $T$ as a function of the time $t$ that has elapsed since the faucet was turned on.

SOLUTION The initial temperature of the running water is close to room temperature because the water has been sitting in the pipes. When the water from the hot-water tank starts flowing from the faucet, $T$ increases quickly. In the next phase, $T$ is constant at the temperature of the heated water in the tank. When the tank is drained, $T$ decreases to the temperature of the water supply. This enables us to make the rough sketch of $T$ as a function of $t$ in Figure 11.

In the following example we start with a verbal description of a function in a physical situation and obtain an explicit algebraic formula. The ability to do this is a useful skill in solving calculus problems that ask for the maximum or minimum values of quantities.

**V** **EXAMPLE 5** A rectangular storage container with an open top has a volume of 10 m³. The length of its base is twice its width. Material for the base costs $10 per square meter; material for the sides costs $6 per square meter. Express the cost of materials as a function of the width of the base.

$h$

$2w$

$w$

**FIGURE 12**

SOLUTION We draw a diagram as in Figure 12 and introduce notation by letting $w$ and $2w$ be the width and length of the base, respectively, and $h$ be the height.

The area of the base is $(2w)w = 2w^2$, so the cost, in dollars, of the material for the base is $10(2w^2)$. Two of the sides have area $wh$ and the other two have area $2wh$, so the cost of the material for the sides is $6[2(wh) + 2(2wh)]$. The total cost is therefore

$$C = 10(2w^2) + 6[2(wh) + 2(2wh)] = 20w^2 + 36wh$$

To express $C$ as a function of $w$ alone, we need to eliminate $h$ and we do so by using the fact that the volume is 10 m³. Thus

$$w(2w)h = 10$$

which gives

$$h = \frac{10}{2w^2} = \frac{5}{w^2}$$

Substituting this into the expression for $C$, we have

$$C = 20w^2 + 36w\left(\frac{5}{w^2}\right) = 20w^2 + \frac{180}{w}$$

Therefore the equation

$$C(w) = 20w^2 + \frac{180}{w} \qquad w > 0$$

expresses $C$ as a function of $w$.

 In setting up applied functions as in Example 5, it may be useful to review the principles of problem solving as discussed on page 97, particularly *Step 1: Understand the Problem.*

**EXAMPLE 6** Find the domain of each function.

(a) $f(x) = \sqrt{x + 2}$

(b) $g(x) = \dfrac{1}{x^2 - x}$

SOLUTION

(a) Because the square root of a negative number is not defined (as a real number), the domain of $f$ consists of all values of $x$ such that $x + 2 \geq 0$. This is equivalent to $x \geq -2$, so the domain is the interval $[-2, \infty)$.

**Domain Convention**

If a function is given by a formula and the domain is not stated explicitly, the convention is that the domain is the set of all numbers for which the formula makes sense and defines a real number.

(b) Since

$$g(x) = \frac{1}{x^2 - x} = \frac{1}{x(x-1)}$$

and division by 0 is not allowed, we see that $g(x)$ is not defined when $x = 0$ or $x = 1$. Thus the domain of $g$ is

$$\{x \mid x \neq 0, x \neq 1\}$$

which could also be written in interval notation as

$$(-\infty, 0) \cup (0, 1) \cup (1, \infty)$$

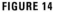

The graph of a function is a curve in the $xy$-plane. But the question arises: Which curves in the $xy$-plane are graphs of functions? This is answered by the following test.

> **The Vertical Line Test**  A curve in the $xy$-plane is the graph of a function of $x$ if and only if no vertical line intersects the curve more than once.

The reason for the truth of the Vertical Line Test can be seen in Figure 13. If each vertical line $x = a$ intersects a curve only once, at $(a, b)$, then exactly one functional value is defined by $f(a) = b$. But if a line $x = a$ intersects the curve twice, at $(a, b)$ and $(a, c)$, then the curve can't represent a function because a function can't assign two different values to $a$.

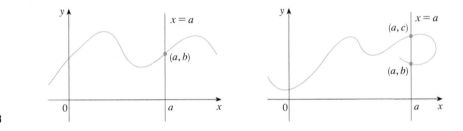

**FIGURE 13**

For example, the parabola $x = y^2 - 2$ shown in Figure 14(a) is not the graph of a function of $x$ because, as you can see, there are vertical lines that intersect the parabola twice. The parabola, however, does contain the graphs of *two* functions of $x$. Notice that the equation $x = y^2 - 2$ implies $y^2 = x + 2$, so $y = \pm\sqrt{x+2}$. Thus the upper and lower halves of the parabola are the graphs of the functions $f(x) = \sqrt{x+2}$ [from Example 6(a)] and $g(x) = -\sqrt{x+2}$. [See Figures 14(b) and (c).] We observe that if we reverse the roles of $x$ and $y$, then the equation $x = h(y) = y^2 - 2$ *does* define $x$ as a function of $y$ (with $y$ as the independent variable and $x$ as the dependent variable) and the parabola now appears as the graph of the function $h$.

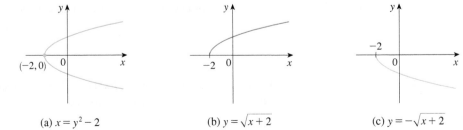

**FIGURE 14**      (a) $x = y^2 - 2$                 (b) $y = \sqrt{x+2}$                 (c) $y = -\sqrt{x+2}$

### Piecewise Defined Functions

The functions in the following four examples are defined by different formulas in different parts of their domains. Such functions are called **piecewise defined functions**.

**ⓋEXAMPLE 7** A function $f$ is defined by

$$f(x) = \begin{cases} 1 - x & \text{if } x \leq -1 \\ x^2 & \text{if } x > -1 \end{cases}$$

Evaluate $f(-2)$, $f(-1)$, and $f(0)$ and sketch the graph.

SOLUTION Remember that a function is a rule. For this particular function the rule is the following: First look at the value of the input $x$. If it happens that $x \leq -1$, then the value of $f(x)$ is $1 - x$. On the other hand, if $x > -1$, then the value of $f(x)$ is $x^2$.

$$\text{Since } -2 \leq -1, \text{ we have } f(-2) = 1 - (-2) = 3.$$

$$\text{Since } -1 \leq -1, \text{ we have } f(-1) = 1 - (-1) = 2.$$

$$\text{Since } 0 > -1, \text{ we have } f(0) = 0^2 = 0.$$

How do we draw the graph of $f$? We observe that if $x \leq -1$, then $f(x) = 1 - x$, so the part of the graph of $f$ that lies to the left of the vertical line $x = -1$ must coincide with the line $y = 1 - x$, which has slope $-1$ and $y$-intercept 1. If $x > -1$, then $f(x) = x^2$, so the part of the graph of $f$ that lies to the right of the line $x = -1$ must coincide with the graph of $y = x^2$, which is a parabola. This enables us to sketch the graph in Figure 15. The solid dot indicates that the point $(-1, 2)$ is included on the graph; the open dot indicates that the point $(-1, 1)$ is excluded from the graph.

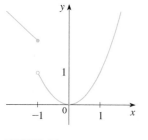

**FIGURE 15**

The next example of a piecewise defined function is the absolute value function. Recall that the **absolute value** of a number $a$, denoted by $|a|$, is the distance from $a$ to 0 on the real number line. Distances are always positive or 0, so we have

For a more extensive review of absolute values, see Appendix A.

$$|a| \geq 0 \qquad \text{for every number } a$$

For example,

$$|3| = 3 \qquad |-3| = 3 \qquad |0| = 0 \qquad |\sqrt{2} - 1| = \sqrt{2} - 1 \qquad |3 - \pi| = \pi - 3$$

In general, we have

$$\boxed{\begin{aligned} |a| &= a && \text{if } a \geq 0 \\ |a| &= -a && \text{if } a < 0 \end{aligned}}$$

(Remember that if $a$ is negative, then $-a$ is positive.)

**EXAMPLE 8** Sketch the graph of the absolute value function $f(x) = |x|$.

SOLUTION From the preceding discussion we know that

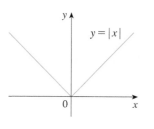

$$|x| = \begin{cases} x & \text{if } x \geq 0 \\ -x & \text{if } x < 0 \end{cases}$$

Using the same method as in Example 7, we see that the graph of $f$ coincides with the line $y = x$ to the right of the $y$-axis and coincides with the line $y = -x$ to the left of the $y$-axis (see Figure 16).

**FIGURE 16**

**EXAMPLE 9** Find a formula for the function $f$ graphed in Figure 17.

**FIGURE 17**

SOLUTION The line through $(0, 0)$ and $(1, 1)$ has slope $m = 1$ and $y$-intercept $b = 0$, so its equation is $y = x$. Thus, for the part of the graph of $f$ that joins $(0, 0)$ to $(1, 1)$, we have

$$f(x) = x \quad \text{if } 0 \leq x \leq 1$$

Point-slope form of the equation of a line:
$$y - y_1 = m(x - x_1)$$
See Appendix B.

The line through $(1, 1)$ and $(2, 0)$ has slope $m = -1$, so its point-slope form is

$$y - 0 = (-1)(x - 2) \quad \text{or} \quad y = 2 - x$$

So we have $\qquad f(x) = 2 - x \quad \text{if } 1 < x \leq 2$

We also see that the graph of $f$ coincides with the $x$-axis for $x > 2$. Putting this information together, we have the following three-piece formula for $f$:

$$f(x) = \begin{cases} x & \text{if } 0 \leq x \leq 1 \\ 2 - x & \text{if } 1 < x \leq 2 \\ 0 & \text{if } x > 2 \end{cases}$$

**EXAMPLE 10** In Example C at the beginning of this section we considered the cost $C(w)$ of mailing a large envelope with weight $w$. In effect, this is a piecewise defined function because, from the table of values on page 13, we have

$$C(w) = \begin{cases} 0.88 & \text{if } 0 < w \leq 1 \\ 1.05 & \text{if } 1 < w \leq 2 \\ 1.22 & \text{if } 2 < w \leq 3 \\ 1.39 & \text{if } 3 < w \leq 4 \\ \vdots & \\ \vdots & \end{cases}$$

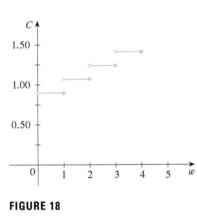

**FIGURE 18**

The graph is shown in Figure 18. You can see why functions similar to this one are called **step functions**—they jump from one value to the next. Such functions will be studied in Chapter 2.

### Symmetry

If a function $f$ satisfies $f(-x) = f(x)$ for every number $x$ in its domain, then $f$ is called an **even function**. For instance, the function $f(x) = x^2$ is even because

$$f(-x) = (-x)^2 = x^2 = f(x)$$

The geometric significance of an even function is that its graph is symmetric with respect to the $y$-axis (see Figure 19). This means that if we have plotted the graph of $f$ for $x \geqslant 0$, we obtain the entire graph simply by reflecting this portion about the $y$-axis.

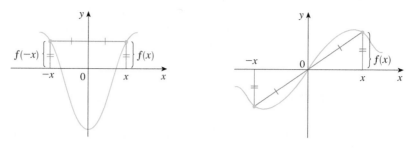

**FIGURE 19**  An even function          **FIGURE 20**  An odd function

If $f$ satisfies $f(-x) = -f(x)$ for every number $x$ in its domain, then $f$ is called an **odd function**. For example, the function $f(x) = x^3$ is odd because

$$f(-x) = (-x)^3 = -x^3 = -f(x)$$

The graph of an odd function is symmetric about the origin (see Figure 20). If we already have the graph of $f$ for $x \geqslant 0$, we can obtain the entire graph by rotating this portion through $180°$ about the origin.

**V   EXAMPLE 11**   Determine whether each of the following functions is even, odd, or neither even nor odd.
(a)  $f(x) = x^5 + x$          (b)  $g(x) = 1 - x^4$          (c)  $h(x) = 2x - x^2$

SOLUTION

(a)
$$f(-x) = (-x)^5 + (-x) = (-1)^5 x^5 + (-x)$$
$$= -x^5 - x = -(x^5 + x)$$
$$= -f(x)$$

Therefore $f$ is an odd function.

(b)
$$g(-x) = 1 - (-x)^4 = 1 - x^4 = g(x)$$

So $g$ is even.

(c)
$$h(-x) = 2(-x) - (-x)^2 = -2x - x^2$$

Since $h(-x) \neq h(x)$ and $h(-x) \neq -h(x)$, we conclude that $h$ is neither even nor odd.

The graphs of the functions in Example 11 are shown in Figure 21. Notice that the graph of $h$ is symmetric neither about the $y$-axis nor about the origin.

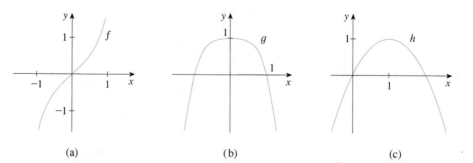

**FIGURE 21**              (a)                          (b)                          (c)

**FIGURE 22**

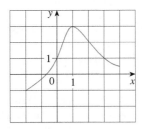

**FIGURE 23**

## Increasing and Decreasing Functions

The graph shown in Figure 22 rises from $A$ to $B$, falls from $B$ to $C$, and rises again from $C$ to $D$. The function $f$ is said to be increasing on the interval $[a, b]$, decreasing on $[b, c]$, and increasing again on $[c, d]$. Notice that if $x_1$ and $x_2$ are any two numbers between $a$ and $b$ with $x_1 < x_2$, then $f(x_1) < f(x_2)$. We use this as the defining property of an increasing function.

---

A function $f$ is called **increasing** on an interval $I$ if

$$f(x_1) < f(x_2) \qquad \text{whenever } x_1 < x_2 \text{ in } I$$

It is called **decreasing** on $I$ if

$$f(x_1) > f(x_2) \qquad \text{whenever } x_1 < x_2 \text{ in } I$$

---

In the definition of an increasing function it is important to realize that the inequality $f(x_1) < f(x_2)$ must be satisfied for *every* pair of numbers $x_1$ and $x_2$ in $I$ with $x_1 < x_2$.

You can see from Figure 23 that the function $f(x) = x^2$ is decreasing on the interval $(-\infty, 0]$ and increasing on the interval $[0, \infty)$.

## 1.1 Exercises

**1.** If $f(x) = x + \sqrt{2 - x}$ and $g(u) = u + \sqrt{2 - u}$, is it true that $f = g$?

**2.** If

$$f(x) = \frac{x^2 - x}{x - 1} \qquad \text{and} \qquad g(x) = x$$

is it true that $f = g$?

**3.** The graph of a function $f$ is given.
(a) State the value of $f(1)$.
(b) Estimate the value of $f(-1)$.
(c) For what values of $x$ is $f(x) = 1$?
(d) Estimate the value of $x$ such that $f(x) = 0$.
(e) State the domain and range of $f$.
(f) On what interval is $f$ increasing?

**4.** The graphs of $f$ and $g$ are given.
(a) State the values of $f(-4)$ and $g(3)$.
(b) For what values of $x$ is $f(x) = g(x)$?

(c) Estimate the solution of the equation $f(x) = -1$.
(d) On what interval is $f$ decreasing?
(e) State the domain and range of $f$.
(f) State the domain and range of $g$.

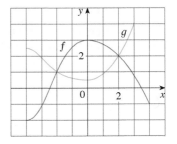

**5.** Figure 1 was recorded by an instrument operated by the California Department of Mines and Geology at the University Hospital of the University of Southern California in Los Angeles. Use it to estimate the range of the vertical ground acceleration function at USC during the Northridge earthquake.

**6.** In this section we discussed examples of ordinary, everyday functions: Population is a function of time, postage cost is a function of weight, water temperature is a function of time. Give three other examples of functions from everyday life that are described verbally. What can you say about the domain and range of each of your functions? If possible, sketch a rough graph of each function.

**1.** Homework Hints available at stewartcalculus.com

**7–10** Determine whether the curve is the graph of a function of $x$. If it is, state the domain and range of the function.

**7.**
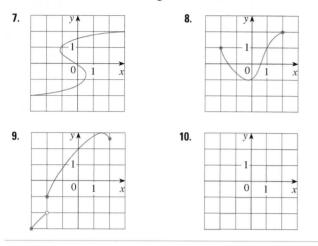

**8.**

**9.**

**10.**

**11.** The graph shown gives the weight of a certain person as a function of age. Describe in words how this person's weight varies over time. What do you think happened when this person was 30 years old?

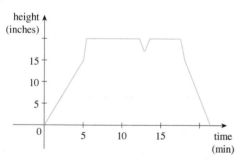

**12.** The graph shows the height of the water in a bathtub as a function of time. Give a verbal description of what you think happened.

**13.** You put some ice cubes in a glass, fill the glass with cold water, and then let the glass sit on a table. Describe how the temperature of the water changes as time passes. Then sketch a rough graph of the temperature of the water as a function of the elapsed time.

**14.** Three runners compete in a 100-meter race. The graph depicts the distance run as a function of time for each runner. Describe in words what the graph tells you about this race. Who won the race? Did each runner finish the race?

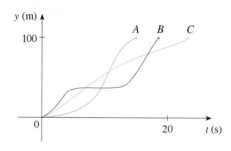

**15.** The graph shows the power consumption for a day in September in San Francisco. ($P$ is measured in megawatts; $t$ is measured in hours starting at midnight.)
   (a) What was the power consumption at 6 AM? At 6 PM?
   (b) When was the power consumption the lowest? When was it the highest? Do these times seem reasonable?

Pacific Gas & Electric

**16.** Sketch a rough graph of the number of hours of daylight as a function of the time of year.

**17.** Sketch a rough graph of the outdoor temperature as a function of time during a typical spring day.

**18.** Sketch a rough graph of the market value of a new car as a function of time for a period of 20 years. Assume the car is well maintained.

**19.** Sketch the graph of the amount of a particular brand of coffee sold by a store as a function of the price of the coffee.

**20.** You place a frozen pie in an oven and bake it for an hour. Then you take it out and let it cool before eating it. Describe how the temperature of the pie changes as time passes. Then sketch a rough graph of the temperature of the pie as a function of time.

**21.** A homeowner mows the lawn every Wednesday afternoon. Sketch a rough graph of the height of the grass as a function of time over the course of a four-week period.

**22.** An airplane takes off from an airport and lands an hour later at another airport, 400 miles away. If $t$ represents the time in minutes since the plane has left the terminal building, let $x(t)$ be

the horizontal distance traveled and $y(t)$ be the altitude of the plane.
(a) Sketch a possible graph of $x(t)$.
(b) Sketch a possible graph of $y(t)$.
(c) Sketch a possible graph of the ground speed.
(d) Sketch a possible graph of the vertical velocity.

**23.** The number $N$ (in millions) of US cellular phone subscribers is shown in the table. (Midyear estimates are given.)

| $t$ | 1996 | 1998 | 2000 | 2002 | 2004 | 2006 |
|---|---|---|---|---|---|---|
| $N$ | 44 | 69 | 109 | 141 | 182 | 233 |

(a) Use the data to sketch a rough graph of $N$ as a function of $t$.
(b) Use your graph to estimate the number of cell-phone subscribers at midyear in 2001 and 2005.

**24.** Temperature readings $T$ (in °F) were recorded every two hours from midnight to 2:00 PM in Phoenix on September 10, 2008. The time $t$ was measured in hours from midnight.

| $t$ | 0 | 2 | 4 | 6 | 8 | 10 | 12 | 14 |
|---|---|---|---|---|---|---|---|---|
| $T$ | 82 | 75 | 74 | 75 | 84 | 90 | 93 | 94 |

(a) Use the readings to sketch a rough graph of $T$ as a function of $t$.
(b) Use your graph to estimate the temperature at 9:00 AM.

**25.** If $f(x) = 3x^2 - x + 2$, find $f(2)$, $f(-2)$, $f(a)$, $f(-a)$, $f(a + 1)$, $2f(a)$, $f(2a)$, $f(a^2)$, $[f(a)]^2$, and $f(a + h)$.

**26.** A spherical balloon with radius $r$ inches has volume $V(r) = \frac{4}{3}\pi r^3$. Find a function that represents the amount of air required to inflate the balloon from a radius of $r$ inches to a radius of $r + 1$ inches.

**27–30** Evaluate the difference quotient for the given function. Simplify your answer.

**27.** $f(x) = 4 + 3x - x^2$,    $\dfrac{f(3 + h) - f(3)}{h}$

**28.** $f(x) = x^3$,    $\dfrac{f(a + h) - f(a)}{h}$

**29.** $f(x) = \dfrac{1}{x}$,    $\dfrac{f(x) - f(a)}{x - a}$

**30.** $f(x) = \dfrac{x + 3}{x + 1}$,    $\dfrac{f(x) - f(1)}{x - 1}$

**31–37** Find the domain of the function.

**31.** $f(x) = \dfrac{x + 4}{x^2 - 9}$

**32.** $f(x) = \dfrac{2x^3 - 5}{x^2 + x - 6}$

**33.** $f(t) = \sqrt[3]{2t - 1}$

**34.** $g(t) = \sqrt{3 - t} - \sqrt{2 + t}$

**35.** $h(x) = \dfrac{1}{\sqrt[4]{x^2 - 5x}}$

**36.** $f(u) = \dfrac{u + 1}{1 + \dfrac{1}{u + 1}}$

**37.** $F(p) = \sqrt{2 - \sqrt{p}}$

**38.** Find the domain and range and sketch the graph of the function $h(x) = \sqrt{4 - x^2}$.

**39–50** Find the domain and sketch the graph of the function.

**39.** $f(x) = 2 - 0.4x$

**40.** $F(x) = x^2 - 2x + 1$

**41.** $f(t) = 2t + t^2$

**42.** $H(t) = \dfrac{4 - t^2}{2 - t}$

**43.** $g(x) = \sqrt{x - 5}$

**44.** $F(x) = |2x + 1|$

**45.** $G(x) = \dfrac{3x + |x|}{x}$

**46.** $g(x) = |x| - x$

**47.** $f(x) = \begin{cases} x + 2 & \text{if } x < 0 \\ 1 - x & \text{if } x \geq 0 \end{cases}$

**48.** $f(x) = \begin{cases} 3 - \frac{1}{2}x & \text{if } x \leq 2 \\ 2x - 5 & \text{if } x > 2 \end{cases}$

**49.** $f(x) = \begin{cases} x + 2 & \text{if } x \leq -1 \\ x^2 & \text{if } x > -1 \end{cases}$

**50.** $f(x) = \begin{cases} x + 9 & \text{if } x < -3 \\ -2x & \text{if } |x| \leq 3 \\ -6 & \text{if } x > 3 \end{cases}$

**51–56** Find an expression for the function whose graph is the given curve.

**51.** The line segment joining the points $(1, -3)$ and $(5, 7)$

**52.** The line segment joining the points $(-5, 10)$ and $(7, -10)$

**53.** The bottom half of the parabola $x + (y - 1)^2 = 0$

**54.** The top half of the circle $x^2 + (y - 2)^2 = 4$

**55.**

**56.**

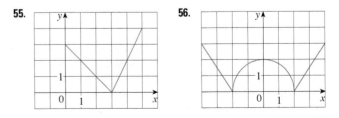

**57–61** Find a formula for the described function and state its domain.

**57.** A rectangle has perimeter 20 m. Express the area of the rectangle as a function of the length of one of its sides.

**58.** A rectangle has area 16 m². Express the perimeter of the rectangle as a function of the length of one of its sides.

**59.** Express the area of an equilateral triangle as a function of the length of a side.

**60.** Express the surface area of a cube as a function of its volume.

**61.** An open rectangular box with volume 2 m³ has a square base. Express the surface area of the box as a function of the length of a side of the base.

**62.** A Norman window has the shape of a rectangle surmounted by a semicircle. If the perimeter of the window is 30 ft, express the area $A$ of the window as a function of the width $x$ of the window.

**63.** A box with an open top is to be constructed from a rectangular piece of cardboard with dimensions 12 in. by 20 in. by cutting out equal squares of side $x$ at each corner and then folding up the sides as in the figure. Express the volume $V$ of the box as a function of $x$.

**64.** A cell phone plan has a basic charge of $35 a month. The plan includes 400 free minutes and charges 10 cents for each additional minute of usage. Write the monthly cost $C$ as a function of the number $x$ of minutes used and graph $C$ as a function of $x$ for $0 \le x \le 600$.

**65.** In a certain state the maximum speed permitted on freeways is 65 mi/h and the minimum speed is 40 mi/h. The fine for violating these limits is $15 for every mile per hour above the maximum speed or below the minimum speed. Express the amount of the fine $F$ as a function of the driving speed $x$ and graph $F(x)$ for $0 \le x \le 100$.

**66.** An electricity company charges its customers a base rate of $10 a month, plus 6 cents per kilowatt-hour (kWh) for the first 1200 kWh and 7 cents per kWh for all usage over 1200 kWh. Express the monthly cost $E$ as a function of the amount $x$ of electricity used. Then graph the function $E$ for $0 \le x \le 2000$.

**67.** In a certain country, income tax is assessed as follows. There is no tax on income up to $10,000. Any income over $10,000 is taxed at a rate of 10%, up to an income of $20,000. Any income over $20,000 is taxed at 15%.
(a) Sketch the graph of the tax rate $R$ as a function of the income $I$.
(b) How much tax is assessed on an income of $14,000? On $26,000?
(c) Sketch the graph of the total assessed tax $T$ as a function of the income $I$.

**68.** The functions in Example 10 and Exercise 67 are called *step functions* because their graphs look like stairs. Give two other examples of step functions that arise in everyday life.

**69–70** Graphs of $f$ and $g$ are shown. Decide whether each function is even, odd, or neither. Explain your reasoning.

**69.**                          **70.**
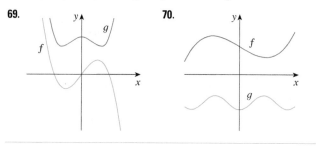

**71.** (a) If the point $(5, 3)$ is on the graph of an even function, what other point must also be on the graph?
(b) If the point $(5, 3)$ is on the graph of an odd function, what other point must also be on the graph?

**72.** A function $f$ has domain $[-5, 5]$ and a portion of its graph is shown.
(a) Complete the graph of $f$ if it is known that $f$ is even.
(b) Complete the graph of $f$ if it is known that $f$ is odd.

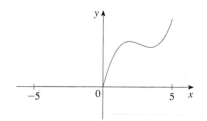

**73–78** Determine whether $f$ is even, odd, or neither. If you have a graphing calculator, use it to check your answer visually.

**73.** $f(x) = \dfrac{x}{x^2 + 1}$     **74.** $f(x) = \dfrac{x^2}{x^4 + 1}$

**75.** $f(x) = \dfrac{x}{x + 1}$     **76.** $f(x) = x|x|$

**77.** $f(x) = 1 + 3x^2 - x^4$     **78.** $f(x) = 1 + 3x^3 - x^5$

**79.** If $f$ and $g$ are both even functions, is $f + g$ even? If $f$ and $g$ are both odd functions, is $f + g$ odd? What if $f$ is even and $g$ is odd? Justify your answers.

**80.** If $f$ and $g$ are both even functions, is the product $fg$ even? If $f$ and $g$ are both odd functions, is $fg$ odd? What if $f$ is even and $g$ is odd? Justify your answers.

## 1.2 Mathematical Models: A Catalog of Essential Functions

A **mathematical model** is a mathematical description (often by means of a function or an equation) of a real-world phenomenon such as the size of a population, the demand for a product, the speed of a falling object, the concentration of a product in a chemical reaction, the life expectancy of a person at birth, or the cost of emission reductions. The purpose of the model is to understand the phenomenon and perhaps to make predictions about future behavior.

Figure 1 illustrates the process of mathematical modeling. Given a real-world problem, our first task is to formulate a mathematical model by identifying and naming the independent and dependent variables and making assumptions that simplify the phenomenon enough to make it mathematically tractable. We use our knowledge of the physical situation and our mathematical skills to obtain equations that relate the variables. In situations where there is no physical law to guide us, we may need to collect data (either from a library or the Internet or by conducting our own experiments) and examine the data in the form of a table in order to discern patterns. From this numerical representation of a function we may wish to obtain a graphical representation by plotting the data. The graph might even suggest a suitable algebraic formula in some cases.

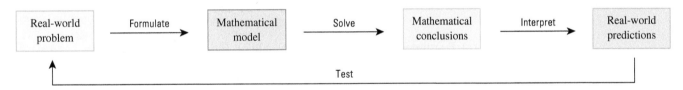

**FIGURE 1** The modeling process

The second stage is to apply the mathematics that we know (such as the calculus that will be developed throughout this book) to the mathematical model that we have formulated in order to derive mathematical conclusions. Then, in the third stage, we take those mathematical conclusions and interpret them as information about the original real-world phenomenon by way of offering explanations or making predictions. The final step is to test our predictions by checking against new real data. If the predictions don't compare well with reality, we need to refine our model or to formulate a new model and start the cycle again.

A mathematical model is never a completely accurate representation of a physical situation—it is an *idealization*. A good model simplifies reality enough to permit mathematical calculations but is accurate enough to provide valuable conclusions. It is important to realize the limitations of the model. In the end, Mother Nature has the final say.

There are many different types of functions that can be used to model relationships observed in the real world. In what follows, we discuss the behavior and graphs of these functions and give examples of situations appropriately modeled by such functions.

### Linear Models

The coordinate geometry of lines is reviewed in Appendix B.

When we say that $y$ is a **linear function** of $x$, we mean that the graph of the function is a line, so we can use the slope-intercept form of the equation of a line to write a formula for the function as

$$y = f(x) = mx + b$$

where $m$ is the slope of the line and $b$ is the $y$-intercept.

A characteristic feature of linear functions is that they grow at a constant rate. For instance, Figure 2 shows a graph of the linear function $f(x) = 3x - 2$ and a table of sample values. Notice that whenever $x$ increases by 0.1, the value of $f(x)$ increases by 0.3. So $f(x)$ increases three times as fast as $x$. Thus the slope of the graph $y = 3x - 2$, namely 3, can be interpreted as the rate of change of $y$ with respect to $x$.

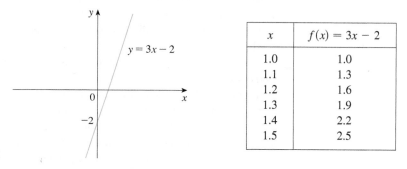

| $x$ | $f(x) = 3x - 2$ |
|-----|-----------------|
| 1.0 | 1.0 |
| 1.1 | 1.3 |
| 1.2 | 1.6 |
| 1.3 | 1.9 |
| 1.4 | 2.2 |
| 1.5 | 2.5 |

**FIGURE 2**

**V  EXAMPLE 1**

(a) As dry air moves upward, it expands and cools. If the ground temperature is 20°C and the temperature at a height of 1 km is 10°C, express the temperature $T$ (in °C) as a function of the height $h$ (in kilometers), assuming that a linear model is appropriate.

(b) Draw the graph of the function in part (a). What does the slope represent?

(c) What is the temperature at a height of 2.5 km?

SOLUTION

(a) Because we are assuming that $T$ is a linear function of $h$, we can write

$$T = mh + b$$

We are given that $T = 20$ when $h = 0$, so

$$20 = m \cdot 0 + b = b$$

In other words, the $y$-intercept is $b = 20$.

We are also given that $T = 10$ when $h = 1$, so

$$10 = m \cdot 1 + 20$$

The slope of the line is therefore $m = 10 - 20 = -10$ and the required linear function is

$$T = -10h + 20$$

(b) The graph is sketched in Figure 3. The slope is $m = -10$°C/km, and this represents the rate of change of temperature with respect to height.

(c) At a height of $h = 2.5$ km, the temperature is

$$T = -10(2.5) + 20 = -5°C$$

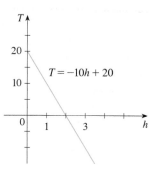

**FIGURE 3**

If there is no physical law or principle to help us formulate a model, we construct an **empirical model**, which is based entirely on collected data. We seek a curve that "fits" the data in the sense that it captures the basic trend of the data points.

**V** **EXAMPLE 2** Table 1 lists the average carbon dioxide level in the atmosphere, measured in parts per million at Mauna Loa Observatory from 1980 to 2008. Use the data in Table 1 to find a model for the carbon dioxide level.

SOLUTION We use the data in Table 1 to make the scatter plot in Figure 4, where $t$ represents time (in years) and $C$ represents the $CO_2$ level (in parts per million, ppm).

**TABLE 1**

| Year | CO₂ level (in ppm) | Year | CO₂ level (in ppm) |
|------|--------------------|------|--------------------|
| 1980 | 338.7 | 1996 | 362.4 |
| 1982 | 341.2 | 1998 | 366.5 |
| 1984 | 344.4 | 2000 | 369.4 |
| 1986 | 347.2 | 2002 | 373.2 |
| 1988 | 351.5 | 2004 | 377.5 |
| 1990 | 354.2 | 2006 | 381.9 |
| 1992 | 356.3 | 2008 | 385.6 |
| 1994 | 358.6 | | |

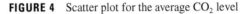

**FIGURE 4**  Scatter plot for the average $CO_2$ level

Notice that the data points appear to lie close to a straight line, so it's natural to choose a linear model in this case. But there are many possible lines that approximate these data points, so which one should we use? One possibility is the line that passes through the first and last data points. The slope of this line is

$$\frac{385.6 - 338.7}{2008 - 1980} = \frac{46.9}{28} = 1.675$$

and its equation is

$$C - 338.7 = 1.675(t - 1980)$$

or

$$\boxed{1} \qquad C = 1.675t - 2977.8$$

Equation 1 gives one possible linear model for the carbon dioxide level; it is graphed in Figure 5.

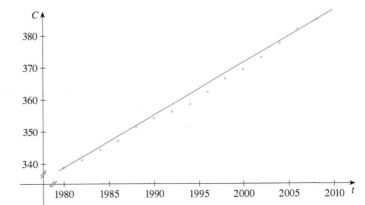

**FIGURE 5**
Linear model through
first and last data points

A computer or graphing calculator finds the regression line by the method of **least squares**, which is to minimize the sum of the squares of the vertical distances between the data points and the line. The details are explained in Section 14.7.

Notice that our model gives values higher than most of the actual $CO_2$ levels. A better linear model is obtained by a procedure from statistics called *linear regression*. If we use a graphing calculator, we enter the data from Table 1 into the data editor and choose the linear regression command. (With Maple we use the fit[leastsquare] command in the stats package; with Mathematica we use the Fit command.) The machine gives the slope and y-intercept of the regression line as

$$m = 1.65429 \qquad b = -2938.07$$

So our least squares model for the $CO_2$ level is

**2**

$$C = 1.65429t - 2938.07$$

In Figure 6 we graph the regression line as well as the data points. Comparing with Figure 5, we see that it gives a better fit than our previous linear model.

**FIGURE 6**
The regression line

**V    EXAMPLE 3**    Use the linear model given by Equation 2 to estimate the average $CO_2$ level for 1987 and to predict the level for the year 2015. According to this model, when will the $CO_2$ level exceed 420 parts per million?

SOLUTION    Using Equation 2 with $t = 1987$, we estimate that the average $CO_2$ level in 1987 was

$$C(1987) = (1.65429)(1987) - 2938.07 \approx 349.00$$

This is an example of *interpolation* because we have estimated a value *between* observed values. (In fact, the Mauna Loa Observatory reported that the average $CO_2$ level in 1987 was 348.93 ppm, so our estimate is quite accurate.)

With $t = 2015$, we get

$$C(2015) = (1.65429)(2015) - 2938.07 \approx 395.32$$

So we predict that the average $CO_2$ level in the year 2015 will be 395.3 ppm. This is an example of *extrapolation* because we have predicted a value *outside* the region of observations. Consequently, we are far less certain about the accuracy of our prediction.

Using Equation 2, we see that the $CO_2$ level exceeds 420 ppm when

$$1.65429t - 2938.07 > 420$$

Solving this inequality, we get

$$t > \frac{3358.07}{1.65429} \approx 2029.92$$

We therefore predict that the $CO_2$ level will exceed 420 ppm by the year 2030. This prediction is risky because it involves a time quite remote from our observations. In fact, we see from Figure 6 that the trend has been for $CO_2$ levels to increase rather more rapidly in recent years, so the level might exceed 420 ppm well before 2030.

## Polynomials

A function $P$ is called a **polynomial** if

$$P(x) = a_n x^n + a_{n-1} x^{n-1} + \cdots + a_2 x^2 + a_1 x + a_0$$

where $n$ is a nonnegative integer and the numbers $a_0, a_1, a_2, \ldots, a_n$ are constants called the **coefficients** of the polynomial. The domain of any polynomial is $\mathbb{R} = (-\infty, \infty)$. If the leading coefficient $a_n \neq 0$, then the **degree** of the polynomial is $n$. For example, the function

$$P(x) = 2x^6 - x^4 + \tfrac{2}{5}x^3 + \sqrt{2}$$

is a polynomial of degree 6.

A polynomial of degree 1 is of the form $P(x) = mx + b$ and so it is a linear function. A polynomial of degree 2 is of the form $P(x) = ax^2 + bx + c$ and is called a **quadratic function**. Its graph is always a parabola obtained by shifting the parabola $y = ax^2$, as we will see in the next section. The parabola opens upward if $a > 0$ and downward if $a < 0$. (See Figure 7.)

**FIGURE 7**

The graphs of quadratic functions are parabolas.

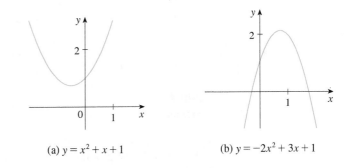

(a) $y = x^2 + x + 1$

(b) $y = -2x^2 + 3x + 1$

A polynomial of degree 3 is of the form

$$P(x) = ax^3 + bx^2 + cx + d \qquad a \neq 0$$

and is called a **cubic function**. Figure 8 shows the graph of a cubic function in part (a) and graphs of polynomials of degrees 4 and 5 in parts (b) and (c). We will see later why the graphs have these shapes.

**FIGURE 8**

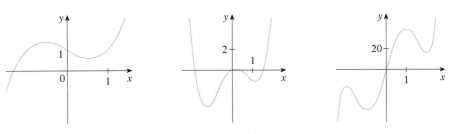

(a) $y = x^3 - x + 1$

(b) $y = x^4 - 3x^2 + x$

(c) $y = 3x^5 - 25x^3 + 60x$

Polynomials are commonly used to model various quantities that occur in the natural and social sciences. For instance, in Section 2.7 we will explain why economists often use a polynomial $P(x)$ to represent the cost of producing $x$ units of a commodity. In the following example we use a quadratic function to model the fall of a ball.

**TABLE 2**

| Time (seconds) | Height (meters) |
|---|---|
| 0 | 450 |
| 1 | 445 |
| 2 | 431 |
| 3 | 408 |
| 4 | 375 |
| 5 | 332 |
| 6 | 279 |
| 7 | 216 |
| 8 | 143 |
| 9 | 61 |

 A ball is dropped from the upper observation deck of the CN Tower, 450 m above the ground, and its height $h$ above the ground is recorded at 1-second intervals in Table 2. Find a model to fit the data and use the model to predict the time at which the ball hits the ground.

SOLUTION   We draw a scatter plot of the data in Figure 9 and observe that a linear model is inappropriate. But it looks as if the data points might lie on a parabola, so we try a quadratic model instead. Using a graphing calculator or computer algebra system (which uses the least squares method), we obtain the following quadratic model:

$$\boxed{3} \qquad h = 449.36 + 0.96t - 4.90t^2$$

**FIGURE 9**
Scatter plot for a falling ball

**FIGURE 10**
Quadratic model for a falling ball

In Figure 10 we plot the graph of Equation 3 together with the data points and see that the quadratic model gives a very good fit.

The ball hits the ground when $h = 0$, so we solve the quadratic equation

$$-4.90t^2 + 0.96t + 449.36 = 0$$

The quadratic formula gives

$$t = \frac{-0.96 \pm \sqrt{(0.96)^2 - 4(-4.90)(449.36)}}{2(-4.90)}$$

The positive root is $t \approx 9.67$, so we predict that the ball will hit the ground after about 9.7 seconds.

## Power Functions

A function of the form $f(x) = x^a$, where $a$ is a constant, is called a **power function**. We consider several cases.

(i) $a = n$, **where $n$ is a positive integer**

The graphs of $f(x) = x^n$ for $n = 1, 2, 3, 4$, and 5 are shown in Figure 11. (These are polynomials with only one term.) We already know the shape of the graphs of $y = x$ (a line through the origin with slope 1) and $y = x^2$ [a parabola, see Example 2(b) in Section 1.1].

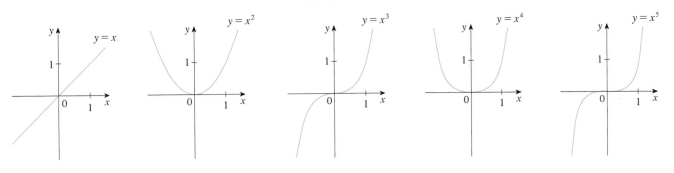

**FIGURE 11**  Graphs of $f(x) = x^n$ for $n = 1, 2, 3, 4, 5$

The general shape of the graph of $f(x) = x^n$ depends on whether $n$ is even or odd. If $n$ is even, then $f(x) = x^n$ is an even function and its graph is similar to the parabola $y = x^2$. If $n$ is odd, then $f(x) = x^n$ is an odd function and its graph is similar to that of $y = x^3$. Notice from Figure 12, however, that as $n$ increases, the graph of $y = x^n$ becomes flatter near 0 and steeper when $|x| \geq 1$. (If $x$ is small, then $x^2$ is smaller, $x^3$ is even smaller, $x^4$ is smaller still, and so on.)

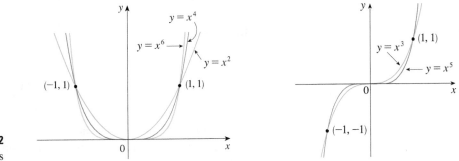

**FIGURE 12**

Families of power functions

(ii) $a = 1/n$, **where $n$ is a positive integer**

The function $f(x) = x^{1/n} = \sqrt[n]{x}$ is a **root function**. For $n = 2$ it is the square root function $f(x) = \sqrt{x}$, whose domain is $[0, \infty)$ and whose graph is the upper half of the parabola $x = y^2$. [See Figure 13(a).] For other even values of $n$, the graph of $y = \sqrt[n]{x}$ is similar to that of $y = \sqrt{x}$. For $n = 3$ we have the cube root function $f(x) = \sqrt[3]{x}$ whose domain is $\mathbb{R}$ (recall that every real number has a cube root) and whose graph is shown in Figure 13(b). The graph of $y = \sqrt[n]{x}$ for $n$ odd $(n > 3)$ is similar to that of $y = \sqrt[3]{x}$.

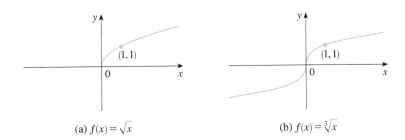

**FIGURE 13**

Graphs of root functions

(a) $f(x) = \sqrt{x}$

(b) $f(x) = \sqrt[3]{x}$

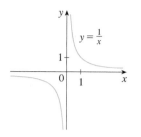

**FIGURE 14**
The reciprocal function

(iii) $a = -1$

The graph of the **reciprocal function** $f(x) = x^{-1} = 1/x$ is shown in Figure 14. Its graph has the equation $y = 1/x$, or $xy = 1$, and is a hyperbola with the coordinate axes as its asymptotes. This function arises in physics and chemistry in connection with Boyle's Law, which says that, when the temperature is constant, the volume $V$ of a gas is inversely proportional to the pressure $P$:

$$V = \frac{C}{P}$$

where $C$ is a constant. Thus the graph of $V$ as a function of $P$ (see Figure 15) has the same general shape as the right half of Figure 14.

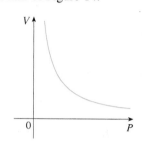

**FIGURE 15**
Volume as a function of pressure
at constant temperature

Power functions are also used to model species-area relationships (Exercises 26–27), illumination as a function of a distance from a light source (Exercise 25), and the period of revolution of a planet as a function of its distance from the sun (Exercise 28).

### Rational Functions

A **rational function** $f$ is a ratio of two polynomials:

$$f(x) = \frac{P(x)}{Q(x)}$$

where $P$ and $Q$ are polynomials. The domain consists of all values of $x$ such that $Q(x) \neq 0$. A simple example of a rational function is the function $f(x) = 1/x$, whose domain is $\{x \mid x \neq 0\}$; this is the reciprocal function graphed in Figure 14. The function

$$f(x) = \frac{2x^4 - x^2 + 1}{x^2 - 4}$$

is a rational function with domain $\{x \mid x \neq \pm 2\}$. Its graph is shown in Figure 16.

**FIGURE 16**
$f(x) = \dfrac{2x^4 - x^2 + 1}{x^2 - 4}$

### Algebraic Functions

A function $f$ is called an **algebraic function** if it can be constructed using algebraic operations (such as addition, subtraction, multiplication, division, and taking roots) starting with polynomials. Any rational function is automatically an algebraic function. Here are two more examples:

$$f(x) = \sqrt{x^2 + 1} \qquad g(x) = \frac{x^4 - 16x^2}{x + \sqrt{x}} + (x - 2)\sqrt[3]{x + 1}$$

When we sketch algebraic functions in Chapter 3, we will see that their graphs can assume a variety of shapes. Figure 17 illustrates some of the possibilities.

**FIGURE 17**  (a) $f(x) = x\sqrt{x+3}$     (b) $g(x) = \sqrt[4]{x^2 - 25}$     (c) $h(x) = x^{2/3}(x-2)^2$

An example of an algebraic function occurs in the theory of relativity. The mass of a particle with velocity $v$ is

$$m = f(v) = \frac{m_0}{\sqrt{1 - v^2/c^2}}$$

where $m_0$ is the rest mass of the particle and $c = 3.0 \times 10^5$ km/s is the speed of light in a vacuum.

## Trigonometric Functions

The Reference Pages are located at the front and back of the book.

Trigonometry and the trigonometric functions are reviewed on Reference Page 2 and also in Appendix D. In calculus the convention is that radian measure is always used (except when otherwise indicated). For example, when we use the function $f(x) = \sin x$, it is understood that $\sin x$ means the sine of the angle whose radian measure is $x$. Thus the graphs of the sine and cosine functions are as shown in Figure 18.

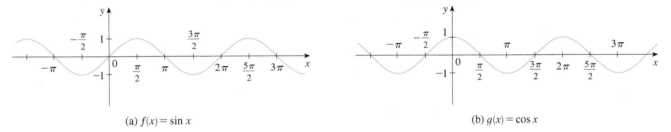

(a) $f(x) = \sin x$                              (b) $g(x) = \cos x$

**FIGURE 18**

Notice that for both the sine and cosine functions the domain is $(-\infty, \infty)$ and the range is the closed interval $[-1, 1]$. Thus, for all values of $x$, we have

$$-1 \leqslant \sin x \leqslant 1 \qquad -1 \leqslant \cos x \leqslant 1$$

or, in terms of absolute values,

$$\left| \sin x \right| \leqslant 1 \qquad \left| \cos x \right| \leqslant 1$$

Also, the zeros of the sine function occur at the integer multiples of $\pi$; that is,

$$\sin x = 0 \qquad \text{when} \qquad x = n\pi \quad n \text{ an integer}$$

An important property of the sine and cosine functions is that they are periodic functions and have period $2\pi$. This means that, for all values of $x$,

$$\sin(x + 2\pi) = \sin x \qquad \cos(x + 2\pi) = \cos x$$

The periodic nature of these functions makes them suitable for modeling repetitive phenomena such as tides, vibrating springs, and sound waves. For instance, in Example 4 in Section 1.3 we will see that a reasonable model for the number of hours of daylight in Philadelphia $t$ days after January 1 is given by the function

$$L(t) = 12 + 2.8 \sin\left[\frac{2\pi}{365}(t - 80)\right]$$

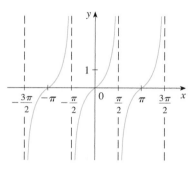

**FIGURE 19**

$y = \tan x$

The tangent function is related to the sine and cosine functions by the equation

$$\tan x = \frac{\sin x}{\cos x}$$

and its graph is shown in Figure 19. It is undefined whenever $\cos x = 0$, that is, when $x = \pm\pi/2, \pm 3\pi/2, \ldots$. Its range is $(-\infty, \infty)$. Notice that the tangent function has period $\pi$:

$$\tan(x + \pi) = \tan x \qquad \text{for all } x$$

The remaining three trigonometric functions (cosecant, secant, and cotangent) are the reciprocals of the sine, cosine, and tangent functions. Their graphs are shown in Appendix D.

## Exponential Functions

The **exponential functions** are the functions of the form $f(x) = a^x$, where the base $a$ is a positive constant. The graphs of $y = 2^x$ and $y = (0.5)^x$ are shown in Figure 20. In both cases the domain is $(-\infty, \infty)$ and the range is $(0, \infty)$.

Exponential functions will be studied in detail in Chapter 6, and we will see that they are useful for modeling many natural phenomena, such as population growth (if $a > 1$) and radioactive decay (if $a < 1$).

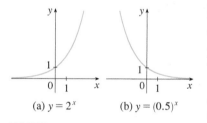

(a) $y = 2^x$      (b) $y = (0.5)^x$

**FIGURE 20**

## Logarithmic Functions

The **logarithmic functions** $f(x) = \log_a x$, where the base $a$ is a positive constant, are the inverse functions of the exponential functions. They will be studied in Chapter 6. Figure 21 shows the graphs of four logarithmic functions with various bases. In each case the domain is $(0, \infty)$, the range is $(-\infty, \infty)$, and the function increases slowly when $x > 1$.

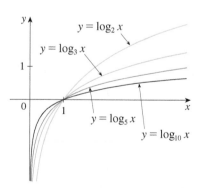

**FIGURE 21**

**EXAMPLE 5** Classify the following functions as one of the types of functions that we have discussed.

(a) $f(x) = 5^x$                     (b) $g(x) = x^5$

(c) $h(x) = \dfrac{1 + x}{1 - \sqrt{x}}$          (d) $u(t) = 1 - t + 5t^4$

SOLUTION

(a) $f(x) = 5^x$ is an exponential function. (The $x$ is the exponent.)

(b) $g(x) = x^5$ is a power function. (The $x$ is the base.) We could also consider it to be a polynomial of degree 5.

(c) $h(x) = \dfrac{1 + x}{1 - \sqrt{x}}$ is an algebraic function.

(d) $u(t) = 1 - t + 5t^4$ is a polynomial of degree 4.

## 1.2 Exercises

**1–2** Classify each function as a power function, root function, polynomial (state its degree), rational function, algebraic function, trigonometric function, exponential function, or logarithmic function.

**1.** (a) $f(x) = \log_2 x$

(b) $g(x) = \sqrt[4]{x}$

(c) $h(x) = \dfrac{2x^3}{1 - x^2}$

(d) $u(t) = 1 - 1.1t + 2.54t^2$

(e) $v(t) = 5^t$

(f) $w(\theta) = \sin\theta\cos^2\theta$

**2.** (a) $y = \pi^x$

(b) $y = x^\pi$

(c) $y = x^2(2 - x^3)$

(d) $y = \tan t - \cos t$

(e) $y = \dfrac{s}{1 + s}$

(f) $y = \dfrac{\sqrt{x^3 - 1}}{1 + \sqrt[3]{x}}$

**3–4** Match each equation with its graph. Explain your choices. (Don't use a computer or graphing calculator.)

**3.** (a) $y = x^2$    (b) $y = x^5$    (c) $y = x^8$

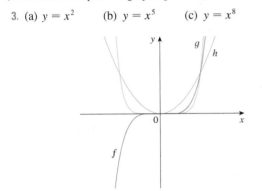

**4.** (a) $y = 3x$

(b) $y = 3^x$

(c) $y = x^3$

(d) $y = \sqrt[3]{x}$

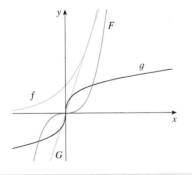

**5.** (a) Find an equation for the family of linear functions with slope 2 and sketch several members of the family.

(b) Find an equation for the family of linear functions such that $f(2) = 1$ and sketch several members of the family.

(c) Which function belongs to both families?

**6.** What do all members of the family of linear functions $f(x) = 1 + m(x + 3)$ have in common? Sketch several members of the family.

**7.** What do all members of the family of linear functions $f(x) = c - x$ have in common? Sketch several members of the family.

**8.** Find expressions for the quadratic functions whose graphs are shown.

**9.** Find an expression for a cubic function $f$ if $f(1) = 6$ and $f(-1) = f(0) = f(2) = 0$.

**10.** Recent studies indicate that the average surface temperature of the earth has been rising steadily. Some scientists have modeled the temperature by the linear function $T = 0.02t + 8.50$, where $T$ is temperature in °C and $t$ represents years since 1900.

(a) What do the slope and $T$-intercept represent?

(b) Use the equation to predict the average global surface temperature in 2100.

**11.** If the recommended adult dosage for a drug is $D$ ( in mg), then to determine the appropriate dosage $c$ for a child of age $a$, pharmacists use the equation $c = 0.0417D(a + 1)$. Suppose the dosage for an adult is 200 mg.

(a) Find the slope of the graph of $c$. What does it represent?

(b) What is the dosage for a newborn?

**12.** The manager of a weekend flea market knows from past experience that if he charges $x$ dollars for a rental space at the market, then the number $y$ of spaces he can rent is given by the equation $y = 200 - 4x$.

(a) Sketch a graph of this linear function. (Remember that the rental charge per space and the number of spaces rented can't be negative quantities.)

(b) What do the slope, the $y$-intercept, and the $x$-intercept of the graph represent?

**13.** The relationship between the Fahrenheit $(F)$ and Celsius $(C)$ temperature scales is given by the linear function $F = \frac{9}{5}C + 32$.

(a) Sketch a graph of this function.

(b) What is the slope of the graph and what does it represent? What is the $F$-intercept and what does it represent?

**14.** Jason leaves Detroit at 2:00 PM and drives at a constant speed west along I-96. He passes Ann Arbor, 40 mi from Detroit, at 2:50 PM.

(a) Express the distance traveled in terms of the time elapsed.

Graphing calculator or computer required    **1.** Homework Hints available at stewartcalculus.com

(b) Draw the graph of the equation in part (a).
(c) What is the slope of this line? What does it represent?

**15.** Biologists have noticed that the chirping rate of crickets of a certain species is related to temperature, and the relationship appears to be very nearly linear. A cricket produces 113 chirps per minute at 70°F and 173 chirps per minute at 80°F.
(a) Find a linear equation that models the temperature $T$ as a function of the number of chirps per minute $N$.
(b) What is the slope of the graph? What does it represent?
(c) If the crickets are chirping at 150 chirps per minute, estimate the temperature.

**16.** The manager of a furniture factory finds that it costs $2200 to manufacture 100 chairs in one day and $4800 to produce 300 chairs in one day.
(a) Express the cost as a function of the number of chairs produced, assuming that it is linear. Then sketch the graph.
(b) What is the slope of the graph and what does it represent?
(c) What is the $y$-intercept of the graph and what does it represent?

**17.** At the surface of the ocean, the water pressure is the same as the air pressure above the water, 15 lb/in². Below the surface, the water pressure increases by 4.34 lb/in² for every 10 ft of descent.
(a) Express the water pressure as a function of the depth below the ocean surface.
(b) At what depth is the pressure 100 lb/in²?

**18.** The monthly cost of driving a car depends on the number of miles driven. Lynn found that in May it cost her $380 to drive 480 mi and in June it cost her $460 to drive 800 mi.
(a) Express the monthly cost $C$ as a function of the distance driven $d$, assuming that a linear relationship gives a suitable model.
(b) Use part (a) to predict the cost of driving 1500 miles per month.
(c) Draw the graph of the linear function. What does the slope represent?
(d) What does the $C$-intercept represent?
(e) Why does a linear function give a suitable model in this situation?

**19–20** For each scatter plot, decide what type of function you might choose as a model for the data. Explain your choices.

**19.** (a)            (b)

**20.** (a)            (b)

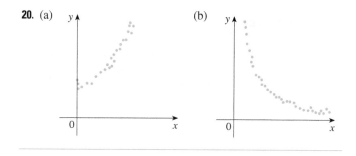

**21.** The table shows (lifetime) peptic ulcer rates (per 100 population) for various family incomes as reported by the National Health Interview Survey.

| Income | Ulcer rate (per 100 population) |
|---|---|
| $4,000 | 14.1 |
| $6,000 | 13.0 |
| $8,000 | 13.4 |
| $12,000 | 12.5 |
| $16,000 | 12.0 |
| $20,000 | 12.4 |
| $30,000 | 10.5 |
| $45,000 | 9.4 |
| $60,000 | 8.2 |

(a) Make a scatter plot of these data and decide whether a linear model is appropriate.
(b) Find and graph a linear model using the first and last data points.
(c) Find and graph the least squares regression line.
(d) Use the linear model in part (c) to estimate the ulcer rate for an income of $25,000.
(e) According to the model, how likely is someone with an income of $80,000 to suffer from peptic ulcers?
(f) Do you think it would be reasonable to apply the model to someone with an income of $200,000?

**22.** Biologists have observed that the chirping rate of crickets of a certain species appears to be related to temperature. The table shows the chirping rates for various temperatures.

| Temperature (°F) | Chirping rate (chirps/min) | Temperature (°F) | Chirping rate (chirps/min) |
|---|---|---|---|
| 50 | 20 | 75 | 140 |
| 55 | 46 | 80 | 173 |
| 60 | 79 | 85 | 198 |
| 65 | 91 | 90 | 211 |
| 70 | 113 | | |

(a) Make a scatter plot of the data.
(b) Find and graph the regression line.
(c) Use the linear model in part (b) to estimate the chirping rate at 100°F.

**23.** The table gives the winning heights for the men's Olympic pole vault competitions up to the year 2004.

| Year | Height (m) | Year | Height (m) |
|------|-----------|------|-----------|
| 1896 | 3.30 | 1960 | 4.70 |
| 1900 | 3.30 | 1964 | 5.10 |
| 1904 | 3.50 | 1968 | 5.40 |
| 1908 | 3.71 | 1972 | 5.64 |
| 1912 | 3.95 | 1976 | 5.64 |
| 1920 | 4.09 | 1980 | 5.78 |
| 1924 | 3.95 | 1984 | 5.75 |
| 1928 | 4.20 | 1988 | 5.90 |
| 1932 | 4.31 | 1992 | 5.87 |
| 1936 | 4.35 | 1996 | 5.92 |
| 1948 | 4.30 | 2000 | 5.90 |
| 1952 | 4.55 | 2004 | 5.95 |
| 1956 | 4.56 |  |  |

(a) Make a scatter plot and decide whether a linear model is appropriate.
(b) Find and graph the regression line.
(c) Use the linear model to predict the height of the winning pole vault at the 2008 Olympics and compare with the actual winning height of 5.96 meters.
(d) Is it reasonable to use the model to predict the winning height at the 2100 Olympics?

**24.** The table shows the percentage of the population of Argentina that has lived in rural areas from 1955 to 2000. Find a model for the data and use it to estimate the rural percentage in 1988 and 2002.

| Year | Percentage rural | Year | Percentage rural |
|------|-----------------|------|-----------------|
| 1955 | 30.4 | 1980 | 17.1 |
| 1960 | 26.4 | 1985 | 15.0 |
| 1965 | 23.6 | 1990 | 13.0 |
| 1970 | 21.1 | 1995 | 11.7 |
| 1975 | 19.0 | 2000 | 10.5 |

**25.** Many physical quantities are connected by *inverse square laws*, that is, by power functions of the form $f(x) = kx^{-2}$. In particular, the illumination of an object by a light source is inversely proportional to the square of the distance from the source. Suppose that after dark you are in a room with just one lamp and you are trying to read a book. The light is too dim and so you move halfway to the lamp. How much brighter is the light?

**26.** It makes sense that the larger the area of a region, the larger the number of species that inhabit the region. Many

ecologists have modeled the species-area relation with a power function and, in particular, the number of species $S$ of bats living in caves in central Mexico has been related to the surface area $A$ of the caves by the equation $S = 0.7A^{0.3}$.
(a) The cave called *Misión Imposible* near Puebla, Mexico, has a surface area of $A = 60$ m$^2$. How many species of bats would you expect to find in that cave?
(b) If you discover that four species of bats live in a cave, estimate the area of the cave.

**27.** The table shows the number $N$ of species of reptiles and amphibians inhabiting Caribbean islands and the area $A$ of the island in square miles.

| Island | $A$ | $N$ |
|--------|-----|-----|
| Saba | 4 | 5 |
| Monserrat | 40 | 9 |
| Puerto Rico | 3,459 | 40 |
| Jamaica | 4,411 | 39 |
| Hispaniola | 29,418 | 84 |
| Cuba | 44,218 | 76 |

(a) Use a power function to model $N$ as a function of $A$.
(b) The Caribbean island of Dominica has area 291 m$^2$. How many species of reptiles and amphibians would you expect to find on Dominica?

**28.** The table shows the mean (average) distances $d$ of the planets from the sun (taking the unit of measurement to be the distance from the earth to the sun) and their periods $T$ (time of revolution in years).

| Planet | $d$ | $T$ |
|--------|-----|-----|
| Mercury | 0.387 | 0.241 |
| Venus | 0.723 | 0.615 |
| Earth | 1.000 | 1.000 |
| Mars | 1.523 | 1.881 |
| Jupiter | 5.203 | 11.861 |
| Saturn | 9.541 | 29.457 |
| Uranus | 19.190 | 84.008 |
| Neptune | 30.086 | 164.784 |

(a) Fit a power model to the data.
(b) Kepler's Third Law of Planetary Motion states that

"The square of the period of revolution of a planet is proportional to the cube of its mean distance from the sun."

Does your model corroborate Kepler's Third Law?

## 1.3   New Functions from Old Functions

In this section we start with the basic functions we discussed in Section 1.2 and obtain new functions by shifting, stretching, and reflecting their graphs. We also show how to combine pairs of functions by the standard arithmetic operations and by composition.

### Transformations of Functions

By applying certain transformations to the graph of a given function we can obtain the graphs of certain related functions. This will give us the ability to sketch the graphs of many functions quickly by hand. It will also enable us to write equations for given graphs. Let's first consider **translations**. If $c$ is a positive number, then the graph of $y = f(x) + c$ is just the graph of $y = f(x)$ shifted upward a distance of $c$ units (because each $y$-coordinate is increased by the same number $c$). Likewise, if $g(x) = f(x - c)$, where $c > 0$, then the value of $g$ at $x$ is the same as the value of $f$ at $x - c$ ($c$ units to the left of $x$). Therefore the graph of $y = f(x - c)$ is just the graph of $y = f(x)$ shifted $c$ units to the right (see Figure 1).

---

**Vertical and Horizontal Shifts**   Suppose $c > 0$. To obtain the graph of

$y = f(x) + c$, shift the graph of $y = f(x)$ a distance $c$ units upward

$y = f(x) - c$, shift the graph of $y = f(x)$ a distance $c$ units downward

$y = f(x - c)$, shift the graph of $y = f(x)$ a distance $c$ units to the right

$y = f(x + c)$, shift the graph of $y = f(x)$ a distance $c$ units to the left

---

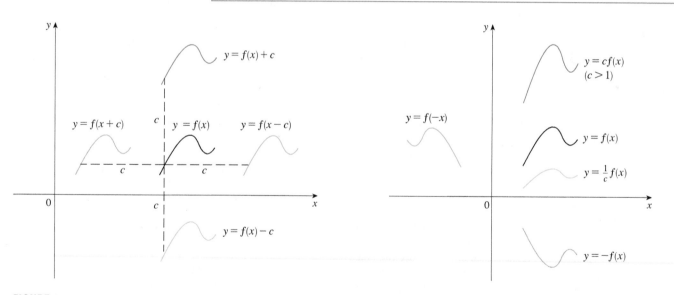

**FIGURE 1**
Translating the graph of $f$

**FIGURE 2**
Stretching and reflecting the graph of $f$

Now let's consider the **stretching** and **reflecting** transformations. If $c > 1$, then the graph of $y = cf(x)$ is the graph of $y = f(x)$ stretched by a factor of $c$ in the vertical direction (because each $y$-coordinate is multiplied by the same number $c$). The graph of $y = -f(x)$ is the graph of $y = f(x)$ reflected about the $x$-axis because the point $(x, y)$ is replaced by the point $(x, -y)$. (See Figure 2 and the following chart, where the results of other stretching, shrinking, and reflecting transformations are also given.)

**Vertical and Horizontal Stretching and Reflecting** Suppose $c > 1$. To obtain the graph of

$y = cf(x)$, stretch the graph of $y = f(x)$ vertically by a factor of $c$

$y = (1/c)f(x)$, shrink the graph of $y = f(x)$ vertically by a factor of $c$

$y = f(cx)$, shrink the graph of $y = f(x)$ horizontally by a factor of $c$

$y = f(x/c)$, stretch the graph of $y = f(x)$ horizontally by a factor of $c$

$y = -f(x)$, reflect the graph of $y = f(x)$ about the $x$-axis

$y = f(-x)$, reflect the graph of $y = f(x)$ about the $y$-axis

Figure 3 illustrates these stretching transformations when applied to the cosine function with $c = 2$. For instance, in order to get the graph of $y = 2 \cos x$ we multiply the $y$-coordinate of each point on the graph of $y = \cos x$ by 2. This means that the graph of $y = \cos x$ gets stretched vertically by a factor of 2.

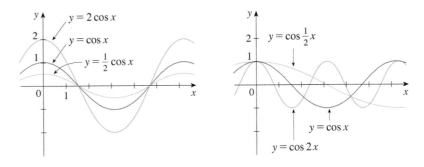

**FIGURE 3**

**V EXAMPLE 1** Given the graph of $y = \sqrt{x}$, use transformations to graph $y = \sqrt{x} - 2$, $y = \sqrt{x - 2}$, $y = -\sqrt{x}$, $y = 2\sqrt{x}$, and $y = \sqrt{-x}$.

SOLUTION The graph of the square root function $y = \sqrt{x}$, obtained from Figure 13(a) in Section 1.2, is shown in Figure 4(a). In the other parts of the figure we sketch $y = \sqrt{x} - 2$ by shifting 2 units downward, $y = \sqrt{x - 2}$ by shifting 2 units to the right, $y = -\sqrt{x}$ by reflecting about the $x$-axis, $y = 2\sqrt{x}$ by stretching vertically by a factor of 2, and $y = \sqrt{-x}$ by reflecting about the $y$-axis.

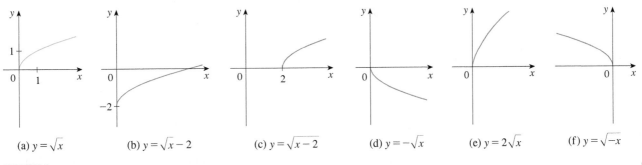

(a) $y = \sqrt{x}$     (b) $y = \sqrt{x} - 2$     (c) $y = \sqrt{x - 2}$     (d) $y = -\sqrt{x}$     (e) $y = 2\sqrt{x}$     (f) $y = \sqrt{-x}$

**FIGURE 4**

**EXAMPLE 2** Sketch the graph of the function $f(x) = x^2 + 6x + 10$.

**SOLUTION** Completing the square, we write the equation of the graph as

$$y = x^2 + 6x + 10 = (x + 3)^2 + 1$$

This means we obtain the desired graph by starting with the parabola $y = x^2$ and shifting 3 units to the left and then 1 unit upward (see Figure 5).

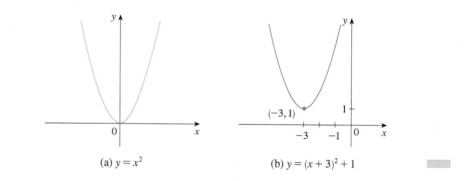

**FIGURE 5**

(a) $y = x^2$        (b) $y = (x + 3)^2 + 1$

**EXAMPLE 3** Sketch the graphs of the following functions.
(a) $y = \sin 2x$           (b) $y = 1 - \sin x$

**SOLUTION**

(a) We obtain the graph of $y = \sin 2x$ from that of $y = \sin x$ by compressing horizontally by a factor of 2. (See Figures 6 and 7.) Thus, whereas the period of $y = \sin x$ is $2\pi$, the period of $y = \sin 2x$ is $2\pi/2 = \pi$.

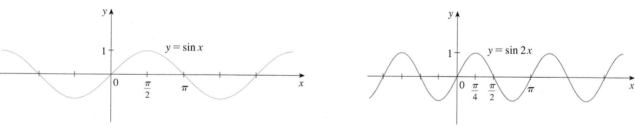

**FIGURE 6**

**FIGURE 7**

(b) To obtain the graph of $y = 1 - \sin x$, we again start with $y = \sin x$. We reflect about the $x$-axis to get the graph of $y = -\sin x$ and then we shift 1 unit upward to get $y = 1 - \sin x$. (See Figure 8.)

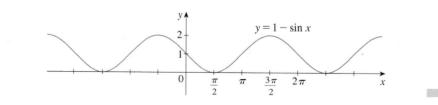

**FIGURE 8**

**EXAMPLE 4** Figure 9 shows graphs of the number of hours of daylight as functions of the time of the year at several latitudes. Given that Philadelphia is located at approximately 40°N latitude, find a function that models the length of daylight at Philadelphia.

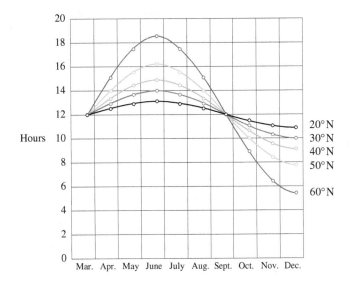

**FIGURE 9**

Graph of the length of daylight
from March 21 through December 21
at various latitudes

Lucia C. Harrison, *Daylight, Twilight, Darkness and Time*
(New York, 1935) page 40.

SOLUTION Notice that each curve resembles a shifted and stretched sine function. By looking at the blue curve we see that, at the latitude of Philadelphia, daylight lasts about 14.8 hours on June 21 and 9.2 hours on December 21, so the amplitude of the curve (the factor by which we have to stretch the sine curve vertically) is $\frac{1}{2}(14.8 - 9.2) = 2.8$.

By what factor do we need to stretch the sine curve horizontally if we measure the time $t$ in days? Because there are about 365 days in a year, the period of our model should be 365. But the period of $y = \sin t$ is $2\pi$, so the horizontal stretching factor is $c = 2\pi/365$.

We also notice that the curve begins its cycle on March 21, the 80th day of the year, so we have to shift the curve 80 units to the right. In addition, we shift it 12 units upward. Therefore we model the length of daylight in Philadelphia on the $t$th day of the year by the function

$$L(t) = 12 + 2.8 \sin\left[\frac{2\pi}{365}(t - 80)\right]$$

Another transformation of some interest is taking the *absolute value* of a function. If $y = |f(x)|$, then according to the definition of absolute value, $y = f(x)$ when $f(x) \geq 0$ and $y = -f(x)$ when $f(x) < 0$. This tells us how to get the graph of $y = |f(x)|$ from the graph of $y = f(x)$: The part of the graph that lies above the $x$-axis remains the same; the part that lies below the $x$-axis is reflected about the $x$-axis.

**V EXAMPLE 5** Sketch the graph of the function $y = |x^2 - 1|$.

SOLUTION We first graph the parabola $y = x^2 - 1$ in Figure 10(a) by shifting the parabola $y = x^2$ downward 1 unit. We see that the graph lies below the $x$-axis when $-1 < x < 1$, so we reflect that part of the graph about the $x$-axis to obtain the graph of $y = |x^2 - 1|$ in Figure 10(b).

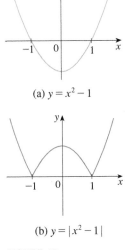

(a) $y = x^2 - 1$

(b) $y = |x^2 - 1|$

**FIGURE 10**

## Combinations of Functions

Two functions $f$ and $g$ can be combined to form new functions $f + g$, $f - g$, $fg$, and $f/g$ in a manner similar to the way we add, subtract, multiply, and divide real numbers. The sum and difference functions are defined by

$$(f + g)(x) = f(x) + g(x) \qquad (f - g)(x) = f(x) - g(x)$$

If the domain of $f$ is $A$ and the domain of $g$ is $B$, then the domain of $f + g$ is the intersection $A \cap B$ because both $f(x)$ and $g(x)$ have to be defined. For example, the domain of $f(x) = \sqrt{x}$ is $A = [0, \infty)$ and the domain of $g(x) = \sqrt{2 - x}$ is $B = (-\infty, 2]$, so the domain of $(f + g)(x) = \sqrt{x} + \sqrt{2 - x}$ is $A \cap B = [0, 2]$.

Similarly, the product and quotient functions are defined by

$$(fg)(x) = f(x)g(x) \qquad \left(\frac{f}{g}\right)(x) = \frac{f(x)}{g(x)}$$

The domain of $fg$ is $A \cap B$, but we can't divide by 0 and so the domain of $f/g$ is $\{x \in A \cap B \mid g(x) \ne 0\}$. For instance, if $f(x) = x^2$ and $g(x) = x - 1$, then the domain of the rational function $(f/g)(x) = x^2/(x - 1)$ is $\{x \mid x \ne 1\}$, or $(-\infty, 1) \cup (1, \infty)$.

There is another way of combining two functions to obtain a new function. For example, suppose that $y = f(u) = \sqrt{u}$ and $u = g(x) = x^2 + 1$. Since $y$ is a function of $u$ and $u$ is, in turn, a function of $x$, it follows that $y$ is ultimately a function of $x$. We compute this by substitution:

$$y = f(u) = f(g(x)) = f(x^2 + 1) = \sqrt{x^2 + 1}$$

The procedure is called *composition* because the new function is *composed* of the two given functions $f$ and $g$.

In general, given any two functions $f$ and $g$, we start with a number $x$ in the domain of $g$ and find its image $g(x)$. If this number $g(x)$ is in the domain of $f$, then we can calculate the value of $f(g(x))$. Notice that the output of one function is used as the input to the next function. The result is a new function $h(x) = f(g(x))$ obtained by substituting $g$ into $f$. It is called the *composition* (or *composite*) of $f$ and $g$ and is denoted by $f \circ g$ ("$f$ circle $g$").

---

> **Definition** Given two functions $f$ and $g$, the **composite function** $f \circ g$ (also called the **composition** of $f$ and $g$) is defined by
>
> $$(f \circ g)(x) = f(g(x))$$

---

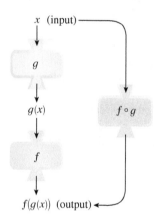

**FIGURE 11**

The $f \circ g$ machine is composed of the $g$ machine (first) and then the $f$ machine.

The domain of $f \circ g$ is the set of all $x$ in the domain of $g$ such that $g(x)$ is in the domain of $f$. In other words, $(f \circ g)(x)$ is defined whenever both $g(x)$ and $f(g(x))$ are defined. Figure 11 shows how to picture $f \circ g$ in terms of machines.

**EXAMPLE 6** If $f(x) = x^2$ and $g(x) = x - 3$, find the composite functions $f \circ g$ and $g \circ f$.

SOLUTION We have

$$(f \circ g)(x) = f(g(x)) = f(x - 3) = (x - 3)^2$$

$$(g \circ f)(x) = g(f(x)) = g(x^2) = x^2 - 3$$

⊘    **NOTE** You can see from Example 6 that, in general, $f \circ g \ne g \circ f$. Remember, the notation $f \circ g$ means that the function $g$ is applied first and then $f$ is applied second. In Example 6, $f \circ g$ is the function that *first* subtracts 3 and *then* squares; $g \circ f$ is the function that *first* squares and *then* subtracts 3.

**V EXAMPLE 7** If $f(x) = \sqrt{x}$ and $g(x) = \sqrt{2-x}$, find each function and its domain.

(a) $f \circ g$     (b) $g \circ f$     (c) $f \circ f$     (d) $g \circ g$

SOLUTION

(a) $$(f \circ g)(x) = f(g(x)) = f\left(\sqrt{2-x}\right) = \sqrt{\sqrt{2-x}} = \sqrt[4]{2-x}$$

The domain of $f \circ g$ is $\{x \mid 2 - x \geqslant 0\} = \{x \mid x \leqslant 2\} = (-\infty, 2]$.

(b) $$(g \circ f)(x) = g(f(x)) = g\left(\sqrt{x}\right) = \sqrt{2 - \sqrt{x}}$$

For $\sqrt{x}$ to be defined we must have $x \geqslant 0$. For $\sqrt{2 - \sqrt{x}}$ to be defined we must have $2 - \sqrt{x} \geqslant 0$, that is, $\sqrt{x} \leqslant 2$, or $x \leqslant 4$. Thus we have $0 \leqslant x \leqslant 4$, so the domain of $g \circ f$ is the closed interval $[0, 4]$.

If $0 \leqslant a \leqslant b$, then $a^2 \leqslant b^2$.

(c) $$(f \circ f)(x) = f(f(x)) = f\left(\sqrt{x}\right) = \sqrt{\sqrt{x}} = \sqrt[4]{x}$$

The domain of $f \circ f$ is $[0, \infty)$.

(d) $$(g \circ g)(x) = g(g(x)) = g\left(\sqrt{2-x}\right) = \sqrt{2 - \sqrt{2-x}}$$

This expression is defined when both $2 - x \geqslant 0$ and $2 - \sqrt{2-x} \geqslant 0$. The first inequality means $x \leqslant 2$, and the second is equivalent to $\sqrt{2-x} \leqslant 2$, or $2 - x \leqslant 4$, or $x \geqslant -2$. Thus $-2 \leqslant x \leqslant 2$, so the domain of $g \circ g$ is the closed interval $[-2, 2]$. ■

It is possible to take the composition of three or more functions. For instance, the composite function $f \circ g \circ h$ is found by first applying $h$, then $g$, and then $f$ as follows:

$$(f \circ g \circ h)(x) = f(g(h(x)))$$

**EXAMPLE 8** Find $f \circ g \circ h$ if $f(x) = x/(x+1)$, $g(x) = x^{10}$, and $h(x) = x + 3$.

SOLUTION
$$(f \circ g \circ h)(x) = f(g(h(x))) = f(g(x+3))$$
$$= f((x+3)^{10}) = \frac{(x+3)^{10}}{(x+3)^{10} + 1} \quad ■$$

So far we have used composition to build complicated functions from simpler ones. But in calculus it is often useful to be able to *decompose* a complicated function into simpler ones, as in the following example.

**EXAMPLE 9** Given $F(x) = \cos^2(x+9)$, find functions $f$, $g$, and $h$ such that $F = f \circ g \circ h$.

SOLUTION Since $F(x) = [\cos(x+9)]^2$, the formula for $F$ says: First add 9, then take the cosine of the result, and finally square. So we let

$$h(x) = x + 9 \qquad g(x) = \cos x \qquad f(x) = x^2$$

Then
$$(f \circ g \circ h)(x) = f(g(h(x))) = f(g(x+9)) = f(\cos(x+9))$$
$$= [\cos(x+9)]^2 = F(x) \quad ■$$

## 1.3  Exercises

1. Suppose the graph of $f$ is given. Write equations for the graphs that are obtained from the graph of $f$ as follows.
   (a) Shift 3 units upward.        (b) Shift 3 units downward.
   (c) Shift 3 units to the right.   (d) Shift 3 units to the left.
   (e) Reflect about the $x$-axis.   (f) Reflect about the $y$-axis.
   (g) Stretch vertically by a factor of 3.
   (h) Shrink vertically by a factor of 3.

2. Explain how each graph is obtained from the graph of $y = f(x)$.
   (a) $y = f(x) + 8$              (b) $y = f(x + 8)$
   (c) $y = 8f(x)$                 (d) $y = f(8x)$
   (e) $y = -f(x) - 1$             (f) $y = 8f\left(\frac{1}{8}x\right)$

3. The graph of $y = f(x)$ is given. Match each equation with its graph and give reasons for your choices.
   (a) $y = f(x - 4)$              (b) $y = f(x) + 3$
   (c) $y = \frac{1}{3}f(x)$       (d) $y = -f(x + 4)$
   (e) $y = 2f(x + 6)$

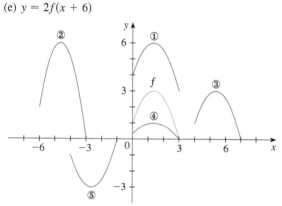

4. The graph of $f$ is given. Draw the graphs of the following functions.
   (a) $y = f(x) - 2$              (b) $y = f(x - 2)$
   (c) $y = -2f(x)$               (d) $y = f\left(\frac{1}{3}x\right) + 1$

5. The graph of $f$ is given. Use it to graph the following functions.
   (a) $y = f(2x)$                 (b) $y = f\left(\frac{1}{2}x\right)$
   (c) $y = f(-x)$                 (d) $y = -f(-x)$

6–7 The graph of $y = \sqrt{3x - x^2}$ is given. Use transformations to create a function whose graph is as shown.

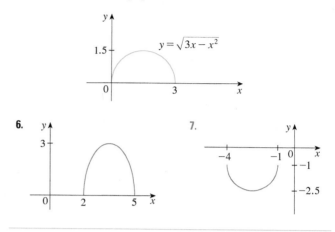

6.

7.

8. (a) How is the graph of $y = 2\sin x$ related to the graph of $y = \sin x$? Use your answer and Figure 6 to sketch the graph of $y = 2\sin x$.
   (b) How is the graph of $y = 1 + \sqrt{x}$ related to the graph of $y = \sqrt{x}$? Use your answer and Figure 4(a) to sketch the graph of $y = 1 + \sqrt{x}$.

9–24 Graph the function by hand, not by plotting points, but by starting with the graph of one of the standard functions given in Section 1.2, and then applying the appropriate transformations.

9. $y = \dfrac{1}{x + 2}$                 10. $y = (x - 1)^3$

11. $y = -\sqrt[3]{x}$                    12. $y = x^2 + 6x + 4$

13. $y = \sqrt{x - 2} - 1$                14. $y = 4\sin 3x$

15. $y = \sin\left(\frac{1}{2}x\right)$   16. $y = \dfrac{2}{x} - 2$

17. $y = \frac{1}{2}(1 - \cos x)$         18. $y = 1 - 2\sqrt{x + 3}$

19. $y = 1 - 2x - x^2$                    20. $y = |x| - 2$

21. $y = |x - 2|$                         22. $y = \dfrac{1}{4}\tan\left(x - \dfrac{\pi}{4}\right)$

23. $y = \left|\sqrt{x} - 1\right|$       24. $y = |\cos \pi x|$

25. The city of New Orleans is located at latitude 30°N. Use Figure 9 to find a function that models the number of hours of daylight at New Orleans as a function of the time of year. To check the accuracy of your model, use the fact that on March 31 the sun rises at 5:51 AM and sets at 6:18 PM in New Orleans.

1. Homework Hints available at stewartcalculus.com

**26.** A variable star is one whose brightness alternately increases and decreases. For the most visible variable star, Delta Cephei, the time between periods of maximum brightness is 5.4 days, the average brightness (or magnitude) of the star is 4.0, and its brightness varies by $\pm 0.35$ magnitude. Find a function that models the brightness of Delta Cephei as a function of time.

**27.** (a) How is the graph of $y = f(|x|)$ related to the graph of $f$?
(b) Sketch the graph of $y = \sin |x|$.
(c) Sketch the graph of $y = \sqrt{|x|}$.

**28.** Use the given graph of $f$ to sketch the graph of $y = 1/f(x)$. Which features of $f$ are the most important in sketching $y = 1/f(x)$? Explain how they are used.

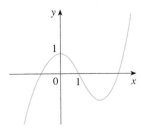

**29–30** Find (a) $f + g$, (b) $f - g$, (c) $fg$, and (d) $f/g$ and state their domains.

**29.** $f(x) = x^3 + 2x^2, \quad g(x) = 3x^2 - 1$

**30.** $f(x) = \sqrt{3 - x}, \quad g(x) = \sqrt{x^2 - 1}$

**31–36** Find the functions (a) $f \circ g$, (b) $g \circ f$, (c) $f \circ f$, and (d) $g \circ g$ and their domains.

**31.** $f(x) = x^2 - 1, \quad g(x) = 2x + 1$

**32.** $f(x) = x - 2, \quad g(x) = x^2 + 3x + 4$

**33.** $f(x) = 1 - 3x, \quad g(x) = \cos x$

**34.** $f(x) = \sqrt{x}, \quad g(x) = \sqrt[3]{1 - x}$

**35.** $f(x) = x + \dfrac{1}{x}, \quad g(x) = \dfrac{x + 1}{x + 2}$

**36.** $f(x) = \dfrac{x}{1 + x}, \quad g(x) = \sin 2x$

**37–40** Find $f \circ g \circ h$.

**37.** $f(x) = 3x - 2, \quad g(x) = \sin x, \quad h(x) = x^2$

**38.** $f(x) = |x - 4|, \quad g(x) = 2^x, \quad h(x) = \sqrt{x}$

**39.** $f(x) = \sqrt{x - 3}, \quad g(x) = x^2, \quad h(x) = x^3 + 2$

**40.** $f(x) = \tan x, \quad g(x) = \dfrac{x}{x - 1}, \quad h(x) = \sqrt[3]{x}$

**41–46** Express the function in the form $f \circ g$.

**41.** $F(x) = (2x + x^2)^4$

**42.** $F(x) = \cos^2 x$

**43.** $F(x) = \dfrac{\sqrt[3]{x}}{1 + \sqrt[3]{x}}$

**44.** $G(x) = \sqrt[3]{\dfrac{x}{1 + x}}$

**45.** $v(t) = \sec(t^2) \tan(t^2)$

**46.** $u(t) = \dfrac{\tan t}{1 + \tan t}$

**47–49** Express the function in the form $f \circ g \circ h$.

**47.** $R(x) = \sqrt{\sqrt{x} - 1}$

**48.** $H(x) = \sqrt[8]{2 + |x|}$

**49.** $H(x) = \sec^4(\sqrt{x})$

**50.** Use the table to evaluate each expression.
(a) $f(g(1))$     (b) $g(f(1))$     (c) $f(f(1))$
(d) $g(g(1))$     (e) $(g \circ f)(3)$     (f) $(f \circ g)(6)$

| $x$ | 1 | 2 | 3 | 4 | 5 | 6 |
|-----|---|---|---|---|---|---|
| $f(x)$ | 3 | 1 | 4 | 2 | 2 | 5 |
| $g(x)$ | 6 | 3 | 2 | 1 | 2 | 3 |

**51.** Use the given graphs of $f$ and $g$ to evaluate each expression, or explain why it is undefined.
(a) $f(g(2))$     (b) $g(f(0))$     (c) $(f \circ g)(0)$
(d) $(g \circ f)(6)$     (e) $(g \circ g)(-2)$     (f) $(f \circ f)(4)$

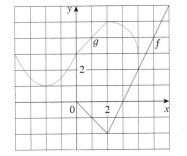

**52.** Use the given graphs of $f$ and $g$ to estimate the value of $f(g(x))$ for $x = -5, -4, -3, \ldots, 5$. Use these estimates to sketch a rough graph of $f \circ g$.

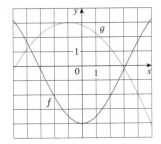

**53.** A stone is dropped into a lake, creating a circular ripple that travels outward at a speed of 60 cm/s.
(a) Express the radius $r$ of this circle as a function of the time $t$ (in seconds).
(b) If $A$ is the area of this circle as a function of the radius, find $A \circ r$ and interpret it.

**54.** A spherical balloon is being inflated and the radius of the balloon is increasing at a rate of 2 cm/s.
(a) Express the radius $r$ of the balloon as a function of the time $t$ (in seconds).
(b) If $V$ is the volume of the balloon as a function of the radius, find $V \circ r$ and interpret it.

**55.** A ship is moving at a speed of 30 km/h parallel to a straight shoreline. The ship is 6 km from shore and it passes a lighthouse at noon.
(a) Express the distance $s$ between the lighthouse and the ship as a function of $d$, the distance the ship has traveled since noon; that is, find $f$ so that $s = f(d)$.
(b) Express $d$ as a function of $t$, the time elapsed since noon; that is, find $g$ so that $d = g(t)$.
(c) Find $f \circ g$. What does this function represent?

**56.** An airplane is flying at a speed of 350 mi/h at an altitude of one mile and passes directly over a radar station at time $t = 0$.
(a) Express the horizontal distance $d$ (in miles) that the plane has flown as a function of $t$.
(b) Express the distance $s$ between the plane and the radar station as a function of $d$.
(c) Use composition to express $s$ as a function of $t$.

**57.** The **Heaviside function** $H$ is defined by

$$H(t) = \begin{cases} 0 & \text{if } t < 0 \\ 1 & \text{if } t \geq 0 \end{cases}$$

It is used in the study of electric circuits to represent the sudden surge of electric current, or voltage, when a switch is instantaneously turned on.
(a) Sketch the graph of the Heaviside function.
(b) Sketch the graph of the voltage $V(t)$ in a circuit if the switch is turned on at time $t = 0$ and 120 volts are applied instantaneously to the circuit. Write a formula for $V(t)$ in terms of $H(t)$.

(c) Sketch the graph of the voltage $V(t)$ in a circuit if the switch is turned on at time $t = 5$ seconds and 240 volts are applied instantaneously to the circuit. Write a formula for $V(t)$ in terms of $H(t)$. (Note that starting at $t = 5$ corresponds to a translation.)

**58.** The Heaviside function defined in Exercise 57 can also be used to define the **ramp function** $y = ctH(t)$, which represents a gradual increase in voltage or current in a circuit.
(a) Sketch the graph of the ramp function $y = tH(t)$.
(b) Sketch the graph of the voltage $V(t)$ in a circuit if the switch is turned on at time $t = 0$ and the voltage is gradually increased to 120 volts over a 60-second time interval. Write a formula for $V(t)$ in terms of $H(t)$ for $t \leq 60$.
(c) Sketch the graph of the voltage $V(t)$ in a circuit if the switch is turned on at time $t = 7$ seconds and the voltage is gradually increased to 100 volts over a period of 25 seconds. Write a formula for $V(t)$ in terms of $H(t)$ for $t \leq 32$.

**59.** Let $f$ and $g$ be linear functions with equations $f(x) = m_1 x + b_1$ and $g(x) = m_2 x + b_2$. Is $f \circ g$ also a linear function? If so, what is the slope of its graph?

**60.** If you invest $x$ dollars at 4% interest compounded annually, then the amount $A(x)$ of the investment after one year is $A(x) = 1.04x$. Find $A \circ A$, $A \circ A \circ A$, and $A \circ A \circ A \circ A$. What do these compositions represent? Find a formula for the composition of $n$ copies of $A$.

**61.** (a) If $g(x) = 2x + 1$ and $h(x) = 4x^2 + 4x + 7$, find a function $f$ such that $f \circ g = h$. (Think about what operations you would have to perform on the formula for $g$ to end up with the formula for $h$.)
(b) If $f(x) = 3x + 5$ and $h(x) = 3x^2 + 3x + 2$, find a function $g$ such that $f \circ g = h$.

**62.** If $f(x) = x + 4$ and $h(x) = 4x - 1$, find a function $g$ such that $g \circ f = h$.

**63.** Suppose $g$ is an even function and let $h = f \circ g$. Is $h$ always an even function?

**64.** Suppose $g$ is an odd function and let $h = f \circ g$. Is $h$ always an odd function? What if $f$ is odd? What if $f$ is even?

**1.4**  **The Tangent and Velocity Problems**

AROG

In this section we see how limits arise when we attempt to find the tangent to a curve or the velocity of an object.

### The Tangent Problem

The word *tangent* is derived from the Latin word *tangens*, which means "touching." Thus a tangent to a curve is a line that touches the curve. In other words, a tangent line should have the same direction as the curve at the point of contact. How can this idea be made precise?

For a circle we could simply follow Euclid and say that a tangent is a line that intersects the circle once and only once, as in Figure 1(a). For more complicated curves this definition is inadequate. Figure l(b) shows two lines $l$ and $t$ passing through a point $P$ on a curve $C$. The line $l$ intersects $C$ only once, but it certainly does not look like what we think of as a tangent. The line $t$, on the other hand, looks like a tangent but it intersects $C$ twice.

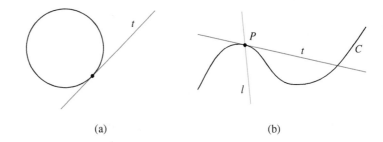

**FIGURE 1**

(a)                                                    (b)

To be specific, let's look at the problem of trying to find a tangent line $t$ to the parabola $y = x^2$ in the following example.

**V** **EXAMPLE 1**  Find an equation of the tangent line to the parabola $y = x^2$ at the point $P(1, 1)$.

SOLUTION   We will be able to find an equation of the tangent line $t$ as soon as we know its slope $m$. The difficulty is that we know only one point, $P$, on $t$, whereas we need two points to compute the slope. But observe that we can compute an approximation to $m$ by choosing a nearby point $Q(x, x^2)$ on the parabola (as in Figure 2) and computing the slope $m_{PQ}$ of the secant line $PQ$. [A **secant line**, from the Latin word *secans*, meaning cutting, is a line that cuts (intersects) a curve more than once.]

We choose $x \neq 1$ so that $Q \neq P$. Then

$$m_{PQ} = \frac{x^2 - 1}{x - 1}$$

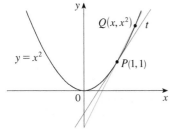

**FIGURE 2**

For instance, for the point $Q(1.5, 2.25)$ we have

$$m_{PQ} = \frac{2.25 - 1}{1.5 - 1} = \frac{1.25}{0.5} = 2.5$$

The tables in the margin show the values of $m_{PQ}$ for several values of $x$ close to 1. The closer $Q$ is to $P$, the closer $x$ is to 1 and, it appears from the tables, the closer $m_{PQ}$ is to 2. This suggests that the slope of the tangent line $t$ should be $m = 2$.

We say that the slope of the tangent line is the *limit* of the slopes of the secant lines, and we express this symbolically by writing

$$\lim_{Q \to P} m_{PQ} = m \qquad \text{and} \qquad \lim_{x \to 1} \frac{x^2 - 1}{x - 1} = 2$$

Assuming that the slope of the tangent line is indeed 2, we use the point-slope form of the equation of a line (see Appendix B) to write the equation of the tangent line through $(1, 1)$ as

$$y - 1 = 2(x - 1) \qquad \text{or} \qquad y = 2x - 1$$

| $x$ | $m_{PQ}$ |
|-----|----------|
| 2 | 3 |
| 1.5 | 2.5 |
| 1.1 | 2.1 |
| 1.01 | 2.01 |
| 1.001 | 2.001 |

| $x$ | $m_{PQ}$ |
|-----|----------|
| 0 | 1 |
| 0.5 | 1.5 |
| 0.9 | 1.9 |
| 0.99 | 1.99 |
| 0.999 | 1.999 |

Figure 3 illustrates the limiting process that occurs in this example. As $Q$ approaches $P$ along the parabola, the corresponding secant lines rotate about $P$ and approach the tangent line $t$.

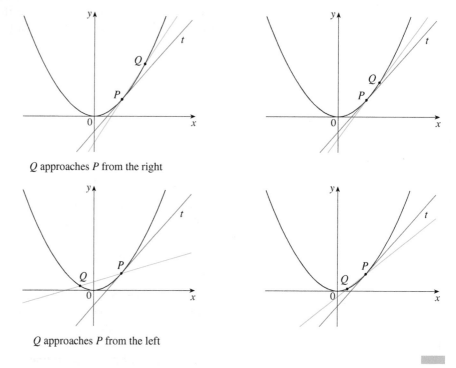

$Q$ approaches $P$ from the right

$Q$ approaches $P$ from the left

**FIGURE 3**

**TEC** In Visual 1.4 you can see how the process in Figure 3 works for additional functions.

| $t$ | $Q$ |
|------|--------|
| 0.00 | 100.00 |
| 0.02 | 81.87 |
| 0.04 | 67.03 |
| 0.06 | 54.88 |
| 0.08 | 44.93 |
| 0.10 | 36.76 |

Many functions that occur in science are not described by explicit equations; they are defined by experimental data. The next example shows how to estimate the slope of the tangent line to the graph of such a function.

**V** **EXAMPLE 2** The flash unit on a camera operates by storing charge on a capacitor and releasing it suddenly when the flash is set off. The data in the table describe the charge $Q$ remaining on the capacitor (measured in microcoulombs) at time $t$ (measured in seconds after the flash goes off). Use the data to draw the graph of this function and estimate the slope of the tangent line at the point where $t = 0.04$. [*Note:* The slope of the tangent line represents the electric current flowing from the capacitor to the flash bulb (measured in microamperes).]

SOLUTION In Figure 4 we plot the given data and use them to sketch a curve that approximates the graph of the function.

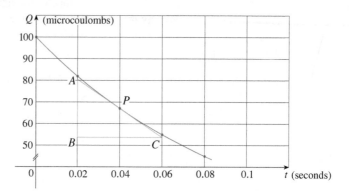

**FIGURE 4**

Given the points $P(0.04, 67.03)$ and $R(0.00, 100.00)$ on the graph, we find that the slope of the secant line $PR$ is

$$m_{PR} = \frac{100.00 - 67.03}{0.00 - 0.04} = -824.25$$

| $R$ | $m_{PR}$ |
|---|---|
| (0.00, 100.00) | −824.25 |
| (0.02, 81.87) | −742.00 |
| (0.06, 54.88) | −607.50 |
| (0.08, 44.93) | −552.50 |
| (0.10, 36.76) | −504.50 |

The table at the left shows the results of similar calculations for the slopes of other secant lines. From this table we would expect the slope of the tangent line at $t = 0.04$ to lie somewhere between −742 and −607.5. In fact, the average of the slopes of the two closest secant lines is

$$\tfrac{1}{2}(-742 - 607.5) = -674.75$$

So, by this method, we estimate the slope of the tangent line to be −675.

Another method is to draw an approximation to the tangent line at $P$ and measure the sides of the triangle $ABC$, as in Figure 4. This gives an estimate of the slope of the tangent line as

$$-\frac{|AB|}{|BC|} \approx -\frac{80.4 - 53.6}{0.06 - 0.02} = -670$$

The physical meaning of the answer in Example 2 is that the electric current flowing from the capacitor to the flash bulb after 0.04 second is about –670 microamperes.

### The Velocity Problem

If you watch the speedometer of a car as you travel in city traffic, you see that the needle doesn't stay still for very long; that is, the velocity of the car is not constant. We assume from watching the speedometer that the car has a definite velocity at each moment, but how is the "instantaneous" velocity defined? Let's investigate the example of a falling ball.

**V EXAMPLE 3** Suppose that a ball is dropped from the upper observation deck of the CN Tower in Toronto, 450 m above the ground. Find the velocity of the ball after 5 seconds.

SOLUTION Through experiments carried out four centuries ago, Galileo discovered that the distance fallen by any freely falling body is proportional to the square of the time it has been falling. (This model for free fall neglects air resistance.) If the distance fallen after $t$ seconds is denoted by $s(t)$ and measured in meters, then Galileo's law is expressed by the equation

$$s(t) = 4.9t^2$$

The difficulty in finding the velocity after 5 s is that we are dealing with a single instant of time ($t = 5$), so no time interval is involved. However, we can approximate the desired quantity by computing the average velocity over the brief time interval of a tenth of a second from $t = 5$ to $t = 5.1$:

$$\text{average velocity} = \frac{\text{change in position}}{\text{time elapsed}}$$

$$= \frac{s(5.1) - s(5)}{0.1}$$

$$= \frac{4.9(5.1)^2 - 4.9(5)^2}{0.1} = 49.49 \text{ m/s}$$

The CN Tower in Toronto was the tallest free-standing building in the world for 32 years.

The following table shows the results of similar calculations of the average velocity over successively smaller time periods.

| Time interval | Average velocity (m/s) |
|---|---|
| $5 \leqslant t \leqslant 6$ | 53.9 |
| $5 \leqslant t \leqslant 5.1$ | 49.49 |
| $5 \leqslant t \leqslant 5.05$ | 49.245 |
| $5 \leqslant t \leqslant 5.01$ | 49.049 |
| $5 \leqslant t \leqslant 5.001$ | 49.0049 |

It appears that as we shorten the time period, the average velocity is becoming closer to 49 m/s. The **instantaneous velocity** when $t = 5$ is defined to be the limiting value of these average velocities over shorter and shorter time periods that start at $t = 5$. Thus the (instantaneous) velocity after 5 s is

$$v = 49 \text{ m/s}$$

You may have the feeling that the calculations used in solving this problem are very similar to those used earlier in this section to find tangents. In fact, there is a close connection between the tangent problem and the problem of finding velocities. If we draw the graph of the distance function of the ball (as in Figure 5) and we consider the points $P(a, 4.9a^2)$ and $Q(a + h, 4.9(a + h)^2)$ on the graph, then the slope of the secant line $PQ$ is

$$m_{PQ} = \frac{4.9(a + h)^2 - 4.9a^2}{(a + h) - a}$$

which is the same as the average velocity over the time interval $[a, a + h]$. Therefore the velocity at time $t = a$ (the limit of these average velocities as $h$ approaches 0) must be equal to the slope of the tangent line at $P$ (the limit of the slopes of the secant lines).

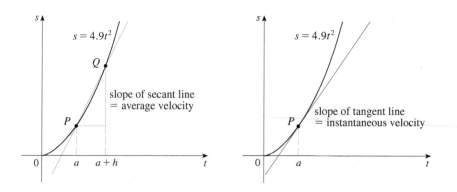

**FIGURE 5**

Examples 1 and 3 show that in order to solve tangent and velocity problems we must be able to find limits. After studying methods for computing limits in the next four sections, we will return to the problems of finding tangents and velocities in Chapter 2.

## 1.4   Exercises

**1.** A tank holds 1000 gallons of water, which drains from the bottom of the tank in half an hour. The values in the table show the volume $V$ of water remaining in the tank (in gallons) after $t$ minutes.

| $t$ (min) | 5 | 10 | 15 | 20 | 25 | 30 |
|-----------|-----|-----|-----|-----|-----|-----|
| $V$ (gal) | 694 | 444 | 250 | 111 | 28 | 0 |

(a) If $P$ is the point $(15, 250)$ on the graph of $V$, find the slopes of the secant lines $PQ$ when $Q$ is the point on the graph with $t = 5, 10, 20, 25,$ and $30$.

(b) Estimate the slope of the tangent line at $P$ by averaging the slopes of two secant lines.

(c) Use a graph of the function to estimate the slope of the tangent line at $P$. (This slope represents the rate at which the water is flowing from the tank after 15 minutes.)

**2.** A cardiac monitor is used to measure the heart rate of a patient after surgery. It compiles the number of heartbeats after $t$ minutes. When the data in the table are graphed, the slope of the tangent line represents the heart rate in beats per minute.

| $t$ (min) | 36 | 38 | 40 | 42 | 44 |
|-----------|------|------|------|------|------|
| Heartbeats | 2530 | 2661 | 2806 | 2948 | 3080 |

The monitor estimates this value by calculating the slope of a secant line. Use the data to estimate the patient's heart rate after 42 minutes using the secant line between the points with the given values of $t$.

(a) $t = 36$  and  $t = 42$     (b) $t = 38$  and  $t = 42$

(c) $t = 40$  and  $t = 42$     (d) $t = 42$  and  $t = 44$

What are your conclusions?

**3.** The point $P(2, -1)$ lies on the curve $y = 1/(1 - x)$.

(a) If $Q$ is the point $(x, 1/(1 - x))$, use your calculator to find the slope of the secant line $PQ$ (correct to six decimal places) for the following values of $x$:

  (i) 1.5    (ii) 1.9    (iii) 1.99    (iv) 1.999
  (v) 2.5    (vi) 2.1    (vii) 2.01    (viii) 2.001

(b) Using the results of part (a), guess the value of the slope of the tangent line to the curve at $P(2, -1)$.

(c) Using the slope from part (b), find an equation of the tangent line to the curve at $P(2, -1)$.

**4.** The point $P(0.5, 0)$ lies on the curve $y = \cos \pi x$.

(a) If $Q$ is the point $(x, \cos \pi x)$, use your calculator to find the slope of the secant line $PQ$ (correct to six decimal places) for the following values of $x$:

  (i) 0    (ii) 0.4    (iii) 0.49    (iv) 0.499
  (v) 1    (vi) 0.6    (vii) 0.51    (viii) 0.501

(b) Using the results of part (a), guess the value of the slope of the tangent line to the curve at $P(0.5, 0)$.

(c) Using the slope from part (b), find an equation of the tangent line to the curve at $P(0.5, 0)$.

(d) Sketch the curve, two of the secant lines, and the tangent line.

**5.** If a ball is thrown into the air with a velocity of 40 ft/s, its height in feet $t$ seconds later is given by $y = 40t - 16t^2$.

(a) Find the average velocity for the time period beginning when $t = 2$ and lasting

  (i) 0.5 second     (ii) 0.1 second
  (iii) 0.05 second    (iv) 0.01 second

(b) Estimate the instantaneous velocity when $t = 2$.

**6.** If a rock is thrown upward on the planet Mars with a velocity of 10 m/s, its height in meters $t$ seconds later is given by $y = 10t - 1.86t^2$.

(a) Find the average velocity over the given time intervals:

  (i) [1, 2]     (ii) [1, 1.5]     (iii) [1, 1.1]
  (iv) [1, 1.01]     (v) [1, 1.001]

(b) Estimate the instantaneous velocity when $t = 1$.

**7.** The table shows the position of a cyclist.

| $t$ (seconds) | 0 | 1 | 2 | 3 | 4 | 5 |
|---------------|---|-----|-----|------|------|------|
| $s$ (meters) | 0 | 1.4 | 5.1 | 10.7 | 17.7 | 25.8 |

(a) Find the average velocity for each time period:

  (i) [1, 3]     (ii) [2, 3]     (iii) [3, 5]     (iv) [3, 4]

(b) Use the graph of $s$ as a function of $t$ to estimate the instantaneous velocity when $t = 3$.

**8.** The displacement (in centimeters) of a particle moving back and forth along a straight line is given by the equation of motion $s = 2 \sin \pi t + 3 \cos \pi t$, where $t$ is measured in seconds.

(a) Find the average velocity during each time period:

  (i) [1, 2]     (ii) [1, 1.1]
  (iii) [1, 1.01]     (iv) [1, 1.001]

(b) Estimate the instantaneous velocity of the particle when $t = 1$.

**9.** The point $P(1, 0)$ lies on the curve $y = \sin(10\pi/x)$.

(a) If $Q$ is the point $(x, \sin(10\pi/x))$, find the slope of the secant line $PQ$ (correct to four decimal places) for $x = 2, 1.5, 1.4, 1.3, 1.2, 1.1, 0.5, 0.6, 0.7, 0.8,$ and $0.9$. Do the slopes appear to be approaching a limit?

(b) Use a graph of the curve to explain why the slopes of the secant lines in part (a) are not close to the slope of the tangent line at $P$.

(c) By choosing appropriate secant lines, estimate the slope of the tangent line at $P$.

---

## 1.5 The Limit of a Function

Having seen in the preceding section how limits arise when we want to find the tangent to a curve or the velocity of an object, we now turn our attention to limits in general and numerical and graphical methods for computing them.

Let's investigate the behavior of the function $f$ defined by $f(x) = x^2 - x + 2$ for values of $x$ near 2. The following table gives values of $f(x)$ for values of $x$ close to 2 but not equal to 2.

FIGURE 1

| $x$ | $f(x)$ | $x$ | $f(x)$ |
|-----|--------|-----|--------|
| 1.0 | 2.000000 | 3.0 | 8.000000 |
| 1.5 | 2.750000 | 2.5 | 5.750000 |
| 1.8 | 3.440000 | 2.2 | 4.640000 |
| 1.9 | 3.710000 | 2.1 | 4.310000 |
| 1.95 | 3.852500 | 2.05 | 4.152500 |
| 1.99 | 3.970100 | 2.01 | 4.030100 |
| 1.995 | 3.985025 | 2.005 | 4.015025 |
| 1.999 | 3.997001 | 2.001 | 4.003001 |

From the table and the graph of $f$ (a parabola) shown in Figure 1 we see that when $x$ is close to 2 (on either side of 2), $f(x)$ is close to 4. In fact, it appears that we can make the values of $f(x)$ as close as we like to 4 by taking $x$ sufficiently close to 2. We express this by saying "the limit of the function $f(x) = x^2 - x + 2$ as $x$ approaches 2 is equal to 4." The notation for this is

$$\lim_{x \to 2} (x^2 - x + 2) = 4$$

In general, we use the following notation.

---

**1** **Definition**   Suppose $f(x)$ is defined when $x$ is near the number $a$. (This means that $f$ is defined on some open interval that contains $a$, except possibly at $a$ itself.) Then we write

$$\lim_{x \to a} f(x) = L$$

and say        "the limit of $f(x)$, as $x$ approaches $a$, equals $L$"

if we can make the values of $f(x)$ arbitrarily close to $L$ (as close to $L$ as we like) by taking $x$ to be sufficiently close to $a$ (on either side of $a$) but not equal to $a$.

---

Roughly speaking, this says that the values of $f(x)$ approach $L$ as $x$ approaches $a$. In other words, the values of $f(x)$ tend to get closer and closer to the number $L$ as $x$ gets closer and closer to the number $a$ (from either side of $a$) but $x \neq a$. (A more precise definition will be given in Section 1.7.)

An alternative notation for

$$\lim_{x \to a} f(x) = L$$

is                            $f(x) \to L$        as        $x \to a$

which is usually read "$f(x)$ approaches $L$ as $x$ approaches $a$."

Notice the phrase "but $x \neq a$" in the definition of limit. This means that in finding the limit of $f(x)$ as $x$ approaches $a$, we never consider $x = a$. In fact, $f(x)$ need not even be defined when $x = a$. The only thing that matters is how $f$ is defined *near a*.

Figure 2 shows the graphs of three functions. Note that in part (c), $f(a)$ is not defined and in part (b), $f(a) \neq L$. But in each case, regardless of what happens at $a$, it is true that $\lim_{x \to a} f(x) = L$.

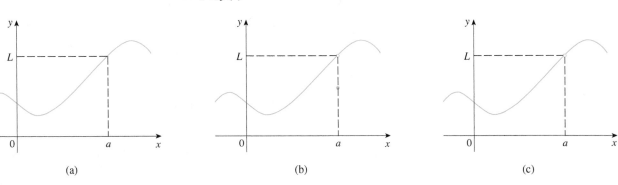

|     |     |     |
|:---:|:---:|:---:|
| (a) | (b) | (c) |

**FIGURE 2**   $\lim_{x \to a} f(x) = L$ in all three cases

**EXAMPLE 1**   Guess the value of $\displaystyle \lim_{x \to 1} \frac{x - 1}{x^2 - 1}$.

SOLUTION   Notice that the function $f(x) = (x - 1)/(x^2 - 1)$ is not defined when $x = 1$, but that doesn't matter because the definition of $\lim_{x \to a} f(x)$ says that we consider values of $x$ that are close to $a$ but not equal to $a$.

The tables at the left give values of $f(x)$ (correct to six decimal places) for values of $x$ that approach 1 (but are not equal to 1). On the basis of the values in the tables, we make the guess that

$$\lim_{x \to 1} \frac{x - 1}{x^2 - 1} = 0.5$$

| $x < 1$ | $f(x)$ |
|:---:|:---:|
| 0.5 | 0.666667 |
| 0.9 | 0.526316 |
| 0.99 | 0.502513 |
| 0.999 | 0.500250 |
| 0.9999 | 0.500025 |

| $x > 1$ | $f(x)$ |
|:---:|:---:|
| 1.5 | 0.400000 |
| 1.1 | 0.476190 |
| 1.01 | 0.497512 |
| 1.001 | 0.499750 |
| 1.0001 | 0.499975 |

Example 1 is illustrated by the graph of $f$ in Figure 3. Now let's change $f$ slightly by giving it the value 2 when $x = 1$ and calling the resulting function $g$:

$$g(x) = \begin{cases} \dfrac{x - 1}{x^2 - 1} & \text{if } x \neq 1 \\ 2 & \text{if } x = 1 \end{cases}$$

This new function $g$ still has the same limit as $x$ approaches 1. (See Figure 4.)

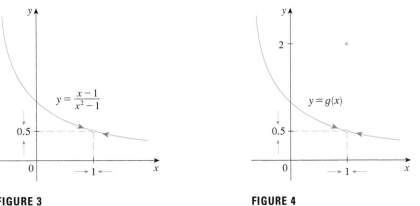

**FIGURE 3**                    **FIGURE 4**

**EXAMPLE 2** Estimate the value of $\displaystyle\lim_{t \to 0} \frac{\sqrt{t^2 + 9} - 3}{t^2}$.

SOLUTION The table lists values of the function for several values of $t$ near 0.

| $t$ | $\dfrac{\sqrt{t^2 + 9} - 3}{t^2}$ |
|---|---|
| $\pm 1.0$ | 0.16228 |
| $\pm 0.5$ | 0.16553 |
| $\pm 0.1$ | 0.16662 |
| $\pm 0.05$ | 0.16666 |
| $\pm 0.01$ | 0.16667 |

As $t$ approaches 0, the values of the function seem to approach $0.1666666\ldots$ and so we guess that

$$\lim_{t \to 0} \frac{\sqrt{t^2 + 9} - 3}{t^2} = \frac{1}{6}$$

| $t$ | $\dfrac{\sqrt{t^2 + 9} - 3}{t^2}$ |
|---|---|
| $\pm 0.0005$ | 0.16800 |
| $\pm 0.0001$ | 0.20000 |
| $\pm 0.00005$ | 0.00000 |
| $\pm 0.00001$ | 0.00000 |

**www.stewartcalculus.com**

For a further explanation of why calculators sometimes give false values, click on *Lies My Calculator and Computer Told Me.* In particular, see the section called *The Perils of Subtraction.*

In Example 2 what would have happened if we had taken even smaller values of $t$? The table in the margin shows the results from one calculator; you can see that something strange seems to be happening.

If you try these calculations on your own calculator you might get different values, but eventually you will get the value 0 if you make $t$ sufficiently small. Does this mean that the answer is really 0 instead of $\frac{1}{6}$? No, the value of the limit is $\frac{1}{6}$, as we will show in the next section. The problem is that the calculator gave false values because $\sqrt{t^2 + 9}$ is very close to 3 when $t$ is small. (In fact, when $t$ is sufficiently small, a calculator's value for $\sqrt{t^2 + 9}$ is $3.000\ldots$ to as many digits as the calculator is capable of carrying.)

Something similar happens when we try to graph the function

$$f(t) = \frac{\sqrt{t^2 + 9} - 3}{t^2}$$

of Example 2 on a graphing calculator or computer. Parts (a) and (b) of Figure 5 show quite accurate graphs of $f$, and when we use the trace mode (if available) we can estimate easily that the limit is about $\frac{1}{6}$. But if we zoom in too much, as in parts (c) and (d), then we get inaccurate graphs, again because of problems with subtraction.

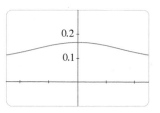

(a) $[-5, 5]$ by $[-0.1, 0.3]$

(b) $[-0.1, 0.1]$ by $[-0.1, 0.3]$

(c) $[-10^{-6}, 10^{-6}]$ by $[-0.1, 0.3]$

(d) $[-10^{-7}, 10^{-7}]$ by $[-0.1, 0.3]$

**FIGURE 5**

**V** **EXAMPLE 3** Guess the value of $\lim\limits_{x \to 0} \dfrac{\sin x}{x}$.

**SOLUTION** The function $f(x) = (\sin x)/x$ is not defined when $x = 0$. Using a calculator (and remembering that, if $x \in \mathbb{R}$, $\sin x$ means the sine of the angle whose *radian* measure is $x$), we construct a table of values correct to eight decimal places. From the table at the left and the graph in Figure 6 we guess that

$$\lim_{x \to 0} \frac{\sin x}{x} = 1$$

This guess is in fact correct, as will be proved in Chapter 2 using a geometric argument.

| $x$ | $\dfrac{\sin x}{x}$ |
|---|---|
| $\pm 1.0$ | 0.84147098 |
| $\pm 0.5$ | 0.95885108 |
| $\pm 0.4$ | 0.97354586 |
| $\pm 0.3$ | 0.98506736 |
| $\pm 0.2$ | 0.99334665 |
| $\pm 0.1$ | 0.99833417 |
| $\pm 0.05$ | 0.99958339 |
| $\pm 0.01$ | 0.99998333 |
| $\pm 0.005$ | 0.99999583 |
| $\pm 0.001$ | 0.99999983 |

**FIGURE 6**

**V** **EXAMPLE 4** Investigate $\lim\limits_{x \to 0} \sin \dfrac{\pi}{x}$.

**SOLUTION** Again the function $f(x) = \sin(\pi/x)$ is undefined at 0. Evaluating the function for some small values of $x$, we get

$$f(1) = \sin \pi = 0 \qquad\qquad f\left(\tfrac{1}{2}\right) = \sin 2\pi = 0$$

$$f\left(\tfrac{1}{3}\right) = \sin 3\pi = 0 \qquad\qquad f\left(\tfrac{1}{4}\right) = \sin 4\pi = 0$$

$$f(0.1) = \sin 10\pi = 0 \qquad\qquad f(0.01) = \sin 100\pi = 0$$

Similarly, $f(0.001) = f(0.0001) = 0$. On the basis of this information we might be tempted to guess that

$$\lim_{x \to 0} \sin \frac{\pi}{x} = 0$$

⊘ but this time our guess is wrong. Note that although $f(1/n) = \sin n\pi = 0$ for any integer $n$, it is also true that $f(x) = 1$ for infinitely many values of $x$ that approach 0. You can see this from the graph of $f$ shown in Figure 7.

**Computer Algebra Systems**

Computer algebra systems (CAS) have commands that compute limits. In order to avoid the types of pitfalls demonstrated in Examples 2, 4, and 5, they don't find limits by numerical experimentation. Instead, they use more sophisticated techniques such as computing infinite series. If you have access to a CAS, use the limit command to compute the limits in the examples of this section and to check your answers in the exercises of this chapter.

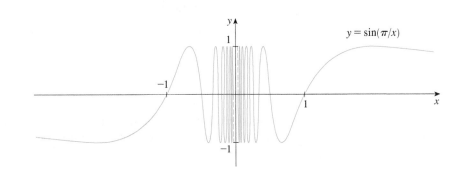

**FIGURE 7**

The dashed lines near the $y$-axis indicate that the values of $\sin(\pi/x)$ oscillate between 1 and $-1$ infinitely often as $x$ approaches 0. (See Exercise 43.)

Since the values of $f(x)$ do not approach a fixed number as $x$ approaches 0,

$$\lim_{x \to 0} \sin \frac{\pi}{x} \quad \text{does not exist}$$

| $x$ | $x^3 + \dfrac{\cos 5x}{10{,}000}$ |
|---|---|
| 1 | 1.000028 |
| 0.5 | 0.124920 |
| 0.1 | 0.001088 |
| 0.05 | 0.000222 |
| 0.01 | 0.000101 |

| $x$ | $x^3 + \dfrac{\cos 5x}{10{,}000}$ |
|---|---|
| 0.005 | 0.00010009 |
| 0.001 | 0.00010000 |

**EXAMPLE 5** Find $\lim\limits_{x \to 0} \left( x^3 + \dfrac{\cos 5x}{10{,}000} \right)$.

SOLUTION As before, we construct a table of values. From the first table in the margin it appears that

$$\lim_{x \to 0} \left( x^3 + \frac{\cos 5x}{10{,}000} \right) = 0$$

But if we persevere with smaller values of $x$, the second table suggests that

$$\lim_{x \to 0} \left( x^3 + \frac{\cos 5x}{10{,}000} \right) = 0.000100 = \frac{1}{10{,}000}$$

Later we will see that $\lim_{x \to 0} \cos 5x = 1$; then it follows that the limit is 0.0001.

⊘ Examples 4 and 5 illustrate some of the pitfalls in guessing the value of a limit. It is easy to guess the wrong value if we use inappropriate values of $x$, but it is difficult to know when to stop calculating values. And, as the discussion after Example 2 shows, sometimes calculators and computers give the wrong values. In the next section, however, we will develop foolproof methods for calculating limits.

**V** **EXAMPLE 6** The Heaviside function $H$ is defined by

$$H(t) = \begin{cases} 0 & \text{if } t < 0 \\ 1 & \text{if } t \geq 0 \end{cases}$$

[This function is named after the electrical engineer Oliver Heaviside (1850–1925) and can be used to describe an electric current that is switched on at time $t = 0$.] Its graph is shown in Figure 8.

As $t$ approaches 0 from the left, $H(t)$ approaches 0. As $t$ approaches 0 from the right, $H(t)$ approaches 1. There is no single number that $H(t)$ approaches as $t$ approaches 0. Therefore $\lim_{t \to 0} H(t)$ does not exist.

**FIGURE 8**
The Heaviside function

### One-Sided Limits

We noticed in Example 6 that $H(t)$ approaches 0 as $t$ approaches 0 from the left and $H(t)$ approaches 1 as $t$ approaches 0 from the right. We indicate this situation symbolically by writing

$$\lim_{t \to 0^-} H(t) = 0 \qquad \text{and} \qquad \lim_{t \to 0^+} H(t) = 1$$

The symbol "$t \to 0^-$" indicates that we consider only values of $t$ that are less than 0. Likewise, "$t \to 0^+$" indicates that we consider only values of $t$ that are greater than 0.

> **2**  **Definition**   We write
>
> $$\lim_{x \to a^-} f(x) = L$$
>
> and say the **left-hand limit of** $f(x)$ **as** $x$ **approaches** $a$ [or the **limit of** $f(x)$ **as** $x$ **approaches** $a$ **from the left**] is equal to $L$ if we can make the values of $f(x)$ arbitrarily close to $L$ by taking $x$ to be sufficiently close to $a$ and $x$ less than $a$.

Notice that Definition 2 differs from Definition 1 only in that we require $x$ to be less than $a$. Similarly, if we require that $x$ be greater than $a$, we get "the **right-hand limit of** $f(x)$ **as** $x$ **approaches** $a$ is equal to $L$" and we write

$$\lim_{x \to a^+} f(x) = L$$

Thus the symbol "$x \to a^+$" means that we consider only $x > a$. These definitions are illustrated in Figure 9.

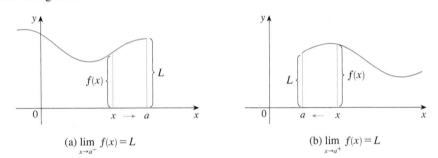

**FIGURE 9**

(a) $\lim\limits_{x \to a^-} f(x) = L$          (b) $\lim\limits_{x \to a^+} f(x) = L$

By comparing Definition 1 with the definitions of one-sided limits, we see that the following is true.

> **3**    $\lim\limits_{x \to a} f(x) = L$   if and only if   $\lim\limits_{x \to a^-} f(x) = L$  and  $\lim\limits_{x \to a^+} f(x) = L$

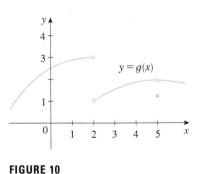

**FIGURE 10**

**V** **EXAMPLE 7**   The graph of a function $g$ is shown in Figure 10. Use it to state the values (if they exist) of the following:

(a) $\lim\limits_{x \to 2^-} g(x)$          (b) $\lim\limits_{x \to 2^+} g(x)$          (c) $\lim\limits_{x \to 2} g(x)$

(d) $\lim\limits_{x \to 5^-} g(x)$          (e) $\lim\limits_{x \to 5^+} g(x)$          (f) $\lim\limits_{x \to 5} g(x)$

SOLUTION  From the graph we see that the values of $g(x)$ approach 3 as $x$ approaches 2 from the left, but they approach 1 as $x$ approaches 2 from the right. Therefore

(a) $\lim\limits_{x \to 2^-} g(x) = 3$     and     (b) $\lim\limits_{x \to 2^+} g(x) = 1$

(c)  Since the left and right limits are different, we conclude from $\boxed{3}$ that $\lim_{x \to 2} g(x)$ does not exist.

The graph also shows that

(d) $\lim\limits_{x \to 5^-} g(x) = 2$     and     (e) $\lim\limits_{x \to 5^+} g(x) = 2$

(f) This time the left and right limits are the same and so, by $\boxed{3}$, we have

$$\lim_{x \to 5} g(x) = 2$$

Despite this fact, notice that $g(5) \neq 2$.

### Infinite Limits

**EXAMPLE 8**  Find $\lim_{x \to 0} \dfrac{1}{x^2}$ if it exists.

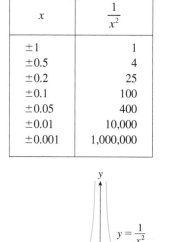

| $x$ | $\dfrac{1}{x^2}$ |
|---|---|
| $\pm 1$ | 1 |
| $\pm 0.5$ | 4 |
| $\pm 0.2$ | 25 |
| $\pm 0.1$ | 100 |
| $\pm 0.05$ | 400 |
| $\pm 0.01$ | 10,000 |
| $\pm 0.001$ | 1,000,000 |

SOLUTION  As $x$ becomes close to 0, $x^2$ also becomes close to 0, and $1/x^2$ becomes very large. (See the table in the margin.) In fact, it appears from the graph of the function $f(x) = 1/x^2$ shown in Figure 11 that the values of $f(x)$ can be made arbitrarily large by taking $x$ close enough to 0. Thus the values of $f(x)$ do not approach a number, so $\lim_{x \to 0} (1/x^2)$ does not exist.

To indicate the kind of behavior exhibited in Example 8, we use the notation

$$\lim_{x \to 0} \frac{1}{x^2} = \infty$$

$y = \dfrac{1}{x^2}$

**FIGURE 11**

⊘ This does not mean that we are regarding $\infty$ as a number. Nor does it mean that the limit exists. It simply expresses the particular way in which the limit does not exist: $1/x^2$ can be made as large as we like by taking $x$ close enough to 0.

In general, we write symbolically

$$\lim_{x \to a} f(x) = \infty$$

to indicate that the values of $f(x)$ tend to become larger and larger (or "increase without bound") as $x$ becomes closer and closer to $a$.

$\boxed{4}$  **Definition**  Let $f$ be a function defined on both sides of $a$, except possibly at $a$ itself. Then

$$\lim_{x \to a} f(x) = \infty$$

means that the values of $f(x)$ can be made arbitrarily large (as large as we please) by taking $x$ sufficiently close to $a$, but not equal to $a$.

$y = f(x)$

$x = a$

**FIGURE 12**

$\lim_{x \to a} f(x) = \infty$

Another notation for $\lim_{x \to a} f(x) = \infty$ is

$$f(x) \to \infty \qquad \text{as} \qquad x \to a$$

Again, the symbol $\infty$ is not a number, but the expression $\lim_{x \to a} f(x) = \infty$ is often read as

"the limit of $f(x)$, as $x$ approaches $a$, is infinity"

or            "$f(x)$ becomes infinite as $x$ approaches $a$"

or            "$f(x)$ increases without bound as $x$ approaches $a$"

This definition is illustrated graphically in Figure 12.

When we say a number is "large negative," we mean that it is negative but its magnitude (absolute value) is large.

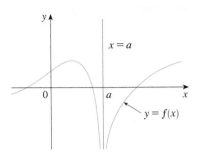

**FIGURE 13**

$$\lim_{x \to a} f(x) = -\infty$$

A similar sort of limit, for functions that become large negative as $x$ gets close to $a$, is defined in Definition 5 and is illustrated in Figure 13.

> **5** **Definition** Let $f$ be defined on both sides of $a$, except possibly at $a$ itself. Then
>
> $$\lim_{x \to a} f(x) = -\infty$$
>
> means that the values of $f(x)$ can be made arbitrarily large negative by taking $x$ sufficiently close to $a$, but not equal to $a$.

The symbol $\lim_{x \to a} f(x) = -\infty$ can be read as "the limit of $f(x)$, as $x$ approaches $a$, is negative infinity" or "$f(x)$ decreases without bound as $x$ approaches $a$." As an example we have

$$\lim_{x \to 0} \left( -\frac{1}{x^2} \right) = -\infty$$

Similar definitions can be given for the one-sided infinite limits

$$\lim_{x \to a^-} f(x) = \infty \qquad\qquad \lim_{x \to a^+} f(x) = \infty$$

$$\lim_{x \to a^-} f(x) = -\infty \qquad\qquad \lim_{x \to a^+} f(x) = -\infty$$

remembering that "$x \to a^-$" means that we consider only values of $x$ that are less than $a$, and similarly "$x \to a^+$" means that we consider only $x > a$. Illustrations of these four cases are given in Figure 14.

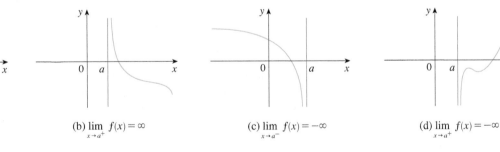

(a) $\lim\limits_{x \to a^-} f(x) = \infty$     (b) $\lim\limits_{x \to a^+} f(x) = \infty$     (c) $\lim\limits_{x \to a^-} f(x) = -\infty$     (d) $\lim\limits_{x \to a^+} f(x) = -\infty$

**FIGURE 14**

> **6** **Definition** The line $x = a$ is called a **vertical asymptote** of the curve $y = f(x)$ if at least one of the following statements is true:
>
> $$\lim_{x \to a} f(x) = \infty \qquad \lim_{x \to a^-} f(x) = \infty \qquad \lim_{x \to a^+} f(x) = \infty$$
>
> $$\lim_{x \to a} f(x) = -\infty \qquad \lim_{x \to a^-} f(x) = -\infty \qquad \lim_{x \to a^+} f(x) = -\infty$$

For instance, the $y$-axis is a vertical asymptote of the curve $y = 1/x^2$ because $\lim_{x \to 0} (1/x^2) = \infty$. In Figure 14 the line $x = a$ is a vertical asymptote in each of the four cases shown. In general, knowledge of vertical asymptotes is very useful in sketching graphs.

**EXAMPLE 9**  Find $\lim\limits_{x \to 3^+} \dfrac{2x}{x-3}$ and $\lim\limits_{x \to 3^-} \dfrac{2x}{x-3}$.

SOLUTION  If $x$ is close to 3 but larger than 3, then the denominator $x-3$ is a small positive number and $2x$ is close to 6. So the quotient $2x/(x-3)$ is a large *positive* number. Thus, intuitively, we see that

$$\lim_{x \to 3^+} \frac{2x}{x-3} = \infty$$

Likewise, if $x$ is close to 3 but smaller than 3, then $x-3$ is a small negative number but $2x$ is still a positive number (close to 6). So $2x/(x-3)$ is a numerically large *negative* number. Thus

$$\lim_{x \to 3^-} \frac{2x}{x-3} = -\infty$$

The graph of the curve $y = 2x/(x-3)$ is given in Figure 15. The line $x = 3$ is a vertical asymptote.

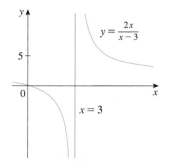

**FIGURE 15**

**EXAMPLE 10**  Find the vertical asymptotes of $f(x) = \tan x$.

SOLUTION  Because

$$\tan x = \frac{\sin x}{\cos x}$$

there are potential vertical asymptotes where $\cos x = 0$. In fact, since $\cos x \to 0^+$ as $x \to (\pi/2)^-$ and $\cos x \to 0^-$ as $x \to (\pi/2)^+$, whereas $\sin x$ is positive when $x$ is near $\pi/2$, we have

$$\lim_{x \to (\pi/2)^-} \tan x = \infty \qquad \text{and} \qquad \lim_{x \to (\pi/2)^+} \tan x = -\infty$$

This shows that the line $x = \pi/2$ is a vertical asymptote. Similar reasoning shows that the lines $x = (2n+1)\pi/2$, where $n$ is an integer, are all vertical asymptotes of $f(x) = \tan x$. The graph in Figure 16 confirms this.

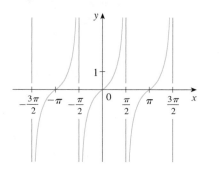

**FIGURE 16**
$y = \tan x$

**1.5**  **Exercises**

**1.** Explain in your own words what is meant by the equation

$$\lim_{x \to 2} f(x) = 5$$

Is it possible for this statement to be true and yet $f(2) = 3$? Explain.

**2.** Explain what it means to say that

$$\lim_{x \to 1^-} f(x) = 3 \quad \text{and} \quad \lim_{x \to 1^+} f(x) = 7$$

In this situation is it possible that $\lim_{x \to 1} f(x)$ exists? Explain.

**3.** Explain the meaning of each of the following.
  (a) $\lim_{x \to -3} f(x) = \infty$       (b) $\lim_{x \to 4^+} f(x) = -\infty$

**4.** Use the given graph of $f$ to state the value of each quantity, if it exists. If it does not exist, explain why.
  (a) $\lim_{x \to 2^-} f(x)$       (b) $\lim_{x \to 2^+} f(x)$       (c) $\lim_{x \to 2} f(x)$

  (d) $f(2)$       (e) $\lim_{x \to 4} f(x)$       (f) $f(4)$

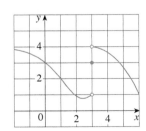

**5.** For the function $f$ whose graph is given, state the value of each quantity, if it exists. If it does not exist, explain why.
  (a) $\lim_{x \to 1} f(x)$       (b) $\lim_{x \to 3^-} f(x)$       (c) $\lim_{x \to 3^+} f(x)$

  (d) $\lim_{x \to 3} f(x)$       (e) $f(3)$

**6.** For the function $h$ whose graph is given, state the value of each quantity, if it exists. If it does not exist, explain why.
  (a) $\lim_{x \to -3^-} h(x)$       (b) $\lim_{x \to -3^+} h(x)$       (c) $\lim_{x \to -3} h(x)$

  (d) $h(-3)$       (e) $\lim_{x \to 0^-} h(x)$       (f) $\lim_{x \to 0^+} h(x)$

  (g) $\lim_{x \to 0} h(x)$       (h) $h(0)$       (i) $\lim_{x \to 2} h(x)$

  (j) $h(2)$       (k) $\lim_{x \to 5^+} h(x)$       (l) $\lim_{x \to 5^-} h(x)$

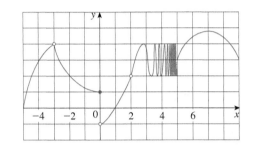

**7.** For the function $g$ whose graph is given, state the value of each quantity, if it exists. If it does not exist, explain why.
  (a) $\lim_{t \to 0^-} g(t)$       (b) $\lim_{t \to 0^+} g(t)$       (c) $\lim_{t \to 0} g(t)$

  (d) $\lim_{t \to 2^-} g(t)$       (e) $\lim_{t \to 2^+} g(t)$       (f) $\lim_{t \to 2} g(t)$

  (g) $g(2)$       (h) $\lim_{t \to 4} g(t)$

**8.** For the function $R$ whose graph is shown, state the following.
  (a) $\lim_{x \to 2} R(x)$                 (b) $\lim_{x \to 5} R(x)$

  (c) $\lim_{x \to -3^-} R(x)$              (d) $\lim_{x \to -3^+} R(x)$

  (e) The equations of the vertical asymptotes.

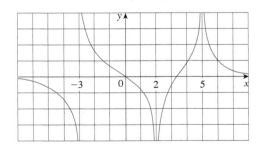

**9.** For the function $f$ whose graph is shown, state the following.

(a) $\lim\limits_{x \to -7} f(x)$    (b) $\lim\limits_{x \to -3} f(x)$    (c) $\lim\limits_{x \to 0} f(x)$

(d) $\lim\limits_{x \to 6^-} f(x)$    (e) $\lim\limits_{x \to 6^+} f(x)$

(f) The equations of the vertical asymptotes.

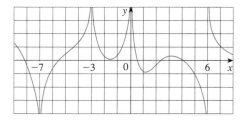

**10.** A patient receives a 150-mg injection of a drug every 4 hours. The graph shows the amount $f(t)$ of the drug in the bloodstream after $t$ hours. Find

$$\lim_{t \to 12^-} f(t) \qquad \text{and} \qquad \lim_{t \to 12^+} f(t)$$

and explain the significance of these one-sided limits.

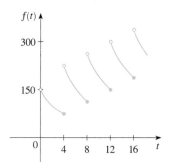

**11–12** Sketch the graph of the function and use it to determine the values of $a$ for which $\lim_{x \to a} f(x)$ exists.

**11.** $f(x) = \begin{cases} 1 + x & \text{if } x < -1 \\ x^2 & \text{if } -1 \le x < 1 \\ 2 - x & \text{if } x \ge 1 \end{cases}$

**12.** $f(x) = \begin{cases} 1 + \sin x & \text{if } x < 0 \\ \cos x & \text{if } 0 \le x \le \pi \\ \sin x & \text{if } x > \pi \end{cases}$

**13–14** Use the graph of the function $f$ to state the value of each limit, if it exists. If it does not exist, explain why.

(a) $\lim\limits_{x \to 0^-} f(x)$    (b) $\lim\limits_{x \to 0^+} f(x)$    (c) $\lim\limits_{x \to 0} f(x)$

**13.** $f(x) = \dfrac{1}{1 + 2^{1/x}}$

**14.** $f(x) = \dfrac{x^2 + x}{\sqrt{x^3 + x^2}}$

**15–18** Sketch the graph of an example of a function $f$ that satisfies all of the given conditions.

**15.** $\lim\limits_{x \to 0^-} f(x) = -1$,    $\lim\limits_{x \to 0^+} f(x) = 2$,    $f(0) = 1$

**16.** $\lim\limits_{x \to 0} f(x) = 1$,    $\lim\limits_{x \to 3^-} f(x) = -2$,    $\lim\limits_{x \to 3^+} f(x) = 2$,
$f(0) = -1$,    $f(3) = 1$

**17.** $\lim\limits_{x \to 3^+} f(x) = 4$,    $\lim\limits_{x \to 3^-} f(x) = 2$,    $\lim\limits_{x \to -2} f(x) = 2$,
$f(3) = 3$,    $f(-2) = 1$

**18.** $\lim\limits_{x \to 0^-} f(x) = 2$,    $\lim\limits_{x \to 0^+} f(x) = 0$,    $\lim\limits_{x \to 4^-} f(x) = 3$,
$\lim\limits_{x \to 4^+} f(x) = 0$,    $f(0) = 2$,    $f(4) = 1$

**19–22** Guess the value of the limit (if it exists) by evaluating the function at the given numbers (correct to six decimal places).

**19.** $\lim\limits_{x \to 2} \dfrac{x^2 - 2x}{x^2 - x - 2}$,
$x = 2.5, 2.1, 2.05, 2.01, 2.005, 2.001,$
$1.9, 1.95, 1.99, 1.995, 1.999$

**20.** $\lim\limits_{x \to -1} \dfrac{x^2 - 2x}{x^2 - x - 2}$,
$x = 0, -0.5, -0.9, -0.95, -0.99, -0.999,$
$-2, -1.5, -1.1, -1.01, -1.001$

**21.** $\lim\limits_{x \to 0} \dfrac{\sin x}{x + \tan x}$,    $x = \pm 1, \pm 0.5, \pm 0.2, \pm 0.1, \pm 0.05, \pm 0.01$

**22.** $\lim\limits_{h \to 0} \dfrac{(2 + h)^5 - 32}{h}$,
$h = \pm 0.5, \pm 0.1, \pm 0.01, \pm 0.001, \pm 0.0001$

**23–26** Use a table of values to estimate the value of the limit. If you have a graphing device, use it to confirm your result graphically.

**23.** $\lim\limits_{x \to 0} \dfrac{\sqrt{x + 4} - 2}{x}$    **24.** $\lim\limits_{x \to 0} \dfrac{\tan 3x}{\tan 5x}$

**25.** $\lim\limits_{x \to 1} \dfrac{x^6 - 1}{x^{10} - 1}$    **26.** $\lim\limits_{x \to 0} \dfrac{9^x - 5^x}{x}$

**27.** (a) By graphing the function $f(x) = (\cos 2x - \cos x)/x^2$ and zooming in toward the point where the graph crosses the $y$-axis, estimate the value of $\lim_{x \to 0} f(x)$.

(b) Check your answer in part (a) by evaluating $f(x)$ for values of $x$ that approach 0.

**28.** (a) Estimate the value of

$$\lim_{x \to 0} \frac{\sin x}{\sin \pi x}$$

by graphing the function $f(x) = (\sin x)/(\sin \pi x)$. State your answer correct to two decimal places.

(b) Check your answer in part (a) by evaluating $f(x)$ for values of $x$ that approach 0.

**29–37** Determine the infinite limit.

**29.** $\displaystyle\lim_{x \to -3^+} \frac{x + 2}{x + 3}$

**30.** $\displaystyle\lim_{x \to -3^-} \frac{x + 2}{x + 3}$

**31.** $\displaystyle\lim_{x \to 1} \frac{2 - x}{(x - 1)^2}$

**32.** $\displaystyle\lim_{x \to 0} \frac{x - 1}{x^2(x + 2)}$

**33.** $\displaystyle\lim_{x \to -2^+} \frac{x - 1}{x^2(x + 2)}$

**34.** $\displaystyle\lim_{x \to \pi^-} \cot x$

**35.** $\displaystyle\lim_{x \to 2\pi^-} x \csc x$

**36.** $\displaystyle\lim_{x \to 2^-} \frac{x^2 - 2x}{x^2 - 4x + 4}$

**37.** $\displaystyle\lim_{x \to 2^+} \frac{x^2 - 2x - 8}{x^2 - 5x + 6}$

**38.** (a) Find the vertical asymptotes of the function

$$y = \frac{x^2 + 1}{3x - 2x^2}$$

(b) Confirm your answer to part (a) by graphing the function.

**39.** Determine $\displaystyle\lim_{x \to 1^-} \frac{1}{x^3 - 1}$ and $\displaystyle\lim_{x \to 1^+} \frac{1}{x^3 - 1}$

(a) by evaluating $f(x) = 1/(x^3 - 1)$ for values of $x$ that approach 1 from the left and from the right,

(b) by reasoning as in Example 9, and

(c) from a graph of $f$.

**40.** (a) By graphing the function $f(x) = (\tan 4x)/x$ and zooming in toward the point where the graph crosses the $y$-axis, estimate the value of $\lim_{x \to 0} f(x)$.

(b) Check your answer in part (a) by evaluating $f(x)$ for values of $x$ that approach 0.

**41.** (a) Evaluate the function $f(x) = x^2 - (2^x/1000)$ for $x = 1$, $0.8, 0.6, 0.4, 0.2, 0.1,$ and $0.05$, and guess the value of

$$\lim_{x \to 0} \left( x^2 - \frac{2^x}{1000} \right)$$

(b) Evaluate $f(x)$ for $x = 0.04, 0.02, 0.01, 0.005, 0.003,$ and $0.001$. Guess again.

**42.** (a) Evaluate $h(x) = (\tan x - x)/x^3$ for $x = 1, 0.5, 0.1, 0.05$, $0.01,$ and $0.005$.

(b) Guess the value of $\displaystyle\lim_{x \to 0} \frac{\tan x - x}{x^3}$.

(c) Evaluate $h(x)$ for successively smaller values of $x$ until you finally reach a value of 0 for $h(x)$. Are you still confident that your guess in part (b) is correct? Explain why you eventually obtained 0 values. (In Section 6.8 a method for evaluating the limit will be explained.)

(d) Graph the function $h$ in the viewing rectangle $[-1, 1]$ by $[0, 1]$. Then zoom in toward the point where the graph crosses the $y$-axis to estimate the limit of $h(x)$ as $x$ approaches 0. Continue to zoom in until you observe distortions in the graph of $h$. Compare with the results of part (c).

**43.** Graph the function $f(x) = \sin(\pi/x)$ of Example 4 in the viewing rectangle $[-1, 1]$ by $[-1, 1]$. Then zoom in toward the origin several times. Comment on the behavior of this function.

**44.** In the theory of relativity, the mass of a particle with velocity $v$ is

$$m = \frac{m_0}{\sqrt{1 - v^2/c^2}}$$

where $m_0$ is the mass of the particle at rest and $c$ is the speed of light. What happens as $v \to c^-$?

**45.** Use a graph to estimate the equations of all the vertical asymptotes of the curve

$$y = \tan(2 \sin x) \qquad -\pi \leqslant x \leqslant \pi$$

Then find the exact equations of these asymptotes.

**46.** (a) Use numerical and graphical evidence to guess the value of the limit

$$\lim_{x \to 1} \frac{x^3 - 1}{\sqrt{x} - 1}$$

(b) How close to 1 does $x$ have to be to ensure that the function in part (a) is within a distance 0.5 of its limit?

## 1.6    Calculating Limits Using the Limit Laws

In Section 1.5 we used calculators and graphs to guess the values of limits, but we saw that such methods don't always lead to the correct answer. In this section we use the following properties of limits, called the *Limit Laws*, to calculate limits.

> **Limit Laws**  Suppose that $c$ is a constant and the limits
>
> $$\lim_{x \to a} f(x) \quad \text{and} \quad \lim_{x \to a} g(x)$$
>
> exist. Then
>
> **1.** $\lim_{x \to a} [f(x) + g(x)] = \lim_{x \to a} f(x) + \lim_{x \to a} g(x)$
>
> **2.** $\lim_{x \to a} [f(x) - g(x)] = \lim_{x \to a} f(x) - \lim_{x \to a} g(x)$
>
> **3.** $\lim_{x \to a} [cf(x)] = c \lim_{x \to a} f(x)$
>
> **4.** $\lim_{x \to a} [f(x) g(x)] = \lim_{x \to a} f(x) \cdot \lim_{x \to a} g(x)$
>
> **5.** $\lim_{x \to a} \dfrac{f(x)}{g(x)} = \dfrac{\lim\limits_{x \to a} f(x)}{\lim\limits_{x \to a} g(x)} \quad \text{if } \lim_{x \to a} g(x) \neq 0$

These five laws can be stated verbally as follows:

Sum Law

**1.** The limit of a sum is the sum of the limits.

Difference Law

**2.** The limit of a difference is the difference of the limits.

Constant Multiple Law

**3.** The limit of a constant times a function is the constant times the limit of the function.

Product Law

**4.** The limit of a product is the product of the limits.

Quotient Law

**5.** The limit of a quotient is the quotient of the limits (provided that the limit of the denominator is not 0).

It is easy to believe that these properties are true. For instance, if $f(x)$ is close to $L$ and $g(x)$ is close to $M$, it is reasonable to conclude that $f(x) + g(x)$ is close to $L + M$. This gives us an intuitive basis for believing that Law 1 is true. In Section 1.7 we give a precise definition of a limit and use it to prove this law. The proofs of the remaining laws are given in Appendix F.

**EXAMPLE 1**  Use the Limit Laws and the graphs of $f$ and $g$ in Figure 1 to evaluate the following limits, if they exist.

(a) $\lim\limits_{x \to -2} [f(x) + 5g(x)]$      (b) $\lim\limits_{x \to 1} [f(x)g(x)]$      (c) $\lim\limits_{x \to 2} \dfrac{f(x)}{g(x)}$

SOLUTION

(a) From the graphs of $f$ and $g$ we see that

$$\lim_{x \to -2} f(x) = 1 \quad \text{and} \quad \lim_{x \to -2} g(x) = -1$$

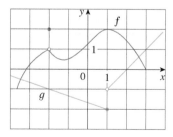

**FIGURE 1**

Therefore we have

$$\lim_{x \to -2} [f(x) + 5g(x)] = \lim_{x \to -2} f(x) + \lim_{x \to -2} [5g(x)] \qquad \text{(by Law 1)}$$

$$= \lim_{x \to -2} f(x) + 5 \lim_{x \to -2} g(x) \qquad \text{(by Law 3)}$$

$$= 1 + 5(-1) = -4$$

(b)  We see that $\lim_{x \to 1} f(x) = 2$. But $\lim_{x \to 1} g(x)$ does not exist because the left and right limits are different:

$$\lim_{x \to 1^-} g(x) = -2 \qquad \lim_{x \to 1^+} g(x) = -1$$

So we can't use Law 4 for the desired limit. But we *can* use Law 4 for the one-sided limits:

$$\lim_{x \to 1^-} [f(x)g(x)] = 2 \cdot (-2) = -4 \qquad \lim_{x \to 1^+} [f(x)g(x)] = 2 \cdot (-1) = -2$$

The left and right limits aren't equal, so $\lim_{x \to 1} [f(x)g(x)]$ does not exist.

(c)  The graphs show that

$$\lim_{x \to 2} f(x) \approx 1.4 \qquad \text{and} \qquad \lim_{x \to 2} g(x) = 0$$

Because the limit of the denominator is 0, we can't use Law 5. The given limit does not exist because the denominator approaches 0 while the numerator approaches a nonzero number.

If we use the Product Law repeatedly with $g(x) = f(x)$, we obtain the following law.

Power Law

> **6.** $\lim_{x \to a} [f(x)]^n = \left[ \lim_{x \to a} f(x) \right]^n \qquad$ where $n$ is a positive integer

In applying these six limit laws, we need to use two special limits:

> **7.** $\lim_{x \to a} c = c$ $\qquad\qquad\qquad\qquad$ **8.** $\lim_{x \to a} x = a$

These limits are obvious from an intuitive point of view (state them in words or draw graphs of $y = c$ and $y = x$), but proofs based on the precise definition are requested in the exercises for Section 1.7.

If we now put $f(x) = x$ in Law 6 and use Law 8, we get another useful special limit.

> **9.** $\lim_{x \to a} x^n = a^n \qquad$ where $n$ is a positive integer

A similar limit holds for roots as follows. (For square roots the proof is outlined in Exercise 37 in Section 1.7.)

> **10.** $\lim_{x \to a} \sqrt[n]{x} = \sqrt[n]{a} \qquad$ where $n$ is a positive integer
>
> (If $n$ is even, we assume that $a > 0$.)

More generally, we have the following law, which is proved in Section 1.8 as a consequence of Law 10.

**11.** $\lim\limits_{x \to a} \sqrt[n]{f(x)} = \sqrt[n]{\lim\limits_{x \to a} f(x)}$     where $n$ is a positive integer

$\left[\text{If } n \text{ is even, we assume that } \lim\limits_{x \to a} f(x) > 0.\right]$

## Newton and Limits

Isaac Newton was born on Christmas Day in 1642, the year of Galileo's death. When he entered Cambridge University in 1661 Newton didn't know much mathematics, but he learned quickly by reading Euclid and Descartes and by attending the lectures of Isaac Barrow. Cambridge was closed because of the plague in 1665 and 1666, and Newton returned home to reflect on what he had learned. Those two years were amazingly productive for at that time he made four of his major discoveries: (1) his representation of functions as sums of infinite series, including the binomial theorem; (2) his work on differential and integral calculus; (3) his laws of motion and law of universal gravitation; and (4) his prism experiments on the nature of light and color. Because of a fear of controversy and criticism, he was reluctant to publish his discoveries and it wasn't until 1687, at the urging of the astronomer Halley, that Newton published *Principia Mathematica*. In this work, the greatest scientific treatise ever written, Newton set forth his version of calculus and used it to investigate mechanics, fluid dynamics, and wave motion, and to explain the motion of planets and comets.

The beginnings of calculus are found in the calculations of areas and volumes by ancient Greek scholars such as Eudoxus and Archimedes. Although aspects of the idea of a limit are implicit in their "method of exhaustion," Eudoxus and Archimedes never explicitly formulated the concept of a limit. Likewise, mathematicians such as Cavalieri, Fermat, and Barrow, the immediate precursors of Newton in the development of calculus, did not actually use limits. It was Isaac Newton who was the first to talk explicitly about limits. He explained that the main idea behind limits is that quantities "approach nearer than by any given difference." Newton stated that the limit was the basic concept in calculus, but it was left to later mathematicians like Cauchy to clarify his ideas about limits.

**EXAMPLE 2** Evaluate the following limits and justify each step.

(a) $\lim\limits_{x \to 5} (2x^2 - 3x + 4)$ 

(b) $\lim\limits_{x \to -2} \dfrac{x^3 + 2x^2 - 1}{5 - 3x}$

SOLUTION

(a) 
$$\lim_{x \to 5} (2x^2 - 3x + 4) = \lim_{x \to 5} (2x^2) - \lim_{x \to 5} (3x) + \lim_{x \to 5} 4 \qquad \text{(by Laws 2 and 1)}$$

$$= 2 \lim_{x \to 5} x^2 - 3 \lim_{x \to 5} x + \lim_{x \to 5} 4 \qquad \text{(by 3)}$$

$$= 2(5^2) - 3(5) + 4 \qquad \text{(by 9, 8, and 7)}$$

$$= 39$$

(b) We start by using Law 5, but its use is fully justified only at the final stage when we see that the limits of the numerator and denominator exist and the limit of the denominator is not 0.

$$\lim_{x \to -2} \frac{x^3 + 2x^2 - 1}{5 - 3x} = \frac{\lim\limits_{x \to -2} (x^3 + 2x^2 - 1)}{\lim\limits_{x \to -2} (5 - 3x)} \qquad \text{(by Law 5)}$$

$$= \frac{\lim\limits_{x \to -2} x^3 + 2 \lim\limits_{x \to -2} x^2 - \lim\limits_{x \to -2} 1}{\lim\limits_{x \to -2} 5 - 3 \lim\limits_{x \to -2} x} \qquad \text{(by 1, 2, and 3)}$$

$$= \frac{(-2)^3 + 2(-2)^2 - 1}{5 - 3(-2)} \qquad \text{(by 9, 8, and 7)}$$

$$= -\frac{1}{11}$$

**NOTE** If we let $f(x) = 2x^2 - 3x + 4$, then $f(5) = 39$. In other words, we would have gotten the correct answer in Example 2(a) by substituting 5 for $x$. Similarly, direct substitution provides the correct answer in part (b). The functions in Example 2 are a polynomial and a rational function, respectively, and similar use of the Limit Laws proves that direct substitution always works for such functions (see Exercises 55 and 56). We state this fact as follows.

**Direct Substitution Property** If $f$ is a polynomial or a rational function and $a$ is in the domain of $f$, then

$$\lim_{x \to a} f(x) = f(a)$$

Functions with the Direct Substitution Property are called *continuous at a* and will be studied in Section 1.8. However, not all limits can be evaluated by direct substitution, as the following examples show.

**EXAMPLE 3**  Find $\lim\limits_{x \to 1} \dfrac{x^2 - 1}{x - 1}$.

SOLUTION  Let $f(x) = (x^2 - 1)/(x - 1)$. We can't find the limit by substituting $x = 1$ because $f(1)$ isn't defined. Nor can we apply the Quotient Law, because the limit of the denominator is 0. Instead, we need to do some preliminary algebra. We factor the numerator as a difference of squares:

$$\frac{x^2 - 1}{x - 1} = \frac{(x - 1)(x + 1)}{x - 1}$$

The numerator and denominator have a common factor of $x - 1$. When we take the limit as $x$ approaches 1, we have $x \neq 1$ and so $x - 1 \neq 0$. Therefore we can cancel the common factor and compute the limit as follows:

$$\lim_{x \to 1} \frac{x^2 - 1}{x - 1} = \lim_{x \to 1} \frac{(x - 1)(x + 1)}{x - 1}$$

$$= \lim_{x \to 1} (x + 1)$$

$$= 1 + 1 = 2$$

The limit in this example arose in Section 1.4 when we were trying to find the tangent to the parabola $y = x^2$ at the point $(1, 1)$.

**NOTE**  In Example 3 we were able to compute the limit by replacing the given function $f(x) = (x^2 - 1)/(x - 1)$ by a simpler function, $g(x) = x + 1$, with the same limit. This is valid because $f(x) = g(x)$ except when $x = 1$, and in computing a limit as $x$ approaches 1 we don't consider what happens when $x$ is actually *equal* to 1. In general, we have the following useful fact.

> If $f(x) = g(x)$ when $x \neq a$, then $\lim\limits_{x \to a} f(x) = \lim\limits_{x \to a} g(x)$, provided the limits exist.

**EXAMPLE 4**  Find $\lim\limits_{x \to 1} g(x)$ where

$$g(x) = \begin{cases} x + 1 & \text{if } x \neq 1 \\ \pi & \text{if } x = 1 \end{cases}$$

SOLUTION  Here $g$ is defined at $x = 1$ and $g(1) = \pi$, but the value of a limit as $x$ approaches 1 does not depend on the value of the function at 1. Since $g(x) = x + 1$ for $x \neq 1$, we have

$$\lim_{x \to 1} g(x) = \lim_{x \to 1} (x + 1) = 2$$

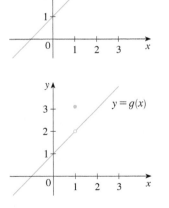

**FIGURE 2**
The graphs of the functions $f$ (from Example 3) and $g$ (from Example 4)

Note that the values of the functions in Examples 3 and 4 are identical except when $x = 1$ (see Figure 2) and so they have the same limit as $x$ approaches 1.

**V EXAMPLE 5** Evaluate $\lim\limits_{h \to 0} \dfrac{(3 + h)^2 - 9}{h}$.

SOLUTION  If we define

$$F(h) = \frac{(3 + h)^2 - 9}{h}$$

then, as in Example 3, we can't compute $\lim_{h \to 0} F(h)$ by letting $h = 0$ since $F(0)$ is undefined. But if we simplify $F(h)$ algebraically, we find that

$$F(h) = \frac{(9 + 6h + h^2) - 9}{h} = \frac{6h + h^2}{h} = 6 + h$$

(Recall that we consider only $h \neq 0$ when letting $h$ approach 0.) Thus

$$\lim_{h \to 0} \frac{(3 + h)^2 - 9}{h} = \lim_{h \to 0} (6 + h) = 6$$

**EXAMPLE 6** Find $\lim\limits_{t \to 0} \dfrac{\sqrt{t^2 + 9} - 3}{t^2}$.

SOLUTION  We can't apply the Quotient Law immediately, since the limit of the denominator is 0. Here the preliminary algebra consists of rationalizing the numerator:

$$\begin{aligned}
\lim_{t \to 0} \frac{\sqrt{t^2 + 9} - 3}{t^2} &= \lim_{t \to 0} \frac{\sqrt{t^2 + 9} - 3}{t^2} \cdot \frac{\sqrt{t^2 + 9} + 3}{\sqrt{t^2 + 9} + 3} \\
&= \lim_{t \to 0} \frac{(t^2 + 9) - 9}{t^2 \left( \sqrt{t^2 + 9} + 3 \right)} \\
&= \lim_{t \to 0} \frac{t^2}{t^2 \left( \sqrt{t^2 + 9} + 3 \right)} \\
&= \lim_{t \to 0} \frac{1}{\sqrt{t^2 + 9} + 3} \\
&= \frac{1}{\sqrt{\lim\limits_{t \to 0} (t^2 + 9)} + 3} \\
&= \frac{1}{3 + 3} = \frac{1}{6}
\end{aligned}$$

This calculation confirms the guess that we made in Example 2 in Section 1.5.

Some limits are best calculated by first finding the left- and right-hand limits. The following theorem is a reminder of what we discovered in Section 1.5. It says that a two-sided limit exists if and only if both of the one-sided limits exist and are equal.

**1  Theorem**    $\lim\limits_{x \to a} f(x) = L$     if and only if     $\lim\limits_{x \to a^-} f(x) = L = \lim\limits_{x \to a^+} f(x)$

When computing one-sided limits, we use the fact that the Limit Laws also hold for one-sided limits.

**EXAMPLE 7** Show that $\lim\limits_{x \to 0} |x| = 0$.

**SOLUTION** Recall that

$$|x| = \begin{cases} x & \text{if } x \geq 0 \\ -x & \text{if } x < 0 \end{cases}$$

Since $|x| = x$ for $x > 0$, we have

$$\lim_{x \to 0^+} |x| = \lim_{x \to 0^+} x = 0$$

For $x < 0$ we have $|x| = -x$ and so

$$\lim_{x \to 0^-} |x| = \lim_{x \to 0^-} (-x) = 0$$

Therefore, by Theorem 1,

$$\lim_{x \to 0} |x| = 0$$

The result of Example 7 looks plausible from Figure 3.

**FIGURE 3**

**V EXAMPLE 8** Prove that $\lim\limits_{x \to 0} \dfrac{|x|}{x}$ does not exist.

**SOLUTION**
$$\lim_{x \to 0^+} \frac{|x|}{x} = \lim_{x \to 0^+} \frac{x}{x} = \lim_{x \to 0^+} 1 = 1$$

$$\lim_{x \to 0^-} \frac{|x|}{x} = \lim_{x \to 0^-} \frac{-x}{x} = \lim_{x \to 0^-} (-1) = -1$$

Since the right- and left-hand limits are different, it follows from Theorem 1 that $\lim_{x \to 0} |x|/x$ does not exist. The graph of the function $f(x) = |x|/x$ is shown in Figure 4 and supports the one-sided limits that we found.

**FIGURE 4**

**EXAMPLE 9** If

$$f(x) = \begin{cases} \sqrt{x - 4} & \text{if } x > 4 \\ 8 - 2x & \text{if } x < 4 \end{cases}$$

determine whether $\lim_{x \to 4} f(x)$ exists.

**SOLUTION** Since $f(x) = \sqrt{x - 4}$ for $x > 4$, we have

$$\lim_{x \to 4^+} f(x) = \lim_{x \to 4^+} \sqrt{x - 4} = \sqrt{4 - 4} = 0$$

Since $f(x) = 8 - 2x$ for $x < 4$, we have

$$\lim_{x \to 4^-} f(x) = \lim_{x \to 4^-} (8 - 2x) = 8 - 2 \cdot 4 = 0$$

The right- and left-hand limits are equal. Thus the limit exists and

$$\lim_{x \to 4} f(x) = 0$$

It is shown in Example 3 in Section 1.7 that $\lim_{x \to 0^+} \sqrt{x} = 0$.

**FIGURE 5**

The graph of $f$ is shown in Figure 5.

Other notations for $[\![x]\!]$ are $[x]$ and $\lfloor x \rfloor$. The greatest integer function is sometimes called the *floor function*.

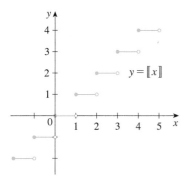

**FIGURE 6**
Greatest integer function

**EXAMPLE 10** The **greatest integer function** is defined by $[\![x]\!]$ = the largest integer that is less than or equal to $x$. (For instance, $[\![4]\!] = 4$, $[\![4.8]\!] = 4$, $[\![\pi]\!] = 3$, $[\![\sqrt{2}\,]\!] = 1$, $[\![-\frac{1}{2}]\!] = -1$.) Show that $\lim_{x \to 3} [\![x]\!]$ does not exist.

SOLUTION The graph of the greatest integer function is shown in Figure 6. Since $[\![x]\!] = 3$ for $3 \le x < 4$, we have

$$\lim_{x \to 3^+} [\![x]\!] = \lim_{x \to 3^+} 3 = 3$$

Since $[\![x]\!] = 2$ for $2 \le x < 3$, we have

$$\lim_{x \to 3^-} [\![x]\!] = \lim_{x \to 3^-} 2 = 2$$

Because these one-sided limits are not equal, $\lim_{x \to 3} [\![x]\!]$ does not exist by Theorem 1.

The next two theorems give two additional properties of limits. Their proofs can be found in Appendix F.

**2** **Theorem** If $f(x) \le g(x)$ when $x$ is near $a$ (except possibly at $a$) and the limits of $f$ and $g$ both exist as $x$ approaches $a$, then

$$\lim_{x \to a} f(x) \le \lim_{x \to a} g(x)$$

**3** **The Squeeze Theorem** If $f(x) \le g(x) \le h(x)$ when $x$ is near $a$ (except possibly at $a$) and

$$\lim_{x \to a} f(x) = \lim_{x \to a} h(x) = L$$

then

$$\lim_{x \to a} g(x) = L$$

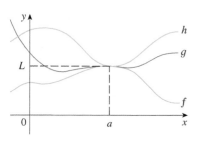

**FIGURE 7**

The Squeeze Theorem, which is sometimes called the Sandwich Theorem or the Pinching Theorem, is illustrated by Figure 7. It says that if $g(x)$ is squeezed between $f(x)$ and $h(x)$ near $a$, and if $f$ and $h$ have the same limit $L$ at $a$, then $g$ is forced to have the same limit $L$ at $a$.

**V** **EXAMPLE 11** Show that $\lim_{x \to 0} x^2 \sin \dfrac{1}{x} = 0$.

SOLUTION First note that we **cannot** use

$$\lim_{x \to 0} x^2 \sin \frac{1}{x} = \lim_{x \to 0} x^2 \cdot \lim_{x \to 0} \sin \frac{1}{x}$$

because $\lim_{x \to 0} \sin(1/x)$ does not exist (see Example 4 in Section 1.5).

Instead we apply the Squeeze Theorem, and so we need to find a function $f$ smaller than $g(x) = x^2 \sin(1/x)$ and a function $h$ bigger than $g$ such that both $f(x)$ and $h(x)$

approach 0. To do this we use our knowledge of the sine function. Because the sine of any number lies between $-1$ and 1, we can write

$$\boxed{4} \qquad -1 \leq \sin\frac{1}{x} \leq 1$$

Any inequality remains true when multiplied by a positive number. We know that $x^2 \geq 0$ for all $x$ and so, multiplying each side of the inequalities in $\boxed{4}$ by $x^2$, we get

$$-x^2 \leq x^2 \sin\frac{1}{x} \leq x^2$$

as illustrated by Figure 8. We know that

$$\lim_{x \to 0} x^2 = 0 \qquad \text{and} \qquad \lim_{x \to 0} (-x^2) = 0$$

Taking $f(x) = -x^2$, $g(x) = x^2 \sin(1/x)$, and $h(x) = x^2$ in the Squeeze Theorem, we obtain

$$\lim_{x \to 0} x^2 \sin\frac{1}{x} = 0$$

**FIGURE 8**

$y = x^2 \sin(1/x)$

## 1.6 Exercises

**1.** Given that

$$\lim_{x \to 2} f(x) = 4 \qquad \lim_{x \to 2} g(x) = -2 \qquad \lim_{x \to 2} h(x) = 0$$

find the limits that exist. If the limit does not exist, explain why.

(a) $\lim_{x \to 2} [f(x) + 5g(x)]$

(b) $\lim_{x \to 2} [g(x)]^3$

(c) $\lim_{x \to 2} \sqrt{f(x)}$

(d) $\lim_{x \to 2} \dfrac{3f(x)}{g(x)}$

(e) $\lim_{x \to 2} \dfrac{g(x)}{h(x)}$

(f) $\lim_{x \to 2} \dfrac{g(x)h(x)}{f(x)}$

**2.** The graphs of $f$ and $g$ are given. Use them to evaluate each limit, if it exists. If the limit does not exist, explain why.

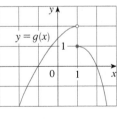

(a) $\lim_{x \to 2} [f(x) + g(x)]$

(b) $\lim_{x \to 1} [f(x) + g(x)]$

(c) $\lim_{x \to 0} [f(x)g(x)]$

(d) $\lim_{x \to -1} \dfrac{f(x)}{g(x)}$

(e) $\lim_{x \to 2} [x^3 f(x)]$

(f) $\lim_{x \to 1} \sqrt{3 + f(x)}$

**3–9** Evaluate the limit and justify each step by indicating the appropriate Limit Law(s).

**3.** $\lim_{x \to 3} (5x^3 - 3x^2 + x - 6)$

**4.** $\lim_{x \to -1} (x^4 - 3x)(x^2 + 5x + 3)$

**5.** $\lim_{t \to -2} \dfrac{t^4 - 2}{2t^2 - 3t + 2}$

**6.** $\lim_{u \to -2} \sqrt{u^4 + 3u + 6}$

**7.** $\lim_{x \to 8} (1 + \sqrt[3]{x})(2 - 6x^2 + x^3)$

**8.** $\lim_{t \to 2} \left(\dfrac{t^2 - 2}{t^3 - 3t + 5}\right)^2$

**9.** $\lim_{x \to 2} \sqrt{\dfrac{2x^2 + 1}{3x - 2}}$

**10.** (a) What is wrong with the following equation?

$$\frac{x^2 + x - 6}{x - 2} = x + 3$$

(b) In view of part (a), explain why the equation

$$\lim_{x \to 2} \frac{x^2 + x - 6}{x - 2} = \lim_{x \to 2} (x + 3)$$

is correct.

---

⌂ Graphing calculator or computer required  **1.** Homework Hints available at stewartcalculus.com

**11–32** Evaluate the limit, if it exists.

**11.** $\lim\limits_{x \to 5} \dfrac{x^2 - 6x + 5}{x - 5}$

**12.** $\lim\limits_{x \to 4} \dfrac{x^2 - 4x}{x^2 - 3x - 4}$

**13.** $\lim\limits_{x \to 5} \dfrac{x^2 - 5x + 6}{x - 5}$

**14.** $\lim\limits_{x \to -1} \dfrac{x^2 - 4x}{x^2 - 3x - 4}$

**15.** $\lim\limits_{t \to -3} \dfrac{t^2 - 9}{2t^2 + 7t + 3}$

**16.** $\lim\limits_{x \to -1} \dfrac{2x^2 + 3x + 1}{x^2 - 2x - 3}$

**17.** $\lim\limits_{h \to 0} \dfrac{(-5 + h)^2 - 25}{h}$

**18.** $\lim\limits_{h \to 0} \dfrac{(2 + h)^3 - 8}{h}$

**19.** $\lim\limits_{x \to -2} \dfrac{x + 2}{x^3 + 8}$

**20.** $\lim\limits_{t \to 1} \dfrac{t^4 - 1}{t^3 - 1}$

**21.** $\lim\limits_{h \to 0} \dfrac{\sqrt{9 + h} - 3}{h}$

**22.** $\lim\limits_{u \to 2} \dfrac{\sqrt{4u + 1} - 3}{u - 2}$

**23.** $\lim\limits_{x \to -4} \dfrac{\frac{1}{4} + \frac{1}{x}}{4 + x}$

**24.** $\lim\limits_{x \to -1} \dfrac{x^2 + 2x + 1}{x^4 - 1}$

**25.** $\lim\limits_{t \to 0} \dfrac{\sqrt{1 + t} - \sqrt{1 - t}}{t}$

**26.** $\lim\limits_{t \to 0} \left( \dfrac{1}{t} - \dfrac{1}{t^2 + t} \right)$

**27.** $\lim\limits_{x \to 16} \dfrac{4 - \sqrt{x}}{16x - x^2}$

**28.** $\lim\limits_{h \to 0} \dfrac{(3 + h)^{-1} - 3^{-1}}{h}$

**29.** $\lim\limits_{t \to 0} \left( \dfrac{1}{t\sqrt{1 + t}} - \dfrac{1}{t} \right)$

**30.** $\lim\limits_{x \to -4} \dfrac{\sqrt{x^2 + 9} - 5}{x + 4}$

**31.** $\lim\limits_{h \to 0} \dfrac{(x + h)^3 - x^3}{h}$

**32.** $\lim\limits_{h \to 0} \dfrac{\frac{1}{(x + h)^2} - \frac{1}{x^2}}{h}$

**33.** (a) Estimate the value of
$$\lim_{x \to 0} \frac{x}{\sqrt{1 + 3x} - 1}$$
by graphing the function $f(x) = x/(\sqrt{1 + 3x} - 1)$.
(b) Make a table of values of $f(x)$ for $x$ close to 0 and guess the value of the limit.
(c) Use the Limit Laws to prove that your guess is correct.

**34.** (a) Use a graph of
$$f(x) = \frac{\sqrt{3 + x} - \sqrt{3}}{x}$$
to estimate the value of $\lim_{x \to 0} f(x)$ to two decimal places.
(b) Use a table of values of $f(x)$ to estimate the limit to four decimal places.
(c) Use the Limit Laws to find the exact value of the limit.

**35.** Use the Squeeze Theorem to show that $\lim_{x \to 0} (x^2 \cos 20\pi x) = 0$. Illustrate by graphing the functions $f(x) = -x^2$, $g(x) = x^2 \cos 20\pi x$, and $h(x) = x^2$ on the same screen.

**36.** Use the Squeeze Theorem to show that
$$\lim_{x \to 0} \sqrt{x^3 + x^2} \, \sin \frac{\pi}{x} = 0$$
Illustrate by graphing the functions $f$, $g$, and $h$ (in the notation of the Squeeze Theorem) on the same screen.

**37.** If $4x - 9 \leqslant f(x) \leqslant x^2 - 4x + 7$ for $x \geqslant 0$, find $\lim\limits_{x \to 4} f(x)$.

**38.** If $2x \leqslant g(x) \leqslant x^4 - x^2 + 2$ for all $x$, evaluate $\lim\limits_{x \to 1} g(x)$.

**39.** Prove that $\lim\limits_{x \to 0} x^4 \cos \dfrac{2}{x} = 0$.

**40.** Prove that $\lim\limits_{x \to 0^+} \sqrt{x}\, [1 + \sin^2(2\pi/x)] = 0$.

**41–46** Find the limit, if it exists. If the limit does not exist, explain why.

**41.** $\lim\limits_{x \to 3} (2x + |x - 3|)$

**42.** $\lim\limits_{x \to -6} \dfrac{2x + 12}{|x + 6|}$

**43.** $\lim\limits_{x \to 0.5^-} \dfrac{2x - 1}{|2x^3 - x^2|}$

**44.** $\lim\limits_{x \to -2} \dfrac{2 - |x|}{2 + x}$

**45.** $\lim\limits_{x \to 0^-} \left( \dfrac{1}{x} - \dfrac{1}{|x|} \right)$

**46.** $\lim\limits_{x \to 0^+} \left( \dfrac{1}{x} - \dfrac{1}{|x|} \right)$

**47.** The *signum* (or *sign*) *function*, denoted by sgn, is defined by
$$\operatorname{sgn} x = \begin{cases} -1 & \text{if } x < 0 \\ 0 & \text{if } x = 0 \\ 1 & \text{if } x > 0 \end{cases}$$
(a) Sketch the graph of this function.
(b) Find each of the following limits or explain why it does not exist.
  (i) $\lim\limits_{x \to 0^+} \operatorname{sgn} x$   (ii) $\lim\limits_{x \to 0^-} \operatorname{sgn} x$
  (iii) $\lim\limits_{x \to 0} \operatorname{sgn} x$   (iv) $\lim\limits_{x \to 0} |\operatorname{sgn} x|$

**48.** Let
$$f(x) = \begin{cases} x^2 + 1 & \text{if } x < 1 \\ (x - 2)^2 & \text{if } x \geqslant 1 \end{cases}$$
(a) Find $\lim_{x \to 1^-} f(x)$ and $\lim_{x \to 1^+} f(x)$.
(b) Does $\lim_{x \to 1} f(x)$ exist?
(c) Sketch the graph of $f$.

**49.** Let $g(x) = \dfrac{x^2 + x - 6}{|x - 2|}$.

(a) Find

(i) $\lim\limits_{x \to 2^+} g(x)$     (ii) $\lim\limits_{x \to 2^-} g(x)$

(b) Does $\lim\limits_{x \to 2} g(x)$ exist?

(c) Sketch the graph of $g$.

**50.** Let

$$g(x) = \begin{cases} x & \text{if } x < 1 \\ 3 & \text{if } x = 1 \\ 2 - x^2 & \text{if } 1 < x \le 2 \\ x - 3 & \text{if } x > 2 \end{cases}$$

(a) Evaluate each of the following, if it exists.

(i) $\lim\limits_{x \to 1^-} g(x)$    (ii) $\lim\limits_{x \to 1} g(x)$    (iii) $g(1)$

(iv) $\lim\limits_{x \to 2^-} g(x)$    (v) $\lim\limits_{x \to 2^+} g(x)$    (vi) $\lim\limits_{x \to 2} g(x)$

(b) Sketch the graph of $g$.

**51.** (a) If the symbol $[\![\ ]\!]$ denotes the greatest integer function defined in Example 10, evaluate

(i) $\lim\limits_{x \to -2^+} [\![x]\!]$    (ii) $\lim\limits_{x \to -2} [\![x]\!]$    (iii) $\lim\limits_{x \to -2.4} [\![x]\!]$

(b) If $n$ is an integer, evaluate

(i) $\lim\limits_{x \to n^-} [\![x]\!]$    (ii) $\lim\limits_{x \to n^+} [\![x]\!]$

(c) For what values of $a$ does $\lim\limits_{x \to a} [\![x]\!]$ exist?

**52.** Let $f(x) = [\![\cos x]\!]$, $-\pi \le x \le \pi$.

(a) Sketch the graph of $f$.

(b) Evaluate each limit, if it exists.

(i) $\lim\limits_{x \to 0} f(x)$      (ii) $\lim\limits_{x \to (\pi/2)^-} f(x)$

(iii) $\lim\limits_{x \to (\pi/2)^+} f(x)$      (iv) $\lim\limits_{x \to \pi/2} f(x)$

(c) For what values of $a$ does $\lim\limits_{x \to a} f(x)$ exist?

**53.** If $f(x) = [\![x]\!] + [\![-x]\!]$, show that $\lim\limits_{x \to 2} f(x)$ exists but is not equal to $f(2)$.

**54.** In the theory of relativity, the Lorentz contraction formula

$$L = L_0 \sqrt{1 - v^2/c^2}$$

expresses the length $L$ of an object as a function of its velocity $v$ with respect to an observer, where $L_0$ is the length of the object at rest and $c$ is the speed of light. Find $\lim\limits_{v \to c^-} L$ and interpret the result. Why is a left-hand limit necessary?

**55.** If $p$ is a polynomial, show that $\lim\limits_{x \to a} p(x) = p(a)$.

**56.** If $r$ is a rational function, use Exercise 55 to show that $\lim\limits_{x \to a} r(x) = r(a)$ for every number $a$ in the domain of $r$.

**57.** If $\lim\limits_{x \to 1} \dfrac{f(x) - 8}{x - 1} = 10$, find $\lim\limits_{x \to 1} f(x)$.

**58.** If $\lim\limits_{x \to 0} \dfrac{f(x)}{x^2} = 5$, find the following limits.

(a) $\lim\limits_{x \to 0} f(x)$      (b) $\lim\limits_{x \to 0} \dfrac{f(x)}{x}$

**59.** If

$$f(x) = \begin{cases} x^2 & \text{if } x \text{ is rational} \\ 0 & \text{if } x \text{ is irrational} \end{cases}$$

prove that $\lim\limits_{x \to 0} f(x) = 0$.

**60.** Show by means of an example that $\lim\limits_{x \to a} [f(x) + g(x)]$ may exist even though neither $\lim\limits_{x \to a} f(x)$ nor $\lim\limits_{x \to a} g(x)$ exists.

**61.** Show by means of an example that $\lim\limits_{x \to a} [f(x) g(x)]$ may exist even though neither $\lim\limits_{x \to a} f(x)$ nor $\lim\limits_{x \to a} g(x)$ exists.

**62.** Evaluate $\lim\limits_{x \to 2} \dfrac{\sqrt{6 - x} - 2}{\sqrt{3 - x} - 1}$.

**63.** Is there a number $a$ such that

$$\lim\limits_{x \to -2} \dfrac{3x^2 + ax + a + 3}{x^2 + x - 2}$$

exists? If so, find the value of $a$ and the value of the limit.

**64.** The figure shows a fixed circle $C_1$ with equation $(x - 1)^2 + y^2 = 1$ and a shrinking circle $C_2$ with radius $r$ and center the origin. $P$ is the point $(0, r)$, $Q$ is the upper point of intersection of the two circles, and $R$ is the point of intersection of the line $PQ$ and the $x$-axis. What happens to $R$ as $C_2$ shrinks, that is, as $r \to 0^+$?

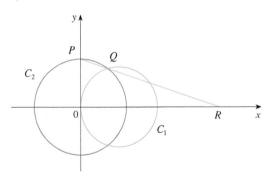

## 1.7  The Precise Definition of a Limit

The intuitive definition of a limit given in Section 1.5 is inadequate for some purposes because such phrases as "$x$ is close to 2" and "$f(x)$ gets closer and closer to $L$" are vague. In order to be able to prove conclusively that

$$\lim_{x \to 0} \left( x^3 + \frac{\cos 5x}{10,000} \right) = 0.0001 \qquad \text{or} \qquad \lim_{x \to 0} \frac{\sin x}{x} = 1$$

we must make the definition of a limit precise.

To motivate the precise definition of a limit, let's consider the function

$$f(x) = \begin{cases} 2x - 1 & \text{if } x \neq 3 \\ 6 & \text{if } x = 3 \end{cases}$$

Intuitively, it is clear that when $x$ is close to 3 but $x \neq 3$, then $f(x)$ is close to 5, and so $\lim_{x \to 3} f(x) = 5$.

To obtain more detailed information about how $f(x)$ varies when $x$ is close to 3, we ask the following question:

How close to 3 does $x$ have to be so that $f(x)$ differs from 5 by less than 0.1?

The distance from $x$ to 3 is $|x - 3|$ and the distance from $f(x)$ to 5 is $|f(x) - 5|$, so our problem is to find a number $\delta$ such that

> It is traditional to use the Greek letter $\delta$ (delta) in this situation.

$$|f(x) - 5| < 0.1 \qquad \text{if} \qquad |x - 3| < \delta \ \text{ but } x \neq 3$$

If $|x - 3| > 0$, then $x \neq 3$, so an equivalent formulation of our problem is to find a number $\delta$ such that

$$|f(x) - 5| < 0.1 \qquad \text{if} \qquad 0 < |x - 3| < \delta$$

Notice that if $0 < |x - 3| < (0.1)/2 = 0.05$, then

$$|f(x) - 5| = |(2x - 1) - 5| = |2x - 6| = 2|x - 3| < 2(0.05) = 0.1$$

that is,

$$|f(x) - 5| < 0.1 \qquad \text{if} \qquad 0 < |x - 3| < 0.05$$

Thus an answer to the problem is given by $\delta = 0.05$; that is, if $x$ is within a distance of 0.05 from 3, then $f(x)$ will be within a distance of 0.1 from 5.

If we change the number 0.1 in our problem to the smaller number 0.01, then by using the same method we find that $f(x)$ will differ from 5 by less than 0.01 provided that $x$ differs from 3 by less than $(0.01)/2 = 0.005$:

$$|f(x) - 5| < 0.01 \qquad \text{if} \qquad 0 < |x - 3| < 0.005$$

Similarly,

$$|f(x) - 5| < 0.001 \qquad \text{if} \qquad 0 < |x - 3| < 0.0005$$

The numbers 0.1, 0.01, and 0.001 that we have considered are *error tolerances* that we might allow. For 5 to be the precise limit of $f(x)$ as $x$ approaches 3, we must not only be able to bring the difference between $f(x)$ and 5 below each of these three numbers; we must be able to bring it below *any* positive number. And, by the same reasoning, we can! If we write $\varepsilon$ (the Greek letter epsilon) for an arbitrary positive number, then we find as before that

$$\boxed{1} \qquad |f(x) - 5| < \varepsilon \quad \text{if} \quad 0 < |x - 3| < \delta = \frac{\varepsilon}{2}$$

This is a precise way of saying that $f(x)$ is close to 5 when $x$ is close to 3 because $\boxed{1}$ says that we can make the values of $f(x)$ within an arbitrary distance $\varepsilon$ from 5 by taking the values of $x$ within a distance $\varepsilon/2$ from 3 (but $x \neq 3$).

Note that $\boxed{1}$ can be rewritten as follows:

$$\text{if} \quad 3 - \delta < x < 3 + \delta \quad (x \neq 3) \qquad \text{then} \qquad 5 - \varepsilon < f(x) < 5 + \varepsilon$$

and this is illustrated in Figure 1. By taking the values of $x$ ($\neq 3$) to lie in the interval $(3 - \delta, 3 + \delta)$ we can make the values of $f(x)$ lie in the interval $(5 - \varepsilon, 5 + \varepsilon)$.

Using $\boxed{1}$ as a model, we give a precise definition of a limit.

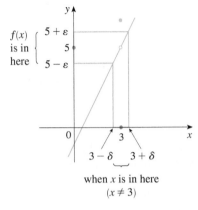

**FIGURE 1**

---

$\boxed{2}$ **Definition** Let $f$ be a function defined on some open interval that contains the number $a$, except possibly at $a$ itself. Then we say that the **limit of $f(x)$ as $x$ approaches $a$ is $L$**, and we write

$$\lim_{x \to a} f(x) = L$$

if for every number $\varepsilon > 0$ there is a number $\delta > 0$ such that

$$\text{if} \quad 0 < |x - a| < \delta \qquad \text{then} \qquad |f(x) - L| < \varepsilon$$

---

Since $|x - a|$ is the distance from $x$ to $a$ and $|f(x) - L|$ is the distance from $f(x)$ to $L$, and since $\varepsilon$ can be arbitrarily small, the definition of a limit can be expressed in words as follows:

$\lim_{x \to a} f(x) = L$ means that the distance between $f(x)$ and $L$ can be made arbitrarily small by taking the distance from $x$ to $a$ sufficiently small (but not 0).

Alternatively,

$\lim_{x \to a} f(x) = L$ means that the values of $f(x)$ can be made as close as we please to $L$ by taking $x$ close enough to $a$ (but not equal to $a$).

We can also reformulate Definition 2 in terms of intervals by observing that the inequality $|x - a| < \delta$ is equivalent to $-\delta < x - a < \delta$, which in turn can be written as $a - \delta < x < a + \delta$. Also $0 < |x - a|$ is true if and only if $x - a \neq 0$, that is, $x \neq a$. Similarly, the inequality $|f(x) - L| < \varepsilon$ is equivalent to the pair of inequalities $L - \varepsilon < f(x) < L + \varepsilon$. Therefore, in terms of intervals, Definition 2 can be stated as follows:

$\lim_{x \to a} f(x) = L$ means that for every $\varepsilon > 0$ (no matter how small $\varepsilon$ is) we can find $\delta > 0$ such that if $x$ lies in the open interval $(a - \delta, a + \delta)$ and $x \neq a$, then $f(x)$ lies in the open interval $(L - \varepsilon, L + \varepsilon)$.

We interpret this statement geometrically by representing a function by an arrow diagram as in Figure 2, where $f$ maps a subset of $\mathbb{R}$ onto another subset of $\mathbb{R}$.

**FIGURE 2**

The definition of limit says that if any small interval $(L - \varepsilon, L + \varepsilon)$ is given around $L$, then we can find an interval $(a - \delta, a + \delta)$ around $a$ such that $f$ maps all the points in $(a - \delta, a + \delta)$ (except possibly $a$) into the interval $(L - \varepsilon, L + \varepsilon)$. (See Figure 3.)

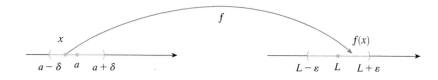

**FIGURE 3**

Another geometric interpretation of limits can be given in terms of the graph of a function. If $\varepsilon > 0$ is given, then we draw the horizontal lines $y = L + \varepsilon$ and $y = L - \varepsilon$ and the graph of $f$. (See Figure 4.) If $\lim_{x \to a} f(x) = L$, then we can find a number $\delta > 0$ such that if we restrict $x$ to lie in the interval $(a - \delta, a + \delta)$ and take $x \neq a$, then the curve $y = f(x)$ lies between the lines $y = L - \varepsilon$ and $y = L + \varepsilon$. (See Figure 5.) You can see that if such a $\delta$ has been found, then any smaller $\delta$ will also work.

It is important to realize that the process illustrated in Figures 4 and 5 must work for *every* positive number $\varepsilon$, no matter how small it is chosen. Figure 6 shows that if a smaller $\varepsilon$ is chosen, then a smaller $\delta$ may be required.

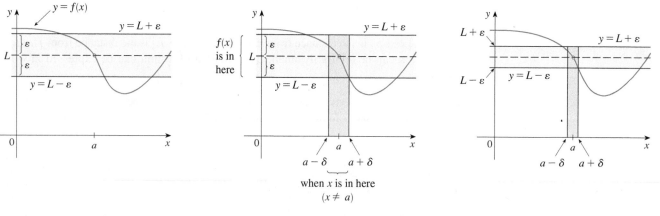

**FIGURE 4**          **FIGURE 5**          **FIGURE 6**

**EXAMPLE 1** Use a graph to find a number $\delta$ such that

$$\text{if} \quad |x - 1| < \delta \quad \text{then} \quad |(x^3 - 5x + 6) - 2| < 0.2$$

In other words, find a number $\delta$ that corresponds to $\varepsilon = 0.2$ in the definition of a limit for the function $f(x) = x^3 - 5x + 6$ with $a = 1$ and $L = 2$.

**FIGURE 7**

**FIGURE 8**

**SOLUTION** A graph of $f$ is shown in Figure 7; we are interested in the region near the point $(1, 2)$. Notice that we can rewrite the inequality

$$|(x^3 - 5x + 6) - 2| < 0.2$$

as

$$1.8 < x^3 - 5x + 6 < 2.2$$

So we need to determine the values of $x$ for which the curve $y = x^3 - 5x + 6$ lies between the horizontal lines $y = 1.8$ and $y = 2.2$. Therefore we graph the curves $y = x^3 - 5x + 6$, $y = 1.8$, and $y = 2.2$ near the point $(1, 2)$ in Figure 8. Then we use the cursor to estimate that the $x$-coordinate of the point of intersection of the line $y = 2.2$ and the curve $y = x^3 - 5x + 6$ is about 0.911. Similarly, $y = x^3 - 5x + 6$ intersects the line $y = 1.8$ when $x \approx 1.124$. So, rounding to be safe, we can say that

$$\text{if} \qquad 0.92 < x < 1.12 \qquad \text{then} \qquad 1.8 < x^3 - 5x + 6 < 2.2$$

This interval $(0.92, 1.12)$ is not symmetric about $x = 1$. The distance from $x = 1$ to the left endpoint is $1 - 0.92 = 0.08$ and the distance to the right endpoint is 0.12. We can choose $\delta$ to be the smaller of these numbers, that is, $\delta = 0.08$. Then we can rewrite our inequalities in terms of distances as follows:

$$\text{if} \qquad |x - 1| < 0.08 \qquad \text{then} \qquad |(x^3 - 5x + 6) - 2| < 0.2$$

This just says that by keeping $x$ within 0.08 of 1, we are able to keep $f(x)$ within 0.2 of 2.

Although we chose $\delta = 0.08$, any smaller positive value of $\delta$ would also have worked. ▪

The graphical procedure in Example 1 gives an illustration of the definition for $\varepsilon = 0.2$, but it does not *prove* that the limit is equal to 2. A proof has to provide a $\delta$ for *every* $\varepsilon$.

In proving limit statements it may be helpful to think of the definition of limit as a challenge. First it challenges you with a number $\varepsilon$. Then you must be able to produce a suitable $\delta$. You have to be able to do this for *every* $\varepsilon > 0$, not just a particular $\varepsilon$.

Imagine a contest between two people, A and B, and imagine yourself to be B. Person A stipulates that the fixed number $L$ should be approximated by the values of $f(x)$ to within a degree of accuracy $\varepsilon$ (say, 0.01). Person B then responds by finding a number $\delta$ such that if $0 < |x - a| < \delta$, then $|f(x) - L| < \varepsilon$. Then A may become more exacting and challenge B with a smaller value of $\varepsilon$ (say, 0.0001). Again B has to respond by finding a corresponding $\delta$. Usually the smaller the value of $\varepsilon$, the smaller the corresponding value of $\delta$ must be. If B always wins, no matter how small A makes $\varepsilon$, then $\lim_{x \to a} f(x) = L$.

**V EXAMPLE 2** Prove that $\lim_{x \to 3} (4x - 5) = 7$.

**SOLUTION**

**1.** *Preliminary analysis of the problem (guessing a value for $\delta$).* Let $\varepsilon$ be a given positive number. We want to find a number $\delta$ such that

$$\text{if} \qquad 0 < |x - 3| < \delta \qquad \text{then} \qquad |(4x - 5) - 7| < \varepsilon$$

But $|(4x - 5) - 7| = |4x - 12| = |4(x - 3)| = 4|x - 3|$. Therefore we want $\delta$ such that

$$\text{if} \qquad 0 < |x - 3| < \delta \qquad \text{then} \qquad 4|x - 3| < \varepsilon$$

that is,

$$\text{if} \qquad 0 < |x - 3| < \delta \qquad \text{then} \qquad |x - 3| < \frac{\varepsilon}{4}$$

This suggests that we should choose $\delta = \varepsilon/4$.

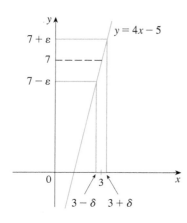

**FIGURE 9**

**2.** *Proof (showing that this δ works).*  Given $\varepsilon > 0$, choose $\delta = \varepsilon/4$. If $0 < |x - 3| < \delta$, then

$$|(4x - 5) - 7| = |4x - 12| = 4|x - 3| < 4\delta = 4\left(\frac{\varepsilon}{4}\right) = \varepsilon$$

Thus

$$\text{if} \qquad 0 < |x - 3| < \delta \qquad \text{then} \qquad |(4x - 5) - 7| < \varepsilon$$

Therefore, by the definition of a limit,

$$\lim_{x \to 3} (4x - 5) = 7$$

This example is illustrated by Figure 9.

Note that in the solution of Example 2 there were two stages—guessing and proving. We made a preliminary analysis that enabled us to guess a value for δ. But then in the second stage we had to go back and prove in a careful, logical fashion that we had made a correct guess. This procedure is typical of much of mathematics. Sometimes it is necessary to first make an intelligent guess about the answer to a problem and then prove that the guess is correct.

The intuitive definitions of one-sided limits that were given in Section 1.5 can be precisely reformulated as follows.

---

**3**  **Definition of Left-Hand Limit**

$$\lim_{x \to a^-} f(x) = L$$

if for every number $\varepsilon > 0$ there is a number $\delta > 0$ such that

$$\text{if} \qquad a - \delta < x < a \qquad \text{then} \qquad |f(x) - L| < \varepsilon$$

---

**4**  **Definition of Right-Hand Limit**

$$\lim_{x \to a^+} f(x) = L$$

if for every number $\varepsilon > 0$ there is a number $\delta > 0$ such that

$$\text{if} \qquad a < x < a + \delta \qquad \text{then} \qquad |f(x) - L| < \varepsilon$$

---

Notice that Definition 3 is the same as Definition 2 except that $x$ is restricted to lie in the *left* half $(a - \delta, a)$ of the interval $(a - \delta, a + \delta)$. In Definition 4, $x$ is restricted to lie in the *right* half $(a, a + \delta)$ of the interval $(a - \delta, a + \delta)$.

**V** **EXAMPLE 3**  Use Definition 4 to prove that $\lim_{x \to 0^+} \sqrt{x} = 0$.

SOLUTION

**1.** *Guessing a value for δ.* Let $\varepsilon$ be a given positive number. Here $a = 0$ and $L = 0$, so we want to find a number δ such that

$$\text{if} \qquad 0 < x < \delta \qquad \text{then} \qquad |\sqrt{x} - 0| < \varepsilon$$

that is,    $$\text{if} \qquad 0 < x < \delta \qquad \text{then} \qquad \sqrt{x} < \varepsilon$$

or, squaring both sides of the inequality $\sqrt{x} < \varepsilon$, we get

$$\text{if} \quad 0 < x < \delta \quad \text{then} \quad x < \varepsilon^2$$

This suggests that we should choose $\delta = \varepsilon^2$.

**2.** *Showing that this $\delta$ works.* Given $\varepsilon > 0$, let $\delta = \varepsilon^2$. If $0 < x < \delta$, then

$$\sqrt{x} < \sqrt{\delta} = \sqrt{\varepsilon^2} = \varepsilon$$

so

$$|\sqrt{x} - 0| < \varepsilon$$

According to Definition 4, this shows that $\lim_{x \to 0^+} \sqrt{x} = 0$.

**EXAMPLE 4** Prove that $\lim_{x \to 3} x^2 = 9$.

SOLUTION

**1.** *Guessing a value for $\delta$.* Let $\varepsilon > 0$ be given. We have to find a number $\delta > 0$ such that

$$\text{if} \quad 0 < |x - 3| < \delta \quad \text{then} \quad |x^2 - 9| < \varepsilon$$

To connect $|x^2 - 9|$ with $|x - 3|$ we write $|x^2 - 9| = |(x + 3)(x - 3)|$. Then we want

$$\text{if} \quad 0 < |x - 3| < \delta \quad \text{then} \quad |x + 3||x - 3| < \varepsilon$$

Notice that if we can find a positive constant $C$ such that $|x + 3| < C$, then

$$|x + 3||x - 3| < C|x - 3|$$

and we can make $C|x - 3| < \varepsilon$ by taking $|x - 3| < \varepsilon/C = \delta$.

We can find such a number $C$ if we restrict $x$ to lie in some interval centered at 3. In fact, since we are interested only in values of $x$ that are close to 3, it is reasonable to assume that $x$ is within a distance 1 from 3, that is, $|x - 3| < 1$. Then $2 < x < 4$, so $5 < x + 3 < 7$. Thus we have $|x + 3| < 7$, and so $C = 7$ is a suitable choice for the constant.

But now there are two restrictions on $|x - 3|$, namely

$$|x - 3| < 1 \quad \text{and} \quad |x - 3| < \frac{\varepsilon}{C} = \frac{\varepsilon}{7}$$

To make sure that both of these inequalities are satisfied, we take $\delta$ to be the smaller of the two numbers 1 and $\varepsilon/7$. The notation for this is $\delta = \min\{1, \varepsilon/7\}$.

**2.** *Showing that this $\delta$ works.* Given $\varepsilon > 0$, let $\delta = \min\{1, \varepsilon/7\}$. If $0 < |x - 3| < \delta$, then $|x - 3| < 1 \Rightarrow 2 < x < 4 \Rightarrow |x + 3| < 7$ (as in part 1). We also have $|x - 3| < \varepsilon/7$, so

$$|x^2 - 9| = |x + 3||x - 3| < 7 \cdot \frac{\varepsilon}{7} = \varepsilon$$

This shows that $\lim_{x \to 3} x^2 = 9$.

As Example 4 shows, it is not always easy to prove that limit statements are true using the $\varepsilon$, $\delta$ definition. In fact, if we had been given a more complicated function such as $f(x) = (6x^2 - 8x + 9)/(2x^2 - 1)$, a proof would require a great deal of ingenuity. Fortunately this is unnecessary because the Limit Laws stated in Section 1.6 can be proved using Definition 2, and then the limits of complicated functions can be found rigorously from the Limit Laws without resorting to the definition directly.

For instance, we prove the Sum Law: If $\lim_{x \to a} f(x) = L$ and $\lim_{x \to a} g(x) = M$ both exist, then

$$\lim_{x \to a} [f(x) + g(x)] = L + M$$

The remaining laws are proved in the exercises and in Appendix F.

PROOF OF THE SUM LAW   Let $\varepsilon > 0$ be given. We must find $\delta > 0$ such that

$$\text{if} \quad 0 < |x - a| < \delta \quad \text{then} \quad |f(x) + g(x) - (L + M)| < \varepsilon$$

Using the Triangle Inequality we can write

Triangle Inequality:

$$|a + b| \le |a| + |b|$$

(See Appendix A.)

$$\boxed{5} \qquad |f(x) + g(x) - (L + M)| = |(f(x) - L) + (g(x) - M)|$$

$$\le |f(x) - L| + |g(x) - M|$$

We make $|f(x) + g(x) - (L + M)|$ less than $\varepsilon$ by making each of the terms $|f(x) - L|$ and $|g(x) - M|$ less than $\varepsilon/2$.

Since $\varepsilon/2 > 0$ and $\lim_{x \to a} f(x) = L$, there exists a number $\delta_1 > 0$ such that

$$\text{if} \quad 0 < |x - a| < \delta_1 \quad \text{then} \quad |f(x) - L| < \frac{\varepsilon}{2}$$

Similarly, since $\lim_{x \to a} g(x) = M$, there exists a number $\delta_2 > 0$ such that

$$\text{if} \quad 0 < |x - a| < \delta_2 \quad \text{then} \quad |g(x) - M| < \frac{\varepsilon}{2}$$

Let $\delta = \min\{\delta_1, \delta_2\}$, the smaller of the numbers $\delta_1$ and $\delta_2$. Notice that

$$\text{if} \quad 0 < |x - a| < \delta \quad \text{then} \quad 0 < |x - a| < \delta_1 \quad \text{and} \quad 0 < |x - a| < \delta_2$$

and so

$$|f(x) - L| < \frac{\varepsilon}{2} \quad \text{and} \quad |g(x) - M| < \frac{\varepsilon}{2}$$

Therefore, by $\boxed{5}$,

$$|f(x) + g(x) - (L + M)| \le |f(x) - L| + |g(x) - M|$$

$$< \frac{\varepsilon}{2} + \frac{\varepsilon}{2} = \varepsilon$$

To summarize,

$$\text{if} \quad 0 < |x - a| < \delta \quad \text{then} \quad |f(x) + g(x) - (L + M)| < \varepsilon$$

Thus, by the definition of a limit,

$$\lim_{x \to a} [f(x) + g(x)] = L + M$$

### Infinite Limits

Infinite limits can also be defined in a precise way. The following is a precise version of Definition 4 in Section 1.5.

---

**6 Definition** Let $f$ be a function defined on some open interval that contains the number $a$, except possibly at $a$ itself. Then

$$\lim_{x \to a} f(x) = \infty$$

means that for every positive number $M$ there is a positive number $\delta$ such that

$$\text{if} \quad 0 < |x - a| < \delta \quad \text{then} \quad f(x) > M$$

---

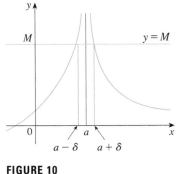

**FIGURE 10**

This says that the values of $f(x)$ can be made arbitrarily large (larger than any given number $M$) by taking $x$ close enough to $a$ (within a distance $\delta$, where $\delta$ depends on $M$, but with $x \neq a$). A geometric illustration is shown in Figure 10.

Given any horizontal line $y = M$, we can find a number $\delta > 0$ such that if we restrict $x$ to lie in the interval $(a - \delta, a + \delta)$ but $x \neq a$, then the curve $y = f(x)$ lies above the line $y = M$. You can see that if a larger $M$ is chosen, then a smaller $\delta$ may be required.

**V EXAMPLE 5** Use Definition 6 to prove that $\lim\limits_{x \to 0} \dfrac{1}{x^2} = \infty$.

SOLUTION Let $M$ be a given positive number. We want to find a number $\delta$ such that

$$\text{if} \quad 0 < |x| < \delta \quad \text{then} \quad 1/x^2 > M$$

But

$$\frac{1}{x^2} > M \quad \Longleftrightarrow \quad x^2 < \frac{1}{M} \quad \Longleftrightarrow \quad |x| < \frac{1}{\sqrt{M}}$$

So if we choose $\delta = 1/\sqrt{M}$ and $0 < |x| < \delta = 1/\sqrt{M}$, then $1/x^2 > M$. This shows that $1/x^2 \to \infty$ as $x \to 0$.

Similarly, the following is a precise version of Definition 5 in Section 1.5. It is illustrated by Figure 11.

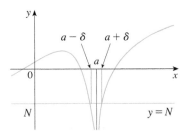

**FIGURE 11**

---

**7 Definition** Let $f$ be a function defined on some open interval that contains the number $a$, except possibly at $a$ itself. Then

$$\lim_{x \to a} f(x) = -\infty$$

means that for every negative number $N$ there is a positive number $\delta$ such that

$$\text{if} \quad 0 < |x - a| < \delta \quad \text{then} \quad f(x) < N$$

---

1.7 **Exercises**

**1.** Use the given graph of $f$ to find a number $\delta$ such that

$$\text{if} \quad |x - 1| < \delta \quad \text{then} \quad |f(x) - 1| < 0.2$$

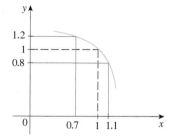

**2.** Use the given graph of $f$ to find a number $\delta$ such that

$$\text{if} \quad 0 < |x - 3| < \delta \quad \text{then} \quad |f(x) - 2| < 0.5$$

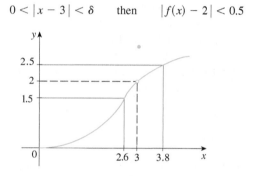

**3.** Use the given graph of $f(x) = \sqrt{x}$ to find a number $\delta$ such that

$$\text{if} \quad |x - 4| < \delta \quad \text{then} \quad |\sqrt{x} - 2| < 0.4$$

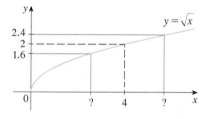

**4.** Use the given graph of $f(x) = x^2$ to find a number $\delta$ such that

$$\text{if} \quad |x - 1| < \delta \quad \text{then} \quad |x^2 - 1| < \tfrac{1}{2}$$

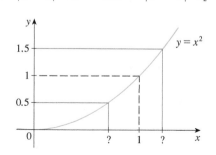

**5.** Use a graph to find a number $\delta$ such that

$$\text{if} \quad \left| x - \frac{\pi}{4} \right| < \delta \quad \text{then} \quad |\tan x - 1| < 0.2$$

**6.** Use a graph to find a number $\delta$ such that

$$\text{if} \quad |x - 1| < \delta \quad \text{then} \quad \left| \frac{2x}{x^2 + 4} - 0.4 \right| < 0.1$$

**7.** For the limit

$$\lim_{x \to 2} (x^3 - 3x + 4) = 6$$

illustrate Definition 2 by finding values of $\delta$ that correspond to $\varepsilon = 0.2$ and $\varepsilon = 0.1$.

**8.** For the limit

$$\lim_{x \to 2} \frac{4x + 1}{3x - 4} = 4.5$$

illustrate Definition 2 by finding values of $\delta$ that correspond to $\varepsilon = 0.5$ and $\varepsilon = 0.1$.

**9.** Given that $\lim_{x \to \pi/2} \tan^2 x = \infty$, illustrate Definition 6 by finding values of $\delta$ that correspond to (a) $M = 1000$ and (b) $M = 10,000$.

**10.** Use a graph to find a number $\delta$ such that

$$\text{if} \quad 5 < x < 5 + \delta \quad \text{then} \quad \frac{x^2}{\sqrt{x - 5}} > 100$$

**11.** A machinist is required to manufacture a circular metal disk with area 1000 cm².
   (a) What radius produces such a disk?
   (b) If the machinist is allowed an error tolerance of ±5 cm² in the area of the disk, how close to the ideal radius in part (a) must the machinist control the radius?
   (c) In terms of the $\varepsilon$, $\delta$ definition of $\lim_{x \to a} f(x) = L$, what is $x$? What is $f(x)$? What is $a$? What is $L$? What value of $\varepsilon$ is given? What is the corresponding value of $\delta$?

**12.** A crystal growth furnace is used in research to determine how best to manufacture crystals used in electronic components for the space shuttle. For proper growth of the crystal, the temperature must be controlled accurately by adjusting the input power. Suppose the relationship is given by

$$T(w) = 0.1w^2 + 2.155w + 20$$

where $T$ is the temperature in degrees Celsius and $w$ is the power input in watts.
   (a) How much power is needed to maintain the temperature at 200°C?
   (b) If the temperature is allowed to vary from 200°C by up to ±1°C, what range of wattage is allowed for the input power?

(c) In terms of the $\varepsilon$, $\delta$ definition of $\lim_{x \to a} f(x) = L$, what is $x$? What is $f(x)$? What is $a$? What is $L$? What value of $\varepsilon$ is given? What is the corresponding value of $\delta$?

**13.** (a) Find a number $\delta$ such that if $|x - 2| < \delta$, then $|4x - 8| < \varepsilon$, where $\varepsilon = 0.1$.
(b) Repeat part (a) with $\varepsilon = 0.01$.

**14.** Given that $\lim_{x \to 2}(5x - 7) = 3$, illustrate Definition 2 by finding values of $\delta$ that correspond to $\varepsilon = 0.1$, $\varepsilon = 0.05$, and $\varepsilon = 0.01$.

**15–18** Prove the statement using the $\varepsilon$, $\delta$ definition of a limit and illustrate with a diagram like Figure 9.

**15.** $\lim_{x \to 3} \left(1 + \frac{1}{3}x\right) = 2$

**16.** $\lim_{x \to 4} (2x - 5) = 3$

**17.** $\lim_{x \to -3} (1 - 4x) = 13$

**18.** $\lim_{x \to -2} (3x + 5) = -1$

**19–32** Prove the statement using the $\varepsilon$, $\delta$ definition of a limit.

**19.** $\lim_{x \to 1} \dfrac{2 + 4x}{3} = 2$

**20.** $\lim_{x \to 10} \left(3 - \frac{4}{5}x\right) = -5$

**21.** $\lim_{x \to 2} \dfrac{x^2 + x - 6}{x - 2} = 5$

**22.** $\lim_{x \to -1.5} \dfrac{9 - 4x^2}{3 + 2x} = 6$

**23.** $\lim_{x \to a} x = a$

**24.** $\lim_{x \to a} c = c$

**25.** $\lim_{x \to 0} x^2 = 0$

**26.** $\lim_{x \to 0} x^3 = 0$

**27.** $\lim_{x \to 0} |x| = 0$

**28.** $\lim_{x \to -6^+} \sqrt[8]{6 + x} = 0$

**29.** $\lim_{x \to 2} (x^2 - 4x + 5) = 1$

**30.** $\lim_{x \to 2} (x^2 + 2x - 7) = 1$

**31.** $\lim_{x \to -2} (x^2 - 1) = 3$

**32.** $\lim_{x \to 2} x^3 = 8$

**33.** Verify that another possible choice of $\delta$ for showing that $\lim_{x \to 3} x^2 = 9$ in Example 4 is $\delta = \min\{2, \varepsilon/8\}$.

**34.** Verify, by a geometric argument, that the largest possible choice of $\delta$ for showing that $\lim_{x \to 3} x^2 = 9$ is $\delta = \sqrt{9 + \varepsilon} - 3$.

CAS **35.** (a) For the limit $\lim_{x \to 1} (x^3 + x + 1) = 3$, use a graph to find a value of $\delta$ that corresponds to $\varepsilon = 0.4$.

(b) By using a computer algebra system to solve the cubic equation $x^3 + x + 1 = 3 + \varepsilon$, find the largest possible value of $\delta$ that works for any given $\varepsilon > 0$.
(c) Put $\varepsilon = 0.4$ in your answer to part (b) and compare with your answer to part (a).

**36.** Prove that $\lim_{x \to 2} \dfrac{1}{x} = \dfrac{1}{2}$.

**37.** Prove that $\lim_{x \to a} \sqrt{x} = \sqrt{a}$ if $a > 0$.

$$\left[ \text{Hint: Use } \left|\sqrt{x} - \sqrt{a}\right| = \frac{|x - a|}{\sqrt{x} + \sqrt{a}}. \right]$$

**38.** If $H$ is the Heaviside function defined in Example 6 in Section 1.5, prove, using Definition 2, that $\lim_{t \to 0} H(t)$ does not exist. [*Hint:* Use an indirect proof as follows. Suppose that the limit is $L$. Take $\varepsilon = \frac{1}{2}$ in the definition of a limit and try to arrive at a contradiction.]

**39.** If the function $f$ is defined by

$$f(x) = \begin{cases} 0 & \text{if } x \text{ is rational} \\ 1 & \text{if } x \text{ is irrational} \end{cases}$$

prove that $\lim_{x \to 0} f(x)$ does not exist.

**40.** By comparing Definitions 2, 3, and 4, prove Theorem 1 in Section 1.6.

**41.** How close to $-3$ do we have to take $x$ so that

$$\frac{1}{(x + 3)^4} > 10,000$$

**42.** Prove, using Definition 6, that $\lim_{x \to -3} \dfrac{1}{(x + 3)^4} = \infty$.

**43.** Prove that $\lim_{x \to -1^-} \dfrac{5}{(x + 1)^3} = -\infty$.

**44.** Suppose that $\lim_{x \to a} f(x) = \infty$ and $\lim_{x \to a} g(x) = c$, where $c$ is a real number. Prove each statement.
(a) $\lim_{x \to a} [f(x) + g(x)] = \infty$
(b) $\lim_{x \to a} [f(x)g(x)] = \infty$   if $c > 0$
(c) $\lim_{x \to a} [f(x)g(x)] = -\infty$   if $c < 0$

## 1.8   Continuity

We noticed in Section 1.6 that the limit of a function as $x$ approaches $a$ can often be found simply by calculating the value of the function at $a$. Functions with this property are called *continuous at a*. We will see that the mathematical definition of continuity corresponds closely with the meaning of the word *continuity* in everyday language. (A continuous process is one that takes place gradually, without interruption or abrupt change.)

As illustrated in Figure 1, if $f$ is continuous, then the points $(x, f(x))$ on the graph of $f$ approach the point $(a, f(a))$ on the graph. So there is no gap in the curve.

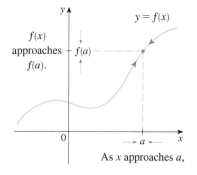

$f(x)$ approaches $f(a)$.

As $x$ approaches $a$,

**FIGURE 1**

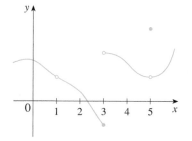

**FIGURE 2**

> **1** **Definition** A function $f$ is **continuous at a number $a$** if
> $$\lim_{x \to a} f(x) = f(a)$$

Notice that Definition 1 implicitly requires three things if $f$ is continuous at $a$:

**1.** $f(a)$ is defined (that is, $a$ is in the domain of $f$)

**2.** $\lim_{x \to a} f(x)$ exists

**3.** $\lim_{x \to a} f(x) = f(a)$

The definition says that $f$ is continuous at $a$ if $f(x)$ approaches $f(a)$ as $x$ approaches $a$. Thus a continuous function $f$ has the property that a small change in $x$ produces only a small change in $f(x)$. In fact, the change in $f(x)$ can be kept as small as we please by keeping the change in $x$ sufficiently small.

If $f$ is defined near $a$ (in other words, $f$ is defined on an open interval containing $a$, except perhaps at $a$), we say that $f$ is **discontinuous at $a$** (or $f$ has a **discontinuity** at $a$) if $f$ is not continuous at $a$.

Physical phenomena are usually continuous. For instance, the displacement or velocity of a vehicle varies continuously with time, as does a person's height. But discontinuities do occur in such situations as electric currents. [See Example 6 in Section 1.5, where the Heaviside function is discontinuous at 0 because $\lim_{t \to 0} H(t)$ does not exist.]

Geometrically, you can think of a function that is continuous at every number in an interval as a function whose graph has no break in it. The graph can be drawn without removing your pen from the paper.

**EXAMPLE 1** Figure 2 shows the graph of a function $f$. At which numbers is $f$ discontinuous? Why?

SOLUTION It looks as if there is a discontinuity when $a = 1$ because the graph has a break there. The official reason that $f$ is discontinuous at 1 is that $f(1)$ is not defined.

The graph also has a break when $a = 3$, but the reason for the discontinuity is different. Here, $f(3)$ is defined, but $\lim_{x \to 3} f(x)$ does not exist (because the left and right limits are different). So $f$ is discontinuous at 3.

What about $a = 5$? Here, $f(5)$ is defined and $\lim_{x \to 5} f(x)$ exists (because the left and right limits are the same). But

$$\lim_{x \to 5} f(x) \neq f(5)$$

So $f$ is discontinuous at 5.

Now let's see how to detect discontinuities when a function is defined by a formula.

**EXAMPLE 2** Where are each of the following functions discontinuous?

(a) $f(x) = \dfrac{x^2 - x - 2}{x - 2}$

(b) $f(x) = \begin{cases} \dfrac{1}{x^2} & \text{if } x \neq 0 \\ 1 & \text{if } x = 0 \end{cases}$

(c) $f(x) = \begin{cases} \dfrac{x^2 - x - 2}{x - 2} & \text{if } x \neq 2 \\ 1 & \text{if } x = 2 \end{cases}$

(d) $f(x) = [\![x]\!]$

SOLUTION

(a) Notice that $f(2)$ is not defined, so $f$ is discontinuous at 2. Later we'll see why $f$ is continuous at all other numbers.

(b) Here $f(0) = 1$ is defined but

$$\lim_{x \to 0} f(x) = \lim_{x \to 0} \frac{1}{x^2}$$

does not exist. (See Example 8 in Section 1.5.) So $f$ is discontinuous at 0.

(c) Here $f(2) = 1$ is defined and

$$\lim_{x \to 2} f(x) = \lim_{x \to 2} \frac{x^2 - x - 2}{x - 2} = \lim_{x \to 2} \frac{(x - 2)(x + 1)}{x - 2} = \lim_{x \to 2} (x + 1) = 3$$

exists. But

$$\lim_{x \to 2} f(x) \neq f(2)$$

so $f$ is not continuous at 2.

(d) The greatest integer function $f(x) = [\![x]\!]$ has discontinuities at all of the integers because $\lim_{x \to n} [\![x]\!]$ does not exist if $n$ is an integer. (See Example 10 and Exercise 51 in Section 1.6.)

Figure 3 shows the graphs of the functions in Example 2. In each case the graph can't be drawn without lifting the pen from the paper because a hole or break or jump occurs in the graph. The kind of discontinuity illustrated in parts (a) and (c) is called **removable** because we could remove the discontinuity by redefining $f$ at just the single number 2. [The function $g(x) = x + 1$ is continuous.] The discontinuity in part (b) is called an **infinite discontinuity**. The discontinuities in part (d) are called **jump discontinuities** because the function "jumps" from one value to another.

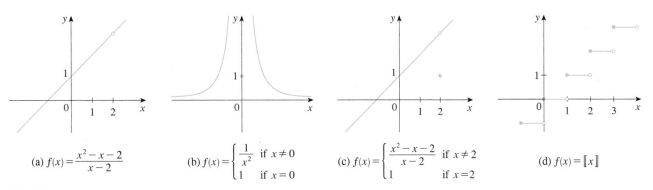

(a) $f(x) = \dfrac{x^2 - x - 2}{x - 2}$

(b) $f(x) = \begin{cases} \dfrac{1}{x^2} & \text{if } x \neq 0 \\ 1 & \text{if } x = 0 \end{cases}$

(c) $f(x) = \begin{cases} \dfrac{x^2 - x - 2}{x - 2} & \text{if } x \neq 2 \\ 1 & \text{if } x = 2 \end{cases}$

(d) $f(x) = [\![x]\!]$

**FIGURE 3**
Graphs of the functions in Example 2

---

$\boxed{2}$ **Definition** A function $f$ is **continuous from the right at a number** $a$ if

$$\lim_{x \to a^+} f(x) = f(a)$$

and $f$ is **continuous from the left at** $a$ if

$$\lim_{x \to a^-} f(x) = f(a)$$

**EXAMPLE 3** At each integer $n$, the function $f(x) = [\![x]\!]$ [see Figure 3(d)] is continuous from the right but discontinuous from the left because

$$\lim_{x \to n^+} f(x) = \lim_{x \to n^+} [\![x]\!] = n = f(n)$$

but

$$\lim_{x \to n^-} f(x) = \lim_{x \to n^-} [\![x]\!] = n - 1 \neq f(n)$$

> **3** **Definition** A function $f$ is **continuous on an interval** if it is continuous at every number in the interval. (If $f$ is defined only on one side of an endpoint of the interval, we understand *continuous* at the endpoint to mean *continuous from the right* or *continuous from the left*.)

**EXAMPLE 4** Show that the function $f(x) = 1 - \sqrt{1 - x^2}$ is continuous on the interval $[-1, 1]$.

SOLUTION If $-1 < a < 1$, then using the Limit Laws, we have

$$\lim_{x \to a} f(x) = \lim_{x \to a} \left(1 - \sqrt{1 - x^2}\right)$$

$$= 1 - \lim_{x \to a} \sqrt{1 - x^2} \qquad \text{(by Laws 2 and 7)}$$

$$= 1 - \sqrt{\lim_{x \to a} (1 - x^2)} \qquad \text{(by 11)}$$

$$= 1 - \sqrt{1 - a^2} \qquad \text{(by 2, 7, and 9)}$$

$$= f(a)$$

Thus, by Definition 1, $f$ is continuous at $a$ if $-1 < a < 1$. Similar calculations show that

$$\lim_{x \to -1^+} f(x) = 1 = f(-1) \qquad \text{and} \qquad \lim_{x \to 1^-} f(x) = 1 = f(1)$$

so $f$ is continuous from the right at $-1$ and continuous from the left at 1. Therefore, according to Definition 3, $f$ is continuous on $[-1, 1]$.

The graph of $f$ is sketched in Figure 4. It is the lower half of the circle

$$x^2 + (y - 1)^2 = 1$$

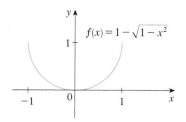

$f(x) = 1 - \sqrt{1 - x^2}$

**FIGURE 4**

Instead of always using Definitions 1, 2, and 3 to verify the continuity of a function as we did in Example 4, it is often convenient to use the next theorem, which shows how to build up complicated continuous functions from simple ones.

> **4** **Theorem** If $f$ and $g$ are continuous at $a$ and $c$ is a constant, then the following functions are also continuous at $a$:
>
> **1.** $f + g$         **2.** $f - g$         **3.** $cf$
>
> **4.** $fg$         **5.** $\dfrac{f}{g}$   if $g(a) \neq 0$

PROOF Each of the five parts of this theorem follows from the corresponding Limit Law in Section 1.6. For instance, we give the proof of part 1. Since $f$ and $g$ are continuous at $a$, we have

$$\lim_{x \to a} f(x) = f(a) \qquad \text{and} \qquad \lim_{x \to a} g(x) = g(a)$$

Therefore

$$\lim_{x \to a} (f + g)(x) = \lim_{x \to a} [f(x) + g(x)]$$

$$= \lim_{x \to a} f(x) + \lim_{x \to a} g(x) \qquad \text{(by Law 1)}$$

$$= f(a) + g(a)$$

$$= (f + g)(a)$$

This shows that $f + g$ is continuous at $a$. ▪

It follows from Theorem 4 and Definition 3 that if $f$ and $g$ are continuous on an interval, then so are the functions $f + g$, $f - g$, $cf$, $fg$, and (if $g$ is never 0) $f/g$. The following theorem was stated in Section 1.6 as the Direct Substitution Property.

---

**5** | **Theorem**

(a) Any polynomial is continuous everywhere; that is, it is continuous on $\mathbb{R} = (-\infty, \infty)$.

(b) Any rational function is continuous wherever it is defined; that is, it is continuous on its domain.

---

PROOF

(a) A polynomial is a function of the form

$$P(x) = c_n x^n + c_{n-1} x^{n-1} + \cdots + c_1 x + c_0$$

where $c_0, c_1, \ldots, c_n$ are constants. We know that

$$\lim_{x \to a} c_0 = c_0 \qquad \text{(by Law 7)}$$

and

$$\lim_{x \to a} x^m = a^m \qquad m = 1, 2, \ldots, n \qquad \text{(by 9)}$$

This equation is precisely the statement that the function $f(x) = x^m$ is a continuous function. Thus, by part 3 of Theorem 4, the function $g(x) = cx^m$ is continuous. Since $P$ is a sum of functions of this form and a constant function, it follows from part 1 of Theorem 4 that $P$ is continuous.

(b) A rational function is a function of the form

$$f(x) = \frac{P(x)}{Q(x)}$$

where $P$ and $Q$ are polynomials. The domain of $f$ is $D = \{x \in \mathbb{R} \mid Q(x) \neq 0\}$. We know from part (a) that $P$ and $Q$ are continuous everywhere. Thus, by part 5 of Theorem 4, $f$ is continuous at every number in $D$. ▪

As an illustration of Theorem 5, observe that the volume of a sphere varies continuously with its radius because the formula $V(r) = \frac{4}{3}\pi r^3$ shows that $V$ is a polynomial function of $r$. Likewise, if a ball is thrown vertically into the air with a velocity of 50 ft/s, then the height of the ball in feet $t$ seconds later is given by the formula $h = 50t - 16t^2$. Again this is a polynomial function, so the height is a continuous function of the elapsed time.

Knowledge of which functions are continuous enables us to evaluate some limits very quickly, as the following example shows. Compare it with Example 2(b) in Section 1.6.

**EXAMPLE 5**   Find $\lim\limits_{x \to -2} \dfrac{x^3 + 2x^2 - 1}{5 - 3x}$.

SOLUTION   The function

$$f(x) = \frac{x^3 + 2x^2 - 1}{5 - 3x}$$

is rational, so by Theorem 5 it is continuous on its domain, which is $\left\{ x \mid x \neq \frac{5}{3} \right\}$. Therefore

$$\lim_{x \to -2} \frac{x^3 + 2x^2 - 1}{5 - 3x} = \lim_{x \to -2} f(x) = f(-2)$$

$$= \frac{(-2)^3 + 2(-2)^2 - 1}{5 - 3(-2)} = -\frac{1}{11}$$

It turns out that most of the familiar functions are continuous at every number in their domains. For instance, Limit Law 10 (page 63) is exactly the statement that root functions are continuous.

From the appearance of the graphs of the sine and cosine functions (Figure 18 in Section 1.2), we would certainly guess that they are continuous. We know from the definitions of $\sin \theta$ and $\cos \theta$ that the coordinates of the point $P$ in Figure 5 are $(\cos \theta, \sin \theta)$. As $\theta \to 0$, we see that $P$ approaches the point $(1, 0)$ and so $\cos \theta \to 1$ and $\sin \theta \to 0$. Thus

**FIGURE 5**

Another way to establish the limits in ⑥ is to use the Squeeze Theorem with the inequality $\sin \theta < \theta$ (for $\theta > 0$), which is proved in Section 2.4.

$$\boxed{6} \qquad \lim_{\theta \to 0} \cos \theta = 1 \qquad \lim_{\theta \to 0} \sin \theta = 0$$

Since $\cos 0 = 1$ and $\sin 0 = 0$, the equations in ⑥ assert that the cosine and sine functions are continuous at 0. The addition formulas for cosine and sine can then be used to deduce that these functions are continuous everywhere (see Exercises 60 and 61).

It follows from part 5 of Theorem 4 that

$$\tan x = \frac{\sin x}{\cos x}$$

is continuous except where $\cos x = 0$. This happens when $x$ is an odd integer multiple of $\pi/2$, so $y = \tan x$ has infinite discontinuities when $x = \pm\pi/2, \pm 3\pi/2, \pm 5\pi/2$, and so on (see Figure 6).

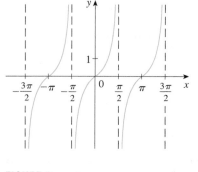

**FIGURE 6**   $y = \tan x$

> ⑦ **Theorem**   The following types of functions are continuous at every number in their domains:
>
> | polynomials | rational functions |
> |---|---|
> | root functions | trigonometric functions |

**EXAMPLE 6** On what intervals is each function continuous?

(a) $f(x) = x^{100} - 2x^{37} + 75$ (b) $g(x) = \dfrac{x^2 + 2x + 17}{x^2 - 1}$

(c) $h(x) = \sqrt{x} + \dfrac{x + 1}{x - 1} - \dfrac{x + 1}{x^2 + 1}$

SOLUTION

(a) $f$ is a polynomial, so it is continuous on $(-\infty, \infty)$ by Theorem 5(a).

(b) $g$ is a rational function, so by Theorem 5(b), it is continuous on its domain, which is $D = \{x \mid x^2 - 1 \neq 0\} = \{x \mid x \neq \pm 1\}$. Thus $g$ is continuous on the intervals $(-\infty, -1)$, $(-1, 1)$, and $(1, \infty)$.

(c) We can write $h(x) = F(x) + G(x) - H(x)$, where

$$F(x) = \sqrt{x} \qquad G(x) = \frac{x + 1}{x - 1} \qquad H(x) = \frac{x + 1}{x^2 + 1}$$

$F$ is continuous on $[0, \infty)$ by Theorem 7. $G$ is a rational function, so it is continuous everywhere except when $x - 1 = 0$, that is, $x = 1$. $H$ is also a rational function, but its denominator is never 0, so $H$ is continuous everywhere. Thus, by parts 1 and 2 of Theorem 4, $h$ is continuous on the intervals $[0, 1)$ and $(1, \infty)$.

**EXAMPLE 7** Evaluate $\lim\limits_{x \to \pi} \dfrac{\sin x}{2 + \cos x}$.

SOLUTION Theorem 7 tells us that $y = \sin x$ is continuous. The function in the denominator, $y = 2 + \cos x$, is the sum of two continuous functions and is therefore continuous. Notice that this function is never 0 because $\cos x \geq -1$ for all $x$ and so $2 + \cos x > 0$ everywhere. Thus the ratio

$$f(x) = \frac{\sin x}{2 + \cos x}$$

is continuous everywhere. Hence, by the definition of a continuous function,

$$\lim_{x \to \pi} \frac{\sin x}{2 + \cos x} = \lim_{x \to \pi} f(x) = f(\pi) = \frac{\sin \pi}{2 + \cos \pi} = \frac{0}{2 - 1} = 0$$

Another way of combining continuous functions $f$ and $g$ to get a new continuous function is to form the composite function $f \circ g$. This fact is a consequence of the following theorem.

This theorem says that a limit symbol can be moved through a function symbol if the function is continuous and the limit exists. In other words, the order of these two symbols can be reversed.

> **8 Theorem** If $f$ is continuous at $b$ and $\lim\limits_{x \to a} g(x) = b$, then $\lim\limits_{x \to a} f(g(x)) = f(b)$.
> In other words,
> $$\lim_{x \to a} f(g(x)) = f\left(\lim_{x \to a} g(x)\right)$$

Intuitively, Theorem 8 is reasonable because if $x$ is close to $a$, then $g(x)$ is close to $b$, and since $f$ is continuous at $b$, if $g(x)$ is close to $b$, then $f(g(x))$ is close to $f(b)$. A proof of Theorem 8 is given in Appendix F.

Let's now apply Theorem 8 in the special case where $f(x) = \sqrt[n]{x}$, with $n$ being a positive integer. Then

$$f(g(x)) = \sqrt[n]{g(x)}$$

and

$$f\left(\lim_{x \to a} g(x)\right) = \sqrt[n]{\lim_{x \to a} g(x)}$$

If we put these expressions into Theorem 8, we get

$$\lim_{x \to a} \sqrt[n]{g(x)} = \sqrt[n]{\lim_{x \to a} g(x)}$$

and so Limit Law 11 has now been proved. (We assume that the roots exist.)

---

**9  Theorem**   If $g$ is continuous at $a$ and $f$ is continuous at $g(a)$, then the composite function $f \circ g$ given by $(f \circ g)(x) = f(g(x))$ is continuous at $a$.

---

This theorem is often expressed informally by saying "a continuous function of a continuous function is a continuous function."

PROOF   Since $g$ is continuous at $a$, we have

$$\lim_{x \to a} g(x) = g(a)$$

Since $f$ is continuous at $b = g(a)$, we can apply Theorem 8 to obtain

$$\lim_{x \to a} f(g(x)) = f(g(a))$$

which is precisely the statement that the function $h(x) = f(g(x))$ is continuous at $a$; that is, $f \circ g$ is continuous at $a$.    ▬

**V  EXAMPLE 8**   Where are the following functions continuous?

(a) $h(x) = \sin(x^2)$ 

(b) $F(x) = \dfrac{1}{\sqrt{x^2 + 7} - 4}$

SOLUTION

(a) We have $h(x) = f(g(x))$, where

$$g(x) = x^2 \qquad \text{and} \qquad f(x) = \sin x$$

Now $g$ is continuous on $\mathbb{R}$ since it is a polynomial, and $f$ is also continuous everywhere. Thus $h = f \circ g$ is continuous on $\mathbb{R}$ by Theorem 9.

(b) Notice that $F$ can be broken up as the composition of four continuous functions:

$$F = f \circ g \circ h \circ k \qquad \text{or} \qquad F(x) = f(g(h(k(x))))$$

where   $f(x) = \dfrac{1}{x}$      $g(x) = x - 4$      $h(x) = \sqrt{x}$      $k(x) = x^2 + 7$

We know that each of these functions is continuous on its domain (by Theorems 5 and 7), so by Theorem 9, $F$ is continuous on its domain, which is

$$\{x \in \mathbb{R} \mid \sqrt{x^2 + 7} \neq 4\} = \{x \mid x \neq \pm 3\} = (-\infty, -3) \cup (-3, 3) \cup (3, \infty)$$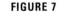

An important property of continuous functions is expressed by the following theorem, whose proof is found in more advanced books on calculus.

---

**10** **The Intermediate Value Theorem**   Suppose that $f$ is continuous on the closed interval $[a, b]$ and let $N$ be any number between $f(a)$ and $f(b)$, where $f(a) \neq f(b)$. Then there exists a number $c$ in $(a, b)$ such that $f(c) = N$.

---

The Intermediate Value Theorem states that a continuous function takes on every intermediate value between the function values $f(a)$ and $f(b)$. It is illustrated by Figure 7. Note that the value $N$ can be taken on once [as in part (a)] or more than once [as in part (b)].

**FIGURE 7**

(a)

(b)

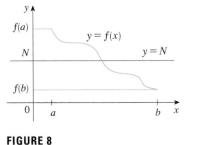

**FIGURE 8**

If we think of a continuous function as a function whose graph has no hole or break, then it is easy to believe that the Intermediate Value Theorem is true. In geometric terms it says that if any horizontal line $y = N$ is given between $y = f(a)$ and $y = f(b)$ as in Figure 8, then the graph of $f$ can't jump over the line. It must intersect $y = N$ somewhere.

It is important that the function $f$ in Theorem 10 be continuous. The Intermediate Value Theorem is not true in general for discontinuous functions (see Exercise 48).

One use of the Intermediate Value Theorem is in locating roots of equations as in the following example.

**V** **EXAMPLE 9** Show that there is a root of the equation

$$4x^3 - 6x^2 + 3x - 2 = 0$$

between 1 and 2.

SOLUTION Let $f(x) = 4x^3 - 6x^2 + 3x - 2$. We are looking for a solution of the given equation, that is, a number $c$ between 1 and 2 such that $f(c) = 0$. Therefore we take $a = 1$, $b = 2$, and $N = 0$ in Theorem 10. We have

$$f(1) = 4 - 6 + 3 - 2 = -1 < 0$$

and

$$f(2) = 32 - 24 + 6 - 2 = 12 > 0$$

Thus $f(1) < 0 < f(2)$; that is, $N = 0$ is a number between $f(1)$ and $f(2)$. Now $f$ is continuous since it is a polynomial, so the Intermediate Value Theorem says there is a number $c$ between 1 and 2 such that $f(c) = 0$. In other words, the equation $4x^3 - 6x^2 + 3x - 2 = 0$ has at least one root $c$ in the interval $(1, 2)$.

In fact, we can locate a root more precisely by using the Intermediate Value Theorem again. Since

$$f(1.2) = -0.128 < 0 \qquad \text{and} \qquad f(1.3) = 0.548 > 0$$

a root must lie between 1.2 and 1.3. A calculator gives, by trial and error,

$$f(1.22) = -0.007008 < 0 \qquad \text{and} \qquad f(1.23) = 0.056068 > 0$$

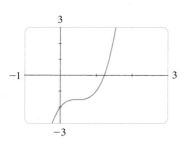

so a root lies in the interval $(1.22, 1.23)$.

**FIGURE 9**

We can use a graphing calculator or computer to illustrate the use of the Intermediate Value Theorem in Example 9. Figure 9 shows the graph of $f$ in the viewing rectangle $[-1, 3]$ by $[-3, 3]$ and you can see that the graph crosses the $x$-axis between 1 and 2. Figure 10 shows the result of zooming in to the viewing rectangle $[1.2, 1.3]$ by $[-0.2, 0.2]$.

In fact, the Intermediate Value Theorem plays a role in the very way these graphing devices work. A computer calculates a finite number of points on the graph and turns on the pixels that contain these calculated points. It assumes that the function is continuous and takes on all the intermediate values between two consecutive points. The computer therefore connects the pixels by turning on the intermediate pixels.

**FIGURE 10**

---

## 1.8 Exercises

**1.** Write an equation that expresses the fact that a function $f$ is continuous at the number 4.

**2.** If $f$ is continuous on $(-\infty, \infty)$, what can you say about its graph?

**3.** (a) From the graph of $f$, state the numbers at which $f$ is discontinuous and explain why.
   (b) For each of the numbers stated in part (a), determine whether $f$ is continuous from the right, or from the left, or neither.

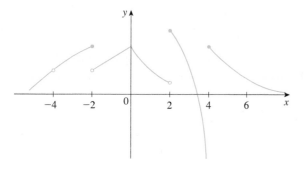

**4.** From the graph of $g$, state the intervals on which $g$ is continuous.

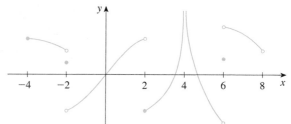

**5–8** Sketch the graph of a function $f$ that is continuous except for the stated discontinuity.

**5.** Discontinuous, but continuous from the right, at 2

**6.** Discontinuities at $-1$ and 4, but continuous from the left at $-1$ and from the right at 4

**7.** Removable discontinuity at 3, jump discontinuity at 5

**8.** Neither left nor right continuous at $-2$, continuous only from the left at 2

---

9. The toll $T$ charged for driving on a certain stretch of a toll road is \$5 except during rush hours (between 7 AM and 10 AM and between 4 PM and 7 PM) when the toll is \$7.
   (a) Sketch a graph of $T$ as a function of the time $t$, measured in hours past midnight.
   (b) Discuss the discontinuities of this function and their significance to someone who uses the road.

10. Explain why each function is continuous or discontinuous.
    (a) The temperature at a specific location as a function of time
    (b) The temperature at a specific time as a function of the distance due west from New York City
    (c) The altitude above sea level as a function of the distance due west from New York City
    (d) The cost of a taxi ride as a function of the distance traveled
    (e) The current in the circuit for the lights in a room as a function of time

11. Suppose $f$ and $g$ are continuous functions such that $g(2) = 6$ and $\lim_{x \to 2} [3f(x) + f(x)g(x)] = 36$. Find $f(2)$.

**12–14** Use the definition of continuity and the properties of limits to show that the function is continuous at the given number $a$.

12. $f(x) = 3x^4 - 5x + \sqrt[3]{x^2 + 4}$, $a = 2$

13. $f(x) = (x + 2x^3)^4$, $a = -1$

14. $h(t) = \dfrac{2t - 3t^2}{1 + t^3}$, $a = 1$

**15–16** Use the definition of continuity and the properties of limits to show that the function is continuous on the given interval.

15. $f(x) = \dfrac{2x + 3}{x - 2}$, $(2, \infty)$

16. $g(x) = 2\sqrt{3 - x}$, $(-\infty, 3]$

**17–22** Explain why the function is discontinuous at the given number $a$. Sketch the graph of the function.

17. $f(x) = \dfrac{1}{x + 2}$ $\qquad\qquad a = -2$

18. $f(x) = \begin{cases} \dfrac{1}{x + 2} & \text{if } x \neq -2 \\ 1 & \text{if } x = -2 \end{cases}$ $\qquad a = -2$

19. $f(x) = \begin{cases} 1 - x^2 & \text{if } x < 1 \\ 1/x & \text{if } x \geq 1 \end{cases}$ $\qquad a = 1$

20. $f(x) = \begin{cases} \dfrac{x^2 - x}{x^2 - 1} & \text{if } x \neq 1 \\ 1 & \text{if } x = 1 \end{cases}$ $\qquad a = 1$

21. $f(x) = \begin{cases} \cos x & \text{if } x < 0 \\ 0 & \text{if } x = 0 \\ 1 - x^2 & \text{if } x > 0 \end{cases}$ $\qquad a = 0$

22. $f(x) = \begin{cases} \dfrac{2x^2 - 5x - 3}{x - 3} & \text{if } x \neq 3 \\ 6 & \text{if } x = 3 \end{cases}$ $\qquad a = 3$

**23–24** How would you "remove the discontinuity" of $f$? In other words, how would you define $f(2)$ in order to make $f$ continuous at 2?

23. $f(x) = \dfrac{x^2 - x - 2}{x - 2}$

24. $f(x) = \dfrac{x^3 - 8}{x^2 - 4}$

**25–32** Explain, using Theorems 4, 5, 7, and 9, why the function is continuous at every number in its domain. State the domain.

25. $F(x) = \dfrac{2x^2 - x - 1}{x^2 + 1}$

26. $G(x) = \dfrac{x^2 + 1}{2x^2 - x - 1}$

27. $Q(x) = \dfrac{\sqrt[3]{x - 2}}{x^3 - 2}$

28. $h(x) = \dfrac{\sin x}{x + 1}$

29. $h(x) = \cos(1 - x^2)$

30. $B(x) = \dfrac{\tan x}{\sqrt{4 - x^2}}$

31. $M(x) = \sqrt{1 + \dfrac{1}{x}}$

32. $F(x) = \sin(\cos(\sin x))$

**33–34** Locate the discontinuities of the function and illustrate by graphing.

33. $y = \dfrac{1}{1 + \sin x}$

34. $y = \tan \sqrt{x}$

**35–38** Use continuity to evaluate the limit.

35. $\lim\limits_{x \to 4} \dfrac{5 + \sqrt{x}}{\sqrt{5 + x}}$

36. $\lim\limits_{x \to \pi} \sin(x + \sin x)$

37. $\lim\limits_{x \to \pi/4} x \cos^2 x$

38. $\lim\limits_{x \to 2} (x^3 - 3x + 1)^{-3}$

**39–40** Show that $f$ is continuous on $(-\infty, \infty)$.

**39.** $f(x) = \begin{cases} x^2 & \text{if } x < 1 \\ \sqrt{x} & \text{if } x \geqslant 1 \end{cases}$

**40.** $f(x) = \begin{cases} \sin x & \text{if } x < \pi/4 \\ \cos x & \text{if } x \geqslant \pi/4 \end{cases}$

**41–43** Find the numbers at which $f$ is discontinuous. At which of these numbers is $f$ continuous from the right, from the left, or neither? Sketch the graph of $f$.

**41.** $f(x) = \begin{cases} 1 + x^2 & \text{if } x \leqslant 0 \\ 2 - x & \text{if } 0 < x \leqslant 2 \\ (x - 2)^2 & \text{if } x > 2 \end{cases}$

**42.** $f(x) = \begin{cases} x + 1 & \text{if } x \leqslant 1 \\ 1/x & \text{if } 1 < x < 3 \\ \sqrt{x - 3} & \text{if } x \geqslant 3 \end{cases}$

**43.** $f(x) = \begin{cases} x + 2 & \text{if } x < 0 \\ 2x^2 & \text{if } 0 \leqslant x \leqslant 1 \\ 2 - x & \text{if } x > 1 \end{cases}$

**44.** The gravitational force exerted by the planet Earth on a unit mass at a distance $r$ from the center of the planet is

$$F(r) = \begin{cases} \dfrac{GMr}{R^3} & \text{if } r < R \\[2ex] \dfrac{GM}{r^2} & \text{if } r \geqslant R \end{cases}$$

where $M$ is the mass of Earth, $R$ is its radius, and $G$ is the gravitational constant. Is $F$ a continuous function of $r$?

**45.** For what value of the constant $c$ is the function $f$ continuous on $(-\infty, \infty)$?

$$f(x) = \begin{cases} cx^2 + 2x & \text{if } x < 2 \\ x^3 - cx & \text{if } x \geqslant 2 \end{cases}$$

**46.** Find the values of $a$ and $b$ that make $f$ continuous everywhere.

$$f(x) = \begin{cases} \dfrac{x^2 - 4}{x - 2} & \text{if } x < 2 \\[2ex] ax^2 - bx + 3 & \text{if } 2 \leqslant x < 3 \\[1ex] 2x - a + b & \text{if } x \geqslant 3 \end{cases}$$

**47.** Which of the following functions $f$ has a removable discontinuity at $a$? If the discontinuity is removable, find a function $g$ that agrees with $f$ for $x \neq a$ and is continuous at $a$.

(a) $f(x) = \dfrac{x^4 - 1}{x - 1}, \quad a = 1$

(b) $f(x) = \dfrac{x^3 - x^2 - 2x}{x - 2}, \quad a = 2$

(c) $f(x) = [\![ \sin x ]\!], \quad a = \pi$

**48.** Suppose that a function $f$ is continuous on $[0, 1]$ except at 0.25 and that $f(0) = 1$ and $f(1) = 3$. Let $N = 2$. Sketch two possible graphs of $f$, one showing that $f$ might not satisfy the conclusion of the Intermediate Value Theorem and one showing that $f$ might still satisfy the conclusion of the Intermediate Value Theorem (even though it doesn't satisfy the hypothesis).

**49.** If $f(x) = x^2 + 10 \sin x$, show that there is a number $c$ such that $f(c) = 1000$.

**50.** Suppose $f$ is continuous on $[1, 5]$ and the only solutions of the equation $f(x) = 6$ are $x = 1$ and $x = 4$. If $f(2) = 8$, explain why $f(3) > 6$.

**51–54** Use the Intermediate Value Theorem to show that there is a root of the given equation in the specified interval.

**51.** $x^4 + x - 3 = 0, \quad (1, 2)$      **52.** $\sqrt[3]{x} = 1 - x, \quad (0, 1)$

**53.** $\cos x = x, \quad (0, 1)$      **54.** $\sin x = x^2 - x, \quad (1, 2)$

**55–56** (a) Prove that the equation has at least one real root.
(b) Use your calculator to find an interval of length 0.01 that contains a root.

**55.** $\cos x = x^3$      **56.** $x^5 - x^2 + 2x + 3 = 0$

**57–58** (a) Prove that the equation has at least one real root.
(b) Use your graphing device to find the root correct to three decimal places.

**57.** $x^5 - x^2 - 4 = 0$      **58.** $\sqrt{x - 5} = \dfrac{1}{x + 3}$

**59.** Prove that $f$ is continuous at $a$ if and only if

$$\lim_{h \to 0} f(a + h) = f(a)$$

**60.** To prove that sine is continuous, we need to show that $\lim_{x \to a} \sin x = \sin a$ for every real number $a$. By Exercise 59 an equivalent statement is that

$$\lim_{h \to 0} \sin(a + h) = \sin a$$

Use $\boxed{6}$ to show that this is true.

**61.** Prove that cosine is a continuous function.

**62.** (a) Prove Theorem 4, part 3.
(b) Prove Theorem 4, part 5.

**63.** For what values of $x$ is $f$ continuous?

$$f(x) = \begin{cases} 0 & \text{if } x \text{ is rational} \\ 1 & \text{if } x \text{ is irrational} \end{cases}$$

**64.** For what values of $x$ is $g$ continuous?

$$g(x) = \begin{cases} 0 & \text{if } x \text{ is rational} \\ x & \text{if } x \text{ is irrational} \end{cases}$$

**65.** Is there a number that is exactly 1 more than its cube?

**66.** If $a$ and $b$ are positive numbers, prove that the equation

$$\frac{a}{x^3 + 2x^2 - 1} + \frac{b}{x^3 + x - 2} = 0$$

has at least one solution in the interval $(-1, 1)$.

**67.** Show that the function

$$f(x) = \begin{cases} x^4 \sin(1/x) & \text{if } x \neq 0 \\ 0 & \text{if } x = 0 \end{cases}$$

is continuous on $(-\infty, \infty)$.

**68.** (a) Show that the absolute value function $F(x) = |x|$ is continuous everywhere.
(b) Prove that if $f$ is a continuous function on an interval, then so is $|f|$.
(c) Is the converse of the statement in part (b) also true? In other words, if $|f|$ is continuous, does it follow that $f$ is continuous? If so, prove it. If not, find a counterexample.

**69.** A Tibetan monk leaves the monastery at 7:00 AM and takes his usual path to the top of the mountain, arriving at 7:00 PM. The following morning, he starts at 7:00 AM at the top and takes the same path back, arriving at the monastery at 7:00 PM. Use the Intermediate Value Theorem to show that there is a point on the path that the monk will cross at exactly the same time of day on both days.

---

| 1 | **Review** |
|---|---|

## Concept Check

**1.** (a) What is a function? What are its domain and range?
(b) What is the graph of a function?
(c) How can you tell whether a given curve is the graph of a function?

**2.** Discuss four ways of representing a function. Illustrate your discussion with examples.

**3.** (a) What is an even function? How can you tell if a function is even by looking at its graph? Give three examples of an even function.
(b) What is an odd function? How can you tell if a function is odd by looking at its graph? Give three examples of an odd function.

**4.** What is an increasing function?

**5.** What is a mathematical model?

**6.** Give an example of each type of function.
(a) Linear function       (b) Power function
(c) Exponential function  (d) Quadratic function
(e) Polynomial of degree 5  (f) Rational function

**7.** Sketch by hand, on the same axes, the graphs of the following functions.
(a) $f(x) = x$       (b) $g(x) = x^2$
(c) $h(x) = x^3$     (d) $j(x) = x^4$

**8.** Draw, by hand, a rough sketch of the graph of each function.
(a) $y = \sin x$       (b) $y = \tan x$
(c) $y = 2^x$          (d) $y = 1/x$
(e) $y = |x|$          (f) $y = \sqrt{x}$

**9.** Suppose that $f$ has domain $A$ and $g$ has domain $B$.
(a) What is the domain of $f + g$?
(b) What is the domain of $fg$?
(c) What is the domain of $f/g$?

**10.** How is the composite function $f \circ g$ defined? What is its domain?

**11.** Suppose the graph of $f$ is given. Write an equation for each of the graphs that are obtained from the graph of $f$ as follows.
(a) Shift 2 units upward.
(b) Shift 2 units downward.
(c) Shift 2 units to the right.
(d) Shift 2 units to the left.
(e) Reflect about the $x$-axis.
(f) Reflect about the $y$-axis.
(g) Stretch vertically by a factor of 2.
(h) Shrink vertically by a factor of 2.
(i) Stretch horizontally by a factor of 2.
(j) Shrink horizontally by a factor of 2.

**12.** Explain what each of the following means and illustrate with a sketch.

(a) $\lim_{x \to a} f(x) = L$

(b) $\lim_{x \to a^+} f(x) = L$

(c) $\lim_{x \to a^-} f(x) = L$

(d) $\lim_{x \to a} f(x) = \infty$

(e) $\lim_{x \to a} f(x) = -\infty$

**13.** Describe several ways in which a limit can fail to exist. Illustrate with sketches.

**14.** What does it mean to say that the line $x = a$ is a vertical asymptote of the curve $y = f(x)$? Draw curves to illustrate the various possibilities.

**15.** State the following Limit Laws.

(a) Sum Law

(b) Difference Law

(c) Constant Multiple Law

(d) Product Law

(e) Quotient Law

(f) Power Law

(g) Root Law

**16.** What does the Squeeze Theorem say?

**17.** (a) What does it mean for $f$ to be continuous at $a$?

(b) What does it mean for $f$ to be continuous on the interval $(-\infty, \infty)$? What can you say about the graph of such a function?

**18.** What does the Intermediate Value Theorem say?

## True-False Quiz

Determine whether the statement is true or false. If it is true, explain why. If it is false, explain why or give an example that disproves the statement.

**1.** If $f$ is a function, then $f(s + t) = f(s) + f(t)$.

**2.** If $f(s) = f(t)$, then $s = t$.

**3.** If $f$ is a function, then $f(3x) = 3f(x)$.

**4.** If $x_1 < x_2$ and $f$ is a decreasing function, then $f(x_1) > f(x_2)$.

**5.** A vertical line intersects the graph of a function at most once.

**6.** If $x$ is any real number, then $\sqrt{x^2} = x$.

**7.** $\lim_{x \to 4} \left( \dfrac{2x}{x - 4} - \dfrac{8}{x - 4} \right) = \lim_{x \to 4} \dfrac{2x}{x - 4} - \lim_{x \to 4} \dfrac{8}{x - 4}$

**8.** $\lim_{x \to 1} \dfrac{x^2 + 6x - 7}{x^2 + 5x - 6} = \dfrac{\lim_{x \to 1} (x^2 + 6x - 7)}{\lim_{x \to 1} (x^2 + 5x - 6)}$

**9.** $\lim_{x \to 1} \dfrac{x - 3}{x^2 + 2x - 4} = \dfrac{\lim_{x \to 1} (x - 3)}{\lim_{x \to 1} (x^2 + 2x - 4)}$

**10.** If $\lim_{x \to 5} f(x) = 2$ and $\lim_{x \to 5} g(x) = 0$, then $\lim_{x \to 5} [f(x)/g(x)]$ does not exist.

**11.** If $\lim_{x \to 5} f(x) = 0$ and $\lim_{x \to 5} g(x) = 0$, then $\lim_{x \to 5} [f(x)/g(x)]$ does not exist.

**12.** If neither $\lim_{x \to a} f(x)$ nor $\lim_{x \to a} g(x)$ exists, then $\lim_{x \to a} [f(x) + g(x)]$ does not exist.

**13.** If $\lim_{x \to a} f(x)$ exists but $\lim_{x \to a} g(x)$ does not exist, then $\lim_{x \to a} [f(x) + g(x)]$ does not exist.

**14.** If $\lim_{x \to 6} [f(x) g(x)]$ exists, then the limit must be $f(6) g(6)$.

**15.** If $p$ is a polynomial, then $\lim_{x \to b} p(x) = p(b)$.

**16.** If $\lim_{x \to 0} f(x) = \infty$ and $\lim_{x \to 0} g(x) = \infty$, then $\lim_{x \to 0} [f(x) - g(x)] = 0$.

**17.** If the line $x = 1$ is a vertical asymptote of $y = f(x)$, then $f$ is not defined at 1.

**18.** If $f(1) > 0$ and $f(3) < 0$, then there exists a number $c$ between 1 and 3 such that $f(c) = 0$.

**19.** If $f$ is continuous at 5 and $f(5) = 2$ and $f(4) = 3$, then $\lim_{x \to 2} f(4x^2 - 11) = 2$.

**20.** If $f$ is continuous on $[-1, 1]$ and $f(-1) = 4$ and $f(1) = 3$, then there exists a number $r$ such that $|r| < 1$ and $f(r) = \pi$.

**21.** Let $f$ be a function such that $\lim_{x \to 0} f(x) = 6$. Then there exists a number $\delta$ such that if $0 < |x| < \delta$, then $|f(x) - 6| < 1$.

**22.** If $f(x) > 1$ for all $x$ and $\lim_{x \to 0} f(x)$ exists, then $\lim_{x \to 0} f(x) > 1$.

**23.** The equation $x^{10} - 10x^2 + 5 = 0$ has a root in the interval $(0, 2)$.

**24.** If $f$ is continuous at $a$, so is $|f|$.

**25.** If $|f|$ is continuous at $a$, so is $f$.

## Exercises

**1.** Let $f$ be the function whose graph is given.
   (a) Estimate the value of $f(2)$.
   (b) Estimate the values of $x$ such that $f(x) = 3$.
   (c) State the domain of $f$.
   (d) State the range of $f$.
   (e) On what interval is $f$ increasing?
   (f) Is $f$ even, odd, or neither even nor odd? Explain.

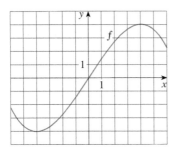

**2.** Determine whether each curve is the graph of a function of $x$. If it is, state the domain and range of the function.

(a)                     (b)

**3.** If $f(x) = x^2 - 2x + 3$, evaluate the difference quotient

$$\frac{f(a + h) - f(a)}{h}$$

**4.** Sketch a rough graph of the yield of a crop as a function of the amount of fertilizer used.

**5–8** Find the domain and range of the function. Write your answer in interval notation.

**5.** $f(x) = 2/(3x - 1)$       **6.** $g(x) = \sqrt{16 - x^4}$

**7.** $y = 1 + \sin x$         **8.** $F(t) = 3 + \cos 2t$

**9.** Suppose that the graph of $f$ is given. Describe how the graphs of the following functions can be obtained from the graph of $f$.
   (a) $y = f(x) + 8$       (b) $y = f(x + 8)$
   (c) $y = 1 + 2f(x)$     (d) $y = f(x - 2) - 2$
   (e) $y = -f(x)$        (f) $y = 3 - f(x)$

**10.** The graph of $f$ is given. Draw the graphs of the following functions.
   (a) $y = f(x - 8)$       (b) $y = -f(x)$

(c) $y = 2 - f(x)$       (d) $y = \frac{1}{2}f(x) - 1$

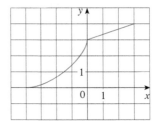

**11–16** Use transformations to sketch the graph of the function.

**11.** $y = -\sin 2x$       **12.** $y = (x - 2)^2$

**13.** $y = 1 + \frac{1}{2}x^3$     **14.** $y = 2 - \sqrt{x}$

**15.** $f(x) = \dfrac{1}{x + 2}$     **16.** $f(x) = \begin{cases} 1 + x & \text{if } x < 0 \\ 1 + x^2 & \text{if } x \geq 0 \end{cases}$

**17.** Determine whether $f$ is even, odd, or neither even nor odd.
   (a) $f(x) = 2x^5 - 3x^2 + 2$
   (b) $f(x) = x^3 - x^7$
   (c) $f(x) = \cos(x^2)$
   (d) $f(x) = 1 + \sin x$

**18.** Find an expression for the function whose graph consists of the line segment from the point $(-2, 2)$ to the point $(-1, 0)$ together with the top half of the circle with center the origin and radius 1.

**19.** If $f(x) = \sqrt{x}$ and $g(x) = \sin x$, find the functions (a) $f \circ g$, (b) $g \circ f$, (c) $f \circ f$, (d) $g \circ g$, and their domains.

**20.** Express the function $F(x) = 1/\sqrt{x + \sqrt{x}}$ as a composition of three functions.

**21.** Life expectancy improved dramatically in the 20th century. The table gives the life expectancy at birth (in years) of males born in the United States. Use a scatter plot to choose an appropriate type of model. Use your model to predict the life span of a male born in the year 2010.

| Birth year | Life expectancy | Birth year | Life expectancy |
|------------|-----------------|------------|-----------------|
| 1900 | 48.3 | 1960 | 66.6 |
| 1910 | 51.1 | 1970 | 67.1 |
| 1920 | 55.2 | 1980 | 70.0 |
| 1930 | 57.4 | 1990 | 71.8 |
| 1940 | 62.5 | 2000 | 73.0 |
| 1950 | 65.6 | | |

**22.** A small-appliance manufacturer finds that it costs \$9000 to produce 1000 toaster ovens a week and \$12,000 to produce 1500 toaster ovens a week.

(a) Express the cost as a function of the number of toaster ovens produced, assuming that it is linear. Then sketch the graph.

(b) What is the slope of the graph and what does it represent?

(c) What is the $y$-intercept of the graph and what does it represent?

**23.** The graph of $f$ is given.

(a) Find each limit, or explain why it does not exist.

    (i) $\displaystyle\lim_{x \to 2^+} f(x)$       (ii) $\displaystyle\lim_{x \to -3^+} f(x)$

    (iii) $\displaystyle\lim_{x \to -3} f(x)$       (iv) $\displaystyle\lim_{x \to 4} f(x)$

    (v) $\displaystyle\lim_{x \to 0} f(x)$       (vi) $\displaystyle\lim_{x \to 2^-} f(x)$

(b) State the equations of the vertical asymptotes.

(c) At what numbers is $f$ discontinuous? Explain.

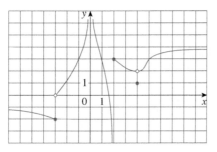

**24.** Sketch the graph of an example of a function $f$ that satisfies all of the following conditions:

$$\lim_{x \to 0^+} f(x) = -2, \quad \lim_{x \to 0^-} f(x) = 1, \quad f(0) = -1,$$

$$\lim_{x \to 2^-} f(x) = \infty, \quad \lim_{x \to 2^+} f(x) = -\infty$$

**25–38** Find the limit.

**25.** $\displaystyle\lim_{x \to 0} \cos(x + \sin x)$

**26.** $\displaystyle\lim_{x \to 3} \frac{x^2 - 9}{x^2 + 2x - 3}$

**27.** $\displaystyle\lim_{x \to -3} \frac{x^2 - 9}{x^2 + 2x - 3}$

**28.** $\displaystyle\lim_{x \to 1^+} \frac{x^2 - 9}{x^2 + 2x - 3}$

**29.** $\displaystyle\lim_{h \to 0} \frac{(h-1)^3 + 1}{h}$

**30.** $\displaystyle\lim_{t \to 2} \frac{t^2 - 4}{t^3 - 8}$

**31.** $\displaystyle\lim_{r \to 9} \frac{\sqrt{r}}{(r-9)^4}$

**32.** $\displaystyle\lim_{v \to 4^+} \frac{4 - v}{|4 - v|}$

**33.** $\displaystyle\lim_{u \to 1} \frac{u^4 - 1}{u^3 + 5u^2 - 6u}$

**34.** $\displaystyle\lim_{x \to 3} \frac{\sqrt{x + 6} - x}{x^3 - 3x^2}$

**35.** $\displaystyle\lim_{s \to 16} \frac{4 - \sqrt{s}}{s - 16}$

**36.** $\displaystyle\lim_{v \to 2} \frac{v^2 + 2v - 8}{v^4 - 16}$

**37.** $\displaystyle\lim_{x \to 0} \frac{1 - \sqrt{1 - x^2}}{x}$

**38.** $\displaystyle\lim_{x \to 1} \left( \frac{1}{x - 1} + \frac{1}{x^2 - 3x + 2} \right)$

**39.** If $2x - 1 \leqslant f(x) \leqslant x^2$ for $0 < x < 3$, find $\lim_{x \to 1} f(x)$.

**40.** Prove that $\lim_{x \to 0} x^2 \cos(1/x^2) = 0$.

**41–44** Prove the statement using the precise definition of a limit.

**41.** $\displaystyle\lim_{x \to 2} (14 - 5x) = 4$

**42.** $\displaystyle\lim_{x \to 0} \sqrt[3]{x} = 0$

**43.** $\displaystyle\lim_{x \to 2} (x^2 - 3x) = -2$

**44.** $\displaystyle\lim_{x \to 4^+} \frac{2}{\sqrt{x - 4}} = \infty$

**45.** Let

$$f(x) = \begin{cases} \sqrt{-x} & \text{if } x < 0 \\ 3 - x & \text{if } 0 \leqslant x < 3 \\ (x - 3)^2 & \text{if } x > 3 \end{cases}$$

(a) Evaluate each limit, if it exists.

    (i) $\displaystyle\lim_{x \to 0^+} f(x)$   (ii) $\displaystyle\lim_{x \to 0^-} f(x)$   (iii) $\displaystyle\lim_{x \to 0} f(x)$

    (iv) $\displaystyle\lim_{x \to 3^-} f(x)$   (v) $\displaystyle\lim_{x \to 3^+} f(x)$   (vi) $\displaystyle\lim_{x \to 3} f(x)$

(b) Where is $f$ discontinuous?

(c) Sketch the graph of $f$.

**46.** Let

$$g(x) = \begin{cases} 2x - x^2 & \text{if } 0 \leqslant x \leqslant 2 \\ 2 - x & \text{if } 2 < x \leqslant 3 \\ x - 4 & \text{if } 3 < x < 4 \\ \pi & \text{if } x \geqslant 4 \end{cases}$$

(a) For each of the numbers 2, 3, and 4, discover whether $g$ is continuous from the left, continuous from the right, or continuous at the number.

(b) Sketch the graph of $g$.

**47–48** Show that the function is continuous on its domain. State the domain.

**47.** $h(x) = \sqrt[4]{x} + x^3 \cos x$

**48.** $g(x) = \dfrac{\sqrt{x^2 - 9}}{x^2 - 2}$

**49–50** Use the Intermediate Value Theorem to show that there is a root of the equation in the given interval.

**49.** $x^5 - x^3 + 3x - 5 = 0, \quad (1, 2)$

**50.** $2 \sin x = 3 - 2x, \quad (0, 1)$

# Principles of Problem Solving

There are no hard and fast rules that will ensure success in solving problems. However, it is possible to outline some general steps in the problem-solving process and to give some principles that may be useful in the solution of certain problems. These steps and principles are just common sense made explicit. They have been adapted from George Polya's book *How To Solve It*.

1 UNDERSTAND THE PROBLEM The first step is to read the problem and make sure that you understand it clearly. Ask yourself the following questions:

> *What is the unknown?*
> *What are the given quantities?*
> *What are the given conditions?*

For many problems it is useful to

> *draw a diagram*

and identify the given and required quantities on the diagram.

Usually it is necessary to

> *introduce suitable notation*

In choosing symbols for the unknown quantities we often use letters such as $a$, $b$, $c$, $m$, $n$, $x$, and $y$, but in some cases it helps to use initials as suggestive symbols; for instance, $V$ for volume or $t$ for time.

2 THINK OF A PLAN Find a connection between the given information and the unknown that will enable you to calculate the unknown. It often helps to ask yourself explicitly: "How can I relate the given to the unknown?" If you don't see a connection immediately, the following ideas may be helpful in devising a plan.

**Try to Recognize Something Familiar** Relate the given situation to previous knowledge. Look at the unknown and try to recall a more familiar problem that has a similar unknown.

**Try to Recognize Patterns** Some problems are solved by recognizing that some kind of pattern is occurring. The pattern could be geometric, or numerical, or algebraic. If you can see regularity or repetition in a problem, you might be able to guess what the continuing pattern is and then prove it.

**Use Analogy** Try to think of an analogous problem, that is, a similar problem, a related problem, but one that is easier than the original problem. If you can solve the similar, simpler problem, then it might give you the clues you need to solve the original, more difficult problem. For instance, if a problem involves very large numbers, you could first try a similar problem with smaller numbers. Or if the problem involves three-dimensional geometry, you could look for a similar problem in two-dimensional geometry. Or if the problem you start with is a general one, you could first try a special case.

**Introduce Something Extra** It may sometimes be necessary to introduce something new, an auxiliary aid, to help make the connection between the given and the unknown. For instance, in a problem where a diagram is useful the auxiliary aid could be a new line drawn in a diagram. In a more algebraic problem it could be a new unknown that is related to the original unknown.

**Take Cases** We may sometimes have to split a problem into several cases and give a different argument for each of the cases. For instance, we often have to use this strategy in dealing with absolute value.

**Work Backward** Sometimes it is useful to imagine that your problem is solved and work backward, step by step, until you arrive at the given data. Then you may be able to reverse your steps and thereby construct a solution to the original problem. This procedure is commonly used in solving equations. For instance, in solving the equation $3x - 5 = 7$, we suppose that $x$ is a number that satisfies $3x - 5 = 7$ and work backward. We add 5 to each side of the equation and then divide each side by 3 to get $x = 4$. Since each of these steps can be reversed, we have solved the problem.

**Establish Subgoals** In a complex problem it is often useful to set subgoals (in which the desired situation is only partially fulfilled). If we can first reach these subgoals, then we may be able to build on them to reach our final goal.

**Indirect Reasoning** Sometimes it is appropriate to attack a problem indirectly. In using proof by contradiction to prove that $P$ implies $Q$, we assume that $P$ is true and $Q$ is false and try to see why this can't happen. Somehow we have to use this information and arrive at a contradiction to what we absolutely know is true.

**Mathematical Induction** In proving statements that involve a positive integer $n$, it is frequently helpful to use the following principle.

---

**Principle of Mathematical Induction** Let $S_n$ be a statement about the positive integer $n$. Suppose that

**1.** $S_1$ is true.

**2.** $S_{k+1}$ is true whenever $S_k$ is true.

Then $S_n$ is true for all positive integers $n$.

---

This is reasonable because, since $S_1$ is true, it follows from condition 2 (with $k = 1$) that $S_2$ is true. Then, using condition 2 with $k = 2$, we see that $S_3$ is true. Again using condition 2, this time with $k = 3$, we have that $S_4$ is true. This procedure can be followed indefinitely.

**3 CARRY OUT THE PLAN**

In Step 2 a plan was devised. In carrying out that plan we have to check each stage of the plan and write the details that prove that each stage is correct.

**4 LOOK BACK**

Having completed our solution, it is wise to look back over it, partly to see if we have made errors in the solution and partly to see if we can think of an easier way to solve the problem. Another reason for looking back is that it will familiarize us with the method of solution and this may be useful for solving a future problem. Descartes said, "Every problem that I solved became a rule which served afterwards to solve other problems."

These principles of problem solving are illustrated in the following examples. Before you look at the solutions, try to solve these problems yourself, referring to these Principles of Problem Solving if you get stuck. You may find it useful to refer to this section from time to time as you solve the exercises in the remaining chapters of this book.

As the first example illustrates, it is often necessary to use the problem-solving principle of *taking cases* when dealing with absolute values.

**EXAMPLE 1** Solve the inequality $|x - 3| + |x + 2| < 11$.

**SOLUTION** Recall the definition of absolute value:

$$|x| = \begin{cases} x & \text{if } x \geq 0 \\ -x & \text{if } x < 0 \end{cases}$$

It follows that

$$|x - 3| = \begin{cases} x - 3 & \text{if } x - 3 \geq 0 \\ -(x - 3) & \text{if } x - 3 < 0 \end{cases}$$

$$= \begin{cases} x - 3 & \text{if } x \geq 3 \\ -x + 3 & \text{if } x < 3 \end{cases}$$

Similarly

$$|x + 2| = \begin{cases} x + 2 & \text{if } x + 2 \geq 0 \\ -(x + 2) & \text{if } x + 2 < 0 \end{cases}$$

$$= \begin{cases} x + 2 & \text{if } x \geq -2 \\ -x - 2 & \text{if } x < -2 \end{cases}$$

PS Take cases

These expressions show that we must consider three cases:

$$x < -2 \qquad -2 \leq x < 3 \qquad x \geq 3$$

**CASE I** If $x < -2$, we have

$$|x - 3| + |x + 2| < 11$$
$$-x + 3 - x - 2 < 11$$
$$-2x < 10$$
$$x > -5$$

**CASE II** If $-2 \leq x < 3$, the given inequality becomes

$$-x + 3 + x + 2 < 11$$
$$5 < 11 \qquad \text{(always true)}$$

**CASE III** If $x \geq 3$, the inequality becomes

$$x - 3 + x + 2 < 11$$
$$2x < 12$$
$$x < 6$$

Combining cases I, II, and III, we see that the inequality is satisfied when $-5 < x < 6$. So the solution is the interval $(-5, 6)$.

**99**

In the following example we first guess the answer by looking at special cases and recognizing a pattern. Then we prove our conjecture by mathematical induction.

In using the Principle of Mathematical Induction, we follow three steps:

**Step 1** Prove that $S_n$ is true when $n = 1$.

**Step 2** Assume that $S_n$ is true when $n = k$ and deduce that $S_n$ is true when $n = k + 1$.

**Step 3** Conclude that $S_n$ is true for all $n$ by the Principle of Mathematical Induction.

**EXAMPLE 2** If $f_0(x) = x/(x + 1)$ and $f_{n+1} = f_0 \circ f_n$ for $n = 0, 1, 2, \ldots$, find a formula for $f_n(x)$.

PS Analogy: Try a similar, simpler problem

SOLUTION We start by finding formulas for $f_n(x)$ for the special cases $n = 1, 2,$ and $3$.

$$f_1(x) = (f_0 \circ f_0)(x) = f_0(f_0(x)) = f_0\left(\frac{x}{x + 1}\right)$$

$$= \frac{\dfrac{x}{x + 1}}{\dfrac{x}{x + 1} + 1} = \frac{\dfrac{x}{x + 1}}{\dfrac{2x + 1}{x + 1}} = \frac{x}{2x + 1}$$

$$f_2(x) = (f_0 \circ f_1)(x) = f_0(f_1(x)) = f_0\left(\frac{x}{2x + 1}\right)$$

$$= \frac{\dfrac{x}{2x + 1}}{\dfrac{x}{2x + 1} + 1} = \frac{\dfrac{x}{2x + 1}}{\dfrac{3x + 1}{2x + 1}} = \frac{x}{3x + 1}$$

$$f_3(x) = (f_0 \circ f_2)(x) = f_0(f_2(x)) = f_0\left(\frac{x}{3x + 1}\right)$$

PS Look for a pattern

$$= \frac{\dfrac{x}{3x + 1}}{\dfrac{x}{3x + 1} + 1} = \frac{\dfrac{x}{3x + 1}}{\dfrac{4x + 1}{3x + 1}} = \frac{x}{4x + 1}$$

We notice a pattern: The coefficient of $x$ in the denominator of $f_n(x)$ is $n + 1$ in the three cases we have computed. So we make the guess that, in general,

$$\boxed{4} \qquad\qquad f_n(x) = \frac{x}{(n + 1)x + 1}$$

To prove this, we use the Principle of Mathematical Induction. We have already verified that $\boxed{4}$ is true for $n = 1$. Assume that it is true for $n = k$, that is,

$$f_k(x) = \frac{x}{(k + 1)x + 1}$$

Then $\qquad f_{k+1}(x) = (f_0 \circ f_k)(x) = f_0(f_k(x)) = f_0\left(\dfrac{x}{(k+1)x+1}\right)$

$$= \dfrac{\dfrac{x}{(k+1)x+1}}{\dfrac{x}{(k+1)x+1}+1} = \dfrac{\dfrac{x}{(k+1)x+1}}{\dfrac{(k+2)x+1}{(k+1)x+1}} = \dfrac{x}{(k+2)x+1}$$

This expression shows that $\boxed{4}$ is true for $n = k + 1$. Therefore, by mathematical induction, it is true for all positive integers $n$.

In the following example we show how the problem solving strategy of *introducing something extra* is sometimes useful when we evaluate limits. The idea is to change the variable—to introduce a new variable that is related to the original variable—in such a way as to make the problem simpler. Later, in Section 4.5, we will make more extensive use of this general idea.

**EXAMPLE 3** Evaluate $\displaystyle\lim_{x \to 0} \dfrac{\sqrt[3]{1+cx}-1}{x}$, where $c$ is a constant.

SOLUTION As it stands, this limit looks challenging. In Section 1.6 we evaluated several limits in which both numerator and denominator approached 0. There our strategy was to perform some sort of algebraic manipulation that led to a simplifying cancellation, but here it's not clear what kind of algebra is necessary.

So we introduce a new variable $t$ by the equation

$$t = \sqrt[3]{1+cx}$$

We also need to express $x$ in terms of $t$, so we solve this equation:

$$t^3 = 1 + cx \qquad x = \dfrac{t^3 - 1}{c} \quad (\text{if } c \neq 0)$$

Notice that $x \to 0$ is equivalent to $t \to 1$. This allows us to convert the given limit into one involving the variable $t$:

$$\lim_{x \to 0} \dfrac{\sqrt[3]{1+cx}-1}{x} = \lim_{t \to 1} \dfrac{t-1}{(t^3-1)/c} = \lim_{t \to 1} \dfrac{c(t-1)}{t^3-1}$$

The change of variable allowed us to replace a relatively complicated limit by a simpler one of a type that we have seen before. Factoring the denominator as a difference of cubes, we get

$$\lim_{t \to 1} \dfrac{c(t-1)}{t^3-1} = \lim_{t \to 1} \dfrac{c(t-1)}{(t-1)(t^2+t+1)}$$

$$= \lim_{t \to 1} \dfrac{c}{t^2+t+1} = \dfrac{c}{3}$$

In making the change of variable we had to rule out the case $c = 0$. But if $c = 0$, the function is 0 for all nonzero $x$ and so its limit is 0. Therefore, in all cases, the limit is $c/3$.

The following problems are meant to test and challenge your problem-solving skills. Some of them require a considerable amount of time to think through, so don't be discouraged if you can't solve them right away. If you get stuck, you might find it helpful to refer to the discussion of the principles of problem solving.

1. Draw the graph of the equation $x + |x| = y + |y|$.

2. Sketch the region in the plane consisting of all points $(x, y)$ such that $|x - y| + |x| - |y| \leq 2$.

3. If $f_0(x) = x^2$ and $f_{n+1}(x) = f_0(f_n(x))$ for $n = 0, 1, 2, \ldots$, find a formula for $f_n(x)$.

4. (a) If $f_0(x) = \dfrac{1}{2 - x}$ and $f_{n+1} = f_0 \circ f_n$ for $n = 0, 1, 2, \ldots$, find an expression for $f_n(x)$ and use mathematical induction to prove it.

   (b) Graph $f_0, f_1, f_2, f_3$ on the same screen and describe the effects of repeated composition.

5. Evaluate $\displaystyle\lim_{x \to 1} \frac{\sqrt[3]{x} - 1}{\sqrt{x} - 1}$.

6. Find numbers $a$ and $b$ such that $\displaystyle\lim_{x \to 0} \frac{\sqrt{ax + b} - 2}{x} = 1$.

7. Evaluate $\displaystyle\lim_{x \to 0} \frac{|2x - 1| - |2x + 1|}{x}$.

8. The figure shows a point $P$ on the parabola $y = x^2$ and the point $Q$ where the perpendicular bisector of $OP$ intersects the $y$-axis. As $P$ approaches the origin along the parabola, what happens to $Q$? Does it have a limiting position? If so, find it.

9. Evaluate the following limits, if they exist, where $[\![x]\!]$ denotes the greatest integer function.

   (a) $\displaystyle\lim_{x \to 0} \frac{[\![x]\!]}{x}$   (b) $\displaystyle\lim_{x \to 0} x [\![1/x]\!]$

10. Sketch the region in the plane defined by each of the following equations.

    (a) $[\![x]\!]^2 + [\![y]\!]^2 = 1$   (b) $[\![x]\!]^2 - [\![y]\!]^2 = 3$   (c) $[\![x + y]\!]^2 = 1$   (d) $[\![x]\!] + [\![y]\!] = 1$

11. Find all values of $a$ such that $f$ is continuous on $\mathbb{R}$:

$$f(x) = \begin{cases} x + 1 & \text{if } x \leq a \\ x^2 & \text{if } x > a \end{cases}$$

12. A **fixed point** of a function $f$ is a number $c$ in its domain such that $f(c) = c$. (The function doesn't move $c$; it stays fixed.)

    (a) Sketch the graph of a continuous function with domain $[0, 1]$ whose range also lies in $[0, 1]$. Locate a fixed point of $f$.

    (b) Try to draw the graph of a continuous function with domain $[0, 1]$ and range in $[0, 1]$ that does *not* have a fixed point. What is the obstacle?

    (c) Use the Intermediate Value Theorem to prove that any continuous function with domain $[0, 1]$ and range in $[0, 1]$ must have a fixed point.

13. If $\lim_{x \to a} [f(x) + g(x)] = 2$ and $\lim_{x \to a} [f(x) - g(x)] = 1$, find $\lim_{x \to a} [f(x) g(x)]$.

14. (a) The figure shows an isosceles triangle $ABC$ with $\angle B = \angle C$. The bisector of angle $B$ intersects the side $AC$ at the point $P$. Suppose that the base $BC$ remains fixed but the altitude $|AM|$ of the triangle approaches 0, so $A$ approaches the midpoint $M$ of $BC$. What happens to $P$ during this process? Does it have a limiting position? If so, find it.

    (b) Try to sketch the path traced out by $P$ during this process. Then find an equation of this curve and use this equation to sketch the curve.

15. (a) If we start from $0°$ latitude and proceed in a westerly direction, we can let $T(x)$ denote the temperature at the point $x$ at any given time. Assuming that $T$ is a continuous function of $x$, show that at any fixed time there are at least two diametrically opposite points on the equator that have exactly the same temperature.

    (b) Does the result in part (a) hold for points lying on any circle on the earth's surface?

    (c) Does the result in part (a) hold for barometric pressure and for altitude above sea level?

**FIGURE FOR PROBLEM 8**

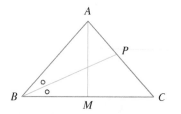

**FIGURE FOR PROBLEM 14**

---

⊞ Graphing calculator or computer required

# 2 Derivatives

For a roller coaster ride to be smooth, the straight stretches of the track need to be connected to the curved segments so that there are no abrupt changes in direction. In the project on page 140 you will see how to design the first ascent and drop of a new coaster for a smooth ride.

© Brett Mulcahy / Shutterstock

In this chapter we begin our study of differential calculus, which is concerned with how one quantity changes in relation to another quantity. The central concept of differential calculus is the *derivative*, which is an outgrowth of the velocities and slopes of tangents that we considered in Chapter 1. After learning how to calculate derivatives, we use them to solve problems involving rates of change and the approximation of functions.

**Derivatives and Rates of Change**

The problem of finding the tangent line to a curve and the problem of finding the velocity of an object both involve finding the same type of limit, as we saw in Section 1.4. This special type of limit is called a *derivative* and we will see that it can be interpreted as a rate of change in any of the sciences or engineering.

### Tangents

If a curve $C$ has equation $y = f(x)$ and we want to find the tangent line to $C$ at the point $P(a, f(a))$, then we consider a nearby point $Q(x, f(x))$, where $x \neq a$, and compute the slope of the secant line $PQ$:

$$m_{PQ} = \frac{f(x) - f(a)}{x - a}$$

Then we let $Q$ approach $P$ along the curve $C$ by letting $x$ approach $a$. If $m_{PQ}$ approaches a number $m$, then we define the *tangent* $t$ to be the line through $P$ with slope $m$. (This amounts to saying that the tangent line is the limiting position of the secant line $PQ$ as $Q$ approaches $P$. See Figure 1.)

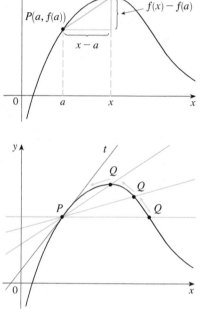

---

**1** **Definition** The **tangent line** to the curve $y = f(x)$ at the point $P(a, f(a))$ is the line through $P$ with slope

$$m = \lim_{x \to a} \frac{f(x) - f(a)}{x - a}$$

provided that this limit exists.

---

**FIGURE 1**

In our first example we confirm the guess we made in Example 1 in Section 1.4.

**V** **EXAMPLE 1** Find an equation of the tangent line to the parabola $y = x^2$ at the point $P(1, 1)$.

SOLUTION Here we have $a = 1$ and $f(x) = x^2$, so the slope is

$$m = \lim_{x \to 1} \frac{f(x) - f(1)}{x - 1} = \lim_{x \to 1} \frac{x^2 - 1}{x - 1}$$

$$= \lim_{x \to 1} \frac{(x - 1)(x + 1)}{x - 1}$$

$$= \lim_{x \to 1} (x + 1) = 1 + 1 = 2$$

Point-slope form for a line through the point $(x_1, y_1)$ with slope $m$:

$$y - y_1 = m(x - x_1)$$

Using the point-slope form of the equation of a line, we find that an equation of the tangent line at $(1, 1)$ is

$$y - 1 = 2(x - 1) \qquad \text{or} \qquad y = 2x - 1$$

We sometimes refer to the slope of the tangent line to a curve at a point as the **slope of the curve** at the point. The idea is that if we zoom in far enough toward the point, the curve looks almost like a straight line. Figure 2 illustrates this procedure for the curve $y = x^2$ in

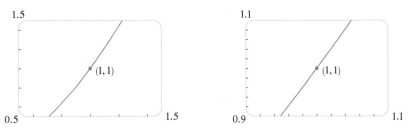

**TEC** Visual 2.1 shows an animation of Figure 2.

Example 1. The more we zoom in, the more the parabola looks like a line. In other words, the curve becomes almost indistinguishable from its tangent line.

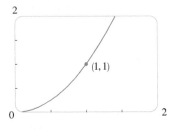

**FIGURE 2**   Zooming in toward the point $(1, 1)$ on the parabola $y = x^2$

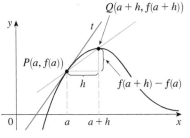

**FIGURE 3**

There is another expression for the slope of a tangent line that is sometimes easier to use. If $h = x - a$, then $x = a + h$ and so the slope of the secant line $PQ$ is

$$m_{PQ} = \frac{f(a + h) - f(a)}{h}$$

(See Figure 3 where the case $h > 0$ is illustrated and $Q$ is to the right of $P$. If it happened that $h < 0$, however, $Q$ would be to the left of $P$.)

Notice that as $x$ approaches $a$, $h$ approaches 0 (because $h = x - a$) and so the expression for the slope of the tangent line in Definition 1 becomes

$$\boxed{2} \qquad m = \lim_{h \to 0} \frac{f(a + h) - f(a)}{h}$$

**EXAMPLE 2**   Find an equation of the tangent line to the hyperbola $y = 3/x$ at the point $(3, 1)$.

SOLUTION   Let $f(x) = 3/x$. Then the slope of the tangent at $(3, 1)$ is

$$m = \lim_{h \to 0} \frac{f(3 + h) - f(3)}{h} = \lim_{h \to 0} \frac{\dfrac{3}{3 + h} - 1}{h} = \lim_{h \to 0} \frac{\dfrac{3 - (3 + h)}{3 + h}}{h}$$

$$= \lim_{h \to 0} \frac{-h}{h(3 + h)} = \lim_{h \to 0} -\frac{1}{3 + h} = -\frac{1}{3}$$

Therefore an equation of the tangent at the point $(3, 1)$ is

$$y - 1 = -\tfrac{1}{3}(x - 3)$$

which simplifies to

$$x + 3y - 6 = 0$$

The hyperbola and its tangent are shown in Figure 4.

**FIGURE 4**

## Velocities

In Section 1.4 we investigated the motion of a ball dropped from the CN Tower and defined its velocity to be the limiting value of average velocities over shorter and shorter time periods.

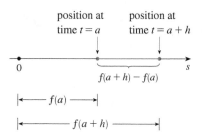

position at          position at
time $t = a$         time $t = a + h$

$f(a + h) - f(a)$

$f(a)$

$f(a + h)$

**FIGURE 5**

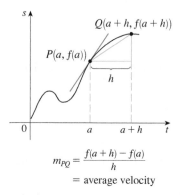

$m_{PQ} = \dfrac{f(a + h) - f(a)}{h}$

$= $ average velocity

**FIGURE 6**

Recall from Section 1.4: The distance
(in meters) fallen after $t$ seconds is $4.9t^2$.

In general, suppose an object moves along a straight line according to an equation of motion $s = f(t)$, where $s$ is the displacement (directed distance) of the object from the origin at time $t$. The function $f$ that describes the motion is called the **position function** of the object. In the time interval from $t = a$ to $t = a + h$ the change in position is $f(a + h) - f(a)$. (See Figure 5.) The average velocity over this time interval is

$$\text{average velocity} = \frac{\text{displacement}}{\text{time}} = \frac{f(a + h) - f(a)}{h}$$

which is the same as the slope of the secant line $PQ$ in Figure 6.

Now suppose we compute the average velocities over shorter and shorter time intervals $[a, a + h]$. In other words, we let $h$ approach 0. As in the example of the falling ball, we define the **velocity** (or **instantaneous velocity**) $v(a)$ at time $t = a$ to be the limit of these average velocities:

**3** $$v(a) = \lim_{h \to 0} \frac{f(a + h) - f(a)}{h}$$

This means that the velocity at time $t = a$ is equal to the slope of the tangent line at $P$ (compare Equations 2 and 3).

Now that we know how to compute limits, let's reconsider the problem of the falling ball.

**V** **EXAMPLE 3**  Suppose that a ball is dropped from the upper observation deck of the CN Tower, 450 m above the ground.
(a) What is the velocity of the ball after 5 seconds?
(b) How fast is the ball traveling when it hits the ground?

SOLUTION  We will need to find the velocity both when $t = 5$ and when the ball hits the ground, so it's efficient to start by finding the velocity at a general time $t = a$. Using the equation of motion $s = f(t) = 4.9t^2$, we have

$$v(a) = \lim_{h \to 0} \frac{f(a + h) - f(a)}{h} = \lim_{h \to 0} \frac{4.9(a + h)^2 - 4.9a^2}{h}$$

$$= \lim_{h \to 0} \frac{4.9(a^2 + 2ah + h^2 - a^2)}{h} = \lim_{h \to 0} \frac{4.9(2ah + h^2)}{h}$$

$$= \lim_{h \to 0} 4.9(2a + h) = 9.8a$$

(a) The velocity after 5 s is $v(5) = (9.8)(5) = 49$ m/s.

(b) Since the observation deck is 450 m above the ground, the ball will hit the ground at the time $t_1$ when $s(t_1) = 450$, that is,

$$4.9t_1^2 = 450$$

This gives

$$t_1^2 = \frac{450}{4.9} \qquad \text{and} \qquad t_1 = \sqrt{\frac{450}{4.9}} \approx 9.6 \text{ s}$$

The velocity of the ball as it hits the ground is therefore

$$v(t_1) = 9.8t_1 = 9.8\sqrt{\frac{450}{4.9}} \approx 94 \text{ m/s}$$

## Derivatives

We have seen that the same type of limit arises in finding the slope of a tangent line (Equation 2) or the velocity of an object (Equation 3). In fact, limits of the form

$$\lim_{h \to 0} \frac{f(a + h) - f(a)}{h}$$

arise whenever we calculate a rate of change in any of the sciences or engineering, such as a rate of reaction in chemistry or a marginal cost in economics. Since this type of limit occurs so widely, it is given a special name and notation.

> **4** **Definition** The **derivative of a function $f$ at a number $a$**, denoted by $f'(a)$, is
>
> $$f'(a) = \lim_{h \to 0} \frac{f(a + h) - f(a)}{h}$$
>
> if this limit exists.

$f'(a)$ is read "$f$ prime of $a$."

If we write $x = a + h$, then we have $h = x - a$ and $h$ approaches 0 if and only if $x$ approaches $a$. Therefore an equivalent way of stating the definition of the derivative, as we saw in finding tangent lines, is

> **5**
>
> $$f'(a) = \lim_{x \to a} \frac{f(x) - f(a)}{x - a}$$

**V** **EXAMPLE 4** Find the derivative of the function $f(x) = x^2 - 8x + 9$ at the number $a$.

SOLUTION From Definition 4 we have

$$
\begin{aligned}
f'(a) &= \lim_{h \to 0} \frac{f(a + h) - f(a)}{h} \\
&= \lim_{h \to 0} \frac{[(a + h)^2 - 8(a + h) + 9] - [a^2 - 8a + 9]}{h} \\
&= \lim_{h \to 0} \frac{a^2 + 2ah + h^2 - 8a - 8h + 9 - a^2 + 8a - 9}{h} \\
&= \lim_{h \to 0} \frac{2ah + h^2 - 8h}{h} = \lim_{h \to 0} (2a + h - 8) \\
&= 2a - 8
\end{aligned}
$$

We defined the tangent line to the curve $y = f(x)$ at the point $P(a, f(a))$ to be the line that passes through $P$ and has slope $m$ given by Equation 1 or 2. Since, by Definition 4, this is the same as the derivative $f'(a)$, we can now say the following.

> The tangent line to $y = f(x)$ at $(a, f(a))$ is the line through $(a, f(a))$ whose slope is equal to $f'(a)$, the derivative of $f$ at $a$.

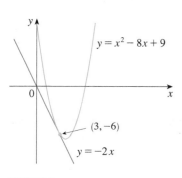

**FIGURE 7**

If we use the point-slope form of the equation of a line, we can write an equation of the tangent line to the curve $y = f(x)$ at the point $(a, f(a))$:

$$y - f(a) = f'(a)(x - a)$$

**V EXAMPLE 5** Find an equation of the tangent line to the parabola $y = x^2 - 8x + 9$ at the point $(3, -6)$.

SOLUTION From Example 4 we know that the derivative of $f(x) = x^2 - 8x + 9$ at the number $a$ is $f'(a) = 2a - 8$. Therefore the slope of the tangent line at $(3, -6)$ is $f'(3) = 2(3) - 8 = -2$. Thus an equation of the tangent line, shown in Figure 7, is

$$y - (-6) = (-2)(x - 3) \quad \text{or} \quad y = -2x$$

## Rates of Change

Suppose $y$ is a quantity that depends on another quantity $x$. Thus $y$ is a function of $x$ and we write $y = f(x)$. If $x$ changes from $x_1$ to $x_2$, then the change in $x$ (also called the **increment** of $x$) is

$$\Delta x = x_2 - x_1$$

and the corresponding change in $y$ is

$$\Delta y = f(x_2) - f(x_1)$$

The difference quotient

$$\frac{\Delta y}{\Delta x} = \frac{f(x_2) - f(x_1)}{x_2 - x_1}$$

is called the **average rate of change of $y$ with respect to $x$** over the interval $[x_1, x_2]$ and can be interpreted as the slope of the secant line $PQ$ in Figure 8.

By analogy with velocity, we consider the average rate of change over smaller and smaller intervals by letting $x_2$ approach $x_1$ and therefore letting $\Delta x$ approach 0. The limit of these average rates of change is called the **(instantaneous) rate of change of $y$ with respect to $x$ at $x = x_1$**, which is interpreted as the slope of the tangent to the curve $y = f(x)$ at $P(x_1, f(x_1))$:

average rate of change $= m_{PQ}$

instantaneous rate of change $=$
         slope of tangent at $P$

**FIGURE 8**

$$\boxed{6} \qquad \text{instantaneous rate of change} = \lim_{\Delta x \to 0} \frac{\Delta y}{\Delta x} = \lim_{x_2 \to x_1} \frac{f(x_2) - f(x_1)}{x_2 - x_1}$$

We recognize this limit as being the derivative $f'(x_1)$.

We know that one interpretation of the derivative $f'(a)$ is as the slope of the tangent line to the curve $y = f(x)$ when $x = a$. We now have a second interpretation:

The derivative $f'(a)$ is the instantaneous rate of change of $y = f(x)$ with respect to $x$ when $x = a$.

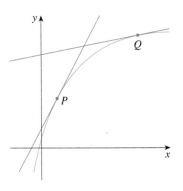

**FIGURE 9**

The $y$-values are changing rapidly at $P$ and slowly at $Q$.

The connection with the first interpretation is that if we sketch the curve $y = f(x)$, then the instantaneous rate of change is the slope of the tangent to this curve at the point where $x = a$. This means that when the derivative is large (and therefore the curve is steep, as at the point $P$ in Figure 9), the $y$-values change rapidly. When the derivative is small, the curve is relatively flat (as at point $Q$) and the $y$-values change slowly.

In particular, if $s = f(t)$ is the position function of a particle that moves along a straight line, then $f'(a)$ is the rate of change of the displacement $s$ with respect to the time $t$. In other words, $f'(a)$ *is the velocity of the particle at time* $t = a$. The **speed** of the particle is the absolute value of the velocity, that is, $|f'(a)|$.

In the next example we discuss the meaning of the derivative of a function that is defined verbally.

**V EXAMPLE 6** A manufacturer produces bolts of a fabric with a fixed width. The cost of producing $x$ yards of this fabric is $C = f(x)$ dollars.
(a) What is the meaning of the derivative $f'(x)$? What are its units?
(b) In practical terms, what does it mean to say that $f'(1000) = 9$?
(c) Which do you think is greater, $f'(50)$ or $f'(500)$? What about $f'(5000)$?

SOLUTION

(a) The derivative $f'(x)$ is the instantaneous rate of change of $C$ with respect to $x$; that is, $f'(x)$ means the rate of change of the production cost with respect to the number of yards produced. (Economists call this rate of change the *marginal cost*. This idea is discussed in more detail in Sections 2.7 and 3.7.)

Because

$$f'(x) = \lim_{\Delta x \to 0} \frac{\Delta C}{\Delta x}$$

the units for $f'(x)$ are the same as the units for the difference quotient $\Delta C / \Delta x$. Since $\Delta C$ is measured in dollars and $\Delta x$ in yards, it follows that the units for $f'(x)$ are dollars per yard.

(b) The statement that $f'(1000) = 9$ means that, after 1000 yards of fabric have been manufactured, the rate at which the production cost is increasing is \$9/yard. (When $x = 1000$, $C$ is increasing 9 times as fast as $x$.)

Since $\Delta x = 1$ is small compared with $x = 1000$, we could use the approximation

Here we are assuming that the cost function is well behaved; in other words, $C(x)$ doesn't oscillate rapidly near $x = 1000$.

$$f'(1000) \approx \frac{\Delta C}{\Delta x} = \frac{\Delta C}{1} = \Delta C$$

and say that the cost of manufacturing the 1000th yard (or the 1001st) is about \$9.

(c) The rate at which the production cost is increasing (per yard) is probably lower when $x = 500$ than when $x = 50$ (the cost of making the 500th yard is less than the cost of the 50th yard) because of economies of scale. (The manufacturer makes more efficient use of the fixed costs of production.) So

$$f'(50) > f'(500)$$

But, as production expands, the resulting large-scale operation might become inefficient and there might be overtime costs. Thus it is possible that the rate of increase of costs will eventually start to rise. So it may happen that

$$f'(5000) > f'(500)$$

In the following example we estimate the rate of change of the national debt with respect to time. Here the function is defined not by a formula but by a table of values.

**V** **EXAMPLE 7**  Let $D(t)$ be the US national debt at time $t$. The table in the margin gives approximate values of this function by providing end of year estimates, in billions of dollars, from 1980 to 2005. Interpret and estimate the value of $D'(1990)$.

SOLUTION  The derivative $D'(1990)$ means the rate of change of $D$ with respect to $t$ when $t = 1990$, that is, the rate of increase of the national debt in 1990.

According to Equation 5,

$$D'(1990) = \lim_{t \to 1990} \frac{D(t) - D(1990)}{t - 1990}$$

| $t$ | $D(t)$ |
|------|--------|
| 1980 | 930.2 |
| 1985 | 1945.9 |
| 1990 | 3233.3 |
| 1995 | 4974.0 |
| 2000 | 5674.2 |
| 2005 | 7932.7 |

So we compute and tabulate values of the difference quotient (the average rates of change) as shown in the table at the left. From this table we see that $D'(1990)$ lies somewhere between 257.48 and 348.14 billion dollars per year. [Here we are making the reasonable assumption that the debt didn't fluctuate wildly between 1980 and 2000.] We estimate that the rate of increase of the national debt of the United States in 1990 was the average of these two numbers, namely

| $t$ | $\dfrac{D(t) - D(1990)}{t - 1990}$ |
|------|-----------|
| 1980 | 230.31 |
| 1985 | 257.48 |
| 1995 | 348.14 |
| 2000 | 244.09 |
| 2005 | 313.29 |

$$D'(1990) \approx 303 \text{ billion dollars per year}$$

Another method would be to plot the debt function and estimate the slope of the tangent line when $t = 1990$.    ▪

**A Note on Units**

The units for the average rate of change $\Delta D / \Delta t$ are the units for $\Delta D$ divided by the units for $\Delta t$, namely, billions of dollars per year. The instantaneous rate of change is the limit of the average rates of change, so it is measured in the same units: billions of dollars per year.

In Examples 3, 6, and 7 we saw three specific examples of rates of change: the velocity of an object is the rate of change of displacement with respect to time; marginal cost is the rate of change of production cost with respect to the number of items produced; the rate of change of the debt with respect to time is of interest in economics. Here is a small sample of other rates of change: In physics, the rate of change of work with respect to time is called *power*. Chemists who study a chemical reaction are interested in the rate of change in the concentration of a reactant with respect to time (called the *rate of reaction*). A biologist is interested in the rate of change of the population of a colony of bacteria with respect to time. In fact, the computation of rates of change is important in all of the natural sciences, in engineering, and even in the social sciences. Further examples will be given in Section 2.7.

All these rates of change are derivatives and can therefore be interpreted as slopes of tangents. This gives added significance to the solution of the tangent problem. Whenever we solve a problem involving tangent lines, we are not just solving a problem in geometry. We are also implicitly solving a great variety of problems involving rates of change in science and engineering.

**1.** A curve has equation $y = f(x)$.
  (a) Write an expression for the slope of the secant line through the points $P(3, f(3))$ and $Q(x, f(x))$.
  (b) Write an expression for the slope of the tangent line at $P$.

**2.** Graph the curve $y = \sin x$ in the viewing rectangles $[-2, 2]$ by $[-2, 2]$, $[-1, 1]$ by $[-1, 1]$, and $[-0.5, 0.5]$ by

$[-0.5, 0.5]$. What do you notice about the curve as you zoom in toward the origin?

**3.** (a) Find the slope of the tangent line to the parabola $y = 4x - x^2$ at the point $(1, 3)$
    (i)  using Definition 1    (ii)  using Equation 2
  (b) Find an equation of the tangent line in part (a).

▱ Graphing calculator or computer required    **1.** Homework Hints available at stewartcalculus.com

(c) Graph the parabola and the tangent line. As a check on your work, zoom in toward the point $(1, 3)$ until the parabola and the tangent line are indistinguishable.

**4.** (a) Find the slope of the tangent line to the curve $y = x - x^3$ at the point $(1, 0)$
   (i) using Definition 1      (ii) using Equation 2
   (b) Find an equation of the tangent line in part (a).
   (c) Graph the curve and the tangent line in successively smaller viewing rectangles centered at $(1, 0)$ until the curve and the line appear to coincide.

**5–8** Find an equation of the tangent line to the curve at the given point.

**5.** $y = 4x - 3x^2$,   $(2, -4)$     **6.** $y = x^3 - 3x + 1$,   $(2, 3)$

**7.** $y = \sqrt{x}$,   $(1, 1)$     **8.** $y = \dfrac{2x + 1}{x + 2}$,   $(1, 1)$

**9.** (a) Find the slope of the tangent to the curve $y = 3 + 4x^2 - 2x^3$ at the point where $x = a$.
   (b) Find equations of the tangent lines at the points $(1, 5)$ and $(2, 3)$.
   (c) Graph the curve and both tangents on a common screen.

**10.** (a) Find the slope of the tangent to the curve $y = 1/\sqrt{x}$ at the point where $x = a$.
   (b) Find equations of the tangent lines at the points $(1, 1)$ and $\left(4, \tfrac{1}{2}\right)$.
   (c) Graph the curve and both tangents on a common screen.

**11.** (a) A particle starts by moving to the right along a horizontal line; the graph of its position function is shown. When is the particle moving to the right? Moving to the left? Standing still?
   (b) Draw a graph of the velocity function.

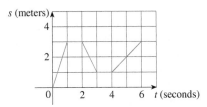

**12.** Shown are graphs of the position functions of two runners, A and B, who run a 100-m race and finish in a tie.

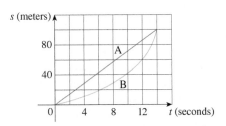

   (a) Describe and compare how the runners run the race.

(b) At what time is the distance between the runners the greatest?
(c) At what time do they have the same velocity?

**13.** If a ball is thrown into the air with a velocity of 40 ft/s, its height (in feet) after $t$ seconds is given by $y = 40t - 16t^2$. Find the velocity when $t = 2$.

**14.** If a rock is thrown upward on the planet Mars with a velocity of 10 m/s, its height (in meters) after $t$ seconds is given by $H = 10t - 1.86t^2$.
   (a) Find the velocity of the rock after one second.
   (b) Find the velocity of the rock when $t = a$.
   (c) When will the rock hit the surface?
   (d) With what velocity will the rock hit the surface?

**15.** The displacement (in meters) of a particle moving in a straight line is given by the equation of motion $s = 1/t^2$, where $t$ is measured in seconds. Find the velocity of the particle at times $t = a, t = 1, t = 2$, and $t = 3$.

**16.** The displacement (in meters) of a particle moving in a straight line is given by $s = t^2 - 8t + 18$, where $t$ is measured in seconds.
   (a) Find the average velocity over each time interval:
      (i) $[3, 4]$      (ii) $[3.5, 4]$
      (iii) $[4, 5]$      (iv) $[4, 4.5]$
   (b) Find the instantaneous velocity when $t = 4$.
   (c) Draw the graph of $s$ as a function of $t$ and draw the secant lines whose slopes are the average velocities in part (a) and the tangent line whose slope is the instantaneous velocity in part (b).

**17.** For the function $g$ whose graph is given, arrange the following numbers in increasing order and explain your reasoning:

$$0 \qquad g'(-2) \qquad g'(0) \qquad g'(2) \qquad g'(4)$$

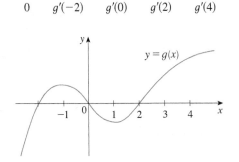

**18.** Find an equation of the tangent line to the graph of $y = g(x)$ at $x = 5$ if $g(5) = -3$ and $g'(5) = 4$.

**19.** If an equation of the tangent line to the curve $y = f(x)$ at the point where $a = 2$ is $y = 4x - 5$, find $f(2)$ and $f'(2)$.

**20.** If the tangent line to $y = f(x)$ at $(4, 3)$ passes through the point $(0, 2)$, find $f(4)$ and $f'(4)$.

**21.** Sketch the graph of a function $f$ for which $f(0) = 0$, $f'(0) = 3$, $f'(1) = 0$, and $f'(2) = -1$.

**22.** Sketch the graph of a function $g$ for which $g(0) = g(2) = g(4) = 0$, $g'(1) = g'(3) = 0$, $g'(0) = g'(4) = 1$, $g'(2) = -1$, $\lim_{x \to 5^-} g(x) = \infty$, and $\lim_{x \to -1^+} g(x) = -\infty$.

**23.** If $f(x) = 3x^2 - x^3$, find $f'(1)$ and use it to find an equation of the tangent line to the curve $y = 3x^2 - x^3$ at the point $(1, 2)$.

**24.** If $g(x) = x^4 - 2$, find $g'(1)$ and use it to find an equation of the tangent line to the curve $y = x^4 - 2$ at the point $(1, -1)$.

**25.** (a) If $F(x) = 5x/(1 + x^2)$, find $F'(2)$ and use it to find an equation of the tangent line to the curve $y = 5x/(1 + x^2)$ at the point $(2, 2)$.
(b) Illustrate part (a) by graphing the curve and the tangent line on the same screen.

**26.** (a) If $G(x) = 4x^2 - x^3$, find $G'(a)$ and use it to find equations of the tangent lines to the curve $y = 4x^2 - x^3$ at the points $(2, 8)$ and $(3, 9)$.
(b) Illustrate part (a) by graphing the curve and the tangent lines on the same screen.

**27–32** Find $f'(a)$.

**27.** $f(x) = 3x^2 - 4x + 1$

**28.** $f(t) = 2t^3 + t$

**29.** $f(t) = \dfrac{2t + 1}{t + 3}$

**30.** $f(x) = x^{-2}$

**31.** $f(x) = \sqrt{1 - 2x}$

**32.** $f(x) = \dfrac{4}{\sqrt{1 - x}}$

**33–38** Each limit represents the derivative of some function $f$ at some number $a$. State such an $f$ and $a$ in each case.

**33.** $\lim\limits_{h \to 0} \dfrac{(1 + h)^{10} - 1}{h}$

**34.** $\lim\limits_{h \to 0} \dfrac{\sqrt[4]{16 + h} - 2}{h}$

**35.** $\lim\limits_{x \to 5} \dfrac{2^x - 32}{x - 5}$

**36.** $\lim\limits_{x \to \pi/4} \dfrac{\tan x - 1}{x - \pi/4}$

**37.** $\lim\limits_{h \to 0} \dfrac{\cos(\pi + h) + 1}{h}$

**38.** $\lim\limits_{t \to 1} \dfrac{t^4 + t - 2}{t - 1}$

**39–40** A particle moves along a straight line with equation of motion $s = f(t)$, where $s$ is measured in meters and $t$ in seconds. Find the velocity and the speed when $t = 5$.

**39.** $f(t) = 100 + 50t - 4.9t^2$

**40.** $f(t) = t^{-1} - t$

**41.** A warm can of soda is placed in a cold refrigerator. Sketch the graph of the temperature of the soda as a function of time. Is the initial rate of change of temperature greater or less than the rate of change after an hour?

**42.** A roast turkey is taken from an oven when its temperature has reached 185°F and is placed on a table in a room where the temperature is 75°F. The graph shows how the temperature of the turkey decreases and eventually approaches room temperature. By measuring the slope of the tangent, estimate the rate of change of the temperature after an hour.

**43.** The number $N$ of US cellular phone subscribers (in millions) is shown in the table. (Midyear estimates are given.)

| $t$ | 1996 | 1998 | 2000 | 2002 | 2004 | 2006 |
|-----|------|------|------|------|------|------|
| $N$ | 44 | 69 | 109 | 141 | 182 | 233 |

(a) Find the average rate of cell phone growth
　(i) from 2002 to 2006　　(ii) from 2002 to 2004
　(iii) from 2000 to 2002
　In each case, include the units.
(b) Estimate the instantaneous rate of growth in 2002 by taking the average of two average rates of change. What are its units?
(c) Estimate the instantaneous rate of growth in 2002 by measuring the slope of a tangent.

**44.** The number $N$ of locations of a popular coffeehouse chain is given in the table. (The numbers of locations as of October 1 are given.)

| Year | 2004 | 2005 | 2006 | 2007 | 2008 |
|------|------|------|------|------|------|
| $N$ | 8569 | 10,241 | 12,440 | 15,011 | 16,680 |

(a) Find the average rate of growth
　(i) from 2006 to 2008　　(ii) from 2006 to 2007
　(iii) from 2005 to 2006
　In each case, include the units.
(b) Estimate the instantaneous rate of growth in 2006 by taking the average of two average rates of change. What are its units?
(c) Estimate the instantaneous rate of growth in 2006 by measuring the slope of a tangent.
(d) Estimate the intantaneous rate of growth in 2007 and compare it with the growth rate in 2006. What do you conclude?

**45.** The cost (in dollars) of producing $x$ units of a certain commodity is $C(x) = 5000 + 10x + 0.05x^2$.
(a) Find the average rate of change of $C$ with respect to $x$ when the production level is changed
　(i) from $x = 100$ to $x = 105$
　(ii) from $x = 100$ to $x = 101$
(b) Find the instantaneous rate of change of $C$ with respect to $x$ when $x = 100$. (This is called the *marginal cost*. Its significance will be explained in Section 2.7.)

**46.** If a cylindrical tank holds 100,000 gallons of water, which can be drained from the bottom of the tank in an hour, then Torricelli's Law gives the volume $V$ of water remaining in the tank after $t$ minutes as

$$V(t) = 100,000\left(1 - \tfrac{1}{60}t\right)^2 \qquad 0 \le t \le 60$$

Find the rate at which the water is flowing out of the tank (the instantaneous rate of change of $V$ with respect to $t$) as a function of $t$. What are its units? For times $t = 0, 10, 20, 30, 40, 50$, and 60 min, find the flow rate and the amount of water remaining in the tank. Summarize your findings in a sentence or two. At what time is the flow rate the greatest? The least?

**47.** The cost of producing $x$ ounces of gold from a new gold mine is $C = f(x)$ dollars.
  (a) What is the meaning of the derivative $f'(x)$? What are its units?
  (b) What does the statement $f'(800) = 17$ mean?
  (c) Do you think the values of $f'(x)$ will increase or decrease in the short term? What about the long term? Explain.

**48.** The number of bacteria after $t$ hours in a controlled laboratory experiment is $n = f(t)$.
  (a) What is the meaning of the derivative $f'(5)$? What are its units?
  (b) Suppose there is an unlimited amount of space and nutrients for the bacteria. Which do you think is larger, $f'(5)$ or $f'(10)$? If the supply of nutrients is limited, would that affect your conclusion? Explain.

**49.** Let $T(t)$ be the temperature (in °F) in Phoenix $t$ hours after midnight on September 10, 2008. The table shows values of this function recorded every two hours. What is the meaning of $T'(8)$? Estimate its value.

| $t$ | 0 | 2 | 4 | 6 | 8 | 10 | 12 | 14 |
|---|---|---|---|---|---|---|---|---|
| $T$ | 82 | 75 | 74 | 75 | 84 | 90 | 93 | 94 |

**50.** The quantity (in pounds) of a gourmet ground coffee that is sold by a coffee company at a price of $p$ dollars per pound is $Q = f(p)$.
  (a) What is the meaning of the derivative $f'(8)$? What are its units?
  (b) Is $f'(8)$ positive or negative? Explain.

**51.** The quantity of oxygen that can dissolve in water depends on the temperature of the water. (So thermal pollution influences the oxygen content of water.) The graph shows how oxygen solubility $S$ varies as a function of the water temperature $T$.
  (a) What is the meaning of the derivative $S'(T)$? What are its units?
  (b) Estimate the value of $S'(16)$ and interpret it.

Adapted from *Environmental Science: Living Within the System of Nature*, 2d ed.; by Charles E. Kupchella, © 1989. Reprinted by permission of Prentice-Hall, Inc., Upper Saddle River, NJ.

**52.** The graph shows the influence of the temperature $T$ on the maximum sustainable swimming speed $S$ of Coho salmon.
  (a) What is the meaning of the derivative $S'(T)$? What are its units?
  (b) Estimate the values of $S'(15)$ and $S'(25)$ and interpret them.

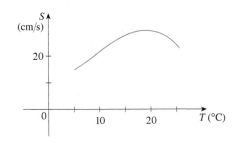

**53–54** Determine whether $f'(0)$ exists.

**53.** $f(x) = \begin{cases} x \sin \dfrac{1}{x} & \text{if } x \ne 0 \\ 0 & \text{if } x = 0 \end{cases}$

**54.** $f(x) = \begin{cases} x^2 \sin \dfrac{1}{x} & \text{if } x \ne 0 \\ 0 & \text{if } x = 0 \end{cases}$

**WRITING PROJECT**  **EARLY METHODS FOR FINDING TANGENTS**

The first person to formulate explicitly the ideas of limits and derivatives was Sir Isaac Newton in the 1660s. But Newton acknowledged that "If I have seen further than other men, it is because I have stood on the shoulders of giants." Two of those giants were Pierre Fermat (1601–1665) and Newton's mentor at Cambridge, Isaac Barrow (1630–1677). Newton was familiar with the methods that these men used to find tangent lines, and their methods played a role in Newton's eventual formulation of calculus.

The following references contain explanations of these methods. Read one or more of the references and write a report comparing the methods of either Fermat or Barrow to modern methods. In particular, use the method of Section 2.1 to find an equation of the tangent line to the curve $y = x^3 + 2x$ at the point $(1, 3)$ and show how either Fermat or Barrow would have solved the same problem. Although you used derivatives and they did not, point out similarities between the methods.

1. Carl Boyer and Uta Merzbach, *A History of Mathematics* (New York: Wiley, 1989), pp. 389, 432.
2. C. H. Edwards, *The Historical Development of the Calculus* (New York: Springer-Verlag, 1979), pp. 124, 132.
3. Howard Eves, *An Introduction to the History of Mathematics*, 6th ed. (New York: Saunders, 1990), pp. 391, 395.
4. Morris Kline, *Mathematical Thought from Ancient to Modern Times* (New York: Oxford University Press, 1972), pp. 344, 346.

## 2.2 The Derivative as a Function

In the preceding section we considered the derivative of a function $f$ at a fixed number $a$:

$$\boxed{1} \qquad f'(a) = \lim_{h \to 0} \frac{f(a + h) - f(a)}{h}$$

Here we change our point of view and let the number $a$ vary. If we replace $a$ in Equation 1 by a variable $x$, we obtain

$$\boxed{2} \qquad f'(x) = \lim_{h \to 0} \frac{f(x + h) - f(x)}{h}$$

Given any number $x$ for which this limit exists, we assign to $x$ the number $f'(x)$. So we can regard $f'$ as a new function, called the **derivative of $f$** and defined by Equation 2. We know that the value of $f'$ at $x$, $f'(x)$, can be interpreted geometrically as the slope of the tangent line to the graph of $f$ at the point $(x, f(x))$.

The function $f'$ is called the derivative of $f$ because it has been "derived" from $f$ by the limiting operation in Equation 2. The domain of $f'$ is the set $\{x \mid f'(x) \text{ exists}\}$ and may be smaller than the domain of $f$.

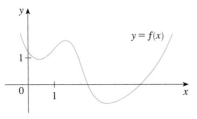

**FIGURE 1**

**V EXAMPLE 1**  The graph of a function $f$ is given in Figure 1. Use it to sketch the graph of the derivative $f'$.

SOLUTION  We can estimate the value of the derivative at any value of $x$ by drawing the tangent at the point $(x, f(x))$ and estimating its slope. For instance, for $x = 5$ we draw the tangent at $P$ in Figure 2(a) and estimate its slope to be about $\frac{3}{2}$, so $f'(5) \approx 1.5$. This allows us to plot the point $P'(5, 1.5)$ on the graph of $f'$ directly beneath $P$. Repeating this procedure at several points, we get the graph shown in Figure 2(b). Notice that the tangents at $A$, $B$, and $C$ are horizontal, so the derivative is 0 there and the graph of $f'$ crosses the $x$-axis at the points $A'$, $B'$, and $C'$, directly beneath $A$, $B$, and $C$. Between $A$ and $B$ the tangents have positive slope, so $f'(x)$ is positive there. But between $B$ and $C$ the tangents have negative slope, so $f'(x)$ is negative there.

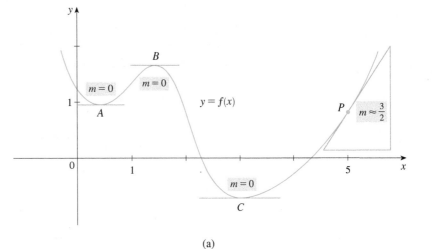

(a)

**TEC** Visual 2.2 shows an animation of Figure 2 for several functions.

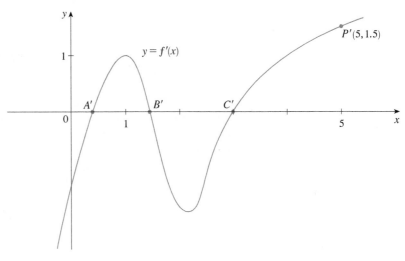

**FIGURE 2**

(b)

**V EXAMPLE 2**
(a) If $f(x) = x^3 - x$, find a formula for $f'(x)$.
(b) Illustrate by comparing the graphs of $f$ and $f'$.

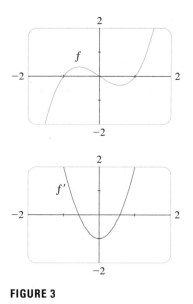

**FIGURE 3**

SOLUTION

(a) When using Equation 2 to compute a derivative, we must remember that the variable is $h$ and that $x$ is temporarily regarded as a constant during the calculation of the limit.

$$f'(x) = \lim_{h \to 0} \frac{f(x+h) - f(x)}{h} = \lim_{h \to 0} \frac{[(x+h)^3 - (x+h)] - [x^3 - x]}{h}$$

$$= \lim_{h \to 0} \frac{x^3 + 3x^2 h + 3xh^2 + h^3 - x - h - x^3 + x}{h}$$

$$= \lim_{h \to 0} \frac{3x^2 h + 3xh^2 + h^3 - h}{h}$$

$$= \lim_{h \to 0} (3x^2 + 3xh + h^2 - 1) = 3x^2 - 1$$

(b) We use a graphing device to graph $f$ and $f'$ in Figure 3. Notice that $f'(x) = 0$ when $f$ has horizontal tangents and $f'(x)$ is positive when the tangents have positive slope. So these graphs serve as a check on our work in part (a).

**EXAMPLE 3**  If $f(x) = \sqrt{x}$, find the derivative of $f$. State the domain of $f'$.

SOLUTION

$$f'(x) = \lim_{h \to 0} \frac{f(x+h) - f(x)}{h} = \lim_{h \to 0} \frac{\sqrt{x+h} - \sqrt{x}}{h}$$

$$= \lim_{h \to 0} \left( \frac{\sqrt{x+h} - \sqrt{x}}{h} \cdot \frac{\sqrt{x+h} + \sqrt{x}}{\sqrt{x+h} + \sqrt{x}} \right)$$

$$= \lim_{h \to 0} \frac{(x+h) - x}{h(\sqrt{x+h} + \sqrt{x})} = \lim_{h \to 0} \frac{1}{\sqrt{x+h} + \sqrt{x}}$$

$$= \frac{1}{\sqrt{x} + \sqrt{x}} = \frac{1}{2\sqrt{x}}$$

We see that $f'(x)$ exists if $x > 0$, so the domain of $f'$ is $(0, \infty)$. This is smaller than the domain of $f$, which is $[0, \infty)$.

Let's check to see that the result of Example 3 is reasonable by looking at the graphs of $f$ and $f'$ in Figure 4. When $x$ is close to 0, $\sqrt{x}$ is also close to 0, so $f'(x) = 1/(2\sqrt{x})$ is very large and this corresponds to the steep tangent lines near $(0, 0)$ in Figure 4(a) and the large values of $f'(x)$ just to the right of 0 in Figure 4(b). When $x$ is large, $f'(x)$ is very small and this corresponds to the flatter tangent lines at the far right of the graph of $f$ and the horizontal asymptote of the graph of $f'$.

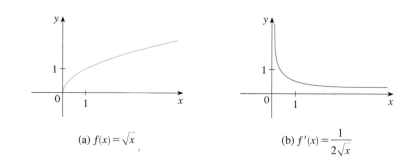

**FIGURE 4**    (a) $f(x) = \sqrt{x}$          (b) $f'(x) = \dfrac{1}{2\sqrt{x}}$

**EXAMPLE 4** Find $f'$ if $f(x) = \dfrac{1 - x}{2 + x}$.

SOLUTION

$$
f'(x) = \lim_{h \to 0} \frac{f(x + h) - f(x)}{h} = \lim_{h \to 0} \frac{\dfrac{1 - (x + h)}{2 + (x + h)} - \dfrac{1 - x}{2 + x}}{h}
$$

$$
= \lim_{h \to 0} \frac{(1 - x - h)(2 + x) - (1 - x)(2 + x + h)}{h(2 + x + h)(2 + x)}
$$

$$
= \lim_{h \to 0} \frac{(2 - x - 2h - x^2 - xh) - (2 - x + h - x^2 - xh)}{h(2 + x + h)(2 + x)}
$$

$$
= \lim_{h \to 0} \frac{-3h}{h(2 + x + h)(2 + x)}
$$

$$
= \lim_{h \to 0} \frac{-3}{(2 + x + h)(2 + x)} = -\frac{3}{(2 + x)^2}
$$

$$
\frac{\dfrac{a}{b} - \dfrac{c}{d}}{e} = \frac{ad - bc}{bd} \cdot \frac{1}{e}
$$

## Other Notations

If we use the traditional notation $y = f(x)$ to indicate that the independent variable is $x$ and the dependent variable is $y$, then some common alternative notations for the derivative are as follows:

$$
f'(x) = y' = \frac{dy}{dx} = \frac{df}{dx} = \frac{d}{dx} f(x) = Df(x) = D_x f(x)
$$

The symbols $D$ and $d/dx$ are called **differentiation operators** because they indicate the operation of **differentiation**, which is the process of calculating a derivative.

The symbol $dy/dx$, which was introduced by Leibniz, should not be regarded as a ratio (for the time being); it is simply a synonym for $f'(x)$. Nonetheless, it is a very useful and suggestive notation, especially when used in conjunction with increment notation. Referring to Equation 2.1.6, we can rewrite the definition of derivative in Leibniz notation in the form

$$
\frac{dy}{dx} = \lim_{\Delta x \to 0} \frac{\Delta y}{\Delta x}
$$

If we want to indicate the value of a derivative $dy/dx$ in Leibniz notation at a specific number $a$, we use the notation

$$
\frac{dy}{dx}\bigg|_{x=a} \qquad \text{or} \qquad \frac{dy}{dx}\bigg]_{x=a}
$$

which is a synonym for $f'(a)$.

> **3 Definition** A function $f$ is **differentiable at $a$** if $f'(a)$ exists. It is **differentiable on an open interval** $(a, b)$ [or $(a, \infty)$ or $(-\infty, a)$ or $(-\infty, \infty)$] if it is differentiable at every number in the interval.

**Leibniz**

Gottfried Wilhelm Leibniz was born in Leipzig in 1646 and studied law, theology, philosophy, and mathematics at the university there, graduating with a bachelor's degree at age 17. After earning his doctorate in law at age 20, Leibniz entered the diplomatic service and spent most of his life traveling to the capitals of Europe on political missions. In particular, he worked to avert a French military threat against Germany and attempted to reconcile the Catholic and Protestant churches.

His serious study of mathematics did not begin until 1672 while he was on a diplomatic mission in Paris. There he built a calculating machine and met scientists, like Huygens, who directed his attention to the latest developments in mathematics and science. Leibniz sought to develop a symbolic logic and system of notation that would simplify logical reasoning. In particular, the version of calculus that he published in 1684 established the notation and the rules for finding derivatives that we use today.

Unfortunately, a dreadful priority dispute arose in the 1690s between the followers of Newton and those of Leibniz as to who had invented calculus first. Leibniz was even accused of plagiarism by members of the Royal Society in England. The truth is that each man invented calculus independently. Newton arrived at his version of calculus first but, because of his fear of controversy, did not publish it immediately. So Leibniz's 1684 account of calculus was the first to be published.

**EXAMPLE 5**  Where is the function $f(x) = |x|$ differentiable?

SOLUTION  If $x > 0$, then $|x| = x$ and we can choose $h$ small enough that $x + h > 0$ and hence $|x + h| = x + h$. Therefore, for $x > 0$, we have

$$f'(x) = \lim_{h \to 0} \frac{|x + h| - |x|}{h} = \lim_{h \to 0} \frac{(x + h) - x}{h}$$

$$= \lim_{h \to 0} \frac{h}{h} = \lim_{h \to 0} 1 = 1$$

and so $f$ is differentiable for any $x > 0$.

Similarly, for $x < 0$ we have $|x| = -x$ and $h$ can be chosen small enough that $x + h < 0$ and so $|x + h| = -(x + h)$. Therefore, for $x < 0$,

$$f'(x) = \lim_{h \to 0} \frac{|x + h| - |x|}{h} = \lim_{h \to 0} \frac{-(x + h) - (-x)}{h}$$

$$= \lim_{h \to 0} \frac{-h}{h} = \lim_{h \to 0} (-1) = -1$$

and so $f$ is differentiable for any $x < 0$.

For $x = 0$ we have to investigate

$$f'(0) = \lim_{h \to 0} \frac{f(0 + h) - f(0)}{h}$$

$$= \lim_{h \to 0} \frac{|0 + h| - |0|}{h} \qquad \text{(if it exists)}$$

Let's compute the left and right limits separately:

$$\lim_{h \to 0^+} \frac{|0 + h| - |0|}{h} = \lim_{h \to 0^+} \frac{|h|}{h} = \lim_{h \to 0^+} \frac{h}{h} = \lim_{h \to 0^+} 1 = 1$$

and

$$\lim_{h \to 0^-} \frac{|0 + h| - |0|}{h} = \lim_{h \to 0^-} \frac{|h|}{h} = \lim_{h \to 0^-} \frac{-h}{h} = \lim_{h \to 0^-} (-1) = -1$$

Since these limits are different, $f'(0)$ does not exist. Thus $f$ is differentiable at all $x$ except 0.

A formula for $f'$ is given by

$$f'(x) = \begin{cases} 1 & \text{if } x > 0 \\ -1 & \text{if } x < 0 \end{cases}$$

and its graph is shown in Figure 5(b). The fact that $f'(0)$ does not exist is reflected geometrically in the fact that the curve $y = |x|$ does not have a tangent line at $(0, 0)$. [See Figure 5(a).]

Both continuity and differentiability are desirable properties for a function to have. The following theorem shows how these properties are related.

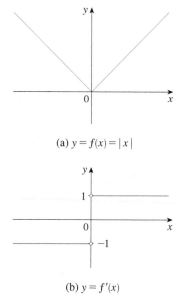

(a) $y = f(x) = |x|$

(b) $y = f'(x)$

**FIGURE 5**

> **4**   **Theorem**   If $f$ is differentiable at $a$, then $f$ is continuous at $a$.

PROOF   To prove that $f$ is continuous at $a$, we have to show that $\lim_{x \to a} f(x) = f(a)$. We do this by showing that the difference $f(x) - f(a)$ approaches 0.

The given information is that $f$ is differentiable at $a$, that is,

$$f'(a) = \lim_{x \to a} \frac{f(x) - f(a)}{x - a}$$

PS An important aspect of problem solving is trying to find a connection between the given and the unknown. See Step 2 (Think of a Plan) in *Principles of Problem Solving* on page 97.

exists (see Equation 2.1.5). To connect the given and the unknown, we divide and multiply $f(x) - f(a)$ by $x - a$ (which we can do when $x \neq a$):

$$f(x) - f(a) = \frac{f(x) - f(a)}{x - a} (x - a)$$

Thus, using the Product Law and (2.1.5), we can write

$$\lim_{x \to a} [f(x) - f(a)] = \lim_{x \to a} \frac{f(x) - f(a)}{x - a} (x - a)$$

$$= \lim_{x \to a} \frac{f(x) - f(a)}{x - a} \cdot \lim_{x \to a} (x - a)$$

$$= f'(a) \cdot 0 = 0$$

To use what we have just proved, we start with $f(x)$ and add and subtract $f(a)$:

$$\lim_{x \to a} f(x) = \lim_{x \to a} [f(a) + (f(x) - f(a))]$$

$$= \lim_{x \to a} f(a) + \lim_{x \to a} [f(x) - f(a)]$$

$$= f(a) + 0 = f(a)$$

Therefore $f$ is continuous at $a$.   ▪

⊘   **NOTE**   The converse of Theorem 4 is false; that is, there are functions that are continuous but not differentiable. For instance, the function $f(x) = |x|$ is continuous at 0 because

$$\lim_{x \to 0} f(x) = \lim_{x \to 0} |x| = 0 = f(0)$$

(See Example 7 in Section 1.6.) But in Example 5 we showed that $f$ is not differentiable at 0.

### How Can a Function Fail to Be Differentiable?

We saw that the function $y = |x|$ in Example 5 is not differentiable at 0 and Figure 5(a) shows that its graph changes direction abruptly when $x = 0$. In general, if the graph of a function $f$ has a "corner" or "kink" in it, then the graph of $f$ has no tangent at this point and $f$ is not differentiable there. [In trying to compute $f'(a)$, we find that the left and right limits are different.]

**FIGURE 6**

Theorem 4 gives another way for a function not to have a derivative. It says that if $f$ is not continuous at $a$, then $f$ is not differentiable at $a$. So at any discontinuity (for instance, a jump discontinuity) $f$ fails to be differentiable.

A third possibility is that the curve has a **vertical tangent line** when $x = a$; that is, $f$ is continuous at $a$ and

$$\lim_{x \to a} \left| f'(x) \right| = \infty$$

This means that the tangent lines become steeper and steeper as $x \to a$. Figure 6 shows one way that this can happen; Figure 7(c) shows another. Figure 7 illustrates the three possibilities that we have discussed.

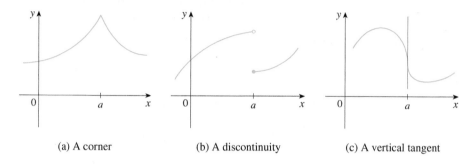

**FIGURE 7**

Three ways for $f$ not to be differentiable at $a$

(a) A corner          (b) A discontinuity          (c) A vertical tangent

A graphing calculator or computer provides another way of looking at differentiability. If $f$ is differentiable at $a$, then when we zoom in toward the point $(a, f(a))$ the graph straightens out and appears more and more like a line. (See Figure 8. We saw a specific example of this in Figure 2 in Section 2.1.) But no matter how much we zoom in toward a point like the ones in Figures 6 and 7(a), we can't eliminate the sharp point or corner (see Figure 9).

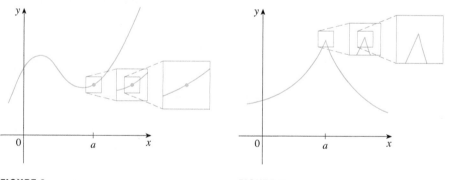

**FIGURE 8**
$f$ is differentiable at $a$.

**FIGURE 9**
$f$ is not differentiable at $a$.

## Higher Derivatives

If $f$ is a differentiable function, then its derivative $f'$ is also a function, so $f'$ may have a derivative of its own, denoted by $(f')' = f''$. This new function $f''$ is called the **second derivative** of $f$ because it is the derivative of the derivative of $f$. Using Leibniz notation, we write the second derivative of $y = f(x)$ as

$$\frac{d}{dx}\left(\frac{dy}{dx}\right) = \frac{d^2 y}{dx^2}$$

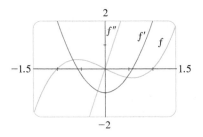

**FIGURE 10**

**EXAMPLE 6** If $f(x) = x^3 - x$, find and interpret $f''(x)$.

SOLUTION  In Example 2 we found that the first derivative is $f'(x) = 3x^2 - 1$. So the second derivative is

$$f''(x) = (f')'(x) = \lim_{h \to 0} \frac{f'(x + h) - f'(x)}{h} = \lim_{h \to 0} \frac{[3(x + h)^2 - 1] - [3x^2 - 1]}{h}$$

$$= \lim_{h \to 0} \frac{3x^2 + 6xh + 3h^2 - 1 - 3x^2 + 1}{h} = \lim_{h \to 0} (6x + 3h) = 6x$$

The graphs of $f$, $f'$, and $f''$ are shown in Figure 10.

We can interpret $f''(x)$ as the slope of the curve $y = f'(x)$ at the point $(x, f'(x))$. In other words, it is the rate of change of the slope of the original curve $y = f(x)$.

Notice from Figure 10 that $f''(x)$ is negative when $y = f'(x)$ has negative slope and positive when $y = f'(x)$ has positive slope. So the graphs serve as a check on our calculations.

In general, we can interpret a second derivative as a rate of change of a rate of change. The most familiar example of this is *acceleration,* which we define as follows.

If $s = s(t)$ is the position function of an object that moves in a straight line, we know that its first derivative represents the velocity $v(t)$ of the object as a function of time:

$$v(t) = s'(t) = \frac{ds}{dt}$$

The instantaneous rate of change of velocity with respect to time is called the **acceleration** $a(t)$ of the object. Thus the acceleration function is the derivative of the velocity function and is therefore the second derivative of the position function:

$$a(t) = v'(t) = s''(t)$$

or, in Leibniz notation,

$$a = \frac{dv}{dt} = \frac{d^2s}{dt^2}$$

The **third derivative** $f'''$ is the derivative of the second derivative: $f''' = (f'')'$. So $f'''(x)$ can be interpreted as the slope of the curve $y = f''(x)$ or as the rate of change of $f''(x)$. If $y = f(x)$, then alternative notations for the third derivative are

$$y''' = f'''(x) = \frac{d}{dx}\left(\frac{d^2y}{dx^2}\right) = \frac{d^3y}{dx^3}$$

The process can be continued. The fourth derivative $f''''$ is usually denoted by $f^{(4)}$. In general, the $n$th derivative of $f$ is denoted by $f^{(n)}$ and is obtained from $f$ by differentiating $n$ times. If $y = f(x)$, we write

$$y^{(n)} = f^{(n)}(x) = \frac{d^ny}{dx^n}$$

**EXAMPLE 7** If $f(x) = x^3 - x$, find $f'''(x)$ and $f^{(4)}(x)$.

SOLUTION  In Example 6 we found that $f''(x) = 6x$. The graph of the second derivative has equation $y = 6x$ and so it is a straight line with slope 6. Since the derivative $f'''(x)$ is the slope of $f''(x)$, we have

$$f'''(x) = 6$$

for all values of $x$. So $f'''$ is a constant function and its graph is a horizontal line. Therefore, for all values of $x$,

$$f^{(4)}(x) = 0$$

We can also interpret the third derivative physically in the case where the function is the position function $s = s(t)$ of an object that moves along a straight line. Because $s''' = (s'')' = a'$, the third derivative of the position function is the derivative of the acceleration function and is called the **jerk**:

$$j = \frac{da}{dt} = \frac{d^3 s}{dt^3}$$

Thus the jerk $j$ is the rate of change of acceleration. It is aptly named because a large jerk means a sudden change in acceleration, which causes an abrupt movement in a vehicle.

We have seen that one application of second and third derivatives occurs in analyzing the motion of objects using acceleration and jerk. We will investigate another application of second derivatives in Section 3.3, where we show how knowledge of $f''$ gives us information about the shape of the graph of $f$. In Chapter 11 we will see how second and higher derivatives enable us to represent functions as sums of infinite series.

## 2.2  Exercises

1–2 Use the given graph to estimate the value of each derivative. Then sketch the graph of $f'$.

**1.** (a) $f'(-3)$     (b) $f'(-2)$     (c) $f'(-1)$
(d) $f'(0)$     (e) $f'(1)$     (f) $f'(2)$
(g) $f'(3)$

**2.** (a) $f'(0)$     (b) $f'(1)$     (c) $f'(2)$
(d) $f'(3)$     (e) $f'(4)$     (f) $f'(5)$
(g) $f'(6)$     (h) $f'(7)$

**3.** Match the graph of each function in (a)–(d) with the graph of its derivative in I–IV. Give reasons for your choices.

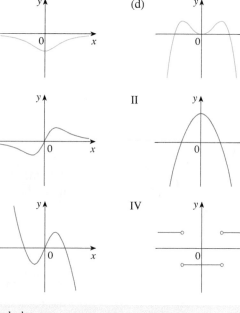

**4–11** Trace or copy the graph of the given function $f$. (Assume that the axes have equal scales.) Then use the method of Example 1 to sketch the graph of $f'$ below it.

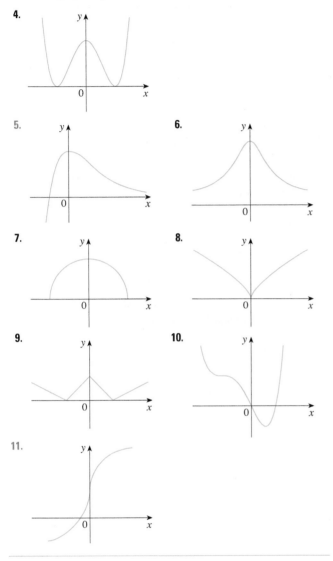

**4.**

**5.**

**6.**

**7.**

**8.**

**9.**

**10.**

**11.**

**12.** Shown is the graph of the population function $P(t)$ for yeast cells in a laboratory culture. Use the method of Example 1 to graph the derivative $P'(t)$. What does the graph of $P'$ tell us about the yeast population?

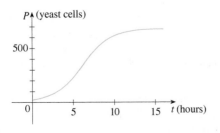

**13.** A rechargeable battery is plugged into a charger. The graph shows $C(t)$, the percentage of full capacity that the battery reaches as a function of time $t$ elapsed (in hours).
(a) What is the meaning of the derivative $C'(t)$?
(b) Sketch the graph of $C'(t)$. What does the graph tell you?

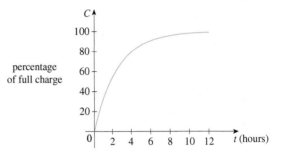

**14.** The graph (from the US Department of Energy) shows how driving speed affects gas mileage. Fuel economy $F$ is measured in miles per gallon and speed $v$ is measured in miles per hour.
(a) What is the meaning of the derivative $F'(v)$?
(b) Sketch the graph of $F'(v)$.
(c) At what speed should you drive if you want to save on gas?

**15.** The graph shows how the average age of first marriage of Japanese men varied in the last half of the 20th century. Sketch the graph of the derivative function $M'(t)$. During which years was the derivative negative?

**16.** Make a careful sketch of the graph of the sine function and below it sketch the graph of its derivative in the same manner as in Exercises 4–11. Can you guess what the derivative of the sine function is from its graph?

**17.** Let $f(x) = x^2$.
 (a) Estimate the values of $f'(0), f'(\frac{1}{2}), f'(1),$ and $f'(2)$ by using a graphing device to zoom in on the graph of $f$.
 (b) Use symmetry to deduce the values of $f'(-\frac{1}{2}), f'(-1),$ and $f'(-2)$.
 (c) Use the results from parts (a) and (b) to guess a formula for $f'(x)$.
 (d) Use the definition of derivative to prove that your guess in part (c) is correct.

**18.** Let $f(x) = x^3$.
 (a) Estimate the values of $f'(0), f'(\frac{1}{2}), f'(1), f'(2),$ and $f'(3)$ by using a graphing device to zoom in on the graph of $f$.
 (b) Use symmetry to deduce the values of $f'(-\frac{1}{2}), f'(-1),$ $f'(-2),$ and $f'(-3)$.
 (c) Use the values from parts (a) and (b) to graph $f'$.
 (d) Guess a formula for $f'(x)$.
 (e) Use the definition of derivative to prove that your guess in part (d) is correct.

**19–29** Find the derivative of the function using the definition of derivative. State the domain of the function and the domain of its derivative.

**19.** $f(x) = \frac{1}{2}x - \frac{1}{3}$

**20.** $f(x) = mx + b$

**21.** $f(t) = 5t - 9t^2$

**22.** $f(x) = 1.5x^2 - x + 3.7$

**23.** $f(x) = x^2 - 2x^3$

**24.** $g(t) = \dfrac{1}{\sqrt{t}}$

**25.** $g(x) = \sqrt{9 - x}$

**26.** $f(x) = \dfrac{x^2 - 1}{2x - 3}$

**27.** $G(t) = \dfrac{1 - 2t}{3 + t}$

**28.** $f(x) = x^{3/2}$

**29.** $f(x) = x^4$

**30.** (a) Sketch the graph of $f(x) = \sqrt{6 - x}$ by starting with the graph of $y = \sqrt{x}$ and using the transformations of Section 1.3.
 (b) Use the graph from part (a) to sketch the graph of $f'$.
 (c) Use the definition of a derivative to find $f'(x)$. What are the domains of $f$ and $f'$?
 (d) Use a graphing device to graph $f'$ and compare with your sketch in part (b).

**31.** (a) If $f(x) = x^4 + 2x$, find $f'(x)$.
 (b) Check to see that your answer to part (a) is reasonable by comparing the graphs of $f$ and $f'$.

**32.** (a) If $f(x) = x + 1/x$, find $f'(x)$.
 (b) Check to see that your answer to part (a) is reasonable by comparing the graphs of $f$ and $f'$.

**33.** The unemployment rate $U(t)$ varies with time. The table (from the Bureau of Labor Statistics) gives the percentage of unemployed in the US labor force from 1999 to 2008.

| $t$ | $U(t)$ | $t$ | $U(t)$ |
|------|--------|------|--------|
| 1999 | 4.2 | 2004 | 5.5 |
| 2000 | 4.0 | 2005 | 5.1 |
| 2001 | 4.7 | 2006 | 4.6 |
| 2002 | 5.8 | 2007 | 4.6 |
| 2003 | 6.0 | 2008 | 5.8 |

 (a) What is the meaning of $U'(t)$? What are its units?
 (b) Construct a table of estimated values for $U'(t)$.

**34.** Let $P(t)$ be the percentage of Americans under the age of 18 at time $t$. The table gives values of this function in census years from 1950 to 2000.

| $t$ | $P(t)$ | $t$ | $P(t)$ |
|------|--------|------|--------|
| 1950 | 31.1 | 1980 | 28.0 |
| 1960 | 35.7 | 1990 | 25.7 |
| 1970 | 34.0 | 2000 | 25.7 |

 (a) What is the meaning of $P'(t)$? What are its units?
 (b) Construct a table of estimated values for $P'(t)$.
 (c) Graph $P$ and $P'$.
 (d) How would it be possible to get more accurate values for $P'(t)$?

**35–38** The graph of $f$ is given. State, with reasons, the numbers at which $f$ is not differentiable.

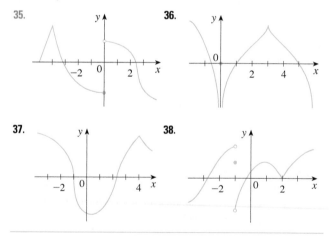

**39.** Graph the function $f(x) = x + \sqrt{|x|}$. Zoom in repeatedly, first toward the point $(-1, 0)$ and then toward the origin. What is different about the behavior of $f$ in the vicinity of these two points? What do you conclude about the differentiability of $f$?

**40.** Zoom in toward the points $(1, 0), (0, 1),$ and $(-1, 0)$ on the graph of the function $g(x) = (x^2 - 1)^{2/3}$. What do you notice? Account for what you see in terms of the differentiability of $g$.

**41.** The figure shows the graphs of $f$, $f'$, and $f''$. Identify each curve, and explain your choices.

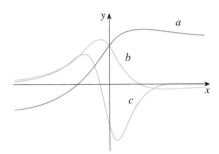

**42.** The figure shows graphs of $f$, $f'$, $f''$, and $f'''$. Identify each curve, and explain your choices.

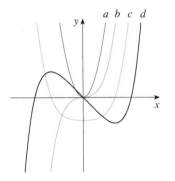

**43.** The figure shows the graphs of three functions. One is the position function of a car, one is the velocity of the car, and one is its acceleration. Identify each curve, and explain your choices.

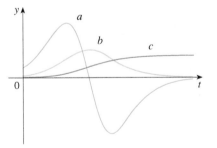

**44.** The figure shows the graphs of four functions. One is the position function of a car, one is the velocity of the car, one is its acceleration, and one is its jerk. Identify each curve, and explain your choices.

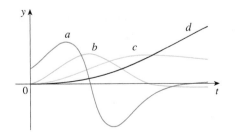

**45–46** Use the definition of a derivative to find $f'(x)$ and $f''(x)$. Then graph $f$, $f'$, and $f''$ on a common screen and check to see if your answers are reasonable.

**45.** $f(x) = 3x^2 + 2x + 1$

**46.** $f(x) = x^3 - 3x$

---

**47.** If $f(x) = 2x^2 - x^3$, find $f'(x)$, $f''(x)$, $f'''(x)$, and $f^{(4)}(x)$. Graph $f$, $f'$, $f''$, and $f'''$ on a common screen. Are the graphs consistent with the geometric interpretations of these derivatives?

**48.** (a) The graph of a position function of a car is shown, where $s$ is measured in feet and $t$ in seconds. Use it to graph the velocity and acceleration of the car. What is the acceleration at $t = 10$ seconds?

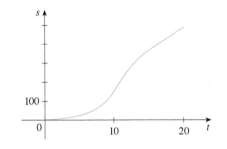

(b) Use the acceleration curve from part (a) to estimate the jerk at $t = 10$ seconds. What are the units for jerk?

**49.** Let $f(x) = \sqrt[3]{x}$.
  (a) If $a \neq 0$, use Equation 2.1.5 to find $f'(a)$.
  (b) Show that $f'(0)$ does not exist.
  (c) Show that $y = \sqrt[3]{x}$ has a vertical tangent line at $(0, 0)$. (Recall the shape of the graph of $f$. See Figure 13 in Section 1.2.)

**50.** (a) If $g(x) = x^{2/3}$, show that $g'(0)$ does not exist.
  (b) If $a \neq 0$, find $g'(a)$.
  (c) Show that $y = x^{2/3}$ has a vertical tangent line at $(0, 0)$.
  (d) Illustrate part (c) by graphing $y = x^{2/3}$.

**51.** Show that the function $f(x) = |x - 6|$ is not differentiable at 6. Find a formula for $f'$ and sketch its graph.

**52.** Where is the greatest integer function $f(x) = [\![x]\!]$ not differentiable? Find a formula for $f'$ and sketch its graph.

**53.** (a) Sketch the graph of the function $f(x) = x|x|$.
  (b) For what values of $x$ is $f$ differentiable?
  (c) Find a formula for $f'$.

**54.** The **left-hand** and **right-hand derivatives** of $f$ at $a$ are defined by

$$f'_-(a) = \lim_{h \to 0^-} \frac{f(a+h) - f(a)}{h}$$

and

$$f'_+(a) = \lim_{h \to 0^+} \frac{f(a+h) - f(a)}{h}$$

if these limits exist. Then $f'(a)$ exists if and only if these one-sided derivatives exist and are equal.

(a) Find $f'_-(4)$ and $f'_+(4)$ for the function

$$f(x) = \begin{cases} 0 & \text{if } x \le 0 \\ 5 - x & \text{if } 0 < x < 4 \\ \dfrac{1}{5 - x} & \text{if } x \ge 4 \end{cases}$$

(b) Sketch the graph of $f$.

(c) Where is $f$ discontinuous?

(d) Where is $f$ not differentiable?

**55.** Recall that a function $f$ is called *even* if $f(-x) = f(x)$ for all $x$ in its domain and *odd* if $f(-x) = -f(x)$ for all such $x$. Prove each of the following.

(a) The derivative of an even function is an odd function.

(b) The derivative of an odd function is an even function.

**56.** When you turn on a hot-water faucet, the temperature $T$ of the water depends on how long the water has been running.

(a) Sketch a possible graph of $T$ as a function of the time $t$ that has elapsed since the faucet was turned on.

(b) Describe how the rate of change of $T$ with respect to $t$ varies as $t$ increases.

(c) Sketch a graph of the derivative of $T$.

**57.** Let $\ell$ be the tangent line to the parabola $y = x^2$ at the point $(1, 1)$. The *angle of inclination* of $\ell$ is the angle $\phi$ that $\ell$ makes with the positive direction of the $x$-axis. Calculate $\phi$ correct to the nearest degree.

## 2.3 Differentiation Formulas

If it were always necessary to compute derivatives directly from the definition, as we did in the preceding section, such computations would be tedious and the evaluation of some limits would require ingenuity. Fortunately, several rules have been developed for finding derivatives without having to use the definition directly. These formulas greatly simplify the task of differentiation.

Let's start with the simplest of all functions, the constant function $f(x) = c$. The graph of this function is the horizontal line $y = c$, which has slope 0, so we must have $f'(x) = 0$. (See Figure 1.) A formal proof, from the definition of a derivative, is also easy:

$$f'(x) = \lim_{h \to 0} \frac{f(x+h) - f(x)}{h} = \lim_{h \to 0} \frac{c - c}{h} = \lim_{h \to 0} 0 = 0$$

**FIGURE 1**
The graph of $f(x) = c$ is the line $y = c$, so $f'(x) = 0$.

In Leibniz notation, we write this rule as follows.

> **Derivative of a Constant Function**
>
> $$\frac{d}{dx}(c) = 0$$

### Power Functions

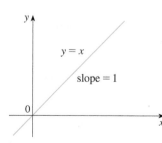

**FIGURE 2**
The graph of $f(x) = x$ is the line $y = x$, so $f'(x) = 1$.

We next look at the functions $f(x) = x^n$, where $n$ is a positive integer. If $n = 1$, the graph of $f(x) = x$ is the line $y = x$, which has slope 1. (See Figure 2.) So

$$\boxed{1} \qquad \frac{d}{dx}(x) = 1$$

(You can also verify Equation 1 from the definition of a derivative.) We have already investigated the cases $n = 2$ and $n = 3$. In fact, in Section 2.2 (Exercises 17 and 18) we found that

$$\boxed{2} \qquad \frac{d}{dx}(x^2) = 2x \qquad \frac{d}{dx}(x^3) = 3x^2$$

For $n = 4$ we find the derivative of $f(x) = x^4$ as follows:

$$f'(x) = \lim_{h \to 0} \frac{f(x+h) - f(x)}{h} = \lim_{h \to 0} \frac{(x+h)^4 - x^4}{h}$$

$$= \lim_{h \to 0} \frac{x^4 + 4x^3h + 6x^2h^2 + 4xh^3 + h^4 - x^4}{h}$$

$$= \lim_{h \to 0} \frac{4x^3h + 6x^2h^2 + 4xh^3 + h^4}{h}$$

$$= \lim_{h \to 0} (4x^3 + 6x^2h + 4xh^2 + h^3) = 4x^3$$

Thus

$$\boxed{3} \qquad \frac{d}{dx}(x^4) = 4x^3$$

Comparing the equations in $\boxed{1}$, $\boxed{2}$, and $\boxed{3}$, we see a pattern emerging. It seems to be a reasonable guess that, when $n$ is a positive integer, $(d/dx)(x^n) = nx^{n-1}$. This turns out to be true. We prove it in two ways; the second proof uses the Binomial Theorem.

---

**The Power Rule** If $n$ is a positive integer, then

$$\frac{d}{dx}(x^n) = nx^{n-1}$$

---

FIRST PROOF  The formula

$$x^n - a^n = (x - a)(x^{n-1} + x^{n-2}a + \cdots + xa^{n-2} + a^{n-1})$$

can be verified simply by multiplying out the right-hand side (or by summing the second factor as a geometric series). If $f(x) = x^n$, we can use Equation 2.1.5 for $f'(a)$ and the equation above to write

$$f'(a) = \lim_{x \to a} \frac{f(x) - f(a)}{x - a} = \lim_{x \to a} \frac{x^n - a^n}{x - a}$$

$$= \lim_{x \to a} (x^{n-1} + x^{n-2}a + \cdots + xa^{n-2} + a^{n-1})$$

$$= a^{n-1} + a^{n-2}a + \cdots + aa^{n-2} + a^{n-1}$$

$$= na^{n-1}$$

SECOND PROOF

$$f'(x) = \lim_{h \to 0} \frac{f(x + h) - f(x)}{h} = \lim_{h \to 0} \frac{(x + h)^n - x^n}{h}$$

The Binomial Theorem is given on Reference Page 1.

In finding the derivative of $x^4$ we had to expand $(x + h)^4$. Here we need to expand $(x + h)^n$ and we use the Binomial Theorem to do so:

$$f'(x) = \lim_{h \to 0} \frac{\left[ x^n + nx^{n-1}h + \dfrac{n(n - 1)}{2}x^{n-2}h^2 + \cdots + nxh^{n-1} + h^n \right] - x^n}{h}$$

$$= \lim_{h \to 0} \frac{nx^{n-1}h + \dfrac{n(n - 1)}{2}x^{n-2}h^2 + \cdots + nxh^{n-1} + h^n}{h}$$

$$= \lim_{h \to 0} \left[ nx^{n-1} + \dfrac{n(n - 1)}{2}x^{n-2}h + \cdots + nxh^{n-2} + h^{n-1} \right]$$

$$= nx^{n-1}$$

because every term except the first has $h$ as a factor and therefore approaches 0.

We illustrate the Power Rule using various notations in Example 1.

**EXAMPLE 1**

(a) If $f(x) = x^6$, then $f'(x) = 6x^5$.     (b) If $y = x^{1000}$, then $y' = 1000x^{999}$.

(c) If $y = t^4$, then $\dfrac{dy}{dt} = 4t^3$.     (d) $\dfrac{d}{dr}(r^3) = 3r^2$

## New Derivatives from Old

When new functions are formed from old functions by addition, subtraction, or multiplication by a constant, their derivatives can be calculated in terms of derivatives of the old functions. In particular, the following formula says that *the derivative of a constant times a function is the constant times the derivative of the function.*

GEOMETRIC INTERPRETATION OF THE CONSTANT MULTIPLE RULE

Multiplying by $c = 2$ stretches the graph vertically by a factor of 2. All the rises have been doubled but the runs stay the same. So the slopes are doubled too.

---

**The Constant Multiple Rule** If $c$ is a constant and $f$ is a differentiable function, then

$$\frac{d}{dx}[cf(x)] = c\frac{d}{dx}f(x)$$

---

PROOF Let $g(x) = cf(x)$. Then

$$g'(x) = \lim_{h \to 0} \frac{g(x + h) - g(x)}{h} = \lim_{h \to 0} \frac{cf(x + h) - cf(x)}{h}$$

$$= \lim_{h \to 0} c\left[ \frac{f(x + h) - f(x)}{h} \right]$$

$$= c \lim_{h \to 0} \frac{f(x + h) - f(x)}{h} \qquad \text{(by Law 3 of limits)}$$

$$= cf'(x)$$

EXAMPLE 2

(a) $\dfrac{d}{dx}(3x^4) = 3\,\dfrac{d}{dx}(x^4) = 3(4x^3) = 12x^3$

(b) $\dfrac{d}{dx}(-x) = \dfrac{d}{dx}[(-1)x] = (-1)\dfrac{d}{dx}(x) = -1(1) = -1$ ▬

The next rule tells us that *the derivative of a sum of functions is the sum of the derivatives.*

---

Using prime notation, we can write the Sum Rule as

$$(f + g)' = f' + g'$$

**The Sum Rule** If $f$ and $g$ are both differentiable, then

$$\frac{d}{dx}[f(x) + g(x)] = \frac{d}{dx}f(x) + \frac{d}{dx}g(x)$$

---

PROOF Let $F(x) = f(x) + g(x)$. Then

$$F'(x) = \lim_{h \to 0} \frac{F(x + h) - F(x)}{h}$$

$$= \lim_{h \to 0} \frac{[f(x + h) + g(x + h)] - [f(x) + g(x)]}{h}$$

$$= \lim_{h \to 0} \left[ \frac{f(x + h) - f(x)}{h} + \frac{g(x + h) - g(x)}{h} \right]$$

$$= \lim_{h \to 0} \frac{f(x + h) - f(x)}{h} + \lim_{h \to 0} \frac{g(x + h) - g(x)}{h} \qquad \text{(by Law 1)}$$

$$= f'(x) + g'(x) \qquad\qquad\qquad ▬$$

The Sum Rule can be extended to the sum of any number of functions. For instance, using this theorem twice, we get

$$(f + g + h)' = [(f + g) + h]' = (f + g)' + h' = f' + g' + h'$$

By writing $f - g$ as $f + (-1)g$ and applying the Sum Rule and the Constant Multiple Rule, we get the following formula.

**The Difference Rule** If $f$ and $g$ are both differentiable, then

$$\frac{d}{dx}[f(x) - g(x)] = \frac{d}{dx}f(x) - \frac{d}{dx}g(x)$$

The Constant Multiple Rule, the Sum Rule, and the Difference Rule can be combined with the Power Rule to differentiate any polynomial, as the following examples demonstrate.

**EXAMPLE 3**

$$\frac{d}{dx}(x^8 + 12x^5 - 4x^4 + 10x^3 - 6x + 5)$$

$$= \frac{d}{dx}(x^8) + 12\frac{d}{dx}(x^5) - 4\frac{d}{dx}(x^4) + 10\frac{d}{dx}(x^3) - 6\frac{d}{dx}(x) + \frac{d}{dx}(5)$$

$$= 8x^7 + 12(5x^4) - 4(4x^3) + 10(3x^2) - 6(1) + 0$$

$$= 8x^7 + 60x^4 - 16x^3 + 30x^2 - 6$$

**V   EXAMPLE 4**  Find the points on the curve $y = x^4 - 6x^2 + 4$ where the tangent line is horizontal.

SOLUTION  Horizontal tangents occur where the derivative is zero. We have

$$\frac{dy}{dx} = \frac{d}{dx}(x^4) - 6\frac{d}{dx}(x^2) + \frac{d}{dx}(4)$$

$$= 4x^3 - 12x + 0 = 4x(x^2 - 3)$$

Thus $dy/dx = 0$ if $x = 0$ or $x^2 - 3 = 0$, that is, $x = \pm\sqrt{3}$. So the given curve has horizontal tangents when $x = 0$, $\sqrt{3}$, and $-\sqrt{3}$. The corresponding points are $(0, 4)$, $(\sqrt{3}, -5)$, and $(-\sqrt{3}, -5)$. (See Figure 3.)

**EXAMPLE 5**  The equation of motion of a particle is $s = 2t^3 - 5t^2 + 3t + 4$, where $s$ is measured in centimeters and $t$ in seconds. Find the acceleration as a function of time. What is the acceleration after 2 seconds?

SOLUTION  The velocity and acceleration are

$$v(t) = \frac{ds}{dt} = 6t^2 - 10t + 3$$

$$a(t) = \frac{dv}{dt} = 12t - 10$$

The acceleration after 2 s is $a(2) = 14 \text{ cm/s}^2$.

Next we need a formula for the derivative of a product of two functions. By analogy with the Sum and Difference Rules, one might be tempted to guess, as Leibniz did three centuries ago, that the derivative of a product is the product of the derivatives. We can see, however, that this guess is wrong by looking at a particular example. Let $f(x) = x$ and $g(x) = x^2$. Then the Power Rule gives $f'(x) = 1$ and $g'(x) = 2x$. But $(fg)(x) = x^3$, so $(fg)'(x) = 3x^2$. Thus $(fg)' \neq f'g'$. The correct formula was discovered by Leibniz (soon after his false start) and is called the Product Rule.

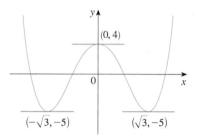

**FIGURE 3**

The curve $y = x^4 - 6x^2 + 4$ and its horizontal tangents

We can write the Product Rule in prime notation as

$$(fg)' = fg' + gf'$$

---

**The Product Rule**  If $f$ and $g$ are both differentiable, then

$$\frac{d}{dx}[f(x)g(x)] = f(x)\frac{d}{dx}[g(x)] + g(x)\frac{d}{dx}[f(x)]$$

PROOF Let $F(x) = f(x)g(x)$. Then

$$F'(x) = \lim_{h \to 0} \frac{F(x+h) - F(x)}{h}$$

$$= \lim_{h \to 0} \frac{f(x+h)g(x+h) - f(x)g(x)}{h}$$

In order to evaluate this limit, we would like to separate the functions $f$ and $g$ as in the proof of the Sum Rule. We can achieve this separation by subtracting and adding the term $f(x+h)g(x)$ in the numerator:

$$F'(x) = \lim_{h \to 0} \frac{f(x+h)g(x+h) - f(x+h)g(x) + f(x+h)g(x) - f(x)g(x)}{h}$$

$$= \lim_{h \to 0} \left[ f(x+h) \frac{g(x+h) - g(x)}{h} + g(x) \frac{f(x+h) - f(x)}{h} \right]$$

$$= \lim_{h \to 0} f(x+h) \cdot \lim_{h \to 0} \frac{g(x+h) - g(x)}{h} + \lim_{h \to 0} g(x) \cdot \lim_{h \to 0} \frac{f(x+h) - f(x)}{h}$$

$$= f(x)g'(x) + g(x)f'(x)$$

Note that $\lim_{h \to 0} g(x) = g(x)$ because $g(x)$ is a constant with respect to the variable $h$. Also, since $f$ is differentiable at $x$, it is continuous at $x$ by Theorem 2.2.4, and so $\lim_{h \to 0} f(x+h) = f(x)$. (See Exercise 59 in Section 1.8.)

In words, the Product Rule says that *the derivative of a product of two functions is the first function times the derivative of the second function plus the second function times the derivative of the first function.*

**EXAMPLE 6** Find $F'(x)$ if $F(x) = (6x^3)(7x^4)$.

SOLUTION By the Product Rule, we have

$$F'(x) = (6x^3) \frac{d}{dx}(7x^4) + (7x^4) \frac{d}{dx}(6x^3)$$

$$= (6x^3)(28x^3) + (7x^4)(18x^2)$$

$$= 168x^6 + 126x^6 = 294x^6$$

Notice that we could verify the answer to Example 6 directly by first multiplying the factors:

$$F(x) = (6x^3)(7x^4) = 42x^7 \quad \Rightarrow \quad F'(x) = 42(7x^6) = 294x^6$$

But later we will meet functions, such as $y = x^2 \sin x$, for which the Product Rule is the only possible method.

**V** **EXAMPLE 7** If $h(x) = xg(x)$ and it is known that $g(3) = 5$ and $g'(3) = 2$, find $h'(3)$.

SOLUTION Applying the Product Rule, we get

$$h'(x) = \frac{d}{dx}[xg(x)] = x\frac{d}{dx}[g(x)] + g(x)\frac{d}{dx}[x]$$

$$= xg'(x) + g(x)$$

Therefore
$$h'(3) = 3g'(3) + g(3) = 3 \cdot 2 + 5 = 11$$

In prime notation we can write the Quotient Rule as

$$\left(\frac{f}{g}\right)' = \frac{gf' - fg'}{g^2}$$

> **The Quotient Rule** If $f$ and $g$ are differentiable, then
>
> $$\frac{d}{dx}\left[\frac{f(x)}{g(x)}\right] = \frac{g(x)\frac{d}{dx}[f(x)] - f(x)\frac{d}{dx}[g(x)]}{[g(x)]^2}$$

PROOF Let $F(x) = f(x)/g(x)$. Then

$$F'(x) = \lim_{h \to 0} \frac{F(x + h) - F(x)}{h} = \lim_{h \to 0} \frac{\dfrac{f(x + h)}{g(x + h)} - \dfrac{f(x)}{g(x)}}{h}$$

$$= \lim_{h \to 0} \frac{f(x + h)g(x) - f(x)g(x + h)}{hg(x + h)g(x)}$$

We can separate $f$ and $g$ in this expression by subtracting and adding the term $f(x)g(x)$ in the numerator:

$$F'(x) = \lim_{h \to 0} \frac{f(x + h)g(x) - f(x)g(x) + f(x)g(x) - f(x)g(x + h)}{hg(x + h)g(x)}$$

$$= \lim_{h \to 0} \frac{g(x)\dfrac{f(x + h) - f(x)}{h} - f(x)\dfrac{g(x + h) - g(x)}{h}}{g(x + h)g(x)}$$

$$= \frac{\displaystyle\lim_{h \to 0} g(x) \cdot \lim_{h \to 0} \frac{f(x + h) - f(x)}{h} - \lim_{h \to 0} f(x) \cdot \lim_{h \to 0} \frac{g(x + h) - g(x)}{h}}{\displaystyle\lim_{h \to 0} g(x + h) \cdot \lim_{h \to 0} g(x)}$$

$$= \frac{g(x)f'(x) - f(x)g'(x)}{[g(x)]^2}$$

Again $g$ is continuous by Theorem 2.2.4, so $\lim_{h \to 0} g(x + h) = g(x)$.

In words, the Quotient Rule says that the *derivative of a quotient is the denominator times the derivative of the numerator minus the numerator times the derivative of the denominator, all divided by the square of the denominator.*

The theorems of this section show that any polynomial is differentiable on $\mathbb{R}$ and any rational function is differentiable on its domain. Furthermore, the Quotient Rule and the other differentiation formulas enable us to compute the derivative of any rational function, as the next example illustrates.

We can use a graphing device to check that the answer to Example 8 is plausible. Figure 4 shows the graphs of the function of Example 8 and its derivative. Notice that when $y$ grows rapidly (near $-2$), $y'$ is large. And when $y$ grows slowly, $y'$ is near 0.

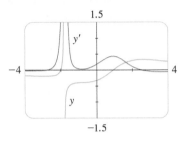

**FIGURE 4**

**V EXAMPLE 8** Let $y = \dfrac{x^2 + x - 2}{x^3 + 6}$. Then

$$y' = \frac{(x^3 + 6)\dfrac{d}{dx}(x^2 + x - 2) - (x^2 + x - 2)\dfrac{d}{dx}(x^3 + 6)}{(x^3 + 6)^2}$$

$$= \frac{(x^3 + 6)(2x + 1) - (x^2 + x - 2)(3x^2)}{(x^3 + 6)^2}$$

$$= \frac{(2x^4 + x^3 + 12x + 6) - (3x^4 + 3x^3 - 6x^2)}{(x^3 + 6)^2}$$

$$= \frac{-x^4 - 2x^3 + 6x^2 + 12x + 6}{(x^3 + 6)^2}$$

**NOTE** Don't use the Quotient Rule *every* time you see a quotient. Sometimes it's easier to rewrite a quotient first to put it in a form that is simpler for the purpose of differentiation. For instance, although it is possible to differentiate the function

$$F(x) = \frac{3x^2 + 2\sqrt{x}}{x}$$

using the Quotient Rule, it is much easier to perform the division first and write the function as

$$F(x) = 3x + 2x^{-1/2}$$

before differentiating.

## General Power Functions

The Quotient Rule can be used to extend the Power Rule to the case where the exponent is a negative integer.

---

If $n$ is a positive integer, then

$$\frac{d}{dx}(x^{-n}) = -nx^{-n-1}$$

---

PROOF
$$\frac{d}{dx}(x^{-n}) = \frac{d}{dx}\left(\frac{1}{x^n}\right)$$

$$= \frac{x^n \dfrac{d}{dx}(1) - 1 \cdot \dfrac{d}{dx}(x^n)}{(x^n)^2} = \frac{x^n \cdot 0 - 1 \cdot nx^{n-1}}{x^{2n}}$$

$$= \frac{-nx^{n-1}}{x^{2n}} = -nx^{n-1-2n} = -nx^{-n-1}$$

**EXAMPLE 9**

(a) If $y = \dfrac{1}{x}$, then $\dfrac{dy}{dx} = \dfrac{d}{dx}(x^{-1}) = -x^{-2} = -\dfrac{1}{x^2}$

(b) $\dfrac{d}{dt}\left(\dfrac{6}{t^3}\right) = 6\dfrac{d}{dt}(t^{-3}) = 6(-3)t^{-4} = -\dfrac{18}{t^4}$

So far we know that the Power Rule holds if the exponent $n$ is a positive or negative integer. If $n = 0$, then $x^0 = 1$, which we know has a derivative of 0. Thus the Power Rule holds for any integer $n$. What if the exponent is a fraction? In Example 3 in Section 2.2 we found that

$$\frac{d}{dx}\sqrt{x} = \frac{1}{2\sqrt{x}}$$

which can be written as

$$\frac{d}{dx}(x^{1/2}) = \tfrac{1}{2}x^{-1/2}$$

This shows that the Power Rule is true even when $n = \tfrac{1}{2}$. In fact, it also holds for *any real number* $n$, as we will prove in Chapter 6. (A proof for rational values of $n$ is indicated in Exercise 48 in Section 2.6.) In the meantime we state the general version and use it in the examples and exercises.

> **The Power Rule (General Version)**   If $n$ is any real number, then
>
> $$\frac{d}{dx}(x^n) = nx^{n-1}$$

**EXAMPLE 10**

(a) If $f(x) = x^\pi$, then $f'(x) = \pi x^{\pi-1}$.

(b) Let
$$y = \frac{1}{\sqrt[3]{x^2}}$$

Then
$$\frac{dy}{dx} = \frac{d}{dx}(x^{-2/3}) = -\tfrac{2}{3}x^{-(2/3)-1}$$
$$= -\tfrac{2}{3}x^{-5/3}$$

In Example 11, $a$ and $b$ are constants. It is customary in mathematics to use letters near the beginning of the alphabet to represent constants and letters near the end of the alphabet to represent variables.

**EXAMPLE 11**  Differentiate the function $f(t) = \sqrt{t}\,(a + bt)$.

SOLUTION 1  Using the Product Rule, we have

$$f'(t) = \sqrt{t}\,\frac{d}{dt}(a + bt) + (a + bt)\frac{d}{dt}(\sqrt{t})$$

$$= \sqrt{t}\cdot b + (a + bt)\cdot\tfrac{1}{2}t^{-1/2}$$

$$= b\sqrt{t} + \frac{a + bt}{2\sqrt{t}} = \frac{a + 3bt}{2\sqrt{t}}$$

SOLUTION 2 If we first use the laws of exponents to rewrite $f(t)$, then we can proceed directly without using the Product Rule.

$$f(t) = a\sqrt{t} + bt\sqrt{t} = at^{1/2} + bt^{3/2}$$

$$f'(t) = \tfrac{1}{2}at^{-1/2} + \tfrac{3}{2}bt^{1/2}$$

which is equivalent to the answer given in Solution 1.

The differentiation rules enable us to find tangent lines without having to resort to the definition of a derivative. They also enable us to find *normal lines*. The **normal line** to a curve $C$ at point $P$ is the line through $P$ that is perpendicular to the tangent line at $P$. (In the study of optics, one needs to consider the angle between a light ray and the normal line to a lens.)

**EXAMPLE 12** Find equations of the tangent line and normal line to the curve $y = \sqrt{x}/(1 + x^2)$ at the point $\left(1, \tfrac{1}{2}\right)$.

SOLUTION According to the Quotient Rule, we have

$$\frac{dy}{dx} = \frac{(1 + x^2)\dfrac{d}{dx}\left(\sqrt{x}\right) - \sqrt{x}\,\dfrac{d}{dx}(1 + x^2)}{(1 + x^2)^2}$$

$$= \frac{(1 + x^2)\dfrac{1}{2\sqrt{x}} - \sqrt{x}\,(2x)}{(1 + x^2)^2}$$

$$= \frac{(1 + x^2) - 4x^2}{2\sqrt{x}\,(1 + x^2)^2} = \frac{1 - 3x^2}{2\sqrt{x}\,(1 + x^2)^2}$$

So the slope of the tangent line at $\left(1, \tfrac{1}{2}\right)$ is

$$\frac{dy}{dx}\bigg|_{x=1} = \frac{1 - 3 \cdot 1^2}{2\sqrt{1}(1 + 1^2)^2} = -\frac{1}{4}$$

We use the point-slope form to write an equation of the tangent line at $\left(1, \tfrac{1}{2}\right)$:

$$y - \tfrac{1}{2} = -\tfrac{1}{4}(x - 1) \qquad \text{or} \qquad y = -\tfrac{1}{4}x + \tfrac{3}{4}$$

The slope of the normal line at $\left(1, \tfrac{1}{2}\right)$ is the negative reciprocal of $-\tfrac{1}{4}$, namely 4, so an equation is

$$y - \tfrac{1}{2} = 4(x - 1) \qquad \text{or} \qquad y = 4x - \tfrac{7}{2}$$

The curve and its tangent and normal lines are graphed in Figure 5.

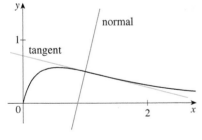

**FIGURE 5**

**EXAMPLE 13** At what points on the hyperbola $xy = 12$ is the tangent line parallel to the line $3x + y = 0$?

SOLUTION Since $xy = 12$ can be written as $y = 12/x$, we have

$$\frac{dy}{dx} = 12\frac{d}{dx}\left(x^{-1}\right) = 12(-x^{-2}) = -\frac{12}{x^2}$$

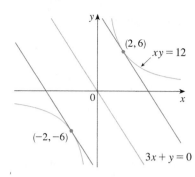

**FIGURE 6**

Let the $x$-coordinate of one of the points in question be $a$. Then the slope of the tangent line at that point is $-12/a^2$. This tangent line will be parallel to the line $3x + y = 0$, or $y = -3x$, if it has the same slope, that is, $-3$. Equating slopes, we get

$$-\frac{12}{a^2} = -3 \quad \text{or} \quad a^2 = 4 \quad \text{or} \quad a = \pm 2$$

Therefore the required points are $(2, 6)$ and $(-2, -6)$. The hyperbola and the tangents are shown in Figure 6.

We summarize the differentiation formulas we have learned so far as follows.

**Table of Differentiation Formulas**

$$\frac{d}{dx}(c) = 0 \qquad \frac{d}{dx}(x^n) = nx^{n-1}$$

$$(cf)' = cf' \qquad (f+g)' = f' + g' \qquad (f-g)' = f' - g'$$

$$(fg)' = fg' + gf' \qquad \left(\frac{f}{g}\right)' = \frac{gf' - fg'}{g^2}$$

---

## 2.3 Exercises

**1–22** Differentiate the function.

**1.** $f(x) = 2^{40}$

**2.** $f(x) = \pi^2$

**3.** $f(t) = 2 - \frac{2}{3}t$

**4.** $F(x) = \frac{3}{4}x^8$

**5.** $f(x) = x^3 - 4x + 6$

**6.** $f(t) = \frac{1}{2}t^6 - 3t^4 + t$

**7.** $g(x) = x^2(1 - 2x)$

**8.** $h(x) = (x - 2)(2x + 3)$

**9.** $g(t) = 2t^{-3/4}$

**10.** $B(y) = cy^{-6}$

**11.** $A(s) = -\dfrac{12}{s^5}$

**12.** $y = x^{5/3} - x^{2/3}$

**13.** $S(p) = \sqrt{p} - p$

**14.** $y = \sqrt{x}\,(x - 1)$

**15.** $R(a) = (3a + 1)^2$

**16.** $S(R) = 4\pi R^2$

**17.** $y = \dfrac{x^2 + 4x + 3}{\sqrt{x}}$

**18.** $y = \dfrac{\sqrt{x} + x}{x^2}$

**19.** $H(x) = (x + x^{-1})^3$

**20.** $g(u) = \sqrt{2}\,u + \sqrt{3u}$

**21.** $u = \sqrt[5]{t} + 4\sqrt{t^5}$

**22.** $v = \left(\sqrt{x} + \dfrac{1}{\sqrt[3]{x}}\right)^2$

**23.** Find the derivative of $f(x) = (1 + 2x^2)(x - x^2)$ in two ways: by using the Product Rule and by performing the multiplication first. Do your answers agree?

**24.** Find the derivative of the function

$$F(x) = \frac{x^4 - 5x^3 + \sqrt{x}}{x^2}$$

in two ways: by using the Quotient Rule and by simplifying first. Show that your answers are equivalent. Which method do you prefer?

**25–44** Differentiate.

**25.** $V(x) = (2x^3 + 3)(x^4 - 2x)$

**26.** $L(x) = (1 + x + x^2)(2 - x^4)$

**27.** $F(y) = \left(\dfrac{1}{y^2} - \dfrac{3}{y^4}\right)(y + 5y^3)$

**28.** $J(v) = (v^3 - 2v)(v^{-4} + v^{-2})$

**29.** $g(x) = \dfrac{1 + 2x}{3 - 4x}$

**30.** $f(x) = \dfrac{x - 3}{x + 3}$

---

Graphing calculator or computer required    **1.** Homework Hints available at stewartcalculus.com

**31.** $y = \dfrac{x^3}{1 - x^2}$

**32.** $y = \dfrac{x + 1}{x^3 + x - 2}$

**33.** $y = \dfrac{v^3 - 2v\sqrt{v}}{v}$

**34.** $y = \dfrac{t}{(t - 1)^2}$

**35.** $y = \dfrac{t^2 + 2}{t^4 - 3t^2 + 1}$

**36.** $g(t) = \dfrac{t - \sqrt{t}}{t^{1/3}}$

**37.** $y = ax^2 + bx + c$

**38.** $y = A + \dfrac{B}{x} + \dfrac{C}{x^2}$

**39.** $f(t) = \dfrac{2t}{2 + \sqrt{t}}$

**40.** $y = \dfrac{cx}{1 + cx}$

**41.** $y = \sqrt[3]{t}\,(t^2 + t + t^{-1})$

**42.** $y = \dfrac{u^6 - 2u^3 + 5}{u^2}$

**43.** $f(x) = \dfrac{x}{x + \dfrac{c}{x}}$

**44.** $f(x) = \dfrac{ax + b}{cx + d}$

**45.** The general polynomial of degree $n$ has the form

$$P(x) = a_n x^n + a_{n-1} x^{n-1} + \cdots + a_2 x^2 + a_1 x + a_0$$

where $a_n \neq 0$. Find the derivative of $P$.

 **46–48** Find $f'(x)$. Compare the graphs of $f$ and $f'$ and use them to explain why your answer is reasonable.

**46.** $f(x) = x/(x^2 - 1)$

**47.** $f(x) = 3x^{15} - 5x^3 + 3$     **48.** $f(x) = x + \dfrac{1}{x}$

**49.** (a) Use a graphing calculator or computer to graph the function $f(x) = x^4 - 3x^3 - 6x^2 + 7x + 30$ in the viewing rectangle $[-3, 5]$ by $[-10, 50]$.
 (b) Using the graph in part (a) to estimate slopes, make a rough sketch, by hand, of the graph of $f'$. (See Example 1 in Section 2.2.)
 (c) Calculate $f'(x)$ and use this expression, with a graphing device, to graph $f'$. Compare with your sketch in part (b).

**50.** (a) Use a graphing calculator or computer to graph the function $g(x) = x^2/(x^2 + 1)$ in the viewing rectangle $[-4, 4]$ by $[-1, 1.5]$.
 (b) Using the graph in part (a) to estimate slopes, make a rough sketch, by hand, of the graph of $g'$. (See Example 1 in Section 2.2.)
 (c) Calculate $g'(x)$ and use this expression, with a graphing device, to graph $g'$. Compare with your sketch in part (b).

**51–52** Find an equation of the tangent line to the curve at the given point.

**51.** $y = \dfrac{2x}{x + 1}$,  $(1, 1)$

**52.** $y = x^4 + 2x^2 - x$,  $(1, 2)$

**53.** (a) The curve $y = 1/(1 + x^2)$ is called a **witch of Maria Agnesi**. Find an equation of the tangent line to this curve at the point $\left(-1, \frac{1}{2}\right)$.
 (b) Illustrate part (a) by graphing the curve and the tangent line on the same screen.

**54.** (a) The curve $y = x/(1 + x^2)$ is called a **serpentine**. Find an equation of the tangent line to this curve at the point $(3, 0.3)$.
 (b) Illustrate part (a) by graphing the curve and the tangent line on the same screen.

**55–58** Find equations of the tangent line and normal line to the curve at the given point.

**55.** $y = x + \sqrt{x}$,  $(1, 2)$     **56.** $y = (1 + 2x)^2$,  $(1, 9)$

**57.** $y = \dfrac{3x + 1}{x^2 + 1}$,  $(1, 2)$     **58.** $y = \dfrac{\sqrt{x}}{x + 1}$,  $(4, 0.4)$

**59–62** Find the first and second derivatives of the function.

**59.** $f(x) = x^4 - 3x^3 + 16x$     **60.** $G(r) = \sqrt{r} + \sqrt[3]{r}$

**61.** $f(x) = \dfrac{x^2}{1 + 2x}$     **62.** $f(x) = \dfrac{1}{3 - x}$

**63.** The equation of motion of a particle is $s = t^3 - 3t$, where $s$ is in meters and $t$ is in seconds. Find
 (a) the velocity and acceleration as functions of $t$,
 (b) the acceleration after 2 s, and
 (c) the acceleration when the velocity is 0.

**64.** The equation of motion of a particle is

$$s = t^4 - 2t^3 + t^2 - t$$

where $s$ is in meters and $t$ is in seconds.
 (a) Find the velocity and acceleration as functions of $t$.
 (b) Find the acceleration after 1 s.
 (c) Graph the position, velocity, and acceleration functions on the same screen.

**65.** Boyle's Law states that when a sample of gas is compressed at a constant pressure, the pressure $P$ of the gas is inversely proportional to the volume $V$ of the gas.
(a) Suppose that the pressure of a sample of air that occupies $0.106 \text{ m}^3$ at $25°C$ is 50 kPa. Write $V$ as a function of $P$.
(b) Calculate $dV/dP$ when $P = 50$ kPa. What is the meaning of the derivative? What are its units?

**66.** Car tires need to be inflated properly because overinflation or underinflation can cause premature treadware. The data in the table show tire life $L$ (in thousands of miles) for a certain type of tire at various pressures $P$ (in lb/in$^2$).

| $P$ | 26 | 28 | 31 | 35 | 38 | 42 | 45 |
|---|---|---|---|---|---|---|---|
| $L$ | 50 | 66 | 78 | 81 | 74 | 70 | 59 |

(a) Use a graphing calculator or computer to model tire life with a quadratic function of the pressure.
(b) Use the model to estimate $dL/dP$ when $P = 30$ and when $P = 40$. What is the meaning of the derivative? What are the units? What is the significance of the signs of the derivatives?

**67.** Suppose that $f(5) = 1$, $f'(5) = 6$, $g(5) = -3$, and $g'(5) = 2$. Find the following values.
(a) $(fg)'(5)$     (b) $(f/g)'(5)$
(c) $(g/f)'(5)$

**68.** Find $h'(2)$, given that $f(2) = -3$, $g(2) = 4$, $f'(2) = -2$, and $g'(2) = 7$.
(a) $h(x) = 5f(x) - 4g(x)$     (b) $h(x) = f(x)g(x)$
(c) $h(x) = \dfrac{f(x)}{g(x)}$     (d) $h(x) = \dfrac{g(x)}{1 + f(x)}$

**69.** If $f(x) = \sqrt{x}\, g(x)$, where $g(4) = 8$ and $g'(4) = 7$, find $f'(4)$.

**70.** If $h(2) = 4$ and $h'(2) = -3$, find

$$\frac{d}{dx}\left(\frac{h(x)}{x}\right)\bigg|_{x=2}$$

**71.** If $f$ and $g$ are the functions whose graphs are shown, let $u(x) = f(x)g(x)$ and $v(x) = f(x)/g(x)$.
(a) Find $u'(1)$.     (b) Find $v'(5)$.

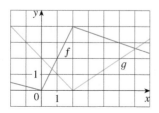

**72.** Let $P(x) = F(x)\, G(x)$ and $Q(x) = F(x)/G(x)$, where $F$ and $G$ are the functions whose graphs are shown.
(a) Find $P'(2)$.     (b) Find $Q'(7)$.

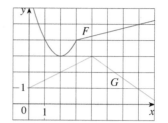

**73.** If $g$ is a differentiable function, find an expression for the derivative of each of the following functions.
(a) $y = xg(x)$     (b) $y = \dfrac{x}{g(x)}$     (c) $y = \dfrac{g(x)}{x}$

**74.** If $f$ is a differentiable function, find an expression for the derivative of each of the following functions.
(a) $y = x^2 f(x)$     (b) $y = \dfrac{f(x)}{x^2}$
(c) $y = \dfrac{x^2}{f(x)}$     (d) $y = \dfrac{1 + xf(x)}{\sqrt{x}}$

**75.** Find the points on the curve $y = 2x^3 + 3x^2 - 12x + 1$ where the tangent is horizontal.

**76.** For what values of $x$ does the graph of $f(x) = x^3 + 3x^2 + x + 3$ have a horizontal tangent?

**77.** Show that the curve $y = 6x^3 + 5x - 3$ has no tangent line with slope 4.

**78.** Find an equation of the tangent line to the curve $y = x\sqrt{x}$ that is parallel to the line $y = 1 + 3x$.

**79.** Find equations of both lines that are tangent to the curve $y = 1 + x^3$ and are parallel to the line $12x - y = 1$.

**80.** Find equations of the tangent lines to the curve

$$y = \frac{x - 1}{x + 1}$$

that are parallel to the line $x - 2y = 2$.

**81.** Find an equation of the normal line to the parabola $y = x^2 - 5x + 4$ that is parallel to the line $x - 3y = 5$.

**82.** Where does the normal line to the parabola $y = x - x^2$ at the point $(1, 0)$ intersect the parabola a second time? Illustrate with a sketch.

**83.** Draw a diagram to show that there are two tangent lines to the parabola $y = x^2$ that pass through the point $(0, -4)$. Find the coordinates of the points where these tangent lines intersect the parabola.

**84.** (a) Find equations of both lines through the point $(2, -3)$ that are tangent to the parabola $y = x^2 + x$.

(b) Show that there is no line through the point $(2, 7)$ that is tangent to the parabola. Then draw a diagram to see why.

**85.** (a) Use the Product Rule twice to prove that if $f$, $g$, and $h$ are differentiable, then $(fgh)' = f'gh + fg'h + fgh'$.

(b) Taking $f = g = h$ in part (a), show that

$$\frac{d}{dx}[f(x)]^3 = 3[f(x)]^2 f'(x)$$

(c) Use part (b) to differentiate $y = (x^4 + 3x^3 + 17x + 82)^3$.

**86.** Find the $n$th derivative of each function by calculating the first few derivatives and observing the pattern that occurs.
(a) $f(x) = x^n$        (b) $f(x) = 1/x$

**87.** Find a second-degree polynomial $P$ such that $P(2) = 5$, $P'(2) = 3$, and $P''(2) = 2$.

**88.** The equation $y'' + y' - 2y = x^2$ is called a **differential equation** because it involves an unknown function $y$ and its derivatives $y'$ and $y''$. Find constants $A$, $B$, and $C$ such that the function $y = Ax^2 + Bx + C$ satisfies this equation. (Differential equations will be studied in detail in Chapter 9.)

**89.** Find a cubic function $y = ax^3 + bx^2 + cx + d$ whose graph has horizontal tangents at the points $(-2, 6)$ and $(2, 0)$.

**90.** Find a parabola with equation $y = ax^2 + bx + c$ that has slope 4 at $x = 1$, slope $-8$ at $x = -1$, and passes through the point $(2, 15)$.

**91.** In this exercise we estimate the rate at which the total personal income is rising in the Richmond-Petersburg, Virginia, metropolitan area. In 1999, the population of this area was 961,400, and the population was increasing at roughly 9200 people per year. The average annual income was \$30,593 per capita, and this average was increasing at about \$1400 per year (a little above the national average of about \$1225 yearly). Use the Product Rule and these figures to estimate the rate at which total personal income was rising in the Richmond-Petersburg area in 1999. Explain the meaning of each term in the Product Rule.

**92.** A manufacturer produces bolts of a fabric with a fixed width. The quantity $q$ of this fabric (measured in yards) that is sold is a function of the selling price $p$ (in dollars per yard), so we can write $q = f(p)$. Then the total revenue earned with selling price $p$ is $R(p) = pf(p)$.
(a) What does it mean to say that $f(20) = 10,000$ and $f'(20) = -350$?
(b) Assuming the values in part (a), find $R'(20)$ and interpret your answer.

**93.** Let

$$f(x) = \begin{cases} x^2 + 1 & \text{if } x < 1 \\ x + 1 & \text{if } x \geqslant 1 \end{cases}$$

Is $f$ differentiable at 1? Sketch the graphs of $f$ and $f'$.

**94.** At what numbers is the following function $g$ differentiable?

$$g(x) = \begin{cases} 2x & \text{if } x \leqslant 0 \\ 2x - x^2 & \text{if } 0 < x < 2 \\ 2 - x & \text{if } x \geqslant 2 \end{cases}$$

Give a formula for $g'$ and sketch the graphs of $g$ and $g'$.

**95.** (a) For what values of $x$ is the function $f(x) = |x^2 - 9|$ differentiable? Find a formula for $f'$.
(b) Sketch the graphs of $f$ and $f'$.

**96.** Where is the function $h(x) = |x - 1| + |x + 2|$ differentiable? Give a formula for $h'$ and sketch the graphs of $h$ and $h'$.

**97.** For what values of $a$ and $b$ is the line $2x + y = b$ tangent to the parabola $y = ax^2$ when $x = 2$?

**98.** (a) If $F(x) = f(x)g(x)$, where $f$ and $g$ have derivatives of all orders, show that $F'' = f''g + 2f'g' + fg''$.
(b) Find similar formulas for $F'''$ and $F^{(4)}$.
(c) Guess a formula for $F^{(n)}$.

**99.** Find the value of $c$ such that the line $y = \frac{3}{2}x + 6$ is tangent to the curve $y = c\sqrt{x}$.

**100.** Let

$$f(x) = \begin{cases} x^2 & \text{if } x \leqslant 2 \\ mx + b & \text{if } x > 2 \end{cases}$$

Find the values of $m$ and $b$ that make $f$ differentiable everywhere.

**101.** An easy proof of the Quotient Rule can be given if we make the prior assumption that $F'(x)$ exists, where $F = f/g$. Write $f = Fg$; then differentiate using the Product Rule and solve the resulting equation for $F'$.

**102.** A tangent line is drawn to the hyperbola $xy = c$ at a point $P$.
(a) Show that the midpoint of the line segment cut from this tangent line by the coordinate axes is $P$.
(b) Show that the triangle formed by the tangent line and the coordinate axes always has the same area, no matter where $P$ is located on the hyperbola.

**103.** Evaluate $\lim\limits_{x \to 1} \dfrac{x^{1000} - 1}{x - 1}$.

**104.** Draw a diagram showing two perpendicular lines that intersect on the $y$-axis and are both tangent to the parabola $y = x^2$. Where do these lines intersect?

**105.** If $c > \frac{1}{2}$, how many lines through the point $(0, c)$ are normal lines to the parabola $y = x^2$? What if $c \leqslant \frac{1}{2}$?

**106.** Sketch the parabolas $y = x^2$ and $y = x^2 - 2x + 2$. Do you think there is a line that is tangent to both curves? If so, find its equation. If not, why not?

## BUILDING A BETTER ROLLER COASTER

Suppose you are asked to design the first ascent and drop for a new roller coaster. By studying photographs of your favorite coasters, you decide to make the slope of the ascent 0.8 and the slope of the drop $-1.6$. You decide to connect these two straight stretches $y = L_1(x)$ and $y = L_2(x)$ with part of a parabola $y = f(x) = ax^2 + bx + c$, where $x$ and $f(x)$ are measured in feet. For the track to be smooth there can't be abrupt changes in direction, so you want the linear segments $L_1$ and $L_2$ to be tangent to the parabola at the transition points $P$ and $Q$. (See the figure.) To simplify the equations, you decide to place the origin at $P$.

1. (a) Suppose the horizontal distance between $P$ and $Q$ is 100 ft. Write equations in $a$, $b$, and $c$ that will ensure that the track is smooth at the transition points.
   (b) Solve the equations in part (a) for $a$, $b$, and $c$ to find a formula for $f(x)$.
   (c) Plot $L_1$, $f$, and $L_2$ to verify graphically that the transitions are smooth.
   (d) Find the difference in elevation between $P$ and $Q$.

2. The solution in Problem 1 might *look* smooth, but it might not *feel* smooth because the piecewise defined function [consisting of $L_1(x)$ for $x < 0$, $f(x)$ for $0 \le x \le 100$, and $L_2(x)$ for $x > 100$] doesn't have a continuous second derivative. So you decide to improve the design by using a quadratic function $q(x) = ax^2 + bx + c$ only on the interval $10 \le x \le 90$ and connecting it to the linear functions by means of two cubic functions:

$$g(x) = kx^3 + lx^2 + mx + n \qquad 0 \le x < 10$$

$$h(x) = px^3 + qx^2 + rx + s \qquad 90 < x \le 100$$

   (a) Write a system of equations in 11 unknowns that ensure that the functions and their first two derivatives agree at the transition points.
   (b) Solve the equations in part (a) with a computer algebra system to find formulas for $q(x)$, $g(x)$, and $h(x)$.
   (c) Plot $L_1$, $g$, $q$, $h$, and $L_2$, and compare with the plot in Problem 1(c).

---

Graphing calculator or computer required

Computer algebra system required

## 2.4   Derivatives of Trigonometric Functions

A review of trigonometric functions is given in Appendix D.

Before starting this section, you might need to review the trigonometric functions. In particular, it is important to remember that when we talk about the function $f$ defined for all real numbers $x$ by

$$f(x) = \sin x$$

it is understood that $\sin x$ means the sine of the angle whose *radian* measure is $x$. A similar convention holds for the other trigonometric functions cos, tan, csc, sec, and cot. Recall from Section 1.8 that all of the trigonometric functions are continuous at every number in their domains.

If we sketch the graph of the function $f(x) = \sin x$ and use the interpretation of $f'(x)$ as the slope of the tangent to the sine curve in order to sketch the graph of $f'$ (see Exercise 16 in Section 2.2), then it looks as if the graph of $f'$ may be the same as the cosine curve (see Figure 1).

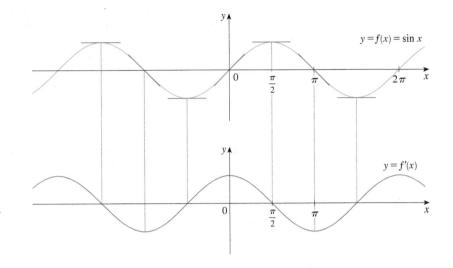

**TEC** Visual 2.4 shows an animation of Figure 1.

**FIGURE 1**

Let's try to confirm our guess that if $f(x) = \sin x$, then $f'(x) = \cos x$. From the definition of a derivative, we have

$$f'(x) = \lim_{h \to 0} \frac{f(x + h) - f(x)}{h} = \lim_{h \to 0} \frac{\sin(x + h) - \sin x}{h}$$

We have used the addition formula for sine. See Appendix D.

$$= \lim_{h \to 0} \frac{\sin x \cos h + \cos x \sin h - \sin x}{h}$$

$$= \lim_{h \to 0} \left[ \frac{\sin x \cos h - \sin x}{h} + \frac{\cos x \sin h}{h} \right]$$

$$= \lim_{h \to 0} \left[ \sin x \left( \frac{\cos h - 1}{h} \right) + \cos x \left( \frac{\sin h}{h} \right) \right]$$

$$\boxed{1} \qquad = \lim_{h \to 0} \sin x \cdot \lim_{h \to 0} \frac{\cos h - 1}{h} + \lim_{h \to 0} \cos x \cdot \lim_{h \to 0} \frac{\sin h}{h}$$

Two of these four limits are easy to evaluate. Since we regard $x$ as a constant when computing a limit as $h \to 0$, we have

$$\lim_{h \to 0} \sin x = \sin x \qquad \text{and} \qquad \lim_{h \to 0} \cos x = \cos x$$

The limit of $(\sin h)/h$ is not so obvious. In Example 3 in Section 1.5 we made the guess, on the basis of numerical and graphical evidence, that

$$\boxed{2} \qquad \lim_{\theta \to 0} \frac{\sin \theta}{\theta} = 1$$

We now use a geometric argument to prove Equation 2. Assume first that $\theta$ lies between 0 and $\pi/2$. Figure 2(a) shows a sector of a circle with center $O$, central angle $\theta$, and

(a)

(b)

**FIGURE 2**

radius 1. $BC$ is drawn perpendicular to $OA$. By the definition of radian measure, we have arc $AB = \theta$. Also $|BC| = |OB| \sin \theta = \sin \theta$. From the diagram we see that

$$|BC| < |AB| < \text{arc } AB$$

Therefore
$$\sin \theta < \theta \qquad \text{so} \qquad \frac{\sin \theta}{\theta} < 1$$

Let the tangent lines at $A$ and $B$ intersect at $E$. You can see from Figure 2(b) that the circumference of a circle is smaller than the length of a circumscribed polygon, and so arc $AB < |AE| + |EB|$. Thus

$$\theta = \text{arc } AB < |AE| + |EB|$$
$$< |AE| + |ED|$$
$$= |AD| = |OA| \tan \theta$$
$$= \tan \theta$$

(In Appendix F the inequality $\theta \le \tan \theta$ is proved directly from the definition of the length of an arc without resorting to geometric intuition as we did here.) Therefore we have

$$\theta < \frac{\sin \theta}{\cos \theta}$$

so
$$\cos \theta < \frac{\sin \theta}{\theta} < 1$$

We know that $\lim_{\theta \to 0} 1 = 1$ and $\lim_{\theta \to 0} \cos \theta = 1$, so by the Squeeze Theorem, we have

$$\lim_{\theta \to 0^+} \frac{\sin \theta}{\theta} = 1$$

But the function $(\sin \theta)/\theta$ is an even function, so its right and left limits must be equal. Hence, we have

$$\lim_{\theta \to 0} \frac{\sin \theta}{\theta} = 1$$

so we have proved Equation 2.

We can deduce the value of the remaining limit in $\boxed{1}$ as follows:

We multiply numerator and denominator by $\cos \theta + 1$ in order to put the function in a form in which we can use the limits we know.

$$\lim_{\theta \to 0} \frac{\cos \theta - 1}{\theta} = \lim_{\theta \to 0} \left( \frac{\cos \theta - 1}{\theta} \cdot \frac{\cos \theta + 1}{\cos \theta + 1} \right) = \lim_{\theta \to 0} \frac{\cos^2 \theta - 1}{\theta (\cos \theta + 1)}$$

$$= \lim_{\theta \to 0} \frac{-\sin^2 \theta}{\theta (\cos \theta + 1)} = -\lim_{\theta \to 0} \left( \frac{\sin \theta}{\theta} \cdot \frac{\sin \theta}{\cos \theta + 1} \right)$$

$$= -\lim_{\theta \to 0} \frac{\sin \theta}{\theta} \cdot \lim_{\theta \to 0} \frac{\sin \theta}{\cos \theta + 1}$$

$$= -1 \cdot \left( \frac{0}{1 + 1} \right) = 0 \qquad \text{(by Equation 2)}$$

$$\boxed{3} \qquad \lim_{\theta \to 0} \frac{\cos \theta - 1}{\theta} = 0$$

If we now put the limits $\boxed{2}$ and $\boxed{3}$ in $\boxed{1}$, we get

$$f'(x) = \lim_{h \to 0} \sin x \cdot \lim_{h \to 0} \frac{\cos h - 1}{h} + \lim_{h \to 0} \cos x \cdot \lim_{h \to 0} \frac{\sin h}{h}$$

$$= (\sin x) \cdot 0 + (\cos x) \cdot 1 = \cos x$$

So we have proved the formula for the derivative of the sine function:

$$\boxed{4} \qquad \frac{d}{dx} (\sin x) = \cos x$$

Figure 3 shows the graphs of the function of Example 1 and its derivative. Notice that $y' = 0$ whenever $y$ has a horizontal tangent.

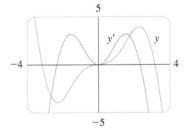

**FIGURE 3**

**V EXAMPLE 1** Differentiate $y = x^2 \sin x$.

SOLUTION Using the Product Rule and Formula 4, we have

$$\frac{dy}{dx} = x^2 \frac{d}{dx} (\sin x) + \sin x \frac{d}{dx} (x^2)$$

$$= x^2 \cos x + 2x \sin x$$

Using the same methods as in the proof of Formula 4, one can prove (see Exercise 20) that

$$\boxed{5} \qquad \frac{d}{dx} (\cos x) = -\sin x$$

The tangent function can also be differentiated by using the definition of a derivative, but it is easier to use the Quotient Rule together with Formulas 4 and 5:

$$\frac{d}{dx} (\tan x) = \frac{d}{dx} \left( \frac{\sin x}{\cos x} \right)$$

$$= \frac{\cos x \dfrac{d}{dx} (\sin x) - \sin x \dfrac{d}{dx} (\cos x)}{\cos^2 x}$$

$$= \frac{\cos x \cdot \cos x - \sin x (-\sin x)}{\cos^2 x}$$

$$= \frac{\cos^2 x + \sin^2 x}{\cos^2 x}$$

$$= \frac{1}{\cos^2 x} = \sec^2 x$$

$$\boxed{6} \qquad \frac{d}{dx}(\tan x) = \sec^2 x$$

The derivatives of the remaining trigonometric functions, csc, sec , and cot , can also be found easily using the Quotient Rule (see Exercises 17–19). We collect all the differentiation formulas for trigonometric functions in the following table. Remember that they are valid only when $x$ is measured in radians.

When you memorize this table, it is helpful to notice that the minus signs go with the derivatives of the "cofunctions," that is, cosine, cosecant, and cotangent.

---

**Derivatives of Trigonometric Functions**

$$\frac{d}{dx}(\sin x) = \cos x \qquad\qquad \frac{d}{dx}(\csc x) = -\csc x \cot x$$

$$\frac{d}{dx}(\cos x) = -\sin x \qquad\qquad \frac{d}{dx}(\sec x) = \sec x \tan x$$

$$\frac{d}{dx}(\tan x) = \sec^2 x \qquad\qquad \frac{d}{dx}(\cot x) = -\csc^2 x$$

---

**EXAMPLE 2** Differentiate $f(x) = \dfrac{\sec x}{1 + \tan x}$. For what values of $x$ does the graph of $f$ have a horizontal tangent?

SOLUTION The Quotient Rule gives

$$f'(x) = \frac{(1 + \tan x)\dfrac{d}{dx}(\sec x) - \sec x \dfrac{d}{dx}(1 + \tan x)}{(1 + \tan x)^2}$$

$$= \frac{(1 + \tan x)\sec x \tan x - \sec x \cdot \sec^2 x}{(1 + \tan x)^2}$$

$$= \frac{\sec x (\tan x + \tan^2 x - \sec^2 x)}{(1 + \tan x)^2}$$

$$= \frac{\sec x (\tan x - 1)}{(1 + \tan x)^2}$$

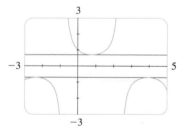

**FIGURE 4**
The horizontal tangents in Example 2

In simplifying the answer we have used the identity $\tan^2 x + 1 = \sec^2 x$.

Since $\sec x$ is never 0, we see that $f'(x) = 0$ when $\tan x = 1$, and this occurs when $x = n\pi + \pi/4$, where $n$ is an integer (see Figure 4).

Trigonometric functions are often used in modeling real-world phenomena. In particular, vibrations, waves, elastic motions, and other quantities that vary in a periodic manner can be described using trigonometric functions. In the following example we discuss an instance of simple harmonic motion.

**FIGURE 5**

**FIGURE 6**

PS Look for a pattern.

V EXAMPLE 3 An object at the end of a vertical spring is stretched 4 cm beyond its rest position and released at time $t = 0$. (See Figure 5 and note that the downward direction is positive.) Its position at time $t$ is

$$s = f(t) = 4 \cos t$$

Find the velocity and acceleration at time $t$ and use them to analyze the motion of the object.

SOLUTION The velocity and acceleration are

$$v = \frac{ds}{dt} = \frac{d}{dt}(4 \cos t) = 4\frac{d}{dt}(\cos t) = -4 \sin t$$

$$a = \frac{dv}{dt} = \frac{d}{dt}(-4 \sin t) = -4\frac{d}{dt}(\sin t) = -4 \cos t$$

The object oscillates from the lowest point ($s = 4$ cm) to the highest point ($s = -4$ cm). The period of the oscillation is $2\pi$, the period of $\cos t$.

The speed is $|v| = 4|\sin t|$, which is greatest when $|\sin t| = 1$, that is, when $\cos t = 0$. So the object moves fastest as it passes through its equilibrium position ($s = 0$). Its speed is 0 when $\sin t = 0$, that is, at the high and low points.

The acceleration $a = -4 \cos t = 0$ when $s = 0$. It has greatest magnitude at the high and low points. See the graphs in Figure 6. ▬

EXAMPLE 4 Find the 27th derivative of $\cos x$.

SOLUTION The first few derivatives of $f(x) = \cos x$ are as follows:

$$f'(x) = -\sin x$$
$$f''(x) = -\cos x$$
$$f'''(x) = \sin x$$
$$f^{(4)}(x) = \cos x$$
$$f^{(5)}(x) = -\sin x$$

We see that the successive derivatives occur in a cycle of length 4 and, in particular, $f^{(n)}(x) = \cos x$ whenever $n$ is a multiple of 4. Therefore

$$f^{(24)}(x) = \cos x$$

and, differentiating three more times, we have

$$f^{(27)}(x) = \sin x$$ ▬

Our main use for the limit in Equation 2 has been to prove the differentiation formula for the sine function. But this limit is also useful in finding certain other trigonometric limits, as the following two examples show.

EXAMPLE 5 Find $\lim\limits_{x \to 0} \dfrac{\sin 7x}{4x}$.

SOLUTION In order to apply Equation 2, we first rewrite the function by multiplying and dividing by 7:

$$\frac{\sin 7x}{4x} = \frac{7}{4}\left(\frac{\sin 7x}{7x}\right)$$

Note that $\sin 7x \neq 7 \sin x$.

If we let $\theta = 7x$, then $\theta \to 0$ as $x \to 0$, so by Equation 2 we have

$$\lim_{x \to 0} \frac{\sin 7x}{4x} = \frac{7}{4} \lim_{x \to 0} \left( \frac{\sin 7x}{7x} \right)$$

$$= \frac{7}{4} \lim_{\theta \to 0} \frac{\sin \theta}{\theta} = \frac{7}{4} \cdot 1 = \frac{7}{4}$$

**V** **EXAMPLE 6**  Calculate $\lim_{x \to 0} x \cot x$.

**SOLUTION**  Here we divide numerator and denominator by $x$:

$$\lim_{x \to 0} x \cot x = \lim_{x \to 0} \frac{x \cos x}{\sin x}$$

$$= \lim_{x \to 0} \frac{\cos x}{\dfrac{\sin x}{x}} = \frac{\displaystyle \lim_{x \to 0} \cos x}{\displaystyle \lim_{x \to 0} \frac{\sin x}{x}}$$

$$= \frac{\cos 0}{1} \qquad \text{(by the continuity of cosine and Equation 2)}$$

$$= 1$$

## 2.4 Exercises

**1–16** Differentiate.

**1.** $f(x) = 3x^2 - 2 \cos x$

**2.** $f(x) = \sqrt{x} \sin x$

**3.** $f(x) = \sin x + \frac{1}{2} \cot x$

**4.** $y = 2 \sec x - \csc x$

**5.** $y = \sec \theta \tan \theta$

**6.** $g(t) = 4 \sec t + \tan t$

**7.** $y = c \cos t + t^2 \sin t$

**8.** $y = u(a \cos u + b \cot u)$

**9.** $y = \dfrac{x}{2 - \tan x}$

**10.** $y = \sin \theta \cos \theta$

**11.** $f(\theta) = \dfrac{\sec \theta}{1 + \sec \theta}$

**12.** $y = \dfrac{\cos x}{1 - \sin x}$

**13.** $y = \dfrac{t \sin t}{1 + t}$

**14.** $y = \dfrac{1 - \sec x}{\tan x}$

**15.** $h(\theta) = \theta \csc \theta - \cot \theta$

**16.** $y = x^2 \sin x \tan x$

**17.** Prove that $\dfrac{d}{dx} (\csc x) = -\csc x \cot x$.

**18.** Prove that $\dfrac{d}{dx} (\sec x) = \sec x \tan x$.

**19.** Prove that $\dfrac{d}{dx} (\cot x) = -\csc^2 x$.

**20.** Prove, using the definition of derivative, that if $f(x) = \cos x$, then $f'(x) = -\sin x$.

**21–24** Find an equation of the tangent line to the curve at the given point.

**21.** $y = \sec x$,  $(\pi/3, 2)$

**22.** $y = (1 + x) \cos x$,  $(0, 1)$

**23.** $y = \cos x - \sin x$,  $(\pi, -1)$

**24.** $y = x + \tan x$,  $(\pi, \pi)$

**25.** (a) Find an equation of the tangent line to the curve $y = 2x \sin x$ at the point $(\pi/2, \pi)$.

(b) Illustrate part (a) by graphing the curve and the tangent line on the same screen.

**26.** (a) Find an equation of the tangent line to the curve $y = 3x + 6 \cos x$ at the point $(\pi/3, \pi + 3)$.

(b) Illustrate part (a) by graphing the curve and the tangent line on the same screen.

**27.** (a) If $f(x) = \sec x - x$, find $f'(x)$.

(b) Check to see that your answer to part (a) is reasonable by graphing both $f$ and $f'$ for $|x| < \pi/2$.

**28.** (a) If $f(x) = \sqrt{x} \sin x$, find $f'(x)$.

(b) Check to see that your answer to part (a) is reasonable by graphing both $f$ and $f'$ for $0 \le x \le 2\pi$.

Graphing calculator or computer required    **1.** Homework Hints available at stewartcalculus.com

**29.** If $H(\theta) = \theta \sin \theta$, find $H'(\theta)$ and $H''(\theta)$.

**30.** If $f(t) = \csc t$, find $f''(\pi/6)$.

**31.** (a) Use the Quotient Rule to differentiate the function

$$f(x) = \frac{\tan x - 1}{\sec x}$$

(b) Simplify the expression for $f(x)$ by writing it in terms of $\sin x$ and $\cos x$, and then find $f'(x)$.

(c) Show that your answers to parts (a) and (b) are equivalent.

**32.** Suppose $f(\pi/3) = 4$ and $f'(\pi/3) = -2$, and let $g(x) = f(x) \sin x$ and $h(x) = (\cos x)/f(x)$. Find
(a) $g'(\pi/3)$       (b) $h'(\pi/3)$

**33.** For what values of $x$ does the graph of $f(x) = x + 2 \sin x$ have a horizontal tangent?

**34.** Find the points on the curve $y = (\cos x)/(2 + \sin x)$ at which the tangent is horizontal.

**35.** A mass on a spring vibrates horizontally on a smooth level surface (see the figure). Its equation of motion is $x(t) = 8 \sin t$, where $t$ is in seconds and $x$ in centimeters.
(a) Find the velocity and acceleration at time $t$.
(b) Find the position, velocity, and acceleration of the mass at time $t = 2\pi/3$. In what direction is it moving at that time?

equilibrium
position

$0$        $x$        $x$

**36.** An elastic band is hung on a hook and a mass is hung on the lower end of the band. When the mass is pulled downward and then released, it vibrates vertically. The equation of motion is $s = 2 \cos t + 3 \sin t$, $t \geq 0$, where $s$ is measured in centimeters and $t$ in seconds. (Take the positive direction to be downward.)
(a) Find the velocity and acceleration at time $t$.
(b) Graph the velocity and acceleration functions.
(c) When does the mass pass through the equilibrium position for the first time?
(d) How far from its equilibrium position does the mass travel?
(e) When is the speed the greatest?

**37.** A ladder 10 ft long rests against a vertical wall. Let $\theta$ be the angle between the top of the ladder and the wall and let $x$ be the distance from the bottom of the ladder to the wall. If the bottom of the ladder slides away from the wall, how fast does $x$ change with respect to $\theta$ when $\theta = \pi/3$?

**38.** An object with weight $W$ is dragged along a horizontal plane by a force acting along a rope attached to the object. If the rope makes an angle $\theta$ with the plane, then the magnitude of the force is

$$F = \frac{\mu W}{\mu \sin \theta + \cos \theta}$$

where $\mu$ is a constant called the *coefficient of friction*.
(a) Find the rate of change of $F$ with respect to $\theta$.
(b) When is this rate of change equal to 0?
(c) If $W = 50$ lb and $\mu = 0.6$, draw the graph of $F$ as a function of $\theta$ and use it to locate the value of $\theta$ for which $dF/d\theta = 0$. Is the value consistent with your answer to part (b)?

**39–48** Find the limit.

**39.** $\displaystyle\lim_{x \to 0} \frac{\sin 3x}{x}$

**40.** $\displaystyle\lim_{x \to 0} \frac{\sin 4x}{\sin 6x}$

**41.** $\displaystyle\lim_{t \to 0} \frac{\tan 6t}{\sin 2t}$

**42.** $\displaystyle\lim_{\theta \to 0} \frac{\cos \theta - 1}{\sin \theta}$

**43.** $\displaystyle\lim_{x \to 0} \frac{\sin 3x}{5x^3 - 4x}$

**44.** $\displaystyle\lim_{x \to 0} \frac{\sin 3x \sin 5x}{x^2}$

**45.** $\displaystyle\lim_{\theta \to 0} \frac{\sin \theta}{\theta + \tan \theta}$

**46.** $\displaystyle\lim_{x \to 0} \frac{\sin(x^2)}{x}$

**47.** $\displaystyle\lim_{x \to \pi/4} \frac{1 - \tan x}{\sin x - \cos x}$

**48.** $\displaystyle\lim_{x \to 1} \frac{\sin(x - 1)}{x^2 + x - 2}$

**49–50** Find the given derivative by finding the first few derivatives and observing the pattern that occurs.

**49.** $\dfrac{d^{99}}{dx^{99}} (\sin x)$

**50.** $\dfrac{d^{35}}{dx^{35}} (x \sin x)$

**51.** Find constants $A$ and $B$ such that the function $y = A \sin x + B \cos x$ satisfies the differential equation $y'' + y' - 2y = \sin x$.

**52.** (a) Evaluate $\displaystyle\lim_{x \to \infty} x \sin \dfrac{1}{x}$.

(b) Evaluate $\displaystyle\lim_{x \to 0} x \sin \dfrac{1}{x}$.

(c) Illustrate parts (a) and (b) by graphing $y = x \sin(1/x)$.

**53.** Differentiate each trigonometric identity to obtain a new (or familiar) identity.

(a) $\tan x = \dfrac{\sin x}{\cos x}$

(b) $\sec x = \dfrac{1}{\cos x}$

(c) $\sin x + \cos x = \dfrac{1 + \cot x}{\csc x}$

**54.** A semicircle with diameter $PQ$ sits on an isosceles triangle $PQR$ to form a region shaped like a two-dimensional ice-cream cone, as shown in the figure. If $A(\theta)$ is the area of the semicircle and $B(\theta)$ is the area of the triangle, find

$$\lim_{\theta \to 0^+} \frac{A(\theta)}{B(\theta)}$$

**55.** The figure shows a circular arc of length $s$ and a chord of length $d$, both subtended by a central angle $\theta$. Find

$$\lim_{\theta \to 0^+} \frac{s}{d}$$

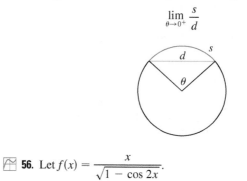

**56.** Let $f(x) = \dfrac{x}{\sqrt{1 - \cos 2x}}$.

(a) Graph $f$. What type of discontinuity does it appear to have at 0?

(b) Calculate the left and right limits of $f$ at 0. Do these values confirm your answer to part (a)?

## 2.5    The Chain Rule

Suppose you are asked to differentiate the function

$$F(x) = \sqrt{x^2 + 1}$$

The differentiation formulas you learned in the previous sections of this chapter do not enable you to calculate $F'(x)$.

*See Section 1.3 for a review of composite functions.*

Observe that $F$ is a composite function. In fact, if we let $y = f(u) = \sqrt{u}$ and let $u = g(x) = x^2 + 1$, then we can write $y = F(x) = f(g(x))$, that is, $F = f \circ g$. We know how to differentiate both $f$ and $g$, so it would be useful to have a rule that tells us how to find the derivative of $F = f \circ g$ in terms of the derivatives of $f$ and $g$.

It turns out that the derivative of the composite function $f \circ g$ is the product of the derivatives of $f$ and $g$. This fact is one of the most important of the differentiation rules and is called the *Chain Rule*. It seems plausible if we interpret derivatives as rates of change. Regard $du/dx$ as the rate of change of $u$ with respect to $x$, $dy/du$ as the rate of change of $y$ with respect to $u$, and $dy/dx$ as the rate of change of $y$ with respect to $x$. If $u$ changes twice as fast as $x$ and $y$ changes three times as fast as $u$, then it seems reasonable that $y$ changes six times as fast as $x$, and so we expect that

$$\frac{dy}{dx} = \frac{dy}{du} \frac{du}{dx}$$

---

**The Chain Rule**  If $g$ is differentiable at $x$ and $f$ is differentiable at $g(x)$, then the composite function $F = f \circ g$ defined by $F(x) = f(g(x))$ is differentiable at $x$ and $F'$ is given by the product

$$F'(x) = f'(g(x)) \cdot g'(x)$$

In Leibniz notation, if $y = f(u)$ and $u = g(x)$ are both differentiable functions, then

$$\frac{dy}{dx} = \frac{dy}{du} \frac{du}{dx}$$

---

**COMMENTS ON THE PROOF OF THE CHAIN RULE**   Let $\Delta u$ be the change in $u$ corresponding to a change of $\Delta x$ in $x$, that is,

$$\Delta u = g(x + \Delta x) - g(x)$$

Then the corresponding change in $y$ is

$$\Delta y = f(u + \Delta u) - f(u)$$

It is tempting to write

$$\frac{dy}{dx} = \lim_{\Delta x \to 0} \frac{\Delta y}{\Delta x}$$

1
$$= \lim_{\Delta x \to 0} \frac{\Delta y}{\Delta u} \cdot \frac{\Delta u}{\Delta x}$$

$$= \lim_{\Delta x \to 0} \frac{\Delta y}{\Delta u} \cdot \lim_{\Delta x \to 0} \frac{\Delta u}{\Delta x}$$

$$= \lim_{\Delta u \to 0} \frac{\Delta y}{\Delta u} \cdot \lim_{\Delta x \to 0} \frac{\Delta u}{\Delta x} \qquad \text{(Note that } \Delta u \to 0 \text{ as } \Delta x \to 0 \text{ since } g \text{ is continuous.)}$$

$$= \frac{dy}{du} \frac{du}{dx}$$

The only flaw in this reasoning is that in $\boxed{1}$ it might happen that $\Delta u = 0$ (even when $\Delta x \neq 0$) and, of course, we can't divide by 0. Nonetheless, this reasoning does at least *suggest* that the Chain Rule is true. A full proof of the Chain Rule is given at the end of this section.

The Chain Rule can be written either in the prime notation

2
$$(f \circ g)'(x) = f'(g(x)) \cdot g'(x)$$

or, if $y = f(u)$ and $u = g(x)$, in Leibniz notation:

3
$$\frac{dy}{dx} = \frac{dy}{du} \frac{du}{dx}$$

Equation 3 is easy to remember because if $dy/du$ and $du/dx$ were quotients, then we could cancel $du$. Remember, however, that $du$ has not been defined and $du/dx$ should not be thought of as an actual quotient.

**EXAMPLE 1**   Find $F'(x)$ if $F(x) = \sqrt{x^2 + 1}$.

SOLUTION 1   (using Equation 2):   At the beginning of this section we expressed $F$ as $F(x) = (f \circ g)(x) = f(g(x))$ where $f(u) = \sqrt{u}$ and $g(x) = x^2 + 1$. Since

$$f'(u) = \tfrac{1}{2}u^{-1/2} = \frac{1}{2\sqrt{u}} \qquad \text{and} \qquad g'(x) = 2x$$

we have
$$F'(x) = f'(g(x)) \cdot g'(x)$$

$$= \frac{1}{2\sqrt{x^2 + 1}} \cdot 2x = \frac{x}{\sqrt{x^2 + 1}}$$

SOLUTION 2 (using Equation 3): If we let $u = x^2 + 1$ and $y = \sqrt{u}$, then

$$F'(x) = \frac{dy}{du}\frac{du}{dx} = \frac{1}{2\sqrt{u}}(2x) = \frac{1}{2\sqrt{x^2+1}}(2x) = \frac{x}{\sqrt{x^2+1}}$$

When using Formula 3 we should bear in mind that $dy/dx$ refers to the derivative of $y$ when $y$ is considered as a function of $x$ (called the *derivative of y with respect to x*), whereas $dy/du$ refers to the derivative of $y$ when considered as a function of $u$ (the derivative of $y$ with respect to $u$). For instance, in Example 1, $y$ can be considered as a function of $x$ $\left( y = \sqrt{x^2+1} \right)$ and also as a function of $u$ $\left( y = \sqrt{u} \right)$. Note that

$$\frac{dy}{dx} = F'(x) = \frac{x}{\sqrt{x^2+1}} \qquad \text{whereas} \qquad \frac{dy}{du} = f'(u) = \frac{1}{2\sqrt{u}}$$

**NOTE** In using the Chain Rule we work from the outside to the inside. Formula 2 says that *we differentiate the outer function $f$ [at the inner function $g(x)$] and then we multiply by the derivative of the inner function.*

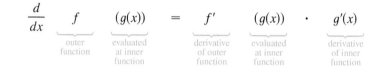

**EXAMPLE 2** Differentiate (a) $y = \sin(x^2)$ and (b) $y = \sin^2 x$.

SOLUTION

(a) If $y = \sin(x^2)$, then the outer function is the sine function and the inner function is the squaring function, so the Chain Rule gives

$$= 2x \cos(x^2)$$

(b) Note that $\sin^2 x = (\sin x)^2$. Here the outer function is the squaring function and the inner function is the sine function. So

The answer can be left as $2 \sin x \cos x$ or written as $\sin 2x$ (by a trigonometric identity known as the double-angle formula).

See Reference Page 2 or Appendix D.

In Example 2(a) we combined the Chain Rule with the rule for differentiating the sine function. In general, if $y = \sin u$, where $u$ is a differentiable function of $x$, then, by the Chain Rule,

$$\frac{dy}{dx} = \frac{dy}{du}\frac{du}{dx} = \cos u \, \frac{du}{dx}$$

Thus

$$\frac{d}{dx}(\sin u) = \cos u \, \frac{du}{dx}$$

In a similar fashion, all of the formulas for differentiating trigonometric functions can be combined with the Chain Rule.

Let's make explicit the special case of the Chain Rule where the outer function $f$ is a power function. If $y = [g(x)]^n$, then we can write $y = f(u) = u^n$ where $u = g(x)$. By using the Chain Rule and then the Power Rule, we get

$$\frac{dy}{dx} = \frac{dy}{du}\frac{du}{dx} = nu^{n-1}\frac{du}{dx} = n[g(x)]^{n-1}g'(x)$$

---

**4**   **The Power Rule Combined with the Chain Rule**   If $n$ is any real number and $u = g(x)$ is differentiable, then

$$\frac{d}{dx}(u^n) = nu^{n-1}\frac{du}{dx}$$

Alternatively,     $\dfrac{d}{dx}[g(x)]^n = n[g(x)]^{n-1} \cdot g'(x)$

---

Notice that the derivative in Example 1 could be calculated by taking $n = \frac{1}{2}$ in Rule 4.

**EXAMPLE 3**   Differentiate $y = (x^3 - 1)^{100}$.

SOLUTION   Taking $u = g(x) = x^3 - 1$ and $n = 100$ in $\boxed{4}$, we have

$$\frac{dy}{dx} = \frac{d}{dx}(x^3 - 1)^{100} = 100(x^3 - 1)^{99}\frac{d}{dx}(x^3 - 1)$$

$$= 100(x^3 - 1)^{99} \cdot 3x^2 = 300x^2(x^3 - 1)^{99}$$

**V   EXAMPLE 4**   Find $f'(x)$ if $f(x) = \dfrac{1}{\sqrt[3]{x^2 + x + 1}}$.

SOLUTION   First rewrite $f$:     $f(x) = (x^2 + x + 1)^{-1/3}$

Thus     $f'(x) = -\frac{1}{3}(x^2 + x + 1)^{-4/3}\dfrac{d}{dx}(x^2 + x + 1)$

$$= -\tfrac{1}{3}(x^2 + x + 1)^{-4/3}(2x + 1)$$

**EXAMPLE 5**   Find the derivative of the function

$$g(t) = \left(\frac{t - 2}{2t + 1}\right)^9$$

SOLUTION   Combining the Power Rule, Chain Rule, and Quotient Rule, we get

$$g'(t) = 9\left(\frac{t - 2}{2t + 1}\right)^8\frac{d}{dt}\left(\frac{t - 2}{2t + 1}\right)$$

$$= 9\left(\frac{t - 2}{2t + 1}\right)^8\frac{(2t + 1) \cdot 1 - 2(t - 2)}{(2t + 1)^2} = \frac{45(t - 2)^8}{(2t + 1)^{10}}$$

The graphs of the functions $y$ and $y'$ in Example 6 are shown in Figure 1. Notice that $y'$ is large when $y$ increases rapidly and $y' = 0$ when $y$ has a horizontal tangent. So our answer appears to be reasonable.

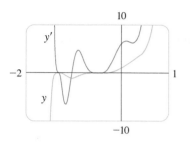

**FIGURE 1**

**EXAMPLE 6** Differentiate $y = (2x + 1)^5(x^3 - x + 1)^4$.

SOLUTION In this example we must use the Product Rule before using the Chain Rule:

$$\frac{dy}{dx} = (2x + 1)^5 \frac{d}{dx}(x^3 - x + 1)^4 + (x^3 - x + 1)^4 \frac{d}{dx}(2x + 1)^5$$

$$= (2x + 1)^5 \cdot 4(x^3 - x + 1)^3 \frac{d}{dx}(x^3 - x + 1)$$

$$+ (x^3 - x + 1)^4 \cdot 5(2x + 1)^4 \frac{d}{dx}(2x + 1)$$

$$= 4(2x + 1)^5(x^3 - x + 1)^3(3x^2 - 1) + 5(x^3 - x + 1)^4(2x + 1)^4 \cdot 2$$

Noticing that each term has the common factor $2(2x + 1)^4(x^3 - x + 1)^3$, we could factor it out and write the answer as

$$\frac{dy}{dx} = 2(2x + 1)^4(x^3 - x + 1)^3(17x^3 + 6x^2 - 9x + 3) \qquad \blacksquare$$

The reason for the name "Chain Rule" becomes clear when we make a longer chain by adding another link. Suppose that $y = f(u)$, $u = g(x)$, and $x = h(t)$, where $f$, $g$, and $h$ are differentiable functions. Then, to compute the derivative of $y$ with respect to $t$, we use the Chain Rule twice:

$$\frac{dy}{dt} = \frac{dy}{dx}\frac{dx}{dt} = \frac{dy}{du}\frac{du}{dx}\frac{dx}{dt}$$

**V EXAMPLE 7** If $f(x) = \sin(\cos(\tan x))$, then

$$f'(x) = \cos(\cos(\tan x)) \frac{d}{dx}\cos(\tan x)$$

$$= \cos(\cos(\tan x))[-\sin(\tan x)]\frac{d}{dx}(\tan x)$$

$$= -\cos(\cos(\tan x))\sin(\tan x)\sec^2 x$$

Notice that we used the Chain Rule twice.   $\blacksquare$

**EXAMPLE 8** Differentiate $y = \sqrt{\sec x^3}$.

SOLUTION Here the outer function is the square root function, the middle function is the secant function, and the inner function is the cubing function. So we have

$$\frac{dy}{dx} = \frac{1}{2\sqrt{\sec x^3}}\frac{d}{dx}(\sec x^3)$$

$$= \frac{1}{2\sqrt{\sec x^3}}\sec x^3 \tan x^3 \frac{d}{dx}(x^3)$$

$$= \frac{3x^2 \sec x^3 \tan x^3}{2\sqrt{\sec x^3}} \qquad \blacksquare$$

### How to Prove the Chain Rule

Recall that if $y = f(x)$ and $x$ changes from $a$ to $a + \Delta x$, we define the increment of $y$ as

$$\Delta y = f(a + \Delta x) - f(a)$$

According to the definition of a derivative, we have

$$\lim_{\Delta x \to 0} \frac{\Delta y}{\Delta x} = f'(a)$$

So if we denote by $\varepsilon$ the difference between the difference quotient and the derivative, we obtain

$$\lim_{\Delta x \to 0} \varepsilon = \lim_{\Delta x \to 0} \left( \frac{\Delta y}{\Delta x} - f'(a) \right) = f'(a) - f'(a) = 0$$

But

$$\varepsilon = \frac{\Delta y}{\Delta x} - f'(a) \quad \Rightarrow \quad \Delta y = f'(a)\, \Delta x + \varepsilon\, \Delta x$$

If we define $\varepsilon$ to be 0 when $\Delta x = 0$, then $\varepsilon$ becomes a continuous function of $\Delta x$. Thus, for a differentiable function $f$, we can write

$$\boxed{5} \qquad \Delta y = f'(a)\, \Delta x + \varepsilon\, \Delta x \qquad \text{where} \quad \varepsilon \to 0 \text{ as } \Delta x \to 0$$

and $\varepsilon$ is a continuous function of $\Delta x$. This property of differentiable functions is what enables us to prove the Chain Rule.

PROOF OF THE CHAIN RULE   Suppose $u = g(x)$ is differentiable at $a$ and $y = f(u)$ is differentiable at $b = g(a)$. If $\Delta x$ is an increment in $x$ and $\Delta u$ and $\Delta y$ are the corresponding increments in $u$ and $y$, then we can use Equation 5 to write

$$\boxed{6} \qquad \Delta u = g'(a)\, \Delta x + \varepsilon_1\, \Delta x = [g'(a) + \varepsilon_1]\, \Delta x$$

where $\varepsilon_1 \to 0$ as $\Delta x \to 0$. Similarly

$$\boxed{7} \qquad \Delta y = f'(b)\, \Delta u + \varepsilon_2\, \Delta u = [f'(b) + \varepsilon_2]\, \Delta u$$

where $\varepsilon_2 \to 0$ as $\Delta u \to 0$. If we now substitute the expression for $\Delta u$ from Equation 6 into Equation 7, we get

$$\Delta y = [f'(b) + \varepsilon_2][g'(a) + \varepsilon_1]\, \Delta x$$

so

$$\frac{\Delta y}{\Delta x} = [f'(b) + \varepsilon_2][g'(a) + \varepsilon_1]$$

As $\Delta x \to 0$, Equation 6 shows that $\Delta u \to 0$. So both $\varepsilon_1 \to 0$ and $\varepsilon_2 \to 0$ as $\Delta x \to 0$. Therefore

$$\frac{dy}{dx} = \lim_{\Delta x \to 0} \frac{\Delta y}{\Delta x} = \lim_{\Delta x \to 0} [f'(b) + \varepsilon_2][g'(a) + \varepsilon_1]$$

$$= f'(b)\, g'(a) = f'(g(a))\, g'(a)$$

This proves the Chain Rule.

**Exercises**

**1–6** Write the composite function in the form $f(g(x))$. [Identify the inner function $u = g(x)$ and the outer function $y = f(u)$.] Then find the derivative $dy/dx$.

**1.** $y = \sqrt[3]{1 + 4x}$

**2.** $y = (2x^3 + 5)^4$

**3.** $y = \tan \pi x$

**4.** $y = \sin(\cot x)$

**5.** $y = \sqrt{\sin x}$

**6.** $y = \sin \sqrt{x}$

**7–46** Find the derivative of the function.

**7.** $F(x) = (x^4 + 3x^2 - 2)^5$

**8.** $F(x) = (4x - x^2)^{100}$

**9.** $F(x) = \sqrt{1 - 2x}$

**10.** $f(x) = \dfrac{1}{(1 + \sec x)^2}$

**11.** $f(z) = \dfrac{1}{z^2 + 1}$

**12.** $f(t) = \sqrt[3]{1 + \tan t}$

**13.** $y = \cos(a^3 + x^3)$

**14.** $y = a^3 + \cos^3 x$

**15.** $y = x \sec kx$

**16.** $y = 3 \cot n\theta$

**17.** $f(x) = (2x - 3)^4 (x^2 + x + 1)^5$

**18.** $g(x) = (x^2 + 1)^3 (x^2 + 2)^6$

**19.** $h(t) = (t + 1)^{2/3} (2t^2 - 1)^3$

**20.** $F(t) = (3t - 1)^4 (2t + 1)^{-3}$

**21.** $y = \left(\dfrac{x^2 + 1}{x^2 - 1}\right)^3$

**22.** $f(s) = \sqrt{\dfrac{s^2 + 1}{s^2 + 4}}$

**23.** $y = \sin(x \cos x)$

**24.** $f(x) = \dfrac{x}{\sqrt{7 - 3x}}$

**25.** $F(z) = \sqrt{\dfrac{z - 1}{z + 1}}$

**26.** $G(y) = \dfrac{(y - 1)^4}{(y^2 + 2y)^5}$

**27.** $y = \dfrac{r}{\sqrt{r^2 + 1}}$

**28.** $y = \dfrac{\cos \pi x}{\sin \pi x + \cos \pi x}$

**29.** $y = \sin \sqrt{1 + x^2}$

**30.** $F(v) = \left(\dfrac{v}{v^3 + 1}\right)^6$

**31.** $y = \sin(\tan 2x)$

**32.** $y = \sec^2(m\theta)$

**33.** $y = \sec^2 x + \tan^2 x$

**34.** $y = x \sin \dfrac{1}{x}$

**35.** $y = \left(\dfrac{1 - \cos 2x}{1 + \cos 2x}\right)^4$

**36.** $f(t) = \sqrt{\dfrac{t}{t^2 + 4}}$

**37.** $y = \cot^2(\sin \theta)$

**38.** $y = \left(ax + \sqrt{x^2 + b^2}\right)^{-2}$

**39.** $y = [x^2 + (1 - 3x)^5]^3$

**40.** $y = \sin(\sin(\sin x))$

**41.** $y = \sqrt{x + \sqrt{x}}$

**42.** $y = \sqrt{x + \sqrt{x + \sqrt{x}}}$

**43.** $g(x) = (2r \sin rx + n)^p$

**44.** $y = \cos^4(\sin^3 x)$

**45.** $y = \cos \sqrt{\sin(\tan \pi x)}$

**46.** $y = [x + (x + \sin^2 x)^3]^4$

**47–50** Find the first and second derivatives of the function.

**47.** $y = \cos(x^2)$

**48.** $y = \cos^2 x$

**49.** $H(t) = \tan 3t$

**50.** $y = \dfrac{4x}{\sqrt{x + 1}}$

**51–54** Find an equation of the tangent line to the curve at the given point.

**51.** $y = (1 + 2x)^{10}$,   $(0, 1)$

**52.** $y = \sqrt{1 + x^3}$,   $(2, 3)$

**53.** $y = \sin(\sin x)$,   $(\pi, 0)$

**54.** $y = \sin x + \sin^2 x$,   $(0, 0)$

**55.** (a) Find an equation of the tangent line to the curve $y = \tan(\pi x^2/4)$ at the point $(1, 1)$.

(b) Illustrate part (a) by graphing the curve and the tangent line on the same screen.

**56.** (a) The curve $y = |x|/\sqrt{2 - x^2}$ is called a *bullet-nose curve*. Find an equation of the tangent line to this curve at the point $(1, 1)$.

(b) Illustrate part (a) by graphing the curve and the tangent line on the same screen.

**57.** (a) If $f(x) = x\sqrt{2 - x^2}$, find $f'(x)$.

(b) Check to see that your answer to part (a) is reasonable by comparing the graphs of $f$ and $f'$.

**58.** The function $f(x) = \sin(x + \sin 2x)$, $0 \le x \le \pi$, arises in applications to frequency modulation (FM) synthesis.

(a) Use a graph of $f$ produced by a graphing device to make a rough sketch of the graph of $f'$.

(b) Calculate $f'(x)$ and use this expression, with a graphing device, to graph $f'$. Compare with your sketch in part (a).

**59.** Find all points on the graph of the function $f(x) = 2 \sin x + \sin^2 x$ at which the tangent line is horizontal.

**60.** Find the $x$-coordinates of all points on the curve $y = \sin 2x - 2 \sin x$ at which the tangent line is horizontal.

**61.** If $F(x) = f(g(x))$, where $f(-2) = 8$, $f'(-2) = 4$, $f'(5) = 3$, $g(5) = -2$, and $g'(5) = 6$, find $F'(5)$.

**62.** If $h(x) = \sqrt{4 + 3f(x)}$, where $f(1) = 7$ and $f'(1) = 4$, find $h'(1)$.

**63.** A table of values for $f$, $g$, $f'$, and $g'$ is given.

| $x$ | $f(x)$ | $g(x)$ | $f'(x)$ | $g'(x)$ |
|---|---|---|---|---|
| 1 | 3 | 2 | 4 | 6 |
| 2 | 1 | 8 | 5 | 7 |
| 3 | 7 | 2 | 7 | 9 |

(a) If $h(x) = f(g(x))$, find $h'(1)$.

(b) If $H(x) = g(f(x))$, find $H'(1)$.

---

Graphing calculator or computer required      CAS Computer algebra system required      **1.** Homework Hints available at stewartcalculus.com

**64.** Let $f$ and $g$ be the functions in Exercise 63.
(a) If $F(x) = f(f(x))$, find $F'(2)$.
(b) If $G(x) = g(g(x))$, find $G'(3)$.

**65.** If $f$ and $g$ are the functions whose graphs are shown, let $u(x) = f(g(x))$, $v(x) = g(f(x))$, and $w(x) = g(g(x))$. Find each derivative, if it exists. If it does not exist, explain why.
(a) $u'(1)$     (b) $v'(1)$     (c) $w'(1)$

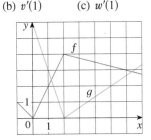

**66.** If $f$ is the function whose graph is shown, let $h(x) = f(f(x))$ and $g(x) = f(x^2)$. Use the graph of $f$ to estimate the value of each derivative.
(a) $h'(2)$     (b) $g'(2)$

**67.** If $g(x) = \sqrt{f(x)}$, where the graph of $f$ is shown, evaluate $g'(3)$.

**68.** Suppose $f$ is differentiable on $\mathbb{R}$ and $\alpha$ is a real number. Let $F(x) = f(x^\alpha)$ and $G(x) = [f(x)]^\alpha$. Find expressions for (a) $F'(x)$ and (b) $G'(x)$.

**69.** Let $r(x) = f(g(h(x)))$, where $h(1) = 2$, $g(2) = 3$, $h'(1) = 4$, $g'(2) = 5$, and $f'(3) = 6$. Find $r'(1)$.

**70.** If $g$ is a twice differentiable function and $f(x) = xg(x^2)$, find $f''$ in terms of $g$, $g'$, and $g''$.

**71.** If $F(x) = f(3f(4f(x)))$, where $f(0) = 0$ and $f'(0) = 2$, find $F'(0)$.

**72.** If $F(x) = f(xf(xf(x)))$, where $f(1) = 2$, $f(2) = 3$, $f'(1) = 4$, $f'(2) = 5$, and $f'(3) = 6$, find $F'(1)$.

**73–74** Find the given derivative by finding the first few derivatives and observing the pattern that occurs.

**73.** $D^{103}\cos 2x$

**74.** $D^{35} x \sin \pi x$

**75.** The displacement of a particle on a vibrating string is given by the equation $s(t) = 10 + \frac{1}{4}\sin(10\pi t)$ where $s$ is measured in centimeters and $t$ in seconds. Find the velocity of the particle after $t$ seconds.

**76.** If the equation of motion of a particle is given by $s = A\cos(\omega t + \delta)$, the particle is said to undergo *simple harmonic motion.*
(a) Find the velocity of the particle at time $t$.
(b) When is the velocity 0?

**77.** A Cepheid variable star is a star whose brightness alternately increases and decreases. The most easily visible such star is Delta Cephei, for which the interval between times of maximum brightness is 5.4 days. The average brightness of this star is 4.0 and its brightness changes by $\pm 0.35$. In view of these data, the brightness of Delta Cephei at time $t$, where $t$ is measured in days, has been modeled by the function

$$B(t) = 4.0 + 0.35\sin\left(\frac{2\pi t}{5.4}\right)$$

(a) Find the rate of change of the brightness after $t$ days.
(b) Find, correct to two decimal places, the rate of increase after one day.

**78.** In Example 4 in Section 1.3 we arrived at a model for the length of daylight (in hours) in Philadelphia on the $t$th day of the year:

$$L(t) = 12 + 2.8\sin\left[\frac{2\pi}{365}(t - 80)\right]$$

Use this model to compare how the number of hours of daylight is increasing in Philadelphia on March 21 and May 21.

**79.** A particle moves along a straight line with displacement $s(t)$, velocity $v(t)$, and acceleration $a(t)$. Show that

$$a(t) = v(t)\frac{dv}{ds}$$

Explain the difference between the meanings of the derivatives $dv/dt$ and $dv/ds$.

**80.** Air is being pumped into a spherical weather balloon. At any time $t$, the volume of the balloon is $V(t)$ and its radius is $r(t)$.
(a) What do the derivatives $dV/dr$ and $dV/dt$ represent?
(b) Express $dV/dt$ in terms of $dr/dt$.

**[CAS] 81.** Computer algebra systems have commands that differentiate functions, but the form of the answer may not be convenient and so further commands may be necessary to simplify the answer.
(a) Use a CAS to find the derivative in Example 5 and compare with the answer in that example. Then use the simplify command and compare again.
(b) Use a CAS to find the derivative in Example 6. What happens if you use the simplify command? What happens if you use the factor command? Which form of the answer would be best for locating horizontal tangents?

CAS **82.** (a) Use a CAS to differentiate the function

$$f(x) = \sqrt{\frac{x^4 - x + 1}{x^4 + x + 1}}$$

and to simplify the result.
(b) Where does the graph of $f$ have horizontal tangents?
(c) Graph $f$ and $f'$ on the same screen. Are the graphs consistent with your answer to part (b)?

**83.** Use the Chain Rule to prove the following.
(a) The derivative of an even function is an odd function.
(b) The derivative of an odd function is an even function.

**84.** Use the Chain Rule and the Product Rule to give an alternative proof of the Quotient Rule.
[*Hint:* Write $f(x)/g(x) = f(x)[g(x)]^{-1}$.]

**85.** (a) If $n$ is a positive integer, prove that

$$\frac{d}{dx}(\sin^n x \cos nx) = n \sin^{n-1}x \cos(n + 1)x$$

(b) Find a formula for the derivative of $y = \cos^n x \cos nx$ that is similar to the one in part (a).

**86.** Suppose $y = f(x)$ is a curve that always lies above the $x$-axis and never has a horizontal tangent, where $f$ is differentiable everywhere. For what value of $y$ is the rate of change of $y^5$ with respect to $x$ eighty times the rate of change of $y$ with respect to $x$?

**87.** Use the Chain Rule to show that if $\theta$ is measured in degrees, then

$$\frac{d}{d\theta}(\sin\theta) = \frac{\pi}{180}\cos\theta$$

(This gives one reason for the convention that radian measure is always used when dealing with trigonometric functions in calculus: The differentiation formulas would not be as simple if we used degree measure.)

**88.** (a) Write $|x| = \sqrt{x^2}$ and use the Chain Rule to show that

$$\frac{d}{dx}|x| = \frac{x}{|x|}$$

(b) If $f(x) = |\sin x|$, find $f'(x)$ and sketch the graphs of $f$ and $f'$. Where is $f$ not differentiable?
(c) If $g(x) = \sin|x|$, find $g'(x)$ and sketch the graphs of $g$ and $g'$. Where is $g$ not differentiable?

**89.** If $y = f(u)$ and $u = g(x)$, where $f$ and $g$ are twice differentiable functions, show that

$$\frac{d^2y}{dx^2} = \frac{d^2y}{du^2}\left(\frac{du}{dx}\right)^2 + \frac{dy}{du}\frac{d^2u}{dx^2}$$

**90.** If $y = f(u)$ and $u = g(x)$, where $f$ and $g$ possess third derivatives, find a formula for $d^3y/dx^3$ similar to the one given in Exercise 89.

---

APPLIED PROJECT   **WHERE SHOULD A PILOT START DESCENT?**

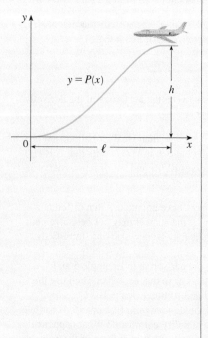

$y = P(x)$

An approach path for an aircraft landing is shown in the figure and satisfies the following conditions:

(i) The cruising altitude is $h$ when descent starts at a horizontal distance $\ell$ from touchdown at the origin.

(ii) The pilot must maintain a constant horizontal speed $v$ throughout descent.

(iii) The absolute value of the vertical acceleration should not exceed a constant $k$ (which is much less than the acceleration due to gravity).

**1.** Find a cubic polynomial $P(x) = ax^3 + bx^2 + cx + d$ that satisfies condition (i) by imposing suitable conditions on $P(x)$ and $P'(x)$ at the start of descent and at touchdown.

**2.** Use conditions (ii) and (iii) to show that

$$\frac{6hv^2}{\ell^2} \le k$$

**3.** Suppose that an airline decides not to allow vertical acceleration of a plane to exceed $k = 860$ mi/h$^2$. If the cruising altitude of a plane is 35,000 ft and the speed is 300 mi/h, how far away from the airport should the pilot start descent?

📈 **4.** Graph the approach path if the conditions stated in Problem 3 are satisfied.

---

📈 Graphing calculator or computer required

## 2.6  Implicit Differentiation

The functions that we have met so far can be described by expressing one variable explicitly in terms of another variable—for example,

$$y = \sqrt{x^3 + 1} \qquad \text{or} \qquad y = x \sin x$$

or, in general, $y = f(x)$. Some functions, however, are defined implicitly by a relation between $x$ and $y$ such as

$$\boxed{1} \qquad\qquad x^2 + y^2 = 25$$

or

$$\boxed{2} \qquad\qquad x^3 + y^3 = 6xy$$

In some cases it is possible to solve such an equation for $y$ as an explicit function (or several functions) of $x$. For instance, if we solve Equation 1 for $y$, we get $y = \pm\sqrt{25 - x^2}$, so two of the functions determined by the implicit Equation 1 are $f(x) = \sqrt{25 - x^2}$ and $g(x) = -\sqrt{25 - x^2}$. The graphs of $f$ and $g$ are the upper and lower semicircles of the circle $x^2 + y^2 = 25$. (See Figure 1.)

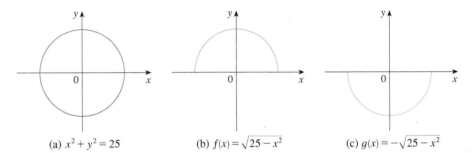

**FIGURE 1**    (a) $x^2 + y^2 = 25$        (b) $f(x) = \sqrt{25 - x^2}$        (c) $g(x) = -\sqrt{25 - x^2}$

It's not easy to solve Equation 2 for $y$ explicitly as a function of $x$ by hand. (A computer algebra system has no trouble, but the expressions it obtains are very complicated.) Nonetheless, $\boxed{2}$ is the equation of a curve called the **folium of Descartes** shown in Figure 2 and it implicitly defines $y$ as several functions of $x$. The graphs of three such functions are shown in Figure 3. When we say that $f$ is a function defined implicitly by Equation 2, we mean that the equation

$$x^3 + [f(x)]^3 = 6xf(x)$$

is true for all values of $x$ in the domain of $f$.

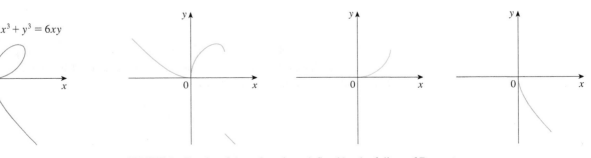

**FIGURE 2**  The folium of Descartes        **FIGURE 3**  Graphs of three functions defined by the folium of Descartes

Fortunately, we don't need to solve an equation for $y$ in terms of $x$ in order to find the derivative of $y$. Instead we can use the method of **implicit differentiation**. This consists of differentiating both sides of the equation with respect to $x$ and then solving the resulting equation for $y'$. In the examples and exercises of this section it is always assumed that the given equation determines $y$ implicitly as a differentiable function of $x$ so that the method of implicit differentiation can be applied.

**V   EXAMPLE 1**

(a) If $x^2 + y^2 = 25$, find $\dfrac{dy}{dx}$.

(b) Find an equation of the tangent to the circle $x^2 + y^2 = 25$ at the point $(3, 4)$.

SOLUTION 1

(a) Differentiate both sides of the equation $x^2 + y^2 = 25$:

$$\frac{d}{dx}(x^2 + y^2) = \frac{d}{dx}(25)$$

$$\frac{d}{dx}(x^2) + \frac{d}{dx}(y^2) = 0$$

Remembering that $y$ is a function of $x$ and using the Chain Rule, we have

$$\frac{d}{dx}(y^2) = \frac{d}{dy}(y^2)\frac{dy}{dx} = 2y\frac{dy}{dx}$$

Thus                              $2x + 2y\dfrac{dy}{dx} = 0$

Now we solve this equation for $dy/dx$:

$$\frac{dy}{dx} = -\frac{x}{y}$$

(b) At the point $(3, 4)$ we have $x = 3$ and $y = 4$, so

$$\frac{dy}{dx} = -\frac{3}{4}$$

An equation of the tangent to the circle at $(3, 4)$ is therefore

$$y - 4 = -\tfrac{3}{4}(x - 3) \qquad \text{or} \qquad 3x + 4y = 25$$

SOLUTION 2

(b) Solving the equation $x^2 + y^2 = 25$, we get $y = \pm\sqrt{25 - x^2}$. The point $(3, 4)$ lies on the upper semicircle $y = \sqrt{25 - x^2}$ and so we consider the function $f(x) = \sqrt{25 - x^2}$. Differentiating $f$ using the Chain Rule, we have

$$f'(x) = \tfrac{1}{2}(25 - x^2)^{-1/2}\frac{d}{dx}(25 - x^2)$$

$$= \tfrac{1}{2}(25 - x^2)^{-1/2}(-2x) = -\frac{x}{\sqrt{25 - x^2}}$$

Example 1 illustrates that even when it is possible to solve an equation explicitly for $y$ in terms of $x$, it may be easier to use implicit differentiation.

So

$$f'(3) = -\frac{3}{\sqrt{25 - 3^2}} = -\frac{3}{4}$$

and, as in Solution 1, an equation of the tangent is $3x + 4y = 25$.

**NOTE 1** The expression $dy/dx = -x/y$ in Solution 1 gives the derivative in terms of both $x$ and $y$. It is correct no matter which function $y$ is determined by the given equation. For instance, for $y = f(x) = \sqrt{25 - x^2}$ we have

$$\frac{dy}{dx} = -\frac{x}{y} = -\frac{x}{\sqrt{25 - x^2}}$$

whereas for $y = g(x) = -\sqrt{25 - x^2}$ we have

$$\frac{dy}{dx} = -\frac{x}{y} = -\frac{x}{-\sqrt{25 - x^2}} = \frac{x}{\sqrt{25 - x^2}}$$

**V EXAMPLE 2**

(a)  Find $y'$ if $x^3 + y^3 = 6xy$.
(b)  Find the tangent to the folium of Descartes $x^3 + y^3 = 6xy$ at the point $(3, 3)$.
(c)  At what point in the first quadrant is the tangent line horizontal?

SOLUTION

(a)  Differentiating both sides of $x^3 + y^3 = 6xy$ with respect to $x$, regarding $y$ as a function of $x$, and using the Chain Rule on the term $y^3$ and the Product Rule on the term $6xy$, we get

$$3x^2 + 3y^2y' = 6xy' + 6y$$

or

$$x^2 + y^2y' = 2xy' + 2y$$

We now solve for $y'$:

$$y^2y' - 2xy' = 2y - x^2$$

$$(y^2 - 2x)y' = 2y - x^2$$

$$y' = \frac{2y - x^2}{y^2 - 2x}$$

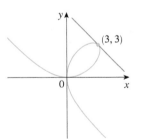

(b)  When $x = y = 3$,

$$y' = \frac{2 \cdot 3 - 3^2}{3^2 - 2 \cdot 3} = -1$$

**FIGURE 4**

and a glance at Figure 4 confirms that this is a reasonable value for the slope at $(3, 3)$. So an equation of the tangent to the folium at $(3, 3)$ is

$$y - 3 = -1(x - 3) \qquad \text{or} \qquad x + y = 6$$

(c)  The tangent line is horizontal if $y' = 0$. Using the expression for $y'$ from part (a), we see that $y' = 0$ when $2y - x^2 = 0$ (provided that $y^2 - 2x \neq 0$). Substituting $y = \frac{1}{2}x^2$ in the equation of the curve, we get

$$x^3 + \left(\tfrac{1}{2}x^2\right)^3 = 6x\left(\tfrac{1}{2}x^2\right)$$

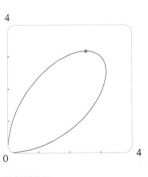

which simplifies to $x^6 = 16x^3$. Since $x \neq 0$ in the first quadrant, we have $x^3 = 16$. If $x = 16^{1/3} = 2^{4/3}$, then $y = \frac{1}{2}(2^{8/3}) = 2^{5/3}$. Thus the tangent is horizontal at $(2^{4/3}, 2^{5/3})$, which is approximately $(2.5198, 3.1748)$. Looking at Figure 5, we see that our answer is reasonable.

**FIGURE 5**

**NOTE 2** There is a formula for the three roots of a cubic equation that is like the quadratic formula but much more complicated. If we use this formula (or a computer algebra system) to solve the equation $x^3 + y^3 = 6xy$ for $y$ in terms of $x$, we get three functions determined by the equation:

$$y = f(x) = \sqrt[3]{-\tfrac{1}{2}x^3 + \sqrt{\tfrac{1}{4}x^6 - 8x^3}} + \sqrt[3]{-\tfrac{1}{2}x^3 - \sqrt{\tfrac{1}{4}x^6 - 8x^3}}$$

and

$$y = \tfrac{1}{2}\left[ -f(x) \pm \sqrt{-3}\left( \sqrt[3]{-\tfrac{1}{2}x^3 + \sqrt{\tfrac{1}{4}x^6 - 8x^3}} - \sqrt[3]{-\tfrac{1}{2}x^3 - \sqrt{\tfrac{1}{4}x^6 - 8x^3}} \right) \right]$$

(These are the three functions whose graphs are shown in Figure 3.) You can see that the method of implicit differentiation saves an enormous amount of work in cases such as this. Moreover, implicit differentiation works just as easily for equations such as

$$y^5 + 3x^2y^2 + 5x^4 = 12$$

for which it is *impossible* to find a similar expression for $y$ in terms of $x$.

**EXAMPLE 3** Find $y'$ if $\sin(x + y) = y^2 \cos x$.

SOLUTION Differentiating implicitly with respect to $x$ and remembering that $y$ is a function of $x$, we get

$$\cos(x + y) \cdot (1 + y') = y^2(-\sin x) + (\cos x)(2yy')$$

(Note that we have used the Chain Rule on the left side and the Product Rule and Chain Rule on the right side.) If we collect the terms that involve $y'$, we get

$$\cos(x + y) + y^2 \sin x = (2y \cos x)y' - \cos(x + y) \cdot y'$$

So

$$y' = \frac{y^2 \sin x + \cos(x + y)}{2y \cos x - \cos(x + y)}$$

Figure 6, drawn with the implicit-plotting command of a computer algebra system, shows part of the curve $\sin(x + y) = y^2 \cos x$. As a check on our calculation, notice that $y' = -1$ when $x = y = 0$ and it appears from the graph that the slope is approximately $-1$ at the origin.

Figures 7, 8, and 9 show three more curves produced by a computer algebra system with an implicit-plotting command. In Exercises 41–42 you will have an opportunity to create and examine unusual curves of this nature.

**Abel and Galois**

The Norwegian mathematician Niels Abel proved in 1824 that no general formula can be given for the roots of a fifth-degree equation in terms of radicals. Later the French mathematician Evariste Galois proved that it is impossible to find a general formula for the roots of an $n$th-degree equation (in terms of algebraic operations on the coefficients) if $n$ is any integer larger than 4.

**FIGURE 6**

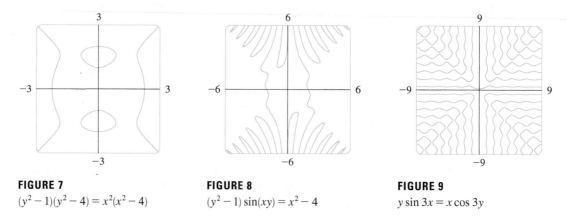

**FIGURE 7**
$(y^2 - 1)(y^2 - 4) = x^2(x^2 - 4)$

**FIGURE 8**
$(y^2 - 1)\sin(xy) = x^2 - 4$

**FIGURE 9**
$y \sin 3x = x \cos 3y$

The following example shows how to find the second derivative of a function that is defined implicitly.

**EXAMPLE 4** Find $y''$ if $x^4 + y^4 = 16$.

SOLUTION Differentiating the equation implicitly with respect to $x$, we get

$$4x^3 + 4y^3 y' = 0$$

Solving for $y'$ gives

$$\boxed{3} \qquad y' = -\frac{x^3}{y^3}$$

Figure 10 shows the graph of the curve $x^4 + y^4 = 16$ of Example 4. Notice that it's a stretched and flattened version of the circle $x^2 + y^2 = 4$. For this reason it's sometimes called a *fat circle*. It starts out very steep on the left but quickly becomes very flat. This can be seen from the expression

$$y' = -\frac{x^3}{y^3} = -\left(\frac{x}{y}\right)^3$$

To find $y''$ we differentiate this expression for $y'$ using the Quotient Rule and remembering that $y$ is a function of $x$:

$$y'' = \frac{d}{dx}\left(-\frac{x^3}{y^3}\right) = -\frac{y^3\,(d/dx)(x^3) - x^3\,(d/dx)(y^3)}{(y^3)^2}$$

$$= -\frac{y^3 \cdot 3x^2 - x^3(3y^2 y')}{y^6}$$

If we now substitute Equation 3 into this expression, we get

$$y'' = -\frac{3x^2 y^3 - 3x^3 y^2\left(-\dfrac{x^3}{y^3}\right)}{y^6}$$

$$= -\frac{3(x^2 y^4 + x^6)}{y^7} = -\frac{3x^2(y^4 + x^4)}{y^7}$$

But the values of $x$ and $y$ must satisfy the original equation $x^4 + y^4 = 16$. So the answer simplifies to

$$y'' = -\frac{3x^2(16)}{y^7} = -48\,\frac{x^2}{y^7}$$

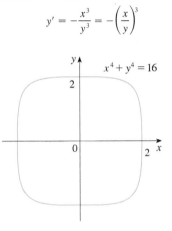

**FIGURE 10**

---

## 2.6 Exercises

**1–4**

(a) Find $y'$ by implicit differentiation.
(b) Solve the equation explicitly for $y$ and differentiate to get $y'$ in terms of $x$.
(c) Check that your solutions to parts (a) and (b) are consistent by substituting the expression for $y$ into your solution for part (a).

**1.** $9x^2 - y^2 = 1$      **2.** $2x^2 + x + xy = 1$

**3.** $\dfrac{1}{x} + \dfrac{1}{y} = 1$      **4.** $\cos x + \sqrt{y} = 5$

**5–20** Find $dy/dx$ by implicit differentiation.

**5.** $x^3 + y^3 = 1$      **6.** $2\sqrt{x} + \sqrt{y} = 3$

**7.** $x^2 + xy - y^2 = 4$      **8.** $2x^3 + x^2 y - xy^3 = 2$

**9.** $x^4(x + y) = y^2(3x - y)$      **10.** $y^5 + x^2 y^3 = 1 + x^4 y$

**11.** $y \cos x = x^2 + y^2$      **12.** $\cos(xy) = 1 + \sin y$

**13.** $4 \cos x \sin y = 1$      **14.** $y \sin(x^2) = x \sin(y^2)$

**15.** $\tan(x/y) = x + y$      **16.** $\sqrt{x + y} = 1 + x^2 y^2$

**17.** $\sqrt{xy} = 1 + x^2 y$      **18.** $x \sin y + y \sin x = 1$

**19.** $y \cos x = 1 + \sin(xy)$      **20.** $\tan(x - y) = \dfrac{y}{1 + x^2}$

**21.** If $f(x) + x^2[f(x)]^3 = 10$ and $f(1) = 2$, find $f'(1)$.

**22.** If $g(x) + x \sin g(x) = x^2$, find $g'(0)$.

---

⌁ Graphing calculator or computer required     CAS Computer algebra system required     **1.** Homework Hints available at stewartcalculus.com

**23–24** Regard $y$ as the independent variable and $x$ as the dependent variable and use implicit differentiation to find $dx/dy$.

**23.** $x^4y^2 - x^3y + 2xy^3 = 0$       **24.** $y \sec x = x \tan y$

**25–32** Use implicit differentiation to find an equation of the tangent line to the curve at the given point.

**25.** $y \sin 2x = x \cos 2y$,   $(\pi/2, \pi/4)$

**26.** $\sin(x + y) = 2x - 2y$,   $(\pi, \pi)$

**27.** $x^2 + xy + y^2 = 3$,   $(1, 1)$   (ellipse)

**28.** $x^2 + 2xy - y^2 + x = 2$,   $(1, 2)$   (hyperbola)

**29.** $x^2 + y^2 = (2x^2 + 2y^2 - x)^2$       **30.** $x^{2/3} + y^{2/3} = 4$
$\left(0, \frac{1}{2}\right)$              $(-3\sqrt{3}, 1)$
(cardioid)                 (astroid)

**31.** $2(x^2 + y^2)^2 = 25(x^2 - y^2)$   **32.** $y^2(y^2 - 4) = x^2(x^2 - 5)$
(3, 1)                     (0, −2)
(lemniscate)               (devil's curve)

**33.** (a) The curve with equation $y^2 = 5x^4 - x^2$ is called a **kampyle of Eudoxus**. Find an equation of the tangent line to this curve at the point $(1, 2)$.
(b) Illustrate part (a) by graphing the curve and the tangent line on a common screen. (If your graphing device will graph implicitly defined curves, then use that capability. If not, you can still graph this curve by graphing its upper and lower halves separately.)

**34.** (a) The curve with equation $y^2 = x^3 + 3x^2$ is called the **Tschirnhausen cubic**. Find an equation of the tangent line to this curve at the point $(1, -2)$.
(b) At what points does this curve have horizontal tangents?
(c) Illustrate parts (a) and (b) by graphing the curve and the tangent lines on a common screen.

**35–38** Find $y''$ by implicit differentiation.

**35.** $9x^2 + y^2 = 9$           **36.** $\sqrt{x} + \sqrt{y} = 1$

**37.** $x^3 + y^3 = 1$           **38.** $x^4 + y^4 = a^4$

**39.** If $xy + y^3 = 1$, find the value of $y''$ at the point where $x = 0$.

**40.** If $x^2 + xy + y^3 = 1$, find the value of $y'''$ at the point where $x = 1$.

**41.** Fanciful shapes can be created by using the implicit plotting capabilities of computer algebra systems.
(a) Graph the curve with equation

$$y(y^2 - 1)(y - 2) = x(x - 1)(x - 2)$$

At how many points does this curve have horizontal tangents? Estimate the $x$-coordinates of these points.
(b) Find equations of the tangent lines at the points $(0, 1)$ and $(0, 2)$.
(c) Find the exact $x$-coordinates of the points in part (a).
(d) Create even more fanciful curves by modifying the equation in part (a).

**42.** (a) The curve with equation

$$2y^3 + y^2 - y^5 = x^4 - 2x^3 + x^2$$

has been likened to a bouncing wagon. Use a computer algebra system to graph this curve and discover why.
(b) At how many points does this curve have horizontal tangent lines? Find the $x$-coordinates of these points.

**43.** Find the points on the lemniscate in Exercise 31 where the tangent is horizontal.

**44.** Show by implicit differentiation that the tangent to the ellipse

$$\frac{x^2}{a^2} + \frac{y^2}{b^2} = 1$$

at the point $(x_0, y_0)$ is

$$\frac{x_0 x}{a^2} + \frac{y_0 y}{b^2} = 1$$

**45.** Find an equation of the tangent line to the hyperbola

$$\frac{x^2}{a^2} - \frac{y^2}{b^2} = 1$$

at the point $(x_0, y_0)$.

**46.** Show that the sum of the $x$- and $y$-intercepts of any tangent line to the curve $\sqrt{x} + \sqrt{y} = \sqrt{c}$ is equal to $c$.

**47.** Show, using implicit differentiation, that any tangent line at a point $P$ to a circle with center $O$ is perpendicular to the radius $OP$.

**48.** The Power Rule can be proved using implicit differentiation for the case where $n$ is a rational number, $n = p/q$, and $y = f(x) = x^n$ is assumed beforehand to be a differentiable function. If $y = x^{p/q}$, then $y^q = x^p$. Use implicit differentiation to show that

$$y' = \frac{p}{q} x^{(p/q)-1}$$

**49–52** Two curves are **orthogonal** if their tangent lines are perpendicular at each point of intersection. Show that the given families of curves are **orthogonal trajectories** of each other; that is, every curve in one family is orthogonal to every curve in the other family. Sketch both families of curves on the same axes.

**49.** $x^2 + y^2 = r^2$,    $ax + by = 0$

**50.** $x^2 + y^2 = ax$,    $x^2 + y^2 = by$

**51.** $y = cx^2$,    $x^2 + 2y^2 = k$

**52.** $y = ax^3$,    $x^2 + 3y^2 = b$

**53.** Show that the ellipse $x^2/a^2 + y^2/b^2 = 1$ and the hyperbola $x^2/A^2 - y^2/B^2 = 1$ are orthogonal trajectories if $A^2 < a^2$ and $a^2 - b^2 = A^2 + B^2$ (so the ellipse and hyperbola have the same foci).

**54.** Find the value of the number $a$ such that the families of curves $y = (x + c)^{-1}$ and $y = a(x + k)^{1/3}$ are orthogonal trajectories.

**55.** (a) The *van der Waals equation* for $n$ moles of a gas is

$$\left(P + \frac{n^2 a}{V^2}\right)(V - nb) = nRT$$

where $P$ is the pressure, $V$ is the volume, and $T$ is the temperature of the gas. The constant $R$ is the universal gas constant and $a$ and $b$ are positive constants that are characteristic of a particular gas. If $T$ remains constant, use implicit differentiation to find $dV/dP$.

(b) Find the rate of change of volume with respect to pressure of 1 mole of carbon dioxide at a volume of $V = 10$ L and a pressure of $P = 2.5$ atm. Use $a = 3.592$ L$^2$-atm/mole$^2$ and $b = 0.04267$ L/mole.

**56.** (a) Use implicit differentiation to find $y'$ if
$x^2 + xy + y^2 + 1 = 0$.

(b) Plot the curve in part (a). What do you see? Prove that what you see is correct.

(c) In view of part (b), what can you say about the expression for $y'$ that you found in part (a)?

**57.** The equation $x^2 - xy + y^2 = 3$ represents a "rotated ellipse," that is, an ellipse whose axes are not parallel to the coordinate axes. Find the points at which this ellipse crosses the $x$-axis and show that the tangent lines at these points are parallel.

**58.** (a) Where does the normal line to the ellipse $x^2 - xy + y^2 = 3$ at the point $(-1, 1)$ intersect the ellipse a second time?

(b) Illustrate part (a) by graphing the ellipse and the normal line.

**59.** Find all points on the curve $x^2y^2 + xy = 2$ where the slope of the tangent line is $-1$.

**60.** Find equations of both the tangent lines to the ellipse $x^2 + 4y^2 = 36$ that pass through the point $(12, 3)$.

**61.** The **Bessel function** of order 0, $y = J(x)$, satisfies the differential equation $xy'' + y' + xy = 0$ for all values of $x$ and its value at 0 is $J(0) = 1$.
(a) Find $J'(0)$.
(b) Use implicit differentiation to find $J''(0)$.

**62.** The figure shows a lamp located three units to the right of the $y$-axis and a shadow created by the elliptical region $x^2 + 4y^2 \leq 5$. If the point $(-5, 0)$ is on the edge of the shadow, how far above the $x$-axis is the lamp located?

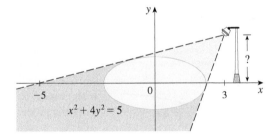

## LABORATORY PROJECT    CAS **FAMILIES OF IMPLICIT CURVES**

In this project you will explore the changing shapes of implicitly defined curves as you vary the constants in a family, and determine which features are common to all members of the family.

**1.** Consider the family of curves

$$y^2 - 2x^2(x + 8) = c[(y + 1)^2(y + 9) - x^2]$$

(a) By graphing the curves with $c = 0$ and $c = 2$, determine how many points of intersection there are. (You might have to zoom in to find all of them.)

(b) Now add the curves with $c = 5$ and $c = 10$ to your graphs in part (a). What do you notice? What about other values of $c$?

CAS Computer algebra system required

**2.** (a) Graph several members of the family of curves

$$x^2 + y^2 + cx^2y^2 = 1$$

Describe how the graph changes as you change the value of $c$.

(b) What happens to the curve when $c = -1$? Describe what appears on the screen. Can you prove it algebraically?

(c) Find $y'$ by implicit differentiation. For the case $c = -1$, is your expression for $y'$ consistent with what you discovered in part (b)?

## 2.7    Rates of Change in the Natural and Social Sciences

We know that if $y = f(x)$, then the derivative $dy/dx$ can be interpreted as the rate of change of $y$ with respect to $x$. In this section we examine some of the applications of this idea to physics, chemistry, biology, economics, and other sciences.

Let's recall from Section 2.1 the basic idea behind rates of change. If $x$ changes from $x_1$ to $x_2$, then the change in $x$ is

$$\Delta x = x_2 - x_1$$

and the corresponding change in $y$ is

$$\Delta y = f(x_2) - f(x_1)$$

The difference quotient

$$\frac{\Delta y}{\Delta x} = \frac{f(x_2) - f(x_1)}{x_2 - x_1}$$

is the **average rate of change of $y$ with respect to $x$** over the interval $[x_1, x_2]$ and can be interpreted as the slope of the secant line $PQ$ in Figure 1. Its limit as $\Delta x \to 0$ is the derivative $f'(x_1)$, which can therefore be interpreted as the **instantaneous rate of change of $y$ with respect to $x$** or the slope of the tangent line at $P(x_1, f(x_1))$. Using Leibniz notation, we write the process in the form

$$\frac{dy}{dx} = \lim_{\Delta x \to 0} \frac{\Delta y}{\Delta x}$$

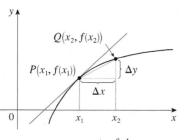

$m_{PQ}$ = average rate of change
$m = f'(x_1)$ = instantaneous rate of change

**FIGURE 1**

Whenever the function $y = f(x)$ has a specific interpretation in one of the sciences, its derivative will have a specific interpretation as a rate of change. (As we discussed in Section 2.1, the units for $dy/dx$ are the units for $y$ divided by the units for $x$.) We now look at some of these interpretations in the natural and social sciences.

### Physics

If $s = f(t)$ is the position function of a particle that is moving in a straight line, then $\Delta s/\Delta t$ represents the average velocity over a time period $\Delta t$, and $v = ds/dt$ represents the instantaneous **velocity** (the rate of change of displacement with respect to time). The instantaneous rate of change of velocity with respect to time is **acceleration**: $a(t) = v'(t) = s''(t)$. This was discussed in Sections 2.1 and 2.2, but now that we know the differentiation formulas, we are able to solve problems involving the motion of objects more easily.

**V EXAMPLE 1** The position of a particle is given by the equation

$$s = f(t) = t^3 - 6t^2 + 9t$$

where $t$ is measured in seconds and $s$ in meters.
(a) Find the velocity at time $t$.
(b) What is the velocity after 2 s? After 4 s?
(c) When is the particle at rest?
(d) When is the particle moving forward (that is, in the positive direction)?
(e) Draw a diagram to represent the motion of the particle.
(f) Find the total distance traveled by the particle during the first five seconds.
(g) Find the acceleration at time $t$ and after 4 s.
(h) Graph the position, velocity, and acceleration functions for $0 \leqslant t \leqslant 5$.
(i) When is the particle speeding up? When is it slowing down?

SOLUTION
(a) The velocity function is the derivative of the position function.

$$s = f(t) = t^3 - 6t^2 + 9t$$

$$v(t) = \frac{ds}{dt} = 3t^2 - 12t + 9$$

(b) The velocity after 2 s means the instantaneous velocity when $t = 2$, that is,

$$v(2) = \frac{ds}{dt}\bigg|_{t=2} = 3(2)^2 - 12(2) + 9 = -3 \text{ m/s}$$

The velocity after 4 s is

$$v(4) = 3(4)^2 - 12(4) + 9 = 9 \text{ m/s}$$

(c) The particle is at rest when $v(t) = 0$, that is,

$$3t^2 - 12t + 9 = 3(t^2 - 4t + 3) = 3(t - 1)(t - 3) = 0$$

and this is true when $t = 1$ or $t = 3$. Thus the particle is at rest after 1 s and after 3 s.
(d) The particle moves in the positive direction when $v(t) > 0$, that is,

$$3t^2 - 12t + 9 = 3(t - 1)(t - 3) > 0$$

This inequality is true when both factors are positive ($t > 3$) or when both factors are negative ($t < 1$). Thus the particle moves in the positive direction in the time intervals $t < 1$ and $t > 3$. It moves backward (in the negative direction) when $1 < t < 3$.

(e) Using the information from part (d) we make a schematic sketch in Figure 2 of the motion of the particle back and forth along a line (the $s$-axis).

(f) Because of what we learned in parts (d) and (e), we need to calculate the distances traveled during the time intervals [0, 1], [1, 3], and [3, 5] separately.
   The distance traveled in the first second is

$$|f(1) - f(0)| = |4 - 0| = 4 \text{ m}$$

From $t = 1$ to $t = 3$ the distance traveled is

$$|f(3) - f(1)| = |0 - 4| = 4 \text{ m}$$

$t = 3$
$s = 0$

$t = 0$      $t = 1$      $s$
$s = 0$      $s = 4$

**FIGURE 2**

From $t = 3$ to $t = 5$ the distance traveled is

$$|f(5) - f(3)| = |20 - 0| = 20 \text{ m}$$

The total distance is $4 + 4 + 20 = 28$ m.

(g)  The acceleration is the derivative of the velocity function:

$$a(t) = \frac{d^2s}{dt^2} = \frac{dv}{dt} = 6t - 12$$

$$a(4) = 6(4) - 12 = 12 \text{ m/s}^2$$

(h)  Figure 3 shows the graphs of $s$, $v$, and $a$.

(i)  The particle speeds up when the velocity is positive and increasing ($v$ and $a$ are both positive) and also when the velocity is negative and decreasing ($v$ and $a$ are both negative). In other words, the particle speeds up when the velocity and acceleration have the same sign. (The particle is pushed in the same direction it is moving.) From Figure 3 we see that this happens when $1 < t < 2$ and when $t > 3$. The particle slows down when $v$ and $a$ have opposite signs, that is, when $0 \leqslant t < 1$ and when $2 < t < 3$. Figure 4 summarizes the motion of the particle.

**FIGURE 3**

**TEC**  In Module 2.7 you can see an animation of Figure 4 with an expression for $s$ that you can choose yourself.

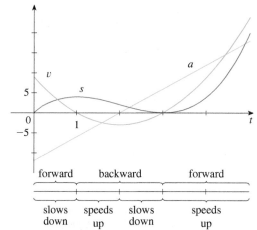

**FIGURE 4**

**EXAMPLE 2**  If a rod or piece of wire is homogeneous, then its linear density is uniform and is defined as the mass per unit length ($\rho = m/l$) and measured in kilograms per meter. Suppose, however, that the rod is not homogeneous but that its mass measured from its left end to a point $x$ is $m = f(x)$, as shown in Figure 5.

**FIGURE 5**    This part of the rod has mass $f(x)$.

The mass of the part of the rod that lies between $x = x_1$ and $x = x_2$ is given by $\Delta m = f(x_2) - f(x_1)$, so the average density of that part of the rod is

$$\text{average density} = \frac{\Delta m}{\Delta x} = \frac{f(x_2) - f(x_1)}{x_2 - x_1}$$

If we now let $\Delta x \to 0$ (that is, $x_2 \to x_1$), we are computing the average density over smaller and smaller intervals. The **linear density** $\rho$ at $x_1$ is the limit of these average densities as $\Delta x \to 0$; that is, the linear density is the rate of change of mass with respect to length. Symbolically,

$$\rho = \lim_{\Delta x \to 0} \frac{\Delta m}{\Delta x} = \frac{dm}{dx}$$

Thus the linear density of the rod is the derivative of mass with respect to length.

For instance, if $m = f(x) = \sqrt{x}$, where $x$ is measured in meters and $m$ in kilograms, then the average density of the part of the rod given by $1 \leqslant x \leqslant 1.2$ is

$$\frac{\Delta m}{\Delta x} = \frac{f(1.2) - f(1)}{1.2 - 1} = \frac{\sqrt{1.2} - 1}{0.2} \approx 0.48 \text{ kg/m}$$

while the density right at $x = 1$ is

$$\rho = \frac{dm}{dx}\bigg|_{x=1} = \frac{1}{2\sqrt{x}}\bigg|_{x=1} = 0.50 \text{ kg/m}$$

**FIGURE 6**

**V EXAMPLE 3** A current exists whenever electric charges move. Figure 6 shows part of a wire and electrons moving through a plane surface, shaded red. If $\Delta Q$ is the net charge that passes through this surface during a time period $\Delta t$, then the average current during this time interval is defined as

$$\text{average current} = \frac{\Delta Q}{\Delta t} = \frac{Q_2 - Q_1}{t_2 - t_1}$$

If we take the limit of this average current over smaller and smaller time intervals, we get what is called the **current** $I$ at a given time $t_1$:

$$I = \lim_{\Delta t \to 0} \frac{\Delta Q}{\Delta t} = \frac{dQ}{dt}$$

Thus the current is the rate at which charge flows through a surface. It is measured in units of charge per unit time (often coulombs per second, called amperes).

Velocity, density, and current are not the only rates of change that are important in physics. Others include power (the rate at which work is done), the rate of heat flow, temperature gradient (the rate of change of temperature with respect to position), and the rate of decay of a radioactive substance in nuclear physics.

## Chemistry

**EXAMPLE 4** A chemical reaction results in the formation of one or more substances (called *products*) from one or more starting materials (called *reactants*). For instance, the "equation"

$$2H_2 + O_2 \to 2H_2O$$

indicates that two molecules of hydrogen and one molecule of oxygen form two molecules of water. Let's consider the reaction

$$A + B \to C$$

where A and B are the reactants and C is the product. The **concentration** of a reactant A is the number of moles (1 mole = $6.022 \times 10^{23}$ molecules) per liter and is denoted by [A]. The concentration varies during a reaction, so [A], [B], and [C] are all functions of time ($t$). The average rate of reaction of the product C over a time interval $t_1 \leqslant t \leqslant t_2$ is

$$\frac{\Delta[\text{C}]}{\Delta t} = \frac{[\text{C}](t_2) - [\text{C}](t_1)}{t_2 - t_1}$$

But chemists are more interested in the **instantaneous rate of reaction**, which is obtained by taking the limit of the average rate of reaction as the time interval $\Delta t$ approaches 0:

$$\text{rate of reaction} = \lim_{\Delta t \to 0} \frac{\Delta[\text{C}]}{\Delta t} = \frac{d[\text{C}]}{dt}$$

Since the concentration of the product increases as the reaction proceeds, the derivative $d[\text{C}]/dt$ will be positive, and so the rate of reaction of C is positive. The concentrations of the reactants, however, decrease during the reaction, so, to make the rates of reaction of A and B positive numbers, we put minus signs in front of the derivatives $d[\text{A}]/dt$ and $d[\text{B}]/dt$. Since [A] and [B] each decrease at the same rate that [C] increases, we have

$$\text{rate of reaction} = \frac{d[\text{C}]}{dt} = -\frac{d[\text{A}]}{dt} = -\frac{d[\text{B}]}{dt}$$

More generally, it turns out that for a reaction of the form

$$a\text{A} + b\text{B} \to c\text{C} + d\text{D}$$

we have

$$-\frac{1}{a}\frac{d[\text{A}]}{dt} = -\frac{1}{b}\frac{d[\text{B}]}{dt} = \frac{1}{c}\frac{d[\text{C}]}{dt} = \frac{1}{d}\frac{d[\text{D}]}{dt}$$

The rate of reaction can be determined from data and graphical methods. In some cases there are explicit formulas for the concentrations as functions of time, which enable us to compute the rate of reaction (see Exercise 24).

**EXAMPLE 5**  One of the quantities of interest in thermodynamics is compressibility. If a given substance is kept at a constant temperature, then its volume $V$ depends on its pressure $P$. We can consider the rate of change of volume with respect to pressure—namely, the derivative $dV/dP$. As $P$ increases, $V$ decreases, so $dV/dP < 0$. The **compressibility** is defined by introducing a minus sign and dividing this derivative by the volume $V$:

$$\text{isothermal compressibility} = \beta = -\frac{1}{V}\frac{dV}{dP}$$

Thus $\beta$ measures how fast, per unit volume, the volume of a substance decreases as the pressure on it increases at constant temperature.

For instance, the volume $V$ (in cubic meters) of a sample of air at 25°C was found to be related to the pressure $P$ (in kilopascals) by the equation

$$V = \frac{5.3}{P}$$

The rate of change of $V$ with respect to $P$ when $P = 50$ kPa is

$$\left.\frac{dV}{dP}\right|_{P=50} = \left.-\frac{5.3}{P^2}\right|_{P=50}$$

$$= -\frac{5.3}{2500} = -0.00212 \text{ m}^3/\text{kPa}$$

The compressibility at that pressure is

$$\beta = \left.-\frac{1}{V}\frac{dV}{dP}\right|_{P=50} = \frac{0.00212}{\dfrac{5.3}{50}} = 0.02 \text{ (m}^3/\text{kPa)}/\text{m}^3$$

## Biology

**EXAMPLE 6**  Let $n = f(t)$ be the number of individuals in an animal or plant population at time $t$. The change in the population size between the times $t = t_1$ and $t = t_2$ is $\Delta n = f(t_2) - f(t_1)$, and so the average rate of growth during the time period $t_1 \leqslant t \leqslant t_2$ is

$$\text{average rate of growth} = \frac{\Delta n}{\Delta t} = \frac{f(t_2) - f(t_1)}{t_2 - t_1}$$

The **instantaneous rate of growth** is obtained from this average rate of growth by letting the time period $\Delta t$ approach 0:

$$\text{growth rate} = \lim_{\Delta t \to 0} \frac{\Delta n}{\Delta t} = \frac{dn}{dt}$$

Strictly speaking, this is not quite accurate because the actual graph of a population function $n = f(t)$ would be a step function that is discontinuous whenever a birth or death occurs and therefore not differentiable. However, for a large animal or plant population, we can replace the graph by a smooth approximating curve as in Figure 7.

**FIGURE 7**
A smooth curve approximating
a growth function

E. coli bacteria are about 2 micrometers (μm) long and 0.75 μm wide. The image was produced with a scanning electron microscope.

To be more specific, consider a population of bacteria in a homogeneous nutrient medium. Suppose that by sampling the population at certain intervals it is determined that the population doubles every hour. If the initial population is $n_0$ and the time $t$ is measured in hours, then

$$f(1) = 2f(0) = 2n_0$$
$$f(2) = 2f(1) = 2^2 n_0$$
$$f(3) = 2f(2) = 2^3 n_0$$

and, in general,

$$f(t) = 2^t n_0$$

The population function is $n = n_0 2^t$.

This is an example of an exponential function. In Chapter 6 we will discuss exponential functions in general; at that time we will be able to compute their derivatives and thereby determine the rate of growth of the bacteria population.

**EXAMPLE 7** When we consider the flow of blood through a blood vessel, such as a vein or artery, we can model the shape of the blood vessel by a cylindrical tube with radius $R$ and length $l$ as illustrated in Figure 8.

**FIGURE 8**
Blood flow in an artery

Because of friction at the walls of the tube, the velocity $v$ of the blood is greatest along the central axis of the tube and decreases as the distance $r$ from the axis increases until $v$ becomes 0 at the wall. The relationship between $v$ and $r$ is given by the **law of laminar flow** discovered by the French physician Jean-Louis-Marie Poiseuille in 1840. This law states that

$$\boxed{1} \qquad v = \frac{P}{4\eta l}(R^2 - r^2)$$

where $\eta$ is the viscosity of the blood and $P$ is the pressure difference between the ends of the tube. If $P$ and $l$ are constant, then $v$ is a function of $r$ with domain $[0, R]$.

For more detailed information, see W. Nichols and M. O'Rourke (eds.), *McDonald's Blood Flow in Arteries: Theoretical, Experimental, and Clinical Principles,* 5th ed. (New York, 2005).

The average rate of change of the velocity as we move from $r = r_1$ outward to $r = r_2$ is given by

$$\frac{\Delta v}{\Delta r} = \frac{v(r_2) - v(r_1)}{r_2 - r_1}$$

and if we let $\Delta r \to 0$, we obtain the **velocity gradient**, that is, the instantaneous rate of change of velocity with respect to $r$:

$$\text{velocity gradient} = \lim_{\Delta r \to 0} \frac{\Delta v}{\Delta r} = \frac{dv}{dr}$$

Using Equation 1, we obtain

$$\frac{dv}{dr} = \frac{P}{4\eta l}(0 - 2r) = -\frac{Pr}{2\eta l}$$

For one of the smaller human arteries we can take $\eta = 0.027$, $R = 0.008$ cm, $l = 2$ cm, and $P = 4000$ dynes/cm$^2$, which gives

$$v = \frac{4000}{4(0.027)2} (0.000064 - r^2)$$

$$\approx 1.85 \times 10^4 (6.4 \times 10^{-5} - r^2)$$

At $r = 0.002$ cm the blood is flowing at a speed of

$$v(0.002) \approx 1.85 \times 10^4 (64 \times 10^{-6} - 4 \times 10^{-6})$$

$$= 1.11 \text{ cm/s}$$

and the velocity gradient at that point is

$$\left. \frac{dv}{dr} \right|_{r=0.002} = -\frac{4000(0.002)}{2(0.027)2} \approx -74 \text{ (cm/s)/cm}$$

To get a feeling for what this statement means, let's change our units from centimeters to micrometers (1 cm $= 10{,}000$ μm). Then the radius of the artery is 80 μm. The velocity at the central axis is 11,850 μm/s, which decreases to 11,110 μm/s at a distance of $r = 20$ μm. The fact that $dv/dr = -74$ (μm/s)/μm means that, when $r = 20$ μm, the velocity is decreasing at a rate of about 74 μm/s for each micrometer that we proceed away from the center.

## Economics

**V** **EXAMPLE 8** Suppose $C(x)$ is the total cost that a company incurs in producing $x$ units of a certain commodity. The function $C$ is called a **cost function**. If the number of items produced is increased from $x_1$ to $x_2$, then the additional cost is $\Delta C = C(x_2) - C(x_1)$, and the average rate of change of the cost is

$$\frac{\Delta C}{\Delta x} = \frac{C(x_2) - C(x_1)}{x_2 - x_1} = \frac{C(x_1 + \Delta x) - C(x_1)}{\Delta x}$$

The limit of this quantity as $\Delta x \to 0$, that is, the instantaneous rate of change of cost with respect to the number of items produced, is called the **marginal cost** by economists:

$$\text{marginal cost} = \lim_{\Delta x \to 0} \frac{\Delta C}{\Delta x} = \frac{dC}{dx}$$

[Since $x$ often takes on only integer values, it may not make literal sense to let $\Delta x$ approach 0, but we can always replace $C(x)$ by a smooth approximating function as in Example 6.]

Taking $\Delta x = 1$ and $n$ large (so that $\Delta x$ is small compared to $n$), we have

$$C'(n) \approx C(n + 1) - C(n)$$

Thus the marginal cost of producing $n$ units is approximately equal to the cost of producing one more unit [the $(n + 1)$st unit].

It is often appropriate to represent a total cost function by a polynomial

$$C(x) = a + bx + cx^2 + dx^3$$

where $a$ represents the overhead cost (rent, heat, maintenance) and the other terms represent the cost of raw materials, labor, and so on. (The cost of raw materials may be proportional to $x$, but labor costs might depend partly on higher powers of $x$ because of overtime costs and inefficiencies involved in large-scale operations.)

For instance, suppose a company has estimated that the cost (in dollars) of producing $x$ items is

$$C(x) = 10{,}000 + 5x + 0.01x^2$$

Then the marginal cost function is

$$C'(x) = 5 + 0.02x$$

The marginal cost at the production level of 500 items is

$$C'(500) = 5 + 0.02(500) = \$15/\text{item}$$

This gives the rate at which costs are increasing with respect to the production level when $x = 500$ and predicts the cost of the 501st item.

The actual cost of producing the 501st item is

$$C(501) - C(500) = [10{,}000 + 5(501) + 0.01(501)^2]$$
$$- [10{,}000 + 5(500) + 0.01(500)^2]$$
$$= \$15.01$$

Notice that $C'(500) \approx C(501) - C(500)$.

Economists also study marginal demand, marginal revenue, and marginal profit, which are the derivatives of the demand, revenue, and profit functions. These will be considered in Chapter 3 after we have developed techniques for finding the maximum and minimum values of functions.

## Other Sciences

Rates of change occur in all the sciences. A geologist is interested in knowing the rate at which an intruded body of molten rock cools by conduction of heat into surrounding rocks. An engineer wants to know the rate at which water flows into or out of a reservoir. An urban geographer is interested in the rate of change of the population density in a city as the distance from the city center increases. A meteorologist is concerned with the rate of change of atmospheric pressure with respect to height (see Exercise 17 in Section 6.5).

In psychology, those interested in learning theory study the so-called learning curve, which graphs the performance $P(t)$ of someone learning a skill as a function of the training time $t$. Of particular interest is the rate at which performance improves as time passes, that is, $dP/dt$.

In sociology, differential calculus is used in analyzing the spread of rumors (or innovations or fads or fashions). If $p(t)$ denotes the proportion of a population that knows a rumor by time $t$, then the derivative $dp/dt$ represents the rate of spread of the rumor (see Exercise 63 in Section 6.2).

## A Single Idea, Many Interpretations

Velocity, density, current, power, and temperature gradient in physics; rate of reaction and compressibility in chemistry; rate of growth and blood velocity gradient in biology; marginal cost and marginal profit in economics; rate of heat flow in geology; rate of improvement of performance in psychology; rate of spread of a rumor in sociology—these are all special cases of a single mathematical concept, the derivative.

This is an illustration of the fact that part of the power of mathematics lies in its abstractness. A single abstract mathematical concept (such as the derivative) can have different interpretations in each of the sciences. When we develop the properties of the mathematical concept once and for all, we can then turn around and apply these results to all of the sciences. This is much more efficient than developing properties of special concepts in each separate science. The French mathematician Joseph Fourier (1768–1830) put it succinctly: "Mathematics compares the most diverse phenomena and discovers the secret analogies that unite them."

## 2.7 Exercises

**1–4** A particle moves according to a law of motion $s = f(t)$, $t \geq 0$, where $t$ is measured in seconds and $s$ in feet.
(a) Find the velocity at time $t$.
(b) What is the velocity after 3 s?
(c) When is the particle at rest?
(d) When is the particle moving in the positive direction?
(e) Find the total distance traveled during the first 8 s.
(f) Draw a diagram like Figure 2 to illustrate the motion of the particle.
(g) Find the acceleration at time $t$ and after 3 s.
(h) Graph the position, velocity, and acceleration functions for $0 \leq t \leq 8$.
(i) When is the particle speeding up? When is it slowing down?

**1.** $f(t) = t^3 - 12t^2 + 36t$      **2.** $f(t) = 0.01t^4 - 0.04t^3$

**3.** $f(t) = \cos(\pi t/4), \quad t \leq 10$      **4.** $f(t) = t/(1 + t^2)$

**5.** Graphs of the *velocity* functions of two particles are shown, where $t$ is measured in seconds. When is each particle speeding up? When is it slowing down? Explain.

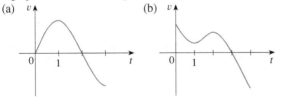

(a)  (b)

**6.** Graphs of the *position* functions of two particles are shown, where $t$ is measured in seconds. When is each particle speeding up? When is it slowing down? Explain.

(a)  (b)

**7.** The height (in meters) of a projectile shot vertically upward from a point 2 m above ground level with an initial velocity of 24.5 m/s is $h = 2 + 24.5t - 4.9t^2$ after $t$ seconds.
(a) Find the velocity after 2 s and after 4 s.
(b) When does the projectile reach its maximum height?
(c) What is the maximum height?
(d) When does it hit the ground?
(e) With what velocity does it hit the ground?

**8.** If a ball is thrown vertically upward with a velocity of 80 ft/s, then its height after $t$ seconds is $s = 80t - 16t^2$.
(a) What is the maximum height reached by the ball?
(b) What is the velocity of the ball when it is 96 ft above the ground on its way up? On its way down?

**9.** If a rock is thrown vertically upward from the surface of Mars with velocity 15 m/s, its height after $t$ seconds is $h = 15t - 1.86t^2$.
(a) What is the velocity of the rock after 2 s?
(b) What is the velocity of the rock when its height is 25 m on its way up? On its way down?

**10.** A particle moves with position function

$$s = t^4 - 4t^3 - 20t^2 + 20t \qquad t \geq 0$$

(a) At what time does the particle have a velocity of 20 m/s?
(b) At what time is the acceleration 0? What is the significance of this value of $t$?

**11.** (a) A company makes computer chips from square wafers of silicon. It wants to keep the side length of a wafer very close to 15 mm and it wants to know how the area $A(x)$ of a wafer changes when the side length $x$ changes. Find $A'(15)$ and explain its meaning in this situation.
(b) Show that the rate of change of the area of a square with respect to its side length is half its perimeter. Try to explain geometrically why this is true by drawing a square whose side length $x$ is increased by an amount $\Delta x$. How can you approximate the resulting change in area $\Delta A$ if $\Delta x$ is small?

**12.** (a) Sodium chlorate crystals are easy to grow in the shape of cubes by allowing a solution of water and sodium chlorate to evaporate slowly. If $V$ is the volume of such a cube with side length $x$, calculate $dV/dx$ when $x = 3$ mm and explain its meaning.

(b) Show that the rate of change of the volume of a cube with respect to its edge length is equal to half the surface area of the cube. Explain geometrically why this result is true by arguing by analogy with Exercise 11(b).

**13.** (a) Find the average rate of change of the area of a circle with respect to its radius $r$ as $r$ changes from
   (i) 2 to 3     (ii) 2 to 2.5     (iii) 2 to 2.1
(b) Find the instantaneous rate of change when $r = 2$.
(c) Show that the rate of change of the area of a circle with respect to its radius (at any $r$) is equal to the circumference of the circle. Try to explain geometrically why this is true by drawing a circle whose radius is increased by an amount $\Delta r$. How can you approximate the resulting change in area $\Delta A$ if $\Delta r$ is small?

**14.** A stone is dropped into a lake, creating a circular ripple that travels outward at a speed of 60 cm/s. Find the rate at which the area within the circle is increasing after (a) 1 s, (b) 3 s, and (c) 5 s. What can you conclude?

**15.** A spherical balloon is being inflated. Find the rate of increase of the surface area ($S = 4\pi r^2$) with respect to the radius $r$ when $r$ is (a) 1 ft, (b) 2 ft, and (c) 3 ft. What conclusion can you make?

**16.** (a) The volume of a growing spherical cell is $V = \frac{4}{3}\pi r^3$, where the radius $r$ is measured in micrometers (1 $\mu$m $= 10^{-6}$ m). Find the average rate of change of $V$ with respect to $r$ when $r$ changes from
   (i) 5 to 8 $\mu$m     (ii) 5 to 6 $\mu$m     (iii) 5 to 5.1 $\mu$m
(b) Find the instantaneous rate of change of $V$ with respect to $r$ when $r = 5$ $\mu$m.
(c) Show that the rate of change of the volume of a sphere with respect to its radius is equal to its surface area. Explain geometrically why this result is true. Argue by analogy with Exercise 13(c).

**17.** The mass of the part of a metal rod that lies between its left end and a point $x$ meters to the right is $3x^2$ kg. Find the linear density (see Example 2) when $x$ is (a) 1 m, (b) 2 m, and (c) 3 m. Where is the density the highest? The lowest?

**18.** If a tank holds 5000 gallons of water, which drains from the bottom of the tank in 40 minutes, then Torricelli's Law gives the volume $V$ of water remaining in the tank after $t$ minutes as

$$V = 5000\left(1 - \tfrac{1}{40}t\right)^2 \qquad 0 \leqslant t \leqslant 40$$

Find the rate at which water is draining from the tank after (a) 5 min, (b) 10 min, (c) 20 min, and (d) 40 min. At what time is the water flowing out the fastest? The slowest? Summarize your findings.

**19.** The quantity of charge $Q$ in coulombs (C) that has passed through a point in a wire up to time $t$ (measured in seconds) is given by $Q(t) = t^3 - 2t^2 + 6t + 2$. Find the current when (a) $t = 0.5$ s and (b) $t = 1$ s. [See Example 3. The unit of current is an ampere (1 A $= 1$ C/s).] At what time is the current lowest?

**20.** Newton's Law of Gravitation says that the magnitude $F$ of the force exerted by a body of mass $m$ on a body of mass $M$ is

$$F = \frac{GmM}{r^2}$$

where $G$ is the gravitational constant and $r$ is the distance between the bodies.
(a) Find $dF/dr$ and explain its meaning. What does the minus sign indicate?
(b) Suppose it is known that the earth attracts an object with a force that decreases at the rate of 2 N/km when $r = 20{,}000$ km. How fast does this force change when $r = 10{,}000$ km?

**21.** The force $F$ acting on a body with mass $m$ and velocity $v$ is the rate of change of momentum: $F = (d/dt)(mv)$. If $m$ is constant, this becomes $F = ma$, where $a = dv/dt$ is the acceleration. But in the theory of relativity the mass of a particle varies with $v$ as follows: $m = m_0/\sqrt{1 - v^2/c^2}$, where $m_0$ is the mass of the particle at rest and $c$ is the speed of light. Show that

$$F = \frac{m_0 a}{(1 - v^2/c^2)^{3/2}}$$

**22.** Some of the highest tides in the world occur in the Bay of Fundy on the Atlantic Coast of Canada. At Hopewell Cape the water depth at low tide is about 2.0 m and at high tide it is about 12.0 m. The natural period of oscillation is a little more than 12 hours and on June 30, 2009, high tide occurred at 6:45 AM. This helps explain the following model for the water depth $D$ (in meters) as a function of the time $t$ (in hours after midnight) on that day:

$$D(t) = 7 + 5\cos[0.503(t - 6.75)]$$

How fast was the tide rising (or falling) at the following times?
(a) 3:00 AM                    (b) 6:00 AM
(c) 9:00 AM                    (d) Noon

**23.** Boyle's Law states that when a sample of gas is compressed at a constant temperature, the product of the pressure and the volume remains constant: $PV = C$.
(a) Find the rate of change of volume with respect to pressure.
(b) A sample of gas is in a container at low pressure and is steadily compressed at constant temperature for 10 minutes. Is the volume decreasing more rapidly at the beginning or the end of the 10 minutes? Explain.
(c) Prove that the isothermal compressibility (see Example 5) is given by $\beta = 1/P$.

**24.** If, in Example 4, one molecule of the product C is formed from one molecule of the reactant A and one molecule of the reactant B, and the initial concentrations of A and B have a common value $[A] = [B] = a$ moles/L, then

$$[C] = a^2kt/(akt + 1)$$

where $k$ is a constant.
(a) Find the rate of reaction at time $t$.
(b) Show that if $x = [C]$, then

$$\frac{dx}{dt} = k(a - x)^2$$

(c) What happens to the concentration as $t \to \infty$?
(d) What happens to the rate of reaction as $t \to \infty$?
(e) What do the results of parts (c) and (d) mean in practical terms?

**25.** The table gives the population of the world in the 20th century.

| Year | Population (in millions) | Year | Population (in millions) |
|------|------|------|------|
| 1900 | 1650 | 1960 | 3040 |
| 1910 | 1750 | 1970 | 3710 |
| 1920 | 1860 | 1980 | 4450 |
| 1930 | 2070 | 1990 | 5280 |
| 1940 | 2300 | 2000 | 6080 |
| 1950 | 2560 | | |

(a) Estimate the rate of population growth in 1920 and in 1980 by averaging the slopes of two secant lines.
(b) Use a graphing calculator or computer to find a cubic function (a third-degree polynomial) that models the data.
(c) Use your model in part (b) to find a model for the rate of population growth in the 20th century.
(d) Use part (c) to estimate the rates of growth in 1920 and 1980. Compare with your estimates in part (a).
(e) Estimate the rate of growth in 1985.

**26.** The table shows how the average age of first marriage of Japanese women varied in the last half of the 20th century.

| $t$ | $A(t)$ | $t$ | $A(t)$ |
|------|------|------|------|
| 1950 | 23.0 | 1980 | 25.2 |
| 1955 | 23.8 | 1985 | 25.5 |
| 1960 | 24.4 | 1990 | 25.9 |
| 1965 | 24.5 | 1995 | 26.3 |
| 1970 | 24.2 | 2000 | 27.0 |
| 1975 | 24.7 | | |

(a) Use a graphing calculator or computer to model these data with a fourth-degree polynomial.
(b) Use part (a) to find a model for $A'(t)$.
(c) Estimate the rate of change of marriage age for women in 1990.
(d) Graph the data points and the models for $A$ and $A'$.

**27.** Refer to the law of laminar flow given in Example 7. Consider a blood vessel with radius 0.01 cm, length 3 cm, pressure difference 3000 dynes/cm², and viscosity $\eta = 0.027$.
(a) Find the velocity of the blood along the centerline $r = 0$, at radius $r = 0.005$ cm, and at the wall $r = R = 0.01$ cm.
(b) Find the velocity gradient at $r = 0$, $r = 0.005$, and $r = 0.01$.
(c) Where is the velocity the greatest? Where is the velocity changing most?

**28.** The frequency of vibrations of a vibrating violin string is given by

$$f = \frac{1}{2L} \sqrt{\frac{T}{\rho}}$$

where $L$ is the length of the string, $T$ is its tension, and $\rho$ is its linear density. [See Chapter 11 in D. E. Hall, *Musical Acoustics,* 3rd ed. (Pacific Grove, CA, 2002).]
(a) Find the rate of change of the frequency with respect to
  (i) the length (when $T$ and $\rho$ are constant),
  (ii) the tension (when $L$ and $\rho$ are constant), and
  (iii) the linear density (when $L$ and $T$ are constant).
(b) The pitch of a note (how high or low the note sounds) is determined by the frequency $f$. (The higher the frequency, the higher the pitch.) Use the signs of the derivatives in part (a) to determine what happens to the pitch of a note
  (i) when the effective length of a string is decreased by placing a finger on the string so a shorter portion of the string vibrates,
  (ii) when the tension is increased by turning a tuning peg,
  (iii) when the linear density is increased by switching to another string.

**29.** The cost, in dollars, of producing $x$ yards of a certain fabric is

$$C(x) = 1200 + 12x - 0.1x^2 + 0.0005x^3$$

(a) Find the marginal cost function.
(b) Find $C'(200)$ and explain its meaning. What does it predict?
(c) Compare $C'(200)$ with the cost of manufacturing the 201st yard of fabric.

**30.** The cost function for production of a commodity is

$$C(x) = 339 + 25x - 0.09x^2 + 0.0004x^3$$

(a) Find and interpret $C'(100)$.
(b) Compare $C'(100)$ with the cost of producing the 101st item.

**31.** If $p(x)$ is the total value of the production when there are $x$ workers in a plant, then the *average productivity* of the workforce at the plant is

$$A(x) = \frac{p(x)}{x}$$

(a) Find $A'(x)$. Why does the company want to hire more workers if $A'(x) > 0$?

(b) Show that $A'(x) > 0$ if $p'(x)$ is greater than the average productivity.

**32.** If $R$ denotes the reaction of the body to some stimulus of strength $x$, the *sensitivity* $S$ is defined to be the rate of change of the reaction with respect to $x$. A particular example is that when the brightness $x$ of a light source is increased, the eye reacts by decreasing the area $R$ of the pupil. The experimental formula

$$R = \frac{40 + 24x^{0.4}}{1 + 4x^{0.4}}$$

has been used to model the dependence of $R$ on $x$ when $R$ is measured in square millimeters and $x$ is measured in appropriate units of brightness.
(a) Find the sensitivity.
(b) Illustrate part (a) by graphing both $R$ and $S$ as functions of $x$. Comment on the values of $R$ and $S$ at low levels of brightness. Is this what you would expect?

**33.** The gas law for an ideal gas at absolute temperature $T$ (in kelvins), pressure $P$ (in atmospheres), and volume $V$ (in liters) is $PV = nRT$, where $n$ is the number of moles of the gas and $R = 0.0821$ is the gas constant. Suppose that, at a certain instant, $P = 8.0$ atm and is increasing at a rate of $0.10$ atm/min and $V = 10$ L and is decreasing at a rate of $0.15$ L/min. Find the rate of change of $T$ with respect to time at that instant if $n = 10$ mol.

**34.** In a fish farm, a population of fish is introduced into a pond and harvested regularly. A model for the rate of change of the fish population is given by the equation

$$\frac{dP}{dt} = r_0\left(1 - \frac{P(t)}{P_c}\right)P(t) - \beta P(t)$$

where $r_0$ is the birth rate of the fish, $P_c$ is the maximum population that the pond can sustain (called the *carrying capacity*), and $\beta$ is the percentage of the population that is harvested.
(a) What value of $dP/dt$ corresponds to a stable population?
(b) If the pond can sustain 10,000 fish, the birth rate is 5%, and the harvesting rate is 4%, find the stable population level.
(c) What happens if $\beta$ is raised to 5%?

**35.** In the study of ecosystems, *predator-prey models* are often used to study the interaction between species. Consider populations of tundra wolves, given by $W(t)$, and caribou, given by $C(t)$, in northern Canada. The interaction has been modeled by the equations

$$\frac{dC}{dt} = aC - bCW \qquad \frac{dW}{dt} = -cW + dCW$$

(a) What values of $dC/dt$ and $dW/dt$ correspond to stable populations?
(b) How would the statement "The caribou go extinct" be represented mathematically?
(c) Suppose that $a = 0.05$, $b = 0.001$, $c = 0.05$, and $d = 0.0001$. Find all population pairs $(C, W)$ that lead to stable populations. According to this model, is it possible for the two species to live in balance or will one or both species become extinct?

**PS** According to the Principles of Problem Solving discussed on page 97, the first step is to understand the problem. This includes reading the problem carefully, identifying the given and the unknown, and introducing suitable notation.

## 2.8 Related Rates

If we are pumping air into a balloon, both the volume and the radius of the balloon are increasing and their rates of increase are related to each other. But it is much easier to measure directly the rate of increase of the volume than the rate of increase of the radius.

In a related rates problem the idea is to compute the rate of change of one quantity in terms of the rate of change of another quantity (which may be more easily measured). The procedure is to find an equation that relates the two quantities and then use the Chain Rule to differentiate both sides with respect to time.

**V EXAMPLE 1** Air is being pumped into a spherical balloon so that its volume increases at a rate of 100 cm$^3$/s. How fast is the radius of the balloon increasing when the diameter is 50 cm?

SOLUTION We start by identifying two things:

the *given information:*

the rate of increase of the volume of air is 100 cm$^3$/s

and the *unknown:*

the rate of increase of the radius when the diameter is 50 cm

In order to express these quantities mathematically, we introduce some suggestive *notation*:

Let $V$ be the volume of the balloon and let $r$ be its radius.

The key thing to remember is that rates of change are derivatives. In this problem, the volume and the radius are both functions of the time $t$. The rate of increase of the volume with respect to time is the derivative $dV/dt$, and the rate of increase of the radius is $dr/dt$. We can therefore restate the given and the unknown as follows:

$$\text{Given:} \qquad \frac{dV}{dt} = 100 \text{ cm}^3/\text{s}$$

$$\text{Unknown:} \qquad \frac{dr}{dt} \quad \text{when } r = 25 \text{ cm}$$

PS The second stage of problem solving is to think of a plan for connecting the given and the unknown.

In order to connect $dV/dt$ and $dr/dt$, we first relate $V$ and $r$ by the formula for the volume of a sphere:

$$V = \tfrac{4}{3}\pi r^3$$

In order to use the given information, we differentiate each side of this equation with respect to $t$. To differentiate the right side, we need to use the Chain Rule:

$$\frac{dV}{dt} = \frac{dV}{dr}\frac{dr}{dt} = 4\pi r^2 \frac{dr}{dt}$$

Now we solve for the unknown quantity:

$$\frac{dr}{dt} = \frac{1}{4\pi r^2}\frac{dV}{dt}$$

Notice that, although $dV/dt$ is constant, $dr/dt$ is *not* constant.

If we put $r = 25$ and $dV/dt = 100$ in this equation, we obtain

$$\frac{dr}{dt} = \frac{1}{4\pi(25)^2}100 = \frac{1}{25\pi}$$

The radius of the balloon is increasing at the rate of $1/(25\pi) \approx 0.0127$ cm/s.

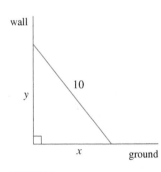

**FIGURE 1**

**EXAMPLE 2**  A ladder 10 ft long rests against a vertical wall. If the bottom of the ladder slides away from the wall at a rate of 1 ft/s, how fast is the top of the ladder sliding down the wall when the bottom of the ladder is 6 ft from the wall?

SOLUTION  We first draw a diagram and label it as in Figure 1. Let $x$ feet be the distance from the bottom of the ladder to the wall and $y$ feet the distance from the top of the ladder to the ground. Note that $x$ and $y$ are both functions of $t$ (time, measured in seconds).

We are given that $dx/dt = 1$ ft/s and we are asked to find $dy/dt$ when $x = 6$ ft (see Figure 2). In this problem, the relationship between $x$ and $y$ is given by the Pythagorean Theorem:

$$x^2 + y^2 = 100$$

Differentiating each side with respect to $t$ using the Chain Rule, we have

$$2x\frac{dx}{dt} + 2y\frac{dy}{dt} = 0$$

and solving this equation for the desired rate, we obtain

$$\frac{dy}{dt} = -\frac{x}{y}\frac{dx}{dt}$$

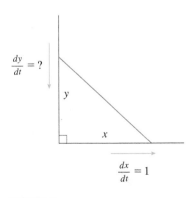

**FIGURE 2**

When $x = 6$, the Pythagorean Theorem gives $y = 8$ and so, substituting these values and $dx/dt = 1$, we have

$$\frac{dy}{dt} = -\frac{6}{8}(1) = -\frac{3}{4} \text{ ft/s}$$

The fact that $dy/dt$ is negative means that the distance from the top of the ladder to the ground is *decreasing* at a rate of $\frac{3}{4}$ ft/s. In other words, the top of the ladder is sliding down the wall at a rate of $\frac{3}{4}$ ft/s.

**EXAMPLE 3** A water tank has the shape of an inverted circular cone with base radius 2 m and height 4 m. If water is being pumped into the tank at a rate of 2 m³/min, find the rate at which the water level is rising when the water is 3 m deep.

SOLUTION We first sketch the cone and label it as in Figure 3. Let $V$, $r$, and $h$ be the volume of the water, the radius of the surface, and the height of the water at time $t$, where $t$ is measured in minutes.

We are given that $dV/dt = 2$ m³/min and we are asked to find $dh/dt$ when $h$ is 3 m. The quantities $V$ and $h$ are related by the equation

$$V = \tfrac{1}{3}\pi r^2 h$$

**FIGURE 3**

but it is very useful to express $V$ as a function of $h$ alone. In order to eliminate $r$, we use the similar triangles in Figure 3 to write

$$\frac{r}{h} = \frac{2}{4} \qquad r = \frac{h}{2}$$

and the expression for $V$ becomes

$$V = \frac{1}{3}\pi\left(\frac{h}{2}\right)^2 h = \frac{\pi}{12}h^3$$

Now we can differentiate each side with respect to $t$:

$$\frac{dV}{dt} = \frac{\pi}{4}h^2\frac{dh}{dt}$$

so

$$\frac{dh}{dt} = \frac{4}{\pi h^2}\frac{dV}{dt}$$

Substituting $h = 3$ m and $dV/dt = 2$ m³/min, we have

$$\frac{dh}{dt} = \frac{4}{\pi(3)^2} \cdot 2 = \frac{8}{9\pi}$$

The water level is rising at a rate of $8/(9\pi) \approx 0.28$ m/min.

PS Look back: What have we learned from Examples 1–3 that will help us solve future problems?

⊘ **WARNING** A common error is to substitute the given numerical information (for quantities that vary with time) too early. This should be done only *after* the differentiation. (Step 7 follows Step 6.) For instance, in Example 3 we dealt with general values of $h$ until we finally substituted $h = 3$ at the last stage. (If we had put $h = 3$ earlier, we would have gotten $dV/dt = 0$, which is clearly wrong.)

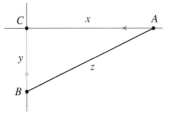

**FIGURE 4**

**Problem Solving Strategy** It is useful to recall some of the problem-solving principles from page 97 and adapt them to related rates in light of our experience in Examples 1–3:

1. Read the problem carefully.
2. Draw a diagram if possible.
3. Introduce notation. Assign symbols to all quantities that are functions of time.
4. Express the given information and the required rate in terms of derivatives.
5. Write an equation that relates the various quantities of the problem. If necessary, use the geometry of the situation to eliminate one of the variables by substitution (as in Example 3).
6. Use the Chain Rule to differentiate both sides of the equation with respect to $t$.
7. Substitute the given information into the resulting equation and solve for the unknown rate.

The following examples are further illustrations of the strategy.

**V** **EXAMPLE 4** Car A is traveling west at 50 mi/h and car B is traveling north at 60 mi/h. Both are headed for the intersection of the two roads. At what rate are the cars approaching each other when car A is 0.3 mi and car B is 0.4 mi from the intersection?

SOLUTION We draw Figure 4, where $C$ is the intersection of the roads. At a given time $t$, let $x$ be the distance from car A to $C$, let $y$ be the distance from car B to $C$, and let $z$ be the distance between the cars, where $x$, $y$, and $z$ are measured in miles.

We are given that $dx/dt = -50$ mi/h and $dy/dt = -60$ mi/h. (The derivatives are negative because $x$ and $y$ are decreasing.) We are asked to find $dz/dt$. The equation that relates $x$, $y$, and $z$ is given by the Pythagorean Theorem:

$$z^2 = x^2 + y^2$$

Differentiating each side with respect to $t$, we have

$$2z \frac{dz}{dt} = 2x \frac{dx}{dt} + 2y \frac{dy}{dt}$$

$$\frac{dz}{dt} = \frac{1}{z}\left(x \frac{dx}{dt} + y \frac{dy}{dt}\right)$$

When $x = 0.3$ mi and $y = 0.4$ mi, the Pythagorean Theorem gives $z = 0.5$ mi, so

$$\frac{dz}{dt} = \frac{1}{0.5}[0.3(-50) + 0.4(-60)]$$

$$= -78 \text{ mi/h}$$

The cars are approaching each other at a rate of 78 mi/h. ▬

**V** **EXAMPLE 5** A man walks along a straight path at a speed of 4 ft/s. A searchlight is located on the ground 20 ft from the path and is kept focused on the man. At what rate is the searchlight rotating when the man is 15 ft from the point on the path closest to the searchlight?

SOLUTION We draw Figure 5 and let $x$ be the distance from the man to the point on the path closest to the searchlight. We let $\theta$ be the angle between the beam of the searchlight and the perpendicular to the path.

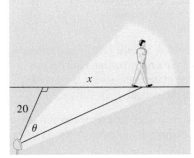

**FIGURE 5**

We are given that $dx/dt = 4$ ft/s and are asked to find $d\theta/dt$ when $x = 15$. The equation that relates $x$ and $\theta$ can be written from Figure 5:

$$\frac{x}{20} = \tan\theta \qquad x = 20\tan\theta$$

Differentiating each side with respect to $t$, we get

$$\frac{dx}{dt} = 20\sec^2\theta\,\frac{d\theta}{dt}$$

so

$$\frac{d\theta}{dt} = \frac{1}{20}\cos^2\theta\,\frac{dx}{dt}$$

$$= \frac{1}{20}\cos^2\theta\,(4) = \frac{1}{5}\cos^2\theta$$

When $x = 15$, the length of the beam is 25, so $\cos\theta = \frac{4}{5}$ and

$$\frac{d\theta}{dt} = \frac{1}{5}\left(\frac{4}{5}\right)^2 = \frac{16}{125} = 0.128$$

The searchlight is rotating at a rate of 0.128 rad/s.

## 2.8 Exercises

1. If $V$ is the volume of a cube with edge length $x$ and the cube expands as time passes, find $dV/dt$ in terms of $dx/dt$.

2. (a) If $A$ is the area of a circle with radius $r$ and the circle expands as time passes, find $dA/dt$ in terms of $dr/dt$.
   (b) Suppose oil spills from a ruptured tanker and spreads in a circular pattern. If the radius of the oil spill increases at a constant rate of 1 m/s, how fast is the area of the spill increasing when the radius is 30 m?

3. Each side of a square is increasing at a rate of 6 cm/s. At what rate is the area of the square increasing when the area of the square is 16 cm$^2$?

4. The length of a rectangle is increasing at a rate of 8 cm/s and its width is increasing at a rate of 3 cm/s. When the length is 20 cm and the width is 10 cm, how fast is the area of the rectangle increasing?

5. A cylindrical tank with radius 5 m is being filled with water at a rate of 3 m$^3$/min. How fast is the height of the water increasing?

6. The radius of a sphere is increasing at a rate of 4 mm/s. How fast is the volume increasing when the diameter is 80 mm?

7. Suppose $y = \sqrt{2x + 1}$, where $x$ and $y$ are functions of $t$.
   (a) If $dx/dt = 3$, find $dy/dt$ when $x = 4$.
   (b) If $dy/dt = 5$, find $dx/dt$ when $x = 12$.

8. Suppose $4x^2 + 9y^2 = 36$, where $x$ and $y$ are functions of $t$.
   (a) If $dy/dt = \frac{1}{3}$, find $dx/dt$ when $x = 2$ and $y = \frac{2}{3}\sqrt{5}$.
   (b) If $dx/dt = 3$, find $dy/dt$ when $x = -2$ and $y = \frac{2}{3}\sqrt{5}$.

9. If $x^2 + y^2 + z^2 = 9$, $dx/dt = 5$, and $dy/dt = 4$, find $dz/dt$ when $(x, y, z) = (2, 2, 1)$.

10. A particle is moving along a hyperbola $xy = 8$. As it reaches the point $(4, 2)$, the $y$-coordinate is decreasing at a rate of 3 cm/s. How fast is the $x$-coordinate of the point changing at that instant?

**11–14**

(a) What quantities are given in the problem?
(b) What is the unknown?
(c) Draw a picture of the situation for any time $t$.
(d) Write an equation that relates the quantities.
(e) Finish solving the problem.

11. A plane flying horizontally at an altitude of 1 mi and a speed of 500 mi/h passes directly over a radar station. Find the rate at which the distance from the plane to the station is increasing when it is 2 mi away from the station.

12. If a snowball melts so that its surface area decreases at a rate of 1 cm$^2$/min, find the rate at which the diameter decreases when the diameter is 10 cm.

---

🖩 Graphing calculator or computer required    **1.** Homework Hints available at stewartcalculus.com

**13.** A street light is mounted at the top of a 15-ft-tall pole. A man 6 ft tall walks away from the pole with a speed of 5 ft/s along a straight path. How fast is the tip of his shadow moving when he is 40 ft from the pole?

**14.** At noon, ship A is 150 km west of ship B. Ship A is sailing east at 35 km/h and ship B is sailing north at 25 km/h. How fast is the distance between the ships changing at 4:00 PM?

**15.** Two cars start moving from the same point. One travels south at 60 mi/h and the other travels west at 25 mi/h. At what rate is the distance between the cars increasing two hours later?

**16.** A spotlight on the ground shines on a wall 12 m away. If a man 2 m tall walks from the spotlight toward the building at a speed of 1.6 m/s, how fast is the length of his shadow on the building decreasing when he is 4 m from the building?

**17.** A man starts walking north at 4 ft/s from a point P. Five minutes later a woman starts walking south at 5 ft/s from a point 500 ft due east of P. At what rate are the people moving apart 15 min after the woman starts walking?

**18.** A baseball diamond is a square with side 90 ft. A batter hits the ball and runs toward first base with a speed of 24 ft/s.
(a) At what rate is his distance from second base decreasing when he is halfway to first base?
(b) At what rate is his distance from third base increasing at the same moment?

90 ft

**19.** The altitude of a triangle is increasing at a rate of 1 cm/min while the area of the triangle is increasing at a rate of 2 cm²/min. At what rate is the base of the triangle changing when the altitude is 10 cm and the area is 100 cm²?

**20.** A boat is pulled into a dock by a rope attached to the bow of the boat and passing through a pulley on the dock that is 1 m higher than the bow of the boat. If the rope is pulled in at a rate of 1 m/s, how fast is the boat approaching the dock when it is 8 m from the dock?

**21.** At noon, ship A is 100 km west of ship B. Ship A is sailing south at 35 km/h and ship B is sailing north at 25 km/h. How fast is the distance between the ships changing at 4:00 PM?

**22.** A particle moves along the curve $y = 2\sin(\pi x/2)$. As the particle passes through the point $\left(\frac{1}{3}, 1\right)$, its $x$-coordinate increases at a rate of $\sqrt{10}$ cm/s. How fast is the distance from the particle to the origin changing at this instant?

**23.** Water is leaking out of an inverted conical tank at a rate of 10,000 cm³/min at the same time that water is being pumped into the tank at a constant rate. The tank has height 6 m and the diameter at the top is 4 m. If the water level is rising at a rate of 20 cm/min when the height of the water is 2 m, find the rate at which water is being pumped into the tank.

**24.** A trough is 10 ft long and its ends have the shape of isosceles triangles that are 3 ft across at the top and have a height of 1 ft. If the trough is being filled with water at a rate of 12 ft³/min, how fast is the water level rising when the water is 6 inches deep?

**25.** A water trough is 10 m long and a cross-section has the shape of an isosceles trapezoid that is 30 cm wide at the bottom, 80 cm wide at the top, and has height 50 cm. If the trough is being filled with water at the rate of 0.2 m³/min, how fast is the water level rising when the water is 30 cm deep?

**26.** A swimming pool is 20 ft wide, 40 ft long, 3 ft deep at the shallow end, and 9 ft deep at its deepest point. A cross-section is shown in the figure. If the pool is being filled at a rate of 0.8 ft³/min, how fast is the water level rising when the depth at the deepest point is 5 ft?

6    12    16    6     3   6

**27.** Gravel is being dumped from a conveyor belt at a rate of 30 ft³/min, and its coarseness is such that it forms a pile in the shape of a cone whose base diameter and height are always equal. How fast is the height of the pile increasing when the pile is 10 ft high?

**28.** A kite 100 ft above the ground moves horizontally at a speed of 8 ft/s. At what rate is the angle between the string and the horizontal decreasing when 200 ft of string has been let out?

**29.** Two sides of a triangle are 4 m and 5 m in length and the angle between them is increasing at a rate of 0.06 rad/s. Find the rate at which the area of the triangle is increasing when the angle between the sides of fixed length is $\pi/3$.

**30.** How fast is the angle between the ladder and the ground changing in Example 2 when the bottom of the ladder is 6 ft from the wall?

**31.** The top of a ladder slides down a vertical wall at a rate of 0.15 m/s. At the moment when the bottom of the ladder is 3 m from the wall, it slides away from the wall at a rate of 0.2 m/s. How long is the ladder?

**32.** A faucet is filling a hemispherical basin of diameter 60 cm with water at a rate of 2 L/min. Find the rate at which the water is rising in the basin when it is half full. [Use the following facts: 1 L is 1000 cm³. The volume of the portion of a sphere with radius $r$ from the bottom to a height $h$ is $V = \pi\left(rh^2 - \frac{1}{3}h^3\right)$, as we will show in Chapter 5.]

**33.** Boyle's Law states that when a sample of gas is compressed at a constant temperature, the pressure $P$ and volume $V$ satisfy the equation $PV = C$, where $C$ is a constant. Suppose that at a certain instant the volume is 600 cm³, the pressure is 150 kPa, and the pressure is increasing at a rate of 20 kPa/min. At what rate is the volume decreasing at this instant?

**34.** When air expands adiabatically (without gaining or losing heat), its pressure $P$ and volume $V$ are related by the equation $PV^{1.4} = C$, where $C$ is a constant. Suppose that at a certain instant the volume is 400 cm³ and the pressure is 80 kPa and is decreasing at a rate of 10 kPa/min. At what rate is the volume increasing at this instant?

**35.** If two resistors with resistances $R_1$ and $R_2$ are connected in parallel, as in the figure, then the total resistance $R$, measured in ohms ($\Omega$), is given by

$$\frac{1}{R} = \frac{1}{R_1} + \frac{1}{R_2}$$

If $R_1$ and $R_2$ are increasing at rates of 0.3 $\Omega$/s and 0.2 $\Omega$/s, respectively, how fast is $R$ changing when $R_1 = 80\ \Omega$ and $R_2 = 100\ \Omega$?

**36.** Brain weight $B$ as a function of body weight $W$ in fish has been modeled by the power function $B = 0.007W^{2/3}$, where $B$ and $W$ are measured in grams. A model for body weight as a function of body length $L$ (measured in centimeters) is $W = 0.12L^{2.53}$. If, over 10 million years, the average length of a certain species of fish evolved from 15 cm to 20 cm at a constant rate, how fast was this species' brain growing when the average length was 18 cm?

**37.** Two sides of a triangle have lengths 12 m and 15 m. The angle between them is increasing at a rate of 2°/min. How fast is the length of the third side increasing when the angle between the sides of fixed length is 60°?

**38.** Two carts, A and B, are connected by a rope 39 ft long that passes over a pulley $P$ (see the figure). The point $Q$ is on the floor 12 ft directly beneath $P$ and between the carts. Cart A is being pulled away from $Q$ at a speed of 2 ft/s. How fast is cart B moving toward $Q$ at the instant when cart A is 5 ft from $Q$?

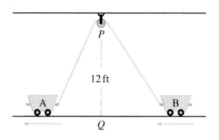

**39.** A television camera is positioned 4000 ft from the base of a rocket launching pad. The angle of elevation of the camera has to change at the correct rate in order to keep the rocket in sight. Also, the mechanism for focusing the camera has to take into account the increasing distance from the camera to the rising rocket. Let's assume the rocket rises vertically and its speed is 600 ft/s when it has risen 3000 ft.
(a) How fast is the distance from the television camera to the rocket changing at that moment?
(b) If the television camera is always kept aimed at the rocket, how fast is the camera's angle of elevation changing at that same moment?

**40.** A lighthouse is located on a small island 3 km away from the nearest point $P$ on a straight shoreline and its light makes four revolutions per minute. How fast is the beam of light moving along the shoreline when it is 1 km from $P$?

**41.** A plane flies horizontally at an altitude of 5 km and passes directly over a tracking telescope on the ground. When the angle of elevation is $\pi/3$, this angle is decreasing at a rate of $\pi/6$ rad/min. How fast is the plane traveling at that time?

**42.** A Ferris wheel with a radius of 10 m is rotating at a rate of one revolution every 2 minutes. How fast is a rider rising when his seat is 16 m above ground level?

**43.** A plane flying with a constant speed of 300 km/h passes over a ground radar station at an altitude of 1 km and climbs at an angle of 30°. At what rate is the distance from the plane to the radar station increasing a minute later?

**44.** Two people start from the same point. One walks east at 3 mi/h and the other walks northeast at 2 mi/h. How fast is the distance between the people changing after 15 minutes?

**45.** A runner sprints around a circular track of radius 100 m at a constant speed of 7 m/s. The runner's friend is standing at a distance 200 m from the center of the track. How fast is the distance between the friends changing when the distance between them is 200 m?

**46.** The minute hand on a watch is 8 mm long and the hour hand is 4 mm long. How fast is the distance between the tips of the hands changing at one o'clock?

## 2.9 Linear Approximations and Differentials

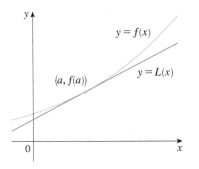

**FIGURE 1**

We have seen that a curve lies very close to its tangent line near the point of tangency. In fact, by zooming in toward a point on the graph of a differentiable function, we noticed that the graph looks more and more like its tangent line. (See Figure 2 in Section 2.1.) This observation is the basis for a method of finding approximate values of functions.

The idea is that it might be easy to calculate a value $f(a)$ of a function, but difficult (or even impossible) to compute nearby values of $f$. So we settle for the easily computed values of the linear function $L$ whose graph is the tangent line of $f$ at $(a, f(a))$. (See Figure 1.)

In other words, we use the tangent line at $(a, f(a))$ as an approximation to the curve $y = f(x)$ when $x$ is near $a$. An equation of this tangent line is

$$y = f(a) + f'(a)(x - a)$$

and the approximation

$$\boxed{1} \qquad f(x) \approx f(a) + f'(a)(x - a)$$

is called the **linear approximation** or **tangent line approximation** of $f$ at $a$. The linear function whose graph is this tangent line, that is,

$$\boxed{2} \qquad L(x) = f(a) + f'(a)(x - a)$$

is called the **linearization** of $f$ at $a$.

**V** **EXAMPLE 1** Find the linearization of the function $f(x) = \sqrt{x + 3}$ at $a = 1$ and use it to approximate the numbers $\sqrt{3.98}$ and $\sqrt{4.05}$. Are these approximations overestimates or underestimates?

SOLUTION The derivative of $f(x) = (x + 3)^{1/2}$ is

$$f'(x) = \tfrac{1}{2}(x + 3)^{-1/2} = \frac{1}{2\sqrt{x + 3}}$$

and so we have $f(1) = 2$ and $f'(1) = \tfrac{1}{4}$. Putting these values into Equation 2, we see that the linearization is

$$L(x) = f(1) + f'(1)(x - 1) = 2 + \tfrac{1}{4}(x - 1) = \frac{7}{4} + \frac{x}{4}$$

The corresponding linear approximation $\boxed{1}$ is

$$\sqrt{x + 3} \approx \frac{7}{4} + \frac{x}{4} \qquad \text{(when } x \text{ is near 1)}$$

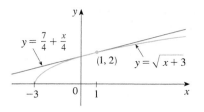

**FIGURE 2**

In particular, we have

$$\sqrt{3.98} \approx \tfrac{7}{4} + \tfrac{0.98}{4} = 1.995 \qquad \text{and} \qquad \sqrt{4.05} \approx \tfrac{7}{4} + \tfrac{1.05}{4} = 2.0125$$

The linear approximation is illustrated in Figure 2. We see that, indeed, the tangent line approximation is a good approximation to the given function when $x$ is near 1. We also see that our approximations are overestimates because the tangent line lies above the curve.

Of course, a calculator could give us approximations for $\sqrt{3.98}$ and $\sqrt{4.05}$, but the linear approximation gives an approximation *over an entire interval.*

In the following table we compare the estimates from the linear approximation in Example 1 with the true values. Notice from this table, and also from Figure 2, that the tangent line approximation gives good estimates when $x$ is close to 1 but the accuracy of the approximation deteriorates when $x$ is farther away from 1.

|  | $x$ | From $L(x)$ | Actual value |
|---|---|---|---|
| $\sqrt{3.9}$ | 0.9 | 1.975 | 1.97484176 . . . |
| $\sqrt{3.98}$ | 0.98 | 1.995 | 1.99499373 . . . |
| $\sqrt{4}$ | 1 | 2 | 2.00000000 . . . |
| $\sqrt{4.05}$ | 1.05 | 2.0125 | 2.01246117 . . . |
| $\sqrt{4.1}$ | 1.1 | 2.025 | 2.02484567 . . . |
| $\sqrt{5}$ | 2 | 2.25 | 2.23606797 . . . |
| $\sqrt{6}$ | 3 | 2.5 | 2.44948974 . . . |

How good is the approximation that we obtained in Example 1? The next example shows that by using a graphing calculator or computer we can determine an interval throughout which a linear approximation provides a specified accuracy.

**EXAMPLE 2** For what values of $x$ is the linear approximation

$$\sqrt{x+3} \approx \frac{7}{4} + \frac{x}{4}$$

accurate to within 0.5? What about accuracy to within 0.1?

SOLUTION Accuracy to within 0.5 means that the functions should differ by less than 0.5:

$$\left| \sqrt{x+3} - \left( \frac{7}{4} + \frac{x}{4} \right) \right| < 0.5$$

Equivalently, we could write

$$\sqrt{x+3} - 0.5 < \frac{7}{4} + \frac{x}{4} < \sqrt{x+3} + 0.5$$

This says that the linear approximation should lie between the curves obtained by shifting the curve $y = \sqrt{x+3}$ upward and downward by an amount 0.5. Figure 3 shows the tangent line $y = (7 + x)/4$ intersecting the upper curve $y = \sqrt{x+3} + 0.5$ at $P$

**FIGURE 3**

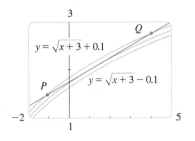

$y = \sqrt{x+3} + 0.1$

$y = \sqrt{x+3} - 0.1$

**FIGURE 4**

and $Q$. Zooming in and using the cursor, we estimate that the $x$-coordinate of $P$ is about $-2.66$ and the $x$-coordinate of $Q$ is about $8.66$. Thus we see from the graph that the approximation

$$\sqrt{x+3} \approx \frac{7}{4} + \frac{x}{4}$$

is accurate to within 0.5 when $-2.6 < x < 8.6$. (We have rounded to be safe.)

Similarly, from Figure 4 we see that the approximation is accurate to within 0.1 when $-1.1 < x < 3.9$.

## Applications to Physics

Linear approximations are often used in physics. In analyzing the consequences of an equation, a physicist sometimes needs to simplify a function by replacing it with its linear approximation. For instance, in deriving a formula for the period of a pendulum, physics textbooks obtain the expression $a_T = -g \sin \theta$ for tangential acceleration and then replace $\sin \theta$ by $\theta$ with the remark that $\sin \theta$ is very close to $\theta$ if $\theta$ is not too large. [See, for example, *Physics: Calculus*, 2d ed., by Eugene Hecht (Pacific Grove, CA, 2000), p. 431.] You can verify that the linearization of the function $f(x) = \sin x$ at $a = 0$ is $L(x) = x$ and so the linear approximation at 0 is

$$\sin x \approx x$$

(see Exercise 40). So, in effect, the derivation of the formula for the period of a pendulum uses the tangent line approximation for the sine function.

Another example occurs in the theory of optics, where light rays that arrive at shallow angles relative to the optical axis are called *paraxial rays*. In paraxial (or Gaussian) optics, both $\sin \theta$ and $\cos \theta$ are replaced by their linearizations. In other words, the linear approximations

$$\sin \theta \approx \theta \qquad \text{and} \qquad \cos \theta \approx 1$$

are used because $\theta$ is close to 0. The results of calculations made with these approximations became the basic theoretical tool used to design lenses. [See *Optics*, 4th ed., by Eugene Hecht (San Francisco, 2002), p. 154.]

In Section 11.11 we will present several other applications of the idea of linear approximations to physics and engineering.

## Differentials

The ideas behind linear approximations are sometimes formulated in the terminology and notation of *differentials*. If $y = f(x)$, where $f$ is a differentiable function, then the **differential** $dx$ is an independent variable; that is, $dx$ can be given the value of any real number. The **differential** $dy$ is then defined in terms of $dx$ by the equation

If $dx \neq 0$, we can divide both sides of Equation 3 by $dx$ to obtain

$$\frac{dy}{dx} = f'(x)$$

We have seen similar equations before, but now the left side can genuinely be interpreted as a ratio of differentials.

$$\boxed{3} \qquad dy = f'(x)\, dx$$

So $dy$ is a dependent variable; it depends on the values of $x$ and $dx$. If $dx$ is given a specific value and $x$ is taken to be some specific number in the domain of $f$, then the numerical value of $dy$ is determined.

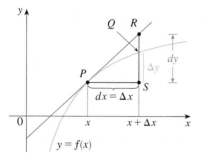

**FIGURE 5**

The geometric meaning of differentials is shown in Figure 5. Let $P(x, f(x))$ and $Q(x + \Delta x, f(x + \Delta x))$ be points on the graph of $f$ and let $dx = \Delta x$. The corresponding change in $y$ is

$$\Delta y = f(x + \Delta x) - f(x)$$

The slope of the tangent line $PR$ is the derivative $f'(x)$. Thus the directed distance from $S$ to $R$ is $f'(x) \, dx = dy$. Therefore $dy$ represents the amount that the tangent line rises or falls (the change in the linearization), whereas $\Delta y$ represents the amount that the curve $y = f(x)$ rises or falls when $x$ changes by an amount $dx$.

**EXAMPLE 3** Compare the values of $\Delta y$ and $dy$ if $y = f(x) = x^3 + x^2 - 2x + 1$ and $x$ changes (a) from 2 to 2.05 and (b) from 2 to 2.01.

SOLUTION

(a) We have

$$f(2) = 2^3 + 2^2 - 2(2) + 1 = 9$$

$$f(2.05) = (2.05)^3 + (2.05)^2 - 2(2.05) + 1 = 9.717625$$

$$\Delta y = f(2.05) - f(2) = 0.717625$$

Figure 6 shows the function in Example 3 and a comparison of $dy$ and $\Delta y$ when $a = 2$. The viewing rectangle is $[1.8, 2.5]$ by $[6, 18]$.

In general,

$$dy = f'(x) \, dx = (3x^2 + 2x - 2) \, dx$$

When $x = 2$ and $dx = \Delta x = 0.05$, this becomes

$$dy = [3(2)^2 + 2(2) - 2]0.05 = 0.7$$

**FIGURE 6**

(b)

$$f(2.01) = (2.01)^3 + (2.01)^2 - 2(2.01) + 1 = 9.140701$$

$$\Delta y = f(2.01) - f(2) = 0.140701$$

When $dx = \Delta x = 0.01$,

$$dy = [3(2)^2 + 2(2) - 2]0.01 = 0.14$$

Notice that the approximation $\Delta y \approx dy$ becomes better as $\Delta x$ becomes smaller in Example 3. Notice also that $dy$ was easier to compute than $\Delta y$. For more complicated functions it may be impossible to compute $\Delta y$ exactly. In such cases the approximation by differentials is especially useful.

In the notation of differentials, the linear approximation $\boxed{1}$ can be written as

$$f(a + dx) \approx f(a) + dy$$

For instance, for the function $f(x) = \sqrt{x + 3}$ in Example 1, we have

$$dy = f'(x) \, dx = \frac{dx}{2\sqrt{x + 3}}$$

If $a = 1$ and $dx = \Delta x = 0.05$, then

$$dy = \frac{0.05}{2\sqrt{1 + 3}} = 0.0125$$

and

$$\sqrt{4.05} = f(1.05) \approx f(1) + dy = 2.0125$$

just as we found in Example 1.

Our final example illustrates the use of differentials in estimating the errors that occur because of approximate measurements.

**V EXAMPLE 4** The radius of a sphere was measured and found to be 21 cm with a possible error in measurement of at most 0.05 cm. What is the maximum error in using this value of the radius to compute the volume of the sphere?

SOLUTION If the radius of the sphere is $r$, then its volume is $V = \frac{4}{3}\pi r^3$. If the error in the measured value of $r$ is denoted by $dr = \Delta r$, then the corresponding error in the calculated value of $V$ is $\Delta V$, which can be approximated by the differential

$$dV = 4\pi r^2 \, dr$$

When $r = 21$ and $dr = 0.05$, this becomes

$$dV = 4\pi(21)^2 0.05 \approx 277$$

The maximum error in the calculated volume is about 277 cm³.

**NOTE** Although the possible error in Example 4 may appear to be rather large, a better picture of the error is given by the **relative error**, which is computed by dividing the error by the total volume:

$$\frac{\Delta V}{V} \approx \frac{dV}{V} = \frac{4\pi r^2 \, dr}{\frac{4}{3}\pi r^3} = 3\frac{dr}{r}$$

Thus the relative error in the volume is about three times the relative error in the radius. In Example 4 the relative error in the radius is approximately $dr/r = 0.05/21 \approx 0.0024$ and it produces a relative error of about 0.007 in the volume. The errors could also be expressed as **percentage errors** of 0.24% in the radius and 0.7% in the volume.

## 2.9 Exercises

**1–4** Find the linearization $L(x)$ of the function at $a$.

**1.** $f(x) = x^4 + 3x^2$, $a = -1$

**2.** $f(x) = \sin x$, $a = \pi/6$

**3.** $f(x) = \sqrt{x}$, $a = 4$

**4.** $f(x) = x^{3/4}$, $a = 16$

**5.** Find the linear approximation of the function $f(x) = \sqrt{1-x}$ at $a = 0$ and use it to approximate the numbers $\sqrt{0.9}$ and $\sqrt{0.99}$. Illustrate by graphing $f$ and the tangent line.

**6.** Find the linear approximation of the function $g(x) = \sqrt[3]{1+x}$ at $a = 0$ and use it to approximate the numbers $\sqrt[3]{0.95}$ and $\sqrt[3]{1.1}$. Illustrate by graphing $g$ and the tangent line.

**7–10** Verify the given linear approximation at $a = 0$. Then determine the values of $x$ for which the linear approximation is accurate to within 0.1.

**7.** $\sqrt[4]{1 + 2x} \approx 1 + \frac{1}{2}x$

**8.** $(1 + x)^{-3} \approx 1 - 3x$

**9.** $1/(1 + 2x)^4 \approx 1 - 8x$

**10.** $\tan x \approx x$

**11–14** Find the differential of each function.

**11.** (a) $y = x^2 \sin 2x$      (b) $y = \sqrt{1 + t^2}$

**12.** (a) $y = s/(1 + 2s)$      (b) $y = u \cos u$

**13.** (a) $y = \tan \sqrt{t}$      (b) $y = \dfrac{1 - v^2}{1 + v^2}$

**14.** (a) $y = (t + \tan t)^5$      (b) $y = \sqrt{z + 1/z}$

**15–18** (a) Find the differential $dy$ and (b) evaluate $dy$ for the given values of $x$ and $dx$.

**15.** $y = \tan x$, $x = \pi/4$, $dx = -0.1$

**16.** $y = \cos \pi x$, $x = \frac{1}{3}$, $dx = -0.02$

**17.** $y = \sqrt{3 + x^2}$, $x = 1$, $dx = -0.1$

**18.** $y = \dfrac{x + 1}{x - 1}$, $x = 2$, $dx = 0.05$

---

Graphing calculator or computer required      **1.** Homework Hints available at stewartcalculus.com

19–22 Compute $\Delta y$ and $dy$ for the given values of $x$ and $dx = \Delta x$. Then sketch a diagram like Figure 5 showing the line segments with lengths $dx$, $dy$, and $\Delta y$.

**19.** $y = 2x - x^2$,   $x = 2$,   $\Delta x = -0.4$

**20.** $y = \sqrt{x}$,   $x = 1$,   $\Delta x = 1$

**21.** $y = 2/x$,   $x = 4$,   $\Delta x = 1$

**22.** $y = x^3$,   $x = 1$,   $\Delta x = 0.5$

23–28 Use a linear approximation (or differentials) to estimate the given number.

**23.** $(1.999)^4$

**24.** $\sin 1°$

**25.** $\sqrt[3]{1001}$

**26.** $1/4.002$

**27.** $\tan 44°$

**28.** $\sqrt{99.8}$

29–30 Explain, in terms of linear approximations or differentials, why the approximation is reasonable.

**29.** $\sec 0.08 \approx 1$

**30.** $(1.01)^6 \approx 1.06$

**31.** The edge of a cube was found to be 30 cm with a possible error in measurement of 0.1 cm. Use differentials to estimate the maximum possible error, relative error, and percentage error in computing (a) the volume of the cube and (b) the surface area of the cube.

**32.** The radius of a circular disk is given as 24 cm with a maximum error in measurement of 0.2 cm.
(a) Use differentials to estimate the maximum error in the calculated area of the disk.
(b) What is the relative error? What is the percentage error?

**33.** The circumference of a sphere was measured to be 84 cm with a possible error of 0.5 cm.
(a) Use differentials to estimate the maximum error in the calculated surface area. What is the relative error?
(b) Use differentials to estimate the maximum error in the calculated volume. What is the relative error?

**34.** Use differentials to estimate the amount of paint needed to apply a coat of paint 0.05 cm thick to a hemispherical dome with diameter 50 m.

**35.** (a) Use differentials to find a formula for the approximate volume of a thin cylindrical shell with height $h$, inner radius $r$, and thickness $\Delta r$.
(b) What is the error involved in using the formula from part (a)?

**36.** One side of a right triangle is known to be 20 cm long and the opposite angle is measured as 30°, with a possible error of $\pm 1°$.
(a) Use differentials to estimate the error in computing the length of the hypotenuse.
(b) What is the percentage error?

**37.** If a current $I$ passes through a resistor with resistance $R$, Ohm's Law states that the voltage drop is $V = RI$. If $V$ is constant and

$R$ is measured with a certain error, use differentials to show that the relative error in calculating $I$ is approximately the same (in magnitude) as the relative error in $R$.

**38.** When blood flows along a blood vessel, the flux $F$ (the volume of blood per unit time that flows past a given point) is proportional to the fourth power of the radius $R$ of the blood vessel:

$$F = kR^4$$

(This is known as Poiseuille's Law; we will show why it is true in Section 8.4.) A partially clogged artery can be expanded by an operation called angioplasty, in which a balloon-tipped catheter is inflated inside the artery in order to widen it and restore the normal blood flow.

Show that the relative change in $F$ is about four times the relative change in $R$. How will a 5% increase in the radius affect the flow of blood?

**39.** Establish the following rules for working with differentials (where $c$ denotes a constant and $u$ and $v$ are functions of $x$).
(a) $dc = 0$  (b) $d(cu) = c\,du$
(c) $d(u + v) = du + dv$  (d) $d(uv) = u\,dv + v\,du$
(e) $d\left(\dfrac{u}{v}\right) = \dfrac{v\,du - u\,dv}{v^2}$  (f) $d(x^n) = nx^{n-1}\,dx$

**40.** On page 431 of *Physics: Calculus*, 2d ed., by Eugene Hecht (Pacific Grove, CA, 2000), in the course of deriving the formula $T = 2\pi\sqrt{L/g}$ for the period of a pendulum of length $L$, the author obtains the equation $a_T = -g\sin\theta$ for the tangential acceleration of the bob of the pendulum. He then says, "for small angles, the value of $\theta$ in radians is very nearly the value of $\sin\theta$; they differ by less than 2% out to about 20°."
(a) Verify the linear approximation at 0 for the sine function:

$$\sin x \approx x$$

(b) Use a graphing device to determine the values of $x$ for which $\sin x$ and $x$ differ by less than 2%. Then verify Hecht's statement by converting from radians to degrees.

**41.** Suppose that the only information we have about a function $f$ is that $f(1) = 5$ and the graph of its *derivative* is as shown.
(a) Use a linear approximation to estimate $f(0.9)$ and $f(1.1)$.
(b) Are your estimates in part (a) too large or too small? Explain.

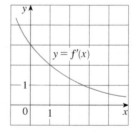

**42.** Suppose that we don't have a formula for $g(x)$ but we know that $g(2) = -4$ and $g'(x) = \sqrt{x^2 + 5}$ for all $x$.
(a) Use a linear approximation to estimate $g(1.95)$ and $g(2.05)$.
(b) Are your estimates in part (a) too large or too small? Explain.

◫ **TAYLOR POLYNOMIALS**

The tangent line approximation $L(x)$ is the best first-degree (linear) approximation to $f(x)$ near $x = a$ because $f(x)$ and $L(x)$ have the same rate of change (derivative) at $a$. For a better approximation than a linear one, let's try a second-degree (quadratic) approximation $P(x)$. In other words, we approximate a curve by a parabola instead of by a straight line. To make sure that the approximation is a good one, we stipulate the following:

(i) $P(a) = f(a)$      ($P$ and $f$ should have the same value at $a$.)

(ii) $P'(a) = f'(a)$      ($P$ and $f$ should have the same rate of change at $a$.)

(iii) $P''(a) = f''(a)$      (The slopes of $P$ and $f$ should change at the same rate at $a$.)

1. Find the quadratic approximation $P(x) = A + Bx + Cx^2$ to the function $f(x) = \cos x$ that satisfies conditions (i), (ii), and (iii) with $a = 0$. Graph $P, f$, and the linear approximation $L(x) = 1$ on a common screen. Comment on how well the functions $P$ and $L$ approximate $f$.

2. Determine the values of $x$ for which the quadratic approximation $f(x) \approx P(x)$ in Problem 1 is accurate to within 0.1. [*Hint:* Graph $y = P(x)$, $y = \cos x - 0.1$, and $y = \cos x + 0.1$ on a common screen.]

3. To approximate a function $f$ by a quadratic function $P$ near a number $a$, it is best to write $P$ in the form

$$P(x) = A + B(x - a) + C(x - a)^2$$

Show that the quadratic function that satisfies conditions (i), (ii), and (iii) is

$$P(x) = f(a) + f'(a)(x - a) + \tfrac{1}{2}f''(a)(x - a)^2$$

4. Find the quadratic approximation to $f(x) = \sqrt{x + 3}$ near $a = 1$. Graph $f$, the quadratic approximation, and the linear approximation from Example 2 in Section 2.9 on a common screen. What do you conclude?

5. Instead of being satisfied with a linear or quadratic approximation to $f(x)$ near $x = a$, let's try to find better approximations with higher-degree polynomials. We look for an $n$th-degree polynomial

$$T_n(x) = c_0 + c_1(x - a) + c_2(x - a)^2 + c_3(x - a)^3 + \cdots + c_n(x - a)^n$$

such that $T_n$ and its first $n$ derivatives have the same values at $x = a$ as $f$ and its first $n$ derivatives. By differentiating repeatedly and setting $x = a$, show that these conditions are satisfied if $c_0 = f(a)$, $c_1 = f'(a)$, $c_2 = \tfrac{1}{2}f''(a)$, and in general

$$c_k = \frac{f^{(k)}(a)}{k!}$$

where $k! = 1 \cdot 2 \cdot 3 \cdot 4 \cdot \cdots \cdot k$. The resulting polynomial

$$T_n(x) = f(a) + f'(a)(x - a) + \frac{f''(a)}{2!}(x - a)^2 + \cdots + \frac{f^{(n)}(a)}{n!}(x - a)^n$$

is called the **$n$th-degree Taylor polynomial of $f$ centered at $a$.**

6. Find the 8th-degree Taylor polynomial centered at $a = 0$ for the function $f(x) = \cos x$. Graph $f$ together with the Taylor polynomials $T_2, T_4, T_6, T_8$ in the viewing rectangle $[-5, 5]$ by $[-1.4, 1.4]$ and comment on how well they approximate $f$.

---

◫ Graphing calculator or computer required

## Concept Check

1. Write an expression for the slope of the tangent line to the curve $y = f(x)$ at the point $(a, f(a))$.

2. Suppose an object moves along a straight line with position $f(t)$ at time $t$. Write an expression for the instantaneous velocity of the object at time $t = a$. How can you interpret this velocity in terms of the graph of $f$?

3. If $y = f(x)$ and $x$ changes from $x_1$ to $x_2$, write expressions for the following.
   (a) The average rate of change of $y$ with respect to $x$ over the interval $[x_1, x_2]$.
   (b) The instantaneous rate of change of $y$ with respect to $x$ at $x = x_1$.

4. Define the derivative $f'(a)$. Discuss two ways of interpreting this number.

5. (a) What does it mean for $f$ to be differentiable at $a$?
   (b) What is the relation between the differentiability and continuity of a function?
   (c) Sketch the graph of a function that is continuous but not differentiable at $a = 2$.

6. Describe several ways in which a function can fail to be differentiable. Illustrate with sketches.

7. What are the second and third derivatives of a function $f$? If $f$ is the position function of an object, how can you interpret $f''$ and $f'''$?

8. State each differentiation rule both in symbols and in words.
   (a) The Power Rule
   (b) The Constant Multiple Rule
   (c) The Sum Rule
   (d) The Difference Rule
   (e) The Product Rule
   (f) The Quotient Rule
   (g) The Chain Rule

9. State the derivative of each function.
   (a) $y = x^n$
   (b) $y = \sin x$
   (c) $y = \cos x$
   (d) $y = \tan x$
   (e) $y = \csc x$
   (f) $y = \sec x$
   (g) $y = \cot x$

10. Explain how implicit differentiation works.

11. Give several examples of how the derivative can be interpreted as a rate of change in physics, chemistry, biology, economics, or other sciences.

12. (a) Write an expression for the linearization of $f$ at $a$.
    (b) If $y = f(x)$, write an expression for the differential $dy$.
    (c) If $dx = \Delta x$, draw a picture showing the geometric meanings of $\Delta y$ and $dy$.

## True-False Quiz

Determine whether the statement is true or false. If it is true, explain why. If it is false, explain why or give an example that disproves the statement.

1. If $f$ is continuous at $a$, then $f$ is differentiable at $a$.

2. If $f$ and $g$ are differentiable, then
$$\frac{d}{dx}[f(x) + g(x)] = f'(x) + g'(x)$$

3. If $f$ and $g$ are differentiable, then
$$\frac{d}{dx}[f(x)g(x)] = f'(x)g'(x)$$

4. If $f$ and $g$ are differentiable, then
$$\frac{d}{dx}[f(g(x))] = f'(g(x))g'(x)$$

5. If $f$ is differentiable, then $\dfrac{d}{dx}\sqrt{f(x)} = \dfrac{f'(x)}{2\sqrt{f(x)}}$.

6. If $f$ is differentiable, then $\dfrac{d}{dx}f(\sqrt{x}) = \dfrac{f'(x)}{2\sqrt{x}}$.

7. $\dfrac{d}{dx}|x^2 + x| = |2x + 1|$

8. If $f'(r)$ exists, then $\lim_{x \to r} f(x) = f(r)$.

9. If $g(x) = x^5$, then $\lim\limits_{x \to 2} \dfrac{g(x) - g(2)}{x - 2} = 80$.

10. $\dfrac{d^2y}{dx^2} = \left(\dfrac{dy}{dx}\right)^2$

11. An equation of the tangent line to the parabola $y = x^2$ at $(-2, 4)$ is $y - 4 = 2x(x + 2)$.

12. $\dfrac{d}{dx}(\tan^2 x) = \dfrac{d}{dx}(\sec^2 x)$

## Exercises

**1.** The displacement (in meters) of an object moving in a straight line is given by $s = 1 + 2t + \frac{1}{4}t^2$, where $t$ is measured in seconds.

(a) Find the average velocity over each time period.
  (i) $[1, 3]$     (ii) $[1, 2]$
  (iii) $[1, 1.5]$     (iv) $[1, 1.1]$

(b) Find the instantaneous velocity when $t = 1$.

**2.** The graph of $f$ is shown. State, with reasons, the numbers at which $f$ is not differentiable.

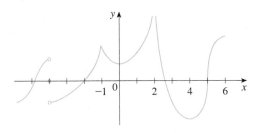

**3–4** Trace or copy the graph of the function. Then sketch a graph of its derivative directly beneath.

**3.**        **4.**

**5.** The figure shows the graphs of $f$, $f'$, and $f''$. Identify each curve, and explain your choices.

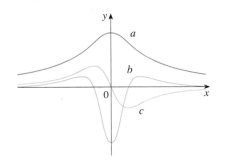

**6.** Find a function $f$ and a number $a$ such that

$$\lim_{h \to 0} \frac{(2 + h)^6 - 64}{h} = f'(a)$$

**7.** The total cost of repaying a student loan at an interest rate of $r\%$ per year is $C = f(r)$.

(a) What is the meaning of the derivative $f'(r)$? What are its units?

(b) What does the statement $f'(10) = 1200$ mean?

(c) Is $f'(r)$ always positive or does it change sign?

**8.** The *total fertility rate* at time $t$, denoted by $F(t)$, is an estimate of the average number of children born to each woman (assuming that current birth rates remain constant). The graph of the total fertility rate in the United States shows the fluctuations from 1940 to 1990.

(a) Estimate the values of $F'(1950)$, $F'(1965)$, and $F'(1987)$.

(b) What are the meanings of these derivatives?

(c) Can you suggest reasons for the values of these derivatives?

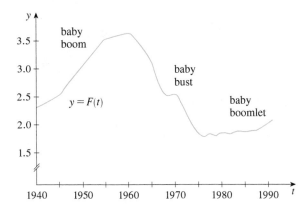

**9.** Let $C(t)$ be the total value of US currency (coins and banknotes) in circulation at time $t$. The table gives values of this function from 1980 to 2000, as of September 30, in billions of dollars. Interpret and estimate the value of $C'(1990)$.

| $t$ | 1980 | 1985 | 1990 | 1995 | 2000 |
|---|---|---|---|---|---|
| $C(t)$ | 129.9 | 187.3 | 271.9 | 409.3 | 568.6 |

**10–11** Find $f'(x)$ from first principles, that is, directly from the definition of a derivative.

**10.** $f(x) = \dfrac{4 - x}{3 + x}$

**11.** $f(x) = x^3 + 5x + 4$

**12.** (a) If $f(x) = \sqrt{3 - 5x}$, use the definition of a derivative to find $f'(x)$.

(b) Find the domains of $f$ and $f'$.

(c) Graph $f$ and $f'$ on a common screen. Compare the graphs to see whether your answer to part (a) is reasonable.

**13–40** Calculate $y'$.

**13.** $y = (x^2 + x^3)^4$

**14.** $y = \dfrac{1}{\sqrt{x}} - \dfrac{1}{\sqrt[5]{x^3}}$

**15.** $y = \dfrac{x^2 - x + 2}{\sqrt{x}}$

**16.** $y = \dfrac{\tan x}{1 + \cos x}$

**17.** $y = x^2 \sin \pi x$

**18.** $y = \left(x + \dfrac{1}{x^2}\right)^{\sqrt{7}}$

Graphing calculator or computer required

**19.** $y = \dfrac{t^4 - 1}{t^4 + 1}$

**20.** $y = \sin(\cos x)$

**21.** $y = \tan \sqrt{1 - x}$

**22.** $y = \dfrac{1}{\sin(x - \sin x)}$

**23.** $xy^4 + x^2 y = x + 3y$

**24.** $y = \sec(1 + x^2)$

**25.** $y = \dfrac{\sec 2\theta}{1 + \tan 2\theta}$

**26.** $x^2 \cos y + \sin 2y = xy$

**27.** $y = (1 - x^{-1})^{-1}$

**28.** $y = 1/\sqrt[3]{x + \sqrt{x}}$

**29.** $\sin(xy) = x^2 - y$

**30.** $y = \sqrt{\sin \sqrt{x}}$

**31.** $y = \cot(3x^2 + 5)$

**32.** $y = \dfrac{(x + \lambda)^4}{x^4 + \lambda^4}$

**33.** $y = \sqrt{x} \cos \sqrt{x}$

**34.** $y = \dfrac{\sin mx}{x}$

**35.** $y = \tan^2(\sin \theta)$

**36.** $x \tan y = y - 1$

**37.** $y = \sqrt[5]{x \tan x}$

**38.** $y = \dfrac{(x - 1)(x - 4)}{(x - 2)(x - 3)}$

**39.** $y = \sin(\tan \sqrt{1 + x^3})$

**40.** $y = \sin^2(\cos \sqrt{\sin \pi x})$

**41.** If $f(t) = \sqrt{4t + 1}$, find $f''(2)$.

**42.** If $g(\theta) = \theta \sin \theta$, find $g''(\pi/6)$.

**43.** Find $y''$ if $x^6 + y^6 = 1$.

**44.** Find $f^{(n)}(x)$ if $f(x) = 1/(2 - x)$.

**45–46** Find the limit.

**45.** $\displaystyle\lim_{x \to 0} \dfrac{\sec x}{1 - \sin x}$

**46.** $\displaystyle\lim_{t \to 0} \dfrac{t^3}{\tan^3 2t}$

**47–48** Find an equation of the tangent to the curve at the given point.

**47.** $y = 4 \sin^2 x$,   $(\pi/6, 1)$

**48.** $y = \dfrac{x^2 - 1}{x^2 + 1}$,   $(0, -1)$

**49–50** Find equations of the tangent line and normal line to the curve at the given point.

**49.** $y = \sqrt{1 + 4 \sin x}$,   $(0, 1)$

**50.** $x^2 + 4xy + y^2 = 13$,   $(2, 1)$

**51.** (a) If $f(x) = x\sqrt{5 - x}$, find $f'(x)$.
    (b) Find equations of the tangent lines to the curve
      $y = x\sqrt{5 - x}$ at the points $(1, 2)$ and $(4, 4)$.
    (c) Illustrate part (b) by graphing the curve and tangent lines
      on the same screen.
    (d) Check to see that your answer to part (a) is reasonable by
      comparing the graphs of $f$ and $f'$.

**52.** (a) If $f(x) = 4x - \tan x$, $-\pi/2 < x < \pi/2$, find $f'$ and $f''$.
    (b) Check to see that your answers to part (a) are reasonable
      by comparing the graphs of $f$, $f'$, and $f''$.

**53.** At what points on the curve $y = \sin x + \cos x$, $0 \le x \le 2\pi$, is the tangent line horizontal?

**54.** Find the points on the ellipse $x^2 + 2y^2 = 1$ where the tangent line has slope 1.

**55.** Find a parabola $y = ax^2 + bx + c$ that passes through the point $(1, 4)$ and whose tangent lines at $x = -1$ and $x = 5$ have slopes 6 and $-2$, respectively.

**56.** How many tangent lines to the curve $y = x/(x + 1)$ pass through the point $(1, 2)$? At which points do these tangent lines touch the curve?

**57.** If $f(x) = (x - a)(x - b)(x - c)$, show that

$$\frac{f'(x)}{f(x)} = \frac{1}{x - a} + \frac{1}{x - b} + \frac{1}{x - c}$$

**58.** (a) By differentiating the double-angle formula

$$\cos 2x = \cos^2 x - \sin^2 x$$

     obtain the double-angle formula for the sine function.
    (b) By differentiating the addition formula

$$\sin(x + a) = \sin x \cos a + \cos x \sin a$$

     obtain the addition formula for the cosine function.

**59.** Suppose that $h(x) = f(x)g(x)$ and $F(x) = f(g(x))$, where $f(2) = 3$, $g(2) = 5$, $g'(2) = 4$, $f'(2) = -2$, and $f'(5) = 11$. Find (a) $h'(2)$ and (b) $F'(2)$.

**60.** If $f$ and $g$ are the functions whose graphs are shown, let $P(x) = f(x)g(x)$, $Q(x) = f(x)/g(x)$, and $C(x) = f(g(x))$. Find (a) $P'(2)$, (b) $Q'(2)$, and (c) $C'(2)$.

**61–68** Find $f'$ in terms of $g'$.

**61.** $f(x) = x^2 g(x)$

**62.** $f(x) = g(x^2)$

**63.** $f(x) = [g(x)]^2$

**64.** $f(x) = x^a g(x^b)$

**65.** $f(x) = g(g(x))$

**66.** $f(x) = \sin(g(x))$

**67.** $f(x) = g(\sin x)$

**68.** $f(x) = g(\tan \sqrt{x})$

**69–71** Find $h'$ in terms of $f'$ and $g'$.

**69.** $h(x) = \dfrac{f(x)g(x)}{f(x) + g(x)}$

**70.** $h(x) = \sqrt{\dfrac{f(x)}{g(x)}}$

**71.** $h(x) = f(g(\sin 4x))$

**72.** A particle moves along a horizontal line so that its coordinate at time $t$ is $x = \sqrt{b^2 + c^2t^2}$, $t \geqslant 0$, where $b$ and $c$ are positive constants.
(a) Find the velocity and acceleration functions.
(b) Show that the particle always moves in the positive direction.

**73.** A particle moves on a vertical line so that its coordinate at time $t$ is $y = t^3 - 12t + 3$, $t \geqslant 0$.
(a) Find the velocity and acceleration functions.
(b) When is the particle moving upward and when is it moving downward?
(c) Find the distance that the particle travels in the time interval $0 \leqslant t \leqslant 3$.
(d) Graph the position, velocity, and acceleration functions for $0 \leqslant t \leqslant 3$.
(e) When is the particle speeding up? When is it slowing down?

**74.** The volume of a right circular cone is $V = \frac{1}{3}\pi r^2 h$, where $r$ is the radius of the base and $h$ is the height.
(a) Find the rate of change of the volume with respect to the height if the radius is constant.
(b) Find the rate of change of the volume with respect to the radius if the height is constant.

**75.** The mass of part of a wire is $x\left(1 + \sqrt{x}\right)$ kilograms, where $x$ is measured in meters from one end of the wire. Find the linear density of the wire when $x = 4$ m.

**76.** The cost, in dollars, of producing $x$ units of a certain commodity is
$$C(x) = 920 + 2x - 0.02x^2 + 0.00007x^3$$
(a) Find the marginal cost function.
(b) Find $C'(100)$ and explain its meaning.
(c) Compare $C'(100)$ with the cost of producing the 101st item.

**77.** The volume of a cube is increasing at a rate of 10 cm³/min. How fast is the surface area increasing when the length of an edge is 30 cm?

**78.** A paper cup has the shape of a cone with height 10 cm and radius 3 cm (at the top). If water is poured into the cup at a rate of 2 cm³/s, how fast is the water level rising when the water is 5 cm deep?

**79.** A balloon is rising at a constant speed of 5 ft/s. A boy is cycling along a straight road at a speed of 15 ft/s. When he passes under the balloon, it is 45 ft above him. How fast is the distance between the boy and the balloon increasing 3 s later?

**80.** A waterskier skis over the ramp shown in the figure at a speed of 30 ft/s. How fast is she rising as she leaves the ramp?

**81.** The angle of elevation of the sun is decreasing at a rate of 0.25 rad/h. How fast is the shadow cast by a 400-ft-tall building increasing when the angle of elevation of the sun is $\pi/6$?

**82.** (a) Find the linear approximation to $f(x) = \sqrt{25 - x^2}$ near 3.
(b) Illustrate part (a) by graphing $f$ and the linear approximation.
(c) For what values of $x$ is the linear approximation accurate to within 0.1?

**83.** (a) Find the linearization of $f(x) = \sqrt[3]{1 + 3x}$ at $a = 0$. State the corresponding linear approximation and use it to give an approximate value for $\sqrt[3]{1.03}$.
(b) Determine the values of $x$ for which the linear approximation given in part (a) is accurate to within 0.1.

**84.** Evaluate $dy$ if $y = x^3 - 2x^2 + 1$, $x = 2$, and $dx = 0.2$.

**85.** A window has the shape of a square surmounted by a semicircle. The base of the window is measured as having width 60 cm with a possible error in measurement of 0.1 cm. Use differentials to estimate the maximum error possible in computing the area of the window.

**86–88** Express the limit as a derivative and evaluate.

**86.** $\displaystyle\lim_{x \to 1} \frac{x^{17} - 1}{x - 1}$

**87.** $\displaystyle\lim_{h \to 0} \frac{\sqrt[4]{16 + h} - 2}{h}$

**88.** $\displaystyle\lim_{\theta \to \pi/3} \frac{\cos\theta - 0.5}{\theta - \pi/3}$

**89.** Evaluate $\displaystyle\lim_{x \to 0} \frac{\sqrt{1 + \tan x} - \sqrt{1 + \sin x}}{x^3}$.

**90.** Suppose $f$ is a differentiable function such that $f(g(x)) = x$ and $f'(x) = 1 + [f(x)]^2$. Show that $g'(x) = 1/(1 + x^2)$.

**91.** Find $f'(x)$ if it is known that
$$\frac{d}{dx}[f(2x)] = x^2$$

**92.** Show that the length of the portion of any tangent line to the astroid $x^{2/3} + y^{2/3} = a^{2/3}$ cut off by the coordinate axes is constant.

# Problems Plus

Before you look at the example, cover up the solution and try it yourself first.

**EXAMPLE 1** How many lines are tangent to both of the parabolas $y = -1 - x^2$ and $y = 1 + x^2$? Find the coordinates of the points at which these tangents touch the parabolas.

**SOLUTION** To gain insight into this problem, it is essential to draw a diagram. So we sketch the parabolas $y = 1 + x^2$ (which is the standard parabola $y = x^2$ shifted 1 unit upward) and $y = -1 - x^2$ (which is obtained by reflecting the first parabola about the $x$-axis). If we try to draw a line tangent to both parabolas, we soon discover that there are only two possibilities, as illustrated in Figure 1.

Let $P$ be a point at which one of these tangents touches the upper parabola and let $a$ be its $x$-coordinate. (The choice of notation for the unknown is important. Of course we could have used $b$ or $c$ or $x_0$ or $x_1$ instead of $a$. However, it's not advisable to use $x$ in place of $a$ because that $x$ could be confused with the variable $x$ in the equation of the parabola.) Then, since $P$ lies on the parabola $y = 1 + x^2$, its $y$-coordinate must be $1 + a^2$. Because of the symmetry shown in Figure 1, the coordinates of the point $Q$ where the tangent touches the lower parabola must be $(-a, -(1 + a^2))$.

To use the given information that the line is a tangent, we equate the slope of the line $PQ$ to the slope of the tangent line at $P$. We have

$$m_{PQ} = \frac{1 + a^2 - (-1 - a^2)}{a - (-a)} = \frac{1 + a^2}{a}$$

If $f(x) = 1 + x^2$, then the slope of the tangent line at $P$ is $f'(a) = 2a$. Thus the condition that we need to use is that

$$\frac{1 + a^2}{a} = 2a$$

Solving this equation, we get $1 + a^2 = 2a^2$, so $a^2 = 1$ and $a = \pm 1$. Therefore the points are $(1, 2)$ and $(-1, -2)$. By symmetry, the two remaining points are $(-1, 2)$ and $(1, -2)$.

**FIGURE 1**

1. Find points $P$ and $Q$ on the parabola $y = 1 - x^2$ so that the triangle $ABC$ formed by the $x$-axis and the tangent lines at $P$ and $Q$ is an equilateral triangle (see the figure).

2. Find the point where the curves $y = x^3 - 3x + 4$ and $y = 3(x^2 - x)$ are tangent to each other, that is, have a common tangent line. Illustrate by sketching both curves and the common tangent.

3. Show that the tangent lines to the parabola $y = ax^2 + bx + c$ at any two points with $x$-coordinates $p$ and $q$ must intersect at a point whose $x$-coordinate is halfway between $p$ and $q$.

4. Show that

$$\frac{d}{dx}\left(\frac{\sin^2 x}{1 + \cot x} + \frac{\cos^2 x}{1 + \tan x}\right) = -\cos 2x$$

5. If $f(x) = \lim_{t \to x} \frac{\sec t - \sec x}{t - x}$, find the value of $f'(\pi/4)$.

**FIGURE FOR PROBLEM 1**

---

 Graphing calculator or computer required

CAS Computer algebra system required

**FIGURE FOR PROBLEM 9**

**FIGURE FOR PROBLEM 11**

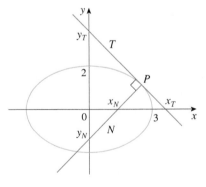

**FIGURE FOR PROBLEM 13**

**6.** Find the values of the constants $a$ and $b$ such that

$$\lim_{x \to 0} \frac{\sqrt[3]{ax + b} - 2}{x} = \frac{5}{12}$$

**7.** Prove that $\dfrac{d^n}{dx^n} (\sin^4 x + \cos^4 x) = 4^{n-1} \cos(4x + n\pi/2)$.

**8.** Find the $n$th derivative of the function $f(x) = x^n/(1 - x)$.

**9.** The figure shows a circle with radius 1 inscribed in the parabola $y = x^2$. Find the center of the circle.

**10.** If $f$ is differentiable at $a$, where $a > 0$, evaluate the following limit in terms of $f'(a)$:

$$\lim_{x \to a} \frac{f(x) - f(a)}{\sqrt{x} - \sqrt{a}}$$

**11.** The figure shows a rotating wheel with radius 40 cm and a connecting rod $AP$ with length 1.2 m. The pin $P$ slides back and forth along the $x$-axis as the wheel rotates counterclockwise at a rate of 360 revolutions per minute.
(a) Find the angular velocity of the connecting rod, $d\alpha/dt$, in radians per second, when $\theta = \pi/3$.
(b) Express the distance $x = |OP|$ in terms of $\theta$.
(c) Find an expression for the velocity of the pin $P$ in terms of $\theta$.

**12.** Tangent lines $T_1$ and $T_2$ are drawn at two points $P_1$ and $P_2$ on the parabola $y = x^2$ and they intersect at a point $P$. Another tangent line $T$ is drawn at a point between $P_1$ and $P_2$; it intersects $T_1$ at $Q_1$ and $T_2$ at $Q_2$. Show that

$$\frac{|PQ_1|}{|PP_1|} + \frac{|PQ_2|}{|PP_2|} = 1$$

**13.** Let $T$ and $N$ be the tangent and normal lines to the ellipse $x^2/9 + y^2/4 = 1$ at any point $P$ on the ellipse in the first quadrant. Let $x_T$ and $y_T$ be the $x$- and $y$-intercepts of $T$ and $x_N$ and $y_N$ be the intercepts of $N$. As $P$ moves along the ellipse in the first quadrant (but not on the axes), what values can $x_T$, $y_T$, $x_N$, and $y_N$ take on? First try to guess the answers just by looking at the figure. Then use calculus to solve the problem and see how good your intuition is.

**14.** Evaluate $\displaystyle\lim_{x \to 0} \frac{\sin(3 + x)^2 - \sin 9}{x}$.

**15.** (a) Use the identity for $\tan(x - y)$ (see Equation 14b in Appendix D) to show that if two lines $L_1$ and $L_2$ intersect at an angle $\alpha$, then

$$\tan \alpha = \frac{m_2 - m_1}{1 + m_1 m_2}$$

where $m_1$ and $m_2$ are the slopes of $L_1$ and $L_2$, respectively.
(b) The **angle between the curves** $C_1$ and $C_2$ at a point of intersection $P$ is defined to be the angle between the tangent lines to $C_1$ and $C_2$ at $P$ (if these tangent lines exist). Use part (a) to find, correct to the nearest degree, the angle between each pair of curves at each point of intersection.
(i) $y = x^2$ and $y = (x - 2)^2$
(ii) $x^2 - y^2 = 3$ and $x^2 - 4x + y^2 + 3 = 0$

**195**

**FIGURE FOR PROBLEM 16**

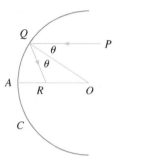

**FIGURE FOR PROBLEM 17**

**16.** Let $P(x_1, y_1)$ be a point on the parabola $y^2 = 4px$ with focus $F(p, 0)$. Let $\alpha$ be the angle between the parabola and the line segment $FP$, and let $\beta$ be the angle between the horizontal line $y = y_1$ and the parabola as in the figure. Prove that $\alpha = \beta$. (Thus, by a principle of geometrical optics, light from a source placed at $F$ will be reflected along a line parallel to the $x$-axis. This explains why *paraboloids*, the surfaces obtained by rotating parabolas about their axes, are used as the shape of some automobile headlights and mirrors for telescopes.)

**17.** Suppose that we replace the parabolic mirror of Problem 16 by a spherical mirror. Although the mirror has no focus, we can show the existence of an *approximate* focus. In the figure, $C$ is a semicircle with center $O$. A ray of light coming in toward the mirror parallel to the axis along the line $PQ$ will be reflected to the point $R$ on the axis so that $\angle PQO = \angle OQR$ (the angle of incidence is equal to the angle of reflection). What happens to the point $R$ as $P$ is taken closer and closer to the axis?

**18.** If $f$ and $g$ are differentiable functions with $f(0) = g(0) = 0$ and $g'(0) \neq 0$, show that

$$\lim_{x \to 0} \frac{f(x)}{g(x)} = \frac{f'(0)}{g'(0)}$$

**19.** Evaluate $\displaystyle\lim_{x \to 0} \frac{\sin(a + 2x) - 2\sin(a + x) + \sin a}{x^2}$.

**20.** Given an ellipse $x^2/a^2 + y^2/b^2 = 1$, where $a \neq b$, find the equation of the set of all points from which there are two tangents to the curve whose slopes are (a) reciprocals and (b) negative reciprocals.

**21.** Find the two points on the curve $y = x^4 - 2x^2 - x$ that have a common tangent line.

**22.** Suppose that three points on the parabola $y = x^2$ have the property that their normal lines intersect at a common point. Show that the sum of their $x$-coordinates is 0.

**23.** A *lattice point* in the plane is a point with integer coordinates. Suppose that circles with radius $r$ are drawn using all lattice points as centers. Find the smallest value of $r$ such that any line with slope $\frac{2}{5}$ intersects some of these circles.

**24.** A cone of radius $r$ centimeters and height $h$ centimeters is lowered point first at a rate of 1 cm/s into a tall cylinder of radius $R$ centimeters that is partially filled with water. How fast is the water level rising at the instant the cone is completely submerged?

**25.** A container in the shape of an inverted cone has height 16 cm and radius 5 cm at the top. It is partially filled with a liquid that oozes through the sides at a rate proportional to the area of the container that is in contact with the liquid. (The surface area of a cone is $\pi r l$, where $r$ is the radius and $l$ is the slant height.) If we pour the liquid into the container at a rate of 2 cm³/min, then the height of the liquid decreases at a rate of 0.3 cm/min when the height is 10 cm. If our goal is to keep the liquid at a constant height of 10 cm, at what rate should we pour the liquid into the container?

CAS **26.** (a) The cubic function $f(x) = x(x - 2)(x - 6)$ has three distinct zeros: 0, 2, and 6. Graph $f$ and its tangent lines at the *average* of each pair of zeros. What do you notice?
(b) Suppose the cubic function $f(x) = (x - a)(x - b)(x - c)$ has three distinct zeros: $a$, $b$, and $c$. Prove, with the help of a computer algebra system, that a tangent line drawn at the average of the zeros $a$ and $b$ intersects the graph of $f$ at the third zero.

# 3 Applications of Differentiation

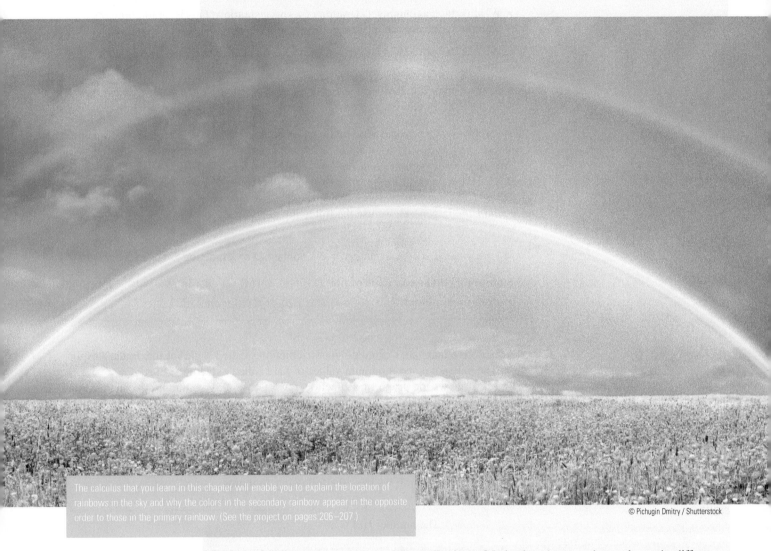

The calculus that you learn in this chapter will enable you to explain the location of rainbows in the sky and why the colors in the secondary rainbow appear in the opposite order to those in the primary rainbow. (See the project on pages 206–207.)

© Pichugin Dmitry / Shutterstock

We have already investigated some of the applications of derivatives, but now that we know the differentiation rules we are in a better position to pursue the applications of differentiation in greater depth. Here we learn how derivatives affect the shape of a graph of a function and, in particular, how they help us locate maximum and minimum values of functions. Many practical problems require us to minimize a cost or maximize an area or somehow find the best possible outcome of a situation. In particular, we will be able to investigate the optimal shape of a can and to explain the location of rainbows in the sky.

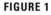

# Maximum and Minimum Values

Some of the most important applications of differential calculus are *optimization problems*, in which we are required to find the optimal (best) way of doing something. Here are examples of such problems that we will solve in this chapter:

- What is the shape of a can that minimizes manufacturing costs?

- What is the maximum acceleration of a space shuttle? (This is an important question to the astronauts who have to withstand the effects of acceleration.)

- What is the radius of a contracted windpipe that expels air most rapidly during a cough?

- At what angle should blood vessels branch so as to minimize the energy expended by the heart in pumping blood?

These problems can be reduced to finding the maximum or minimum values of a function. Let's first explain exactly what we mean by maximum and minimum values.

We see that the highest point on the graph of the function $f$ shown in Figure 1 is the point $(3, 5)$. In other words, the largest value of $f$ is $f(3) = 5$. Likewise, the smallest value is $f(6) = 2$. We say that $f(3) = 5$ is the *absolute maximum* of $f$ and $f(6) = 2$ is the *absolute minimum*. In general, we use the following definition.

**FIGURE 1**

> **1 Definition** Let $c$ be a number in the domain $D$ of a function $f$. Then $f(c)$ is the
>
> - **absolute maximum** value of $f$ on $D$ if $f(c) \geq f(x)$ for all $x$ in $D$.
> - **absolute minimum** value of $f$ on $D$ if $f(c) \leq f(x)$ for all $x$ in $D$.

An absolute maximum or minimum is sometimes called a **global** maximum or minimum. The maximum and minimum values of $f$ are called **extreme values** of $f$.

Figure 2 shows the graph of a function $f$ with absolute maximum at $d$ and absolute minimum at $a$. Note that $(d, f(d))$ is the highest point on the graph and $(a, f(a))$ is the lowest point. In Figure 2, if we consider only values of $x$ near $b$ [for instance, if we restrict our attention to the interval $(a, c)$], then $f(b)$ is the largest of those values of $f(x)$ and is called a *local maximum value* of $f$. Likewise, $f(c)$ is called a *local minimum value* of $f$ because $f(c) \leq f(x)$ for $x$ near $c$ [in the interval $(b, d)$, for instance]. The function $f$ also has a local minimum at $e$. In general, we have the following definition.

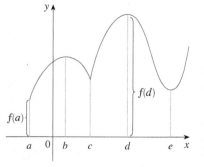

**FIGURE 2**
Abs min $f(a)$, abs max $f(d)$,
loc min $f(c)$, $f(e)$, loc max $f(b)$, $f(d)$

> **2 Definition** The number $f(c)$ is a
>
> - **local maximum** value of $f$ if $f(c) \geq f(x)$ when $x$ is near $c$.
> - **local minimum** value of $f$ if $f(c) \leq f(x)$ when $x$ is near $c$.

In Definition 2 (and elsewhere), if we say that something is true **near** $c$, we mean that it is true on some open interval containing $c$. For instance, in Figure 3 we see that $f(4) = 5$ is a local minimum because it's the smallest value of $f$ on the interval $I$. It's not the absolute minimum because $f(x)$ takes smaller values when $x$ is near 12 (in the interval $K$, for instance). In fact $f(12) = 3$ is both a local minimum and the absolute minimum. Similarly, $f(8) = 7$ is a local maximum, but not the absolute maximum because $f$ takes larger values near 1.

**FIGURE 3**

**EXAMPLE 1** The function $f(x) = \cos x$ takes on its (local and absolute) maximum value of 1 infinitely many times, since $\cos 2n\pi = 1$ for any integer $n$ and $-1 \le \cos x \le 1$ for all $x$. Likewise, $\cos(2n + 1)\pi = -1$ is its minimum value, where $n$ is any integer. ▬

**EXAMPLE 2** If $f(x) = x^2$, then $f(x) \ge f(0)$ because $x^2 \ge 0$ for all $x$. Therefore $f(0) = 0$ is the absolute (and local) minimum value of $f$. This corresponds to the fact that the origin is the lowest point on the parabola $y = x^2$. (See Figure 4.) However, there is no highest point on the parabola and so this function has no maximum value. ▬

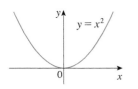

**FIGURE 4**
Minimum value 0, no maximum

**EXAMPLE 3** From the graph of the function $f(x) = x^3$, shown in Figure 5, we see that this function has neither an absolute maximum value nor an absolute minimum value. In fact, it has no local extreme values either.

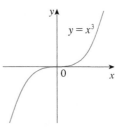

**FIGURE 5**
No minimum, no maximum

**V** **EXAMPLE 4** The graph of the function

$$f(x) = 3x^4 - 16x^3 + 18x^2 \qquad -1 \le x \le 4$$

is shown in Figure 6. You can see that $f(1) = 5$ is a local maximum, whereas the absolute maximum is $f(-1) = 37$. (This absolute maximum is not a local maximum because it occurs at an endpoint.) Also, $f(0) = 0$ is a local minimum and $f(3) = -27$ is both a local and an absolute minimum. Note that $f$ has neither a local nor an absolute maximum at $x = 4$. ▬

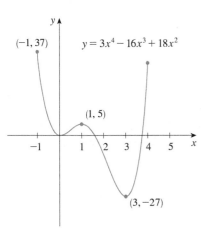

**FIGURE 6**

We have seen that some functions have extreme values, whereas others do not. The following theorem gives conditions under which a function is guaranteed to possess extreme values.

**3** **The Extreme Value Theorem** If $f$ is continuous on a closed interval $[a, b]$, then $f$ attains an absolute maximum value $f(c)$ and an absolute minimum value $f(d)$ at some numbers $c$ and $d$ in $[a, b]$.

The Extreme Value Theorem is illustrated in Figure 7. Note that an extreme value can be taken on more than once. Although the Extreme Value Theorem is intuitively very plausible, it is difficult to prove and so we omit the proof.

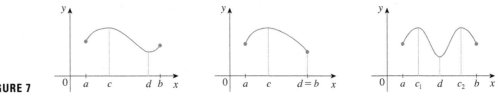

**FIGURE 7**

Figures 8 and 9 show that a function need not possess extreme values if either hypothesis (continuity or closed interval) is omitted from the Extreme Value Theorem.

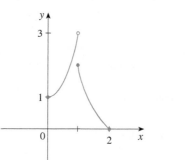

**FIGURE 8**
This function has minimum value $f(2) = 0$, but no maximum value.

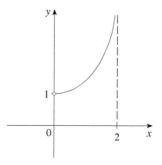

**FIGURE 9**
This continuous function $g$ has no maximum or minimum.

The function $f$ whose graph is shown in Figure 8 is defined on the closed interval [0, 2] but has no maximum value. (Notice that the range of $f$ is [0, 3). The function takes on values arbitrarily close to 3, but never actually attains the value 3.) This does not contradict the Extreme Value Theorem because $f$ is not continuous. [Nonetheless, a discontinuous function *could* have maximum and minimum values. See Exercise 13(b).]

The function $g$ shown in Figure 9 is continuous on the open interval (0, 2) but has neither a maximum nor minimum value. [The range of $g$ is $(1, \infty)$. The function takes on arbitrarily large values.] This does not contradict the Extreme Value Theorem because the interval (0, 2) is not closed.

The Extreme Value Theorem says that a continuous function on a closed interval has a maximum value and a minimum value, but it does not tell us how to find these extreme values. We start by looking for local extreme values.

Figure 10 shows the graph of a function $f$ with a local maximum at $c$ and a local minimum at $d$. It appears that at the maximum and minimum points the tangent lines are horizontal and therefore each has slope 0. We know that the derivative is the slope of the tangent line, so it appears that $f'(c) = 0$ and $f'(d) = 0$. The following theorem says that this is always true for differentiable functions.

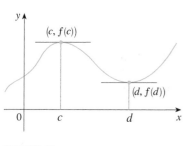

**FIGURE 10**

**Fermat**

Fermat's Theorem is named after Pierre Fermat (1601–1665), a French lawyer who took up mathematics as a hobby. Despite his amateur status, Fermat was one of the two inventors of analytic geometry (Descartes was the other). His methods for finding tangents to curves and maximum and minimum values (before the invention of limits and derivatives) made him a forerunner of Newton in the creation of differential calculus.

> **4**   **Fermat's Theorem**   If $f$ has a local maximum or minimum at $c$, and if $f'(c)$ exists, then $f'(c) = 0$.

**PROOF**   Suppose, for the sake of definiteness, that $f$ has a local maximum at $c$. Then, according to Definition 2, $f(c) \geq f(x)$ if $x$ is sufficiently close to $c$. This implies that if $h$ is sufficiently close to 0, with $h$ being positive or negative, then

$$f(c) \geq f(c + h)$$

and therefore

**5**    $$f(c + h) - f(c) \leq 0$$

We can divide both sides of an inequality by a positive number. Thus, if $h > 0$ and $h$ is sufficiently small, we have

$$\frac{f(c + h) - f(c)}{h} \leq 0$$

Taking the right-hand limit of both sides of this inequality (using Theorem 1.6.2), we get

$$\lim_{h \to 0^+} \frac{f(c + h) - f(c)}{h} \leqslant \lim_{h \to 0^+} 0 = 0$$

But since $f'(c)$ exists, we have

$$f'(c) = \lim_{h \to 0} \frac{f(c + h) - f(c)}{h} = \lim_{h \to 0^+} \frac{f(c + h) - f(c)}{h}$$

and so we have shown that $f'(c) \leqslant 0$.

If $h < 0$, then the direction of the inequality $\boxed{5}$ is reversed when we divide by $h$:

$$\frac{f(c + h) - f(c)}{h} \geqslant 0 \qquad h < 0$$

So, taking the left-hand limit, we have

$$f'(c) = \lim_{h \to 0} \frac{f(c + h) - f(c)}{h} = \lim_{h \to 0^-} \frac{f(c + h) - f(c)}{h} \geqslant 0$$

We have shown that $f'(c) \geqslant 0$ and also that $f'(c) \leqslant 0$. Since both of these inequalities must be true, the only possibility is that $f'(c) = 0$.

We have proved Fermat's Theorem for the case of a local maximum. The case of a local minimum can be proved in a similar manner, or we could use Exercise 70 to deduce it from the case we have just proved (see Exercise 71). ▪

The following examples caution us against reading too much into Fermat's Theorem: We can't expect to locate extreme values simply by setting $f'(x) = 0$ and solving for $x$.

**EXAMPLE 5** If $f(x) = x^3$, then $f'(x) = 3x^2$, so $f'(0) = 0$. But $f$ has no maximum or minimum at 0, as you can see from its graph in Figure 11. (Or observe that $x^3 > 0$ for $x > 0$ but $x^3 < 0$ for $x < 0$.) The fact that $f'(0) = 0$ simply means that the curve $y = x^3$ has a horizontal tangent at $(0, 0)$. Instead of having a maximum or minimum at $(0, 0)$, the curve crosses its horizontal tangent there. ▪

**EXAMPLE 6** The function $f(x) = |x|$ has its (local and absolute) minimum value at 0, but that value can't be found by setting $f'(x) = 0$ because, as was shown in Example 5 in Section 2.2, $f'(0)$ does not exist. (See Figure 12.) ▪

⊘ **WARNING** Examples 5 and 6 show that we must be careful when using Fermat's Theorem. **Example 5** demonstrates that even when $f'(c) = 0$ there need not be a maximum or minimum at $c$. (In other words, the converse of Fermat's Theorem is false in general.) Furthermore, there may be an extreme value even when $f'(c)$ does not exist (as in Example 6).

Fermat's Theorem does suggest that we should at least *start* looking for extreme values of $f$ at the numbers $c$ where $f'(c) = 0$ or where $f'(c)$ does not exist. Such numbers are given a special name.

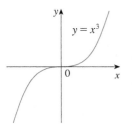

**FIGURE 11**
If $f(x) = x^3$, then $f'(0) = 0$ but $f$ has no maximum or minimum.

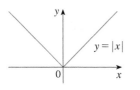

**FIGURE 12**
If $f(x) = |x|$, then $f(0) = 0$ is a minimum value, but $f'(0)$ does not exist.

> $\boxed{6}$ **Definition** A **critical number** of a function $f$ is a number $c$ in the domain of $f$ such that either $f'(c) = 0$ or $f'(c)$ does not exist.

Figure 13 shows a graph of the function $f$ in Example 7. It supports our answer because there is a horizontal tangent when $x = 1.5$ and a vertical tangent when $x = 0$.

**FIGURE 13**

 **EXAMPLE 7**    Find the critical numbers of $f(x) = x^{3/5}(4 - x)$.

SOLUTION    The Product Rule gives

$$f'(x) = x^{3/5}(-1) + (4 - x)(\tfrac{3}{5}x^{-2/5}) = -x^{3/5} + \frac{3(4 - x)}{5x^{2/5}}$$

$$= \frac{-5x + 3(4 - x)}{5x^{2/5}} = \frac{12 - 8x}{5x^{2/5}}$$

[The same result could be obtained by first writing $f(x) = 4x^{3/5} - x^{8/5}$.] Therefore $f'(x) = 0$ if $12 - 8x = 0$, that is, $x = \tfrac{3}{2}$, and $f'(x)$ does not exist when $x = 0$. Thus the critical numbers are $\tfrac{3}{2}$ and $0$.

In terms of critical numbers, Fermat's Theorem can be rephrased as follows (compare Definition 6 with Theorem 4):

> **7**    If $f$ has a local maximum or minimum at $c$, then $c$ is a critical number of $f$.

To find an absolute maximum or minimum of a continuous function on a closed interval, we note that either it is local [in which case it occurs at a critical number by **7**] or it occurs at an endpoint of the interval. Thus the following three-step procedure always works.

> **The Closed Interval Method**    To find the *absolute* maximum and minimum values of a continuous function $f$ on a closed interval $[a, b]$:
>
> **1.** Find the values of $f$ at the critical numbers of $f$ in $(a, b)$.
>
> **2.** Find the values of $f$ at the endpoints of the interval.
>
> **3.** The largest of the values from Steps 1 and 2 is the absolute maximum value; the smallest of these values is the absolute minimum value.

 **EXAMPLE 8**    Find the absolute maximum and minimum values of the function

$$f(x) = x^3 - 3x^2 + 1 \qquad -\tfrac{1}{2} \le x \le 4$$

SOLUTION    Since $f$ is continuous on $\left[-\tfrac{1}{2}, 4\right]$, we can use the Closed Interval Method:

$$f(x) = x^3 - 3x^2 + 1$$

$$f'(x) = 3x^2 - 6x = 3x(x - 2)$$

Since $f'(x)$ exists for all $x$, the only critical numbers of $f$ occur when $f'(x) = 0$, that is, $x = 0$ or $x = 2$. Notice that each of these critical numbers lies in the interval $\left(-\tfrac{1}{2}, 4\right)$. The values of $f$ at these critical numbers are

$$f(0) = 1 \qquad f(2) = -3$$

The values of $f$ at the endpoints of the interval are

$$f\left(-\tfrac{1}{2}\right) = \tfrac{1}{8} \qquad f(4) = 17$$

Comparing these four numbers, we see that the absolute maximum value is $f(4) = 17$ and the absolute minimum value is $f(2) = -3$.

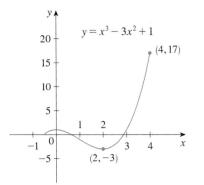

$y = x^3 - 3x^2 + 1$

(4, 17)

(2, −3)

**FIGURE 14**

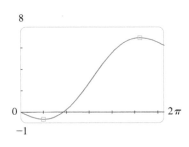

8

0                           $2\pi$

−1

**FIGURE 15**

Note that in this example the absolute maximum occurs at an endpoint, whereas the absolute minimum occurs at a critical number. The graph of $f$ is sketched in Figure 14.

If you have a graphing calculator or a computer with graphing software, it is possible to estimate maximum and minimum values very easily. But, as the next example shows, calculus is needed to find the *exact* values.

**EXAMPLE 9**

(a) Use a graphing device to estimate the absolute minimum and maximum values of the function $f(x) = x - 2 \sin x$, $0 \le x \le 2\pi$.
(b) Use calculus to find the exact minimum and maximum values.

SOLUTION

(a) Figure 15 shows a graph of $f$ in the viewing rectangle $[0, 2\pi]$ by $[-1, 8]$. By moving the cursor close to the maximum point, we see that the $y$-coordinates don't change very much in the vicinity of the maximum. The absolute maximum value is about 6.97 and it occurs when $x \approx 5.2$. Similarly, by moving the cursor close to the minimum point, we see that the absolute minimum value is about $-0.68$ and it occurs when $x \approx 1.0$. It is possible to get more accurate estimates by zooming in toward the maximum and minimum points, but instead let's use calculus.

(b) The function $f(x) = x - 2 \sin x$ is continuous on $[0, 2\pi]$. Since $f'(x) = 1 - 2 \cos x$, we have $f'(x) = 0$ when $\cos x = \frac{1}{2}$ and this occurs when $x = \pi/3$ or $5\pi/3$. The values of $f$ at these critical numbers are

$$f(\pi/3) = \frac{\pi}{3} - 2 \sin \frac{\pi}{3} = \frac{\pi}{3} - \sqrt{3} \approx -0.684853$$

and

$$f(5\pi/3) = \frac{5\pi}{3} - 2 \sin \frac{5\pi}{3} = \frac{5\pi}{3} + \sqrt{3} \approx 6.968039$$

The values of $f$ at the endpoints are

$$f(0) = 0 \quad \text{and} \quad f(2\pi) = 2\pi \approx 6.28$$

Comparing these four numbers and using the Closed Interval Method, we see that the absolute minimum value is $f(\pi/3) = \pi/3 - \sqrt{3}$ and the absolute maximum value is $f(5\pi/3) = 5\pi/3 + \sqrt{3}$. The values from part (a) serve as a check on our work.

**EXAMPLE 10**  The Hubble Space Telescope was deployed on April 24, 1990, by the space shuttle *Discovery*. A model for the velocity of the shuttle during this mission, from liftoff at $t = 0$ until the solid rocket boosters were jettisoned at $t = 126$ s, is given by

$$v(t) = 0.001302t^3 - 0.09029t^2 + 23.61t - 3.083$$

(in feet per second). Using this model, estimate the absolute maximum and minimum values of the *acceleration* of the shuttle between liftoff and the jettisoning of the boosters.

SOLUTION  We are asked for the extreme values not of the given velocity function, but rather of the acceleration function. So we first need to differentiate to find the acceleration:

$$a(t) = v'(t) = \frac{d}{dt}(0.001302t^3 - 0.09029t^2 + 23.61t - 3.083)$$

$$= 0.003906t^2 - 0.18058t + 23.61$$

NASA

We now apply the Closed Interval Method to the continuous function $a$ on the interval $0 \leq t \leq 126$. Its derivative is

$$a'(t) = 0.007812t - 0.18058$$

The only critical number occurs when $a'(t) = 0$:

$$t_1 = \frac{0.18058}{0.007812} \approx 23.12$$

Evaluating $a(t)$ at the critical number and at the endpoints, we have

$$a(0) = 23.61 \qquad a(t_1) \approx 21.52 \qquad a(126) \approx 62.87$$

So the maximum acceleration is about 62.87 ft/s$^2$ and the minimum acceleration is about 21.52 ft/s$^2$.

## 3.1  Exercises

**1.** Explain the difference between an absolute minimum and a local minimum.

**2.** Suppose $f$ is a continuous function defined on a closed interval $[a, b]$.
(a) What theorem guarantees the existence of an absolute maximum value and an absolute minimum value for $f$?
(b) What steps would you take to find those maximum and minimum values?

**3–4** For each of the numbers $a$, $b$, $c$, $d$, $r$, and $s$, state whether the function whose graph is shown has an absolute maximum or minimum, a local maximum or minimum, or neither a maximum nor a minimum.

**3.** **4.**

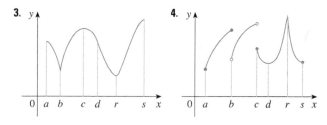

**5–6** Use the graph to state the absolute and local maximum and minimum values of the function.

**5.** **6.**

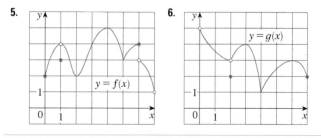

**7–10** Sketch the graph of a function $f$ that is continuous on $[1, 5]$ and has the given properties.

**7.** Absolute minimum at 2, absolute maximum at 3, local minimum at 4

**8.** Absolute minimum at 1, absolute maximum at 5, local maximum at 2, local minimum at 4

**9.** Absolute maximum at 5, absolute minimum at 2, local maximum at 3, local minima at 2 and 4

**10.** $f$ has no local maximum or minimum, but 2 and 4 are critical numbers

**11.** (a) Sketch the graph of a function that has a local maximum at 2 and is differentiable at 2.
(b) Sketch the graph of a function that has a local maximum at 2 and is continuous but not differentiable at 2.
(c) Sketch the graph of a function that has a local maximum at 2 and is not continuous at 2.

**12.** (a) Sketch the graph of a function on $[-1, 2]$ that has an absolute maximum but no local maximum.
(b) Sketch the graph of a function on $[-1, 2]$ that has a local maximum but no absolute maximum.

**13.** (a) Sketch the graph of a function on $[-1, 2]$ that has an absolute maximum but no absolute minimum.
(b) Sketch the graph of a function on $[-1, 2]$ that is discontinuous but has both an absolute maximum and an absolute minimum.

**14.** (a) Sketch the graph of a function that has two local maxima, one local minimum, and no absolute minimum.

(b) Sketch the graph of a function that has three local minima, two local maxima, and seven critical numbers.

**15–28** Sketch the graph of $f$ by hand and use your sketch to find the absolute and local maximum and minimum values of $f$. (Use the graphs and transformations of Sections 1.2 and 1.3.)

**15.** $f(x) = \frac{1}{2}(3x - 1), \quad x \le 3$

**16.** $f(x) = 2 - \frac{1}{3}x, \quad x \ge -2$

**17.** $f(x) = 1/x, \quad x \ge 1$

**18.** $f(x) = 1/x, \quad 1 < x < 3$

**19.** $f(x) = \sin x, \quad 0 \le x < \pi/2$

**20.** $f(x) = \sin x, \quad 0 < x \le \pi/2$

**21.** $f(x) = \sin x, \quad -\pi/2 \le x \le \pi/2$

**22.** $f(t) = \cos t, \quad -3\pi/2 \le t \le 3\pi/2$

**23.** $f(x) = 1 + (x + 1)^2, \quad -2 \le x < 5$

**24.** $f(x) = |x|$

**25.** $f(x) = 1 - \sqrt{x}$

**26.** $f(x) = 1 - x^3$

**27.** $f(x) = \begin{cases} 1 - x & \text{if } 0 \le x < 2 \\ 2x - 4 & \text{if } 2 \le x \le 3 \end{cases}$

**28.** $f(x) = \begin{cases} 4 - x^2 & \text{if } -2 \le x < 0 \\ 2x - 1 & \text{if } 0 \le x \le 2 \end{cases}$

**29–42** Find the critical numbers of the function.

**29.** $f(x) = 4 + \frac{1}{3}x - \frac{1}{2}x^2$

**30.** $f(x) = x^3 + 6x^2 - 15x$

**31.** $f(x) = 2x^3 - 3x^2 - 36x$

**32.** $f(x) = 2x^3 + x^2 + 2x$

**33.** $g(t) = t^4 + t^3 + t^2 + 1$

**34.** $g(t) = |3t - 4|$

**35.** $g(y) = \dfrac{y - 1}{y^2 - y + 1}$

**36.** $h(p) = \dfrac{p - 1}{p^2 + 4}$

**37.** $h(t) = t^{3/4} - 2t^{1/4}$

**38.** $g(x) = x^{1/3} - x^{-2/3}$

**39.** $F(x) = x^{4/5}(x - 4)^2$

**40.** $g(\theta) = 4\theta - \tan\theta$

**41.** $f(\theta) = 2\cos\theta + \sin^2\theta$

**42.** $g(x) = \sqrt{1 - x^2}$

**43–44** A formula for the *derivative* of a function $f$ is given. How many critical numbers does $f$ have?

**43.** $f'(x) = 1 + \dfrac{210\sin x}{x^2 - 6x + 10}$

**44.** $f'(x) = \dfrac{100\cos^2 x}{10 + x^2} - 1$

**45–56** Find the absolute maximum and absolute minimum values of $f$ on the given interval.

**45.** $f(x) = 12 + 4x - x^2, \quad [0, 5]$

**46.** $f(x) = 5 + 54x - 2x^3, \quad [0, 4]$

**47.** $f(x) = 2x^3 - 3x^2 - 12x + 1, \quad [-2, 3]$

**48.** $f(x) = x^3 - 6x^2 + 5, \quad [-3, 5]$

**49.** $f(x) = 3x^4 - 4x^3 - 12x^2 + 1, \quad [-2, 3]$

**50.** $f(x) = (x^2 - 1)^3, \quad [-1, 2]$

**51.** $f(x) = x + \dfrac{1}{x}, \quad [0.2, 4]$

**52.** $f(x) = \dfrac{x}{x^2 - x + 1}, \quad [0, 3]$

**53.** $f(t) = t\sqrt{4 - t^2}, \quad [-1, 2]$

**54.** $f(t) = \sqrt[3]{t}\,(8 - t), \quad [0, 8]$

**55.** $f(t) = 2\cos t + \sin 2t, \quad [0, \pi/2]$

**56.** $f(t) = t + \cot(t/2), \quad [\pi/4, 7\pi/4]$

**57.** If $a$ and $b$ are positive numbers, find the maximum value of $f(x) = x^a(1 - x)^b, 0 \le x \le 1$.

**58.** Use a graph to estimate the critical numbers of $f(x) = |x^3 - 3x^2 + 2|$ correct to one decimal place.

**59–62**
(a) Use a graph to estimate the absolute maximum and minimum values of the function to two decimal places.
(b) Use calculus to find the exact maximum and minimum values.

**59.** $f(x) = x^5 - x^3 + 2, \quad -1 \le x \le 1$

**60.** $f(x) = x^4 - 3x^3 + 3x^2 - x, \quad 0 \le x \le 2$

**61.** $f(x) = x\sqrt{x - x^2}$

**62.** $f(x) = x - 2\cos x, \quad -2 \le x \le 0$

**63.** Between 0°C and 30°C, the volume $V$ (in cubic centimeters) of 1 kg of water at a temperature $T$ is given approximately by the formula

$$V = 999.87 - 0.06426T + 0.0085043T^2 - 0.0000679T^3$$

Find the temperature at which water has its maximum density.

**64.** An object with weight $W$ is dragged along a horizontal plane by a force acting along a rope attached to the object. If the rope makes an angle $\theta$ with the plane, then the magnitude of the force is

$$F = \frac{\mu W}{\mu \sin\theta + \cos\theta}$$

where $\mu$ is a positive constant called the *coefficient(s) of friction* and where $0 \le \theta \le \pi/2$. Show that $F$ is minimized when $\tan\theta = \mu$.

**65.** A model for the US average price of a pound of white sugar from 1993 to 2003 is given by the function

$$S(t) = -0.00003237t^5 + 0.0009037t^4 - 0.008956t^3$$
$$+ 0.03629t^2 - 0.04458t + 0.4074$$

where $t$ is measured in years since August of 1993. Estimate the times when sugar was cheapest and most expensive during the period 1993–2003.

**66.** On May 7, 1992, the space shuttle *Endeavour* was launched on mission STS-49, the purpose of which was to install a new perigee kick motor in an Intelsat communications satellite. The table gives the velocity data for the shuttle between liftoff and the jettisoning of the solid rocket boosters.

| Event | Time (s) | Velocity (ft/s) |
|---|---|---|
| Launch | 0 | 0 |
| Begin roll maneuver | 10 | 185 |
| End roll maneuver | 15 | 319 |
| Throttle to 89% | 20 | 447 |
| Throttle to 67% | 32 | 742 |
| Throttle to 104% | 59 | 1325 |
| Maximum dynamic pressure | 62 | 1445 |
| Solid rocket booster separation | 125 | 4151 |

(a) Use a graphing calculator or computer to find the cubic polynomial that best models the velocity of the shuttle for the time interval $t \in [0, 125]$. Then graph this polynomial.

(b) Find a model for the acceleration of the shuttle and use it to estimate the maximum and minimum values of the acceleration during the first 125 seconds.

**67.** When a foreign object lodged in the trachea (windpipe) forces a person to cough, the diaphragm thrusts upward causing an increase in pressure in the lungs. This is accompanied by a contraction of the trachea, making a narrower channel for the expelled air to flow through. For a given amount of

air to escape in a fixed time, it must move faster through the narrower channel than the wider one. The greater the velocity of the airstream, the greater the force on the foreign object. X rays show that the radius of the circular tracheal tube contracts to about two-thirds of its normal radius during a cough. According to a mathematical model of coughing, the velocity $v$ of the airstream is related to the radius $r$ of the trachea by the equation

$$v(r) = k(r_0 - r)r^2 \qquad \tfrac{1}{2}r_0 \le r \le r_0$$

where $k$ is a constant and $r_0$ is the normal radius of the trachea. The restriction on $r$ is due to the fact that the tracheal wall stiffens under pressure and a contraction greater than $\frac{1}{2}r_0$ is prevented (otherwise the person would suffocate).

(a) Determine the value of $r$ in the interval $\left[\frac{1}{2}r_0, r_0\right]$ at which $v$ has an absolute maximum. How does this compare with experimental evidence?

(b) What is the absolute maximum value of $v$ on the interval?

(c) Sketch the graph of $v$ on the interval $[0, r_0]$.

**68.** Show that 5 is a critical number of the function

$$g(x) = 2 + (x - 5)^3$$

but $g$ does not have a local extreme value at 5.

**69.** Prove that the function

$$f(x) = x^{101} + x^{51} + x + 1$$

has neither a local maximum nor a local minimum.

**70.** If $f$ has a local minimum value at $c$, show that the function $g(x) = -f(x)$ has a local maximum value at $c$.

**71.** Prove Fermat's Theorem for the case in which $f$ has a local minimum at $c$.

**72.** A cubic function is a polynomial of degree 3; that is, it has the form $f(x) = ax^3 + bx^2 + cx + d$, where $a \ne 0$.

(a) Show that a cubic function can have two, one, or no critical number(s). Give examples and sketches to illustrate the three possibilities.

(b) How many local extreme values can a cubic function have?

---

**APPLIED PROJECT**　　**THE CALCULUS OF RAINBOWS**

Rainbows are created when raindrops scatter sunlight. They have fascinated mankind since ancient times and have inspired attempts at scientific explanation since the time of Aristotle. In this project we use the ideas of Descartes and Newton to explain the shape, location, and colors of rainbows.

Formation of the primary rainbow

**1.** The figure shows a ray of sunlight entering a spherical raindrop at $A$. Some of the light is reflected, but the line $AB$ shows the path of the part that enters the drop. Notice that the light is refracted toward the normal line $AO$ and in fact Snell's Law says that $\sin \alpha = k \sin \beta$, where $\alpha$ is the angle of incidence, $\beta$ is the angle of refraction, and $k \approx \frac{4}{3}$ is the index of refraction for water. At $B$ some of the light passes through the drop and is refracted into the air, but the line $BC$ shows the part that is reflected. (The angle of incidence equals the angle of reflection.) When the ray reaches $C$, part of it is reflected, but for the time being we are

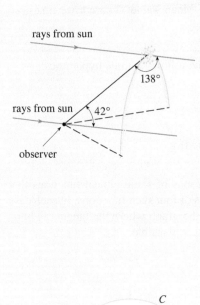

rays from sun

138°

rays from sun   42°

observer

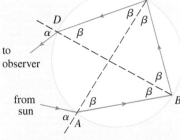

C

D

$\beta$ $\beta$

$\alpha$ $\beta$

$\beta$

to observer

$\beta$

$\beta$

from sun   $\beta$   B

$\alpha$ / A

Formation of the secondary rainbow

42° 51°

more interested in the part that leaves the raindrop at $C$. (Notice that it is refracted away from the normal line.) The *angle of deviation* $D(\alpha)$ is the amount of clockwise rotation that the ray has undergone during this three-stage process. Thus

$$D(\alpha) = (\alpha - \beta) + (\pi - 2\beta) + (\alpha - \beta) = \pi + 2\alpha - 4\beta$$

Show that the minimum value of the deviation is $D(\alpha) \approx 138°$ and occurs when $\alpha \approx 59.4°$.

The significance of the minimum deviation is that when $\alpha \approx 59.4°$ we have $D'(\alpha) \approx 0$, so $\Delta D/\Delta \alpha \approx 0$. This means that many rays with $\alpha \approx 59.4°$ become deviated by approximately the same amount. It is the *concentration* of rays coming from near the direction of minimum deviation that creates the brightness of the primary rainbow. The figure at the left shows that the angle of elevation from the observer up to the highest point on the rainbow is $180° - 138° = 42°$. (This angle is called the *rainbow angle*.)

2. Problem 1 explains the location of the primary rainbow, but how do we explain the colors? Sunlight comprises a range of wavelengths, from the red range through orange, yellow, green, blue, indigo, and violet. As Newton discovered in his prism experiments of 1666, the index of refraction is different for each color. (The effect is called *dispersion*.) For red light the refractive index is $k \approx 1.3318$ whereas for violet light it is $k \approx 1.3435$. By repeating the calculation of Problem 1 for these values of $k$, show that the rainbow angle is about 42.3° for the red bow and 40.6° for the violet bow. So the rainbow really consists of seven individual bows corresponding to the seven colors.

3. Perhaps you have seen a fainter secondary rainbow above the primary bow. That results from the part of a ray that enters a raindrop and is refracted at $A$, reflected twice (at $B$ and $C$), and refracted as it leaves the drop at $D$ (see the figure at the left). This time the deviation angle $D(\alpha)$ is the total amount of counterclockwise rotation that the ray undergoes in this four-stage process. Show that

$$D(\alpha) = 2\alpha - 6\beta + 2\pi$$

and $D(\alpha)$ has a minimum value when

$$\cos \alpha = \sqrt{\frac{k^2 - 1}{8}}$$

Taking $k = \frac{4}{3}$, show that the minimum deviation is about 129° and so the rainbow angle for the secondary rainbow is about 51°, as shown in the figure at the left.

4. Show that the colors in the secondary rainbow appear in the opposite order from those in the primary rainbow.

## The Mean Value Theorem

We will see that many of the results of this chapter depend on one central fact, which is called the Mean Value Theorem. But to arrive at the Mean Value Theorem we first need the following result.

**Rolle**

Rolle's Theorem was first published in 1691 by the French mathematician Michel Rolle (1652–1719) in a book entitled *Méthode pour resoudre les Egalitez.* He was a vocal critic of the methods of his day and attacked calculus as being a "collection of ingenious fallacies." Later, however, he became convinced of the essential correctness of the methods of calculus.

> **Rolle's Theorem**  Let $f$ be a function that satisfies the following three hypotheses:
> 1. $f$ is continuous on the closed interval $[a, b]$.
> 2. $f$ is differentiable on the open interval $(a, b)$.
> 3. $f(a) = f(b)$
>
> Then there is a number $c$ in $(a, b)$ such that $f'(c) = 0$.

Before giving the proof let's take a look at the graphs of some typical functions that satisfy the three hypotheses. Figure 1 shows the graphs of four such functions. In each case it appears that there is at least one point $(c, f(c))$ on the graph where the tangent is horizontal and therefore $f'(c) = 0$. Thus Rolle's Theorem is plausible.

(a)              (b)              (c)              (d)

**FIGURE 1**

PS Take cases

PROOF   There are three cases:

CASE I   $f(x) = k$, **a constant**
Then $f'(x) = 0$, so the number $c$ can be taken to be *any* number in $(a, b)$.

CASE II   $f(x) > f(a)$ **for some $x$ in $(a, b)$**  [as in Figure 1(b) or (c)]
By the Extreme Value Theorem (which we can apply by hypothesis 1), $f$ has a maximum value somewhere in $[a, b]$. Since $f(a) = f(b)$, it must attain this maximum value at a number $c$ in the open interval $(a, b)$. Then $f$ has a *local* maximum at $c$ and, by hypothesis 2, $f$ is differentiable at $c$. Therefore $f'(c) = 0$ by Fermat's Theorem.

CASE III   $f(x) < f(a)$ **for some $x$ in $(a, b)$**  [as in Figure 1(c) or (d)]
By the Extreme Value Theorem, $f$ has a minimum value in $[a, b]$ and, since $f(a) = f(b)$, it attains this minimum value at a number $c$ in $(a, b)$. Again $f'(c) = 0$ by Fermat's Theorem.

EXAMPLE 1   Let's apply Rolle's Theorem to the position function $s = f(t)$ of a moving object. If the object is in the same place at two different instants $t = a$ and $t = b$, then $f(a) = f(b)$. Rolle's Theorem says that there is some instant of time $t = c$ between $a$ and $b$ when $f'(c) = 0$; that is, the velocity is 0. (In particular, you can see that this is true when a ball is thrown directly upward.)

EXAMPLE 2   Prove that the equation $x^3 + x - 1 = 0$ has exactly one real root.

SOLUTION   First we use the Intermediate Value Theorem (1.8.10) to show that a root exists. Let $f(x) = x^3 + x - 1$. Then $f(0) = -1 < 0$ and $f(1) = 1 > 0$. Since $f$ is a

Figure 2 shows a graph of the function $f(x) = x^3 + x - 1$ discussed in Example 2. Rolle's Theorem shows that, no matter how much we enlarge the viewing rectangle, we can never find a second $x$-intercept.

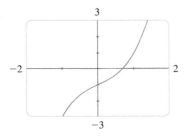

**FIGURE 2**

The Mean Value Theorem is an example of what is called an existence theorem. Like the Intermediate Value Theorem, the Extreme Value Theorem, and Rolle's Theorem, it guarantees that there *exists* a number with a certain property, but it doesn't tell us how to find the number.

polynomial, it is continuous, so the Intermediate Value Theorem states that there is a number $c$ between 0 and 1 such that $f(c) = 0$. Thus the given equation has a root.

To show that the equation has no other real root, we use Rolle's Theorem and argue by contradiction. Suppose that it had two roots $a$ and $b$. Then $f(a) = 0 = f(b)$ and, since $f$ is a polynomial, it is differentiable on $(a, b)$ and continuous on $[a, b]$. Thus, by Rolle's Theorem, there is a number $c$ between $a$ and $b$ such that $f'(c) = 0$. But

$$f'(x) = 3x^2 + 1 \geq 1 \qquad \text{for all } x$$

(since $x^2 \geq 0$) so $f'(x)$ can never be 0. This gives a contradiction. Therefore the equation can't have two real roots.

Our main use of Rolle's Theorem is in proving the following important theorem, which was first stated by another French mathematician, Joseph-Louis Lagrange.

---

**The Mean Value Theorem** Let $f$ be a function that satisfies the following hypotheses:

**1.** $f$ is continuous on the closed interval $[a, b]$.

**2.** $f$ is differentiable on the open interval $(a, b)$.

Then there is a number $c$ in $(a, b)$ such that

$$\boxed{1} \qquad f'(c) = \frac{f(b) - f(a)}{b - a}$$

or, equivalently,

$$\boxed{2} \qquad f(b) - f(a) = f'(c)(b - a)$$

---

Before proving this theorem, we can see that it is reasonable by interpreting it geometrically. Figures 3 and 4 show the points $A(a, f(a))$ and $B(b, f(b))$ on the graphs of two differentiable functions. The slope of the secant line $AB$ is

$$\boxed{3} \qquad m_{AB} = \frac{f(b) - f(a)}{b - a}$$

which is the same expression as on the right side of Equation 1. Since $f'(c)$ is the slope of the tangent line at the point $(c, f(c))$, the Mean Value Theorem, in the form given by Equation 1, says that there is at least one point $P(c, f(c))$ on the graph where the slope of the tangent line is the same as the slope of the secant line $AB$. In other words, there is a point $P$ where the tangent line is parallel to the secant line $AB$. (Imagine a line parallel to $AB$, starting far away and moving parallel to itself until it touches the graph for the first time.)

**FIGURE 3**          **FIGURE 4**

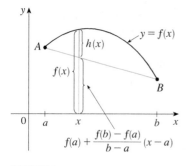

**FIGURE 5**

**Lagrange and the
Mean Value Theorem**

The Mean Value Theorem was first formulated by Joseph-Louis Lagrange (1736–1813), born in Italy of a French father and an Italian mother. He was a child prodigy and became a professor in Turin at the tender age of 19. Lagrange made great contributions to number theory, theory of functions, theory of equations, and analytical and celestial mechanics. In particular, he applied calculus to the analysis of the stability of the solar system. At the invitation of Frederick the Great, he succeeded Euler at the Berlin Academy and, when Frederick died, Lagrange accepted King Louis XVI's invitation to Paris, where he was given apartments in the Louvre and became a professor at the Ecole Polytechnique. Despite all the trappings of luxury and fame, he was a kind and quiet man, living only for science.

**PROOF**   We apply Rolle's Theorem to a new function $h$ defined as the difference between $f$ and the function whose graph is the secant line $AB$. Using Equation 3, we see that the equation of the line $AB$ can be written as

$$y - f(a) = \frac{f(b) - f(a)}{b - a}(x - a)$$

or as

$$y = f(a) + \frac{f(b) - f(a)}{b - a}(x - a)$$

So, as shown in Figure 5,

$$\boxed{4} \qquad h(x) = f(x) - f(a) - \frac{f(b) - f(a)}{b - a}(x - a)$$

First we must verify that $h$ satisfies the three hypotheses of Rolle's Theorem.

1. The function $h$ is continuous on $[a, b]$ because it is the sum of $f$ and a first-degree polynomial, both of which are continuous.

2. The function $h$ is differentiable on $(a, b)$ because both $f$ and the first-degree polynomial are differentiable. In fact, we can compute $h'$ directly from Equation 4:

$$h'(x) = f'(x) - \frac{f(b) - f(a)}{b - a}$$

(Note that $f(a)$ and $[f(b) - f(a)]/(b - a)$ are constants.)

3.
$$h(a) = f(a) - f(a) - \frac{f(b) - f(a)}{b - a}(a - a) = 0$$

$$h(b) = f(b) - f(a) - \frac{f(b) - f(a)}{b - a}(b - a)$$

$$= f(b) - f(a) - [f(b) - f(a)] = 0$$

Therefore $h(a) = h(b)$.

Since $h$ satisfies the hypotheses of Rolle's Theorem, that theorem says there is a number $c$ in $(a, b)$ such that $h'(c) = 0$. Therefore

$$0 = h'(c) = f'(c) - \frac{f(b) - f(a)}{b - a}$$

and so

$$f'(c) = \frac{f(b) - f(a)}{b - a} \qquad \blacksquare$$

**V   EXAMPLE 3**   To illustrate the Mean Value Theorem with a specific function, let's consider $f(x) = x^3 - x$, $a = 0$, $b = 2$. Since $f$ is a polynomial, it is continuous and differentiable for all $x$, so it is certainly continuous on $[0, 2]$ and differentiable on $(0, 2)$. Therefore, by the Mean Value Theorem, there is a number $c$ in $(0, 2)$ such that

$$f(2) - f(0) = f'(c)(2 - 0)$$

Now $f(2) = 6$, $f(0) = 0$, and $f'(x) = 3x^2 - 1$, so this equation becomes

$$6 = (3c^2 - 1)2 = 6c^2 - 2$$

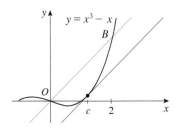

**FIGURE 6**

which gives $c^2 = \frac{4}{3}$, that is, $c = \pm 2/\sqrt{3}$. But $c$ must lie in $(0, 2)$, so $c = 2/\sqrt{3}$. Figure 6 illustrates this calculation: The tangent line at this value of $c$ is parallel to the secant line $OB$.

**V**  If an object moves in a straight line with position function $s = f(t)$, then the average velocity between $t = a$ and $t = b$ is

$$\frac{f(b) - f(a)}{b - a}$$

and the velocity at $t = c$ is $f'(c)$. Thus the Mean Value Theorem (in the form of Equation 1) tells us that at some time $t = c$ between $a$ and $b$ the instantaneous velocity $f'(c)$ is equal to that average velocity. For instance, if a car traveled 180 km in 2 hours, then the speedometer must have read 90 km/h at least once.

In general, the Mean Value Theorem can be interpreted as saying that there is a number at which the instantaneous rate of change is equal to the average rate of change over an interval.

The main significance of the Mean Value Theorem is that it enables us to obtain information about a function from information about its derivative. The next example provides an instance of this principle.

**V EXAMPLE 5** Suppose that $f(0) = -3$ and $f'(x) \le 5$ for all values of $x$. How large can $f(2)$ possibly be?

SOLUTION We are given that $f$ is differentiable (and therefore continuous) everywhere. In particular, we can apply the Mean Value Theorem on the interval $[0, 2]$. There exists a number $c$ such that

$$f(2) - f(0) = f'(c)(2 - 0)$$

so $$f(2) = f(0) + 2f'(c) = -3 + 2f'(c)$$

We are given that $f'(x) \le 5$ for all $x$, so in particular we know that $f'(c) \le 5$. Multiplying both sides of this inequality by 2, we have $2f'(c) \le 10$, so

$$f(2) = -3 + 2f'(c) \le -3 + 10 = 7$$

The largest possible value for $f(2)$ is 7.

The Mean Value Theorem can be used to establish some of the basic facts of differential calculus. One of these basic facts is the following theorem. Others will be found in the following sections.

---

**5 Theorem** If $f'(x) = 0$ for all $x$ in an interval $(a, b)$, then $f$ is constant on $(a, b)$.

---

PROOF Let $x_1$ and $x_2$ be any two numbers in $(a, b)$ with $x_1 < x_2$. Since $f$ is differentiable on $(a, b)$, it must be differentiable on $(x_1, x_2)$ and continuous on $[x_1, x_2]$. By applying the Mean Value Theorem to $f$ on the interval $[x_1, x_2]$, we get a number $c$ such that $x_1 < c < x_2$ and

**6** $$f(x_2) - f(x_1) = f'(c)(x_2 - x_1)$$

Since $f'(x) = 0$ for all $x$, we have $f'(c) = 0$, and so Equation 6 becomes

$$f(x_2) - f(x_1) = 0 \quad \text{or} \quad f(x_2) = f(x_1)$$

Therefore $f$ has the same value at *any* two numbers $x_1$ and $x_2$ in $(a, b)$. This means that $f$ is constant on $(a, b)$.

---

> **7**   **Corollary**   If $f'(x) = g'(x)$ for all $x$ in an interval $(a, b)$, then $f - g$ is constant on $(a, b)$; that is, $f(x) = g(x) + c$ where $c$ is a constant.

---

PROOF   Let $F(x) = f(x) - g(x)$. Then

$$F'(x) = f'(x) - g'(x) = 0$$

for all $x$ in $(a, b)$. Thus, by Theorem 5, $F$ is constant; that is, $f - g$ is constant.

NOTE   Care must be taken in applying Theorem 5. Let

$$f(x) = \frac{x}{|x|} = \begin{cases} 1 & \text{if } x > 0 \\ -1 & \text{if } x < 0 \end{cases}$$

The domain of $f$ is $D = \{x \mid x \neq 0\}$ and $f'(x) = 0$ for all $x$ in $D$. But $f$ is obviously not a constant function. This does not contradict Theorem 5 because $D$ is not an interval. Notice that $f$ is constant on the interval $(0, \infty)$ and also on the interval $(-\infty, 0)$.

## 3.2   Exercises

**1–4** Verify that the function satisfies the three hypotheses of Rolle's Theorem on the given interval. Then find all numbers $c$ that satisfy the conclusion of Rolle's Theorem.

**1.** $f(x) = 5 - 12x + 3x^2, \quad [1, 3]$

**2.** $f(x) = x^3 - x^2 - 6x + 2, \quad [0, 3]$

**3.** $f(x) = \sqrt{x} - \frac{1}{3}x, \quad [0, 9]$

**4.** $f(x) = \cos 2x, \quad [\pi/8, 7\pi/8]$

---

**5.** Let $f(x) = 1 - x^{2/3}$. Show that $f(-1) = f(1)$ but there is no number $c$ in $(-1, 1)$ such that $f'(c) = 0$. Why does this not contradict Rolle's Theorem?

**6.** Let $f(x) = \tan x$. Show that $f(0) = f(\pi)$ but there is no number $c$ in $(0, \pi)$ such that $f'(c) = 0$. Why does this not contradict Rolle's Theorem?

**7.** Use the graph of $f$ to estimate the values of $c$ that satisfy the conclusion of the Mean Value Theorem for the interval $[0, 8]$.

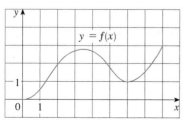

**8.** Use the graph of $f$ given in Exercise 7 to estimate the values of $c$ that satisfy the conclusion of the Mean Value Theorem for the interval $[1, 7]$.

**9–12** Verify that the function satisfies the hypotheses of the Mean Value Theorem on the given interval. Then find all numbers $c$ that satisfy the conclusion of the Mean Value Theorem.

**9.** $f(x) = 2x^2 - 3x + 1, \quad [0, 2]$

**10.** $f(x) = x^3 - 3x + 2, \quad [-2, 2]$

**11.** $f(x) = \sqrt[3]{x}, \quad [0, 1]$

**12.** $f(x) = 1/x, \quad [1, 3]$

---

**13–14** Find the number $c$ that satisfies the conclusion of the Mean Value Theorem on the given interval. Graph the function, the secant line through the endpoints, and the tangent line at $(c, f(c))$. Are the secant line and the tangent line(s) parallel?

**13.** $f(x) = \sqrt{x}, \quad [0, 4]$

**14.** $f(x) = x^3 - 2x, \quad [-2, 2]$

---

**15.** Let $f(x) = (x - 3)^{-2}$. Show that there is no value of $c$ in $(1, 4)$ such that $f(4) - f(1) = f'(c)(4 - 1)$. Why does this not contradict the Mean Value Theorem?

---

Graphing calculator or computer required   **1.** Homework Hints available at stewartcalculus.com

**16.** Let $f(x) = 2 - |2x - 1|$. Show that there is no value of $c$ such that $f(3) - f(0) = f'(c)(3 - 0)$. Why does this not contradict the Mean Value Theorem?

**17–18** Show that the equation has exactly one real root.

**17.** $2x + \cos x = 0$        **18.** $2x - 1 - \sin x = 0$

**19.** Show that the equation $x^3 - 15x + c = 0$ has at most one root in the interval $[-2, 2]$.

**20.** Show that the equation $x^4 + 4x + c = 0$ has at most two real roots.

**21.** (a) Show that a polynomial of degree 3 has at most three real roots.
(b) Show that a polynomial of degree $n$ has at most $n$ real roots.

**22.** (a) Suppose that $f$ is differentiable on $\mathbb{R}$ and has two roots. Show that $f'$ has at least one root.
(b) Suppose $f$ is twice differentiable on $\mathbb{R}$ and has three roots. Show that $f''$ has at least one real root.
(c) Can you generalize parts (a) and (b)?

**23.** If $f(1) = 10$ and $f'(x) \geqslant 2$ for $1 \leqslant x \leqslant 4$, how small can $f(4)$ possibly be?

**24.** Suppose that $3 \leqslant f'(x) \leqslant 5$ for all values of $x$. Show that $18 \leqslant f(8) - f(2) \leqslant 30$.

**25.** Does there exist a function $f$ such that $f(0) = -1$, $f(2) = 4$, and $f'(x) \leqslant 2$ for all $x$?

**26.** Suppose that $f$ and $g$ are continuous on $[a, b]$ and differentiable on $(a, b)$. Suppose also that $f(a) = g(a)$ and

$f'(x) < g'(x)$ for $a < x < b$. Prove that $f(b) < g(b)$. [*Hint:* Apply the Mean Value Theorem to the function $h = f - g$.]

**27.** Show that $\sqrt{1 + x} < 1 + \frac{1}{2}x$ if $x > 0$.

**28.** Suppose $f$ is an odd function and is differentiable everywhere. Prove that for every positive number $b$, there exists a number $c$ in $(-b, b)$ such that $f'(c) = f(b)/b$.

**29.** Use the Mean Value Theorem to prove the inequality

$$|\sin a - \sin b| \leqslant |a - b| \qquad \text{for all } a \text{ and } b$$

**30.** If $f'(x) = c$ ($c$ a constant) for all $x$, use Corollary 7 to show that $f(x) = cx + d$ for some constant $d$.

**31.** Let $f(x) = 1/x$ and

$$g(x) = \begin{cases} \dfrac{1}{x} & \text{if } x > 0 \\[2mm] 1 + \dfrac{1}{x} & \text{if } x < 0 \end{cases}$$

Show that $f'(x) = g'(x)$ for all $x$ in their domains. Can we conclude from Corollary 7 that $f - g$ is constant?

**32.** At 2:00 PM a car's speedometer reads 30 mi/h. At 2:10 PM it reads 50 mi/h. Show that at some time between 2:00 and 2:10 the acceleration is exactly 120 mi/h$^2$.

**33.** Two runners start a race at the same time and finish in a tie. Prove that at some time during the race they have the same speed. [*Hint:* Consider $f(t) = g(t) - h(t)$, where $g$ and $h$ are the position functions of the two runners.]

**34.** A number $a$ is called a **fixed point** of a function $f$ if $f(a) = a$. Prove that if $f'(x) \neq 1$ for all real numbers $x$, then $f$ has at most one fixed point.

---

## 3.3 How Derivatives Affect the Shape of a Graph

Many of the applications of calculus depend on our ability to deduce facts about a function $f$ from information concerning its derivatives. Because $f'(x)$ represents the slope of the curve $y = f(x)$ at the point $(x, f(x))$, it tells us the direction in which the curve proceeds at each point. So it is reasonable to expect that information about $f'(x)$ will provide us with information about $f(x)$.

### What Does $f'$ Say About $f$?

To see how the derivative of $f$ can tell us where a function is increasing or decreasing, look at Figure 1. (Increasing functions and decreasing functions were defined in Section 1.1.) Between $A$ and $B$ and between $C$ and $D$, the tangent lines have positive slope and so $f'(x) > 0$. Between $B$ and $C$, the tangent lines have negative slope and so $f'(x) < 0$. Thus it appears that $f$ increases when $f'(x)$ is positive and decreases when $f'(x)$ is negative. To prove that this is always the case, we use the Mean Value Theorem.

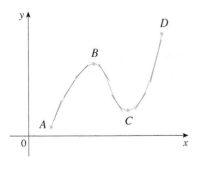

**FIGURE 1**

Let's abbreviate the name of this test to the I/D Test.

---

**Increasing/Decreasing Test**

(a) If $f'(x) > 0$ on an interval, then $f$ is increasing on that interval.

(b) If $f'(x) < 0$ on an interval, then $f$ is decreasing on that interval.

---

PROOF

(a) Let $x_1$ and $x_2$ be any two numbers in the interval with $x_1 < x_2$. According to the definition of an increasing function (page 19), we have to show that $f(x_1) < f(x_2)$.

Because we are given that $f'(x) > 0$, we know that $f$ is differentiable on $[x_1, x_2]$. So, by the Mean Value Theorem, there is a number $c$ between $x_1$ and $x_2$ such that

$$\boxed{1} \qquad\qquad f(x_2) - f(x_1) = f'(c)(x_2 - x_1)$$

Now $f'(c) > 0$ by assumption and $x_2 - x_1 > 0$ because $x_1 < x_2$. Thus the right side of Equation 1 is positive, and so

$$f(x_2) - f(x_1) > 0 \qquad \text{or} \qquad f(x_1) < f(x_2)$$

This shows that $f$ is increasing.

Part (b) is proved similarly.

**▼ EXAMPLE 1** Find where the function $f(x) = 3x^4 - 4x^3 - 12x^2 + 5$ is increasing and where it is decreasing.

SOLUTION $\qquad f'(x) = 12x^3 - 12x^2 - 24x = 12x(x - 2)(x + 1)$

To use the I/D Test we have to know where $f'(x) > 0$ and where $f'(x) < 0$. This depends on the signs of the three factors of $f'(x)$, namely, $12x$, $x - 2$, and $x + 1$. We divide the real line into intervals whose endpoints are the critical numbers $-1$, $0$, and $2$ and arrange our work in a chart. A plus sign indicates that the given expression is positive, and a minus sign indicates that it is negative. The last column of the chart gives the conclusion based on the I/D Test. For instance, $f'(x) < 0$ for $0 < x < 2$, so $f$ is decreasing on $(0, 2)$. (It would also be true to say that $f$ is decreasing on the closed interval $[0, 2]$.)

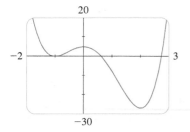

| Interval | $12x$ | $x - 2$ | $x + 1$ | $f'(x)$ | $f$ |
|---|---|---|---|---|---|
| $x < -1$ | $-$ | $-$ | $-$ | $-$ | decreasing on $(-\infty, -1)$ |
| $-1 < x < 0$ | $-$ | $-$ | $+$ | $+$ | increasing on $(-1, 0)$ |
| $0 < x < 2$ | $+$ | $-$ | $+$ | $-$ | decreasing on $(0, 2)$ |
| $x > 2$ | $+$ | $+$ | $+$ | $+$ | increasing on $(2, \infty)$ |

**FIGURE 2**

The graph of $f$ shown in Figure 2 confirms the information in the chart.

Recall from Section 3.1 that if $f$ has a local maximum or minimum at $c$, then $c$ must be a critical number of $f$ (by Fermat's Theorem), but not every critical number gives rise to a maximum or a minimum. We therefore need a test that will tell us whether or not $f$ has a local maximum or minimum at a critical number.

You can see from Figure 2 that $f(0) = 5$ is a local maximum value of $f$ because $f$ increases on $(-1, 0)$ and decreases on $(0, 2)$. Or, in terms of derivatives, $f'(x) > 0$ for $-1 < x < 0$ and $f'(x) < 0$ for $0 < x < 2$. In other words, the sign of $f'(x)$ changes from positive to negative at $0$. This observation is the basis of the following test.

**The First Derivative Test**   Suppose that $c$ is a critical number of a continuous function $f$.

(a) If $f'$ changes from positive to negative at $c$, then $f$ has a local maximum at $c$.

(b) If $f'$ changes from negative to positive at $c$, then $f$ has a local minimum at $c$.

(c) If $f'$ does not change sign at $c$ (for example, if $f'$ is positive on both sides of $c$ or negative on both sides), then $f$ has no local maximum or minimum at $c$.

The First Derivative Test is a consequence of the I/D Test. In part (a), for instance, since the sign of $f'(x)$ changes from positive to negative at $c$, $f$ is increasing to the left of $c$ and decreasing to the right of $c$. It follows that $f$ has a local maximum at $c$.

It is easy to remember the First Derivative Test by visualizing diagrams such as those in Figure 3.

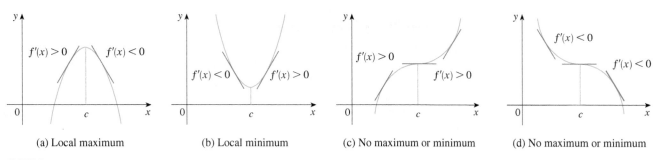

(a) Local maximum         (b) Local minimum         (c) No maximum or minimum         (d) No maximum or minimum

**FIGURE 3**

**V   EXAMPLE 2**   Find the local minimum and maximum values of the function $f$ in Example 1.

SOLUTION   From the chart in the solution to Example 1 we see that $f'(x)$ changes from negative to positive at $-1$, so $f(-1) = 0$ is a local minimum value by the First Derivative Test. Similarly, $f'$ changes from negative to positive at $2$, so $f(2) = -27$ is also a local minimum value. As previously noted, $f(0) = 5$ is a local maximum value because $f'(x)$ changes from positive to negative at $0$. ◼

**EXAMPLE 3**   Find the local maximum and minimum values of the function

$$g(x) = x + 2 \sin x \qquad 0 \le x \le 2\pi$$

SOLUTION   To find the critical numbers of $g$, we differentiate:

$$g'(x) = 1 + 2 \cos x$$

So $g'(x) = 0$ when $\cos x = -\frac{1}{2}$. The solutions of this equation are $2\pi/3$ and $4\pi/3$. Because $g$ is differentiable everywhere, the only critical numbers are $2\pi/3$ and $4\pi/3$ and so we analyze $g$ in the following table.

The + signs in the table come from the fact that $g'(x) > 0$ when $\cos x > -\frac{1}{2}$. From the graph of $y = \cos x$, this is true in the indicated intervals.

| Interval | $g'(x) = 1 + 2 \cos x$ | $g$ |
|---|---|---|
| $0 < x < 2\pi/3$ | $+$ | increasing on $(0, 2\pi/3)$ |
| $2\pi/3 < x < 4\pi/3$ | $-$ | decreasing on $(2\pi/3, 4\pi/3)$ |
| $4\pi/3 < x < 2\pi$ | $+$ | increasing on $(4\pi/3, 2\pi)$ |

Because $g'(x)$ changes from positive to negative at $2\pi/3$, the First Derivative Test tells us that there is a local maximum at $2\pi/3$ and the local maximum value is

$$g(2\pi/3) = \frac{2\pi}{3} + 2\sin\frac{2\pi}{3} = \frac{2\pi}{3} + 2\left(\frac{\sqrt{3}}{2}\right) = \frac{2\pi}{3} + \sqrt{3} \approx 3.83$$

Likewise, $g'(x)$ changes from negative to positive at $4\pi/3$ and so

$$g(4\pi/3) = \frac{4\pi}{3} + 2\sin\frac{4\pi}{3} = \frac{4\pi}{3} + 2\left(-\frac{\sqrt{3}}{2}\right) = \frac{4\pi}{3} - \sqrt{3} \approx 2.46$$

is a local minimum value. The graph of $g$ in Figure 4 supports our conclusion.

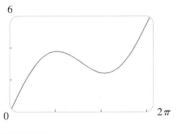

**FIGURE 4**
$g(x) = x + 2\sin x$

## What Does $f''$ Say About $f$?

Figure 5 shows the graphs of two increasing functions on $(a, b)$. Both graphs join point $A$ to point $B$ but they look different because they bend in different directions. How can we distinguish between these two types of behavior? In Figure 6 tangents to these curves have been drawn at several points. In (a) the curve lies above the tangents and $f$ is called *concave upward* on $(a, b)$. In (b) the curve lies below the tangents and $g$ is called *concave downward* on $(a, b)$.

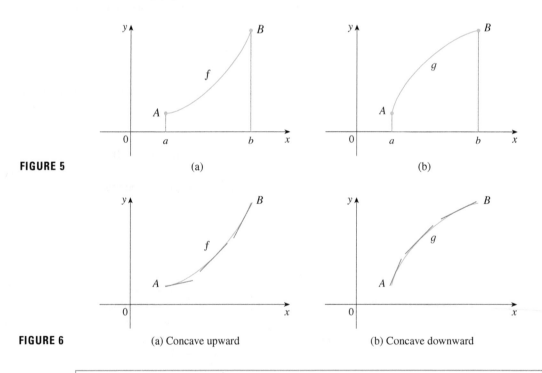

**FIGURE 5**                                      (a)                                      (b)

**FIGURE 6**                    (a) Concave upward                    (b) Concave downward

---

**Definition** If the graph of $f$ lies above all of its tangents on an interval $I$, then it is called **concave upward** on $I$. If the graph of $f$ lies below all of its tangents on $I$, it is called **concave downward** on $I$.

Figure 7 shows the graph of a function that is concave upward (abbreviated CU) on the intervals $(b, c)$, $(d, e)$, and $(e, p)$ and concave downward (CD) on the intervals $(a, b)$, $(c, d)$, and $(p, q)$.

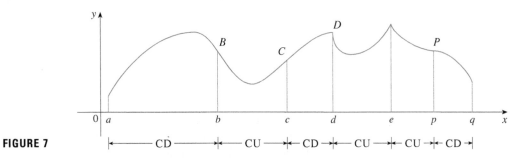

**FIGURE 7**

Let's see how the second derivative helps determine the intervals of concavity. Looking at Figure 6(a), you can see that, going from left to right, the slope of the tangent increases. This means that the derivative $f'$ is an increasing function and therefore its derivative $f''$ is positive. Likewise, in Figure 6(b) the slope of the tangent decreases from left to right, so $f'$ decreases and therefore $f''$ is negative. This reasoning can be reversed and suggests that the following theorem is true. A proof is given in Appendix F with the help of the Mean Value Theorem.

---

**Concavity Test**

(a) If $f''(x) > 0$ for all $x$ in $I$, then the graph of $f$ is concave upward on $I$.

(b) If $f''(x) < 0$ for all $x$ in $I$, then the graph of $f$ is concave downward on $I$.

---

**EXAMPLE 4** Figure 8 shows a population graph for Cyprian honeybees raised in an apiary. How does the rate of population increase change over time? When is this rate highest? Over what intervals is $P$ concave upward or concave downward?

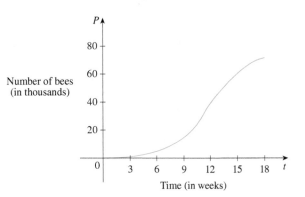

**FIGURE 8**

SOLUTION By looking at the slope of the curve as $t$ increases, we see that the rate of increase of the population is initially very small, then gets larger until it reaches a maximum at about $t = 12$ weeks, and decreases as the population begins to level off. As the population approaches its maximum value of about 75,000 (called the *carrying capacity*), the rate of increase, $P'(t)$, approaches 0. The curve appears to be concave upward on $(0, 12)$ and concave downward on $(12, 18)$.

In Example 4, the population curve changed from concave upward to concave downward at approximately the point (12, 38,000). This point is called an *inflection point* of the curve. The significance of this point is that the rate of population increase has its maximum value there. In general, an inflection point is a point where a curve changes its direction of concavity.

> **Definition**   A point *P* on a curve $y = f(x)$ is called an **inflection point** if $f$ is continuous there and the curve changes from concave upward to concave downward or from concave downward to concave upward at *P*.

For instance, in Figure 7, *B*, *C*, *D*, and *P* are the points of inflection. Notice that if a curve has a tangent at a point of inflection, then the curve crosses its tangent there.

In view of the Concavity Test, there is a point of inflection at any point where the second derivative changes sign.

**EXAMPLE 5**   Sketch a possible graph of a function *f* that satisfies the following conditions:

(i)   $f(0) = 0$,   $f(2) = 3$,   $f(4) = 6$,   $f'(0) = f'(4) = 0$

(ii)   $f'(x) > 0$ for $0 < x < 4$,   $f'(x) < 0$ for $x < 0$ and for $x > 4$

(iii)   $f''(x) > 0$ for $x < 2$,   $f''(x) < 0$ for $x > 2$

SOLUTION   Condition ( i ) tells us that the graph has horizontal tangents at the points (0, 0) and (4, 6). Condition ( ii ) says that *f* is increasing on the interval (0, 4) and decreasing on the intervals $(-\infty, 0)$ and $(4, \infty)$. It follows from the I/D Test that $f(0) = 0$ is a local minimum and $f(4) = 6$ is a local maximum.

Condition ( iii ) says that the graph is concave upward on the interval $(-\infty, 2)$ and concave downward on $(2, \infty)$. Because the curve changes from concave upward to concave downward when $x = 2$, the point (2, 3) is an inflection point.

We use this information to sketch the graph of *f* in Figure 9. Notice that we made the curve bend upward when $x < 2$ and bend downward when $x > 2$.

**FIGURE 9**

Another application of the second derivative is the following test for maximum and minimum values. It is a consequence of the Concavity Test.

> **The Second Derivative Test**   Suppose $f''$ is continuous near *c*.
> (a) If $f'(c) = 0$ and $f''(c) > 0$, then *f* has a local minimum at *c*.
> (b) If $f'(c) = 0$ and $f''(c) < 0$, then *f* has a local maximum at *c*.

For instance, part (a) is true because $f''(x) > 0$ near *c* and so *f* is concave upward near *c*. This means that the graph of *f* lies *above* its horizontal tangent at *c* and so *f* has a local minimum at *c*. (See Figure 10.)

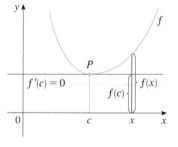

**FIGURE 10**
$f''(c) > 0$, *f* is concave upward

**V  EXAMPLE 6**   Discuss the curve $y = x^4 - 4x^3$ with respect to concavity, points of inflection, and local maxima and minima. Use this information to sketch the curve.

SOLUTION   If $f(x) = x^4 - 4x^3$, then

$$f'(x) = 4x^3 - 12x^2 = 4x^2(x - 3)$$
$$f''(x) = 12x^2 - 24x = 12x(x - 2)$$

To find the critical numbers we set $f'(x) = 0$ and obtain $x = 0$ and $x = 3$. To use the Second Derivative Test we evaluate $f''$ at these critical numbers:

$$f''(0) = 0 \qquad f''(3) = 36 > 0$$

Since $f'(3) = 0$ and $f''(3) > 0$, $f(3) = -27$ is a local minimum. Since $f''(0) = 0$, the Second Derivative Test gives no information about the critical number 0. But since $f'(x) < 0$ for $x < 0$ and also for $0 < x < 3$, the First Derivative Test tells us that $f$ does not have a local maximum or minimum at 0. [In fact, the expression for $f'(x)$ shows that $f$ decreases to the left of 3 and increases to the right of 3.]

Since $f''(x) = 0$ when $x = 0$ or 2, we divide the real line into intervals with these numbers as endpoints and complete the following chart.

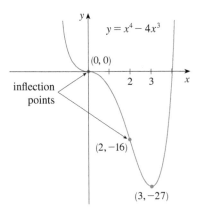

$y = x^4 - 4x^3$

$(0, 0)$

inflection points

$(2, -16)$

$(3, -27)$

**FIGURE 11**

| Interval | $f''(x) = 12x(x - 2)$ | Concavity |
|---|---|---|
| $(-\infty, 0)$ | $+$ | upward |
| $(0, 2)$ | $-$ | downward |
| $(2, \infty)$ | $+$ | upward |

The point $(0, 0)$ is an inflection point since the curve changes from concave upward to concave downward there. Also $(2, -16)$ is an inflection point since the curve changes from concave downward to concave upward there.

Using the local minimum, the intervals of concavity, and the inflection points, we sketch the curve in Figure 11.

**NOTE** The Second Derivative Test is inconclusive when $f''(c) = 0$. In other words, at such a point there might be a maximum, there might be a minimum, or there might be neither (as in Example 6). This test also fails when $f''(c)$ does not exist. In such cases the First Derivative Test must be used. In fact, even when both tests apply, the First Derivative Test is often the easier one to use.

**EXAMPLE 7**   Sketch the graph of the function $f(x) = x^{2/3}(6 - x)^{1/3}$.

SOLUTION   Calculation of the first two derivatives gives

$$f'(x) = \frac{4 - x}{x^{1/3}(6 - x)^{2/3}} \qquad f''(x) = \frac{-8}{x^{4/3}(6 - x)^{5/3}}$$

Use the differentiation rules to check these calculations.

Since $f'(x) = 0$ when $x = 4$ and $f'(x)$ does not exist when $x = 0$ or $x = 6$, the critical numbers are 0, 4, and 6.

| Interval | $4 - x$ | $x^{1/3}$ | $(6 - x)^{2/3}$ | $f'(x)$ | $f$ |
|---|---|---|---|---|---|
| $x < 0$ | $+$ | $-$ | $+$ | $-$ | decreasing on $(-\infty, 0)$ |
| $0 < x < 4$ | $+$ | $+$ | $+$ | $+$ | increasing on $(0, 4)$ |
| $4 < x < 6$ | $-$ | $+$ | $+$ | $-$ | decreasing on $(4, 6)$ |
| $x > 6$ | $-$ | $+$ | $+$ | $-$ | decreasing on $(6, \infty)$ |

To find the local extreme values we use the First Derivative Test. Since $f'$ changes from negative to positive at 0, $f(0) = 0$ is a local minimum. Since $f'$ changes from positive to negative at 4, $f(4) = 2^{5/3}$ is a local maximum. The sign of $f'$ does not change

**TEC** In Module 3.3 you can practice using information about $f'$, $f''$, and asymptotes to determine the shape of the graph of $f$.

Try reproducing the graph in Figure 12 with a graphing calculator or computer. Some machines produce the complete graph, some produce only the portion to the right of the $y$-axis, and some produce only the portion between $x = 0$ and $x = 6$. For an explanation and cure, see Example 7 in Appendix G. An equivalent expression that gives the correct graph is

$$y = (x^2)^{1/3} \cdot \frac{6 - x}{|6 - x|} |6 - x|^{1/3}$$

at 6, so there is no minimum or maximum there. (The Second Derivative Test could be used at 4 but not at 0 or 6 since $f''$ does not exist at either of these numbers.)

Looking at the expression for $f''(x)$ and noting that $x^{4/3} \geq 0$ for all $x$, we have $f''(x) < 0$ for $x < 0$ and for $0 < x < 6$ and $f''(x) > 0$ for $x > 6$. So $f$ is concave downward on $(-\infty, 0)$ and $(0, 6)$ and concave upward on $(6, \infty)$, and the only inflection point is $(6, 0)$. The graph is sketched in Figure 12. Note that the curve has vertical tangents at $(0, 0)$ and $(6, 0)$ because $|f'(x)| \to \infty$ as $x \to 0$ and as $x \to 6$.

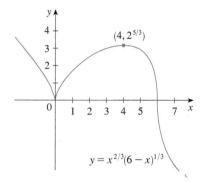

**FIGURE 12**

---

## 3.3   Exercises

**1–2** Use the given graph of $f$ to find the following.
(a) The open intervals on which $f$ is increasing.
(b) The open intervals on which $f$ is decreasing.
(c) The open intervals on which $f$ is concave upward.
(d) The open intervals on which $f$ is concave downward.
(e) The coordinates of the points of inflection.

**1.**      **2.**

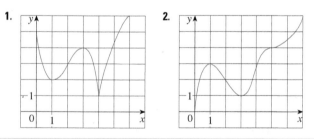

**3.** Suppose you are given a formula for a function $f$.
(a) How do you determine where $f$ is increasing or decreasing?
(b) How do you determine where the graph of $f$ is concave upward or concave downward?
(c) How do you locate inflection points?

**4.** (a) State the First Derivative Test.
(b) State the Second Derivative Test. Under what circumstances is it inconclusive? What do you do if it fails?

**5–6** The graph of the *derivative $f'$* of a function $f$ is shown.
(a) On what intervals is $f$ increasing or decreasing?
(b) At what values of $x$ does $f$ have a local maximum or minimum?

**5.**      **6.**

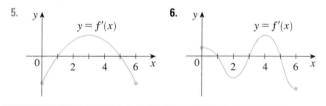

**7.** In each part state the $x$-coordinates of the inflection points of $f$. Give reasons for your answers.
(a) The curve is the graph of $f$.
(b) The curve is the graph of $f'$.
(c) The curve is the graph of $f''$.

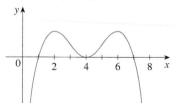

**8.** The graph of the first derivative $f'$ of a function $f$ is shown.
(a) On what intervals is $f$ increasing? Explain.

---

(b) At what values of $x$ does $f$ have a local maximum or minimum? Explain.

(c) On what intervals is $f$ concave upward or concave downward? Explain.

(d) What are the $x$-coordinates of the inflection points of $f$? Why?

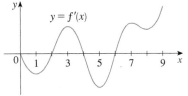

**9–14**

(a) Find the intervals on which $f$ is increasing or decreasing.

(b) Find the local maximum and minimum values of $f$.

(c) Find the intervals of concavity and the inflection points.

**9.** $f(x) = 2x^3 + 3x^2 - 36x$

**10.** $f(x) = 4x^3 + 3x^2 - 6x + 1$

**11.** $f(x) = x^4 - 2x^2 + 3$      **12.** $f(x) = \dfrac{x}{x^2 + 1}$

**13.** $f(x) = \sin x + \cos x, \quad 0 \leqslant x \leqslant 2\pi$

**14.** $f(x) = \cos^2 x - 2\sin x, \quad 0 \leqslant x \leqslant 2\pi$

---

**15–17** Find the local maximum and minimum values of $f$ using both the First and Second Derivative Tests. Which method do you prefer?

**15.** $f(x) = 1 + 3x^2 - 2x^3$      **16.** $f(x) = \dfrac{x^2}{x - 1}$

**17.** $f(x) = \sqrt{x} - \sqrt[4]{x}$

**18.** (a) Find the critical numbers of $f(x) = x^4(x - 1)^3$.

(b) What does the Second Derivative Test tell you about the behavior of $f$ at these critical numbers?

(c) What does the First Derivative Test tell you?

**19.** Suppose $f''$ is continuous on $(-\infty, \infty)$.

(a) If $f'(2) = 0$ and $f''(2) = -5$, what can you say about $f$?

(b) If $f'(6) = 0$ and $f''(6) = 0$, what can you say about $f$?

**20–25** Sketch the graph of a function that satisfies all of the given conditions.

**20.** Vertical asymptote $x = 0$,   $f'(x) > 0$ if $x < -2$,
$f'(x) < 0$ if $x > -2$ $(x \neq 0)$,
$f''(x) < 0$ if $x < 0$,   $f''(x) > 0$ if $x > 0$

**21.** $f'(0) = f'(2) = f'(4) = 0$,
$f'(x) > 0$ if $x < 0$ or $2 < x < 4$,
$f'(x) < 0$ if $0 < x < 2$ or $x > 4$,
$f''(x) > 0$ if $1 < x < 3$,   $f''(x) < 0$ if $x < 1$ or $x > 3$

**22.** $f'(1) = f'(-1) = 0$,   $f'(x) < 0$ if $|x| < 1$,
$f'(x) > 0$ if $1 < |x| < 2$,   $f'(x) = -1$ if $|x| > 2$,
$f''(x) < 0$ if $-2 < x < 0$,   inflection point $(0, 1)$

**23.** $f'(x) > 0$ if $|x| < 2$,   $f'(x) < 0$ if $|x| > 2$,
$f'(-2) = 0$,   $\lim\limits_{x \to 2} |f'(x)| = \infty$,   $f''(x) > 0$ if $x \neq 2$

**24.** $f(0) = f'(0) = f'(2) = f'(4) = f'(6) = 0$,
$f'(x) > 0$ if $0 < x < 2$ or $4 < x < 6$,
$f'(x) < 0$ if $2 < x < 4$ or $x > 6$,
$f''(x) > 0$ if $0 < x < 1$ or $3 < x < 5$,
$f''(x) < 0$ if $1 < x < 3$ or $x > 5$,   $f(-x) = f(x)$

**25.** $f'(x) < 0$ and $f''(x) < 0$ for all $x$

---

**26.** Suppose $f(3) = 2$, $f'(3) = \frac{1}{2}$, and $f'(x) > 0$ and $f''(x) < 0$ for all $x$.

(a) Sketch a possible graph for $f$.

(b) How many solutions does the equation $f(x) = 0$ have? Why?

(c) Is it possible that $f'(2) = \frac{1}{3}$? Why?

**27–28** The graph of the derivative $f'$ of a continuous function $f$ is shown.

(a) On what intervals is $f$ increasing? Decreasing?

(b) At what values of $x$ does $f$ have a local maximum? Local minimum?

(c) On what intervals is $f$ concave upward? Concave downward?

(d) State the $x$-coordinate(s) of the point(s) of inflection.

(e) Assuming that $f(0) = 0$, sketch a graph of $f$.

**27.**

**28.**

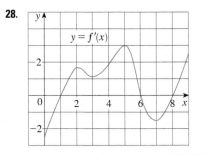

**29–40**
(a) Find the intervals of increase or decrease.
(b) Find the local maximum and minimum values.
(c) Find the intervals of concavity and the inflection points.
(d) Use the information from parts (a)–(c) to sketch the graph. Check your work with a graphing device if you have one.

**29.** $f(x) = x^3 - 12x + 2$     **30.** $f(x) = 36x + 3x^2 - 2x^3$

**31.** $f(x) = 2 + 2x^2 - x^4$     **32.** $g(x) = 200 + 8x^3 + x^4$

**33.** $h(x) = (x + 1)^5 - 5x - 2$     **34.** $h(x) = 5x^3 - 3x^5$

**35.** $F(x) = x\sqrt{6 - x}$     **36.** $G(x) = 5x^{2/3} - 2x^{5/3}$

**37.** $C(x) = x^{1/3}(x + 4)$     **38.** $G(x) = x - 4\sqrt{x}$

**39.** $f(\theta) = 2 \cos \theta + \cos^2\theta, \quad 0 \le \theta \le 2\pi$

**40.** $S(x) = x - \sin x, \quad 0 \le x \le 4\pi$

**41.** Suppose the derivative of a function $f$ is $f'(x) = (x + 1)^2(x - 3)^5(x - 6)^4$. On what interval is $f$ increasing?

**42.** Use the methods of this section to sketch the curve $y = x^3 - 3a^2x + 2a^3$, where $a$ is a positive constant. What do the members of this family of curves have in common? How do they differ from each other?

**43–44**
(a) Use a graph of $f$ to estimate the maximum and minimum values. Then find the exact values.
(b) Estimate the value of $x$ at which $f$ increases most rapidly. Then find the exact value.

**43.** $f(x) = \dfrac{x + 1}{\sqrt{x^2 + 1}}$

**44.** $f(x) = x + 2 \cos x, \quad 0 \le x \le 2\pi$

**45–46**
(a) Use a graph of $f$ to give a rough estimate of the intervals of concavity and the coordinates of the points of inflection.
(b) Use a graph of $f''$ to give better estimates.

**45.** $f(x) = \cos x + \frac{1}{2} \cos 2x, \quad 0 \le x \le 2\pi$

**46.** $f(x) = x^3(x - 2)^4$

CAS **47–48** Estimate the intervals of concavity to one decimal place by using a computer algebra system to compute and graph $f''$.

**47.** $f(x) = \dfrac{x^4 + x^3 + 1}{\sqrt{x^2 + x + 1}}$     **48.** $f(x) = \dfrac{(x + 1)^3(x^2 + 5)}{(x^3 + 1)(x^2 + 4)}$

**49.** A graph of a population of yeast cells in a new laboratory culture as a function of time is shown.
(a) Describe how the rate of population increase varies.
(b) When is this rate highest?
(c) On what intervals is the population function concave upward or downward?
(d) Estimate the coordinates of the inflection point.

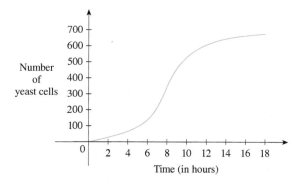

**50.** Let $f(t)$ be the temperature at time $t$ where you live and suppose that at time $t = 3$ you feel uncomfortably hot. How do you feel about the given data in each case?
(a) $f'(3) = 2, \quad f''(3) = 4$
(b) $f'(3) = 2, \quad f''(3) = -4$
(c) $f'(3) = -2, \quad f''(3) = 4$
(d) $f'(3) = -2, \quad f''(3) = -4$

**51.** Let $K(t)$ be a measure of the knowledge you gain by studying for a test for $t$ hours. Which do you think is larger, $K(8) - K(7)$ or $K(3) - K(2)$? Is the graph of $K$ concave upward or concave downward? Why?

**52.** Coffee is being poured into the mug shown in the figure at a constant rate (measured in volume per unit time). Sketch a rough graph of the depth of the coffee in the mug as a function of time. Account for the shape of the graph in terms of concavity. What is the significance of the inflection point?

**53.** Find a cubic function $f(x) = ax^3 + bx^2 + cx + d$ that has a local maximum value of 3 at $x = -2$ and a local minimum value of 0 at $x = 1$.

**54.** Show that the curve $y = (1 + x)/(1 + x^2)$ has three points of inflection and they all lie on one straight line.

**55.** (a) If the function $f(x) = x^3 + ax^2 + bx$ has the local minimum value $-\frac{2}{9}\sqrt{3}$ at $x = 1/\sqrt{3}$, what are the values of $a$ and $b$?

(b) Which of the tangent lines to the curve in part (a) has the smallest slope?

**56.** For what values of $a$ and $b$ is $(2, 2.5)$ an inflection point of the curve $x^2y + ax + by = 0$? What additional inflection points does the curve have?

**57.** Show that the inflection points of the curve $y = x \sin x$ lie on the curve $y^2(x^2 + 4) = 4x^2$.

**58–60** Assume that all of the functions are twice differentiable and the second derivatives are never 0.

**58.** (a) If $f$ and $g$ are concave upward on $I$, show that $f + g$ is concave upward on $I$.

(b) If $f$ is positive and concave upward on $I$, show that the function $g(x) = [f(x)]^2$ is concave upward on $I$.

**59.** (a) If $f$ and $g$ are positive, increasing, concave upward functions on $I$, show that the product function $fg$ is concave upward on $I$.

(b) Show that part (a) remains true if $f$ and $g$ are both decreasing.

(c) Suppose $f$ is increasing and $g$ is decreasing. Show, by giving three examples, that $fg$ may be concave upward, concave downward, or linear. Why doesn't the argument in parts (a) and (b) work in this case?

**60.** Suppose $f$ and $g$ are both concave upward on $(-\infty, \infty)$. Under what condition on $f$ will the composite function $h(x) = f(g(x))$ be concave upward?

**61.** Show that $\tan x > x$ for $0 < x < \pi/2$. [*Hint:* Show that $f(x) = \tan x - x$ is increasing on $(0, \pi/2)$.]

**62.** Prove that, for all $x > 1$,

$$2\sqrt{x} > 3 - \frac{1}{x}$$

**63.** Show that a cubic function (a third-degree polynomial) always has exactly one point of inflection. If its graph has

three $x$-intercepts $x_1, x_2,$ and $x_3$, show that the $x$-coordinate of the inflection point is $(x_1 + x_2 + x_3)/3$.

**64.** For what values of $c$ does the polynomial $P(x) = x^4 + cx^3 + x^2$ have two inflection points? One inflection point? None? Illustrate by graphing $P$ for several values of $c$. How does the graph change as $c$ decreases?

**65.** Prove that if $(c, f(c))$ is a point of inflection of the graph of $f$ and $f''$ exists in an open interval that contains $c$, then $f''(c) = 0$. [*Hint:* Apply the First Derivative Test and Fermat's Theorem to the function $g = f'$.]

**66.** Show that if $f(x) = x^4$, then $f''(0) = 0$, but $(0, 0)$ is not an inflection point of the graph of $f$.

**67.** Show that the function $g(x) = x|x|$ has an inflection point at $(0, 0)$ but $g''(0)$ does not exist.

**68.** Suppose that $f'''$ is continuous and $f'(c) = f''(c) = 0$, but $f'''(c) > 0$. Does $f$ have a local maximum or minimum at $c$? Does $f$ have a point of inflection at $c$?

**69.** Suppose $f$ is differentiable on an interval $I$ and $f'(x) > 0$ for all numbers $x$ in $I$ except for a single number $c$. Prove that $f$ is increasing on the entire interval $I$.

**70.** For what values of $c$ is the function

$$f(x) = cx + \frac{1}{x^2 + 3}$$

increasing on $(-\infty, \infty)$?

**71.** The three cases in the First Derivative Test cover the situations one commonly encounters but do not exhaust all possibilities. Consider the functions $f$, $g$, and $h$ whose values at 0 are all 0 and, for $x \neq 0$,

$$f(x) = x^4 \sin \frac{1}{x} \qquad g(x) = x^4\left(2 + \sin \frac{1}{x}\right)$$

$$h(x) = x^4\left(-2 + \sin \frac{1}{x}\right)$$

(a) Show that 0 is a critical number of all three functions but their derivatives change sign infinitely often on both sides of 0.

(b) Show that $f$ has neither a local maximum nor a local minimum at 0, $g$ has a local minimum, and $h$ has a local maximum.

## 3.4 Limits at Infinity; Horizontal Asymptotes

In Sections 1.5 and 1.7 we investigated infinite limits and vertical asymptotes. There we let $x$ approach a number and the result was that the values of $y$ became arbitrarily large (positive or negative). In this section we let $x$ become arbitrarily large (positive or negative) and see what happens to $y$. We will find it very useful to consider this so-called *end behavior* when sketching graphs.

| $x$ | $f(x)$ |
|---|---|
| 0 | $-1$ |
| $\pm 1$ | 0 |
| $\pm 2$ | 0.600000 |
| $\pm 3$ | 0.800000 |
| $\pm 4$ | 0.882353 |
| $\pm 5$ | 0.923077 |
| $\pm 10$ | 0.980198 |
| $\pm 50$ | 0.999200 |
| $\pm 100$ | 0.999800 |
| $\pm 1000$ | 0.999998 |

Let's begin by investigating the behavior of the function $f$ defined by

$$f(x) = \frac{x^2 - 1}{x^2 + 1}$$

as $x$ becomes large. The table at the left gives values of this function correct to six decimal places, and the graph of $f$ has been drawn by a computer in Figure 1.

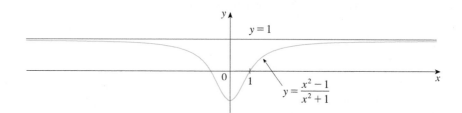

**FIGURE 1**

As $x$ grows larger and larger you can see that the values of $f(x)$ get closer and closer to 1. In fact, it seems that we can make the values of $f(x)$ as close as we like to 1 by taking $x$ sufficiently large. This situation is expressed symbolically by writing

$$\lim_{x \to \infty} \frac{x^2 - 1}{x^2 + 1} = 1$$

In general, we use the notation

$$\lim_{x \to \infty} f(x) = L$$

to indicate that the values of $f(x)$ approach $L$ as $x$ becomes larger and larger.

---

**1**  **Definition**  Let $f$ be a function defined on some interval $(a, \infty)$. Then

$$\lim_{x \to \infty} f(x) = L$$

means that the values of $f(x)$ can be made arbitrarily close to $L$ by taking $x$ sufficiently large.

---

Another notation for $\lim_{x \to \infty} f(x) = L$ is

$$f(x) \to L \quad \text{as} \quad x \to \infty$$

The symbol $\infty$ does not represent a number. Nonetheless, the expression $\lim_{x \to \infty} f(x) = L$ is often read as

"the limit of $f(x)$, as $x$ approaches infinity, is $L$"

or              "the limit of $f(x)$, as $x$ becomes infinite, is $L$"

or              "the limit of $f(x)$, as $x$ increases without bound, is $L$"

The meaning of such phrases is given by Definition 1. A more precise definition, similar to the $\varepsilon$, $\delta$ definition of Section 1.7, is given at the end of this section.

Geometric illustrations of Definition 1 are shown in Figure 2. Notice that there are many ways for the graph of $f$ to approach the line $y = L$ (which is called a *horizontal asymptote*) as we look to the far right of each graph.

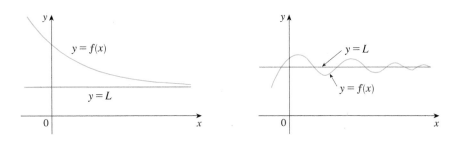

**FIGURE 2**

Examples illustrating $\lim_{x \to \infty} f(x) = L$

Referring back to Figure 1, we see that for numerically large negative values of $x$, the values of $f(x)$ are close to 1. By letting $x$ decrease through negative values without bound, we can make $f(x)$ as close to 1 as we like. This is expressed by writing

$$\lim_{x \to -\infty} \frac{x^2 - 1}{x^2 + 1} = 1$$

The general definition is as follows.

---

**2** **Definition** Let $f$ be a function defined on some interval $(-\infty, a)$. Then

$$\lim_{x \to -\infty} f(x) = L$$

means that the values of $f(x)$ can be made arbitrarily close to $L$ by taking $x$ sufficiently large negative.

---

Again, the symbol $-\infty$ does not represent a number, but the expression $\lim_{x \to -\infty} f(x) = L$ is often read as

"the limit of $f(x)$, as $x$ approaches negative infinity, is $L$"

Definition 2 is illustrated in Figure 3. Notice that the graph approaches the line $y = L$ as we look to the far left of each graph.

---

**3** **Definition** The line $y = L$ is called a **horizontal asymptote** of the curve $y = f(x)$ if either

$$\lim_{x \to \infty} f(x) = L \qquad \text{or} \qquad \lim_{x \to -\infty} f(x) = L$$

---

For instance, the curve illustrated in Figure 1 has the line $y = 1$ as a horizontal asymptote because

$$\lim_{x \to \infty} \frac{x^2 - 1}{x^2 + 1} = 1$$

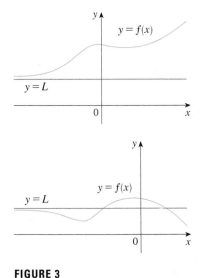

**FIGURE 3**

Examples illustrating $\lim_{x \to -\infty} f(x) = L$

The curve $y = f(x)$ sketched in Figure 4 has both $y = -1$ and $y = 2$ as horizontal asymptotes because

$$\lim_{x \to \infty} f(x) = -1 \qquad \text{and} \qquad \lim_{x \to -\infty} f(x) = 2$$

**FIGURE 4**

**FIGURE 5**

**EXAMPLE 1** Find the infinite limits, limits at infinity, and asymptotes for the function $f$ whose graph is shown in Figure 5.

SOLUTION We see that the values of $f(x)$ become large as $x \to -1$ from both sides, so

$$\lim_{x \to -1} f(x) = \infty$$

Notice that $f(x)$ becomes large negative as $x$ approaches 2 from the left, but large positive as $x$ approaches 2 from the right. So

$$\lim_{x \to 2^-} f(x) = -\infty \qquad \text{and} \qquad \lim_{x \to 2^+} f(x) = \infty$$

Thus both of the lines $x = -1$ and $x = 2$ are vertical asymptotes.

As $x$ becomes large, it appears that $f(x)$ approaches 4. But as $x$ decreases through negative values, $f(x)$ approaches 2. So

$$\lim_{x \to \infty} f(x) = 4 \qquad \text{and} \qquad \lim_{x \to -\infty} f(x) = 2$$

This means that both $y = 4$ and $y = 2$ are horizontal asymptotes.

**EXAMPLE 2** Find $\displaystyle\lim_{x \to \infty} \frac{1}{x}$ and $\displaystyle\lim_{x \to -\infty} \frac{1}{x}$.

SOLUTION Observe that when $x$ is large, $1/x$ is small. For instance,

$$\frac{1}{100} = 0.01 \qquad \frac{1}{10,000} = 0.0001 \qquad \frac{1}{1,000,000} = 0.000001$$

In fact, by taking $x$ large enough, we can make $1/x$ as close to 0 as we please. Therefore, according to Definition 1, we have

$$\lim_{x \to \infty} \frac{1}{x} = 0$$

Similar reasoning shows that when $x$ is large negative, $1/x$ is small negative, so we also have

$$\lim_{x \to -\infty} \frac{1}{x} = 0$$

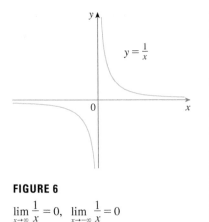

**FIGURE 6**

$$\lim_{x \to \infty} \frac{1}{x} = 0, \quad \lim_{x \to -\infty} \frac{1}{x} = 0$$

It follows that the line $y = 0$ (the $x$-axis) is a horizontal asymptote of the curve $y = 1/x$. (This is an equilateral hyperbola; see Figure 6.)

Most of the Limit Laws that were given in Section 1.6 also hold for limits at infinity. It can be proved that the *Limit Laws listed in Section 1.6 (with the exception of Laws 9 and 10) are also valid if "$x \to a$" is replaced by "$x \to \infty$" or "$x \to -\infty$."* In particular, if we combine Laws 6 and 11 with the results of Example 2, we obtain the following important rule for calculating limits.

---

**4 Theorem** If $r > 0$ is a rational number, then

$$\lim_{x \to \infty} \frac{1}{x^r} = 0$$

If $r > 0$ is a rational number such that $x^r$ is defined for all $x$, then

$$\lim_{x \to -\infty} \frac{1}{x^r} = 0$$

---

**V EXAMPLE 3** Evaluate

$$\lim_{x \to \infty} \frac{3x^2 - x - 2}{5x^2 + 4x + 1}$$

and indicate which properties of limits are used at each stage.

SOLUTION As $x$ becomes large, both numerator and denominator become large, so it isn't obvious what happens to their ratio. We need to do some preliminary algebra.

To evaluate the limit at infinity of any rational function, we first divide both the numerator and denominator by the highest power of $x$ that occurs in the denominator. (We may assume that $x \neq 0$, since we are interested only in large values of $x$.) In this case the highest power of $x$ in the denominator is $x^2$, so we have

$$\lim_{x \to \infty} \frac{3x^2 - x - 2}{5x^2 + 4x + 1} = \lim_{x \to \infty} \frac{\dfrac{3x^2 - x - 2}{x^2}}{\dfrac{5x^2 + 4x + 1}{x^2}} = \lim_{x \to \infty} \frac{3 - \dfrac{1}{x} - \dfrac{2}{x^2}}{5 + \dfrac{4}{x} + \dfrac{1}{x^2}}$$

$$= \frac{\lim_{x \to \infty} \left( 3 - \dfrac{1}{x} - \dfrac{2}{x^2} \right)}{\lim_{x \to \infty} \left( 5 + \dfrac{4}{x} + \dfrac{1}{x^2} \right)} \qquad \text{(by Limit Law 5)}$$

$$= \frac{\lim_{x \to \infty} 3 - \lim_{x \to \infty} \dfrac{1}{x} - 2 \lim_{x \to \infty} \dfrac{1}{x^2}}{\lim_{x \to \infty} 5 + 4 \lim_{x \to \infty} \dfrac{1}{x} + \lim_{x \to \infty} \dfrac{1}{x^2}} \qquad \text{(by 1, 2, and 3)}$$

$$= \frac{3 - 0 - 0}{5 + 0 + 0} \qquad \text{(by 7 and Theorem 4)}$$

$$= \frac{3}{5}$$

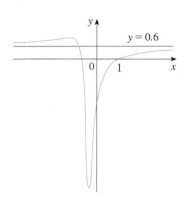

**FIGURE 7**

$$y = \frac{3x^2 - x - 2}{5x^2 + 4x + 1}$$

A similar calculation shows that the limit as $x \to -\infty$ is also $\frac{3}{5}$. Figure 7 illustrates the results of these calculations by showing how the graph of the given rational function approaches the horizontal asymptote $y = \frac{3}{5}$.

**EXAMPLE 4**  Find the horizontal and vertical asymptotes of the graph of the function

$$f(x) = \frac{\sqrt{2x^2 + 1}}{3x - 5}$$

SOLUTION  Dividing both numerator and denominator by $x$ and using the properties of limits, we have

$$\lim_{x \to \infty} \frac{\sqrt{2x^2 + 1}}{3x - 5} = \lim_{x \to \infty} \frac{\sqrt{2 + \dfrac{1}{x^2}}}{3 - \dfrac{5}{x}} \qquad (\text{since } \sqrt{x^2} = x \text{ for } x > 0)$$

$$= \frac{\lim\limits_{x \to \infty} \sqrt{2 + \dfrac{1}{x^2}}}{\lim\limits_{x \to \infty} \left(3 - \dfrac{5}{x}\right)} = \frac{\sqrt{\lim\limits_{x \to \infty} 2 + \lim\limits_{x \to \infty} \dfrac{1}{x^2}}}{\lim\limits_{x \to \infty} 3 - 5 \lim\limits_{x \to \infty} \dfrac{1}{x}}$$

$$= \frac{\sqrt{2 + 0}}{3 - 5 \cdot 0} = \frac{\sqrt{2}}{3}$$

Therefore the line $y = \sqrt{2}/3$ is a horizontal asymptote of the graph of $f$.

In computing the limit as $x \to -\infty$, we must remember that for $x < 0$, we have $\sqrt{x^2} = |x| = -x$. So when we divide the numerator by $x$, for $x < 0$ we get

$$\frac{1}{x} \sqrt{2x^2 + 1} = -\frac{1}{\sqrt{x^2}} \sqrt{2x^2 + 1} = -\sqrt{2 + \frac{1}{x^2}}$$

Therefore

$$\lim_{x \to -\infty} \frac{\sqrt{2x^2 + 1}}{3x - 5} = \lim_{x \to -\infty} \frac{-\sqrt{2 + \dfrac{1}{x^2}}}{3 - \dfrac{5}{x}}$$

$$= \frac{-\sqrt{2 + \lim\limits_{x \to -\infty} \dfrac{1}{x^2}}}{3 - 5 \lim\limits_{x \to -\infty} \dfrac{1}{x}} = -\frac{\sqrt{2}}{3}$$

Thus the line $y = -\sqrt{2}/3$ is also a horizontal asymptote.

A vertical asymptote is likely to occur when the denominator, $3x - 5$, is 0, that is, when $x = \frac{5}{3}$. If $x$ is close to $\frac{5}{3}$ and $x > \frac{5}{3}$, then the denominator is close to 0 and $3x - 5$ is positive. The numerator $\sqrt{2x^2 + 1}$ is always positive, so $f(x)$ is positive. Therefore

$$\lim_{x \to (5/3)^+} \frac{\sqrt{2x^2 + 1}}{3x - 5} = \infty$$

If $x$ is close to $\frac{5}{3}$ but $x < \frac{5}{3}$, then $3x - 5 < 0$ and so $f(x)$ is large negative. Thus

$$\lim_{x \to (5/3)^-} \frac{\sqrt{2x^2 + 1}}{3x - 5} = -\infty$$

The vertical asymptote is $x = \frac{5}{3}$. All three asymptotes are shown in Figure 8.

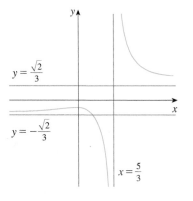

**FIGURE 8**

$$y = \frac{\sqrt{2x^2 + 1}}{3x - 5}$$

**EXAMPLE 5**  Compute $\lim\limits_{x \to \infty} \left( \sqrt{x^2 + 1} - x \right)$.

SOLUTION  Because both $\sqrt{x^2 + 1}$ and $x$ are large when $x$ is large, it's difficult to see what happens to their difference, so we use algebra to rewrite the function. We first multiply numerator and denominator by the conjugate radical:

We can think of the given function as having a denominator of 1.

$$\lim_{x \to \infty} \left( \sqrt{x^2 + 1} - x \right) = \lim_{x \to \infty} \left( \sqrt{x^2 + 1} - x \right) \frac{\sqrt{x^2 + 1} + x}{\sqrt{x^2 + 1} + x}$$

$$= \lim_{x \to \infty} \frac{(x^2 + 1) - x^2}{\sqrt{x^2 + 1} + x} = \lim_{x \to \infty} \frac{1}{\sqrt{x^2 + 1} + x}$$

Notice that the denominator of this last expression $\left( \sqrt{x^2 + 1} + x \right)$ becomes large as $x \to \infty$ ( it's bigger than $x$). So

$$\lim_{x \to \infty} \left( \sqrt{x^2 + 1} - x \right) = \lim_{x \to \infty} \frac{1}{\sqrt{x^2 + 1} + x} = 0$$

$y = \sqrt{x^2 + 1} - x$

**FIGURE 9**

Figure 9 illustrates this result.

**EXAMPLE 6**  Evaluate $\lim\limits_{x \to \infty} \sin \dfrac{1}{x}$.

PS  The problem-solving strategy for Example 6 is *introducing something extra* (see page 97). Here, the something extra, the auxiliary aid, is the new variable $t$.

SOLUTION  If we let $t = 1/x$, then $t \to 0^+$ as $x \to \infty$. Therefore

$$\lim_{x \to \infty} \sin \frac{1}{x} = \lim_{t \to 0^+} \sin t = 0$$

(See Exercise 71.)

**EXAMPLE 7** Evaluate $\lim\limits_{x \to \infty} \sin x$.

SOLUTION As $x$ increases, the values of $\sin x$ oscillate between 1 and $-1$ infinitely often and so they don't approach any definite number. Thus $\lim_{x \to \infty} \sin x$ does not exist.

## Infinite Limits at Infinity

The notation

$$\lim_{x \to \infty} f(x) = \infty$$

is used to indicate that the values of $f(x)$ become large as $x$ becomes large. Similar meanings are attached to the following symbols:

$$\lim_{x \to -\infty} f(x) = \infty \qquad \lim_{x \to \infty} f(x) = -\infty \qquad \lim_{x \to -\infty} f(x) = -\infty$$

**EXAMPLE 8** Find $\lim\limits_{x \to \infty} x^3$ and $\lim\limits_{x \to -\infty} x^3$.

SOLUTION When $x$ becomes large, $x^3$ also becomes large. For instance,

$$10^3 = 1000 \qquad 100^3 = 1,000,000 \qquad 1000^3 = 1,000,000,000$$

In fact, we can make $x^3$ as big as we like by taking $x$ large enough. Therefore we can write

$$\lim_{x \to \infty} x^3 = \infty$$

Similarly, when $x$ is large negative, so is $x^3$. Thus

$$\lim_{x \to -\infty} x^3 = -\infty$$

These limit statements can also be seen from the graph of $y = x^3$ in Figure 10.

**FIGURE 10**
$\lim\limits_{x \to \infty} x^3 = \infty$, $\lim\limits_{x \to -\infty} x^3 = -\infty$

**EXAMPLE 9** Find $\lim\limits_{x \to \infty} (x^2 - x)$.

SOLUTION It would be **wrong** to write

$$\lim_{x \to \infty} (x^2 - x) = \lim_{x \to \infty} x^2 - \lim_{x \to \infty} x = \infty - \infty$$

The Limit Laws can't be applied to infinite limits because $\infty$ is not a number ($\infty - \infty$ can't be defined). However, we *can* write

$$\lim_{x \to \infty} (x^2 - x) = \lim_{x \to \infty} x(x - 1) = \infty$$

because both $x$ and $x - 1$ become arbitrarily large and so their product does too.

**EXAMPLE 10** Find $\lim\limits_{x \to \infty} \dfrac{x^2 + x}{3 - x}$.

SOLUTION As in Example 3, we divide the numerator and denominator by the highest power of $x$ in the denominator, which is just $x$:

$$\lim_{x \to \infty} \frac{x^2 + x}{3 - x} = \lim_{x \to \infty} \frac{x + 1}{\dfrac{3}{x} - 1} = -\infty$$

because $x + 1 \to \infty$ and $3/x - 1 \to -1$ as $x \to \infty$.

The next example shows that by using infinite limits at infinity, together with intercepts, we can get a rough idea of the graph of a polynomial without computing derivatives.

**EXAMPLE 11**  Sketch the graph of $y = (x - 2)^4(x + 1)^3(x - 1)$ by finding its intercepts and its limits as $x \to \infty$ and as $x \to -\infty$.

SOLUTION  The $y$-intercept is $f(0) = (-2)^4(1)^3(-1) = -16$ and the $x$-intercepts are found by setting $y = 0$: $x = 2, -1, 1$. Notice that since $(x - 2)^4$ is positive, the function doesn't change sign at 2; thus the graph doesn't cross the $x$-axis at 2. The graph crosses the axis at $-1$ and 1.

When $x$ is large positive, all three factors are large, so

$$\lim_{x \to \infty} (x - 2)^4(x + 1)^3(x - 1) = \infty$$

When $x$ is large negative, the first factor is large positive and the second and third factors are both large negative, so

$$\lim_{x \to -\infty} (x - 2)^4(x + 1)^3(x - 1) = \infty$$

Combining this information, we give a rough sketch of the graph in Figure 11.

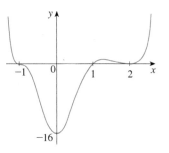

**FIGURE 11**
$y = (x - 2)^4(x + 1)^3(x - 1)$

## Precise Definitions

Definition 1 can be stated precisely as follows.

---

**5**  **Definition**  Let $f$ be a function defined on some interval $(a, \infty)$. Then

$$\lim_{x \to \infty} f(x) = L$$

means that for every $\varepsilon > 0$ there is a corresponding number $N$ such that

$$\text{if} \quad x > N \quad \text{then} \quad |f(x) - L| < \varepsilon$$

---

In words, this says that the values of $f(x)$ can be made arbitrarily close to $L$ (within a distance $\varepsilon$, where $\varepsilon$ is any positive number) by taking $x$ sufficiently large (larger than $N$, where $N$ depends on $\varepsilon$). Graphically it says that by choosing $x$ large enough (larger than some number $N$) we can make the graph of $f$ lie between the given horizontal lines

$y = L - \varepsilon$ and $y = L + \varepsilon$ as in Figure 12. This must be true no matter how small we choose $\varepsilon$. Figure 13 shows that if a smaller value of $\varepsilon$ is chosen, then a larger value of $N$ may be required.

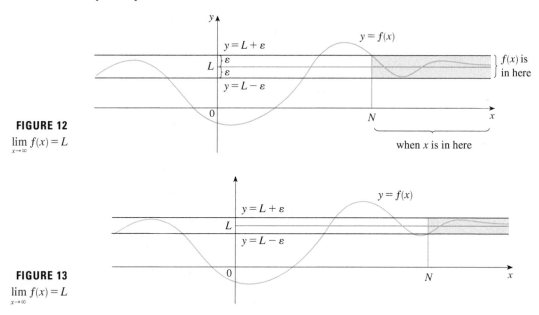

**FIGURE 12**
$\lim\limits_{x \to \infty} f(x) = L$

**FIGURE 13**
$\lim\limits_{x \to \infty} f(x) = L$

Similarly, a precise version of Definition 2 is given by Definition 6, which is illustrated in Figure 14.

---

**6** **Definition** Let $f$ be a function defined on some interval $(-\infty, a)$. Then

$$\lim_{x \to -\infty} f(x) = L$$

means that for every $\varepsilon > 0$ there is a corresponding number $N$ such that

$$\text{if} \quad x < N \quad \text{then} \quad |f(x) - L| < \varepsilon$$

---

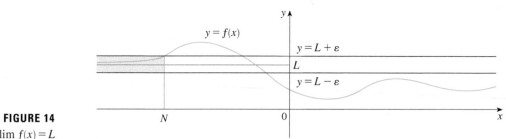

**FIGURE 14**
$\lim\limits_{x \to -\infty} f(x) = L$

In Example 3 we calculated that

$$\lim_{x \to \infty} \frac{3x^2 - x - 2}{5x^2 + 4x + 1} = \frac{3}{5}$$

In the next example we use a graphing device to relate this statement to Definition 5 with $L = \frac{3}{5}$ and $\varepsilon = 0.1$.

**TEC** In Module 1.7/3.4 you can explore the precise definition of a limit both graphically and numerically.

**EXAMPLE 12** Use a graph to find a number $N$ such that

$$\text{if} \quad x > N \quad \text{then} \quad \left| \frac{3x^2 - x - 2}{5x^2 + 4x + 1} - 0.6 \right| < 0.1$$

SOLUTION We rewrite the given inequality as

$$0.5 < \frac{3x^2 - x - 2}{5x^2 + 4x + 1} < 0.7$$

We need to determine the values of $x$ for which the given curve lies between the horizontal lines $y = 0.5$ and $y = 0.7$. So we graph the curve and these lines in Figure 15. Then we use the cursor to estimate that the curve crosses the line $y = 0.5$ when $x \approx 6.7$. To the right of this number it seems that the curve stays between the lines $y = 0.5$ and $y = 0.7$. Rounding to be safe, we can say that

$$\text{if} \quad x > 7 \quad \text{then} \quad \left| \frac{3x^2 - x - 2}{5x^2 + 4x + 1} - 0.6 \right| < 0.1$$

In other words, for $\varepsilon = 0.1$ we can choose $N = 7$ (or any larger number) in Definition 5.

1

$y = 0.7$

$y = 0.5$

$y = \dfrac{3x^2 - x - 2}{5x^2 + 4x + 1}$

0                    15

**FIGURE 15**

**EXAMPLE 13** Use Definition 5 to prove that $\displaystyle\lim_{x \to \infty} \frac{1}{x} = 0$.

SOLUTION Given $\varepsilon > 0$, we want to find $N$ such that

$$\text{if} \quad x > N \quad \text{then} \quad \left| \frac{1}{x} - 0 \right| < \varepsilon$$

In computing the limit we may assume that $x > 0$. Then $1/x < \varepsilon \iff x > 1/\varepsilon$. Let's choose $N = 1/\varepsilon$. So

$$\text{if} \quad x > N = \frac{1}{\varepsilon} \quad \text{then} \quad \left| \frac{1}{x} - 0 \right| = \frac{1}{x} < \varepsilon$$

Therefore, by Definition 5,

$$\lim_{x \to \infty} \frac{1}{x} = 0$$

Figure 16 illustrates the proof by showing some values of $\varepsilon$ and the corresponding values of $N$.

**FIGURE 16**

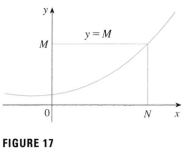

**FIGURE 17**
$\lim\limits_{x \to \infty} f(x) = \infty$

Finally we note that an infinite limit at infinity can be defined as follows. The geometric illustration is given in Figure 17.

---

**7**  **Definition**   Let $f$ be a function defined on some interval $(a, \infty)$. Then

$$\lim_{x \to \infty} f(x) = \infty$$

means that for every positive number $M$ there is a corresponding positive number $N$ such that

$$\text{if} \quad x > N \qquad \text{then} \qquad f(x) > M$$

---

Similar definitions apply when the symbol $\infty$ is replaced by $-\infty$. (See Exercise 72.)

---

## 3.4   Exercises

**1.** Explain in your own words the meaning of each of the following.
   (a) $\lim\limits_{x \to \infty} f(x) = 5$       (b) $\lim\limits_{x \to -\infty} f(x) = 3$

**2.** (a) Can the graph of $y = f(x)$ intersect a vertical asymptote? Can it intersect a horizontal asymptote? Illustrate by sketching graphs.
   (b) How many horizontal asymptotes can the graph of $y = f(x)$ have? Sketch graphs to illustrate the possibilities.

**3.** For the function $f$ whose graph is given, state the following.
   (a) $\lim\limits_{x \to \infty} f(x)$        (b) $\lim\limits_{x \to -\infty} f(x)$
   (c) $\lim\limits_{x \to 1} f(x)$        (d) $\lim\limits_{x \to 3} f(x)$
   (e) The equations of the asymptotes

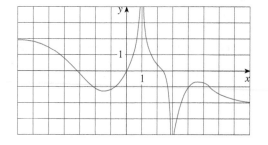

**4.** For the function $g$ whose graph is given, state the following.
   (a) $\lim\limits_{x \to \infty} g(x)$        (b) $\lim\limits_{x \to -\infty} g(x)$
   (c) $\lim\limits_{x \to 0} g(x)$        (d) $\lim\limits_{x \to 2^-} g(x)$
   (e) $\lim\limits_{x \to 2^+} g(x)$        (f) The equations of the asymptotes

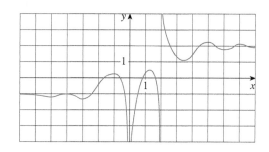

**5.** Guess the value of the limit

$$\lim_{x \to \infty} \frac{x^2}{2^x}$$

by evaluating the function $f(x) = x^2/2^x$ for $x = 0, 1, 2, 3,$ 4, 5, 6, 7, 8, 9, 10, 20, 50, and 100. Then use a graph of $f$ to support your guess.

**6.** (a) Use a graph of

$$f(x) = \left( 1 - \frac{2}{x} \right)^x$$

   to estimate the value of $\lim_{x \to \infty} f(x)$ correct to two decimal places.
   (b) Use a table of values of $f(x)$ to estimate the limit to four decimal places.

**7–8** Evaluate the limit and justify each step by indicating the appropriate properties of limits.

**7.** $\lim\limits_{x \to \infty} \dfrac{3x^2 - x + 4}{2x^2 + 5x - 8}$       **8.** $\lim\limits_{x \to \infty} \sqrt{\dfrac{12x^3 - 5x + 2}{1 + 4x^2 + 3x^3}}$

---

Graphing calculator or computer required      **1.** Homework Hints available at stewartcalculus.com

**9–30** Find the limit or show that it does not exist.

**9.** $\lim\limits_{x \to \infty} \dfrac{3x - 2}{2x + 1}$

**10.** $\lim\limits_{x \to \infty} \dfrac{1 - x^2}{x^3 - x + 1}$

**11.** $\lim\limits_{x \to -\infty} \dfrac{x - 2}{x^2 + 1}$

**12.** $\lim\limits_{x \to -\infty} \dfrac{4x^3 + 6x^2 - 2}{2x^3 - 4x + 5}$

**13.** $\lim\limits_{t \to \infty} \dfrac{\sqrt{t} + t^2}{2t - t^2}$

**14.** $\lim\limits_{t \to \infty} \dfrac{t - t\sqrt{t}}{2t^{3/2} + 3t - 5}$

**15.** $\lim\limits_{x \to \infty} \dfrac{(2x^2 + 1)^2}{(x - 1)^2(x^2 + x)}$

**16.** $\lim\limits_{x \to \infty} \dfrac{x^2}{\sqrt{x^4 + 1}}$

**17.** $\lim\limits_{x \to \infty} \dfrac{\sqrt{9x^6 - x}}{x^3 + 1}$

**18.** $\lim\limits_{x \to -\infty} \dfrac{\sqrt{9x^6 - x}}{x^3 + 1}$

**19.** $\lim\limits_{x \to \infty} \left( \sqrt{9x^2 + x} - 3x \right)$

**20.** $\lim\limits_{x \to -\infty} \left( x + \sqrt{x^2 + 2x} \right)$

**21.** $\lim\limits_{x \to \infty} \left( \sqrt{x^2 + ax} - \sqrt{x^2 + bx} \right)$

**22.** $\lim\limits_{x \to \infty} \cos x$

**23.** $\lim\limits_{x \to \infty} \dfrac{x^4 - 3x^2 + x}{x^3 - x + 2}$

**24.** $\lim\limits_{x \to \infty} \sqrt{x^2 + 1}$

**25.** $\lim\limits_{x \to -\infty} (x^4 + x^5)$

**26.** $\lim\limits_{x \to -\infty} \dfrac{1 + x^6}{x^4 + 1}$

**27.** $\lim\limits_{x \to \infty} \left( x - \sqrt{x} \right)$

**28.** $\lim\limits_{x \to \infty} (x^2 - x^4)$

**29.** $\lim\limits_{x \to \infty} x \sin \dfrac{1}{x}$

**30.** $\lim\limits_{x \to \infty} \sqrt{x} \, \sin \dfrac{1}{x}$

---

**31.** (a) Estimate the value of

$$\lim_{x \to -\infty} \left( \sqrt{x^2 + x + 1} + x \right)$$

by graphing the function $f(x) = \sqrt{x^2 + x + 1} + x$.
(b) Use a table of values of $f(x)$ to guess the value of the limit.
(c) Prove that your guess is correct.

**32.** (a) Use a graph of

$$f(x) = \sqrt{3x^2 + 8x + 6} - \sqrt{3x^2 + 3x + 1}$$

to estimate the value of $\lim_{x \to \infty} f(x)$ to one decimal place.
(b) Use a table of values of $f(x)$ to estimate the limit to four decimal places.
(c) Find the exact value of the limit.

**33–38** Find the horizontal and vertical asymptotes of each curve. If you have a graphing device, check your work by graphing the curve and estimating the asymptotes.

**33.** $y = \dfrac{2x + 1}{x - 2}$

**34.** $y = \dfrac{x^2 + 1}{2x^2 - 3x - 2}$

**35.** $y = \dfrac{2x^2 + x - 1}{x^2 + x - 2}$

**36.** $y = \dfrac{1 + x^4}{x^2 - x^4}$

**37.** $y = \dfrac{x^3 - x}{x^2 - 6x + 5}$

**38.** $F(x) = \dfrac{x - 9}{\sqrt{4x^2 + 3x + 2}}$

**39.** Estimate the horizontal asymptote of the function

$$f(x) = \dfrac{3x^3 + 500x^2}{x^3 + 500x^2 + 100x + 2000}$$

by graphing $f$ for $-10 \leqslant x \leqslant 10$. Then calculate the equation of the asymptote by evaluating the limit. How do you explain the discrepancy?

**40.** (a) Graph the function

$$f(x) = \dfrac{\sqrt{2x^2 + 1}}{3x - 5}$$

How many horizontal and vertical asymptotes do you observe? Use the graph to estimate the values of the limits

$$\lim_{x \to \infty} \dfrac{\sqrt{2x^2 + 1}}{3x - 5} \quad \text{and} \quad \lim_{x \to -\infty} \dfrac{\sqrt{2x^2 + 1}}{3x - 5}$$

(b) By calculating values of $f(x)$, give numerical estimates of the limits in part (a).
(c) Calculate the exact values of the limits in part (a). Did you get the same value or different values for these two limits? [In view of your answer to part (a), you might have to check your calculation for the second limit.]

**41.** Find a formula for a function $f$ that satisfies the following conditions:
$$\lim_{x \to \pm\infty} f(x) = 0, \quad \lim_{x \to 0} f(x) = -\infty, \quad f(2) = 0,$$
$$\lim_{x \to 3^-} f(x) = \infty, \quad \lim_{x \to 3^+} f(x) = -\infty$$

**42.** Find a formula for a function that has vertical asymptotes $x = 1$ and $x = 3$ and horizontal asymptote $y = 1$.

**43.** A function $f$ is a ratio of quadratic functions and has a vertical asymptote $x = 4$ and just one $x$-intercept, $x = 1$. It is known that $f$ has a removable discontinuity at $x = -1$ and $\lim_{x \to -1} f(x) = 2$. Evaluate
(a) $f(0)$
(b) $\lim_{x \to \infty} f(x)$

**44–47** Find the horizontal asymptotes of the curve and use them, together with concavity and intervals of increase and decrease, to sketch the curve.

**44.** $y = \dfrac{1 + 2x^2}{1 + x^2}$

**45.** $y = \dfrac{1 - x}{1 + x}$

**46.** $y = \dfrac{x}{\sqrt{x^2 + 1}}$

**47.** $y = \dfrac{x}{x^2 + 1}$

**48–52** Find the limits as $x \to \infty$ and as $x \to -\infty$. Use this information, together with intercepts, to give a rough sketch of the graph as in Example 11.

**48.** $y = 2x^3 - x^4$

**49.** $y = x^4 - x^6$

**50.** $y = x^3(x + 2)^2(x - 1)$

**51.** $y = (3 - x)(1 + x)^2(1 - x)^4$

**52.** $y = x^2(x^2 - 1)^2(x + 2)$

**53–56** Sketch the graph of a function that satisfies all of the given conditions.

**53.** $f'(2) = 0$, $f(2) = -1$, $f(0) = 0$,
$f'(x) < 0$ if $0 < x < 2$, $f'(x) > 0$ if $x > 2$,
$f''(x) < 0$ if $0 \le x < 1$ or if $x > 4$,
$f''(x) > 0$ if $1 < x < 4$, $\lim_{x \to \infty} f(x) = 1$,
$f(-x) = f(x)$ for all $x$

**54.** $f'(2) = 0$, $f'(0) = 1$, $f'(x) > 0$ if $0 < x < 2$,
$f'(x) < 0$ if $x > 2$, $f''(x) < 0$ if $0 < x < 4$,
$f''(x) > 0$ if $x > 4$, $\lim_{x \to \infty} f(x) = 0$,
$f(-x) = -f(x)$ for all $x$

**55.** $f(1) = f'(1) = 0$, $\lim_{x \to 2^+} f(x) = \infty$, $\lim_{x \to 2^-} f(x) = -\infty$,
$\lim_{x \to 0} f(x) = -\infty$, $\lim_{x \to -\infty} f(x) = \infty$, $\lim_{x \to \infty} f(x) = 0$,
$f''(x) > 0$ for $x > 2$, $f''(x) < 0$ for $x < 0$ and for
$0 < x < 2$

**56.** $g(0) = 0$, $g''(x) < 0$ for $x \ne 0$, $\lim_{x \to -\infty} g(x) = \infty$,
$\lim_{x \to \infty} g(x) = -\infty$, $\lim_{x \to 0^-} g'(x) = -\infty$,
$\lim_{x \to 0^+} g'(x) = \infty$

**57.** (a) Use the Squeeze Theorem to evaluate $\lim_{x \to \infty} \dfrac{\sin x}{x}$.

(b) Graph $f(x) = (\sin x)/x$. How many times does the graph cross the asymptote?

**58.** By the *end behavior* of a function we mean the behavior of its values as $x \to \infty$ and as $x \to -\infty$.

(a) Describe and compare the end behavior of the functions

$$P(x) = 3x^5 - 5x^3 + 2x \qquad Q(x) = 3x^5$$

by graphing both functions in the viewing rectangles $[-2, 2]$ by $[-2, 2]$ and $[-10, 10]$ by $[-10,000, 10,000]$.

(b) Two functions are said to have the *same end behavior* if their ratio approaches 1 as $x \to \infty$. Show that $P$ and $Q$ have the same end behavior.

**59.** Let $P$ and $Q$ be polynomials. Find

$$\lim_{x \to \infty} \frac{P(x)}{Q(x)}$$

if the degree of $P$ is (a) less than the degree of $Q$ and (b) greater than the degree of $Q$.

**60.** Make a rough sketch of the curve $y = x^n$ ($n$ an integer) for the following five cases:

  (i) $n = 0$          (ii) $n > 0$, $n$ odd

  (iii) $n > 0$, $n$ even     (iv) $n < 0$, $n$ odd

  (v) $n < 0$, $n$ even

Then use these sketches to find the following limits.

  (a) $\lim_{x \to 0^+} x^n$         (b) $\lim_{x \to 0^-} x^n$

  (c) $\lim_{x \to \infty} x^n$         (d) $\lim_{x \to -\infty} x^n$

**61.** Find $\lim_{x \to \infty} f(x)$ if

$$\frac{4x - 1}{x} < f(x) < \frac{4x^2 + 3x}{x^2}$$

for all $x > 5$.

**62.** (a) A tank contains 5000 L of pure water. Brine that contains 30 g of salt per liter of water is pumped into the tank at a rate of 25 L/min. Show that the concentration of salt after $t$ minutes (in grams per liter) is

$$C(t) = \frac{30t}{200 + t}$$

(b) What happens to the concentration as $t \to \infty$?

**63.** Use a graph to find a number $N$ such that

$$\text{if} \quad x > N \quad \text{then} \quad \left| \frac{3x^2 + 1}{2x^2 + x + 1} - 1.5 \right| < 0.05$$

**64.** For the limit

$$\lim_{x \to \infty} \frac{\sqrt{4x^2 + 1}}{x + 1} = 2$$

illustrate Definition 5 by finding values of $N$ that correspond to $\varepsilon = 0.5$ and $\varepsilon = 0.1$.

**65.** For the limit

$$\lim_{x \to -\infty} \frac{\sqrt{4x^2 + 1}}{x + 1} = -2$$

illustrate Definition 6 by finding values of $N$ that correspond to $\varepsilon = 0.5$ and $\varepsilon = 0.1$.

**66.** For the limit

$$\lim_{x \to \infty} \frac{2x + 1}{\sqrt{x + 1}} = \infty$$

illustrate Definition 7 by finding a value of $N$ that corresponds to $M = 100$.

**67.** (a) How large do we have to take $x$ so that $1/x^2 < 0.0001$?

(b) Taking $r = 2$ in Theorem 4, we have the statement

$$\lim_{x \to \infty} \frac{1}{x^2} = 0$$

Prove this directly using Definition 5.

**68.** (a) How large do we have to take $x$ so that $1/\sqrt{x} < 0.0001$?

(b) Taking $r = \frac{1}{2}$ in Theorem 4, we have the statement

$$\lim_{x \to \infty} \frac{1}{\sqrt{x}} = 0$$

Prove this directly using Definition 5.

**69.** Use Definition 6 to prove that $\lim\limits_{x \to -\infty} \dfrac{1}{x} = 0$.

**70.** Prove, using Definition 7, that $\lim\limits_{x \to \infty} x^3 = \infty$.

**71.** Prove that

$$\lim_{x \to \infty} f(x) = \lim_{t \to 0^+} f(1/t)$$

and

$$\lim_{x \to -\infty} f(x) = \lim_{t \to 0^-} f(1/t)$$

if these limits exist.

**72.** Formulate a precise definition of

$$\lim_{x \to -\infty} f(x) = -\infty$$

Then use your definition to prove that

$$\lim_{x \to -\infty} (1 + x^3) = -\infty$$

---

## 3.5  Summary of Curve Sketching

So far we have been concerned with some particular aspects of curve sketching: domain, range, symmetry, limits, continuity, and vertical asymptotes in Chapter 1; derivatives and tangents in Chapter 2; and extreme values, intervals of increase and decrease, concavity, points of inflection, and horizontal asymptotes in this chapter. It is now time to put all of this information together to sketch graphs that reveal the important features of functions.

You might ask: Why don't we just use a graphing calculator or computer to graph a curve? Why do we need to use calculus?

It's true that modern technology is capable of producing very accurate graphs. But even the best graphing devices have to be used intelligently. As discussed in Appendix G, it is extremely important to choose an appropriate viewing rectangle to avoid getting a misleading graph. (See especially Examples 1, 3, 4, and 5 in that appendix.) The use of calculus enables us to discover the most interesting aspects of graphs and in many cases to calculate maximum and minimum points and inflection points *exactly* instead of approximately.

For instance, Figure 1 shows the graph of $f(x) = 8x^3 - 21x^2 + 18x + 2$. At first glance it seems reasonable: It has the same shape as cubic curves like $y = x^3$, and it appears to have no maximum or minimum point. But if you compute the derivative, you will see that there is a maximum when $x = 0.75$ and a minimum when $x = 1$. Indeed, if we zoom in to this portion of the graph, we see that behavior exhibited in Figure 2. Without calculus, we could easily have overlooked it.

In the next section we will graph functions by using the interaction between calculus and graphing devices. In this section we draw graphs by first considering the following information. We don't assume that you have a graphing device, but if you do have one you should use it as a check on your work.

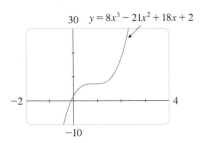

$y = 8x^3 - 21x^2 + 18x + 2$

**FIGURE 1**

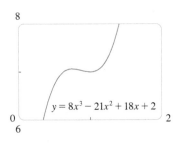

$y = 8x^3 - 21x^2 + 18x + 2$

**FIGURE 2**

### Guidelines for Sketching a Curve

The following checklist is intended as a guide to sketching a curve $y = f(x)$ by hand. Not every item is relevant to every function. (For instance, a given curve might not have an asymptote or possess symmetry.) But the guidelines provide all the information you need to make a sketch that displays the most important aspects of the function.

**A. Domain**   It's often useful to start by determining the domain $D$ of $f$, that is, the set of values of $x$ for which $f(x)$ is defined.

**B. Intercepts** The $y$-intercept is $f(0)$ and this tells us where the curve intersects the $y$-axis. To find the $x$-intercepts, we set $y = 0$ and solve for $x$. (You can omit this step if the equation is difficult to solve.)

**C. Symmetry**

(i) If $f(-x) = f(x)$ for all $x$ in $D$, that is, the equation of the curve is unchanged when $x$ is replaced by $-x$, then $f$ is an **even function** and the curve is symmetric about the $y$-axis. This means that our work is cut in half. If we know what the curve looks like for $x \geqslant 0$, then we need only reflect about the $y$-axis to obtain the complete curve [see Figure 3(a)]. Here are some examples: $y = x^2$, $y = x^4$, $y = |x|$, and $y = \cos x$.

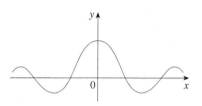

(a) Even function: reflectional symmetry

(ii) If $f(-x) = -f(x)$ for all $x$ in $D$, then $f$ is an **odd function** and the curve is symmetric about the origin. Again we can obtain the complete curve if we know what it looks like for $x \geqslant 0$. [Rotate $180°$ about the origin; see Figure 3(b).] Some simple examples of odd functions are $y = x$, $y = x^3$, $y = x^5$, and $y = \sin x$.

(b) Odd function: rotational symmetry

**FIGURE 3**

(iii) If $f(x + p) = f(x)$ for all $x$ in $D$, where $p$ is a positive constant, then $f$ is called a **periodic function** and the smallest such number $p$ is called the **period.** For instance, $y = \sin x$ has period $2\pi$ and $y = \tan x$ has period $\pi$. If we know what the graph looks like in an interval of length $p$, then we can use translation to sketch the entire graph (see Figure 4).

**FIGURE 4**
Periodic function:
translational symmetry

**D. Asymptotes**

(i) *Horizontal Asymptotes.* Recall from Section 3.4 that if either $\lim_{x \to \infty} f(x) = L$ or $\lim_{x \to -\infty} f(x) = L$, then the line $y = L$ is a horizontal asymptote of the curve $y = f(x)$. If it turns out that $\lim_{x \to \infty} f(x) = \infty$ (or $-\infty$), then we do not have an asymptote to the right, but that is still useful information for sketching the curve.

(ii) *Vertical Asymptotes.* Recall from Section 1.5 that the line $x = a$ is a vertical asymptote if at least one of the following statements is true:

$$\boxed{1} \qquad \lim_{x \to a^+} f(x) = \infty \qquad \lim_{x \to a^-} f(x) = \infty$$

$$\lim_{x \to a^+} f(x) = -\infty \qquad \lim_{x \to a^-} f(x) = -\infty$$

(For rational functions you can locate the vertical asymptotes by equating the denominator to 0 after canceling any common factors. But for other functions this method does not apply.) Furthermore, in sketching the curve it is very useful to know exactly which of the statements in $\boxed{1}$ is true. If $f(a)$ is not defined but $a$ is an endpoint of the domain of $f$, then you should compute $\lim_{x \to a^-} f(x)$ or $\lim_{x \to a^+} f(x)$, whether or not this limit is infinite.

(iii) *Slant Asymptotes.* These are discussed at the end of this section.

**E. Intervals of Increase or Decrease** Use the I/D Test. Compute $f'(x)$ and find the intervals on which $f'(x)$ is positive ($f$ is increasing) and the intervals on which $f'(x)$ is negative ($f$ is decreasing).

**F. Local Maximum and Minimum Values** Find the critical numbers of $f$ [the numbers $c$ where $f'(c) = 0$ or $f'(c)$ does not exist]. Then use the First Derivative Test. If $f'$ changes from positive to negative at a critical number $c$, then $f(c)$ is a local maximum. If $f'$ changes from negative to positive at $c$, then $f(c)$ is a local minimum. Although it is usually preferable to use the First Derivative Test, you can use the Second Derivative Test if $f'(c) = 0$ and $f''(c) \neq 0$. Then $f''(c) > 0$ implies that $f(c)$ is a local minimum, whereas $f''(c) < 0$ implies that $f(c)$ is a local maximum.

**G. Concavity and Points of Inflection** Compute $f''(x)$ and use the Concavity Test. The curve is concave upward where $f''(x) > 0$ and concave downward where $f''(x) < 0$. Inflection points occur where the direction of concavity changes.

**H. Sketch the Curve** Using the information in items A–G, draw the graph. Sketch the asymptotes as dashed lines. Plot the intercepts, maximum and minimum points, and inflection points. Then make the curve pass through these points, rising and falling according to E, with concavity according to G, and approaching the asymptotes. If additional accuracy is desired near any point, you can compute the value of the derivative there. The tangent indicates the direction in which the curve proceeds.

**EXAMPLE 1** Use the guidelines to sketch the curve $y = \dfrac{2x^2}{x^2 - 1}$.

**A.** The domain is

$$\{x \mid x^2 - 1 \neq 0\} = \{x \mid x \neq \pm 1\} = (-\infty, -1) \cup (-1, 1) \cup (1, \infty)$$

**B.** The $x$- and $y$-intercepts are both 0.

**C.** Since $f(-x) = f(x)$, the function $f$ is even. The curve is symmetric about the $y$-axis.

**D.**
$$\lim_{x \to \pm\infty} \frac{2x^2}{x^2 - 1} = \lim_{x \to \pm\infty} \frac{2}{1 - 1/x^2} = 2$$

Therefore the line $y = 2$ is a horizontal asymptote.

Since the denominator is 0 when $x = \pm 1$, we compute the following limits:

$$\lim_{x \to 1^+} \frac{2x^2}{x^2 - 1} = \infty \qquad \lim_{x \to 1^-} \frac{2x^2}{x^2 - 1} = -\infty$$

$$\lim_{x \to -1^+} \frac{2x^2}{x^2 - 1} = -\infty \qquad \lim_{x \to -1^-} \frac{2x^2}{x^2 - 1} = \infty$$

Therefore the lines $x = 1$ and $x = -1$ are vertical asymptotes. This information about limits and asymptotes enables us to draw the preliminary sketch in Figure 5, showing the parts of the curve near the asymptotes.

**E.**
$$f'(x) = \frac{4x(x^2 - 1) - 2x^2 \cdot 2x}{(x^2 - 1)^2} = \frac{-4x}{(x^2 - 1)^2}$$

Since $f'(x) > 0$ when $x < 0$ $(x \neq -1)$ and $f'(x) < 0$ when $x > 0$ $(x \neq 1)$, $f$ is increasing on $(-\infty, -1)$ and $(-1, 0)$ and decreasing on $(0, 1)$ and $(1, \infty)$.

**F.** The only critical number is $x = 0$. Since $f'$ changes from positive to negative at 0, $f(0) = 0$ is a local maximum by the First Derivative Test.

**G.**
$$f''(x) = \frac{-4(x^2 - 1)^2 + 4x \cdot 2(x^2 - 1)2x}{(x^2 - 1)^4} = \frac{12x^2 + 4}{(x^2 - 1)^3}$$

**FIGURE 5**
Preliminary sketch

We have shown the curve approaching its horizontal asymptote from above in Figure 5. This is confirmed by the intervals of increase and decrease.

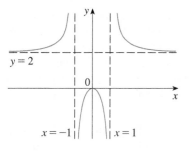

**FIGURE 6**
Finished sketch of $y = \dfrac{2x^2}{x^2 - 1}$

Since $12x^2 + 4 > 0$ for all $x$, we have

$$f''(x) > 0 \iff x^2 - 1 > 0 \iff |x| > 1$$

and $f''(x) < 0 \iff |x| < 1$. Thus the curve is concave upward on the intervals $(-\infty, -1)$ and $(1, \infty)$ and concave downward on $(-1, 1)$. It has no point of inflection since 1 and $-1$ are not in the domain of $f$.

**H.** Using the information in E–G, we finish the sketch in Figure 6.

**EXAMPLE 2** Sketch the graph of $f(x) = \dfrac{x^2}{\sqrt{x + 1}}$.

**A.** Domain $= \{x \mid x + 1 > 0\} = \{x \mid x > -1\} = (-1, \infty)$

**B.** The $x$- and $y$-intercepts are both 0.

**C.** Symmetry: None

**D.** Since

$$\lim_{x \to \infty} \frac{x^2}{\sqrt{x + 1}} = \infty$$

there is no horizontal asymptote. Since $\sqrt{x + 1} \to 0$ as $x \to -1^+$ and $f(x)$ is always positive, we have

$$\lim_{x \to -1^+} \frac{x^2}{\sqrt{x + 1}} = \infty$$

and so the line $x = -1$ is a vertical asymptote.

**E.**
$$f'(x) = \frac{2x\sqrt{x + 1} - x^2 \cdot 1/\left(2\sqrt{x + 1}\right)}{x + 1} = \frac{x(3x + 4)}{2(x + 1)^{3/2}}$$

We see that $f'(x) = 0$ when $x = 0$ $\left(\text{notice that } -\frac{4}{3} \text{ is not in the domain of } f\right)$, so the only critical number is 0. Since $f'(x) < 0$ when $-1 < x < 0$ and $f'(x) > 0$ when $x > 0$, $f$ is decreasing on $(-1, 0)$ and increasing on $(0, \infty)$.

**F.** Since $f'(0) = 0$ and $f'$ changes from negative to positive at 0, $f(0) = 0$ is a local (and absolute) minimum by the First Derivative Test.

**G.**
$$f''(x) = \frac{2(x + 1)^{3/2}(6x + 4) - (3x^2 + 4x)3(x + 1)^{1/2}}{4(x + 1)^3} = \frac{3x^2 + 8x + 8}{4(x + 1)^{5/2}}$$

Note that the denominator is always positive. The numerator is the quadratic $3x^2 + 8x + 8$, which is always positive because its discriminant is $b^2 - 4ac = -32$, which is negative, and the coefficient of $x^2$ is positive. Thus $f''(x) > 0$ for all $x$ in the domain of $f$, which means that $f$ is concave upward on $(-1, \infty)$ and there is no point of inflection.

**H.** The curve is sketched in Figure 7.

**FIGURE 7**

**EXAMPLE 3** Sketch the graph of $f(x) = \dfrac{\cos x}{2 + \sin x}$.

**A.** The domain is $\mathbb{R}$.

**B.** The $y$-intercept is $f(0) = \frac{1}{2}$. The $x$-intercepts occur when $\cos x = 0$, that is, $x = (2n + 1)\pi/2$, where $n$ is an integer.

**C.** $f$ is neither even nor odd, but $f(x + 2\pi) = f(x)$ for all $x$ and so $f$ is periodic and has period $2\pi$. Thus, in what follows, we need to consider only $0 \leq x \leq 2\pi$ and then extend the curve by translation in part H.

**D.** Asymptotes: None

**E.**
$$f'(x) = \frac{(2 + \sin x)(-\sin x) - \cos x\,(\cos x)}{(2 + \sin x)^2} = -\frac{2 \sin x + 1}{(2 + \sin x)^2}$$

Thus $f'(x) > 0$ when $2 \sin x + 1 < 0 \iff \sin x < -\frac{1}{2} \iff$ $7\pi/6 < x < 11\pi/6$. So $f$ is increasing on $(7\pi/6, 11\pi/6)$ and decreasing on $(0, 7\pi/6)$ and $(11\pi/6, 2\pi)$.

**F.** From part E and the First Derivative Test, we see that the local minimum value is $f(7\pi/6) = -1/\sqrt{3}$ and the local maximum value is $f(11\pi/6) = 1/\sqrt{3}$.

**G.** If we use the Quotient Rule again and simplify, we get

$$f''(x) = -\frac{2 \cos x\,(1 - \sin x)}{(2 + \sin x)^3}$$

Because $(2 + \sin x)^3 > 0$ and $1 - \sin x \geq 0$ for all $x$, we know that $f''(x) > 0$ when $\cos x < 0$, that is, $\pi/2 < x < 3\pi/2$. So $f$ is concave upward on $(\pi/2, 3\pi/2)$ and concave downward on $(0, \pi/2)$ and $(3\pi/2, 2\pi)$. The inflection points are $(\pi/2, 0)$ and $(3\pi/2, 0)$.

**H.** The graph of the function restricted to $0 \leq x \leq 2\pi$ is shown in Figure 8. Then we extend it, using periodicity, to the complete graph in Figure 9.

**FIGURE 8**

**FIGURE 9**

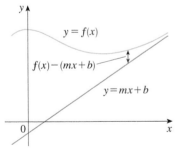

**FIGURE 10**

## Slant Asymptotes

Some curves have asymptotes that are *oblique,* that is, neither horizontal nor vertical. If

$$\lim_{x \to \infty} [f(x) - (mx + b)] = 0$$

then the line $y = mx + b$ is called a **slant asymptote** because the vertical distance between the curve $y = f(x)$ and the line $y = mx + b$ approaches 0, as in Figure 10. (A similar situation exists if we let $x \to -\infty$.) For rational functions, slant asymptotes occur when the degree of the numerator is one more than the degree of the denominator. In such a case the equation of the slant asymptote can be found by long division as in the following example.

**V EXAMPLE 4** Sketch the graph of $f(x) = \dfrac{x^3}{x^2 + 1}$.

**A.** The domain is $\mathbb{R} = (-\infty, \infty)$.

**B.** The $x$- and $y$-intercepts are both 0.

**C.** Since $f(-x) = -f(x)$, $f$ is odd and its graph is symmetric about the origin.

D. Since $x^2 + 1$ is never 0, there is no vertical asymptote. Since $f(x) \to \infty$ as $x \to \infty$ and $f(x) \to -\infty$ as $x \to -\infty$, there is no horizontal asymptote. But long division gives

$$f(x) = \frac{x^3}{x^2 + 1} = x - \frac{x}{x^2 + 1}$$

$$f(x) - x = -\frac{x}{x^2 + 1} = -\frac{\dfrac{1}{x}}{1 + \dfrac{1}{x^2}} \to 0 \quad \text{as} \quad x \to \pm\infty$$

So the line $y = x$ is a slant asymptote.

E.
$$f'(x) = \frac{3x^2(x^2 + 1) - x^3 \cdot 2x}{(x^2 + 1)^2} = \frac{x^2(x^2 + 3)}{(x^2 + 1)^2}$$

Since $f'(x) > 0$ for all $x$ (except 0), $f$ is increasing on $(-\infty, \infty)$.

F. Although $f'(0) = 0$, $f'$ does not change sign at 0, so there is no local maximum or minimum.

G.
$$f''(x) = \frac{(4x^3 + 6x)(x^2 + 1)^2 - (x^4 + 3x^2) \cdot 2(x^2 + 1)2x}{(x^2 + 1)^4} = \frac{2x(3 - x^2)}{(x^2 + 1)^3}$$

Since $f''(x) = 0$ when $x = 0$ or $x = \pm\sqrt{3}$, we set up the following chart:

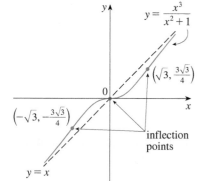

| Interval | $x$ | $3 - x^2$ | $(x^2 + 1)^3$ | $f''(x)$ | $f$ |
|---|---|---|---|---|---|
| $x < -\sqrt{3}$ | $-$ | $-$ | $+$ | $+$ | CU on $\left(-\infty, -\sqrt{3}\right)$ |
| $-\sqrt{3} < x < 0$ | $-$ | $+$ | $+$ | $-$ | CD on $\left(-\sqrt{3}, 0\right)$ |
| $0 < x < \sqrt{3}$ | $+$ | $+$ | $+$ | $+$ | CU on $\left(0, \sqrt{3}\right)$ |
| $x > \sqrt{3}$ | $+$ | $-$ | $+$ | $-$ | CD on $\left(\sqrt{3}, \infty\right)$ |

The points of inflection are $\left(-\sqrt{3}, -\frac{3}{4}\sqrt{3}\right)$, $(0, 0)$, and $\left(\sqrt{3}, \frac{3}{4}\sqrt{3}\right)$.

H. The graph of $f$ is sketched in Figure 11.

**FIGURE 11**

## 3.5 Exercises

1–40 Use the guidelines of this section to sketch the curve.

**1.** $y = x^3 - 12x^2 + 36x$

**2.** $y = 2 + 3x^2 - x^3$

**3.** $y = x^4 - 4x$

**4.** $y = x^4 - 8x^2 + 8$

**5.** $y = x(x - 4)^3$

**6.** $y = x^5 - 5x$

**7.** $y = \frac{1}{5}x^5 - \frac{8}{3}x^3 + 16x$

**8.** $y = (4 - x^2)^5$

**9.** $y = \dfrac{x}{x - 1}$

**10.** $y = \dfrac{x^2 - 4}{x^2 - 2x}$

**11.** $y = \dfrac{x - x^2}{2 - 3x + x^2}$

**12.** $y = \dfrac{x}{x^2 - 9}$

**13.** $y = \dfrac{1}{x^2 - 9}$

**14.** $y = \dfrac{x^2}{x^2 + 9}$

**1.** Homework Hints available at stewartcalculus.com

**15.** $y = \dfrac{x}{x^2 + 9}$

**16.** $y = 1 + \dfrac{1}{x} + \dfrac{1}{x^2}$

**17.** $y = \dfrac{x - 1}{x^2}$

**18.** $y = \dfrac{x}{x^3 - 1}$

**19.** $y = \dfrac{x^2}{x^2 + 3}$

**20.** $y = \dfrac{x^3}{x - 2}$

**21.** $y = (x - 3)\sqrt{x}$

**22.** $y = 2\sqrt{x} - x$

**23.** $y = \sqrt{x^2 + x - 2}$

**24.** $y = \sqrt{x^2 + x} - x$

**25.** $y = \dfrac{x}{\sqrt{x^2 + 1}}$

**26.** $y = x\sqrt{2 - x^2}$

**27.** $y = \dfrac{\sqrt{1 - x^2}}{x}$

**28.** $y = \dfrac{x}{\sqrt{x^2 - 1}}$

**29.** $y = x - 3x^{1/3}$

**30.** $y = x^{5/3} - 5x^{2/3}$

**31.** $y = \sqrt[3]{x^2 - 1}$

**32.** $y = \sqrt[3]{x^3 + 1}$

**33.** $y = \sin^3 x$

**34.** $y = x + \cos x$

**35.** $y = x \tan x, \quad -\pi/2 < x < \pi/2$

**36.** $y = 2x - \tan x, \quad -\pi/2 < x < \pi/2$

**37.** $y = \frac{1}{2}x - \sin x, \quad 0 < x < 3\pi$

**38.** $y = \sec x + \tan x, \quad 0 < x < \pi/2$

**39.** $y = \dfrac{\sin x}{1 + \cos x}$

**40.** $y = \dfrac{\sin x}{2 + \cos x}$

**41.** In the theory of relativity, the mass of a particle is

$$m = \dfrac{m_0}{\sqrt{1 - v^2/c^2}}$$

where $m_0$ is the rest mass of the particle, $m$ is the mass when the particle moves with speed $v$ relative to the observer, and $c$ is the speed of light. Sketch the graph of $m$ as a function of $v$.

**42.** In the theory of relativity, the energy of a particle is

$$E = \sqrt{m_0^2 c^4 + h^2 c^2/\lambda^2}$$

where $m_0$ is the rest mass of the particle, $\lambda$ is its wave length, and $h$ is Planck's constant. Sketch the graph of $E$ as a function of $\lambda$. What does the graph say about the energy?

**43.** The figure shows a beam of length $L$ embedded in concrete walls. If a constant load $W$ is distributed evenly along its length, the beam takes the shape of the deflection curve

$$y = -\dfrac{W}{24EI}x^4 + \dfrac{WL}{12EI}x^3 - \dfrac{WL^2}{24EI}x^2$$

where $E$ and $I$ are positive constants. ($E$ is Young's modulus of elasticity and $I$ is the moment of inertia of a cross-section of the beam.) Sketch the graph of the deflection curve.

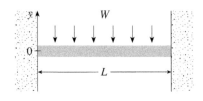

**44.** Coulomb's Law states that the force of attraction between two charged particles is directly proportional to the product of the charges and inversely proportional to the square of the distance between them. The figure shows particles with charge 1 located at positions 0 and 2 on a coordinate line and a particle with charge $-1$ at a position $x$ between them. It follows from Coulomb's Law that the net force acting on the middle particle is

$$F(x) = -\dfrac{k}{x^2} + \dfrac{k}{(x - 2)^2} \qquad 0 < x < 2$$

where $k$ is a positive constant. Sketch the graph of the net force function. What does the graph say about the force?

**45–48** Find an equation of the slant asymptote. Do not sketch the curve.

**45.** $y = \dfrac{x^2 + 1}{x + 1}$

**46.** $y = \dfrac{2x^3 + x^2 + x + 3}{x^2 + 2x}$

**47.** $y = \dfrac{4x^3 - 2x^2 + 5}{2x^2 + x - 3}$

**48.** $y = \dfrac{5x^4 + x^2 + x}{x^3 - x^2 + 2}$

**49–54** Use the guidelines of this section to sketch the curve. In guideline D find an equation of the slant asymptote.

**49.** $y = \dfrac{x^2}{x - 1}$

**50.** $y = \dfrac{1 + 5x - 2x^2}{x - 2}$

**51.** $y = \dfrac{x^3 + 4}{x^2}$

**52.** $y = \dfrac{x^3}{(x + 1)^2}$

**53.** $y = \dfrac{2x^3 + x^2 + 1}{x^2 + 1}$

**54.** $y = \dfrac{(x + 1)^3}{(x - 1)^2}$

**55.** Show that the curve $y = \sqrt{4x^2 + 9}$ has two slant asymptotes: $y = 2x$ and $y = -2x$. Use this fact to help sketch the curve.

**56.** Show that the curve $y = \sqrt{x^2 + 4x}$ has two slant asymptotes: $y = x + 2$ and $y = -x - 2$. Use this fact to help sketch the curve.

**57.** Show that the lines $y = (b/a)x$ and $y = -(b/a)x$ are slant asymptotes of the hyperbola $(x^2/a^2) - (y^2/b^2) = 1$.

**58.** Let $f(x) = (x^3 + 1)/x$. Show that

$$\lim_{x \to \pm\infty} [f(x) - x^2] = 0$$

This shows that the graph of $f$ approaches the graph of $y = x^2$, and we say that the curve $y = f(x)$ is *asymptotic* to the parabola $y = x^2$. Use this fact to help sketch the graph of $f$.

**59.** Discuss the asymptotic behavior of $f(x) = (x^4 + 1)/x$ in the same manner as in Exercise 58. Then use your results to help sketch the graph of $f$.

**60.** Use the asymptotic behavior of $f(x) = \cos x + 1/x^2$ to sketch its graph without going through the curve-sketching procedure of this section.

## 3.6   **Graphing with Calculus *and* Calculators**

If you have not already read Appendix G, you should do so now. In particular, it explains how to avoid some of the pitfalls of graphing devices by choosing appropriate viewing rectangles.

The method we used to sketch curves in the preceding section was a culmination of much of our study of differential calculus. The graph was the final object that we produced. In this section our point of view is completely different. Here we *start* with a graph produced by a graphing calculator or computer and then we refine it. We use calculus to make sure that we reveal all the important aspects of the curve. And with the use of graphing devices we can tackle curves that would be far too complicated to consider without technology. The theme is the *interaction* between calculus and calculators.

**EXAMPLE 1**   Graph the polynomial $f(x) = 2x^6 + 3x^5 + 3x^3 - 2x^2$. Use the graphs of $f'$ and $f''$ to estimate all maximum and minimum points and intervals of concavity.

SOLUTION   If we specify a domain but not a range, many graphing devices will deduce a suitable range from the values computed. Figure 1 shows the plot from one such device if we specify that $-5 \leq x \leq 5$. Although this viewing rectangle is useful for showing that the asymptotic behavior (or end behavior) is the same as for $y = 2x^6$, it is obviously hiding some finer detail. So we change to the viewing rectangle $[-3, 2]$ by $[-50, 100]$ shown in Figure 2.

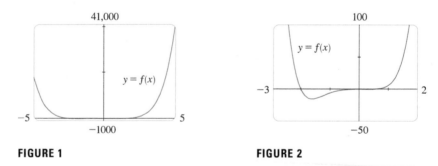

**FIGURE 1**                                       **FIGURE 2**

From this graph it appears that there is an absolute minimum value of about $-15.33$ when $x \approx -1.62$ (by using the cursor) and $f$ is decreasing on $(-\infty, -1.62)$ and increasing on $(-1.62, \infty)$. Also there appears to be a horizontal tangent at the origin and inflection points when $x = 0$ and when $x$ is somewhere between $-2$ and $-1$.

Now let's try to confirm these impressions using calculus. We differentiate and get

$$f'(x) = 12x^5 + 15x^4 + 9x^2 - 4x$$

$$f''(x) = 60x^4 + 60x^3 + 18x - 4$$

When we graph $f'$ in Figure 3 we see that $f'(x)$ changes from negative to positive when $x \approx -1.62$; this confirms (by the First Derivative Test) the minimum value that we found earlier. But, perhaps to our surprise, we also notice that $f'(x)$ changes from positive to negative when $x = 0$ and from negative to positive when $x \approx 0.35$. This means that $f$ has a local maximum at 0 and a local minimum when $x \approx 0.35$, but these were hidden in Figure 2. Indeed, if we now zoom in toward the origin in Figure 4, we see what we missed before: a local maximum value of 0 when $x = 0$ and a local minimum value of about $-0.1$ when $x \approx 0.35$.

**FIGURE 3**                    **FIGURE 4**

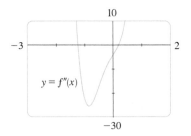

**FIGURE 5**

What about concavity and inflection points? From Figures 2 and 4 there appear to be inflection points when $x$ is a little to the left of $-1$ and when $x$ is a little to the right of 0. But it's difficult to determine inflection points from the graph of $f$, so we graph the second derivative $f''$ in Figure 5. We see that $f''$ changes from positive to negative when $x \approx -1.23$ and from negative to positive when $x \approx 0.19$. So, correct to two decimal places, $f$ is concave upward on $(-\infty, -1.23)$ and $(0.19, \infty)$ and concave downward on $(-1.23, 0.19)$. The inflection points are $(-1.23, -10.18)$ and $(0.19, -0.05)$.

We have discovered that no single graph reveals all the important features of this polynomial. But Figures 2 and 4, when taken together, do provide an accurate picture.

**V** **EXAMPLE 2** Draw the graph of the function

$$f(x) = \frac{x^2 + 7x + 3}{x^2}$$

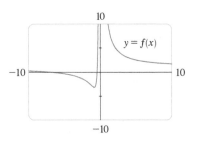

**FIGURE 6**

in a viewing rectangle that contains all the important features of the function. Estimate the maximum and minimum values and the intervals of concavity. Then use calculus to find these quantities exactly.

SOLUTION Figure 6, produced by a computer with automatic scaling, is a disaster. Some graphing calculators use $[-10, 10]$ by $[-10, 10]$ as the default viewing rectangle, so let's try it. We get the graph shown in Figure 7; it's a major improvement.

The $y$-axis appears to be a vertical asymptote and indeed it is because

$$\lim_{x \to 0} \frac{x^2 + 7x + 3}{x^2} = \infty$$

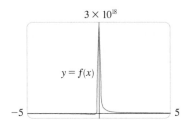

**FIGURE 7**

Figure 7 also allows us to estimate the $x$-intercepts: about $-0.5$ and $-6.5$. The exact values are obtained by using the quadratic formula to solve the equation $x^2 + 7x + 3 = 0$; we get $x = \left(-7 \pm \sqrt{37}\right)/2$.

FIGURE 8

To get a better look at horizontal asymptotes, we change to the viewing rectangle $[-20, 20]$ by $[-5, 10]$ in Figure 8. It appears that $y = 1$ is the horizontal asymptote and this is easily confirmed:

$$\lim_{x \to \pm\infty} \frac{x^2 + 7x + 3}{x^2} = \lim_{x \to \pm\infty} \left( 1 + \frac{7}{x} + \frac{3}{x^2} \right) = 1$$

FIGURE 9

To estimate the minimum value we zoom in to the viewing rectangle $[-3, 0]$ by $[-4, 2]$ in Figure 9. The cursor indicates that the absolute minimum value is about $-3.1$ when $x \approx -0.9$, and we see that the function decreases on $(-\infty, -0.9)$ and $(0, \infty)$ and increases on $(-0.9, 0)$. The exact values are obtained by differentiating:

$$f'(x) = -\frac{7}{x^2} - \frac{6}{x^3} = -\frac{7x + 6}{x^3}$$

This shows that $f'(x) > 0$ when $-\frac{6}{7} < x < 0$ and $f'(x) < 0$ when $x < -\frac{6}{7}$ and when $x > 0$. The exact minimum value is $f\left(-\frac{6}{7}\right) = -\frac{37}{12} \approx -3.08$.

Figure 9 also shows that an inflection point occurs somewhere between $x = -1$ and $x = -2$. We could estimate it much more accurately using the graph of the second derivative, but in this case it's just as easy to find exact values. Since

$$f''(x) = \frac{14}{x^3} + \frac{18}{x^4} = \frac{2(7x + 9)}{x^4}$$

we see that $f''(x) > 0$ when $x > -\frac{9}{7}$ ($x \neq 0$). So $f$ is concave upward on $\left(-\frac{9}{7}, 0\right)$ and $(0, \infty)$ and concave downward on $\left(-\infty, -\frac{9}{7}\right)$. The inflection point is $\left(-\frac{9}{7}, -\frac{71}{27}\right)$.

The analysis using the first two derivatives shows that Figure 8 displays all the major aspects of the curve.

**V  EXAMPLE 3**  Graph the function $f(x) = \dfrac{x^2(x + 1)^3}{(x - 2)^2(x - 4)^4}$.

FIGURE 10

SOLUTION  Drawing on our experience with a rational function in Example 2, let's start by graphing $f$ in the viewing rectangle $[-10, 10]$ by $[-10, 10]$. From Figure 10 we have the feeling that we are going to have to zoom in to see some finer detail and also zoom out to see the larger picture. But, as a guide to intelligent zooming, let's first take a close look at the expression for $f(x)$. Because of the factors $(x - 2)^2$ and $(x - 4)^4$ in the denominator, we expect $x = 2$ and $x = 4$ to be the vertical asymptotes. Indeed

$$\lim_{x \to 2} \frac{x^2(x + 1)^3}{(x - 2)^2(x - 4)^4} = \infty \quad \text{and} \quad \lim_{x \to 4} \frac{x^2(x + 1)^3}{(x - 2)^2(x - 4)^4} = \infty$$

To find the horizontal asymptotes, we divide numerator and denominator by $x^6$:

$$\frac{x^2(x + 1)^3}{(x - 2)^2(x - 4)^4} = \frac{\dfrac{x^2}{x^3} \cdot \dfrac{(x + 1)^3}{x^3}}{\dfrac{(x - 2)^2}{x^2} \cdot \dfrac{(x - 4)^4}{x^4}} = \frac{\dfrac{1}{x}\left(1 + \dfrac{1}{x}\right)^3}{\left(1 - \dfrac{2}{x}\right)^2\left(1 - \dfrac{4}{x}\right)^4}$$

This shows that $f(x) \to 0$ as $x \to \pm\infty$, so the $x$-axis is a horizontal asymptote.

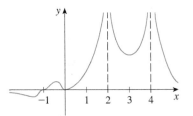

**FIGURE 11**

It is also very useful to consider the behavior of the graph near the $x$-intercepts using an analysis like that in Example 11 in Section 3.4. Since $x^2$ is positive, $f(x)$ does not change sign at 0 and so its graph doesn't cross the $x$-axis at 0. But, because of the factor $(x + 1)^3$, the graph does cross the $x$-axis at $-1$ and has a horizontal tangent there. Putting all this information together, but without using derivatives, we see that the curve has to look something like the one in Figure 11.

Now that we know what to look for, we zoom in (several times) to produce the graphs in Figures 12 and 13 and zoom out (several times) to get Figure 14.

**FIGURE 12**

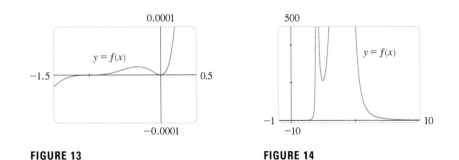

**FIGURE 13**          **FIGURE 14**

We can read from these graphs that the absolute minimum is about $-0.02$ and occurs when $x \approx -20$. There is also a local maximum $\approx 0.00002$ when $x \approx -0.3$ and a local minimum $\approx 211$ when $x \approx 2.5$. These graphs also show three inflection points near $-35$, $-5$, and $-1$ and two between $-1$ and 0. To estimate the inflection points closely we would need to graph $f''$, but to compute $f''$ by hand is an unreasonable chore. If you have a computer algebra system, then it's easy to do (see Exercise 13).

We have seen that, for this particular function, *three* graphs (Figures 12, 13, and 14) are necessary to convey all the useful information. The only way to display all these features of the function on a single graph is to draw it by hand. Despite the exaggerations and distortions, Figure 11 does manage to summarize the essential nature of the function.

The family of functions
$$f(x) = \sin(x + \sin cx)$$
where $c$ is a constant, occurs in applications to frequency modulation (FM) synthesis. A sine wave is modulated by a wave with a different frequency ($\sin cx$). The case where $c = 2$ is studied in Example 4. Exercise 19 explores another special case.

**FIGURE 15**

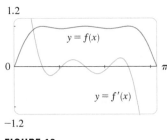

**FIGURE 16**

**EXAMPLE 4** Graph the function $f(x) = \sin(x + \sin 2x)$. For $0 \le x \le \pi$, estimate all maximum and minimum values, intervals of increase and decrease, and inflection points.

SOLUTION We first note that $f$ is periodic with period $2\pi$. Also, $f$ is odd and $|f(x)| \le 1$ for all $x$. So the choice of a viewing rectangle is not a problem for this function: We start with $[0, \pi]$ by $[-1.1, 1.1]$. (See Figure 15.) It appears that there are three local maximum values and two local minimum values in that window. To confirm this and locate them more accurately, we calculate that

$$f'(x) = \cos(x + \sin 2x) \cdot (1 + 2 \cos 2x)$$

and graph both $f$ and $f'$ in Figure 16.

Using zoom-in and the First Derivative Test, we find the following approximate values:

Intervals of increase:     $(0, 0.6),\ (1.0, 1.6),\ (2.1, 2.5)$

Intervals of decrease:     $(0.6, 1.0),\ (1.6, 2.1),\ (2.5, \pi)$

Local maximum values:  $f(0.6) \approx 1,\ f(1.6) \approx 1,\ f(2.5) \approx 1$

Local minimum values:  $f(1.0) \approx 0.94,\ f(2.1) \approx 0.94$

The second derivative is

$$f''(x) = -(1 + 2\cos 2x)^2 \sin(x + \sin 2x) - 4\sin 2x \cos(x + \sin 2x)$$

Graphing both $f$ and $f''$ in Figure 17, we obtain the following approximate values:

Concave upward on:    $(0.8, 1.3)$,  $(1.8, 2.3)$

Concave downward on:  $(0, 0.8)$,  $(1.3, 1.8)$,  $(2.3, \pi)$

Inflection points:    $(0, 0)$,  $(0.8, 0.97)$,  $(1.3, 0.97)$,  $(1.8, 0.97)$,  $(2.3, 0.97)$

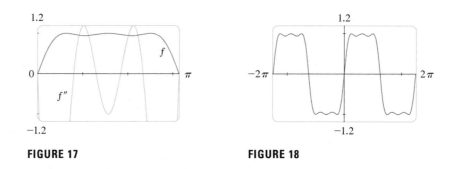

**FIGURE 17**                    **FIGURE 18**

Having checked that Figure 15 does indeed represent $f$ accurately for $0 \le x \le \pi$, we can state that the extended graph in Figure 18 represents $f$ accurately for $-2\pi \le x \le 2\pi$.

Our final example is concerned with *families* of functions. As discussed in Appendix G, this means that the functions in the family are related to each other by a formula that contains one or more arbitrary constants. Each value of the constant gives rise to a member of the family and the idea is to see how the graph of the function changes as the constant changes.

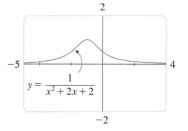

**FIGURE 19**
$c = 2$

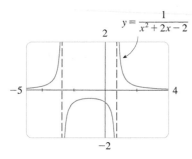

**FIGURE 20**
$c = -2$

**V** **EXAMPLE 5** How does the graph of $f(x) = 1/(x^2 + 2x + c)$ vary as $c$ varies?

SOLUTION The graphs in Figures 19 and 20 (the special cases $c = 2$ and $c = -2$) show two very different-looking curves. Before drawing any more graphs, let's see what members of this family have in common. Since

$$\lim_{x \to \pm\infty} \frac{1}{x^2 + 2x + c} = 0$$

for any value of $c$, they all have the $x$-axis as a horizontal asymptote. A vertical asymptote will occur when $x^2 + 2x + c = 0$. Solving this quadratic equation, we get $x = -1 \pm \sqrt{1 - c}$. When $c > 1$, there is no vertical asymptote (as in Figure 19). When $c = 1$, the graph has a single vertical asymptote $x = -1$ because

$$\lim_{x \to -1} \frac{1}{x^2 + 2x + 1} = \lim_{x \to -1} \frac{1}{(x + 1)^2} = \infty$$

When $c < 1$, there are two vertical asymptotes: $x = -1 \pm \sqrt{1 - c}$ (as in Figure 20). Now we compute the derivative:

$$f'(x) = -\frac{2x + 2}{(x^2 + 2x + c)^2}$$

This shows that $f'(x) = 0$ when $x = -1$ (if $c \neq 1$), $f'(x) > 0$ when $x < -1$, and $f'(x) < 0$ when $x > -1$. For $c \geq 1$, this means that $f$ increases on $(-\infty, -1)$ and decreases on $(-1, \infty)$. For $c > 1$, there is an absolute maximum value $f(-1) = 1/(c - 1)$. For $c < 1$, $f(-1) = 1/(c - 1)$ is a local maximum value and the intervals of increase and decrease are interrupted at the vertical asymptotes.

Figure 21 is a "slide show" displaying five members of the family, all graphed in the viewing rectangle $[-5, 4]$ by $[-2, 2]$. As predicted, $c = 1$ is the value at which a transition takes place from two vertical asymptotes to one, and then to none. As $c$ increases from 1, we see that the maximum point becomes lower; this is explained by the fact that $1/(c - 1) \to 0$ as $c \to \infty$. As $c$ decreases from 1, the vertical asymptotes become more widely separated because the distance between them is $2\sqrt{1 - c}$, which becomes large as $c \to -\infty$. Again, the maximum point approaches the $x$-axis because $1/(c - 1) \to 0$ as $c \to -\infty$.

**TEC** See an animation of Figure 21 in Visual 3.6.

$c = -1$ $\quad\quad\quad$ $c = 0$ $\quad\quad\quad$ $c = 1$ $\quad\quad\quad$ $c = 2$ $\quad\quad\quad$ $c = 3$

**FIGURE 21**   The family of functions $f(x) = 1/(x^2 + 2x + c)$

There is clearly no inflection point when $c \leq 1$. For $c > 1$ we calculate that

$$f''(x) = \frac{2(3x^2 + 6x + 4 - c)}{(x^2 + 2x + c)^3}$$

and deduce that inflection points occur when $x = -1 \pm \sqrt{3(c - 1)}/3$. So the inflection points become more spread out as $c$ increases and this seems plausible from the last two parts of Figure 21.

## 3.6   Exercises

**1–8** Produce graphs of $f$ that reveal all the important aspects of the curve. In particular, you should use graphs of $f'$ and $f''$ to estimate the intervals of increase and decrease, extreme values, intervals of concavity, and inflection points.

**1.** $f(x) = 4x^4 - 32x^3 + 89x^2 - 95x + 29$

**2.** $f(x) = x^6 - 15x^5 + 75x^4 - 125x^3 - x$

**3.** $f(x) = x^6 - 10x^5 - 400x^4 + 2500x^3$

**4.** $f(x) = \dfrac{x^2 - 1}{40x^3 + x + 1}$ $\quad\quad$ **5.** $f(x) = \dfrac{x}{x^3 + x^2 + 1}$

**6.** $f(x) = 6 \sin x - x^2$, $\quad -5 \leq x \leq 3$

**7.** $f(x) = 6 \sin x + \cot x$, $\quad -\pi \leq x \leq \pi$

**8.** $f(x) = \dfrac{\sin x}{x}$, $\quad -2\pi \leq x \leq 2\pi$

**9–10** Produce graphs of $f$ that reveal all the important aspects of the curve. Estimate the intervals of increase and decrease and intervals of concavity, and use calculus to find these intervals exactly.

**9.** $f(x) = 1 + \dfrac{1}{x} + \dfrac{8}{x^2} + \dfrac{1}{x^3}$ $\quad$ **10.** $f(x) = \dfrac{1}{x^8} - \dfrac{2 \times 10^8}{x^4}$

**11–12** Sketch the graph by hand using asymptotes and intercepts, but not derivatives. Then use your sketch as a guide to producing graphs (with a graphing device) that display the major features of the curve. Use these graphs to estimate the maximum and minimum values.

**11.** $f(x) = \dfrac{(x + 4)(x - 3)^2}{x^4(x - 1)}$ $\quad$ **12.** $f(x) = \dfrac{(2x + 3)^2(x - 2)^5}{x^3(x - 5)^2}$

Graphing calculator or computer required    **CAS** Computer algebra system required    **1.** Homework Hints available at stewartcalculus.com

CAS **13.** If $f$ is the function considered in Example 3, use a computer algebra system to calculate $f'$ and then graph it to confirm that all the maximum and minimum values are as given in the example. Calculate $f''$ and use it to estimate the intervals of concavity and inflection points.

CAS **14.** If $f$ is the function of Exercise 12, find $f'$ and $f''$ and use their graphs to estimate the intervals of increase and decrease and concavity of $f$.

CAS **15–18** Use a computer algebra system to graph $f$ and to find $f'$ and $f''$. Use graphs of these derivatives to estimate the intervals of increase and decrease, extreme values, intervals of concavity, and inflection points of $f$.

**15.** $f(x) = \dfrac{x^3 + 5x^2 + 1}{x^4 + x^3 - x^2 + 2}$

**16.** $f(x) = \dfrac{x^{2/3}}{1 + x + x^4}$

**17.** $f(x) = \sqrt{x} + 5 \sin x, \quad x \le 20$

**18.** $f(x) = \dfrac{2x - 1}{\sqrt[4]{x^4 + x + 1}}$

**19.** In Example 4 we considered a member of the family of functions $f(x) = \sin(x + \sin cx)$ that occur in FM synthesis. Here we investigate the function with $c = 3$. Start by graphing $f$ in the viewing rectangle $[0, \pi]$ by $[-1.2, 1.2]$. How many local maximum points do you see? The graph has more than are visible to the naked eye. To discover the hidden maximum and minimum points you will need to examine the graph of $f'$ very carefully. In fact, it helps to look at the graph of $f''$ at the same time. Find all the maximum and minimum values and inflection points. Then graph $f$ in the viewing rectangle $[-2\pi, 2\pi]$ by $[-1.2, 1.2]$ and comment on symmetry.

**20–25** Describe how the graph of $f$ varies as $c$ varies. Graph several members of the family to illustrate the trends that you discover. In particular, you should investigate how maximum and minimum points and inflection points move when $c$ changes. You should also identify any transitional values of $c$ at which the basic shape of the curve changes.

**20.** $f(x) = x^3 + cx$

**21.** $f(x) = \sqrt{x^4 + cx^2}$

**22.** $f(x) = x\sqrt{c^2 - x^2}$

**23.** $f(x) = \dfrac{cx}{1 + c^2x^2}$

**24.** $f(x) = \dfrac{1}{(1 - x^2)^2 + cx^2}$

**25.** $f(x) = cx + \sin x$

**26.** Investigate the family of curves given by the equation $f(x) = x^4 + cx^2 + x$. Start by determining the transitional value of $c$ at which the number of inflection points changes. Then graph several members of the family to see what shapes are possible. There is another transitional value of $c$ at which the number of critical numbers changes. Try to discover it graphically. Then prove what you have discovered.

**27.** (a) Investigate the family of polynomials given by the equation $f(x) = cx^4 - 2x^2 + 1$. For what values of $c$ does the curve have minimum points?
(b) Show that the minimum and maximum points of every curve in the family lie on the parabola $y = 1 - x^2$. Illustrate by graphing this parabola and several members of the family.

**28.** (a) Investigate the family of polynomials given by the equation $f(x) = 2x^3 + cx^2 + 2x$. For what values of $c$ does the curve have maximum and minimum points?
(b) Show that the minimum and maximum points of every curve in the family lie on the curve $y = x - x^3$. Illustrate by graphing this curve and several members of the family.

## 3.7    Optimization Problems

The methods we have learned in this chapter for finding extreme values have practical applications in many areas of life. A businessperson wants to minimize costs and maximize profits. A traveler wants to minimize transportation time. Fermat's Principle in optics states that light follows the path that takes the least time. In this section we solve such problems as maximizing areas, volumes, and profits and minimizing distances, times, and costs.

In solving such practical problems the greatest challenge is often to convert the word problem into a mathematical optimization problem by setting up the function that is to be maximized or minimized. Let's recall the problem-solving principles discussed on page 97 and adapt them to this situation:

**Steps in Solving Optimization Problems**

**1. Understand the Problem**    The first step is to read the problem carefully until it is clearly understood. Ask yourself: What is the unknown? What are the given quantities? What are the given conditions?

2. **Draw a Diagram** In most problems it is useful to draw a diagram and identify the given and required quantities on the diagram.

3. **Introduce Notation** Assign a symbol to the quantity that is to be maximized or minimized (let's call it $Q$ for now). Also select symbols ($a, b, c, \ldots, x, y$) for other unknown quantities and label the diagram with these symbols. It may help to use initials as suggestive symbols—for example, $A$ for area, $h$ for height, $t$ for time.

4. Express $Q$ in terms of some of the other symbols from Step 3.

5. If $Q$ has been expressed as a function of more than one variable in Step 4, use the given information to find relationships (in the form of equations) among these variables. Then use these equations to eliminate all but one of the variables in the expression for $Q$. Thus $Q$ will be expressed as a function of *one* variable $x$, say, $Q = f(x)$. Write the domain of this function.

6. Use the methods of Sections 3.1 and 3.3 to find the *absolute* maximum or minimum value of $f$. In particular, if the domain of $f$ is a closed interval, then the Closed Interval Method in Section 3.1 can be used.

**EXAMPLE 1** A farmer has 2400 ft of fencing and wants to fence off a rectangular field that borders a straight river. He needs no fence along the river. What are the dimensions of the field that has the largest area?

SOLUTION In order to get a feeling for what is happening in this problem, let's experiment with some special cases. Figure 1 (not to scale) shows three possible ways of laying out the 2400 ft of fencing.

PS Understand the problem
PS Analogy: Try special cases
PS Draw diagrams

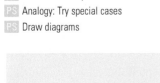

Area = 100 · 2200 = 220,000 ft²          Area = 700 · 1000 = 700,000 ft²          Area = 1000 · 400 = 400,000 ft²

**FIGURE 1**

We see that when we try shallow, wide fields or deep, narrow fields, we get relatively small areas. It seems plausible that there is some intermediate configuration that produces the largest area.

Figure 2 illustrates the general case. We wish to maximize the area $A$ of the rectangle. Let $x$ and $y$ be the depth and width of the rectangle (in feet). Then we express $A$ in terms of $x$ and $y$:

$$A = xy$$

PS Introduce notation

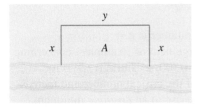

**FIGURE 2**

We want to express $A$ as a function of just one variable, so we eliminate $y$ by expressing it in terms of $x$. To do this we use the given information that the total length of the fencing is 2400 ft. Thus

$$2x + y = 2400$$

From this equation we have $y = 2400 - 2x$, which gives

$$A = x(2400 - 2x) = 2400x - 2x^2$$

Note that $x \geq 0$ and $x \leq 1200$ (otherwise $A < 0$). So the function that we wish to maximize is

$$A(x) = 2400x - 2x^2 \qquad 0 \leq x \leq 1200$$

The derivative is $A'(x) = 2400 - 4x$, so to find the critical numbers we solve the equation

$$2400 - 4x = 0$$

which gives $x = 600$. The maximum value of $A$ must occur either at this critical number or at an endpoint of the interval. Since $A(0) = 0$, $A(600) = 720{,}000$, and $A(1200) = 0$, the Closed Interval Method gives the maximum value as $A(600) = 720{,}000$.

[Alternatively, we could have observed that $A''(x) = -4 < 0$ for all $x$, so $A$ is always concave downward and the local maximum at $x = 600$ must be an absolute maximum.]

Thus the rectangular field should be 600 ft deep and 1200 ft wide. ▪

**V** **EXAMPLE 2** A cylindrical can is to be made to hold 1 L of oil. Find the dimensions that will minimize the cost of the metal to manufacture the can.

SOLUTION Draw the diagram as in Figure 3, where $r$ is the radius and $h$ the height (both in centimeters). In order to minimize the cost of the metal, we minimize the total surface area of the cylinder (top, bottom, and sides). From Figure 4 we see that the sides are made from a rectangular sheet with dimensions $2\pi r$ and $h$. So the surface area is

$$A = 2\pi r^2 + 2\pi r h$$

**FIGURE 3**

To eliminate $h$ we use the fact that the volume is given as 1 L, which we take to be 1000 cm³. Thus

$$\pi r^2 h = 1000$$

which gives $h = 1000/(\pi r^2)$. Substitution of this into the expression for $A$ gives

$$A = 2\pi r^2 + 2\pi r \left( \frac{1000}{\pi r^2} \right) = 2\pi r^2 + \frac{2000}{r}$$

Area $2(\pi r^2)$     Area $(2\pi r)h$

**FIGURE 4**

Therefore the function that we want to minimize is

$$A(r) = 2\pi r^2 + \frac{2000}{r} \qquad r > 0$$

To find the critical numbers, we differentiate:

$$A'(r) = 4\pi r - \frac{2000}{r^2} = \frac{4(\pi r^3 - 500)}{r^2}$$

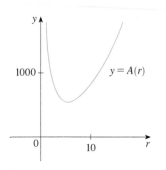

**FIGURE 5**

Then $A'(r) = 0$ when $\pi r^3 = 500$, so the only critical number is $r = \sqrt[3]{500/\pi}$.

Since the domain of $A$ is $(0, \infty)$, we can't use the argument of Example 1 concerning endpoints. But we can observe that $A'(r) < 0$ for $r < \sqrt[3]{500/\pi}$ and $A'(r) > 0$ for $r > \sqrt[3]{500/\pi}$, so $A$ is decreasing for *all* $r$ to the left of the critical number and increasing for *all* $r$ to the right. Thus $r = \sqrt[3]{500/\pi}$ must give rise to an *absolute* minimum.

[Alternatively, we could argue that $A(r) \to \infty$ as $r \to 0^+$ and $A(r) \to \infty$ as $r \to \infty$, so there must be a minimum value of $A(r)$, which must occur at the critical number. See Figure 5.]

In the Applied Project on page 262 we investigate the most economical shape for a can by taking into account other manufacturing costs.

The value of $h$ corresponding to $r = \sqrt[3]{500/\pi}$ is

$$h = \frac{1000}{\pi r^2} = \frac{1000}{\pi (500/\pi)^{2/3}} = 2 \sqrt[3]{\frac{500}{\pi}} = 2r$$

Thus, to minimize the cost of the can, the radius should be $\sqrt[3]{500/\pi}$ cm and the height should be equal to twice the radius, namely, the diameter.

**NOTE 1** The argument used in Example 2 to justify the absolute minimum is a variant of the First Derivative Test (which applies only to *local* maximum or minimum values) and is stated here for future reference.

**TEC** Module 3.7 takes you through six additional optimization problems, including animations of the physical situations.

---

**First Derivative Test for Absolute Extreme Values** Suppose that $c$ is a critical number of a continuous function $f$ defined on an interval.

(a) If $f'(x) > 0$ for all $x < c$ and $f'(x) < 0$ for all $x > c$, then $f(c)$ is the absolute maximum value of $f$.

(b) If $f'(x) < 0$ for all $x < c$ and $f'(x) > 0$ for all $x > c$, then $f(c)$ is the absolute minimum value of $f$.

---

**NOTE 2** An alternative method for solving optimization problems is to use implicit differentiation. Let's look at Example 2 again to illustrate the method. We work with the same equations

$$A = 2\pi r^2 + 2\pi rh \qquad \pi r^2 h = 1000$$

but instead of eliminating $h$, we differentiate both equations implicitly with respect to $r$:

$$A' = 4\pi r + 2\pi h + 2\pi rh' \qquad 2\pi rh + \pi r^2 h' = 0$$

The minimum occurs at a critical number, so we set $A' = 0$, simplify, and arrive at the equations

$$2r + h + rh' = 0 \qquad 2h + rh' = 0$$

and subtraction gives $2r - h = 0$, or $h = 2r$.

**V** **EXAMPLE 3** Find the point on the parabola $y^2 = 2x$ that is closest to the point $(1, 4)$.

SOLUTION The distance between the point $(1, 4)$ and the point $(x, y)$ is

$$d = \sqrt{(x - 1)^2 + (y - 4)^2}$$

(See Figure 6.) But if $(x, y)$ lies on the parabola, then $x = \frac{1}{2}y^2$, so the expression for $d$ becomes

$$d = \sqrt{\left(\tfrac{1}{2}y^2 - 1\right)^2 + (y - 4)^2}$$

(Alternatively, we could have substituted $y = \sqrt{2x}$ to get $d$ in terms of $x$ alone.) Instead of minimizing $d$, we minimize its square:

$$d^2 = f(y) = \left(\tfrac{1}{2}y^2 - 1\right)^2 + (y - 4)^2$$

(You should convince yourself that the minimum of $d$ occurs at the same point as the

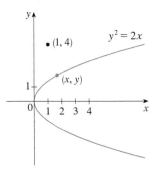

**FIGURE 6**

minimum of $d^2$, but $d^2$ is easier to work with.) Differentiating, we obtain

$$f'(y) = 2\left(\tfrac{1}{2}y^2 - 1\right)y + 2(y - 4) = y^3 - 8$$

so $f'(y) = 0$ when $y = 2$. Observe that $f'(y) < 0$ when $y < 2$ and $f'(y) > 0$ when $y > 2$, so by the First Derivative Test for Absolute Extreme Values, the absolute minimum occurs when $y = 2$. (Or we could simply say that because of the geometric nature of the problem, it's obvious that there is a closest point but not a farthest point.) The corresponding value of $x$ is $x = \tfrac{1}{2}y^2 = 2$. Thus the point on $y^2 = 2x$ closest to $(1, 4)$ is $(2, 2)$.

**EXAMPLE 4** A man launches his boat from point $A$ on a bank of a straight river, 3 km wide, and wants to reach point $B$, 8 km downstream on the opposite bank, as quickly as possible (see Figure 7). He could row his boat directly across the river to point $C$ and then run to $B$, or he could row directly to $B$, or he could row to some point $D$ between $C$ and $B$ and then run to $B$. If he can row 6 km/h and run 8 km/h, where should he land to reach $B$ as soon as possible? (We assume that the speed of the water is negligible compared with the speed at which the man rows.)

SOLUTION If we let $x$ be the distance from $C$ to $D$, then the running distance is $|DB| = 8 - x$ and the Pythagorean Theorem gives the rowing distance as $|AD| = \sqrt{x^2 + 9}$. We use the equation

$$\text{time} = \frac{\text{distance}}{\text{rate}}$$

Then the rowing time is $\sqrt{x^2 + 9}/6$ and the running time is $(8 - x)/8$, so the total time $T$ as a function of $x$ is

$$T(x) = \frac{\sqrt{x^2 + 9}}{6} + \frac{8 - x}{8}$$

The domain of this function $T$ is $[0, 8]$. Notice that if $x = 0$, he rows to $C$ and if $x = 8$, he rows directly to $B$. The derivative of $T$ is

$$T'(x) = \frac{x}{6\sqrt{x^2 + 9}} - \frac{1}{8}$$

Thus, using the fact that $x \geq 0$, we have

$$T'(x) = 0 \iff \frac{x}{6\sqrt{x^2 + 9}} = \frac{1}{8} \iff 4x = 3\sqrt{x^2 + 9}$$

$$\iff 16x^2 = 9(x^2 + 9) \iff 7x^2 = 81$$

$$\iff x = \frac{9}{\sqrt{7}}$$

The only critical number is $x = 9/\sqrt{7}$. To see whether the minimum occurs at this critical number or at an endpoint of the domain $[0, 8]$, we evaluate $T$ at all three points:

$$T(0) = 1.5 \qquad T\left(\frac{9}{\sqrt{7}}\right) = 1 + \frac{\sqrt{7}}{8} \approx 1.33 \qquad T(8) = \frac{\sqrt{73}}{6} \approx 1.42$$

**FIGURE 7**

**FIGURE 8**

**FIGURE 9**

**FIGURE 10**

Since the smallest of these values of $T$ occurs when $x = 9/\sqrt{7}$, the absolute minimum value of $T$ must occur there. Figure 8 illustrates this calculation by showing the graph of $T$.

Thus the man should land the boat at a point $9/\sqrt{7}$ km ($\approx 3.4$ km) downstream from his starting point.

**V EXAMPLE 5** Find the area of the largest rectangle that can be inscribed in a semicircle of radius $r$.

SOLUTION 1 Let's take the semicircle to be the upper half of the circle $x^2 + y^2 = r^2$ with center the origin. Then the word *inscribed* means that the rectangle has two vertices on the semicircle and two vertices on the $x$-axis as shown in Figure 9.

Let $(x, y)$ be the vertex that lies in the first quadrant. Then the rectangle has sides of lengths $2x$ and $y$, so its area is

$$A = 2xy$$

To eliminate $y$ we use the fact that $(x, y)$ lies on the circle $x^2 + y^2 = r^2$ and so $y = \sqrt{r^2 - x^2}$. Thus

$$A = 2x\sqrt{r^2 - x^2}$$

The domain of this function is $0 \leq x \leq r$. Its derivative is

$$A' = 2\sqrt{r^2 - x^2} - \frac{2x^2}{\sqrt{r^2 - x^2}} = \frac{2(r^2 - 2x^2)}{\sqrt{r^2 - x^2}}$$

which is 0 when $2x^2 = r^2$, that is, $x = r/\sqrt{2}$ (since $x \geq 0$). This value of $x$ gives a maximum value of $A$ since $A(0) = 0$ and $A(r) = 0$. Therefore the area of the largest inscribed rectangle is

$$A\left(\frac{r}{\sqrt{2}}\right) = 2\frac{r}{\sqrt{2}}\sqrt{r^2 - \frac{r^2}{2}} = r^2$$

SOLUTION 2 A simpler solution is possible if we think of using an angle as a variable. Let $\theta$ be the angle shown in Figure 10. Then the area of the rectangle is

$$A(\theta) = (2r\cos\theta)(r\sin\theta) = r^2(2\sin\theta\,\cos\theta) = r^2\sin 2\theta$$

We know that $\sin 2\theta$ has a maximum value of 1 and it occurs when $2\theta = \pi/2$. So $A(\theta)$ has a maximum value of $r^2$ and it occurs when $\theta = \pi/4$.

Notice that this trigonometric solution doesn't involve differentiation. In fact, we didn't need to use calculus at all.

## Applications to Business and Economics

In Section 2.7 we introduced the idea of marginal cost. Recall that if $C(x)$, the **cost function**, is the cost of producing $x$ units of a certain product, then the **marginal cost** is the rate of change of $C$ with respect to $x$. In other words, the marginal cost function is the derivative, $C'(x)$, of the cost function.

Now let's consider marketing. Let $p(x)$ be the price per unit that the company can charge if it sells $x$ units. Then $p$ is called the **demand function** (or **price function**) and we would expect it to be a decreasing function of $x$. If $x$ units are sold and the price per unit is $p(x)$, then the total revenue is

$$R(x) = xp(x)$$

and $R$ is called the **revenue function**. The derivative $R'$ of the revenue function is called the **marginal revenue function** and is the rate of change of revenue with respect to the number of units sold.

If $x$ units are sold, then the total profit is

$$P(x) = R(x) - C(x)$$

and $P$ is called the **profit function**. The **marginal profit function** is $P'$, the derivative of the profit function. In Exercises 57–62 you are asked to use the marginal cost, revenue, and profit functions to minimize costs and maximize revenues and profits.

**V  EXAMPLE 6**  A store has been selling 200 Blu-ray disc players a week at \$350 each. A market survey indicates that for each \$10 rebate offered to buyers, the number of units sold will increase by 20 a week. Find the demand function and the revenue function. How large a rebate should the store offer to maximize its revenue?

SOLUTION  If $x$ is the number of Blu-ray players sold per week, then the weekly increase in sales is $x - 200$. For each increase of 20 units sold, the price is decreased by \$10. So for each additional unit sold, the decrease in price will be $\frac{1}{20} \times 10$ and the demand function is

$$p(x) = 350 - \frac{10}{20}(x - 200) = 450 - \frac{1}{2}x$$

The revenue function is

$$R(x) = xp(x) = 450x - \frac{1}{2}x^2$$

Since $R'(x) = 450 - x$, we see that $R'(x) = 0$ when $x = 450$. This value of $x$ gives an absolute maximum by the First Derivative Test (or simply by observing that the graph of $R$ is a parabola that opens downward). The corresponding price is

$$p(450) = 450 - \frac{1}{2}(450) = 225$$

and the rebate is $350 - 225 = 125$. Therefore, to maximize revenue, the store should offer a rebate of \$125.

## 3.7  Exercises

**1.** Consider the following problem: Find two numbers whose sum is 23 and whose product is a maximum.
  (a) Make a table of values, like the following one, so that the sum of the numbers in the first two columns is always 23. On the basis of the evidence in your table, estimate the answer to the problem.

| First number | Second number | Product |
|---|---|---|
| 1 | 22 | 22 |
| 2 | 21 | 42 |
| 3 | 20 | 60 |
| . | . | . |
| . | . | . |
| . | . | . |

  (b) Use calculus to solve the problem and compare with your answer to part (a).

**2.** Find two numbers whose difference is 100 and whose product is a minimum.

**3.** Find two positive numbers whose product is 100 and whose sum is a minimum.

**4.** The sum of two positive numbers is 16. What is the smallest possible value of the sum of their squares?

**5.** What is the maximum vertical distance between the line $y = x + 2$ and the parabola $y = x^2$ for $-1 \le x \le 2$?

**6.** What is the minimum vertical distance between the parabolas $y = x^2 + 1$ and $y = x - x^2$?

Graphing calculator or computer required        CAS  Computer algebra system required        **1.** Homework Hints available at stewartcalculus.com

**7.** Find the dimensions of a rectangle with perimeter 100 m whose area is as large as possible.

**8.** Find the dimensions of a rectangle with area 1000 m$^2$ whose perimeter is as small as possible.

**9.** A model used for the yield $Y$ of an agricultural crop as a function of the nitrogen level $N$ in the soil (measured in appropriate units) is

$$Y = \frac{kN}{1 + N^2}$$

where $k$ is a positive constant. What nitrogen level gives the best yield?

**10.** The rate (in mg carbon/m$^3$/h) at which photosynthesis takes place for a species of phytoplankton is modeled by the function

$$P = \frac{100I}{I^2 + I + 4}$$

where $I$ is the light intensity (measured in thousands of foot-candles). For what light intensity is $P$ a maximum?

**11.** Consider the following problem: A farmer with 750 ft of fencing wants to enclose a rectangular area and then divide it into four pens with fencing parallel to one side of the rectangle. What is the largest possible total area of the four pens?
(a) Draw several diagrams illustrating the situation, some with shallow, wide pens and some with deep, narrow pens. Find the total areas of these configurations. Does it appear that there is a maximum area? If so, estimate it.
(b) Draw a diagram illustrating the general situation. Introduce notation and label the diagram with your symbols.
(c) Write an expression for the total area.
(d) Use the given information to write an equation that relates the variables.
(e) Use part (d) to write the total area as a function of one variable.
(f) Finish solving the problem and compare the answer with your estimate in part (a).

**12.** Consider the following problem: A box with an open top is to be constructed from a square piece of cardboard, 3 ft wide, by cutting out a square from each of the four corners and bending up the sides. Find the largest volume that such a box can have.
(a) Draw several diagrams to illustrate the situation, some short boxes with large bases and some tall boxes with small bases. Find the volumes of several such boxes. Does it appear that there is a maximum volume? If so, estimate it.
(b) Draw a diagram illustrating the general situation. Introduce notation and label the diagram with your symbols.
(c) Write an expression for the volume.
(d) Use the given information to write an equation that relates the variables.
(e) Use part (d) to write the volume as a function of one variable.
(f) Finish solving the problem and compare the answer with your estimate in part (a).

**13.** A farmer wants to fence an area of 1.5 million square feet in a rectangular field and then divide it in half with a fence parallel

to one of the sides of the rectangle. How can he do this so as to minimize the cost of the fence?

**14.** A box with a square base and open top must have a volume of 32,000 cm$^3$. Find the dimensions of the box that minimize the amount of material used.

**15.** If 1200 cm$^2$ of material is available to make a box with a square base and an open top, find the largest possible volume of the box.

**16.** A rectangular storage container with an open top is to have a volume of 10 m$^3$. The length of its base is twice the width. Material for the base costs \$10 per square meter. Material for the sides costs \$6 per square meter. Find the cost of materials for the cheapest such container.

**17.** Do Exercise 16 assuming the container has a lid that is made from the same material as the sides.

**18.** (a) Show that of all the rectangles with a given area, the one with smallest perimeter is a square.
(b) Show that of all the rectangles with a given perimeter, the one with greatest area is a square.

**19.** Find the point on the line $y = 2x + 3$ that is closest to the origin.

**20.** Find the point on the curve $y = \sqrt{x}$ that is closest to the point $(3, 0)$.

**21.** Find the points on the ellipse $4x^2 + y^2 = 4$ that are farthest away from the point $(1, 0)$.

**22.** Find, correct to two decimal places, the coordinates of the point on the curve $y = \sin x$ that is closest to the point $(4, 2)$.

**23.** Find the dimensions of the rectangle of largest area that can be inscribed in a circle of radius $r$.

**24.** Find the area of the largest rectangle that can be inscribed in the ellipse $x^2/a^2 + y^2/b^2 = 1$.

**25.** Find the dimensions of the rectangle of largest area that can be inscribed in an equilateral triangle of side $L$ if one side of the rectangle lies on the base of the triangle.

**26.** Find the area of the largest trapezoid that can be inscribed in a circle of radius 1 and whose base is a diameter of the circle.

**27.** Find the dimensions of the isosceles triangle of largest area that can be inscribed in a circle of radius $r$.

**28.** Find the area of the largest rectangle that can be inscribed in a right triangle with legs of lengths 3 cm and 4 cm if two sides of the rectangle lie along the legs.

**29.** A right circular cylinder is inscribed in a sphere of radius $r$. Find the largest possible volume of such a cylinder.

**30.** A right circular cylinder is inscribed in a cone with height $h$ and base radius $r$. Find the largest possible volume of such a cylinder.

**31.** A right circular cylinder is inscribed in a sphere of radius $r$. Find the largest possible surface area of such a cylinder.

**32.** A Norman window has the shape of a rectangle surmounted by a semicircle. (Thus the diameter of the semicircle is equal to the width of the rectangle. See Exercise 62 on page 22.) If the perimeter of the window is 30 ft, find the dimensions of the window so that the greatest possible amount of light is admitted.

**33.** The top and bottom margins of a poster are each 6 cm and the side margins are each 4 cm. If the area of printed material on the poster is fixed at 384 cm$^2$, find the dimensions of the poster with the smallest area.

**34.** A poster is to have an area of 180 in$^2$ with 1-inch margins at the bottom and sides and a 2-inch margin at the top. What dimensions will give the largest printed area?

**35.** A piece of wire 10 m long is cut into two pieces. One piece is bent into a square and the other is bent into an equilateral triangle. How should the wire be cut so that the total area enclosed is (a) a maximum? (b) A minimum?

**36.** Answer Exercise 35 if one piece is bent into a square and the other into a circle.

**37.** A cylindrical can without a top is made to contain $V$ cm$^3$ of liquid. Find the dimensions that will minimize the cost of the metal to make the can.

**38.** A fence 8 ft tall runs parallel to a tall building at a distance of 4 ft from the building. What is the length of the shortest ladder that will reach from the ground over the fence to the wall of the building?

**39.** A cone-shaped drinking cup is made from a circular piece of paper of radius $R$ by cutting out a sector and joining the edges $CA$ and $CB$. Find the maximum capacity of such a cup.

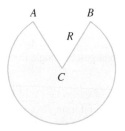

**40.** A cone-shaped paper drinking cup is to be made to hold 27 cm$^3$ of water. Find the height and radius of the cup that will use the smallest amount of paper.

**41.** A cone with height $h$ is inscribed in a larger cone with height $H$ so that its vertex is at the center of the base of the larger cone. Show that the inner cone has maximum volume when $h = \frac{1}{3}H$.

**42.** An object with weight $W$ is dragged along a horizontal plane by a force acting along a rope attached to the object. If the rope makes an angle $\theta$ with a plane, then the magnitude of the force is

$$F = \frac{\mu W}{\mu \sin \theta + \cos \theta}$$

where $\mu$ is a constant called the coefficient of friction. For what value of $\theta$ is $F$ smallest?

**43.** If a resistor of $R$ ohms is connected across a battery of $E$ volts with internal resistance $r$ ohms, then the power (in watts) in the external resistor is

$$P = \frac{E^2 R}{(R + r)^2}$$

If $E$ and $r$ are fixed but $R$ varies, what is the maximum value of the power?

**44.** For a fish swimming at a speed $v$ relative to the water, the energy expenditure per unit time is proportional to $v^3$. It is believed that migrating fish try to minimize the total energy required to swim a fixed distance. If the fish are swimming against a current $u$ ($u < v$), then the time required to swim a distance $L$ is $L/(v - u)$ and the total energy $E$ required to swim the distance is given by

$$E(v) = av^3 \cdot \frac{L}{v - u}$$

where $a$ is the proportionality constant.
(a) Determine the value of $v$ that minimizes $E$.
(b) Sketch the graph of $E$.

*Note:* This result has been verified experimentally; migrating fish swim against a current at a speed 50% greater than the current speed.

**45.** In a beehive, each cell is a regular hexagonal prism, open at one end with a trihedral angle at the other end as in the figure. It is believed that bees form their cells in such a way as to minimize the surface area, thus using the least amount of wax in cell construction. Examination of these cells has shown that the measure of the apex angle $\theta$ is amazingly consistent. Based on the geometry of the cell, it can be shown that the surface area $S$ is given by

$$S = 6sh - \tfrac{3}{2}s^2 \cot \theta + \left(3s^2 \sqrt{3}/2\right) \csc \theta$$

where $s$, the length of the sides of the hexagon, and $h$, the height, are constants.
(a) Calculate $dS/d\theta$.
(b) What angle should the bees prefer?
(c) Determine the minimum surface area of the cell (in terms of $s$ and $h$).

*Note:* Actual measurements of the angle $\theta$ in beehives have been made, and the measures of these angles seldom differ from the calculated value by more than 2°.

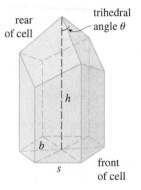

**46.** A boat leaves a dock at 2:00 PM and travels due south at a speed of 20 km/h. Another boat has been heading due east at 15 km/h and reaches the same dock at 3:00 PM. At what time were the two boats closest together?

**47.** Solve the problem in Example 4 if the river is 5 km wide and point $B$ is only 5 km downstream from $A$.

**48.** A woman at a point $A$ on the shore of a circular lake with radius 2 mi wants to arrive at the point $C$ diametrically opposite $A$ on the other side of the lake in the shortest possible time (see the figure). She can walk at the rate of 4 mi/h and row a boat at 2 mi/h. How should she proceed?

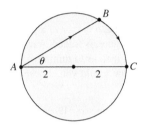

**49.** An oil refinery is located on the north bank of a straight river that is 2 km wide. A pipeline is to be constructed from the refinery to storage tanks located on the south bank of the river 6 km east of the refinery. The cost of laying pipe is $400,000/km over land to a point $P$ on the north bank and $800,000/km under the river to the tanks. To minimize the cost of the pipeline, where should $P$ be located?

**50.** Suppose the refinery in Exercise 49 is located 1 km north of the river. Where should $P$ be located?

**51.** The illumination of an object by a light source is directly proportional to the strength of the source and inversely proportional to the square of the distance from the source. If two light sources, one three times as strong as the other, are placed 10 ft apart, where should an object be placed on the line between the sources so as to receive the least illumination?

**52.** Find an equation of the line through the point $(3, 5)$ that cuts off the least area from the first quadrant.

**53.** Let $a$ and $b$ be positive numbers. Find the length of the shortest line segment that is cut off by the first quadrant and passes through the point $(a, b)$.

**54.** At which points on the curve $y = 1 + 40x^3 - 3x^5$ does the tangent line have the largest slope?

**55.** What is the shortest possible length of the line segment that is cut off by the first quadrant and is tangent to the curve $y = 3/x$ at some point?

**56.** What is the smallest possible area of the triangle that is cut off by the first quadrant and whose hypotenuse is tangent to the parabola $y = 4 - x^2$ at some point?

**57.** (a) If $C(x)$ is the cost of producing $x$ units of a commodity, then the **average cost** per unit is $c(x) = C(x)/x$. Show that if the average cost is a minimum, then the marginal cost equals the average cost.

(b) If $C(x) = 16{,}000 + 200x + 4x^{3/2}$, in dollars, find (i) the cost, average cost, and marginal cost at a production level of 1000 units; (ii) the production level that will minimize the average cost; and (iii) the minimum average cost.

**58.** (a) Show that if the profit $P(x)$ is a maximum, then the marginal revenue equals the marginal cost.

(b) If $C(x) = 16{,}000 + 500x - 1.6x^2 + 0.004x^3$ is the cost function and $p(x) = 1700 - 7x$ is the demand function, find the production level that will maximize profit.

**59.** A baseball team plays in a stadium that holds 55,000 spectators. With ticket prices at $10, the average attendance had been 27,000. When ticket prices were lowered to $8, the average attendance rose to 33,000.

(a) Find the demand function, assuming that it is linear.

(b) How should ticket prices be set to maximize revenue?

**60.** During the summer months Terry makes and sells necklaces on the beach. Last summer he sold the necklaces for $10 each and his sales averaged 20 per day. When he increased the price by $1, he found that the average decreased by two sales per day.

(a) Find the demand function, assuming that it is linear.

(b) If the material for each necklace costs Terry $6, what should the selling price be to maximize his profit?

**61.** A manufacturer has been selling 1000 flat-screen TVs a week at $450 each. A market survey indicates that for each $10 rebate offered to the buyer, the number of TVs sold will increase by 100 per week.

(a) Find the demand function.

(b) How large a rebate should the company offer the buyer in order to maximize its revenue?

(c) If its weekly cost function is $C(x) = 68{,}000 + 150x$, how should the manufacturer set the size of the rebate in order to maximize its profit?

**62.** The manager of a 100-unit apartment complex knows from experience that all units will be occupied if the rent is $800 per month. A market survey suggests that, on average, one additional unit will remain vacant for each $10 increase in rent. What rent should the manager charge to maximize revenue?

**63.** Show that of all the isosceles triangles with a given perimeter, the one with the greatest area is equilateral.

**64.** The frame for a kite is to be made from six pieces of wood. The four exterior pieces have been cut with the lengths indicated in the figure. To maximize the area of the kite, how long should the diagonal pieces be?

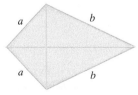

**65.** A point $P$ needs to be located somewhere on the line $AD$ so that the total length $L$ of cables linking $P$ to the points $A$, $B$, and $C$ is minimized (see the figure). Express $L$ as a function of $x = |AP|$ and use the graphs of $L$ and $dL/dx$ to estimate the minimum value of $L$.

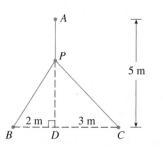

**66.** The graph shows the fuel consumption $c$ of a car (measured in gallons per hour) as a function of the speed $v$ of the car. At very low speeds the engine runs inefficiently, so initially $c$ decreases as the speed increases. But at high speeds the fuel consumption increases. You can see that $c(v)$ is minimized for this car when $v \approx 30$ mi/h. However, for fuel efficiency, what must be minimized is not the consumption in gallons per hour but rather the fuel consumption in gallons *per mile*. Let's call this consumption $G$. Using the graph, estimate the speed at which $G$ has its minimum value.

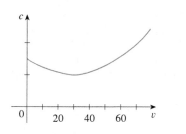

**67.** Let $v_1$ be the velocity of light in air and $v_2$ the velocity of light in water. According to Fermat's Principle, a ray of light will travel from a point $A$ in the air to a point $B$ in the water by a path $ACB$ that minimizes the time taken. Show that

$$\frac{\sin \theta_1}{\sin \theta_2} = \frac{v_1}{v_2}$$

where $\theta_1$ (the angle of incidence) and $\theta_2$ (the angle of refraction) are as shown. This equation is known as Snell's Law.

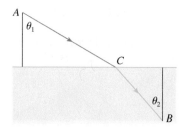

**68.** Two vertical poles $PQ$ and $ST$ are secured by a rope $PRS$ going from the top of the first pole to a point $R$ on the ground

between the poles and then to the top of the second pole as in the figure. Show that the shortest length of such a rope occurs when $\theta_1 = \theta_2$.

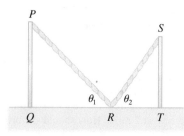

**69.** The upper right-hand corner of a piece of paper, 12 in. by 8 in., as in the figure, is folded over to the bottom edge. How would you fold it so as to minimize the length of the fold? In other words, how would you choose $x$ to minimize $y$?

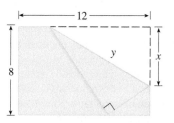

**70.** A steel pipe is being carried down a hallway 9 ft wide. At the end of the hall there is a right-angled turn into a narrower hallway 6 ft wide. What is the length of the longest pipe that can be carried horizontally around the corner?

**71.** An observer stands at a point $P$, one unit away from a track. Two runners start at the point $S$ in the figure and run along the track. One runner runs three times as fast as the other. Find the maximum value of the observer's angle of sight $\theta$ between the runners. [*Hint:* Maximize $\tan \theta$.]

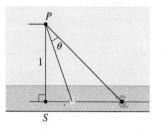

**72.** A rain gutter is to be constructed from a metal sheet of width 30 cm by bending up one-third of the sheet on each side through an angle $\theta$. How should $\theta$ be chosen so that the gutter will carry the maximum amount of water?

**73.** Find the maximum area of a rectangle that can be circumscribed about a given rectangle with length $L$ and width $W$. [*Hint:* Express the area as a function of an angle $\theta$.]

**74.** The blood vascular system consists of blood vessels (arteries, arterioles, capillaries, and veins) that convey blood from the heart to the organs and back to the heart. This system should work so as to minimize the energy expended by the heart in pumping the blood. In particular, this energy is reduced when the resistance of the blood is lowered. One of Poiseuille's Laws gives the resistance $R$ of the blood as

$$R = C \frac{L}{r^4}$$

where $L$ is the length of the blood vessel, $r$ is the radius, and $C$ is a positive constant determined by the viscosity of the blood. (Poiseuille established this law experimentally, but it also follows from Equation 8.4.2.) The figure shows a main blood vessel with radius $r_1$ branching at an angle $\theta$ into a smaller vessel with radius $r_2$.

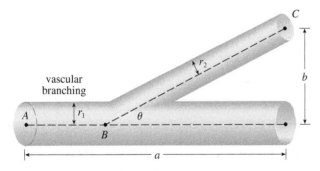

(a) Use Poiseuille's Law to show that the total resistance of the blood along the path $ABC$ is

$$R = C\left(\frac{a - b \cot \theta}{r_1^4} + \frac{b \csc \theta}{r_2^4}\right)$$

where $a$ and $b$ are the distances shown in the figure.
(b) Prove that this resistance is minimized when

$$\cos \theta = \frac{r_2^4}{r_1^4}$$

(c) Find the optimal branching angle (correct to the nearest degree) when the radius of the smaller blood vessel is two-thirds the radius of the larger vessel.

© Manfred Kage / Peter Arnold Images / Photolibrary

**75.** Ornithologists have determined that some species of birds tend to avoid flights over large bodies of water during daylight hours. It is believed that more energy is required to fly over water than over land because air generally rises over land and falls over water during the day. A bird with these tendencies is released from an island that is 5 km from the nearest point $B$ on a straight shoreline, flies to a point $C$ on the shoreline, and then flies along the shoreline to its nesting area $D$. Assume that the bird instinctively chooses a path that will minimize its energy expenditure. Points $B$ and $D$ are 13 km apart.
(a) In general, if it takes 1.4 times as much energy to fly over water as it does over land, to what point $C$ should the bird fly in order to minimize the total energy expended in returning to its nesting area?
(b) Let $W$ and $L$ denote the energy (in joules) per kilometer flown over water and land, respectively. What would a large value of the ratio $W/L$ mean in terms of the bird's flight? What would a small value mean? Determine the ratio $W/L$ corresponding to the minimum expenditure of energy.
(c) What should the value of $W/L$ be in order for the bird to fly directly to its nesting area $D$? What should the value of $W/L$ be for the bird to fly to $B$ and then along the shore to $D$?
(d) If the ornithologists observe that birds of a certain species reach the shore at a point 4 km from $B$, how many times more energy does it take a bird to fly over water than over land?

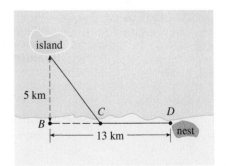

**76.** Two light sources of identical strength are placed 10 m apart. An object is to be placed at a point $P$ on a line $\ell$ parallel to the line joining the light sources and at a distance $d$ meters from it (see the figure). We want to locate $P$ on $\ell$ so that the intensity of illumination is minimized. We need to use the fact that the intensity of illumination for a single source is directly proportional to the strength of the source and inversely proportional to the square of the distance from the source.

(a) Find an expression for the intensity $I(x)$ at the point $P$.

(b) If $d = 5$ m, use graphs of $I(x)$ and $I'(x)$ to show that the intensity is minimized when $x = 5$ m, that is, when $P$ is at the midpoint of $\ell$.

(c) If $d = 10$ m, show that the intensity (perhaps surprisingly) is *not* minimized at the midpoint.

(d) Somewhere between $d = 5$ m and $d = 10$ m there is a transitional value of $d$ at which the point of minimal illumination abruptly changes. Estimate this value of $d$ by graphical methods. Then find the exact value of $d$.

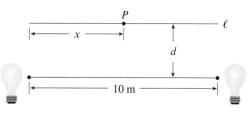

---

## THE SHAPE OF A CAN

In this project we investigate the most economical shape for a can. We first interpret this to mean that the volume $V$ of a cylindrical can is given and we need to find the height $h$ and radius $r$ that minimize the cost of the metal to make the can (see the figure). If we disregard any waste metal in the manufacturing process, then the problem is to minimize the surface area of the cylinder. We solved this problem in Example 2 in Section 3.7 and we found that $h = 2r$; that is, the height should be the same as the diameter. But if you go to your cupboard or your supermarket with a ruler, you will discover that the height is usually greater than the diameter and the ratio $h/r$ varies from 2 up to about 3.8. Let's see if we can explain this phenomenon.

**1.** The material for the cans is cut from sheets of metal. The cylindrical sides are formed by bending rectangles; these rectangles are cut from the sheet with little or no waste. But if the top and bottom discs are cut from squares of side $2r$ (as in the figure), this leaves considerable waste metal, which may be recycled but has little or no value to the can makers. If this is the case, show that the amount of metal used is minimized when

$$\frac{h}{r} = \frac{8}{\pi} \approx 2.55$$

Discs cut from squares

**2.** A more efficient packing of the discs is obtained by dividing the metal sheet into hexagons and cutting the circular lids and bases from the hexagons (see the figure). Show that if this strategy is adopted, then

$$\frac{h}{r} = \frac{4\sqrt{3}}{\pi} \approx 2.21$$

Discs cut from hexagons

**3.** The values of $h/r$ that we found in Problems 1 and 2 are a little closer to the ones that actually occur on supermarket shelves, but they still don't account for everything. If we look more closely at some real cans, we see that the lid and the base are formed from discs with radius larger than $r$ that are bent over the ends of the can. If we allow for this we would increase $h/r$. More significantly, in addition to the cost of the metal we need to incorporate the manufacturing of the can into the cost. Let's assume that most of the expense is incurred in joining the sides to the rims of the cans. If we cut the discs from hexagons as in Problem 2, then the total cost is proportional to

$$4\sqrt{3}\, r^2 + 2\pi r h + k(4\pi r + h)$$

---

Graphing calculator or computer required

where $k$ is the reciprocal of the length that can be joined for the cost of one unit area of metal. Show that this expression is minimized when

$$\frac{\sqrt[3]{V}}{k} = \sqrt{\frac{\pi h}{r}} \cdot \frac{2\pi - h/r}{\pi h/r - 4\sqrt{3}}$$

**4.** Plot $\sqrt[3]{V}/k$ as a function of $x = h/r$ and use your graph to argue that when a can is large or joining is cheap, we should make $h/r$ approximately 2.21 (as in Problem 2). But when the can is small or joining is costly, $h/r$ should be substantially larger.

**5.** Our analysis shows that large cans should be almost square but small cans should be tall and thin. Take a look at the relative shapes of the cans in a supermarket. Is our conclusion usually true in practice? Are there exceptions? Can you suggest reasons why small cans are not always tall and thin?

## 3.8 Newton's Method

Suppose that a car dealer offers to sell you a car for $18,000 or for payments of $375 per month for five years. You would like to know what monthly interest rate the dealer is, in effect, charging you. To find the answer, you have to solve the equation

$$\boxed{1} \qquad 48x(1 + x)^{60} - (1 + x)^{60} + 1 = 0$$

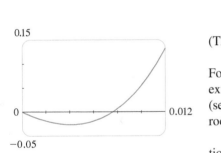

**FIGURE 1**

Try to solve Equation 1 using the numerical rootfinder on your calculator or computer. Some machines are not able to solve it. Others are successful but require you to specify a starting point for the search.

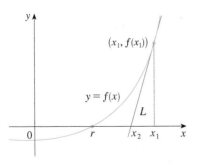

**FIGURE 2**

(The details are explained in Exercise 39.) How would you solve such an equation?

For a quadratic equation $ax^2 + bx + c = 0$ there is a well-known formula for the roots. For third- and fourth-degree equations there are also formulas for the roots, but they are extremely complicated. If $f$ is a polynomial of degree 5 or higher, there is no such formula (see the note on page 160). Likewise, there is no formula that will enable us to find the exact roots of a transcendental equation such as $\cos x = x$.

We can find an *approximate* solution to Equation 1 by plotting the left side of the equation. Using a graphing device, and after experimenting with viewing rectangles, we produce the graph in Figure 1.

We see that in addition to the solution $x = 0$, which doesn't interest us, there is a solution between 0.007 and 0.008. Zooming in shows that the root is approximately 0.0076. If we need more accuracy we could zoom in repeatedly, but that becomes tiresome. A faster alternative is to use a numerical rootfinder on a calculator or computer algebra system. If we do so, we find that the root, correct to nine decimal places, is 0.007628603.

How do those numerical rootfinders work? They use a variety of methods, but most of them make some use of **Newton's method**, also called the **Newton-Raphson method**. We will explain how this method works, partly to show what happens inside a calculator or computer, and partly as an application of the idea of linear approximation.

The geometry behind Newton's method is shown in Figure 2, where the root that we are trying to find is labeled $r$. We start with a first approximation $x_1$, which is obtained by guessing, or from a rough sketch of the graph of $f$, or from a computer-generated graph of $f$. Consider the tangent line $L$ to the curve $y = f(x)$ at the point $(x_1, f(x_1))$ and look at the $x$-intercept of $L$, labeled $x_2$. The idea behind Newton's method is that the tangent line is close to the curve and so its $x$-intercept, $x_2$, is close to the $x$-intercept of the curve (namely, the root $r$ that we are seeking). Because the tangent is a line, we can easily find its $x$-intercept.

To find a formula for $x_2$ in terms of $x_1$ we use the fact that the slope of $L$ is $f'(x_1)$, so its equation is

$$y - f(x_1) = f'(x_1)(x - x_1)$$

Since the $x$-intercept of $L$ is $x_2$, we set $y = 0$ and obtain

$$0 - f(x_1) = f'(x_1)(x_2 - x_1)$$

If $f'(x_1) \neq 0$, we can solve this equation for $x_2$:

$$x_2 = x_1 - \frac{f(x_1)}{f'(x_1)}$$

We use $x_2$ as a second approximation to $r$.

Next we repeat this procedure with $x_1$ replaced by the second approximation $x_2$, using the tangent line at $(x_2, f(x_2))$. This gives a third approximation:

$$x_3 = x_2 - \frac{f(x_2)}{f'(x_2)}$$

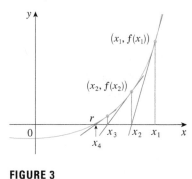

**FIGURE 3**

If we keep repeating this process, we obtain a sequence of approximations $x_1, x_2, x_3, x_4, \ldots$ as shown in Figure 3. In general, if the $n$th approximation is $x_n$ and $f'(x_n) \neq 0$, then the next approximation is given by

$$\boxed{2} \qquad \boxed{\quad x_{n+1} = x_n - \frac{f(x_n)}{f'(x_n)} \quad}$$

Sequences were briefly introduced in *A Preview of Calculus* on page 5. A more thorough discussion starts in Section 11.1.

If the numbers $x_n$ become closer and closer to $r$ as $n$ becomes large, then we say that the sequence *converges* to $r$ and we write

$$\lim_{n \to \infty} x_n = r$$

⊘ Although the sequence of successive approximations converges to the desired root for functions of the type illustrated in Figure 3, in certain circumstances the sequence may not converge. For example, consider the situation shown in Figure 4. You can see that $x_2$ is a worse approximation than $x_1$. This is likely to be the case when $f'(x_1)$ is close to 0. It might even happen that an approximation (such as $x_3$ in Figure 4) falls outside the domain of $f$. Then Newton's method fails and a better initial approximation $x_1$ should be chosen. See Exercises 29–32 for specific examples in which Newton's method works very slowly or does not work at all.

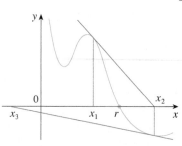

**FIGURE 4**

**V EXAMPLE 1** Starting with $x_1 = 2$, find the third approximation $x_3$ to the root of the equation $x^3 - 2x - 5 = 0$.

SOLUTION  We apply Newton's method with

$$f(x) = x^3 - 2x - 5 \qquad \text{and} \qquad f'(x) = 3x^2 - 2$$

TEC In Module 3.8 you can investigate how Newton's Method works for several functions and what happens when you change $x_1$.

Newton himself used this equation to illustrate his method and he chose $x_1 = 2$ after some experimentation because $f(1) = -6$, $f(2) = -1$, and $f(3) = 16$. Equation 2 becomes

$$x_{n+1} = x_n - \frac{x_n^3 - 2x_n - 5}{3x_n^2 - 2}$$

With $n = 1$ we have

$$x_2 = x_1 - \frac{x_1^3 - 2x_1 - 5}{3x_1^2 - 2}$$

$$= 2 - \frac{2^3 - 2(2) - 5}{3(2)^2 - 2} = 2.1$$

Figure 5 shows the geometry behind the first step in Newton's method in Example 1. Since $f'(2) = 10$, the tangent line to $y = x^3 - 2x - 5$ at $(2, -1)$ has equation $y = 10x - 21$ so its $x$-intercept is $x_2 = 2.1$.

Then with $n = 2$ we obtain

$$x_3 = x_2 - \frac{x_2^3 - 2x_2 - 5}{3x_2^2 - 2}$$

$$= 2.1 - \frac{(2.1)^3 - 2(2.1) - 5}{3(2.1)^2 - 2} \approx 2.0946$$

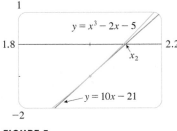

**FIGURE 5**

It turns out that this third approximation $x_3 \approx 2.0946$ is accurate to four decimal places.

Suppose that we want to achieve a given accuracy, say to eight decimal places, using Newton's method. How do we know when to stop? The rule of thumb that is generally used is that we can stop when successive approximations $x_n$ and $x_{n+1}$ agree to eight decimal places. (A precise statement concerning accuracy in Newton's method will be given in Exercise 39 in Section 11.11.)

Notice that the procedure in going from $n$ to $n + 1$ is the same for all values of $n$. (It is called an *iterative* process.) This means that Newton's method is particularly convenient for use with a programmable calculator or a computer.

**V EXAMPLE 2** Use Newton's method to find $\sqrt[6]{2}$ correct to eight decimal places.

SOLUTION First we observe that finding $\sqrt[6]{2}$ is equivalent to finding the positive root of the equation

$$x^6 - 2 = 0$$

so we take $f(x) = x^6 - 2$. Then $f'(x) = 6x^5$ and Formula 2 (Newton's method) becomes

$$x_{n+1} = x_n - \frac{x_n^6 - 2}{6x_n^5}$$

If we choose $x_1 = 1$ as the initial approximation, then we obtain

$$x_2 \approx 1.16666667$$
$$x_3 \approx 1.12644368$$
$$x_4 \approx 1.12249707$$
$$x_5 \approx 1.12246205$$
$$x_6 \approx 1.12246205$$

Since $x_5$ and $x_6$ agree to eight decimal places, we conclude that

$$\sqrt[6]{2} \approx 1.12246205$$

to eight decimal places.

**V** **EXAMPLE 3** Find, correct to six decimal places, the root of the equation $\cos x = x$.

SOLUTION  We first rewrite the equation in standard form:

$$\cos x - x = 0$$

Therefore we let $f(x) = \cos x - x$. Then $f'(x) = -\sin x - 1$, so Formula 2 becomes

$$x_{n+1} = x_n - \frac{\cos x_n - x_n}{-\sin x_n - 1} = x_n + \frac{\cos x_n - x_n}{\sin x_n + 1}$$

**FIGURE 6**

In order to guess a suitable value for $x_1$ we sketch the graphs of $y = \cos x$ and $y = x$ in Figure 6. It appears that they intersect at a point whose $x$-coordinate is somewhat less than 1, so let's take $x_1 = 1$ as a convenient first approximation. Then, remembering to put our calculator in radian mode, we get

$$x_2 \approx 0.75036387$$
$$x_3 \approx 0.73911289$$
$$x_4 \approx 0.73908513$$
$$x_5 \approx 0.73908513$$

Since $x_4$ and $x_5$ agree to six decimal places (eight, in fact), we conclude that the root of the equation, correct to six decimal places, is 0.739085.

**FIGURE 7**

Instead of using the rough sketch in Figure 6 to get a starting approximation for Newton's method in Example 3, we could have used the more accurate graph that a calculator or computer provides. Figure 7 suggests that we use $x_1 = 0.75$ as the initial approximation. Then Newton's method gives

$$x_2 \approx 0.73911114$$
$$x_3 \approx 0.73908513$$
$$x_4 \approx 0.73908513$$

and so we obtain the same answer as before, but with one fewer step.

You might wonder why we bother at all with Newton's method if a graphing device is available. Isn't it easier to zoom in repeatedly and find the roots as in Appendix G? If only one or two decimal places of accuracy are required, then indeed Newton's method is inappropriate and a graphing device suffices. But if six or eight decimal places are required, then repeated zooming becomes tiresome. It is usually faster and more efficient to use a computer and Newton's method in tandem—the graphing device to get started and Newton's method to finish.

**3.8**   **Exercises**

**1.** The figure shows the graph of a function $f$. Suppose that New-ton's method is used to approximate the root $r$ of the equation $f(x) = 0$ with initial approximation $x_1 = 1$.
   (a) Draw the tangent lines that are used to find $x_2$ and $x_3$, and estimate the numerical values of $x_2$ and $x_3$.
   (b) Would $x_1 = 5$ be a better first approximation? Explain.

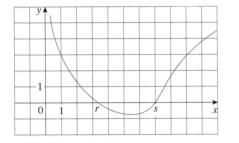

**2.** Follow the instructions for Exercise 1(a) but use $x_1 = 9$ as the starting approximation for finding the root $s$.

**3.** Suppose the tangent line to the curve $y = f(x)$ at the point $(2, 5)$ has the equation $y = 9 - 2x$. If Newton's method is used to locate a root of the equation $f(x) = 0$ and the initial approximation is $x_1 = 2$, find the second approximation $x_2$.

**4.** For each initial approximation, determine graphically what happens if Newton's method is used for the function whose graph is shown.
   (a) $x_1 = 0$    (b) $x_1 = 1$    (c) $x_1 = 3$
   (d) $x_1 = 4$    (e) $x_1 = 5$

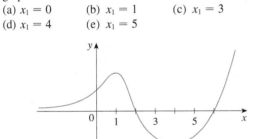

**5.** For which of the initial approximations $x_1 = a, b, c$, and $d$ do you think Newton's method will work and lead to the root of the equation $f(x) = 0$?

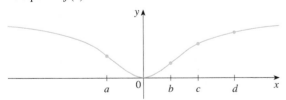

**6–8** Use Newton's method with the specified initial approximation $x_1$ to find $x_3$, the third approximation to the root of the given equation. (Give your answer to four decimal places.)

**6.** $\frac{1}{3}x^3 + \frac{1}{2}x^2 + 3 = 0$,   $x_1 = -3$

**7.** $x^5 - x - 1 = 0$,   $x_1 = 1$       **8.** $x^7 + 4 = 0$,   $x_1 = -1$

**9.** Use Newton's method with initial approximation $x_1 = -1$ to find $x_2$, the second approximation to the root of the equation $x^3 + x + 3 = 0$. Explain how the method works by first graphing the function and its tangent line at $(-1, 1)$.

**10.** Use Newton's method with initial approximation $x_1 = 1$ to find $x_2$, the second approximation to the root of the equation $x^4 - x - 1 = 0$. Explain how the method works by first graph-ing the function and its tangent line at $(1, -1)$.

**11–12** Use Newton's method to approximate the given number cor-rect to eight decimal places.

**11.** $\sqrt[5]{20}$                          **12.** $\sqrt[100]{100}$

**13–16** Use Newton's method to approximate the indicated root of the equation correct to six decimal places.

**13.** The root of $x^4 - 2x^3 + 5x^2 - 6 = 0$ in the interval $[1, 2]$

**14.** The root of $2.2x^5 - 4.4x^3 + 1.3x^2 - 0.9x - 4.0 = 0$ in the interval $[-2, -1]$

**15.** The positive root of $\sin x = x^2$

**16.** The positive root of $3 \sin x = x$

**17–22** Use Newton's method to find all roots of the equation cor-rect to six decimal places.

**17.** $3 \cos x = x + 1$              **18.** $\sqrt{x + 1} = x^2 - x$

**19.** $\sqrt[3]{x} = x^2 - 1$              **20.** $\frac{1}{x} = 1 + x^3$

**21.** $\cos x = \sqrt{x}$                   **22.** $\sin x = x^2 - 2$

**23–26** Use Newton's method to find all the roots of the equation correct to eight decimal places. Start by drawing a graph to find initial approximations.

**23.** $x^6 - x^5 - 6x^4 - x^2 + x + 10 = 0$

**24.** $x^5 - 3x^4 + x^3 - x^2 - x + 6 = 0$

**25.** $\frac{x}{x^2 + 1} = \sqrt{1 - x}$              **26.** $\cos(x^2 - x) = x^4$

**27.** (a) Apply Newton's method to the equation $x^2 - a = 0$ to derive the following square-root algorithm (used by the ancient Babylonians to compute $\sqrt{a}$):

$$x_{n+1} = \frac{1}{2}\left(x_n + \frac{a}{x_n}\right)$$

(b) Use part (a) to compute $\sqrt{1000}$ correct to six decimal places.

**28.** (a) Apply Newton's method to the equation $1/x - a = 0$ to derive the following reciprocal algorithm:

$$x_{n+1} = 2x_n - ax_n^2$$

(This algorithm enables a computer to find reciprocals without actually dividing.)

(b) Use part (a) to compute $1/1.6984$ correct to six decimal places.

**29.** Explain why Newton's method doesn't work for finding the root of the equation $x^3 - 3x + 6 = 0$ if the initial approximation is chosen to be $x_1 = 1$.

**30.** (a) Use Newton's method with $x_1 = 1$ to find the root of the equation $x^3 - x = 1$ correct to six decimal places.
(b) Solve the equation in part (a) using $x_1 = 0.6$ as the initial approximation.
(c) Solve the equation in part (a) using $x_1 = 0.57$. (You definitely need a programmable calculator for this part.)
(d) Graph $f(x) = x^3 - x - 1$ and its tangent lines at $x_1 = 1$, 0.6, and 0.57 to explain why Newton's method is so sensitive to the value of the initial approximation.

**31.** Explain why Newton's method fails when applied to the equation $\sqrt[3]{x} = 0$ with any initial approximation $x_1 \neq 0$. Illustrate your explanation with a sketch.

**32.** If

$$f(x) = \begin{cases} \sqrt{x} & \text{if } x \geq 0 \\ -\sqrt{-x} & \text{if } x < 0 \end{cases}$$

then the root of the equation $f(x) = 0$ is $x = 0$. Explain why Newton's method fails to find the root no matter which initial approximation $x_1 \neq 0$ is used. Illustrate your explanation with a sketch.

**33.** (a) Use Newton's method to find the critical numbers of the function $f(x) = x^6 - x^4 + 3x^3 - 2x$ correct to six decimal places.
(b) Find the absolute minimum value of $f$ correct to four decimal places.

**34.** Use Newton's method to find the absolute maximum value of the function $f(x) = x \cos x$, $0 \leq x \leq \pi$, correct to six decimal places.

**35.** Use Newton's method to find the coordinates of the inflection point of the curve $y = x^2 \sin x$, $0 \leq x \leq \pi$, correct to six decimal places.

**36.** Of the infinitely many lines that are tangent to the curve $y = -\sin x$ and pass through the origin, there is one that has the largest slope. Use Newton's method to find the slope of that line correct to six decimal places.

**37.** Use Newton's method to find the coordinates, correct to six decimal places, of the point on the parabola $y = (x - 1)^2$ that is closest to the origin.

**38.** In the figure, the length of the chord $AB$ is 4 cm and the length of the arc $AB$ is 5 cm. Find the central angle $\theta$, in radians, correct to four decimal places. Then give the answer to the nearest degree.

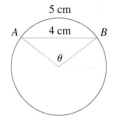

**39.** A car dealer sells a new car for $18,000. He also offers to sell the same car for payments of $375 per month for five years. What monthly interest rate is this dealer charging?

To solve this problem you will need to use the formula for the present value $A$ of an annuity consisting of $n$ equal payments of size $R$ with interest rate $i$ per time period:

$$A = \frac{R}{i}[1 - (1 + i)^{-n}]$$

Replacing $i$ by $x$, show that

$$48x(1 + x)^{60} - (1 + x)^{60} + 1 = 0$$

Use Newton's method to solve this equation.

**40.** The figure shows the sun located at the origin and the earth at the point $(1, 0)$. (The unit here is the distance between the centers of the earth and the sun, called an *astronomical unit*: 1 AU $\approx 1.496 \times 10^8$ km.) There are five locations $L_1$, $L_2$, $L_3$, $L_4$, and $L_5$ in this plane of rotation of the earth about the sun where a satellite remains motionless with respect to the earth because the forces acting on the satellite (including the gravitational attractions of the earth and the sun) balance each other. These locations are called *libration points*. (A solar research satellite has been placed at one of these libration points.) If $m_1$ is the mass of the sun, $m_2$ is the mass of the earth, and $r = m_2/(m_1 + m_2)$, it turns out that the $x$-coordinate of $L_1$ is the unique root of the fifth-degree equation

$$p(x) = x^5 - (2 + r)x^4 + (1 + 2r)x^3 - (1 - r)x^2$$
$$+ 2(1 - r)x + r - 1 = 0$$

and the $x$-coordinate of $L_2$ is the root of the equation

$$p(x) - 2rx^2 = 0$$

Using the value $r \approx 3.04042 \times 10^{-6}$, find the locations of the libration points (a) $L_1$ and (b) $L_2$.

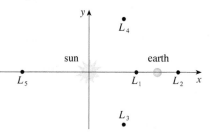

## 3.9 Antiderivatives

A physicist who knows the velocity of a particle might wish to know its position at a given time. An engineer who can measure the variable rate at which water is leaking from a tank wants to know the amount leaked over a certain time period. A biologist who knows the rate at which a bacteria population is increasing might want to deduce what the size of the population will be at some future time. In each case, the problem is to find a function $F$ whose derivative is a known function $f$. If such a function $F$ exists, it is called an *antiderivative* of $f$.

> **Definition** A function $F$ is called an **antiderivative** of $f$ on an interval $I$ if $F'(x) = f(x)$ for all $x$ in $I$.

For instance, let $f(x) = x^2$. It isn't difficult to discover an antiderivative of $f$ if we keep the Power Rule in mind. In fact, if $F(x) = \frac{1}{3}x^3$, then $F'(x) = x^2 = f(x)$. But the function $G(x) = \frac{1}{3}x^3 + 100$ also satisfies $G'(x) = x^2$. Therefore both $F$ and $G$ are antiderivatives of $f$. Indeed, any function of the form $H(x) = \frac{1}{3}x^3 + C$, where $C$ is a constant, is an antiderivative of $f$. The question arises: Are there any others?

To answer this question, recall that in Section 3.2 we used the Mean Value Theorem to prove that if two functions have identical derivatives on an interval, then they must differ by a constant (Corollary 3.2.7). Thus if $F$ and $G$ are any two antiderivatives of $f$, then

$$F'(x) = f(x) = G'(x)$$

so $G(x) - F(x) = C$, where $C$ is a constant. We can write this as $G(x) = F(x) + C$, so we have the following result.

> **1 Theorem** If $F$ is an antiderivative of $f$ on an interval $I$, then the most general antiderivative of $f$ on $I$ is
> $$F(x) + C$$
> where $C$ is an arbitrary constant.

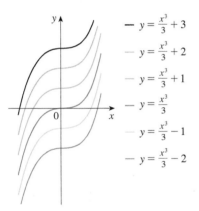

$$— \; y = \frac{x^3}{3} + 3$$
$$— \; y = \frac{x^3}{3} + 2$$
$$— \; y = \frac{x^3}{3} + 1$$
$$— \; y = \frac{x^3}{3}$$
$$— \; y = \frac{x^3}{3} - 1$$
$$— \; y = \frac{x^3}{3} - 2$$

**FIGURE 1**

Members of the family of antiderivatives of $f(x) = x^2$

Going back to the function $f(x) = x^2$, we see that the general antiderivative of $f$ is $\frac{1}{3}x^3 + C$. By assigning specific values to the constant $C$, we obtain a family of functions whose graphs are vertical translates of one another (see Figure 1). This makes sense because each curve must have the same slope at any given value of $x$.

**EXAMPLE 1** Find the most general antiderivative of each of the following functions.
(a) $f(x) = \sin x$      (b) $f(x) = x^n$,   $n \geq 0$      (c) $f(x) = x^{-3}$

SOLUTION

(a) If $F(x) = -\cos x$, then $F'(x) = \sin x$, so an antiderivative of $\sin x$ is $-\cos x$. By Theorem 1, the most general antiderivative is $G(x) = -\cos x + C$.

(b) We use the Power Rule to discover an antiderivative of $x^n$:

$$\frac{d}{dx}\left(\frac{x^{n+1}}{n+1}\right) = \frac{(n+1)x^n}{n+1} = x^n$$

Thus the general antiderivative of $f(x) = x^n$ is

$$F(x) = \frac{x^{n+1}}{n+1} + C$$

This is valid for $n \geq 0$ because then $f(x) = x^n$ is defined on an interval.

(c) If we put $n = -3$ in part (b) we get the particular antiderivative $F(x) = x^{-2}/(-2)$ by the same calculation. But notice that $f(x) = x^{-3}$ is not defined at $x = 0$. Thus Theorem 1 tells us only that the general antiderivative of $f$ is $x^{-2}/(-2) + C$ on any interval that does not contain 0. So the general antiderivative of $f(x) = 1/x^3$ is

$$F(x) = \begin{cases} -\dfrac{1}{2x^2} + C_1 & \text{if } x > 0 \\ -\dfrac{1}{2x^2} + C_2 & \text{if } x < 0 \end{cases}$$

As in Example 1, every differentiation formula, when read from right to left, gives rise to an antidifferentiation formula. In Table 2 we list some particular antiderivatives. Each formula in the table is true because the derivative of the function in the right column appears in the left column. In particular, the first formula says that the antiderivative of a constant times a function is the constant times the antiderivative of the function. The second formula says that the antiderivative of a sum is the sum of the antiderivatives. (We use the notation $F' = f$, $G' = g$.)

**2** Table of Antidifferentiation Formulas

To obtain the most general antiderivative from the particular ones in Table 2, we have to add a constant (or constants), as in Example 1.

| Function | Particular antiderivative | Function | Particular antiderivative |
|---|---|---|---|
| $cf(x)$ | $cF(x)$ | $\cos x$ | $\sin x$ |
| $f(x) + g(x)$ | $F(x) + G(x)$ | $\sin x$ | $-\cos x$ |
| $x^n \ (n \neq -1)$ | $\dfrac{x^{n+1}}{n+1}$ | $\sec^2 x$ | $\tan x$ |
| | | $\sec x \tan x$ | $\sec x$ |

**EXAMPLE 2** Find all functions $g$ such that

$$g'(x) = 4 \sin x + \frac{2x^5 - \sqrt{x}}{x}$$

SOLUTION  We first rewrite the given function as follows:

$$g'(x) = 4 \sin x + \frac{2x^5}{x} - \frac{\sqrt{x}}{x} = 4 \sin x + 2x^4 - \frac{1}{\sqrt{x}}$$

Thus we want to find an antiderivative of

$$g'(x) = 4 \sin x + 2x^4 - x^{-1/2}$$

Using the formulas in Table 2 together with Theorem 1, we obtain

$$g(x) = 4(-\cos x) + 2\frac{x^5}{5} - \frac{x^{1/2}}{\frac{1}{2}} + C$$

$$= -4 \cos x + \tfrac{2}{5} x^5 - 2\sqrt{x} + C$$

In applications of calculus it is very common to have a situation as in Example 2, where it is required to find a function, given knowledge about its derivatives. An equation that involves the derivatives of a function is called a **differential equation**. Such equations will be studied in some detail in Chapter 9, but for the present we can solve some elementary differential equations. The general solution of a differential equation involves an arbitrary constant (or constants) as in Example 2. However, there may be some extra conditions given that will determine the constants and therefore uniquely specify the solution.

**EXAMPLE 3**  Find $f$ if $f'(x) = x\sqrt{x}$ and $f(1) = 2$.

SOLUTION  The general antiderivative of

$$f'(x) = x^{3/2}$$

is

$$f(x) = \frac{x^{5/2}}{\frac{5}{2}} + C = \tfrac{2}{5}x^{5/2} + C$$

To determine $C$ we use the fact that $f(1) = 2$:

$$f(1) = \tfrac{2}{5} + C = 2$$

Solving for $C$, we get $C = 2 - \tfrac{2}{5} = \tfrac{8}{5}$, so the particular solution is

$$f(x) = \frac{2x^{5/2} + 8}{5}$$

**EXAMPLE 4**  Find $f$ if $f''(x) = 12x^2 + 6x - 4$, $f(0) = 4$, and $f(1) = 1$.

SOLUTION  The general antiderivative of $f''(x) = 12x^2 + 6x - 4$ is

$$f'(x) = 12\frac{x^3}{3} + 6\frac{x^2}{2} - 4x + C = 4x^3 + 3x^2 - 4x + C$$

Using the antidifferentiation rules once more, we find that

$$f(x) = 4\frac{x^4}{4} + 3\frac{x^3}{3} - 4\frac{x^2}{2} + Cx + D = x^4 + x^3 - 2x^2 + Cx + D$$

To determine $C$ and $D$ we use the given conditions that $f(0) = 4$ and $f(1) = 1$. Since $f(0) = 0 + D = 4$, we have $D = 4$. Since

$$f(1) = 1 + 1 - 2 + C + 4 = 1$$

we have $C = -3$. Therefore the required function is

$$f(x) = x^4 + x^3 - 2x^2 - 3x + 4$$

If we are given the graph of a function $f$, it seems reasonable that we should be able to sketch the graph of an antiderivative $F$. Suppose, for instance, that we are given that $F(0) = 1$. Then we have a place to start, the point $(0, 1)$, and the direction in which we move our pencil is given at each stage by the derivative $F'(x) = f(x)$. In the next example we use the principles of this chapter to show how to graph $F$ even when we don't have a formula for $f$. This would be the case, for instance, when $f(x)$ is determined by experimental data.

**EXAMPLE 5**  The graph of a function $f$ is given in Figure 2. Make a rough sketch of an antiderivative $F$, given that $F(0) = 2$.

SOLUTION  We are guided by the fact that the slope of $y = F(x)$ is $f(x)$. We start at the point $(0, 2)$ and draw $F$ as an initially decreasing function since $f(x)$ is negative when $0 < x < 1$. Notice that $f(1) = f(3) = 0$, so $F$ has horizontal tangents when $x = 1$ and $x = 3$. For $1 < x < 3$, $f(x)$ is positive and so $F$ is increasing. We see that $F$ has a local minimum when $x = 1$ and a local maximum when $x = 3$. For $x > 3$, $f(x)$ is negative and so $F$ is decreasing on $(3, \infty)$. Since $f(x) \to 0$ as $x \to \infty$, the graph of $F$ becomes flat-

**FIGURE 2**

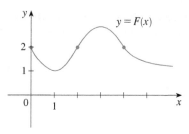

**FIGURE 3**

ter as $x \to \infty$. Also notice that $F''(x) = f'(x)$ changes from positive to negative at $x = 2$ and from negative to positive at $x = 4$, so $F$ has inflection points when $x = 2$ and $x = 4$. We use this information to sketch the graph of the antiderivative in Figure 3.

## Rectilinear Motion

Antidifferentiation is particularly useful in analyzing the motion of an object moving in a straight line. Recall that if the object has position function $s = f(t)$, then the velocity function is $v(t) = s'(t)$. This means that the position function is an antiderivative of the velocity function. Likewise, the acceleration function is $a(t) = v'(t)$, so the velocity function is an antiderivative of the acceleration. If the acceleration and the initial values $s(0)$ and $v(0)$ are known, then the position function can be found by antidifferentiating twice.

**V  EXAMPLE 6**  A particle moves in a straight line and has acceleration given by $a(t) = 6t + 4$. Its initial velocity is $v(0) = -6$ cm/s and its initial displacement is $s(0) = 9$ cm. Find its position function $s(t)$.

SOLUTION  Since $v'(t) = a(t) = 6t + 4$, antidifferentiation gives

$$v(t) = 6\frac{t^2}{2} + 4t + C = 3t^2 + 4t + C$$

Note that $v(0) = C$. But we are given that $v(0) = -6$, so $C = -6$ and

$$v(t) = 3t^2 + 4t - 6$$

Since $v(t) = s'(t)$, $s$ is the antiderivative of $v$:

$$s(t) = 3\frac{t^3}{3} + 4\frac{t^2}{2} - 6t + D = t^3 + 2t^2 - 6t + D$$

This gives $s(0) = D$. We are given that $s(0) = 9$, so $D = 9$ and the required position function is

$$s(t) = t^3 + 2t^2 - 6t + 9$$

An object near the surface of the earth is subject to a gravitational force that produces a downward acceleration denoted by $g$. For motion close to the ground we may assume that $g$ is constant, its value being about 9.8 m/s$^2$ (or 32 ft/s$^2$).

**EXAMPLE 7**  A ball is thrown upward with a speed of 48 ft/s from the edge of a cliff 432 ft above the ground. Find its height above the ground $t$ seconds later. When does it reach its maximum height? When does it hit the ground?

SOLUTION  The motion is vertical and we choose the positive direction to be upward. At time $t$ the distance above the ground is $s(t)$ and the velocity $v(t)$ is decreasing. Therefore the acceleration must be negative and we have

$$a(t) = \frac{dv}{dt} = -32$$

Taking antiderivatives, we have

$$v(t) = -32t + C$$

To determine $C$ we use the given information that $v(0) = 48$. This gives $48 = 0 + C$, so

$$v(t) = -32t + 48$$

The maximum height is reached when $v(t) = 0$, that is, after 1.5 s. Since $s'(t) = v(t)$, we antidifferentiate again and obtain

$$s(t) = -16t^2 + 48t + D$$

Using the fact that $s(0) = 432$, we have $432 = 0 + D$ and so

$$s(t) = -16t^2 + 48t + 432$$

The expression for $s(t)$ is valid until the ball hits the ground. This happens when $s(t) = 0$, that is, when

$$-16t^2 + 48t + 432 = 0$$

or, equivalently,

$$t^2 - 3t - 27 = 0$$

Using the quadratic formula to solve this equation, we get

$$t = \frac{3 \pm 3\sqrt{13}}{2}$$

We reject the solution with the minus sign since it gives a negative value for $t$. Therefore the ball hits the ground after $3(1 + \sqrt{13})/2 \approx 6.9$ s.

Figure 4 shows the position function of the ball in Example 7. The graph corroborates the conclusions we reached: The ball reaches its maximum height after 1.5 s and hits the ground after 6.9 s.

500

0                                        8

**FIGURE 4**

## 3.9 Exercises

**1–18** Find the most general antiderivative of the function. (Check your answer by differentiation.)

**1.** $f(x) = x - 3$

**2.** $f(x) = \frac{1}{2}x^2 - 2x + 6$

**3.** $f(x) = \frac{1}{2} + \frac{3}{4}x^2 - \frac{4}{5}x^3$

**4.** $f(x) = 8x^9 - 3x^6 + 12x^3$

**5.** $f(x) = (x + 1)(2x - 1)$

**6.** $f(x) = x(2 - x)^2$

**7.** $f(x) = 7x^{2/5} + 8x^{-4/5}$

**8.** $f(x) = x^{3.4} - 2x^{\sqrt{2}-1}$

**9.** $f(x) = \sqrt{2}$

**10.** $f(x) = \pi^2$

**11.** $f(x) = \dfrac{10}{x^9}$

**12.** $g(x) = \dfrac{5 - 4x^3 + 2x^6}{x^6}$

**13.** $g(t) = \dfrac{1 + t + t^2}{\sqrt{t}}$

**14.** $f(t) = 3 \cos t - 4 \sin t$

**15.** $h(\theta) = 2 \sin \theta - \sec^2\theta$

**16.** $f(\theta) = 6\theta^2 - 7\sec^2\theta$

**17.** $f(t) = 2 \sec t \tan t + \frac{1}{2}t^{-1/2}$

**18.** $f(x) = 2\sqrt{x} + 6 \cos x$

**19–20** Find the antiderivative $F$ of $f$ that satisfies the given condition. Check your answer by comparing the graphs of $f$ and $F$.

**19.** $f(x) = 5x^4 - 2x^5$,   $F(0) = 4$

**20.** $f(x) = x + 2 \sin x$,   $F(0) = -6$

**21–40** Find $f$.

**21.** $f''(x) = 20x^3 - 12x^2 + 6x$

**22.** $f''(x) = x^6 - 4x^4 + x + 1$

**23.** $f''(x) = \frac{2}{3}x^{2/3}$

**24.** $f''(x) = 6x + \sin x$

**25.** $f'''(t) = \cos t$

**26.** $f'''(t) = t - \sqrt{t}$

**27.** $f'(x) = 1 + 3\sqrt{x}$,   $f(4) = 25$

**28.** $f'(x) = 5x^4 - 3x^2 + 4$,   $f(-1) = 2$

**29.** $f'(x) = \sqrt{x}\,(6 + 5x)$,   $f(1) = 10$

**30.** $f'(t) = t + 1/t^3$,   $t > 0$,   $f(1) = 6$

**31.** $f'(t) = 2 \cos t + \sec^2 t$,   $-\pi/2 < t < \pi/2$,   $f(\pi/3) = 4$

**32.** $f'(x) = x^{-1/3}$,   $f(1) = 1$,   $f(-1) = -1$

**33.** $f''(x) = -2 + 12x - 12x^2$,   $f(0) = 4$,   $f'(0) = 12$

**34.** $f''(x) = 8x^3 + 5$,   $f(1) = 0$,   $f'(1) = 8$

**35.** $f''(\theta) = \sin \theta + \cos \theta$,   $f(0) = 3$,   $f'(0) = 4$

**36.** $f''(t) = 3/\sqrt{t}$,   $f(4) = 20$,   $f'(4) = 7$

**37.** $f''(x) = 4 + 6x + 24x^2$,   $f(0) = 3$,   $f(1) = 10$

**38.** $f''(x) = 20x^3 + 12x^2 + 4$,   $f(0) = 8$,   $f(1) = 5$

**39.** $f''(x) = 2 + \cos x$,   $f(0) = -1$,   $f(\pi/2) = 0$

**40.** $f'''(x) = \cos x$,   $f(0) = 1$,   $f'(0) = 2$,   $f''(0) = 3$

**41.** Given that the graph of $f$ passes through the point $(1, 6)$ and that the slope of its tangent line at $(x, f(x))$ is $2x + 1$, find $f(2)$.

**42.** Find a function $f$ such that $f'(x) = x^3$ and the line $x + y = 0$ is tangent to the graph of $f$.

Graphing calculator or computer required     **1.** Homework Hints available at stewartcalculus.com

**43–44** The graph of a function $f$ is shown. Which graph is an antiderivative of $f$ and why?

**43.**

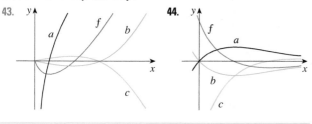

**44.**

**45.** The graph of a function is shown in the figure. Make a rough sketch of an antiderivative $F$, given that $F(0) = 1$.

**46.** The graph of the velocity function of a particle is shown in the figure. Sketch the graph of a position function.

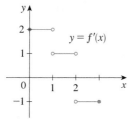

**47.** The graph of $f'$ is shown in the figure. Sketch the graph of $f$ if $f$ is continuous and $f(0) = -1$.

**48.** (a) Use a graphing device to graph $f(x) = 2x - 3\sqrt{x}$.
  (b) Starting with the graph in part (a), sketch a rough graph of the antiderivative $F$ that satisfies $F(0) = 1$.
  (c) Use the rules of this section to find an expression for $F(x)$.
  (d) Graph $F$ using the expression in part (c). Compare with your sketch in part (b).

**49–50** Draw a graph of $f$ and use it to make a rough sketch of the antiderivative that passes through the origin.

**49.** $f(x) = \dfrac{\sin x}{1 + x^2}$,   $-2\pi \leqslant x \leqslant 2\pi$

**50.** $f(x) = \sqrt{x^4 - 2x^2 + 2} - 2$,   $-3 \leqslant x \leqslant 3$

**51–56** A particle is moving with the given data. Find the position of the particle.

**51.** $v(t) = \sin t - \cos t$,   $s(0) = 0$

**52.** $v(t) = 1.5\sqrt{t}$,   $s(4) = 10$

**53.** $a(t) = 2t + 1$,   $s(0) = 3$,   $v(0) = -2$

**54.** $a(t) = 3\cos t - 2\sin t$,   $s(0) = 0$,   $v(0) = 4$

**55.** $a(t) = 10\sin t + 3\cos t$,   $s(0) = 0$,   $s(2\pi) = 12$

**56.** $a(t) = t^2 - 4t + 6$,   $s(0) = 0$,   $s(1) = 20$

**57.** A stone is dropped from the upper observation deck (the Space Deck) of the CN Tower, 450 m above the ground.
  (a) Find the distance of the stone above ground level at time $t$.
  (b) How long does it take the stone to reach the ground?
  (c) With what velocity does it strike the ground?
  (d) If the stone is thrown downward with a speed of 5 m/s, how long does it take to reach the ground?

**58.** Show that for motion in a straight line with constant acceleration $a$, initial velocity $v_0$, and initial displacement $s_0$, the displacement after time $t$ is

$$s = \tfrac{1}{2}at^2 + v_0 t + s_0$$

**59.** An object is projected upward with initial velocity $v_0$ meters per second from a point $s_0$ meters above the ground. Show that

$$[v(t)]^2 = v_0^2 - 19.6[s(t) - s_0]$$

**60.** Two balls are thrown upward from the edge of the cliff in Example 7. The first is thrown with a speed of 48 ft/s and the other is thrown a second later with a speed of 24 ft/s. Do the balls ever pass each other?

**61.** A stone was dropped off a cliff and hit the ground with a speed of 120 ft/s. What is the height of the cliff?

**62.** If a diver of mass $m$ stands at the end of a diving board with length $L$ and linear density $\rho$, then the board takes on the shape of a curve $y = f(x)$, where

$$EIy'' = mg(L - x) + \tfrac{1}{2}\rho g(L - x)^2$$

$E$ and $I$ are positive constants that depend on the material of the board and $g\ (< 0)$ is the acceleration due to gravity.
  (a) Find an expression for the shape of the curve.
  (b) Use $f(L)$ to estimate the distance below the horizontal at the end of the board.

**63.** A company estimates that the marginal cost (in dollars per item) of producing $x$ items is $1.92 - 0.002x$. If the cost of producing one item is $562, find the cost of producing 100 items.

**64.** The linear density of a rod of length 1 m is given by $\rho(x) = 1/\sqrt{x}$, in grams per centimeter, where $x$ is measured in centimeters from one end of the rod. Find the mass of the rod.

**65.** Since raindrops grow as they fall, their surface area increases and therefore the resistance to their falling increases. A raindrop has an initial downward velocity of 10 m/s and its downward acceleration is

$$a = \begin{cases} 9 - 0.9t & \text{if } 0 \le t \le 10 \\ 0 & \text{if } t > 10 \end{cases}$$

If the raindrop is initially 500 m above the ground, how long does it take to fall?

**66.** A car is traveling at 50 mi/h when the brakes are fully applied, producing a constant deceleration of 22 ft/s². What is the distance traveled before the car comes to a stop?

**67.** What constant acceleration is required to increase the speed of a car from 30 mi/h to 50 mi/h in 5 s?

**68.** A car braked with a constant deceleration of 16 ft/s², producing skid marks measuring 200 ft before coming to a stop. How fast was the car traveling when the brakes were first applied?

**69.** A car is traveling at 100 km/h when the driver sees an accident 80 m ahead and slams on the brakes. What constant deceleration is required to stop the car in time to avoid a pileup?

**70.** A model rocket is fired vertically upward from rest. Its acceleration for the first three seconds is $a(t) = 60t$, at which time the fuel is exhausted and it becomes a freely "falling" body. Fourteen seconds later, the rocket's parachute opens, and the (downward) velocity slows linearly to $-18$ ft/s in 5 s. The rocket then "floats" to the ground at that rate.
(a) Determine the position function $s$ and the velocity function $v$ (for all times $t$). Sketch the graphs of $s$ and $v$.
(b) At what time does the rocket reach its maximum height, and what is that height?
(c) At what time does the rocket land?

**71.** A high-speed bullet train accelerates and decelerates at the rate of 4 ft/s². Its maximum cruising speed is 90 mi/h.
(a) What is the maximum distance the train can travel if it accelerates from rest until it reaches its cruising speed and then runs at that speed for 15 minutes?
(b) Suppose that the train starts from rest and must come to a complete stop in 15 minutes. What is the maximum distance it can travel under these conditions?
(c) Find the minimum time that the train takes to travel between two consecutive stations that are 45 miles apart.
(d) The trip from one station to the next takes 37.5 minutes. How far apart are the stations?

# 3 Review

## Concept Check

**1.** Explain the difference between an absolute maximum and a local maximum. Illustrate with a sketch.

**2.** (a) What does the Extreme Value Theorem say?
(b) Explain how the Closed Interval Method works.

**3.** (a) State Fermat's Theorem.
(b) Define a critical number of $f$.

**4.** (a) State Rolle's Theorem.
(b) State the Mean Value Theorem and give a geometric interpretation.

**5.** (a) State the Increasing/Decreasing Test.
(b) What does it mean to say that $f$ is concave upward on an interval $I$?
(c) State the Concavity Test.
(d) What are inflection points? How do you find them?

**6.** (a) State the First Derivative Test.
(b) State the Second Derivative Test.
(c) What are the relative advantages and disadvantages of these tests?

**7.** Explain the meaning of each of the following statements.
(a) $\lim_{x\to\infty} f(x) = L$  (b) $\lim_{x\to-\infty} f(x) = L$  (c) $\lim_{x\to\infty} f(x) = \infty$
(d) The curve $y = f(x)$ has the horizontal asymptote $y = L$.

**8.** If you have a graphing calculator or computer, why do you need calculus to graph a function?

**9.** (a) Given an initial approximation $x_1$ to a root of the equation $f(x) = 0$, explain geometrically, with a diagram, how the second approximation $x_2$ in Newton's method is obtained.
(b) Write an expression for $x_2$ in terms of $x_1$, $f(x_1)$, and $f'(x_1)$.
(c) Write an expression for $x_{n+1}$ in terms of $x_n$, $f(x_n)$, and $f'(x_n)$.
(d) Under what circumstances is Newton's method likely to fail or to work very slowly?

**10.** (a) What is an antiderivative of a function $f$?
(b) Suppose $F_1$ and $F_2$ are both antiderivatives of $f$ on an interval $I$. How are $F_1$ and $F_2$ related?

## True-False Quiz

Determine whether the statement is true or false. If it is true, explain why. If it is false, explain why or give an example that disproves the statement.

**1.** If $f'(c) = 0$, then $f$ has a local maximum or minimum at $c$.

**2.** If $f$ has an absolute minimum value at $c$, then $f'(c) = 0$.

**3.** If $f$ is continuous on $(a, b)$, then $f$ attains an absolute maximum value $f(c)$ and an absolute minimum value $f(d)$ at some numbers $c$ and $d$ in $(a, b)$.

**4.** If $f$ is differentiable and $f(-1) = f(1)$, then there is a number $c$ such that $|c| < 1$ and $f'(c) = 0$.

**5.** If $f'(x) < 0$ for $1 < x < 6$, then $f$ is decreasing on $(1, 6)$.

**6.** If $f''(2) = 0$, then $(2, f(2))$ is an inflection point of the curve $y = f(x)$.

**7.** If $f'(x) = g'(x)$ for $0 < x < 1$, then $f(x) = g(x)$ for $0 < x < 1$.

**8.** There exists a function $f$ such that $f(1) = -2$, $f(3) = 0$, and $f'(x) > 1$ for all $x$.

**9.** There exists a function $f$ such that $f(x) > 0$, $f'(x) < 0$, and $f''(x) > 0$ for all $x$.

**10.** There exists a function $f$ such that $f(x) < 0$, $f'(x) < 0$, and $f''(x) > 0$ for all $x$.

**11.** If $f$ and $g$ are increasing on an interval $I$, then $f + g$ is increasing on $I$.

**12.** If $f$ and $g$ are increasing on an interval $I$, then $f - g$ is increasing on $I$.

**13.** If $f$ and $g$ are increasing on an interval $I$, then $fg$ is increasing on $I$.

**14.** If $f$ and $g$ are positive increasing functions on an interval $I$, then $fg$ is increasing on $I$.

**15.** If $f$ is increasing and $f(x) > 0$ on $I$, then $g(x) = 1/f(x)$ is decreasing on $I$.

**16.** If $f$ is even, then $f'$ is even.

**17.** If $f$ is periodic, then $f'$ is periodic.

**18.** The most general antiderivative of $f(x) = x^{-2}$ is

$$F(x) = -\frac{1}{x} + C$$

**19.** If $f'(x)$ exists and is nonzero for all $x$, then $f(1) \neq f(0)$.

## Exercises

**1–6** Find the local and absolute extreme values of the function on the given interval.

**1.** $f(x) = x^3 - 6x^2 + 9x + 1$, $\quad [2, 4]$

**2.** $f(x) = x\sqrt{1 - x}$, $\quad [-1, 1]$

**3.** $f(x) = \dfrac{3x - 4}{x^2 + 1}$, $\quad [-2, 2]$

**4.** $f(x) = \sqrt{x^2 + x + 1}$, $\quad [-2, 1]$

**5.** $f(x) = x + 2\cos x$, $\quad [-\pi, \pi]$

**6.** $f(x) = \sin x + \cos^2 x$, $\quad [0, \pi]$

**7–12** Find the limit.

**7.** $\displaystyle\lim_{x \to \infty} \frac{3x^4 + x - 5}{6x^4 - 2x^2 + 1}$

**8.** $\displaystyle\lim_{t \to \infty} \frac{t^3 - t + 2}{(2t - 1)(t^2 + t + 1)}$

**9.** $\displaystyle\lim_{x \to -\infty} \frac{\sqrt{4x^2 + 1}}{3x - 1}$

**10.** $\displaystyle\lim_{x \to -\infty} (x^2 + x^3)$

**11.** $\displaystyle\lim_{x \to \infty} \left(\sqrt{4x^2 + 3x} - 2x\right)$

**12.** $\displaystyle\lim_{x \to \infty} \frac{\sin^4 x}{\sqrt{x}}$

**13–15** Sketch the graph of a function that satisfies the given conditions.

**13.** $f(0) = 0$, $\quad f'(-2) = f'(1) = f'(9) = 0$,
$\displaystyle\lim_{x \to \infty} f(x) = 0$, $\quad \displaystyle\lim_{x \to 6} f(x) = -\infty$,
$f'(x) < 0$ on $(-\infty, -2)$, $(1, 6)$, and $(9, \infty)$,
$f'(x) > 0$ on $(-2, 1)$ and $(6, 9)$,
$f''(x) > 0$ on $(-\infty, 0)$ and $(12, \infty)$,
$f''(x) < 0$ on $(0, 6)$ and $(6, 12)$

**14.** $f(0) = 0$, $\quad f$ is continuous and even,
$f'(x) = 2x$ if $0 < x < 1$, $\quad f'(x) = -1$ if $1 < x < 3$,
$f'(x) = 1$ if $x > 3$

**15.** $f$ is odd, $\quad f'(x) < 0$ for $0 < x < 2$,
$f'(x) > 0$ for $x > 2$, $\quad f''(x) > 0$ for $0 < x < 3$,
$f''(x) < 0$ for $x > 3$, $\quad \displaystyle\lim_{x \to \infty} f(x) = -2$

**16.** The figure shows the graph of the *derivative* $f'$ of a function $f$.
(a) On what intervals is $f$ increasing or decreasing?
(b) For what values of $x$ does $f$ have a local maximum or minimum?

⊞ Graphing calculator or computer required

(c) Sketch the graph of $f''$.
(d) Sketch a possible graph of $f$.

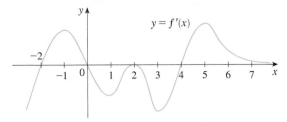

**17–28** Use the guidelines of Section 3.5 to sketch the curve.

**17.** $y = 2 - 2x - x^3$

**18.** $y = x^3 - 6x^2 - 15x + 4$

**19.** $y = x^4 - 3x^3 + 3x^2 - x$

**20.** $y = \dfrac{x}{1 - x^2}$

**21.** $y = \dfrac{1}{x(x - 3)^2}$

**22.** $y = \dfrac{1}{x^2} - \dfrac{1}{(x - 2)^2}$

**23.** $y = x^2/(x + 8)$

**24.** $y = \sqrt{1 - x} + \sqrt{1 + x}$

**25.** $y = x\sqrt{2 + x}$

**26.** $y = \sqrt[3]{x^2 + 1}$

**27.** $y = \sin^2 x - 2\cos x$

**28.** $y = 4x - \tan x, \quad -\pi/2 < x < \pi/2$

**29–32** Produce graphs of $f$ that reveal all the important aspects of the curve. Use graphs of $f'$ and $f''$ to estimate the intervals of increase and decrease, extreme values, intervals of concavity, and inflection points. In Exercise 29 use calculus to find these quantities exactly.

**29.** $f(x) = \dfrac{x^2 - 1}{x^3}$

**30.** $f(x) = \dfrac{x^3 - x}{x^2 + x + 3}$

**31.** $f(x) = 3x^6 - 5x^5 + x^4 - 5x^3 - 2x^2 + 2$

**32.** $f(x) = x^2 + 6.5 \sin x, \quad -5 \leqslant x \leqslant 5$

**33.** Show that the equation $3x + 2\cos x + 5 = 0$ has exactly one real root.

**34.** Suppose that $f$ is continuous on $[0, 4]$, $f(0) = 1$, and $2 \leqslant f'(x) \leqslant 5$ for all $x$ in $(0, 4)$. Show that $9 \leqslant f(4) \leqslant 21$.

**35.** By applying the Mean Value Theorem to the function $f(x) = x^{1/5}$ on the interval $[32, 33]$, show that

$$2 < \sqrt[5]{33} < 2.0125$$

**36.** For what values of the constants $a$ and $b$ is $(1, 3)$ a point of inflection of the curve $y = ax^3 + bx^2$?

**37.** Let $g(x) = f(x^2)$, where $f$ is twice differentiable for all $x$, $f'(x) > 0$ for all $x \neq 0$, and $f$ is concave downward on $(-\infty, 0)$ and concave upward on $(0, \infty)$.
(a) At what numbers does $g$ have an extreme value?
(b) Discuss the concavity of $g$.

**38.** Find two positive integers such that the sum of the first number and four times the second number is 1000 and the product of the numbers is as large as possible.

**39.** Show that the shortest distance from the point $(x_1, y_1)$ to the straight line $Ax + By + C = 0$ is

$$\frac{|Ax_1 + By_1 + C|}{\sqrt{A^2 + B^2}}$$

**40.** Find the point on the hyperbola $xy = 8$ that is closest to the point $(3, 0)$.

**41.** Find the smallest possible area of an isosceles triangle that is circumscribed about a circle of radius $r$.

**42.** Find the volume of the largest circular cone that can be inscribed in a sphere of radius $r$.

**43.** In $\triangle ABC$, $D$ lies on $AB$, $CD \perp AB$, $|AD| = |BD| = 4$ cm, and $|CD| = 5$ cm. Where should a point $P$ be chosen on $CD$ so that the sum $|PA| + |PB| + |PC|$ is a minimum?

**44.** Solve Exercise 43 when $|CD| = 2$ cm.

**45.** The velocity of a wave of length $L$ in deep water is

$$v = K\sqrt{\frac{L}{C} + \frac{C}{L}}$$

where $K$ and $C$ are known positive constants. What is the length of the wave that gives the minimum velocity?

**46.** A metal storage tank with volume $V$ is to be constructed in the shape of a right circular cylinder surmounted by a hemisphere. What dimensions will require the least amount of metal?

**47.** A hockey team plays in an arena with a seating capacity of 15,000 spectators. With the ticket price set at \$12, average attendance at a game has been 11,000. A market survey indicates that for each dollar the ticket price is lowered, average attendance will increase by 1000. How should the owners of the team set the ticket price to maximize their revenue from ticket sales?

**48.** A manufacturer determines that the cost of making $x$ units of a commodity is $C(x) = 1800 + 25x - 0.2x^2 + 0.001x^3$ and the demand function is $p(x) = 48.2 - 0.03x$.
(a) Graph the cost and revenue functions and use the graphs to estimate the production level for maximum profit.
(b) Use calculus to find the production level for maximum profit.
(c) Estimate the production level that minimizes the average cost.

**49.** Use Newton's method to find the root of the equation $x^5 - x^4 + 3x^2 - 3x - 2 = 0$ in the interval $[1, 2]$ correct to six decimal places.

**50.** Use Newton's method to find all roots of the equation $\sin x = x^2 - 3x + 1$ correct to six decimal places.

**51.** Use Newton's method to find the absolute maximum value of the function $f(t) = \cos t + t - t^2$ correct to eight decimal places.

**52.** Use the guidelines in Section 3.5 to sketch the curve $y = x \sin x$, $0 \le x \le 2\pi$. Use Newton's method when necessary.

**53–58** Find $f$.

**53.** $f'(x) = \sqrt{x^3} + \sqrt[3]{x^2}$

**54.** $f'(x) = 8x - 3 \sec^2 x$

**55.** $f'(t) = 2t - 3 \sin t$, $\quad f(0) = 5$

**56.** $f'(u) = \dfrac{u^2 + \sqrt{u}}{u}$, $\quad f(1) = 3$

**57.** $f''(x) = 1 - 6x + 48x^2$, $\quad f(0) = 1$, $\quad f'(0) = 2$

**58.** $f''(x) = 2x^3 + 3x^2 - 4x + 5$, $\quad f(0) = 2$, $\quad f(1) = 0$

**59–60** A particle is moving with the given data. Find the position of the particle.

**59.** $v(t) = 2t - \sin t$, $\quad s(0) = 3$

**60.** $a(t) = \sin t + 3 \cos t$, $\quad s(0) = 0$, $\quad v(0) = 2$

**61.** Use a graphing device to draw a graph of the function $f(x) = x^2 \sin(x^2)$, $0 \le x \le \pi$, and use that graph to sketch the antiderivative $F$ of $f$ that satisfies the initial condition $F(0) = 0$.

**62.** Investigate the family of curves given by
$$f(x) = x^4 + x^3 + cx^2$$
In particular you should determine the transitional value of $c$ at which the number of critical numbers changes and the transitional value at which the number of inflection points changes. Illustrate the various possible shapes with graphs.

**63.** A canister is dropped from a helicopter 500 m above the ground. Its parachute does not open, but the canister has been designed to withstand an impact velocity of 100 m/s. Will it burst?

**64.** In an automobile race along a straight road, car A passed car B twice. Prove that at some time during the race their accelerations were equal. State the assumptions that you make.

**65.** A rectangular beam will be cut from a cylindrical log of radius 10 inches.
(a) Show that the beam of maximal cross-sectional area is a square.
(b) Four rectangular planks will be cut from the four sections of the log that remain after cutting the square beam. Determine the dimensions of the planks that will have maximal cross-sectional area.
(c) Suppose that the strength of a rectangular beam is proportional to the product of its width and the square of its depth. Find the dimensions of the strongest beam that can be cut from the cylindrical log.

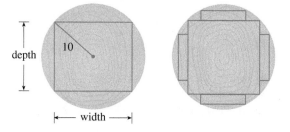

**66.** If a projectile is fired with an initial velocity $v$ at an angle of inclination $\theta$ from the horizontal, then its trajectory, neglecting air resistance, is the parabola
$$y = (\tan \theta)x - \frac{g}{2v^2 \cos^2\theta} x^2 \qquad 0 < \theta < \frac{\pi}{2}$$
(a) Suppose the projectile is fired from the base of a plane that is inclined at an angle $\alpha$, $\alpha > 0$, from the horizontal, as shown in the figure. Show that the range of the projectile, measured up the slope, is given by
$$R(\theta) = \frac{2v^2 \cos \theta \, \sin(\theta - \alpha)}{g \cos^2\alpha}$$
(b) Determine $\theta$ so that $R$ is a maximum.
(c) Suppose the plane is at an angle $\alpha$ *below* the horizontal. Determine the range $R$ in this case, and determine the angle at which the projectile should be fired to maximize $R$

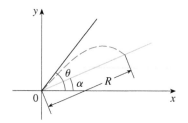

# Problems Plus

One of the most important principles of problem solving is *analogy* (see page 97). If you are having trouble getting started on a problem, it is sometimes helpful to start by solving a similar, but simpler, problem. The following example illustrates the principle. Cover up the solution and try solving it yourself first.

**EXAMPLE 1** If $x$, $y$, and $z$ are positive numbers, prove that

$$\frac{(x^2 + 1)(y^2 + 1)(z^2 + 1)}{xyz} \geq 8$$

**SOLUTION** It may be difficult to get started on this problem. (Some students have tackled it by multiplying out the numerator, but that just creates a mess.) Let's try to think of a similar, simpler problem. When several variables are involved, it's often helpful to think of an analogous problem with fewer variables. In the present case we can reduce the number of variables from three to one and prove the analogous inequality

$$\boxed{1} \qquad \frac{x^2 + 1}{x} \geq 2 \qquad \text{for } x > 0$$

In fact, if we are able to prove $\boxed{1}$, then the desired inequality follows because

$$\frac{(x^2 + 1)(y^2 + 1)(z^2 + 1)}{xyz} = \left(\frac{x^2 + 1}{x}\right)\left(\frac{y^2 + 1}{y}\right)\left(\frac{z^2 + 1}{z}\right) \geq 2 \cdot 2 \cdot 2 = 8$$

The key to proving $\boxed{1}$ is to recognize that it is a disguised version of a minimum problem. If we let

$$f(x) = \frac{x^2 + 1}{x} = x + \frac{1}{x} \qquad x > 0$$

then $f'(x) = 1 - (1/x^2)$, so $f'(x) = 0$ when $x = 1$. Also, $f'(x) < 0$ for $0 < x < 1$ and $f'(x) > 0$ for $x > 1$. Therefore the absolute minimum value of $f$ is $f(1) = 2$. This means that

$$\frac{x^2 + 1}{x} \geq 2 \qquad \text{for all positive values of } x$$

and, as previously mentioned, the given inequality follows by multiplication.

The inequality in $\boxed{1}$ could also be proved without calculus. In fact, if $x > 0$, we have

$$\frac{x^2 + 1}{x} \geq 2 \iff x^2 + 1 \geq 2x \iff x^2 - 2x + 1 \geq 0$$

$$\iff (x - 1)^2 \geq 0$$

Because the last inequality is obviously true, the first one is true too.

**PS LOOK BACK**
What have we learned from the solution to this example?

- To solve a problem involving several variables, it might help to solve a similar problem with just one variable.
- When trying to prove an inequality, it might help to think of it as a maximum or minimum problem.

279

1. Show that $|\sin x - \cos x| \leq \sqrt{2}$ for all $x$.

2. Show that $x^2 y^2 (4 - x^2)(4 - y^2) \leq 16$ for all numbers $x$ and $y$ such that $|x| \leq 2$ and $|y| \leq 2$.

3. Show that the inflection points of the curve $y = (\sin x)/x$ lie on the curve $y^2(x^4 + 4) = 4$.

4. Find the point on the parabola $y = 1 - x^2$ at which the tangent line cuts from the first quadrant the triangle with the smallest area.

5. Find the highest and lowest points on the curve $x^2 + xy + y^2 = 12$.

6. Water is flowing at a constant rate into a spherical tank. Let $V(t)$ be the volume of water in the tank and $H(t)$ be the height of the water in the tank at time $t$.
   (a) What are the meanings of $V'(t)$ and $H'(t)$? Are these derivatives positive, negative, or zero?
   (b) Is $V''(t)$ positive, negative, or zero? Explain.
   (c) Let $t_1$, $t_2$, and $t_3$ be the times when the tank is one-quarter full, half full, and three-quarters full, respectively. Are the values $H''(t_1)$, $H''(t_2)$, and $H''(t_3)$ positive, negative, or zero? Why?

7. Find the absolute maximum value of the function
$$f(x) = \frac{1}{1 + |x|} + \frac{1}{1 + |x - 2|}$$

8. Find a function $f$ such that $f'(-1) = \frac{1}{2}$, $f'(0) = 0$, and $f''(x) > 0$ for all $x$, or prove that such a function cannot exist.

9. The line $y = mx + b$ intersects the parabola $y = x^2$ in points $A$ and $B$. (See the figure.) Find the point $P$ on the arc $AOB$ of the parabola that maximizes the area of the triangle $PAB$.

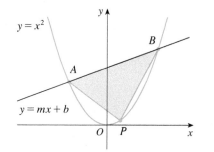

10. Sketch the graph of a function $f$ such that $f'(x) < 0$ for all $x$, $f''(x) > 0$ for $|x| > 1$, $f''(x) < 0$ for $|x| < 1$, and $\lim_{x \to \pm\infty} [f(x) + x] = 0$.

11. Determine the values of the number $a$ for which the function $f$ has no critical number:
$$f(x) = (a^2 + a - 6) \cos 2x + (a - 2)x + \cos 1$$

12. Sketch the region in the plane consisting of all points $(x, y)$ such that
$$2xy \leq |x - y| \leq x^2 + y^2$$

13. Let $ABC$ be a triangle with $\angle BAC = 120°$ and $|AB| \cdot |AC| = 1$.
    (a) Express the length of the angle bisector $AD$ in terms of $x = |AB|$.
    (b) Find the largest possible value of $|AD|$.

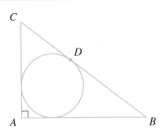

**FIGURE FOR PROBLEM 14**

**14.** (a) Let $ABC$ be a triangle with right angle $A$ and hypotenuse $a = |BC|$. (See the figure.) If the inscribed circle touches the hypotenuse at $D$, show that

$$|CD| = \tfrac{1}{2}(|BC| + |AC| - |AB|)$$

(b) If $\theta = \tfrac{1}{2}\angle C$, express the radius $r$ of the inscribed circle in terms of $a$ and $\theta$.

(c) If $a$ is fixed and $\theta$ varies, find the maximum value of $r$.

**15.** A triangle with sides $a$, $b$, and $c$ varies with time $t$, but its area never changes. Let $\theta$ be the angle opposite the side of length $a$ and suppose $\theta$ always remains acute.

(a) Express $d\theta/dt$ in terms of $b$, $c$, $\theta$, $db/dt$, and $dc/dt$.

(b) Express $da/dt$ in terms of the quantities in part (a).

**16.** $ABCD$ is a square piece of paper with sides of length 1 m. A quarter-circle is drawn from $B$ to $D$ with center $A$. The piece of paper is folded along $EF$, with $E$ on $AB$ and $F$ on $AD$, so that $A$ falls on the quarter-circle. Determine the maximum and minimum areas that the triangle $AEF$ can have.

**17.** The speeds of sound $c_1$ in an upper layer and $c_2$ in a lower layer of rock and the thickness $h$ of the upper layer can be determined by seismic exploration if the speed of sound in the lower layer is greater than the speed in the upper layer. A dynamite charge is detonated at a point $P$ and the transmitted signals are recorded at a point $Q$, which is a distance $D$ from $P$. The first signal to arrive at $Q$ travels along the surface and takes $T_1$ seconds. The next signal travels from $P$ to a point $R$, from $R$ to $S$ in the lower layer, and then to $Q$, taking $T_2$ seconds. The third signal is reflected off the lower layer at the midpoint $O$ of $RS$ and takes $T_3$ seconds to reach $Q$.

(a) Express $T_1$, $T_2$, and $T_3$ in terms of $D$, $h$, $c_1$, $c_2$, and $\theta$.

(b) Show that $T_2$ is a minimum when $\sin\theta = c_1/c_2$.

(c) Suppose that $D = 1$ km, $T_1 = 0.26$ s, $T_2 = 0.32$ s, and $T_3 = 0.34$ s. Find $c_1$, $c_2$, and $h$.

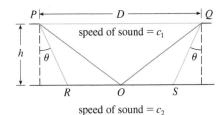

*Note:* Geophysicists use this technique when studying the structure of the earth's crust, whether searching for oil or examining fault lines.

**18.** For what values of $c$ is there a straight line that intersects the curve

$$y = x^4 + cx^3 + 12x^2 - 5x + 2$$

in four distinct points?

**19.** One of the problems posed by the Marquis de l'Hospital in his calculus textbook *Analyse des Infiniment Petits* concerns a pulley that is attached to the ceiling of a room at a point $C$ by a rope of length $r$. At another point $B$ on the ceiling, at a distance $d$ from $C$ (where $d > r$), a rope of length $\ell$ is attached and passed through the pulley at $F$ and connected to a weight $W$. The weight is released and comes to rest at its equilibrium position $D$. As l'Hospital argued, this happens when the distance $|ED|$ is maximized. Show that when the system reaches equilibrium, the value of $x$ is

$$\frac{r}{4d}\left(r + \sqrt{r^2 + 8d^2}\right)$$

Notice that this expression is independent of both $W$ and $\ell$.

**FIGURE FOR PROBLEM 19**

20. Given a sphere with radius $r$, find the height of a pyramid of minimum volume whose base is a square and whose base and triangular faces are all tangent to the sphere. What if the base of the pyramid is a regular $n$-gon? (A regular $n$-gon is a polygon with $n$ equal sides and angles.) (Use the fact that the volume of a pyramid is $\frac{1}{3}Ah$, where $A$ is the area of the base.)

21. Assume that a snowball melts so that its volume decreases at a rate proportional to its surface area. If it takes three hours for the snowball to decrease to half its original volume, how much longer will it take for the snowball to melt completely?

22. A hemispherical bubble is placed on a spherical bubble of radius 1. A smaller hemispherical bubble is then placed on the first one. This process is continued until $n$ chambers, including the sphere, are formed. (The figure shows the case $n = 4$.) Use mathematical induction to prove that the maximum height of any bubble tower with $n$ chambers is $1 + \sqrt{n}$.

# 4 Integrals

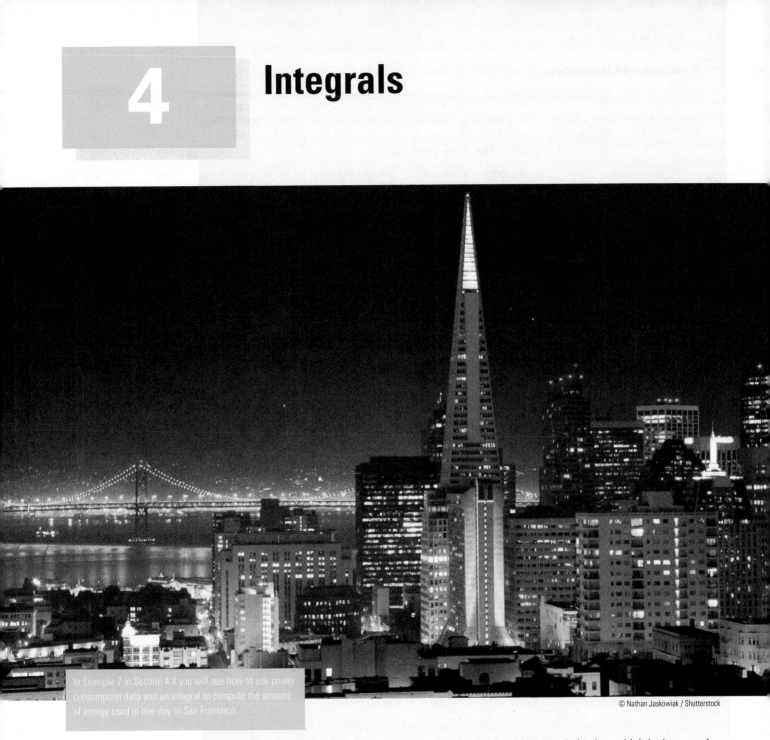

In Example 7 in Section 4.4 you will see how to use power consumption data and an integral to compute the amount of energy used in one day in San Francisco.

In Chapter 2 we used the tangent and velocity problems to introduce the derivative, which is the central idea in differential calculus. In much the same way, this chapter starts with the area and distance problems and uses them to formulate the idea of a definite integral, which is the basic concept of integral calculus. We will see in Chapters 5 and 8 how to use the integral to solve problems concerning volumes, lengths of curves, population predictions, cardiac output, forces on a dam, work, consumer surplus, and baseball, among many others.

There is a connection between integral calculus and differential calculus. The Fundamental Theorem of Calculus relates the integral to the derivative, and we will see in this chapter that it greatly simplifies the solution of many problems.

## 4.1   Areas and Distances

Now is a good time to read (or reread) *A Preview of Calculus* (see page 1). It discusses the unifying ideas of calculus and helps put in perspective where we have been and where we are going.

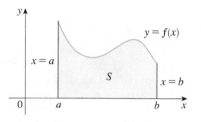

**FIGURE 1**
$S = \{(x, y) \mid a \le x \le b, 0 \le y \le f(x)\}$

In this section we discover that in trying to find the area under a curve or the distance traveled by a car, we end up with the same special type of limit.

### The Area Problem

We begin by attempting to solve the *area problem:* Find the area of the region $S$ that lies under the curve $y = f(x)$ from $a$ to $b$. This means that $S$, illustrated in Figure 1, is bounded by the graph of a continuous function $f$ [where $f(x) \ge 0$], the vertical lines $x = a$ and $x = b$, and the $x$-axis.

In trying to solve the area problem we have to ask ourselves: What is the meaning of the word *area*? This question is easy to answer for regions with straight sides. For a rectangle, the area is defined as the product of the length and the width. The area of a triangle is half the base times the height. The area of a polygon is found by dividing it into triangles (as in Figure 2) and adding the areas of the triangles.

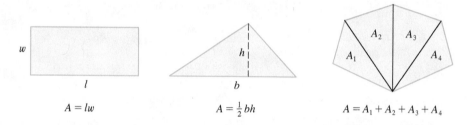

**FIGURE 2**          $A = lw$                $A = \frac{1}{2}bh$              $A = A_1 + A_2 + A_3 + A_4$

However, it isn't so easy to find the area of a region with curved sides. We all have an intuitive idea of what the area of a region is. But part of the area problem is to make this intuitive idea precise by giving an exact definition of area.

Recall that in defining a tangent we first approximated the slope of the tangent line by slopes of secant lines and then we took the limit of these approximations. We pursue a similar idea for areas. We first approximate the region $S$ by rectangles and then we take the limit of the areas of these rectangles as we increase the number of rectangles. The following example illustrates the procedure.

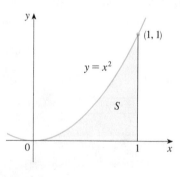

**FIGURE 3**

**V  EXAMPLE 1**   Use rectangles to estimate the area under the parabola $y = x^2$ from 0 to 1 (the parabolic region $S$ illustrated in Figure 3).

SOLUTION   We first notice that the area of $S$ must be somewhere between 0 and 1 because $S$ is contained in a square with side length 1, but we can certainly do better than that. Suppose we divide $S$ into four strips $S_1$, $S_2$, $S_3$, and $S_4$ by drawing the vertical lines $x = \frac{1}{4}$, $x = \frac{1}{2}$, and $x = \frac{3}{4}$ as in Figure 4(a).

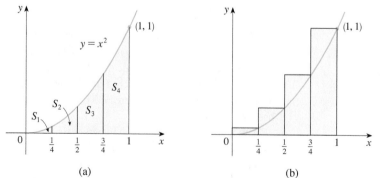

**FIGURE 4**                        (a)                                              (b)

We can approximate each strip by a rectangle that has the same base as the strip and whose height is the same as the right edge of the strip [see Figure 4(b)]. In other words, the heights of these rectangles are the values of the function $f(x) = x^2$ at the *right* endpoints of the subintervals $\left[0, \frac{1}{4}\right]$, $\left[\frac{1}{4}, \frac{1}{2}\right]$, $\left[\frac{1}{2}, \frac{3}{4}\right]$, and $\left[\frac{3}{4}, 1\right]$.

Each rectangle has width $\frac{1}{4}$ and the heights are $\left(\frac{1}{4}\right)^2$, $\left(\frac{1}{2}\right)^2$, $\left(\frac{3}{4}\right)^2$, and $1^2$. If we let $R_4$ be the sum of the areas of these approximating rectangles, we get

$$R_4 = \frac{1}{4} \cdot \left(\frac{1}{4}\right)^2 + \frac{1}{4} \cdot \left(\frac{1}{2}\right)^2 + \frac{1}{4} \cdot \left(\frac{3}{4}\right)^2 + \frac{1}{4} \cdot 1^2 = \frac{15}{32} = 0.46875$$

From Figure 4(b) we see that the area $A$ of $S$ is less than $R_4$, so

$$A < 0.46875$$

Instead of using the rectangles in Figure 4(b) we could use the smaller rectangles in Figure 5 whose heights are the values of $f$ at the *left* endpoints of the subintervals. (The leftmost rectangle has collapsed because its height is 0.) The sum of the areas of these approximating rectangles is

$$L_4 = \frac{1}{4} \cdot 0^2 + \frac{1}{4} \cdot \left(\frac{1}{4}\right)^2 + \frac{1}{4} \cdot \left(\frac{1}{2}\right)^2 + \frac{1}{4} \cdot \left(\frac{3}{4}\right)^2 = \frac{7}{32} = 0.21875$$

We see that the area of $S$ is larger than $L_4$, so we have lower and upper estimates for $A$:

$$0.21875 < A < 0.46875$$

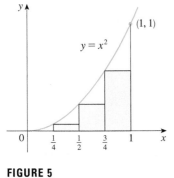

**FIGURE 5**

We can repeat this procedure with a larger number of strips. Figure 6 shows what happens when we divide the region $S$ into eight strips of equal width.

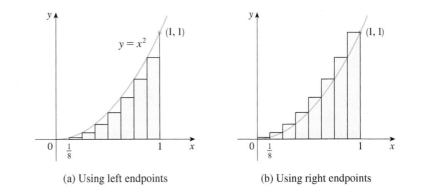

**FIGURE 6**
Approximating $S$ with eight rectangles

(a) Using left endpoints

(b) Using right endpoints

By computing the sum of the areas of the smaller rectangles ($L_8$) and the sum of the areas of the larger rectangles ($R_8$), we obtain better lower and upper estimates for $A$:

$$0.2734375 < A < 0.3984375$$

So one possible answer to the question is to say that the true area of $S$ lies somewhere between 0.2734375 and 0.3984375.

We could obtain better estimates by increasing the number of strips. The table at the left shows the results of similar calculations (with a computer) using $n$ rectangles whose heights are found with left endpoints ($L_n$) or right endpoints ($R_n$). In particular, we see by using 50 strips that the area lies between 0.3234 and 0.3434. With 1000 strips we narrow it down even more: $A$ lies between 0.3328335 and 0.3338335. A good estimate is obtained by averaging these numbers: $A \approx 0.3333335$.

| $n$ | $L_n$ | $R_n$ |
|---|---|---|
| 10 | 0.2850000 | 0.3850000 |
| 20 | 0.3087500 | 0.3587500 |
| 30 | 0.3168519 | 0.3501852 |
| 50 | 0.3234000 | 0.3434000 |
| 100 | 0.3283500 | 0.3383500 |
| 1000 | 0.3328335 | 0.3338335 |

From the values in the table in Example 1, it looks as if $R_n$ is approaching $\frac{1}{3}$ as $n$ increases. We confirm this in the next example.

**V  EXAMPLE 2**  For the region $S$ in Example 1, show that the sum of the areas of the upper approximating rectangles approaches $\frac{1}{3}$, that is,

$$\lim_{n \to \infty} R_n = \tfrac{1}{3}$$

SOLUTION  $R_n$ is the sum of the areas of the $n$ rectangles in Figure 7. Each rectangle has width $1/n$ and the heights are the values of the function $f(x) = x^2$ at the points $1/n, 2/n, 3/n, \ldots, n/n$; that is, the heights are $(1/n)^2, (2/n)^2, (3/n)^2, \ldots, (n/n)^2$. Thus

$$R_n = \frac{1}{n}\left(\frac{1}{n}\right)^2 + \frac{1}{n}\left(\frac{2}{n}\right)^2 + \frac{1}{n}\left(\frac{3}{n}\right)^2 + \cdots + \frac{1}{n}\left(\frac{n}{n}\right)^2$$

$$= \frac{1}{n} \cdot \frac{1}{n^2}(1^2 + 2^2 + 3^2 + \cdots + n^2)$$

$$= \frac{1}{n^3}(1^2 + 2^2 + 3^2 + \cdots + n^2)$$

**FIGURE 7**

Here we need the formula for the sum of the squares of the first $n$ positive integers:

$$\boxed{1} \qquad 1^2 + 2^2 + 3^2 + \cdots + n^2 = \frac{n(n + 1)(2n + 1)}{6}$$

Perhaps you have seen this formula before. It is proved in Example 5 in Appendix E.
Putting Formula 1 into our expression for $R_n$, we get

$$R_n = \frac{1}{n^3} \cdot \frac{n(n + 1)(2n + 1)}{6} = \frac{(n + 1)(2n + 1)}{6n^2}$$

Here we are computing the limit of the sequence $\{R_n\}$. Sequences and their limits were discussed in *A Preview of Calculus* and will be studied in detail in Section 11.1. The idea is very similar to a limit at infinity (Section 3.4) except that in writing $\lim_{n\to\infty}$ we restrict $n$ to be a positive integer. In particular, we know that

$$\lim_{n \to \infty} \frac{1}{n} = 0$$

When we write $\lim_{n\to\infty} R_n = \frac{1}{3}$ we mean that we can make $R_n$ as close to $\frac{1}{3}$ as we like by taking $n$ sufficiently large.

Thus we have

$$\lim_{n \to \infty} R_n = \lim_{n \to \infty} \frac{(n + 1)(2n + 1)}{6n^2}$$

$$= \lim_{n \to \infty} \frac{1}{6}\left(\frac{n + 1}{n}\right)\left(\frac{2n + 1}{n}\right)$$

$$= \lim_{n \to \infty} \frac{1}{6}\left(1 + \frac{1}{n}\right)\left(2 + \frac{1}{n}\right)$$

$$= \frac{1}{6} \cdot 1 \cdot 2 = \frac{1}{3}$$

It can be shown that the lower approximating sums also approach $\frac{1}{3}$, that is,

$$\lim_{n \to \infty} L_n = \tfrac{1}{3}$$

From Figures 8 and 9 it appears that, as $n$ increases, both $L_n$ and $R_n$ become better and better approximations to the area of $S$. Therefore we *define* the area $A$ to be the limit of the sums of the areas of the approximating rectangles, that is,

$$A = \lim_{n \to \infty} R_n = \lim_{n \to \infty} L_n = \tfrac{1}{3}$$

**TEC** In Visual 4.1 you can create pictures like those in Figures 8 and 9 for other values of $n$.

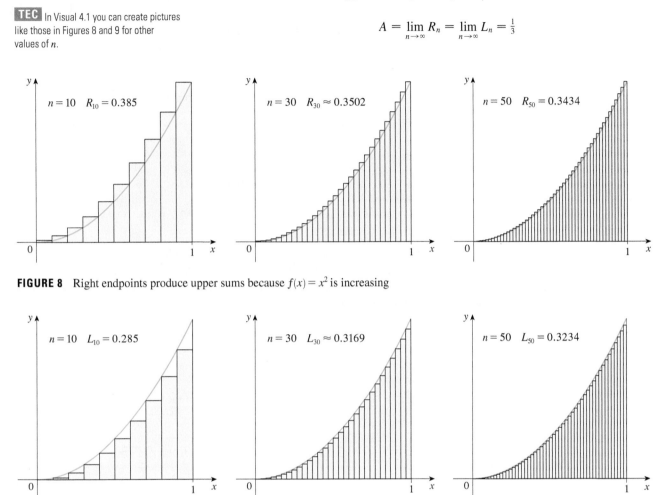

**FIGURE 8** Right endpoints produce upper sums because $f(x) = x^2$ is increasing

**FIGURE 9** Left endpoints produce lower sums because $f(x) = x^2$ is increasing

Let's apply the idea of Examples 1 and 2 to the more general region $S$ of Figure 1. We start by subdividing $S$ into $n$ strips $S_1, S_2, \ldots, S_n$ of equal width as in Figure 10.

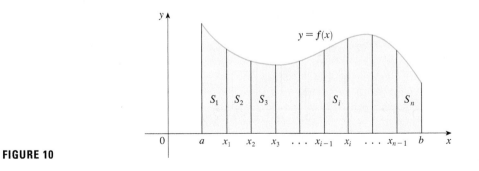

**FIGURE 10**

The width of the interval $[a, b]$ is $b - a$, so the width of each of the $n$ strips is

$$\Delta x = \frac{b - a}{n}$$

These strips divide the interval $[a, b]$ into $n$ subintervals

$$[x_0, x_1], \quad [x_1, x_2], \quad [x_2, x_3], \quad \ldots, \quad [x_{n-1}, x_n]$$

where $x_0 = a$ and $x_n = b$. The right endpoints of the subintervals are

$$x_1 = a + \Delta x,$$

$$x_2 = a + 2\,\Delta x,$$

$$x_3 = a + 3\,\Delta x,$$

$$\vdots$$

Let's approximate the $i$th strip $S_i$ by a rectangle with width $\Delta x$ and height $f(x_i)$, which is the value of $f$ at the right endpoint (see Figure 11). Then the area of the $i$th rectangle is $f(x_i)\,\Delta x$. What we think of intuitively as the area of $S$ is approximated by the sum of the areas of these rectangles, which is

$$R_n = f(x_1)\,\Delta x + f(x_2)\,\Delta x + \cdots + f(x_n)\,\Delta x$$

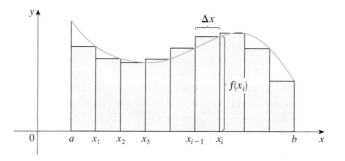

**FIGURE 11**

Figure 12 shows this approximation for $n = 2, 4, 8,$ and $12$. Notice that this approximation appears to become better and better as the number of strips increases, that is, as $n \to \infty$. Therefore we define the area $A$ of the region $S$ in the following way.

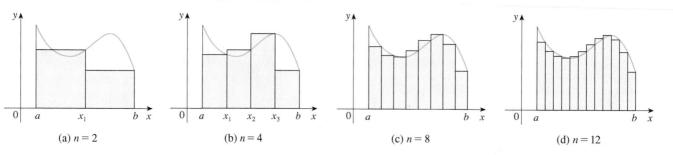

(a) $n = 2$    (b) $n = 4$    (c) $n = 8$    (d) $n = 12$

**FIGURE 12**

> **2** **Definition** The **area** $A$ of the region $S$ that lies under the graph of the continuous function $f$ is the limit of the sum of the areas of approximating rectangles:
>
> $$A = \lim_{n \to \infty} R_n = \lim_{n \to \infty} \left[ f(x_1)\, \Delta x + f(x_2)\, \Delta x + \cdots + f(x_n)\, \Delta x \right]$$

It can be proved that the limit in Definition 2 always exists, since we are assuming that $f$ is continuous. It can also be shown that we get the same value if we use left endpoints:

$$\boxed{3} \qquad A = \lim_{n \to \infty} L_n = \lim_{n \to \infty} \left[ f(x_0)\, \Delta x + f(x_1)\, \Delta x + \cdots + f(x_{n-1})\, \Delta x \right]$$

In fact, instead of using left endpoints or right endpoints, we could take the height of the $i$th rectangle to be the value of $f$ at *any* number $x_i^*$ in the $i$th subinterval $[x_{i-1}, x_i]$. We call the numbers $x_1^*, x_2^*, \ldots, x_n^*$ the **sample points**. Figure 13 shows approximating rectangles when the sample points are not chosen to be endpoints. So a more general expression for the area of $S$ is

$$\boxed{4} \qquad A = \lim_{n \to \infty} \left[ f(x_1^*)\, \Delta x + f(x_2^*)\, \Delta x + \cdots + f(x_n^*)\, \Delta x \right]$$

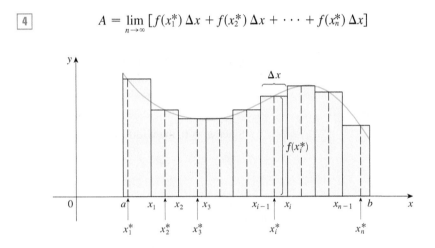

**FIGURE 13**

**NOTE** It can be shown that an equivalent definition of area is the following: *A is the unique number that is smaller than all the upper sums and bigger than all the lower sums.* We saw in Examples 1 and 2, for instance, that the area $\left( A = \frac{1}{3} \right)$ is trapped between all the left approximating sums $L_n$ and all the right approximating sums $R_n$. The function in those examples, $f(x) = x^2$, happens to be increasing on $[0, 1]$ and so the lower sums arise from left endpoints and the upper sums from right endpoints. (See Figures 8 and 9.) In general, we form **lower** (and **upper**) **sums** by choosing the sample points $x_i^*$ so that $f(x_i^*)$ is the minimum (and maximum) value of $f$ on the $i$th subinterval. (See Figure 14 and Exercises 7–8).

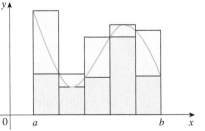

**FIGURE 14**
Lower sums (short rectangles)
and upper sums (tall rectangles)

This tells us to
end with $i = n$.

This tells us
to add.

This tells us to
start with $i = m$.

$$\sum_{i=m}^{n} f(x_i)\,\Delta x$$

If you need practice with sigma notation, look at the examples and try some of the exercises in Appendix E.

We often use **sigma notation** to write sums with many terms more compactly. For instance,

$$\sum_{i=1}^{n} f(x_i)\,\Delta x = f(x_1)\,\Delta x + f(x_2)\,\Delta x + \cdots + f(x_n)\,\Delta x$$

So the expressions for area in Equations 2, 3, and 4 can be written as follows:

$$A = \lim_{n\to\infty} \sum_{i=1}^{n} f(x_i)\,\Delta x$$

$$A = \lim_{n\to\infty} \sum_{i=1}^{n} f(x_{i-1})\,\Delta x$$

$$A = \lim_{n\to\infty} \sum_{i=1}^{n} f(x_i^*)\,\Delta x$$

We can also rewrite Formula 1 in the following way:

$$\sum_{i=1}^{n} i^2 = \frac{n(n+1)(2n+1)}{6}$$

**EXAMPLE 3**    Let $A$ be the area of the region that lies under the graph of $f(x) = \cos x$ between $x = 0$ and $x = b$, where $0 \le b \le \pi/2$.
(a)  Using right endpoints, find an expression for $A$ as a limit. Do not evaluate the limit.
(b)  Estimate the area for the case $b = \pi/2$ by taking the sample points to be midpoints and using four subintervals.

SOLUTION
(a)  Since $a = 0$, the width of a subinterval is

$$\Delta x = \frac{b-0}{n} = \frac{b}{n}$$

So $x_1 = b/n$, $x_2 = 2b/n$, $x_3 = 3b/n$, $x_i = ib/n$, and $x_n = nb/n$. The sum of the areas of the approximating rectangles is

$$R_n = f(x_1)\,\Delta x + f(x_2)\,\Delta x + \cdots + f(x_n)\,\Delta x$$

$$= (\cos x_1)\,\Delta x + (\cos x_2)\,\Delta x + \cdots + (\cos x_n)\,\Delta x$$

$$= \left(\cos\frac{b}{n}\right)\frac{b}{n} + \left(\cos\frac{2b}{n}\right)\frac{b}{n} + \cdots + \left(\cos\frac{nb}{n}\right)\frac{b}{n}$$

According to Definition 2, the area is

$$A = \lim_{n\to\infty} R_n = \lim_{n\to\infty} \frac{b}{n}\left(\cos\frac{b}{n} + \cos\frac{2b}{n} + \cos\frac{3b}{n} + \cdots + \cos\frac{nb}{n}\right)$$

Using sigma notation we could write

$$A = \lim_{n\to\infty} \frac{b}{n} \sum_{i=1}^{n} \cos\frac{ib}{n}$$

It is very difficult to evaluate this limit directly by hand, but with the aid of a computer algebra system it isn't hard (see Exercise 29). In Section 4.3 we will be able to find $A$ more easily using a different method.

**FIGURE 15**

(b) With $n = 4$ and $b = \pi/2$ we have $\Delta x = (\pi/2)/4 = \pi/8$, so the subintervals are $[0, \pi/8]$, $[\pi/8, \pi/4]$, $[\pi/4, 3\pi/8]$, and $[3\pi/8, \pi/2]$. The midpoints of these subintervals are

$$x_1^* = \frac{\pi}{16} \qquad x_2^* = \frac{3\pi}{16} \qquad x_3^* = \frac{5\pi}{16} \qquad x_4^* = \frac{7\pi}{16}$$

and the sum of the areas of the four approximating rectangles (see Figure 15) is

$$
\begin{aligned}
M_4 &= \sum_{i=1}^{4} f(x_i^*) \, \Delta x \\
&= f(\pi/16) \, \Delta x + f(3\pi/16) \, \Delta x + f(5\pi/16) \, \Delta x + f(7\pi/16) \, \Delta x \\
&= \left( \cos \frac{\pi}{16} \right) \frac{\pi}{8} + \left( \cos \frac{3\pi}{16} \right) \frac{\pi}{8} + \left( \cos \frac{5\pi}{16} \right) \frac{\pi}{8} + \left( \cos \frac{7\pi}{16} \right) \frac{\pi}{8} \\
&= \frac{\pi}{8} \left( \cos \frac{\pi}{16} + \cos \frac{3\pi}{16} + \cos \frac{5\pi}{16} + \cos \frac{7\pi}{16} \right) \approx 1.006
\end{aligned}
$$

So an estimate for the area is

$$A \approx 1.006$$

## The Distance Problem

Now let's consider the *distance problem:* Find the distance traveled by an object during a certain time period if the velocity of the object is known at all times. (In a sense this is the inverse problem of the velocity problem that we discussed in Section 1.4.) If the velocity remains constant, then the distance problem is easy to solve by means of the formula

$$\text{distance} = \text{velocity} \times \text{time}$$

But if the velocity varies, it's not so easy to find the distance traveled. We investigate the problem in the following example.

**V EXAMPLE 4** Suppose the odometer on our car is broken and we want to estimate the distance driven over a 30-second time interval. We take speedometer readings every five seconds and record them in the following table:

| Time (s) | 0 | 5 | 10 | 15 | 20 | 25 | 30 |
|---|---|---|---|---|---|---|---|
| Velocity (mi/h) | 17 | 21 | 24 | 29 | 32 | 31 | 28 |

In order to have the time and the velocity in consistent units, let's convert the velocity readings to feet per second (1 mi/h = 5280/3600 ft/s):

| Time (s) | 0 | 5 | 10 | 15 | 20 | 25 | 30 |
|---|---|---|---|---|---|---|---|
| Velocity (ft/s) | 25 | 31 | 35 | 43 | 47 | 46 | 41 |

During the first five seconds the velocity doesn't change very much, so we can estimate the distance traveled during that time by assuming that the velocity is constant. If we take the velocity during that time interval to be the initial velocity (25 ft/s), then we obtain the approximate distance traveled during the first five seconds:

$$25 \text{ ft/s} \times 5 \text{ s} = 125 \text{ ft}$$

Similarly, during the second time interval the velocity is approximately constant and we take it to be the velocity when $t = 5$ s. So our estimate for the distance traveled from $t = 5$ s to $t = 10$ s is

$$31 \text{ ft/s} \times 5 \text{ s} = 155 \text{ ft}$$

If we add similar estimates for the other time intervals, we obtain an estimate for the total distance traveled:

$$(25 \times 5) + (31 \times 5) + (35 \times 5) + (43 \times 5) + (47 \times 5) + (46 \times 5) = 1135 \text{ ft}$$

We could just as well have used the velocity at the *end* of each time period instead of the velocity at the beginning as our assumed constant velocity. Then our estimate becomes

$$(31 \times 5) + (35 \times 5) + (43 \times 5) + (47 \times 5) + (46 \times 5) + (41 \times 5) = 1215 \text{ ft}$$

If we had wanted a more accurate estimate, we could have taken velocity readings every two seconds, or even every second.

Perhaps the calculations in Example 4 remind you of the sums we used earlier to estimate areas. The similarity is explained when we sketch a graph of the velocity function of the car in Figure 16 and draw rectangles whose heights are the initial velocities for each time interval. The area of the first rectangle is $25 \times 5 = 125$, which is also our estimate for the distance traveled in the first five seconds. In fact, the area of each rectangle can be interpreted as a distance because the height represents velocity and the width represents time. The sum of the areas of the rectangles in Figure 16 is $L_6 = 1135$, which is our initial estimate for the total distance traveled.

In general, suppose an object moves with velocity $v = f(t)$, where $a \le t \le b$ and $f(t) \ge 0$ (so the object always moves in the positive direction). We take velocity readings at times $t_0 (= a), t_1, t_2, \ldots, t_n (= b)$ so that the velocity is approximately constant on each subinterval. If these times are equally spaced, then the time between consecutive readings is $\Delta t = (b - a)/n$. During the first time interval the velocity is approximately $f(t_0)$ and so the distance traveled is approximately $f(t_0) \Delta t$. Similarly, the distance traveled during the second time interval is about $f(t_1) \Delta t$ and the total distance traveled during the time interval $[a, b]$ is approximately

$$f(t_0) \, \Delta t + f(t_1) \, \Delta t + \cdots + f(t_{n-1}) \, \Delta t = \sum_{i=1}^{n} f(t_{i-1}) \, \Delta t$$

If we use the velocity at right endpoints instead of left endpoints, our estimate for the total distance becomes

$$f(t_1) \, \Delta t + f(t_2) \, \Delta t + \cdots + f(t_n) \, \Delta t = \sum_{i=1}^{n} f(t_i) \, \Delta t$$

The more frequently we measure the velocity, the more accurate our estimates become, so

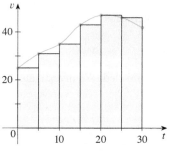

**FIGURE 16**

it seems plausible that the *exact* distance $d$ traveled is the *limit* of such expressions:

$$\boxed{5} \qquad d = \lim_{n \to \infty} \sum_{i=1}^{n} f(t_{i-1}) \, \Delta t = \lim_{n \to \infty} \sum_{i=1}^{n} f(t_i) \, \Delta t$$

We will see in Section 4.4 that this is indeed true.

Because Equation 5 has the same form as our expressions for area in Equations 2 and 3, it follows that the distance traveled is equal to the area under the graph of the velocity function. In Chapter 5 we will see that other quantities of interest in the natural and social sciences—such as the work done by a variable force or the cardiac output of the heart—can also be interpreted as the area under a curve. So when we compute areas in this chapter, bear in mind that they can be interpreted in a variety of practical ways.

## 4.1 Exercises

**1.** (a) By reading values from the given graph of $f$, use four rectangles to find a lower estimate and an upper estimate for the area under the given graph of $f$ from $x = 0$ to $x = 8$. In each case sketch the rectangles that you use.

(b) Find new estimates using eight rectangles in each case.

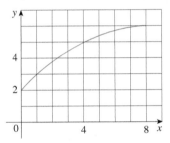

**2.** (a) Use six rectangles to find estimates of each type for the area under the given graph of $f$ from $x = 0$ to $x = 12$.

(i) $L_6$ (sample points are left endpoints)

(ii) $R_6$ (sample points are right endpoints)

(iii) $M_6$ (sample points are midpoints)

(b) Is $L_6$ an underestimate or overestimate of the true area?

(c) Is $R_6$ an underestimate or overestimate of the true area?

(d) Which of the numbers $L_6$, $R_6$, or $M_6$ gives the best estimate? Explain.

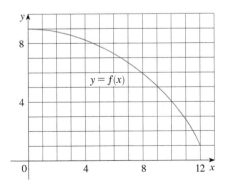

**3.** (a) Estimate the area under the graph of $f(x) = \cos x$ from $x = 0$ to $x = \pi/2$ using four approximating rectangles and right endpoints. Sketch the graph and the rectangles. Is your estimate an underestimate or an overestimate?

(b) Repeat part (a) using left endpoints.

**4.** (a) Estimate the area under the graph of $f(x) = \sqrt{x}$ from $x = 0$ to $x = 4$ using four approximating rectangles and right endpoints. Sketch the graph and the rectangles. Is your estimate an underestimate or an overestimate?

(b) Repeat part (a) using left endpoints.

**5.** (a) Estimate the area under the graph of $f(x) = 1 + x^2$ from $x = -1$ to $x = 2$ using three rectangles and right endpoints. Then improve your estimate by using six rectangles. Sketch the curve and the approximating rectangles.

(b) Repeat part (a) using left endpoints.

(c) Repeat part (a) using midpoints.

(d) From your sketches in parts (a)–(c), which appears to be the best estimate?

**6.** (a) Graph the function

$$f(x) = 1/(1 + x^2) \qquad -2 \le x \le 2$$

(b) Estimate the area under the graph of $f$ using four approximating rectangles and taking the sample points to be (i) right endpoints and (ii) midpoints. In each case sketch the curve and the rectangles.

(c) Improve your estimates in part (b) by using eight rectangles.

**7.** Evaluate the upper and lower sums for $f(x) = 2 + \sin x$, $0 \le x \le \pi$, with $n = 2, 4$, and $8$. Illustrate with diagrams like Figure 14.

**8.** Evaluate the upper and lower sums for $f(x) = 1 + x^2$, $-1 \le x \le 1$, with $n = 3$ and $4$. Illustrate with diagrams like Figure 14.

**9–10** With a programmable calculator (or a computer), it is possible to evaluate the expressions for the sums of areas of approximating rectangles, even for large values of $n$, using looping. (On a TI use the Is> command or a For-EndFor loop, on a Casio use Isz, on an HP or in BASIC use a FOR-NEXT loop.) Compute the sum of the areas of approximating rectangles using equal subintervals and right endpoints for $n = 10, 30, 50$, and $100$. Then guess the value of the exact area.

**9.** The region under $y = x^4$ from 0 to 1

**10.** The region under $y = \cos x$ from 0 to $\pi/2$

CAS **11.** Some computer algebra systems have commands that will draw approximating rectangles and evaluate the sums of their areas, at least if $x_i^*$ is a left or right endpoint. (For instance, in Maple use `leftbox`, `rightbox`, `leftsum`, and `rightsum`.)
   (a) If $f(x) = 1/(x^2 + 1)$, $0 \le x \le 1$, find the left and right sums for $n = 10, 30$, and $50$.
   (b) Illustrate by graphing the rectangles in part (a).
   (c) Show that the exact area under $f$ lies between 0.780 and 0.791.

CAS **12.** (a) If $f(x) = x/(x + 2)$, $1 \le x \le 4$, use the commands discussed in Exercise 11 to find the left and right sums for $n = 10, 30$, and $50$.
   (b) Illustrate by graphing the rectangles in part (a).
   (c) Show that the exact area under $f$ lies between 1.603 and 1.624.

**13.** The speed of a runner increased steadily during the first three seconds of a race. Her speed at half-second intervals is given in the table. Find lower and upper estimates for the distance that she traveled during these three seconds.

| $t$ (s) | 0 | 0.5 | 1.0 | 1.5 | 2.0 | 2.5 | 3.0 |
|---|---|---|---|---|---|---|---|
| $v$ (ft/s) | 0 | 6.2 | 10.8 | 14.9 | 18.1 | 19.4 | 20.2 |

**14.** Speedometer readings for a motorcycle at 12-second intervals are given in the table.
   (a) Estimate the distance traveled by the motorcycle during this time period using the velocities at the beginning of the time intervals.
   (b) Give another estimate using the velocities at the end of the time periods.
   (c) Are your estimates in parts (a) and (b) upper and lower estimates? Explain.

| $t$ (s) | 0 | 12 | 24 | 36 | 48 | 60 |
|---|---|---|---|---|---|---|
| $v$ (ft/s) | 30 | 28 | 25 | 22 | 24 | 27 |

**15.** Oil leaked from a tank at a rate of $r(t)$ liters per hour. The rate decreased as time passed and values of the rate at two-hour time intervals are shown in the table. Find lower and upper estimates for the total amount of oil that leaked out.

| $t$ (h) | 0 | 2 | 4 | 6 | 8 | 10 |
|---|---|---|---|---|---|---|
| $r(t)$ (L/h) | 8.7 | 7.6 | 6.8 | 6.2 | 5.7 | 5.3 |

**16.** When we estimate distances from velocity data, it is sometimes necessary to use times $t_0, t_1, t_2, t_3, \ldots$ that are not equally spaced. We can still estimate distances using the time periods $\Delta t_i = t_i - t_{i-1}$. For example, on May 7, 1992, the space shuttle *Endeavour* was launched on mission STS-49, the purpose of which was to install a new perigee kick motor in an Intelsat communications satellite. The table, provided by NASA, gives the velocity data for the shuttle between liftoff and the jettisoning of the solid rocket boosters. Use these data to estimate the height above the earth's surface of the *Endeavour*, 62 seconds after liftoff.

| Event | Time (s) | Velocity (ft/s) |
|---|---|---|
| Launch | 0 | 0 |
| Begin roll maneuver | 10 | 185 |
| End roll maneuver | 15 | 319 |
| Throttle to 89% | 20 | 447 |
| Throttle to 67% | 32 | 742 |
| Throttle to 104% | 59 | 1325 |
| Maximum dynamic pressure | 62 | 1445 |
| Solid rocket booster separation | 125 | 4151 |

**17.** The velocity graph of a braking car is shown. Use it to estimate the distance traveled by the car while the brakes are applied.

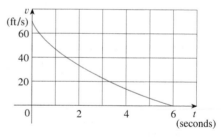

**18.** The velocity graph of a car accelerating from rest to a speed of 120 km/h over a period of 30 seconds is shown. Estimate the distance traveled during this period.

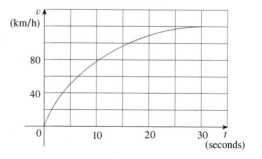

**19–21** Use Definition 2 to find an expression for the area under the graph of $f$ as a limit. Do not evaluate the limit.

**19.** $f(x) = \dfrac{2x}{x^2 + 1}$,   $1 \leqslant x \leqslant 3$

**20.** $f(x) = x^2 + \sqrt{1 + 2x}$,   $4 \leqslant x \leqslant 7$

**21.** $f(x) = \sqrt{\sin x}$,   $0 \leqslant x \leqslant \pi$

---

**22–23** Determine a region whose area is equal to the given limit. Do not evaluate the limit.

**22.** $\displaystyle\lim_{n \to \infty} \sum_{i=1}^{n} \frac{2}{n} \left( 5 + \frac{2i}{n} \right)^{10}$

**23.** $\displaystyle\lim_{n \to \infty} \sum_{i=1}^{n} \frac{\pi}{4n} \tan \frac{i\pi}{4n}$

---

**24.** (a) Use Definition 2 to find an expression for the area under the curve $y = x^3$ from 0 to 1 as a limit.
(b) The following formula for the sum of the cubes of the first $n$ integers is proved in Appendix E. Use it to evaluate the limit in part (a).

$$1^3 + 2^3 + 3^3 + \cdots + n^3 = \left[ \frac{n(n + 1)}{2} \right]^2$$

**25.** Let $A$ be the area under the graph of an increasing continuous function $f$ from $a$ to $b$, and let $L_n$ and $R_n$ be the approximations to $A$ with $n$ subintervals using left and right endpoints, respectively.
(a) How are $A$, $L_n$, and $R_n$ related?
(b) Show that

$$R_n - L_n = \frac{b - a}{n} [f(b) - f(a)]$$

Then draw a diagram to illustrate this equation by showing that the $n$ rectangles representing $R_n - L_n$ can be reassem-

bled to form a single rectangle whose area is the right side of the equation.
(c) Deduce that

$$R_n - A < \frac{b - a}{n} [f(b) - f(a)]$$

**26.** If $A$ is the area under the curve $y = \sin x$ from 0 to $\pi/2$, use Exercise 25 to find a value of $n$ such that $R_n - A < 0.0001$.

**27.** (a) Express the area under the curve $y = x^5$ from 0 to 2 as a limit.
(b) Use a computer algebra system to find the sum in your expression from part (a).
(c) Evaluate the limit in part (a).

**28.** (a) Express the area under the curve $y = x^4 + 5x^2 + x$ from 2 to 7 as a limit.
(b) Use a computer algebra system to evaluate the sum in part (a).
(c) Use a computer algebra system to find the exact area by evaluating the limit of the expression in part (b).

**29.** Find the exact area under the cosine curve $y = \cos x$ from $x = 0$ to $x = b$, where $0 \leqslant b \leqslant \pi/2$. (Use a computer algebra system both to evaluate the sum and compute the limit.) In particular, what is the area if $b = \pi/2$?

**30.** (a) Let $A_n$ be the area of a polygon with $n$ equal sides inscribed in a circle with radius $r$. By dividing the polygon into $n$ congruent triangles with central angle $2\pi/n$, show that

$$A_n = \tfrac{1}{2} nr^2 \sin\left( \frac{2\pi}{n} \right)$$

(b) Show that $\lim_{n \to \infty} A_n = \pi r^2$. [*Hint:* Use Equation 2.4.2 on page 141.]

---

## 4.2    The Definite Integral

We saw in Section 4.1 that a limit of the form

$$\boxed{1} \qquad \lim_{n \to \infty} \sum_{i=1}^{n} f(x_i^*) \, \Delta x = \lim_{n \to \infty} \left[ f(x_1^*) \, \Delta x + f(x_2^*) \, \Delta x + \cdots + f(x_n^*) \, \Delta x \right]$$

arises when we compute an area. We also saw that it arises when we try to find the distance traveled by an object. It turns out that this same type of limit occurs in a wide variety of situations even when $f$ is not necessarily a positive function. In Chapters 5 and 8 we will see that limits of the form $\boxed{1}$ also arise in finding lengths of curves, volumes of solids, centers of mass, force due to water pressure, and work, as well as other quantities. We therefore give this type of limit a special name and notation.

> **2   Definition of a Definite Integral**   If $f$ is a function defined for $a \leq x \leq b$, we divide the interval $[a, b]$ into $n$ subintervals of equal width $\Delta x = (b - a)/n$. We let $x_0 \, (= a), x_1, x_2, \ldots, x_n \, (= b)$ be the endpoints of these subintervals and we let $x_1^*, x_2^*, \ldots, x_n^*$ be any **sample points** in these subintervals, so $x_i^*$ lies in the $i$th subinterval $[x_{i-1}, x_i]$. Then the **definite integral of $f$ from $a$ to $b$** is
>
> $$\int_a^b f(x)\,dx = \lim_{n \to \infty} \sum_{i=1}^{n} f(x_i^*)\,\Delta x$$
>
> provided that this limit exists and gives the same value for all possible choices of sample points. If it does exist, we say that $f$ is **integrable** on $[a, b]$.

The precise meaning of the limit that defines the integral is as follows:

For every number $\varepsilon > 0$ there is an integer $N$ such that

$$\left| \int_a^b f(x)\,dx - \sum_{i=1}^{n} f(x_i^*)\,\Delta x \right| < \varepsilon$$

for every integer $n > N$ and for every choice of $x_i^*$ in $[x_{i-1}, x_i]$.

**NOTE 1**   The symbol $\int$ was introduced by Leibniz and is called an **integral sign**. It is an elongated $S$ and was chosen because an integral is a limit of sums. In the notation $\int_a^b f(x)\,dx$, $f(x)$ is called the **integrand** and $a$ and $b$ are called the **limits of integration**; $a$ is the **lower limit** and $b$ is the **upper limit**. For now, the symbol $dx$ has no meaning by itself; $\int_a^b f(x)\,dx$ is all one symbol. The $dx$ simply indicates that the independent variable is $x$. The procedure of calculating an integral is called **integration**.

**NOTE 2**   The definite integral $\int_a^b f(x)\,dx$ is a number; it does not depend on $x$. In fact, we could use any letter in place of $x$ without changing the value of the integral:

$$\int_a^b f(x)\,dx = \int_a^b f(t)\,dt = \int_a^b f(r)\,dr$$

**NOTE 3**   The sum

$$\sum_{i=1}^{n} f(x_i^*)\,\Delta x$$

that occurs in Definition 2 is called a **Riemann sum** after the German mathematician Bernhard Riemann (1826–1866). So Definition 2 says that the definite integral of an integrable function can be approximated to within any desired degree of accuracy by a Riemann sum.

We know that if $f$ happens to be positive, then the Riemann sum can be interpreted as a sum of areas of approximating rectangles (see Figure 1). By comparing Definition 2 with the definition of area in Section 4.1, we see that the definite integral $\int_a^b f(x)\,dx$ can be interpreted as the area under the curve $y = f(x)$ from $a$ to $b$. (See Figure 2.)

**Riemann**

Bernhard Riemann received his Ph.D. under the direction of the legendary Gauss at the University of Göttingen and remained there to teach. Gauss, who was not in the habit of praising other mathematicians, spoke of Riemann's "creative, active, truly mathematical mind and gloriously fertile originality." The definition [2] of an integral that we use is due to Riemann. He also made major contributions to the theory of functions of a complex variable, mathematical physics, number theory, and the foundations of geometry. Riemann's broad concept of space and geometry turned out to be the right setting, 50 years later, for Einstein's general relativity theory. Riemann's health was poor throughout his life, and he died of tuberculosis at the age of 39.

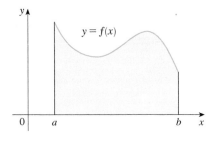

**FIGURE 1**
If $f(x) \geqslant 0$, the Riemann sum $\sum f(x_i^*) \, \Delta x$
is the sum of areas of rectangles.

**FIGURE 2**
If $f(x) \geqslant 0$, the integral $\int_a^b f(x) \, dx$ is the
area under the curve $y = f(x)$ from $a$ to $b$.

**FIGURE 3**
$\sum f(x_i^*) \, \Delta x$ is an approximation to
the net area.

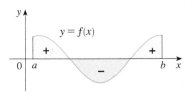

**FIGURE 4**
$\int_a^b f(x) \, dx$ is the net area.

If $f$ takes on both positive and negative values, as in Figure 3, then the Riemann sum is the sum of the areas of the rectangles that lie above the $x$-axis and the *negatives* of the areas of the rectangles that lie below the $x$-axis (the areas of the blue rectangles *minus* the areas of the gold rectangles). When we take the limit of such Riemann sums, we get the situation illustrated in Figure 4. A definite integral can be interpreted as a **net area**, that is, a difference of areas:

$$\int_a^b f(x) \, dx = A_1 - A_2$$

where $A_1$ is the area of the region above the $x$-axis and below the graph of $f$, and $A_2$ is the area of the region below the $x$-axis and above the graph of $f$.

**NOTE 4** Although we have defined $\int_a^b f(x) \, dx$ by dividing $[a, b]$ into subintervals of equal width, there are situations in which it is advantageous to work with subintervals of unequal width. For instance, in Exercise 16 in Section 4.1 NASA provided velocity data at times that were not equally spaced, but we were still able to estimate the distance traveled. And there are methods for numerical integration that take advantage of unequal subintervals.

If the subinterval widths are $\Delta x_1, \Delta x_2, \ldots, \Delta x_n$, we have to ensure that all these widths approach 0 in the limiting process. This happens if the largest width, max $\Delta x_i$, approaches 0. So in this case the definition of a definite integral becomes

$$\int_a^b f(x) \, dx = \lim_{\max \Delta x_i \to 0} \sum_{i=1}^{n} f(x_i^*) \, \Delta x_i$$

**NOTE 5** We have defined the definite integral for an integrable function, but not all functions are integrable (see Exercises 69–70). The following theorem shows that the most commonly occurring functions are in fact integrable. The theorem is proved in more advanced courses.

> **3 Theorem** If $f$ is continuous on $[a, b]$, or if $f$ has only a finite number of jump discontinuities, then $f$ is integrable on $[a, b]$; that is, the definite integral $\int_a^b f(x) \, dx$ exists.

If $f$ is integrable on $[a, b]$, then the limit in Definition 2 exists and gives the same value no matter how we choose the sample points $x_i^*$. To simplify the calculation of the integral we often take the sample points to be right endpoints. Then $x_i^* = x_i$ and the definition of an integral simplifies as follows.

> **4  Theorem**  If $f$ is integrable on $[a, b]$, then
>
> $$\int_a^b f(x)\, dx = \lim_{n \to \infty} \sum_{i=1}^{n} f(x_i)\, \Delta x$$
>
> where $\qquad \Delta x = \dfrac{b - a}{n} \qquad$ and $\qquad x_i = a + i\,\Delta x$

**EXAMPLE 1**  Express

$$\lim_{n \to \infty} \sum_{i=1}^{n} (x_i^3 + x_i \sin x_i)\, \Delta x$$

as an integral on the interval $[0, \pi]$.

SOLUTION  Comparing the given limit with the limit in Theorem 4, we see that they will be identical if we choose $f(x) = x^3 + x \sin x$. We are given that $a = 0$ and $b = \pi$. Therefore, by Theorem 4, we have

$$\lim_{n \to \infty} \sum_{i=1}^{n} (x_i^3 + x_i \sin x_i)\, \Delta x = \int_0^\pi (x^3 + x \sin x)\, dx$$

Later, when we apply the definite integral to physical situations, it will be important to recognize limits of sums as integrals, as we did in Example 1. When Leibniz chose the notation for an integral, he chose the ingredients as reminders of the limiting process. In general, when we write

$$\lim_{n \to \infty} \sum_{i=1}^{n} f(x_i^*)\, \Delta x = \int_a^b f(x)\, dx$$

we replace $\lim \Sigma$ by $\int$, $x_i^*$ by $x$, and $\Delta x$ by $dx$.

### Evaluating Integrals

When we use a limit to evaluate a definite integral, we need to know how to work with sums. The following three equations give formulas for sums of powers of positive integers. Equation 5 may be familiar to you from a course in algebra. Equations 6 and 7 were discussed in Section 4.1 and are proved in Appendix E.

$$\boxed{5} \qquad \sum_{i=1}^{n} i = \frac{n(n + 1)}{2}$$

$$\boxed{6} \qquad \sum_{i=1}^{n} i^2 = \frac{n(n + 1)(2n + 1)}{6}$$

$$\boxed{7} \qquad \sum_{i=1}^{n} i^3 = \left[ \frac{n(n + 1)}{2} \right]^2$$

The remaining formulas are simple rules for working with sigma notation:

Formulas 8–11 are proved by writing out each side in expanded form. The left side of Equation 9 is

$$ca_1 + ca_2 + \cdots + ca_n$$

The right side is

$$c(a_1 + a_2 + \cdots + a_n)$$

These are equal by the distributive property. The other formulas are discussed in Appendix E.

$$\boxed{8} \qquad \sum_{i=1}^{n} c = nc$$

$$\boxed{9} \qquad \sum_{i=1}^{n} ca_i = c \sum_{i=1}^{n} a_i$$

$$\boxed{10} \qquad \sum_{i=1}^{n} (a_i + b_i) = \sum_{i=1}^{n} a_i + \sum_{i=1}^{n} b_i$$

$$\boxed{11} \qquad \sum_{i=1}^{n} (a_i - b_i) = \sum_{i=1}^{n} a_i - \sum_{i=1}^{n} b_i$$

**EXAMPLE 2**

(a) Evaluate the Riemann sum for $f(x) = x^3 - 6x$, taking the sample points to be right endpoints and $a = 0$, $b = 3$, and $n = 6$.

(b) Evaluate $\displaystyle\int_0^3 (x^3 - 6x)\, dx$.

**SOLUTION**

(a) With $n = 6$ the interval width is

$$\Delta x = \frac{b - a}{n} = \frac{3 - 0}{6} = \frac{1}{2}$$

and the right endpoints are $x_1 = 0.5$, $x_2 = 1.0$, $x_3 = 1.5$, $x_4 = 2.0$, $x_5 = 2.5$, and $x_6 = 3.0$. So the Riemann sum is

$$R_6 = \sum_{i=1}^{6} f(x_i)\, \Delta x$$

$$= f(0.5)\, \Delta x + f(1.0)\, \Delta x + f(1.5)\, \Delta x + f(2.0)\, \Delta x + f(2.5)\, \Delta x + f(3.0)\, \Delta x$$

$$= \tfrac{1}{2}(-2.875 - 5 - 5.625 - 4 + 0.625 + 9)$$

$$= -3.9375$$

Notice that $f$ is not a positive function and so the Riemann sum does not represent a sum of areas of rectangles. But it does represent the sum of the areas of the blue rectangles (above the $x$-axis) minus the sum of the areas of the gold rectangles (below the $x$-axis) in Figure 5.

**FIGURE 5**

(b)  With $n$ subintervals we have

$$\Delta x = \frac{b-a}{n} = \frac{3}{n}$$

Thus $x_0 = 0$, $x_1 = 3/n$, $x_2 = 6/n$, $x_3 = 9/n$, and, in general, $x_i = 3i/n$. Since we are using right endpoints, we can use Theorem 4:

$$\int_0^3 (x^3 - 6x)\, dx = \lim_{n\to\infty} \sum_{i=1}^{n} f(x_i)\, \Delta x = \lim_{n\to\infty} \sum_{i=1}^{n} f\!\left(\frac{3i}{n}\right) \frac{3}{n}$$

In the sum, $n$ is a constant (unlike $i$), so we can move $3/n$ in front of the $\Sigma$ sign.

$$= \lim_{n\to\infty} \frac{3}{n} \sum_{i=1}^{n} \left[\left(\frac{3i}{n}\right)^3 - 6\left(\frac{3i}{n}\right)\right] \qquad \text{(Equation 9 with } c = 3/n)$$

$$= \lim_{n\to\infty} \frac{3}{n} \sum_{i=1}^{n} \left[\frac{27}{n^3} i^3 - \frac{18}{n} i\right]$$

$$= \lim_{n\to\infty} \left[\frac{81}{n^4} \sum_{i=1}^{n} i^3 - \frac{54}{n^2} \sum_{i=1}^{n} i\right] \qquad \text{(Equations 11 and 9)}$$

$$= \lim_{n\to\infty} \left\{\frac{81}{n^4} \left[\frac{n(n+1)}{2}\right]^2 - \frac{54}{n^2} \frac{n(n+1)}{2}\right\} \qquad \text{(Equations 7 and 5)}$$

$$= \lim_{n\to\infty} \left[\frac{81}{4}\left(1 + \frac{1}{n}\right)^2 - 27\left(1 + \frac{1}{n}\right)\right]$$

$$= \frac{81}{4} - 27 = -\frac{27}{4} = -6.75$$

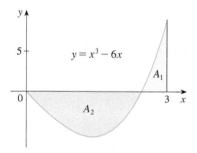

**FIGURE 6**

$\int_0^3 (x^3 - 6x)\, dx = A_1 - A_2 = -6.75$

This integral can't be interpreted as an area because $f$ takes on both positive and negative values. But it can be interpreted as the difference of areas $A_1 - A_2$, where $A_1$ and $A_2$ are shown in Figure 6.

Figure 7 illustrates the calculation by showing the positive and negative terms in the right Riemann sum $R_n$ for $n = 40$. The values in the table show the Riemann sums approaching the exact value of the integral, $-6.75$, as $n \to \infty$.

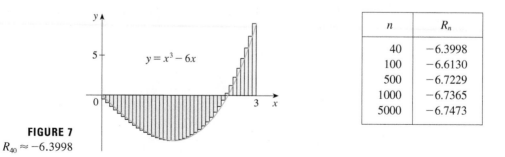

**FIGURE 7**

$R_{40} \approx -6.3998$

| $n$ | $R_n$ |
|---|---|
| 40 | $-6.3998$ |
| 100 | $-6.6130$ |
| 500 | $-6.7229$ |
| 1000 | $-6.7365$ |
| 5000 | $-6.7473$ |

A much simpler method for evaluating the integral in Example 2 will be given in Section 4.4.

Because $f(x) = x^4$ is positive, the integral in Example 3 represents the area shown in Figure 8.

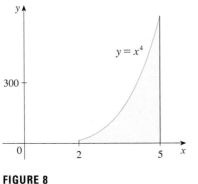

**FIGURE 8**

**EXAMPLE 3**
(a) Set up an expression for $\int_2^5 x^4 \, dx$ as a limit of sums.
(b) Use a computer algebra system to evaluate the expression.

SOLUTION

(a) Here we have $f(x) = x^4$, $a = 2$, $b = 5$, and

$$\Delta x = \frac{b - a}{n} = \frac{3}{n}$$

So $x_0 = 2$, $x_1 = 2 + 3/n$, $x_2 = 2 + 6/n$, $x_3 = 2 + 9/n$, and

$$x_i = 2 + \frac{3i}{n}$$

From Theorem 4, we get

$$\int_2^5 x^4 \, dx = \lim_{n \to \infty} \sum_{i=1}^n f(x_i) \, \Delta x = \lim_{n \to \infty} \sum_{i=1}^n f\left(2 + \frac{3i}{n}\right) \frac{3}{n}$$

$$= \lim_{n \to \infty} \frac{3}{n} \sum_{i=1}^n \left(2 + \frac{3i}{n}\right)^4$$

(b) If we ask a computer algebra system to evaluate the sum and simplify, we obtain

$$\sum_{i=1}^n \left(2 + \frac{3i}{n}\right)^4 = \frac{2062n^4 + 3045n^3 + 1170n^2 - 27}{10n^3}$$

Now we ask the computer algebra system to evaluate the limit:

$$\int_2^5 x^4 \, dx = \lim_{n \to \infty} \frac{3}{n} \sum_{i=1}^n \left(2 + \frac{3i}{n}\right)^4 = \lim_{n \to \infty} \frac{3(2062n^4 + 3045n^3 + 1170n^2 - 27)}{10n^4}$$

$$= \frac{3(2062)}{10} = \frac{3093}{5} = 618.6$$

We will learn a much easier method for the evaluation of integrals in the next section.

**V EXAMPLE 4** Evaluate the following integrals by interpreting each in terms of areas.

(a) $\int_0^1 \sqrt{1 - x^2} \, dx$             (b) $\int_0^3 (x - 1) \, dx$

SOLUTION

(a) Since $f(x) = \sqrt{1 - x^2} \geq 0$, we can interpret this integral as the area under the curve $y = \sqrt{1 - x^2}$ from 0 to 1. But, since $y^2 = 1 - x^2$, we get $x^2 + y^2 = 1$, which shows that the graph of $f$ is the quarter-circle with radius 1 in Figure 9. Therefore

$$\int_0^1 \sqrt{1 - x^2} \, dx = \tfrac{1}{4} \pi (1)^2 = \frac{\pi}{4}$$

(In Section 7.3 we will be able to *prove* that the area of a circle of radius $r$ is $\pi r^2$.)

**FIGURE 9**

(b) The graph of $y = x - 1$ is the line with slope 1 shown in Figure 10. We compute the integral as the difference of the areas of the two triangles:

$$\int_0^3 (x - 1) \, dx = A_1 - A_2 = \tfrac{1}{2}(2 \cdot 2) - \tfrac{1}{2}(1 \cdot 1) = 1.5$$

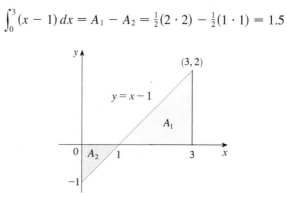

**FIGURE 10**

## The Midpoint Rule

We often choose the sample point $x_i^*$ to be the right endpoint of the $i$th subinterval because it is convenient for computing the limit. But if the purpose is to find an *approximation* to an integral, it is usually better to choose $x_i^*$ to be the midpoint of the interval, which we denote by $\bar{x}_i$. Any Riemann sum is an approximation to an integral, but if we use midpoints we get the following approximation.

**TEC** Module 4.2/7.7 shows how the Midpoint Rule estimates improve as $n$ increases.

---

**Midpoint Rule**

$$\int_a^b f(x) \, dx \approx \sum_{i=1}^{n} f(\bar{x}_i) \, \Delta x = \Delta x \left[ f(\bar{x}_1) + \cdots + f(\bar{x}_n) \right]$$

where

$$\Delta x = \frac{b - a}{n}$$

and

$$\bar{x}_i = \tfrac{1}{2}(x_{i-1} + x_i) = \text{midpoint of } [x_{i-1}, x_i]$$

---

**V  EXAMPLE 5**   Use the Midpoint Rule with $n = 5$ to approximate $\int_1^2 \dfrac{1}{x} \, dx$.

SOLUTION   The endpoints of the five subintervals are 1, 1.2, 1.4, 1.6, 1.8, and 2.0, so the midpoints are 1.1, 1.3, 1.5, 1.7, and 1.9. The width of the subintervals is $\Delta x = (2 - 1)/5 = \tfrac{1}{5}$, so the Midpoint Rule gives

$$\int_1^2 \frac{1}{x} \, dx \approx \Delta x \left[ f(1.1) + f(1.3) + f(1.5) + f(1.7) + f(1.9) \right]$$

$$= \frac{1}{5} \left( \frac{1}{1.1} + \frac{1}{1.3} + \frac{1}{1.5} + \frac{1}{1.7} + \frac{1}{1.9} \right)$$

$$\approx 0.691908$$

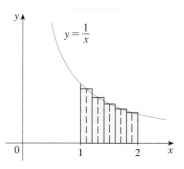

**FIGURE 11**

Since $f(x) = 1/x > 0$ for $1 \le x \le 2$, the integral represents an area, and the approximation given by the Midpoint Rule is the sum of the areas of the rectangles shown in Figure 11.

At the moment we don't know how accurate the approximation in Example 5 is, but in Section 7.7 we will learn a method for estimating the error involved in using the Midpoint Rule. At that time we will discuss other methods for approximating definite integrals.

If we apply the Midpoint Rule to the integral in Example 2, we get the picture in Figure 12. The approximation $M_{40} \approx -6.7563$ is much closer to the true value $-6.75$ than the right endpoint approximation, $R_{40} \approx -6.3998$, shown in Figure 7.

**TEC** In Visual 4.2 you can compare left, right, and midpoint approximations to the integral in Example 2 for different values of $n$.

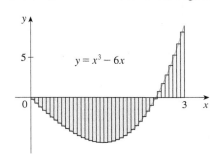

**FIGURE 12**

$M_{40} \approx -6.7563$

## Properties of the Definite Integral

When we defined the definite integral $\int_a^b f(x)\,dx$, we implicitly assumed that $a < b$. But the definition as a limit of Riemann sums makes sense even if $a > b$. Notice that if we reverse $a$ and $b$, then $\Delta x$ changes from $(b - a)/n$ to $(a - b)/n$. Therefore

$$\int_b^a f(x)\,dx = -\int_a^b f(x)\,dx$$

If $a = b$, then $\Delta x = 0$ and so

$$\int_a^a f(x)\,dx = 0$$

We now develop some basic properties of integrals that will help us to evaluate integrals in a simple manner. We assume that $f$ and $g$ are continuous functions.

---

**Properties of the Integral**

**1.** $\displaystyle\int_a^b c\,dx = c(b - a)$,   where $c$ is any constant

**2.** $\displaystyle\int_a^b [f(x) + g(x)]\,dx = \int_a^b f(x)\,dx + \int_a^b g(x)\,dx$

**3.** $\displaystyle\int_a^b cf(x)\,dx = c\int_a^b f(x)\,dx$,   where $c$ is any constant

**4.** $\displaystyle\int_a^b [f(x) - g(x)]\,dx = \int_a^b f(x)\,dx - \int_a^b g(x)\,dx$

---

**FIGURE 13**

$\displaystyle\int_a^b c\,dx = c(b - a)$

Property 1 says that the integral of a constant function $f(x) = c$ is the constant times the length of the interval. If $c > 0$ and $a < b$, this is to be expected because $c(b - a)$ is the area of the shaded rectangle in Figure 13.

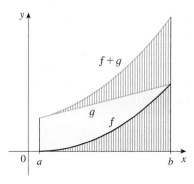

**FIGURE 14**

$$\int_a^b [f(x) + g(x)]\, dx =$$
$$\int_a^b f(x)\, dx + \int_a^b g(x)\, dx$$

Property 3 seems intuitively reasonable because we know that multiplying a function by a positive number $c$ stretches or shrinks its graph vertically by a factor of $c$. So it stretches or shrinks each approximating rectangle by a factor $c$ and therefore it has the effect of multiplying the area by $c$.

Property 2 says that the integral of a sum is the sum of the integrals. For positive functions it says that the area under $f + g$ is the area under $f$ plus the area under $g$. Figure 14 helps us understand why this is true: In view of how graphical addition works, the corresponding vertical line segments have equal height.

In general, Property 2 follows from Theorem 4 and the fact that the limit of a sum is the sum of the limits:

$$\int_a^b [f(x) + g(x)]\, dx = \lim_{n \to \infty} \sum_{i=1}^n [f(x_i) + g(x_i)]\, \Delta x$$

$$= \lim_{n \to \infty} \left[ \sum_{i=1}^n f(x_i)\, \Delta x + \sum_{i=1}^n g(x_i)\, \Delta x \right]$$

$$= \lim_{n \to \infty} \sum_{i=1}^n f(x_i)\, \Delta x + \lim_{n \to \infty} \sum_{i=1}^n g(x_i)\, \Delta x$$

$$= \int_a^b f(x)\, dx + \int_a^b g(x)\, dx$$

Property 3 can be proved in a similar manner and says that the integral of a constant times a function is the constant times the integral of the function. In other words, a constant (but *only* a constant) can be taken in front of an integral sign. Property 4 is proved by writing $f - g = f + (-g)$ and using Properties 2 and 3 with $c = -1$.

**EXAMPLE 6**   Use the properties of integrals to evaluate $\int_0^1 (4 + 3x^2)\, dx$.

SOLUTION   Using Properties 2 and 3 of integrals, we have

$$\int_0^1 (4 + 3x^2)\, dx = \int_0^1 4\, dx + \int_0^1 3x^2\, dx = \int_0^1 4\, dx + 3 \int_0^1 x^2\, dx$$

We know from Property 1 that

$$\int_0^1 4\, dx = 4(1 - 0) = 4$$

and we found in Example 2 in Section 4.1 that $\int_0^1 x^2\, dx = \frac{1}{3}$. So

$$\int_0^1 (4 + 3x^2)\, dx = \int_0^1 4\, dx + 3 \int_0^1 x^2\, dx$$

$$= 4 + 3 \cdot \tfrac{1}{3} = 5$$

The next property tells us how to combine integrals of the same function over adjacent intervals:

**5.**        $$\int_a^c f(x)\, dx + \int_c^b f(x)\, dx = \int_a^b f(x)\, dx$$

This is not easy to prove in general, but for the case where $f(x) \geqslant 0$ and $a < c < b$ Property 5 can be seen from the geometric interpretation in Figure 15: The area under $y = f(x)$ from $a$ to $c$ plus the area from $c$ to $b$ is equal to the total area from $a$ to $b$.

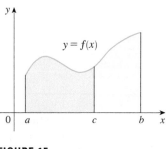

**FIGURE 15**

**V** EXAMPLE 7  If it is known that $\int_0^{10} f(x)\, dx = 17$ and $\int_0^8 f(x)\, dx = 12$, find $\int_8^{10} f(x)\, dx$.

SOLUTION  By Property 5, we have

$$\int_0^8 f(x)\, dx + \int_8^{10} f(x)\, dx = \int_0^{10} f(x)\, dx$$

so $\qquad \int_8^{10} f(x)\, dx = \int_0^{10} f(x)\, dx - \int_0^8 f(x)\, dx = 17 - 12 = 5$

Properties 1–5 are true whether $a < b$, $a = b$, or $a > b$. The following properties, in which we compare sizes of functions and sizes of integrals, are true only if $a \le b$.

---

**Comparison Properties of the Integral**

**6.** If $f(x) \ge 0$ for $a \le x \le b$, then $\int_a^b f(x)\, dx \ge 0$.

**7.** If $f(x) \ge g(x)$ for $a \le x \le b$, then $\int_a^b f(x)\, dx \ge \int_a^b g(x)\, dx$.

**8.** If $m \le f(x) \le M$ for $a \le x \le b$, then

$$m(b - a) \le \int_a^b f(x)\, dx \le M(b - a)$$

---

**FIGURE 16**

If $f(x) \ge 0$, then $\int_a^b f(x)\, dx$ represents the area under the graph of $f$, so the geometric interpretation of Property 6 is simply that areas are positive. (It also follows directly from the definition because all the quantities involved are positive.) Property 7 says that a bigger function has a bigger integral. It follows from Properties 6 and 4 because $f - g \ge 0$.

Property 8 is illustrated by Figure 16 for the case where $f(x) \ge 0$. If $f$ is continuous we could take $m$ and $M$ to be the absolute minimum and maximum values of $f$ on the interval $[a, b]$. In this case Property 8 says that the area under the graph of $f$ is greater than the area of the rectangle with height $m$ and less than the area of the rectangle with height $M$.

PROOF OF PROPERTY 8  Since $m \le f(x) \le M$, Property 7 gives

$$\int_a^b m\, dx \le \int_a^b f(x)\, dx \le \int_a^b M\, dx$$

Using Property 1 to evaluate the integrals on the left and right sides, we obtain

$$m(b - a) \le \int_a^b f(x)\, dx \le M(b - a)$$

Property 8 is useful when all we want is a rough estimate of the size of an integral without going to the bother of using the Midpoint Rule.

EXAMPLE 8  Use Property 8 to estimate $\int_1^4 \sqrt{x}\, dx$.

SOLUTION  Since $f(x) = \sqrt{x}$ is an increasing function, its absolute minimum on $[1, 4]$ is $m = f(1) = 1$ and its absolute maximum on $[1, 4]$ is $M = f(4) = \sqrt{4} = 2$. Thus

**FIGURE 17**

Property 8 gives

$$1(4 - 1) \le \int_1^4 \sqrt{x}\, dx \le 2(4 - 1)$$

or

$$3 \le \int_1^4 \sqrt{x}\, dx \le 6$$

The result of Example 8 is illustrated in Figure 17. The area under $y = \sqrt{x}$ from 1 to 4 is greater than the area of the lower rectangle and less than the area of the large rectangle.

---

## 4.2  Exercises

**1.** Evaluate the Riemann sum for $f(x) = 3 - \frac{1}{2}x$, $2 \le x \le 14$, with six subintervals, taking the sample points to be left endpoints. Explain, with the aid of a diagram, what the Riemann sum represents.

**2.** If $f(x) = x^2 - 2x$, $0 \le x \le 3$, evaluate the Riemann sum with $n = 6$, taking the sample points to be right endpoints. What does the Riemann sum represent? Illustrate with a diagram.

**3.** If $f(x) = \sqrt{x} - 2$, $1 \le x \le 6$, find the Riemann sum with $n = 5$ correct to six decimal places, taking the sample points to be midpoints. What does the Riemann sum represent? Illustrate with a diagram.

**4.** (a) Find the Riemann sum for $f(x) = \sin x$, $0 \le x \le 3\pi/2$, with six terms, taking the sample points to be right endpoints. (Give your answer correct to six decimal places.) Explain what the Riemann sum represents with the aid of a sketch.
   (b) Repeat part (a) with midpoints as the sample points.

**5.** The graph of a function $f$ is given. Estimate $\int_0^{10} f(x)\, dx$ using five subintervals with (a) right endpoints, (b) left endpoints, and (c) midpoints.

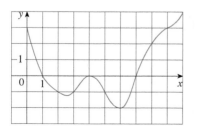

**6.** The graph of $g$ is shown. Estimate $\int_{-2}^{4} g(x)\, dx$ with six subintervals using (a) right endpoints, (b) left endpoints, and (c) midpoints.

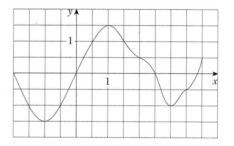

**7.** A table of values of an increasing function $f$ is shown. Use the table to find lower and upper estimates for $\int_{10}^{30} f(x)\, dx$.

| $x$ | 10 | 14 | 18 | 22 | 26 | 30 |
|---|---|---|---|---|---|---|
| $f(x)$ | $-12$ | $-6$ | $-2$ | 1 | 3 | 8 |

**8.** The table gives the values of a function obtained from an experiment. Use them to estimate $\int_3^9 f(x)\, dx$ using three equal subintervals with (a) right endpoints, (b) left endpoints, and (c) midpoints. If the function is known to be an increasing function, can you say whether your estimates are less than or greater than the exact value of the integral?

| $x$ | 3 | 4 | 5 | 6 | 7 | 8 | 9 |
|---|---|---|---|---|---|---|---|
| $f(x)$ | $-3.4$ | $-2.1$ | $-0.6$ | 0.3 | 0.9 | 1.4 | 1.8 |

---

CAS Computer algebra system required    **1.** Homework Hints available at stewartcalculus.com

**9–12** Use the Midpoint Rule with the given value of $n$ to approximate the integral. Round the answer to four decimal places.

**9.** $\int_0^8 \sin \sqrt{x}\, dx, \quad n = 4$

**10.** $\int_0^{\pi/2} \cos^4 x\, dx, \quad n = 4$

**11.** $\int_0^2 \dfrac{x}{x + 1}\, dx, \quad n = 5$

**12.** $\int_1^4 \sqrt{x^3 + 1}\, dx, \quad n = 6$

**CAS** **13.** If you have a CAS that evaluates midpoint approximations and graphs the corresponding rectangles (use `RiemannSum` or `middlesum` and `middlebox` commands in Maple), check the answer to Exercise 11 and illustrate with a graph. Then repeat with $n = 10$ and $n = 20$.

**14.** With a programmable calculator or computer (see the instructions for Exercise 9 in Section 4.1), compute the left and right Riemann sums for the function $f(x) = x/(x + 1)$ on the interval $[0, 2]$ with $n = 100$. Explain why these estimates show that

$$0.8946 < \int_0^2 \dfrac{x}{x + 1}\, dx < 0.9081$$

**15.** Use a calculator or computer to make a table of values of right Riemann sums $R_n$ for the integral $\int_0^\pi \sin x\, dx$ with $n = 5, 10, 50,$ and $100$. What value do these numbers appear to be approaching?

**16.** Use a calculator or computer to make a table of values of left and right Riemann sums $L_n$ and $R_n$ for the integral $\int_0^2 \sqrt{1 + x^4}\, dx$ with $n = 5, 10, 50,$ and $100$. Between what two numbers must the value of the integral lie? Can you make a similar statement for the integral $\int_{-1}^2 \sqrt{1 + x^4}\, dx$? Explain.

**17–20** Express the limit as a definite integral on the given interval.

**17.** $\lim\limits_{n \to \infty} \sum\limits_{i=1}^n \dfrac{1 - x_i^2}{4 + x_i^2}\, \Delta x, \quad [2, 6]$

**18.** $\lim\limits_{n \to \infty} \sum\limits_{i=1}^n \dfrac{\cos x_i}{x_i}\, \Delta x, \quad [\pi, 2\pi]$

**19.** $\lim\limits_{n \to \infty} \sum\limits_{i=1}^n [5(x_i^*)^3 - 4x_i^*]\, \Delta x, \quad [2, 7]$

**20.** $\lim\limits_{n \to \infty} \sum\limits_{i=1}^n \dfrac{x_i^*}{(x_i^*)^2 + 4}\, \Delta x, \quad [1, 3]$

**21–25** Use the form of the definition of the integral given in Theorem 4 to evaluate the integral.

**21.** $\int_2^5 (4 - 2x)\, dx$

**22.** $\int_1^4 (x^2 - 4x + 2)\, dx$

**23.** $\int_{-2}^0 (x^2 + x)\, dx$

**24.** $\int_0^2 (2x - x^3)\, dx$

**25.** $\int_0^1 (x^3 - 3x^2)\, dx$

**26.** (a) Find an approximation to the integral $\int_0^4 (x^2 - 3x)\, dx$ using a Riemann sum with right endpoints and $n = 8$.
(b) Draw a diagram like Figure 3 to illustrate the approximation in part (a).
(c) Use Theorem 4 to evaluate $\int_0^4 (x^2 - 3x)\, dx$.
(d) Interpret the integral in part (c) as a difference of areas and illustrate with a diagram like Figure 4.

**27.** Prove that $\int_a^b x\, dx = \dfrac{b^2 - a^2}{2}$.

**28.** Prove that $\int_a^b x^2\, dx = \dfrac{b^3 - a^3}{3}$.

**29–30** Express the integral as a limit of Riemann sums. Do not evaluate the limit.

**29.** $\int_2^6 \dfrac{x}{1 + x^5}\, dx$

**30.** $\int_0^{2\pi} x^2 \sin x\, dx$

**CAS** **31–32** Express the integral as a limit of sums. Then evaluate, using a computer algebra system to find both the sum and the limit.

**31.** $\int_0^\pi \sin 5x\, dx$

**32.** $\int_2^{10} x^6\, dx$

**33.** The graph of $f$ is shown. Evaluate each integral by interpreting it in terms of areas.

(a) $\int_0^2 f(x)\, dx$

(b) $\int_0^5 f(x)\, dx$

(c) $\int_5^7 f(x)\, dx$

(d) $\int_0^9 f(x)\, dx$

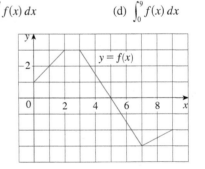

**34.** The graph of $g$ consists of two straight lines and a semicircle. Use it to evaluate each integral.

(a) $\int_0^2 g(x)\, dx$

(b) $\int_2^6 g(x)\, dx$

(c) $\int_0^7 g(x)\, dx$

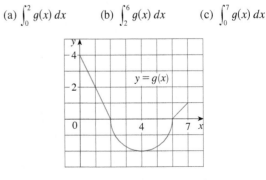

**35–40** Evaluate the integral by interpreting it in terms of areas.

**35.** $\int_{-1}^{2} (1 - x)\, dx$

**36.** $\int_{0}^{9} \left(\frac{1}{3}x - 2\right) dx$

**37.** $\int_{-3}^{0} \left(1 + \sqrt{9 - x^2}\right) dx$

**38.** $\int_{-5}^{5} \left(x - \sqrt{25 - x^2}\right) dx$

**39.** $\int_{-1}^{2} |x|\, dx$

**40.** $\int_{0}^{10} |x - 5|\, dx$

---

**41.** Evaluate $\int_{\pi}^{\pi} \sin^2 x \, \cos^4 x \, dx$.

**42.** Given that $\int_{0}^{1} 3x\sqrt{x^2 + 4}\, dx = 5\sqrt{5} - 8$, what is $\int_{1}^{0} 3u\sqrt{u^2 + 4}\, du$?

**43.** In Example 2 in Section 4.1 we showed that $\int_{0}^{1} x^2\, dx = \frac{1}{3}$. Use this fact and the properties of integrals to evaluate $\int_{0}^{1} (5 - 6x^2)\, dx$.

**44.** Use the properties of integrals and the result of Example 3 to evaluate $\int_{2}^{5} (1 + 3x^4)\, dx$.

**45.** Use the results of Exercises 27 and 28 and the properties of integrals to evaluate $\int_{1}^{4} (2x^2 - 3x + 1)\, dx$.

**46.** Use the result of Exercise 27 and the fact that $\int_{0}^{\pi/2} \cos x\, dx = 1$ (from Exercise 29 in Section 4.1), together with the properties of integrals, to evaluate $\int_{0}^{\pi/2} (2\cos x - 5x)\, dx$.

**47.** Write as a single integral in the form $\int_{a}^{b} f(x)\, dx$:

$$\int_{-2}^{2} f(x)\, dx + \int_{2}^{5} f(x)\, dx - \int_{-2}^{-1} f(x)\, dx$$

**48.** If $\int_{1}^{5} f(x)\, dx = 12$ and $\int_{4}^{5} f(x)\, dx = 3.6$, find $\int_{1}^{4} f(x)\, dx$.

**49.** If $\int_{0}^{9} f(x)\, dx = 37$ and $\int_{0}^{9} g(x)\, dx = 16$, find $\int_{0}^{9} [2f(x) + 3g(x)]\, dx$.

**50.** Find $\int_{0}^{5} f(x)\, dx$ if

$$f(x) = \begin{cases} 3 & \text{for } x < 3 \\ x & \text{for } x \geq 3 \end{cases}$$

**51.** For the function $f$ whose graph is shown, list the following quantities in increasing order, from smallest to largest, and explain your reasoning.

(A) $\int_{0}^{8} f(x)\, dx$

(B) $\int_{0}^{3} f(x)\, dx$

(C) $\int_{3}^{8} f(x)\, dx$

(D) $\int_{4}^{8} f(x)\, dx$

(E) $f'(1)$

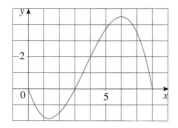

**52.** If $F(x) = \int_{2}^{x} f(t)\, dt$, where $f$ is the function whose graph is given, which of the following values is largest?

(A) $F(0)$           (B) $F(1)$

(C) $F(2)$           (D) $F(3)$

(E) $F(4)$

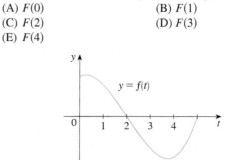

**53.** Each of the regions $A$, $B$, and $C$ bounded by the graph of $f$ and the $x$-axis has area 3. Find the value of

$$\int_{-4}^{2} [f(x) + 2x + 5]\, dx$$

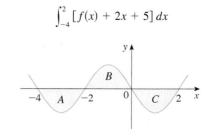

**54.** Suppose $f$ has absolute minimum value $m$ and absolute maximum value $M$. Between what two values must $\int_{0}^{2} f(x)\, dx$ lie? Which property of integrals allows you to make your conclusion?

**55–58** Use the properties of integrals to verify the inequality without evaluating the integrals.

**55.** $\int_{0}^{4} (x^2 - 4x + 4)\, dx \geq 0$

**56.** $\int_{0}^{1} \sqrt{1 + x^2}\, dx \leq \int_{0}^{1} \sqrt{1 + x}\, dx$

**57.** $2 \leq \int_{-1}^{1} \sqrt{1 + x^2}\, dx \leq 2\sqrt{2}$

**58.** $\dfrac{\sqrt{2}\,\pi}{24} \leq \int_{\pi/6}^{\pi/4} \cos x\, dx \leq \dfrac{\sqrt{3}\,\pi}{24}$

---

**59–64** Use Property 8 to estimate the value of the integral.

**59.** $\int_{1}^{4} \sqrt{x}\, dx$

**60.** $\int_{0}^{2} \dfrac{1}{1 + x^2}\, dx$

**61.** $\int_{\pi/4}^{\pi/3} \tan x\, dx$

**62.** $\int_{0}^{2} (x^3 - 3x + 3)\, dx$

**63.** $\int_{-1}^{1} \sqrt{1 + x^4}\, dx$

**64.** $\int_{\pi}^{2\pi} (x - 2\sin x)\, dx$

**65–66** Use properties of integrals, together with Exercises 27 and 28, to prove the inequality.

**65.** $\displaystyle\int_1^3 \sqrt{x^4 + 1}\, dx \geq \frac{26}{3}$

**66.** $\displaystyle\int_0^{\pi/2} x \sin x\, dx \leq \frac{\pi^2}{8}$

**67.** Prove Property 3 of integrals.

**68.** (a) If $f$ is continuous on $[a, b]$, show that

$$\left| \int_a^b f(x)\, dx \right| \leq \int_a^b |f(x)|\, dx$$

[*Hint:* $-|f(x)| \leq f(x) \leq |f(x)|$.]

(b) Use the result of part (a) to show that

$$\left| \int_0^{2\pi} f(x) \sin 2x\, dx \right| \leq \int_0^{2\pi} |f(x)|\, dx$$

**69.** Let $f(x) = 0$ if $x$ is any rational number and $f(x) = 1$ if $x$ is any irrational number. Show that $f$ is not integrable on $[0, 1]$.

**70.** Let $f(0) = 0$ and $f(x) = 1/x$ if $0 < x \leq 1$. Show that $f$ is not integrable on $[0, 1]$. [*Hint:* Show that the first term in the Riemann sum, $f(x_i^*)\,\Delta x$, can be made arbitrarily large.]

**71–72** Express the limit as a definite integral.

**71.** $\displaystyle\lim_{n\to\infty} \sum_{i=1}^n \frac{i^4}{n^5}$   [*Hint:* Consider $f(x) = x^4$.]

**72.** $\displaystyle\lim_{n\to\infty} \frac{1}{n} \sum_{i=1}^n \frac{1}{1 + (i/n)^2}$

**73.** Find $\int_1^2 x^{-2}\, dx$.   *Hint:* Choose $x_i^*$ to be the geometric mean of $x_{i-1}$ and $x_i$ (that is, $x_i^* = \sqrt{x_{i-1}x_i}$) and use the identity

$$\frac{1}{m(m+1)} = \frac{1}{m} - \frac{1}{m+1}$$

---

## DISCOVERY PROJECT   **AREA FUNCTIONS**

**1.** (a) Draw the line $y = 2t + 1$ and use geometry to find the area under this line, above the $t$-axis, and between the vertical lines $t = 1$ and $t = 3$.

(b) If $x > 1$, let $A(x)$ be the area of the region that lies under the line $y = 2t + 1$ between $t = 1$ and $t = x$. Sketch this region and use geometry to find an expression for $A(x)$.

(c) Differentiate the area function $A(x)$. What do you notice?

**2.** (a) If $x \geq -1$, let

$$A(x) = \int_{-1}^x (1 + t^2)\, dt$$

$A(x)$ represents the area of a region. Sketch that region.

(b) Use the result of Exercise 28 in Section 4.2 to find an expression for $A(x)$.

(c) Find $A'(x)$. What do you notice?

(d) If $x \geq -1$ and $h$ is a small positive number, then $A(x + h) - A(x)$ represents the area of a region. Describe and sketch the region.

(e) Draw a rectangle that approximates the region in part (d). By comparing the areas of these two regions, show that

$$\frac{A(x + h) - A(x)}{h} \approx 1 + x^2$$

(f) Use part (e) to give an intuitive explanation for the result of part (c).

**3.** (a) Draw the graph of the function $f(x) = \cos(x^2)$ in the viewing rectangle $[0, 2]$ by $[-1.25, 1.25]$.

(b) If we define a new function $g$ by

$$g(x) = \int_0^x \cos(t^2)\, dt$$

then $g(x)$ is the area under the graph of $f$ from 0 to $x$ [until $f(x)$ becomes negative, at which point $g(x)$ becomes a difference of areas]. Use part (a) to determine the value of

Graphing calculator or computer required

$x$ at which $g(x)$ starts to decrease. [Unlike the integral in Problem 2, it is impossible to evaluate the integral defining $g$ to obtain an explicit expression for $g(x)$.]

(c) Use the integration command on your calculator or computer to estimate $g(0.2)$, $g(0.4)$, $g(0.6)$, ..., $g(1.8)$, $g(2)$. Then use these values to sketch a graph of $g$.

(d) Use your graph of $g$ from part (c) to sketch the graph of $g'$ using the interpretation of $g'(x)$ as the slope of a tangent line. How does the graph of $g'$ compare with the graph of $f$?

**4.** Suppose $f$ is a continuous function on the interval $[a, b]$ and we define a new function $g$ by the equation

$$g(x) = \int_a^x f(t)\, dt$$

Based on your results in Problems 1–3, conjecture an expression for $g'(x)$.

## 4.3    The Fundamental Theorem of Calculus

The Fundamental Theorem of Calculus is appropriately named because it establishes a connection between the two branches of calculus: differential calculus and integral calculus. Differential calculus arose from the tangent problem, whereas integral calculus arose from a seemingly unrelated problem, the area problem. Newton's mentor at Cambridge, Isaac Barrow (1630–1677), discovered that these two problems are actually closely related. In fact, he realized that differentiation and integration are inverse processes. The Fundamental Theorem of Calculus gives the precise inverse relationship between the derivative and the integral. It was Newton and Leibniz who exploited this relationship and used it to develop calculus into a systematic mathematical method. In particular, they saw that the Fundamental Theorem enabled them to compute areas and integrals very easily without having to compute them as limits of sums as we did in Sections 4.1 and 4.2.

The first part of the Fundamental Theorem deals with functions defined by an equation of the form

$$\boxed{1} \qquad\qquad g(x) = \int_a^x f(t)\, dt$$

where $f$ is a continuous function on $[a, b]$ and $x$ varies between $a$ and $b$. Observe that $g$ depends only on $x$, which appears as the variable upper limit in the integral. If $x$ is a fixed number, then the integral $\int_a^x f(t)\, dt$ is a definite number. If we then let $x$ vary, the number $\int_a^x f(t)\, dt$ also varies and defines a function of $x$ denoted by $g(x)$.

If $f$ happens to be a positive function, then $g(x)$ can be interpreted as the area under the graph of $f$ from $a$ to $x$, where $x$ can vary from $a$ to $b$. (Think of $g$ as the "area so far" function; see Figure 1.)

**FIGURE 1**

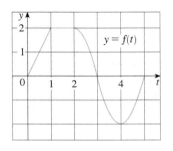

**FIGURE 2**

**V EXAMPLE 1** If $f$ is the function whose graph is shown in Figure 2 and $g(x) = \int_0^x f(t)\, dt$, find the values of $g(0)$, $g(1)$, $g(2)$, $g(3)$, $g(4)$, and $g(5)$. Then sketch a rough graph of $g$.

SOLUTION First we notice that $g(0) = \int_0^0 f(t)\, dt = 0$. From Figure 3 we see that $g(1)$ is the area of a triangle:

$$g(1) = \int_0^1 f(t)\, dt = \tfrac{1}{2}(1 \cdot 2) = 1$$

To find $g(2)$ we add to $g(1)$ the area of a rectangle:

$$g(2) = \int_0^2 f(t)\, dt = \int_0^1 f(t)\, dt + \int_1^2 f(t)\, dt = 1 + (1 \cdot 2) = 3$$

We estimate that the area under $f$ from 2 to 3 is about 1.3, so

$$g(3) = g(2) + \int_2^3 f(t)\, dt \approx 3 + 1.3 = 4.3$$

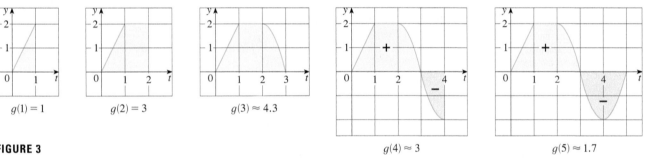

$g(1) = 1$ $g(2) = 3$ $g(3) \approx 4.3$ $g(4) \approx 3$ $g(5) \approx 1.7$

**FIGURE 3**

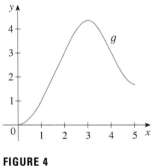

**FIGURE 4**
$g(x) = \int_a^x f(t)\, dt$

For $t > 3$, $f(t)$ is negative and so we start subtracting areas:

$$g(4) = g(3) + \int_3^4 f(t)\, dt \approx 4.3 + (-1.3) = 3.0$$

$$g(5) = g(4) + \int_4^5 f(t)\, dt \approx 3 + (-1.3) = 1.7$$

We use these values to sketch the graph of $g$ in Figure 4. Notice that, because $f(t)$ is positive for $t < 3$, we keep adding area for $t < 3$ and so $g$ is increasing up to $x = 3$, where it attains a maximum value. For $x > 3$, $g$ decreases because $f(t)$ is negative.

If we take $f(t) = t$ and $a = 0$, then, using Exercise 27 in Section 4.2, we have

$$g(x) = \int_0^x t\, dt = \frac{x^2}{2}$$

Notice that $g'(x) = x$, that is, $g' = f$. In other words, if $g$ is defined as the integral of $f$ by Equation 1, then $g$ turns out to be an antiderivative of $f$, at least in this case. And if we sketch the derivative of the function $g$ shown in Figure 4 by estimating slopes of tangents, we get a graph like that of $f$ in Figure 2. So we suspect that $g' = f$ in Example 1 too.

To see why this might be generally true we consider any continuous function $f$ with $f(x) \geq 0$. Then $g(x) = \int_a^x f(t) \, dt$ can be interpreted as the area under the graph of $f$ from $a$ to $x$, as in Figure 1.

In order to compute $g'(x)$ from the definition of a derivative we first observe that, for $h > 0$, $g(x + h) - g(x)$ is obtained by subtracting areas, so it is the area under the graph of $f$ from $x$ to $x + h$ (the blue area in Figure 5). For small $h$ you can see from the figure that this area is approximately equal to the area of the rectangle with height $f(x)$ and width $h$:

$$g(x + h) - g(x) \approx hf(x)$$

so

$$\frac{g(x + h) - g(x)}{h} \approx f(x)$$

Intuitively, we therefore expect that

$$g'(x) = \lim_{h \to 0} \frac{g(x + h) - g(x)}{h} = f(x)$$

The fact that this is true, even when $f$ is not necessarily positive, is the first part of the Fundamental Theorem of Calculus.

**FIGURE 5**

We abbreviate the name of this theorem as FTC1. In words, it says that the derivative of a definite integral with respect to its upper limit is the integrand evaluated at the upper limit.

---

**The Fundamental Theorem of Calculus, Part 1** If $f$ is continuous on $[a, b]$, then the function $g$ defined by

$$g(x) = \int_a^x f(t) \, dt \qquad a \leq x \leq b$$

is continuous on $[a, b]$ and differentiable on $(a, b)$, and $g'(x) = f(x)$.

---

PROOF If $x$ and $x + h$ are in $(a, b)$, then

$$g(x + h) - g(x) = \int_a^{x+h} f(t) \, dt - \int_a^x f(t) \, dt$$

$$= \left( \int_a^x f(t) \, dt + \int_x^{x+h} f(t) \, dt \right) - \int_a^x f(t) \, dt \qquad \text{(by Property 5)}$$

$$= \int_x^{x+h} f(t) \, dt$$

and so, for $h \neq 0$,

$$\boxed{2} \qquad \frac{g(x + h) - g(x)}{h} = \frac{1}{h} \int_x^{x+h} f(t) \, dt$$

For now let's assume that $h > 0$. Since $f$ is continuous on $[x, x + h]$, the Extreme Value Theorem says that there are numbers $u$ and $v$ in $[x, x + h]$ such that $f(u) = m$ and $f(v) = M$, where $m$ and $M$ are the absolute minimum and maximum values of $f$ on $[x, x + h]$. (See Figure 6.)

By Property 8 of integrals, we have

$$mh \leq \int_x^{x+h} f(t) \, dt \leq Mh$$

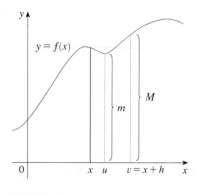

**FIGURE 6**

that is,
$$f(u)h \leq \int_x^{x+h} f(t)\, dt \leq f(v)h$$

Since $h > 0$, we can divide this inequality by $h$:
$$f(u) \leq \frac{1}{h} \int_x^{x+h} f(t)\, dt \leq f(v)$$

Now we use Equation 2 to replace the middle part of this inequality:

$$\boxed{3} \qquad f(u) \leq \frac{g(x+h) - g(x)}{h} \leq f(v)$$

TEC Module 4.3 provides visual evidence for FTC1.

Inequality 3 can be proved in a similar manner for the case where $h < 0$. (See Exercise 63.)

Now we let $h \to 0$. Then $u \to x$ and $v \to x$, since $u$ and $v$ lie between $x$ and $x + h$. Therefore

$$\lim_{h \to 0} f(u) = \lim_{u \to x} f(u) = f(x) \qquad \text{and} \qquad \lim_{h \to 0} f(v) = \lim_{v \to x} f(v) = f(x)$$

because $f$ is continuous at $x$. We conclude, from $\boxed{3}$ and the Squeeze Theorem, that

$$\boxed{4} \qquad g'(x) = \lim_{h \to 0} \frac{g(x+h) - g(x)}{h} = f(x)$$

If $x = a$ or $b$, then Equation 4 can be interpreted as a one-sided limit. Then Theorem 2.2.4 (modified for one-sided limits) shows that $g$ is continuous on $[a, b]$.

Using Leibniz notation for derivatives, we can write FTC1 as

$$\boxed{5} \qquad \frac{d}{dx} \int_a^x f(t)\, dt = f(x)$$

when $f$ is continuous. Roughly speaking, Equation 5 says that if we first integrate $f$ and then differentiate the result, we get back to the original function $f$.

**V EXAMPLE 2** Find the derivative of the function $g(x) = \int_0^x \sqrt{1 + t^2}\, dt$.

SOLUTION Since $f(t) = \sqrt{1 + t^2}$ is continuous, Part 1 of the Fundamental Theorem of Calculus gives
$$g'(x) = \sqrt{1 + x^2}$$

**EXAMPLE 3** Although a formula of the form $g(x) = \int_a^x f(t)\, dt$ may seem like a strange way of defining a function, books on physics, chemistry, and statistics are full of such functions. For instance, the **Fresnel function**

$$S(x) = \int_0^x \sin(\pi t^2 / 2)\, dt$$

is named after the French physicist Augustin Fresnel (1788–1827), who is famous for his works in optics. This function first appeared in Fresnel's theory of the diffraction of light waves, but more recently it has been applied to the design of highways.

Part 1 of the Fundamental Theorem tells us how to differentiate the Fresnel function:

$$S'(x) = \sin(\pi x^2/2)$$

This means that we can apply all the methods of differential calculus to analyze $S$ (see Exercise 57).

Figure 7 shows the graphs of $f(x) = \sin(\pi x^2/2)$ and the Fresnel function $S(x) = \int_0^x f(t)\, dt$. A computer was used to graph $S$ by computing the value of this integral for many values of $x$. It does indeed look as if $S(x)$ is the area under the graph of $f$ from 0 to $x$ [until $x \approx 1.4$ when $S(x)$ becomes a difference of areas]. Figure 8 shows a larger part of the graph of $S$.

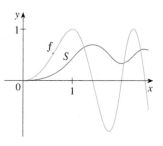

**FIGURE 7**
$f(x) = \sin(\pi x^2/2)$
$S(x) = \int_0^x \sin(\pi t^2/2)\, dt$

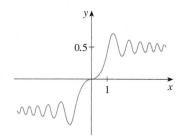

**FIGURE 8**
The Fresnel function $S(x) = \int_0^x \sin(\pi t^2/2)\, dt$

If we now start with the graph of $S$ in Figure 7 and think about what its derivative should look like, it seems reasonable that $S'(x) = f(x)$. [For instance, $S$ is increasing when $f(x) > 0$ and decreasing when $f(x) < 0$.] So this gives a visual confirmation of Part 1 of the Fundamental Theorem of Calculus.

**EXAMPLE 4**  Find $\dfrac{d}{dx} \displaystyle\int_1^{x^4} \sec t\, dt$.

SOLUTION  Here we have to be careful to use the Chain Rule in conjunction with FTC1. Let $u = x^4$. Then

$$\frac{d}{dx} \int_1^{x^4} \sec t\, dt = \frac{d}{dx} \int_1^u \sec t\, dt$$

$$= \frac{d}{du} \left[ \int_1^u \sec t\, dt \right] \frac{du}{dx} \qquad \text{(by the Chain Rule)}$$

$$= \sec u \, \frac{du}{dx} \qquad\qquad \text{(by FTC1)}$$

$$= \sec(x^4) \cdot 4x^3$$

In Section 4.2 we computed integrals from the definition as a limit of Riemann sums and we saw that this procedure is sometimes long and difficult. The second part of the Fundamental Theorem of Calculus, which follows easily from the first part, provides us with a much simpler method for the evaluation of integrals.

**The Fundamental Theorem of Calculus, Part 2**  If $f$ is continuous on $[a, b]$, then

$$\int_a^b f(x)\, dx = F(b) - F(a)$$

where $F$ is any antiderivative of $f$, that is, a function such that $F' = f$.

We abbreviate this theorem as FTC2.

PROOF  Let $g(x) = \int_a^x f(t)\, dt$. We know from Part 1 that $g'(x) = f(x)$; that is, $g$ is an antiderivative of $f$. If $F$ is any other antiderivative of $f$ on $[a, b]$, then we know from Corollary 3.2.7 that $F$ and $g$ differ by a constant:

$$\boxed{6} \qquad F(x) = g(x) + C$$

for $a < x < b$. But both $F$ and $g$ are continuous on $[a, b]$ and so, by taking limits of both sides of Equation 6 (as $x \to a^+$ and $x \to b^-$), we see that it also holds when $x = a$ and $x = b$.

If we put $x = a$ in the formula for $g(x)$, we get

$$g(a) = \int_a^a f(t)\, dt = 0$$

So, using Equation 6 with $x = b$ and $x = a$, we have

$$F(b) - F(a) = [g(b) + C] - [g(a) + C]$$
$$= g(b) - g(a) = g(b) = \int_a^b f(t)\, dt$$

Part 2 of the Fundamental Theorem states that if we know an antiderivative $F$ of $f$, then we can evaluate $\int_a^b f(x)\, dx$ simply by subtracting the values of $F$ at the endpoints of the interval $[a, b]$. It's very surprising that $\int_a^b f(x)\, dx$, which was defined by a complicated procedure involving all of the values of $f(x)$ for $a \le x \le b$, can be found by knowing the values of $F(x)$ at only two points, $a$ and $b$.

Although the theorem may be surprising at first glance, it becomes plausible if we interpret it in physical terms. If $v(t)$ is the velocity of an object and $s(t)$ is its position at time $t$, then $v(t) = s'(t)$, so $s$ is an antiderivative of $v$. In Section 4.1 we considered an object that always moves in the positive direction and made the guess that the area under the velocity curve is equal to the distance traveled. In symbols:

$$\int_a^b v(t)\, dt = s(b) - s(a)$$

That is exactly what FTC2 says in this context.

**V  EXAMPLE 5**  Evaluate the integral $\int_{-2}^1 x^3\, dx$.

SOLUTION  The function $f(x) = x^3$ is continuous on $[-2, 1]$ and we know from Section 3.9 that an antiderivative is $F(x) = \frac{1}{4}x^4$, so Part 2 of the Fundamental Theorem gives

$$\int_{-2}^1 x^3\, dx = F(1) - F(-2) = \tfrac{1}{4}(1)^4 - \tfrac{1}{4}(-2)^4 = -\tfrac{15}{4}$$

Notice that FTC2 says we can use *any* antiderivative $F$ of $f$. So we may as well use the simplest one, namely $F(x) = \frac{1}{4}x^4$, instead of $\frac{1}{4}x^4 + 7$ or $\frac{1}{4}x^4 + C$.

We often use the notation

$$F(x)\Big]_a^b = F(b) - F(a)$$

So the equation of FTC2 can be written as

$$\int_a^b f(x)\,dx = F(x)\Big]_a^b \qquad \text{where} \qquad F' = f$$

Other common notations are $F(x)\big|_a^b$ and $[F(x)]_a^b$.

**EXAMPLE 6** Find the area under the parabola $y = x^2$ from 0 to 1.

SOLUTION An antiderivative of $f(x) = x^2$ is $F(x) = \frac{1}{3}x^3$. The required area $A$ is found using Part 2 of the Fundamental Theorem:

$$A = \int_0^1 x^2\,dx = \frac{x^3}{3}\Bigg]_0^1$$

$$= \frac{1^3}{3} - \frac{0^3}{3} = \frac{1}{3}$$

In applying the Fundamental Theorem we use a particular antiderivative $F$ of $f$. It is not necessary to use the most general antiderivative.

If you compare the calculation in Example 6 with the one in Example 2 in Section 4.1, you will see that the Fundamental Theorem gives a *much* shorter method.

**EXAMPLE 7** Find the area under the cosine curve from 0 to $b$, where $0 \leq b \leq \pi/2$.

SOLUTION Since an antiderivative of $f(x) = \cos x$ is $F(x) = \sin x$, we have

$$A = \int_0^b \cos x\,dx = \sin x\Big]_0^b = \sin b - \sin 0 = \sin b$$

In particular, taking $b = \pi/2$, we have proved that the area under the cosine curve from 0 to $\pi/2$ is $\sin(\pi/2) = 1$. (See Figure 9.)

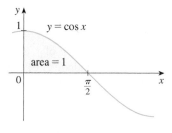

**FIGURE 9**

When the French mathematician Gilles de Roberval first found the area under the sine and cosine curves in 1635, this was a very challenging problem that required a great deal of ingenuity. If we didn't have the benefit of the Fundamental Theorem, we would have to compute a difficult limit of sums using obscure trigonometric identities (or a computer algebra system as in Exercise 29 in Section 4.1). It was even more difficult for Roberval because the apparatus of limits had not been invented in 1635. But in the 1660s and 1670s, when the Fundamental Theorem was discovered by Barrow and exploited by Newton and Leibniz, such problems became very easy, as you can see from Example 7.

**EXAMPLE 8** What is wrong with the following calculation?

$$\oslash \qquad \int_{-1}^3 \frac{1}{x^2}\,dx = \frac{x^{-1}}{-1}\Bigg]_{-1}^3 = -\frac{1}{3} - 1 = -\frac{4}{3}$$

SOLUTION To start, we notice that this calculation must be wrong because the answer is negative but $f(x) = 1/x^2 \geq 0$ and Property 6 of integrals says that $\int_a^b f(x)\,dx \geq 0$ when $f \geq 0$. The Fundamental Theorem of Calculus applies to continuous functions. It can't be applied here because $f(x) = 1/x^2$ is not continuous on $[-1, 3]$. In fact, $f$ has an infinite discontinuity at $x = 0$, so

$$\int_{-1}^{3} \frac{1}{x^2}\,dx \qquad \text{does not exist}$$

### Differentiation and Integration as Inverse Processes

We end this section by bringing together the two parts of the Fundamental Theorem.

---

**The Fundamental Theorem of Calculus** Suppose $f$ is continuous on $[a, b]$.

**1.** If $g(x) = \int_a^x f(t)\,dt$, then $g'(x) = f(x)$.

**2.** $\int_a^b f(x)\,dx = F(b) - F(a)$, where $F$ is any antiderivative of $f$, that is, $F' = f$.

---

We noted that Part 1 can be rewritten as

$$\frac{d}{dx} \int_a^x f(t)\,dt = f(x)$$

which says that if $f$ is integrated and then the result is differentiated, we arrive back at the original function $f$. Since $F'(x) = f(x)$, Part 2 can be rewritten as

$$\int_a^b F'(x)\,dx = F(b) - F(a)$$

This version says that if we take a function $F$, first differentiate it, and then integrate the result, we arrive back at the original function $F$, but in the form $F(b) - F(a)$. Taken together, the two parts of the Fundamental Theorem of Calculus say that differentiation and integration are inverse processes. Each undoes what the other does.

The Fundamental Theorem of Calculus is unquestionably the most important theorem in calculus and, indeed, it ranks as one of the great accomplishments of the human mind. Before it was discovered, from the time of Eudoxus and Archimedes to the time of Galileo and Fermat, problems of finding areas, volumes, and lengths of curves were so difficult that only a genius could meet the challenge. But now, armed with the systematic method that Newton and Leibniz fashioned out of the Fundamental Theorem, we will see in the chapters to come that these challenging problems are accessible to all of us.

## 4.3 Exercises

1. Explain exactly what is meant by the statement that "differentiation and integration are inverse processes."

2. Let $g(x) = \int_0^x f(t)\,dt$, where $f$ is the function whose graph is shown.
   (a) Evaluate $g(x)$ for $x = 0, 1, 2, 3, 4, 5,$ and $6$.
   (b) Estimate $g(7)$.
   (c) Where does $g$ have a maximum value? Where does it have a minimum value?
   (d) Sketch a rough graph of $g$.

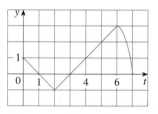

3. Let $g(x) = \int_0^x f(t)\,dt$, where $f$ is the function whose graph is shown.
   (a) Evaluate $g(0), g(1), g(2), g(3),$ and $g(6)$.
   (b) On what interval is $g$ increasing?
   (c) Where does $g$ have a maximum value?
   (d) Sketch a rough graph of $g$.

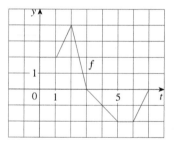

4. Let $g(x) = \int_0^x f(t)\,dt$, where $f$ is the function whose graph is shown.
   (a) Evaluate $g(0)$ and $g(6)$.
   (b) Estimate $g(x)$ for $x = 1, 2, 3, 4,$ and $5$.
   (c) On what interval is $g$ increasing?
   (d) Where does $g$ have a maximum value?
   (e) Sketch a rough graph of $g$.
   (f) Use the graph in part (e) to sketch the graph of $g'(x)$. Compare with the graph of $f$.

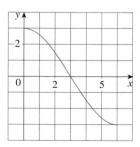

5–6 Sketch the area represented by $g(x)$. Then find $g'(x)$ in two ways: (a) by using Part 1 of the Fundamental Theorem and (b) by evaluating the integral using Part 2 and then differentiating.

5. $g(x) = \int_1^x t^2\,dt$

6. $g(x) = \int_0^x (2 + \sin t)\,dt$

7–18 Use Part 1 of the Fundamental Theorem of Calculus to find the derivative of the function.

7. $g(x) = \int_1^x \dfrac{1}{t^3 + 1}\,dt$

8. $g(x) = \int_1^x (2 + t^4)^5\,dt$

9. $g(s) = \int_5^s (t - t^2)^8\,dt$

10. $g(r) = \int_0^r \sqrt{x^2 + 4}\,dx$

11. $F(x) = \int_x^\pi \sqrt{1 + \sec t}\,dt$

$$\left[\,\text{Hint: } \int_x^\pi \sqrt{1 + \sec t}\,dt = -\int_\pi^x \sqrt{1 + \sec t}\,dt\,\right]$$

12. $G(x) = \int_x^1 \cos\sqrt{t}\,dt$

13. $h(x) = \int_2^{1/x} \sin^4 t\,dt$

14. $h(x) = \int_1^{\sqrt{x}} \dfrac{z^2}{z^4 + 1}\,dz$

15. $y = \int_0^{\tan x} \sqrt{t + \sqrt{t}}\,dt$

16. $y = \int_0^{x^4} \cos^2\theta\,d\theta$

17. $y = \int_{1-3x}^1 \dfrac{u^3}{1 + u^2}\,du$

18. $y = \int_{\sin x}^1 \sqrt{1 + t^2}\,dt$

19–38 Evaluate the integral.

19. $\int_{-1}^2 (x^3 - 2x)\,dx$

20. $\int_{-1}^1 x^{100}\,dx$

21. $\int_1^4 (5 - 2t + 3t^2)\,dt$

22. $\int_0^1 \left(1 + \tfrac{1}{2}u^4 - \tfrac{2}{5}u^9\right)\,du$

23. $\int_1^9 \sqrt{x}\,dx$

24. $\int_1^8 x^{-2/3}\,dx$

25. $\int_{\pi/6}^\pi \sin\theta\,d\theta$

26. $\int_{-5}^5 \pi\,dx$

27. $\int_0^1 (u + 2)(u - 3)\,du$

28. $\int_0^4 (4 - t)\sqrt{t}\,dt$

29. $\int_1^9 \dfrac{x - 1}{\sqrt{x}}\,dx$

30. $\int_0^2 (y - 1)(2y + 1)\,dy$

31. $\int_0^{\pi/4} \sec^2 t\,dt$

32. $\int_0^{\pi/4} \sec\theta\tan\theta\,d\theta$

33. $\int_1^2 (1 + 2y)^2\,dy$

34. $\int_1^2 \dfrac{s^4 + 1}{s^2}\,ds$

**35.** $\displaystyle\int_1^2 \frac{v^5 + 3v^6}{v^4}\, dv$

**36.** $\displaystyle\int_1^{18} \sqrt{\frac{3}{z}}\, dz$

**37.** $\displaystyle\int_0^\pi f(x)\, dx$ where $f(x) = \begin{cases} \sin x & \text{if } 0 \le x < \pi/2 \\ \cos x & \text{if } \pi/2 \le x \le \pi \end{cases}$

**38.** $\displaystyle\int_{-2}^2 f(x)\, dx$ where $f(x) = \begin{cases} 2 & \text{if } -2 \le x \le 0 \\ 4 - x^2 & \text{if } 0 < x \le 2 \end{cases}$

**39–42** What is wrong with the equation?

**39.** $\displaystyle\int_{-2}^1 x^{-4}\, dx = \frac{x^{-3}}{-3}\Big]_{-2}^1 = -\frac{3}{8}$

**40.** $\displaystyle\int_{-1}^2 \frac{4}{x^3}\, dx = -\frac{2}{x^2}\Big]_{-1}^2 = \frac{3}{2}$

**41.** $\displaystyle\int_{\pi/3}^\pi \sec\theta\tan\theta\, d\theta = \sec\theta\Big]_{\pi/3}^\pi = -3$

**42.** $\displaystyle\int_0^\pi \sec^2 x\, dx = \tan x\Big]_0^\pi = 0$

**43–46** Use a graph to give a rough estimate of the area of the region that lies beneath the given curve. Then find the exact area.

**43.** $y = \sqrt[3]{x}, \quad 0 \le x \le 27$

**44.** $y = x^{-4}, \quad 1 \le x \le 6$

**45.** $y = \sin x, \ 0 \le x \le \pi$

**46.** $y = \sec^2 x, \ 0 \le x \le \pi/3$

**47–48** Evaluate the integral and interpret it as a difference of areas. Illustrate with a sketch.

**47.** $\displaystyle\int_{-1}^2 x^3\, dx$

**48.** $\displaystyle\int_{\pi/6}^{2\pi} \cos x\, dx$

**49–52** Find the derivative of the function.

**49.** $g(x) = \displaystyle\int_{2x}^{3x} \frac{u^2 - 1}{u^2 + 1}\, du$

$\left[ \text{Hint: } \displaystyle\int_{2x}^{3x} f(u)\, du = \int_{2x}^0 f(u)\, du + \int_0^{3x} f(u)\, du \right]$

**50.** $g(x) = \displaystyle\int_{1-2x}^{1+2x} t \sin t\, dt$

**51.** $h(x) = \displaystyle\int_{\sqrt{x}}^{x^3} \cos(t^2)\, dt$

**52.** $g(x) = \displaystyle\int_{\tan x}^{x^2} \frac{1}{\sqrt{2 + t^4}}\, dt$

**53.** On what interval is the curve

$$y = \int_0^x \frac{t^2}{t^2 + t + 2}\, dt$$

concave downward?

**54.** If $f(x) = \int_0^x (1 - t^2)\cos^2 t\, dt$, on what interval is $f$ increasing?

**55.** If $f(1) = 12$, $f'$ is continuous, and $\int_1^4 f'(x)\, dx = 17$, what is the value of $f(4)$?

**56.** If $f(x) = \int_0^{\sin x} \sqrt{1 + t^2}\, dt$ and $g(y) = \int_3^y f(x)\, dx$, find $g''(\pi/6)$.

**57.** The Fresnel function $S$ was defined in Example 3 and graphed in Figures 7 and 8.
(a) At what values of $x$ does this function have local maximum values?
(b) On what intervals is the function concave upward?
(c) Use a graph to solve the following equation correct to two decimal places:
$$\int_0^x \sin(\pi t^2/2)\, dt = 0.2$$

**58.** The **sine integral function**
$$\text{Si}(x) = \int_0^x \frac{\sin t}{t}\, dt$$

is important in electrical engineering. [The integrand $f(t) = (\sin t)/t$ is not defined when $t = 0$, but we know that its limit is 1 when $t \to 0$. So we define $f(0) = 1$ and this makes $f$ a continuous function everywhere.]
(a) Draw the graph of Si.
(b) At what values of $x$ does this function have local maximum values?
(c) Find the coordinates of the first inflection point to the right of the origin.
(d) Does this function have horizontal asymptotes?
(e) Solve the following equation correct to one decimal place:
$$\int_0^x \frac{\sin t}{t}\, dt = 1$$

**59–60** Let $g(x) = \int_0^x f(t)\, dt$, where $f$ is the function whose graph is shown.
(a) At what values of $x$ do the local maximum and minimum values of $g$ occur?
(b) Where does $g$ attain its absolute maximum value?
(c) On what intervals is $g$ concave downward?
(d) Sketch the graph of $g$.

**59.**

**60.**

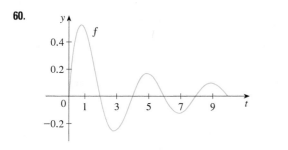

**61–62** Evaluate the limit by first recognizing the sum as a Riemann sum for a function defined on $[0, 1]$.

**61.** $\lim\limits_{n \to \infty} \sum\limits_{i=1}^{n} \dfrac{i^3}{n^4}$

**62.** $\lim\limits_{n \to \infty} \dfrac{1}{n} \left( \sqrt{\dfrac{1}{n}} + \sqrt{\dfrac{2}{n}} + \sqrt{\dfrac{3}{n}} + \cdots + \sqrt{\dfrac{n}{n}} \right)$

**63.** Justify $\boxed{3}$ for the case $h < 0$.

**64.** If $f$ is continuous and $g$ and $h$ are differentiable functions, find a formula for

$$\frac{d}{dx} \int_{g(x)}^{h(x)} f(t)\, dt$$

**65.** (a) Show that $1 \le \sqrt{1 + x^3} \le 1 + x^3$ for $x \ge 0$.
   (b) Show that $1 \le \int_0^1 \sqrt{1 + x^3}\, dx \le 1.25$.

**66.** (a) Show that $\cos(x^2) \ge \cos x$ for $0 \le x \le 1$.
   (b) Deduce that $\int_0^{\pi/6} \cos(x^2)\, dx \ge \frac{1}{2}$.

**67.** Show that

$$0 \le \int_5^{10} \frac{x^2}{x^4 + x^2 + 1}\, dx \le 0.1$$

by comparing the integrand to a simpler function.

**68.** Let

$$f(x) = \begin{cases} 0 & \text{if } x < 0 \\ x & \text{if } 0 \le x \le 1 \\ 2 - x & \text{if } 1 < x \le 2 \\ 0 & \text{if } x > 2 \end{cases}$$

and

$$g(x) = \int_0^x f(t)\, dt$$

(a) Find an expression for $g(x)$ similar to the one for $f(x)$.
(b) Sketch the graphs of $f$ and $g$.
(c) Where is $f$ differentiable? Where is $g$ differentiable?

**69.** Find a function $f$ and a number $a$ such that

$$6 + \int_a^x \frac{f(t)}{t^2}\, dt = 2\sqrt{x} \qquad \text{for all } x > 0$$

**70.** Suppose $h$ is a function such that $h(1) = -2$, $h'(1) = 2$, $h''(1) = 3$, $h(2) = 6$, $h'(2) = 5$, $h''(2) = 13$, and $h''$ is continuous everywhere. Evaluate $\int_1^2 h''(u)\, du$.

**71.** A manufacturing company owns a major piece of equipment that depreciates at the (continuous) rate $f = f(t)$, where $t$ is the time measured in months since its last overhaul. Because a fixed cost $A$ is incurred each time the machine is overhauled, the company wants to determine the optimal time $T$ (in months) between overhauls.
   (a) Explain why $\int_0^t f(s)\, ds$ represents the loss in value of the machine over the period of time $t$ since the last overhaul.
   (b) Let $C = C(t)$ be given by

$$C(t) = \frac{1}{t}\left[ A + \int_0^t f(s)\, ds \right]$$

What does $C$ represent and why would the company want to minimize $C$?
   (c) Show that $C$ has a minimum value at the numbers $t = T$ where $C(T) = f(T)$.

**72.** A high-tech company purchases a new computing system whose initial value is $V$. The system will depreciate at the rate $f = f(t)$ and will accumulate maintenance costs at the rate $g = g(t)$, where $t$ is the time measured in months. The company wants to determine the optimal time to replace the system.
   (a) Let

$$C(t) = \frac{1}{t} \int_0^t [f(s) + g(s)]\, ds$$

Show that the critical numbers of $C$ occur at the numbers $t$ where $C(t) = f(t) + g(t)$.
   (b) Suppose that

$$f(t) = \begin{cases} \dfrac{V}{15} - \dfrac{V}{450}t & \text{if } 0 < t \le 30 \\ 0 & \text{if } t > 30 \end{cases}$$

and $\qquad g(t) = \dfrac{Vt^2}{12{,}900} \qquad t > 0$

Determine the length of time $T$ for the total depreciation $D(t) = \int_0^t f(s)\, ds$ to equal the initial value $V$.
   (c) Determine the absolute minimum of $C$ on $(0, T]$.
   (d) Sketch the graphs of $C$ and $f + g$ in the same coordinate system, and verify the result in part (a) in this case.

---

The following exercises are intended only for those who have already covered Chapter 6.

**73–78** Evaluate the integral.

**73.** $\int_1^9 \dfrac{1}{2x}\, dx$

**74.** $\int_0^1 10^x\, dx$

**75.** $\int_{1/2}^{\sqrt{3}/2} \dfrac{6}{\sqrt{1 - t^2}}\, dt$

**76.** $\int_0^1 \dfrac{4}{t^2 + 1}\, dt$

**77.** $\int_{-1}^1 e^{u+1}\, du$

**78.** $\int_1^2 \dfrac{4 + u^2}{u^3}\, du$

## 4.4  Indefinite Integrals and the Net Change Theorem

We saw in Section 4.3 that the second part of the Fundamental Theorem of Calculus provides a very powerful method for evaluating the definite integral of a function, assuming that we can find an antiderivative of the function. In this section we introduce a notation for antiderivatives, review the formulas for antiderivatives, and use them to evaluate definite integrals. We also reformulate FTC2 in a way that makes it easier to apply to science and engineering problems.

### Indefinite Integrals

Both parts of the Fundamental Theorem establish connections between antiderivatives and definite integrals. Part 1 says that if $f$ is continuous, then $\int_a^x f(t)\, dt$ is an antiderivative of $f$. Part 2 says that $\int_a^b f(x)\, dx$ can be found by evaluating $F(b) - F(a)$, where $F$ is an antiderivative of $f$.

We need a convenient notation for antiderivatives that makes them easy to work with. Because of the relation given by the Fundamental Theorem between antiderivatives and integrals, the notation $\int f(x)\, dx$ is traditionally used for an antiderivative of $f$ and is called an **indefinite integral**. Thus

$$\int f(x)\, dx = F(x) \qquad \text{means} \qquad F'(x) = f(x)$$

For example, we can write

$$\int x^2\, dx = \frac{x^3}{3} + C \qquad \text{because} \qquad \frac{d}{dx}\left(\frac{x^3}{3} + C\right) = x^2$$

So we can regard an indefinite integral as representing an entire *family* of functions (one antiderivative for each value of the constant $C$).

⊘  You should distinguish carefully between definite and indefinite integrals. A definite integral $\int_a^b f(x)\, dx$ is a *number*, whereas an indefinite integral $\int f(x)\, dx$ is a *function* (or family of functions). The connection between them is given by Part 2 of the Fundamental Theorem: If $f$ is continuous on $[a, b]$, then

$$\int_a^b f(x)\, dx = \int f(x)\, dx\,\Big]_a^b$$

The effectiveness of the Fundamental Theorem depends on having a supply of antiderivatives of functions. We therefore restate the Table of Antidifferentiation Formulas from Section 3.9, together with a few others, in the notation of indefinite integrals. Any formula can be verified by differentiating the function on the right side and obtaining the integrand. For instance,

$$\int \sec^2 x\, dx = \tan x + C \qquad \text{because} \qquad \frac{d}{dx}(\tan x + C) = \sec^2 x$$

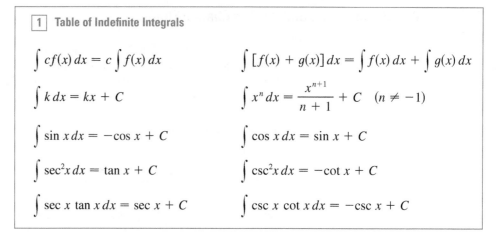

**1** Table of Indefinite Integrals

$$\int cf(x)\, dx = c \int f(x)\, dx \qquad \int [f(x) + g(x)]\, dx = \int f(x)\, dx + \int g(x)\, dx$$

$$\int k\, dx = kx + C \qquad \int x^n\, dx = \frac{x^{n+1}}{n+1} + C \quad (n \neq -1)$$

$$\int \sin x\, dx = -\cos x + C \qquad \int \cos x\, dx = \sin x + C$$

$$\int \sec^2 x\, dx = \tan x + C \qquad \int \csc^2 x\, dx = -\cot x + C$$

$$\int \sec x \tan x\, dx = \sec x + C \qquad \int \csc x \cot x\, dx = -\csc x + C$$

Recall from Theorem 3.9.1 that the most general antiderivative *on a given interval* is obtained by adding a constant to a particular antiderivative. **We adopt the convention that when a formula for a general indefinite integral is given, it is valid only on an interval.** Thus we write

$$\int \frac{1}{x^2}\, dx = -\frac{1}{x} + C$$

with the understanding that it is valid on the interval $(0, \infty)$ or on the interval $(-\infty, 0)$. This is true despite the fact that the general antiderivative of the function $f(x) = 1/x^2$, $x \neq 0$, is

$$F(x) = \begin{cases} -\dfrac{1}{x} + C_1 & \text{if } x < 0 \\[2mm] -\dfrac{1}{x} + C_2 & \text{if } x > 0 \end{cases}$$

**EXAMPLE 1** Find the general indefinite integral

$$\int (10x^4 - 2\sec^2 x)\, dx$$

SOLUTION Using our convention and Table 1, we have

$$\int (10x^4 - 2\sec^2 x)\, dx = 10 \int x^4\, dx - 2 \int \sec^2 x\, dx$$

$$= 10\frac{x^5}{5} - 2\tan x + C = 2x^5 - 2\tan x + C$$

You should check this answer by differentiating it.

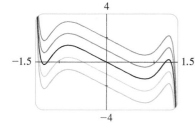

4

−1.5    1.5

−4

**FIGURE 1**

The indefinite integral in Example 1 is graphed in Figure 1 for several values of $C$. Here the value of $C$ is the $y$-intercept.

**V** **EXAMPLE 2** Evaluate $\displaystyle\int \frac{\cos \theta}{\sin^2 \theta}\, d\theta$.

SOLUTION This indefinite integral isn't immediately apparent in Table 1, so we use trigonometric identities to rewrite the function before integrating:

$$\int \frac{\cos \theta}{\sin^2 \theta}\, d\theta = \int \left(\frac{1}{\sin \theta}\right)\left(\frac{\cos \theta}{\sin \theta}\right)\, d\theta$$

$$= \int \csc \theta \cot \theta\, d\theta = -\csc \theta + C$$

**EXAMPLE 3** Evaluate $\int_0^3 (x^3 - 6x)\, dx$.

SOLUTION Using FTC2 and Table 1, we have

$$\int_0^3 (x^3 - 6x)\, dx = \frac{x^4}{4} - 6\frac{x^2}{2}\Bigg]_0^3$$

$$= \left(\tfrac{1}{4} \cdot 3^4 - 3 \cdot 3^2\right) - \left(\tfrac{1}{4} \cdot 0^4 - 3 \cdot 0^2\right)$$

$$= \tfrac{81}{4} - 27 - 0 + 0 = -6.75$$

Compare this calculation with Example 2(b) in Section 4.2.

**EXAMPLE 4** Find $\int_0^{12} (x - 12 \sin x)\, dx$.

SOLUTION The Fundamental Theorem gives

$$\int_0^{12} (x - 12 \sin x)\, dx = \frac{x^2}{2} - 12(-\cos x)\Bigg]_0^{12}$$

$$= \tfrac{1}{2}(12)^2 + 12(\cos 12 - \cos 0)$$

$$= 72 + 12 \cos 12 - 12$$

$$= 60 + 12 \cos 12$$

This is the exact value of the integral. If a decimal approximation is desired, we can use a calculator to approximate $\cos 12$. Doing so, we get

$$\int_0^{12} (x - 12 \sin x)\, dx \approx 70.1262$$

Figure 2 shows the graph of the integrand in Example 4. We know from Section 4.2 that the value of the integral can be interpreted as the sum of the areas labeled with a plus sign minus the areas labeled with a minus sign.

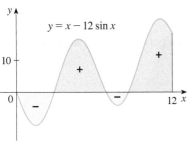

$y = x - 12 \sin x$

**FIGURE 2**

**EXAMPLE 5** Evaluate $\int_1^9 \dfrac{2t^2 + t^2\sqrt{t} - 1}{t^2}\, dt$.

SOLUTION First we need to write the integrand in a simpler form by carrying out the division:

$$\int_1^9 \frac{2t^2 + t^2\sqrt{t} - 1}{t^2}\, dt = \int_1^9 (2 + t^{1/2} - t^{-2})\, dt$$

$$= 2t + \frac{t^{3/2}}{\frac{3}{2}} - \frac{t^{-1}}{-1}\Bigg]_1^9 = 2t + \tfrac{2}{3}t^{3/2} + \frac{1}{t}\Bigg]_1^9$$

$$= \left(2 \cdot 9 + \tfrac{2}{3} \cdot 9^{3/2} + \tfrac{1}{9}\right) - \left(2 \cdot 1 + \tfrac{2}{3} \cdot 1^{3/2} + \tfrac{1}{1}\right)$$

$$= 18 + 18 + \tfrac{1}{9} - 2 - \tfrac{2}{3} - 1 = 32\tfrac{4}{9}$$

## Applications

Part 2 of the Fundamental Theorem says that if $f$ is continuous on $[a, b]$, then

$$\int_a^b f(x)\, dx = F(b) - F(a)$$

where $F$ is any antiderivative of $f$. This means that $F' = f$, so the equation can be rewritten as

$$\int_a^b F'(x)\, dx = F(b) - F(a)$$

We know that $F'(x)$ represents the rate of change of $y = F(x)$ with respect to $x$ and $F(b) - F(a)$ is the change in $y$ when $x$ changes from $a$ to $b$. [Note that $y$ could, for instance, increase, then decrease, then increase again. Although $y$ might change in both directions, $F(b) - F(a)$ represents the *net* change in $y$.] So we can reformulate FTC2 in words as follows.

---

**Net Change Theorem**   The integral of a rate of change is the net change:

$$\int_a^b F'(x)\, dx = F(b) - F(a)$$

---

This principle can be applied to all of the rates of change in the natural and social sciences that we discussed in Section 2.7. Here are a few instances of this idea:

- If $V(t)$ is the volume of water in a reservoir at time $t$, then its derivative $V'(t)$ is the rate at which water flows into the reservoir at time $t$. So

$$\int_{t_1}^{t_2} V'(t)\, dt = V(t_2) - V(t_1)$$

  is the change in the amount of water in the reservoir between time $t_1$ and time $t_2$.

- If $[C](t)$ is the concentration of the product of a chemical reaction at time $t$, then the rate of reaction is the derivative $d[C]/dt$. So

$$\int_{t_1}^{t_2} \frac{d[C]}{dt}\, dt = [C](t_2) - [C](t_1)$$

  is the change in the concentration of C from time $t_1$ to time $t_2$.

- If the mass of a rod measured from the left end to a point $x$ is $m(x)$, then the linear density is $\rho(x) = m'(x)$. So

$$\int_a^b \rho(x)\, dx = m(b) - m(a)$$

  is the mass of the segment of the rod that lies between $x = a$ and $x = b$.

- If the rate of growth of a population is $dn/dt$, then

$$\int_{t_1}^{t_2} \frac{dn}{dt}\, dt = n(t_2) - n(t_1)$$

  is the net change in population during the time period from $t_1$ to $t_2$. (The population increases when births happen and decreases when deaths occur. The net change takes into account both births and deaths.)

- If $C(x)$ is the cost of producing $x$ units of a commodity, then the marginal cost is the derivative $C'(x)$. So

$$\int_{x_1}^{x_2} C'(x)\, dx = C(x_2) - C(x_1)$$

  is the increase in cost when production is increased from $x_1$ units to $x_2$ units.

- If an object moves along a straight line with position function $s(t)$, then its velocity is $v(t) = s'(t)$, so

**2**

$$\int_{t_1}^{t_2} v(t)\, dt = s(t_2) - s(t_1)$$

is the net change of position, or *displacement,* of the particle during the time period from $t_1$ to $t_2$. In Section 4.1 we guessed that this was true for the case where the object moves in the positive direction, but now we have proved that it is always true.

- If we want to calculate the distance the object travels during the time interval, we have to consider the intervals when $v(t) \geq 0$ (the particle moves to the right) and also the intervals when $v(t) \leq 0$ (the particle moves to the left). In both cases the distance is computed by integrating $|v(t)|$, the speed. Therefore

$$\boxed{3} \qquad \int_{t_1}^{t_2} |v(t)| \, dt = \text{total distance traveled}$$

Figure 3 shows how both displacement and distance traveled can be interpreted in terms of areas under a velocity curve.

$$\text{displacement} = \int_{t_1}^{t_2} v(t) \, dt = A_1 - A_2 + A_3$$

$$\text{distance} = \int_{t_1}^{t_2} |v(t)| \, dt = A_1 + A_2 + A_3$$

**FIGURE 3**

- The acceleration of the object is $a(t) = v'(t)$, so

$$\int_{t_1}^{t_2} a(t) \, dt = v(t_2) - v(t_1)$$

is the change in velocity from time $t_1$ to time $t_2$.

**V EXAMPLE 6** A particle moves along a line so that its velocity at time $t$ is $v(t) = t^2 - t - 6$ (measured in meters per second).
(a) Find the displacement of the particle during the time period $1 \leq t \leq 4$.
(b) Find the distance traveled during this time period.

SOLUTION
(a) By Equation 2, the displacement is

$$s(4) - s(1) = \int_1^4 v(t) \, dt = \int_1^4 (t^2 - t - 6) \, dt$$

$$= \left[ \frac{t^3}{3} - \frac{t^2}{2} - 6t \right]_1^4 = -\frac{9}{2}$$

This means that the particle moved 4.5 m toward the left.

(b) Note that $v(t) = t^2 - t - 6 = (t - 3)(t + 2)$ and so $v(t) \leq 0$ on the interval $[1, 3]$ and $v(t) \geq 0$ on $[3, 4]$. Thus, from Equation 3, the distance traveled is

To integrate the absolute value of $v(t)$, we use Property 5 of integrals from Section 4.2 to split the integral into two parts, one where $v(t) \leq 0$ and one where $v(t) \geq 0$.

$$\int_1^4 |v(t)| \, dt = \int_1^3 [-v(t)] \, dt + \int_3^4 v(t) \, dt$$

$$= \int_1^3 (-t^2 + t + 6) \, dt + \int_3^4 (t^2 - t - 6) \, dt$$

$$= \left[ -\frac{t^3}{3} + \frac{t^2}{2} + 6t \right]_1^3 + \left[ \frac{t^3}{3} - \frac{t^2}{2} - 6t \right]_3^4$$

$$= \frac{61}{6} \approx 10.17 \text{ m}$$

**EXAMPLE 7** Figure 4 shows the power consumption in the city of San Francisco for a day in September ($P$ is measured in megawatts; $t$ is measured in hours starting at midnight). Estimate the energy used on that day.

**FIGURE 4**

Pacific Gas & Electric

SOLUTION Power is the rate of change of energy: $P(t) = E'(t)$. So, by the Net Change Theorem,

$$\int_0^{24} P(t)\, dt = \int_0^{24} E'(t)\, dt = E(24) - E(0)$$

is the total amount of energy used on that day. We approximate the value of the integral using the Midpoint Rule with 12 subintervals and $\Delta t = 2$:

$$\int_0^{24} P(t)\, dt \approx [P(1) + P(3) + P(5) + \cdots + P(21) + P(23)]\,\Delta t$$

$$\approx (440 + 400 + 420 + 620 + 790 + 840 + 850$$

$$+ 840 + 810 + 690 + 670 + 550)(2)$$

$$= 15{,}840$$

The energy used was approximately 15,840 megawatt-hours.

A note on units

How did we know what units to use for energy in Example 7? The integral $\int_0^{24} P(t)\, dt$ is defined as the limit of sums of terms of the form $P(t_i^*)\,\Delta t$. Now $P(t_i^*)$ is measured in megawatts and $\Delta t$ is measured in hours, so their product is measured in megawatt-hours. The same is true of the limit. In general, the unit of measurement for $\int_a^b f(x)\, dx$ is the product of the unit for $f(x)$ and the unit for $x$.

## 4.4  Exercises

1–4 Verify by differentiation that the formula is correct.

**1.** $\displaystyle\int \frac{1}{x^2\sqrt{1 + x^2}}\, dx = -\frac{\sqrt{1 + x^2}}{x} + C$

**2.** $\displaystyle\int \cos^2 x\, dx = \tfrac{1}{2}x + \tfrac{1}{4}\sin 2x + C$

**3.** $\displaystyle\int \cos^3 x\, dx = \sin x - \tfrac{1}{3}\sin^3 x + C$

**4.** $\displaystyle\int \frac{x}{\sqrt{a + bx}}\, dx = \frac{2}{3b^2}(bx - 2a)\sqrt{a + bx} + C$

5–16 Find the general indefinite integral.

**5.** $\displaystyle\int (x^2 + x^{-2})\, dx$

**6.** $\displaystyle\int \left(\sqrt{x^3} + \sqrt[3]{x^2}\right) dx$

Graphing calculator or computer required    **1.** Homework Hints available at stewartcalculus.com

**7.** $\int \left( x^4 - \tfrac{1}{2}x^3 + \tfrac{1}{4}x - 2 \right) dx$

**8.** $\int \left( y^3 + 1.8y^2 - 2.4y \right) dy$

**9.** $\int (u + 4)(2u + 1)\, du$

**10.** $\int v(v^2 + 2)^2\, dv$

**11.** $\int \dfrac{x^3 - 2\sqrt{x}}{x}\, dx$

**12.** $\int \left( u^2 + 1 + \dfrac{1}{u^2} \right) du$

**13.** $\int (\theta - \csc\theta \cot\theta)\, d\theta$

**14.** $\int \sec t\, (\sec t + \tan t)\, dt$

**15.** $\int (1 + \tan^2\alpha)\, d\alpha$

**16.** $\int \dfrac{\sin 2x}{\sin x}\, dx$

---

**17–18** Find the general indefinite integral. Illustrate by graphing several members of the family on the same screen.

**17.** $\int \left( \cos x + \tfrac{1}{2}x \right) dx$

**18.** $\int (1 - x^2)^2\, dx$

---

**19–42** Evaluate the integral.

**19.** $\int_{-2}^{3} (x^2 - 3)\, dx$

**20.** $\int_{1}^{2} (4x^3 - 3x^2 + 2x)\, dx$

**21.** $\int_{-2}^{0} \left( \tfrac{1}{2}t^4 + \tfrac{1}{4}t^3 - t \right) dt$

**22.** $\int_{0}^{3} (1 + 6w^2 - 10w^4)\, dw$

**23.** $\int_{0}^{2} (2x - 3)(4x^2 + 1)\, dx$

**24.** $\int_{-1}^{1} t(1 - t)^2\, dt$

**25.** $\int_{0}^{\pi} (4\sin\theta - 3\cos\theta)\, d\theta$

**26.** $\int_{1}^{2} \left( \dfrac{1}{x^2} - \dfrac{4}{x^3} \right) dx$

**27.** $\int_{1}^{4} \left( \dfrac{4 + 6u}{\sqrt{u}} \right) du$

**28.** $\int_{1}^{2} \left( x + \dfrac{1}{x} \right)^2 dx$

**29.** $\int_{1}^{4} \sqrt{\dfrac{5}{x}}\, dx$

**30.** $\int_{1}^{9} \dfrac{3x - 2}{\sqrt{x}}\, dx$

**31.** $\int_{1}^{4} \sqrt{t}\, (1 + t)\, dt$

**32.** $\int_{\pi/4}^{\pi/3} \csc^2\theta\, d\theta$

**33.** $\int_{0}^{\pi/4} \dfrac{1 + \cos^2\theta}{\cos^2\theta}\, d\theta$

**34.** $\int_{0}^{\pi/3} \dfrac{\sin\theta + \sin\theta \tan^2\theta}{\sec^2\theta}\, d\theta$

**35.** $\int_{1}^{64} \dfrac{1 + \sqrt[3]{x}}{\sqrt{x}}\, dx$

**36.** $\int_{1}^{8} \dfrac{x - 1}{\sqrt[3]{x^2}}\, dx$

**37.** $\int_{0}^{1} \left( \sqrt[4]{x^5} + \sqrt[5]{x^4} \right) dx$

**38.** $\int_{0}^{1} (1 + x^2)^3\, dx$

**39.** $\int_{2}^{5} |x - 3|\, dx$

**40.** $\int_{0}^{2} |2x - 1|\, dx$

**41.** $\int_{-1}^{2} (x - 2|x|)\, dx$

**42.** $\int_{0}^{3\pi/2} |\sin x|\, dx$

**43.** Use a graph to estimate the $x$-intercepts of the curve $y = 1 - 2x - 5x^4$. Then use this information to estimate the area of the region that lies under the curve and above the $x$-axis.

**44.** Repeat Exercise 43 for the curve $y = 2x + 3x^4 - 2x^6$.

**45.** The area of the region that lies to the right of the $y$-axis and to the left of the parabola $x = 2y - y^2$ (the shaded region in the figure) is given by the integral $\int_{0}^{2} (2y - y^2)\, dy$. (Turn your head clockwise and think of the region as lying below the curve $x = 2y - y^2$ from $y = 0$ to $y = 2$.) Find the area of the region.

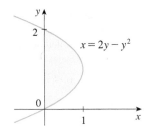

**46.** The boundaries of the shaded region are the $y$-axis, the line $y = 1$, and the curve $y = \sqrt[4]{x}$. Find the area of this region by writing $x$ as a function of $y$ and integrating with respect to $y$ (as in Exercise 45).

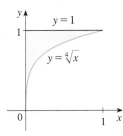

**47.** If $w'(t)$ is the rate of growth of a child in pounds per year, what does $\int_{5}^{10} w'(t)\, dt$ represent?

**48.** The current in a wire is defined as the derivative of the charge: $I(t) = Q'(t)$. (See Example 3 in Section 2.7.) What does $\int_{a}^{b} I(t)\, dt$ represent?

**49.** If oil leaks from a tank at a rate of $r(t)$ gallons per minute at time $t$, what does $\int_{0}^{120} r(t)\, dt$ represent?

**50.** A honeybee population starts with 100 bees and increases at a rate of $n'(t)$ bees per week. What does $100 + \int_{0}^{15} n'(t)\, dt$ represent?

**51.** In Section 3.7 we defined the marginal revenue function $R'(x)$ as the derivative of the revenue function $R(x)$, where $x$ is the number of units sold. What does $\int_{1000}^{5000} R'(x)\, dx$ represent?

**52.** If $f(x)$ is the slope of a trail at a distance of $x$ miles from the start of the trail, what does $\int_{3}^{5} f(x)\, dx$ represent?

**53.** If $x$ is measured in meters and $f(x)$ is measured in newtons, what are the units for $\int_{0}^{100} f(x)\, dx$?

**54.** If the units for $x$ are feet and the units for $a(x)$ are pounds per foot, what are the units for $da/dx$? What units does $\int_2^8 a(x)\, dx$ have?

**55–56** The velocity function (in meters per second) is given for a particle moving along a line. Find (a) the displacement and (b) the distance traveled by the particle during the given time interval.

**55.** $v(t) = 3t - 5$, $\quad 0 \leqslant t \leqslant 3$

**56.** $v(t) = t^2 - 2t - 8$, $\quad 1 \leqslant t \leqslant 6$

**57–58** The acceleration function (in m/s$^2$) and the initial velocity are given for a particle moving along a line. Find (a) the velocity at time $t$ and (b) the distance traveled during the given time interval.

**57.** $a(t) = t + 4$, $\quad v(0) = 5$, $\quad 0 \leqslant t \leqslant 10$

**58.** $a(t) = 2t + 3$, $\quad v(0) = -4$, $\quad 0 \leqslant t \leqslant 3$

**59.** The linear density of a rod of length 4 m is given by $\rho(x) = 9 + 2\sqrt{x}$ measured in kilograms per meter, where $x$ is measured in meters from one end of the rod. Find the total mass of the rod.

**60.** Water flows from the bottom of a storage tank at a rate of $r(t) = 200 - 4t$ liters per minute, where $0 \leqslant t \leqslant 50$. Find the amount of water that flows from the tank during the first 10 minutes.

**61.** The velocity of a car was read from its speedometer at 10-second intervals and recorded in the table. Use the Midpoint Rule to estimate the distance traveled by the car.

| $t$ (s) | $v$ (mi/h) | $t$ (s) | $v$ (mi/h) |
|---|---|---|---|
| 0 | 0 | 60 | 56 |
| 10 | 38 | 70 | 53 |
| 20 | 52 | 80 | 50 |
| 30 | 58 | 90 | 47 |
| 40 | 55 | 100 | 45 |
| 50 | 51 | | |

**62.** Suppose that a volcano is erupting and readings of the rate $r(t)$ at which solid materials are spewed into the atmosphere are given in the table. The time $t$ is measured in seconds and the units for $r(t)$ are tonnes (metric tons) per second.

| $t$ | 0 | 1 | 2 | 3 | 4 | 5 | 6 |
|---|---|---|---|---|---|---|---|
| $r(t)$ | 2 | 10 | 24 | 36 | 46 | 54 | 60 |

(a) Give upper and lower estimates for the total quantity $Q(6)$ of erupted materials after 6 seconds.
(b) Use the Midpoint Rule to estimate $Q(6)$.

**63.** Water flows into and out of a storage tank. A graph of the rate of change $r(t)$ of the volume of water in the tank, in liters per day, is shown. If the amount of water in the tank at time $t = 0$ is 25,000 L, use the Midpoint Rule to estimate the amount of water in the tank four days later.

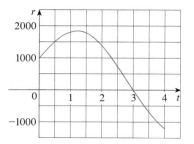

**64.** Shown is the graph of traffic on an Internet service provider's T1 data line from midnight to 8:00 AM. $D$ is the data throughput, measured in megabits per second. Use the Midpoint Rule to estimate the total amount of data transmitted during that time period.

**65.** Shown is the power consumption in the province of Ontario, Canada, for December 9, 2004 ($P$ is measured in megawatts; $t$ is measured in hours starting at midnight). Using the fact that power is the rate of change of energy, estimate the energy used on that day.

Independent Electricity Market Operator

**66.** On May 7, 1992, the space shuttle *Endeavour* was launched on mission STS-49, the purpose of which was to install a new perigee kick motor in an Intelsat communications satellite. The table gives the velocity data for the shuttle between liftoff and the jettisoning of the solid rocket boosters.

| Event | Time (s) | Velocity (ft/s) |
|---|---|---|
| Launch | 0 | 0 |
| Begin roll maneuver | 10 | 185 |
| End roll maneuver | 15 | 319 |
| Throttle to 89% | 20 | 447 |
| Throttle to 67% | 32 | 742 |
| Throttle to 104% | 59 | 1325 |
| Maximum dynamic pressure | 62 | 1445 |
| Solid rocket booster separation | 125 | 4151 |

(a) Use a graphing calculator or computer to model these data by a third-degree polynomial.

(b) Use the model in part (a) to estimate the height reached by the *Endeavour*, 125 seconds after liftoff.

The following exercises are intended only for those who have already covered Chapter 6.

**67–71** Evaluate the integral.

**67.** $\int (\sin x + \sinh x)\, dx$

**68.** $\int_{-10}^{10} \dfrac{2e^x}{\sinh x + \cosh x}\, dx$

**69.** $\int \left( x^2 + 1 + \dfrac{1}{x^2 + 1} \right) dx$

**70.** $\int_{1}^{2} \dfrac{(x-1)^3}{x^2}\, dx$

**71.** $\int_{0}^{1/\sqrt{3}} \dfrac{t^2 - 1}{t^4 - 1}\, dt$

**72.** The area labeled $B$ is three times the area labeled $A$. Express $b$ in terms of $a$.

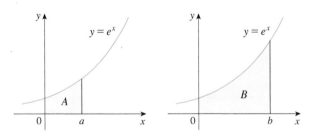

---

## WRITING PROJECT

### NEWTON, LEIBNIZ, AND THE INVENTION OF CALCULUS

We sometimes read that the inventors of calculus were Sir Isaac Newton (1642–1727) and Gottfried Wilhelm Leibniz (1646–1716). But we know that the basic ideas behind integration were investigated 2500 years ago by ancient Greeks such as Eudoxus and Archimedes, and methods for finding tangents were pioneered by Pierre Fermat (1601–1665), Isaac Barrow (1630–1677), and others. Barrow—who taught at Cambridge and was a major influence on Newton—was the first to understand the inverse relationship between differentiation and integration. What Newton and Leibniz did was to use this relationship, in the form of the Fundamental Theorem of Calculus, in order to develop calculus into a systematic mathematical discipline. It is in this sense that Newton and Leibniz are credited with the invention of calculus.

Read about the contributions of these men in one or more of the given references and write a report on one of the following three topics. You can include biographical details, but the main thrust of your report should be a description, in some detail, of their methods and notations. In particular, you should consult one of the sourcebooks, which give excerpts from the original publications of Newton and Leibniz, translated from Latin to English.

- The Role of Newton in the Development of Calculus

- The Role of Leibniz in the Development of Calculus

- The Controversy between the Followers of Newton and Leibniz over Priority in the Invention of Calculus

**References**

**1.** Carl Boyer and Uta Merzbach, *A History of Mathematics* (New York: Wiley, 1987), Chapter 19.

**2.** Carl Boyer, *The History of the Calculus and Its Conceptual Development* (New York: Dover, 1959), Chapter V.

**3.** C. H. Edwards, *The Historical Development of the Calculus* (New York: Springer-Verlag, 1979), Chapters 8 and 9.

**4.** Howard Eves, *An Introduction to the History of Mathematics,* 6th ed. (New York: Saunders, 1990), Chapter 11.

**5.** C. C. Gillispie, ed., *Dictionary of Scientific Biography* (New York: Scribner's, 1974). See the article on Leibniz by Joseph Hofmann in Volume VIII and the article on Newton by I. B. Cohen in Volume X.

**6.** Victor Katz, *A History of Mathematics: An Introduction* (New York: HarperCollins, 1993), Chapter 12.

**7.** Morris Kline, *Mathematical Thought from Ancient to Modern Times* (New York: Oxford University Press, 1972), Chapter 17.

**Sourcebooks**

**1.** John Fauvel and Jeremy Gray, eds., *The History of Mathematics: A Reader* (London: MacMillan Press, 1987), Chapters 12 and 13.

**2.** D. E. Smith, ed., *A Sourcebook in Mathematics* (New York: Dover, 1959), Chapter V.

**3.** D. J. Struik, ed., *A Sourcebook in Mathematics, 1200–1800* (Princeton, NJ: Princeton University Press, 1969), Chapter V.

## 4.5   The Substitution Rule

Because of the Fundamental Theorem, it's important to be able to find antiderivatives. But our antidifferentiation formulas don't tell us how to evaluate integrals such as

$$\boxed{1} \qquad \int 2x\sqrt{1 + x^2}\, dx$$

To find this integral we use the problem-solving strategy of *introducing something extra.* Here the "something extra" is a new variable; we change from the variable $x$ to a new variable $u$. Suppose that we let $u$ be the quantity under the root sign in $\boxed{1}$, $u = 1 + x^2$. Then the differential of $u$ is $du = 2x\, dx$. Notice that if the $dx$ in the notation for an integral were to be interpreted as a differential, then the differential $2x\, dx$ would occur in $\boxed{1}$ and so, formally, without justifying our calculation, we could write

Differentials were defined in Section 2.9. If $u = f(x)$, then

$$du = f'(x)\, dx$$

$$\boxed{2} \qquad \int 2x\sqrt{1 + x^2}\, dx = \int \sqrt{1 + x^2}\, 2x\, dx = \int \sqrt{u}\, du$$

$$= \tfrac{2}{3}u^{3/2} + C = \tfrac{2}{3}(x^2 + 1)^{3/2} + C$$

But now we can check that we have the correct answer by using the Chain Rule to differentiate the final function of Equation 2:

$$\frac{d}{dx}\left[\tfrac{2}{3}(x^2 + 1)^{3/2} + C\right] = \tfrac{2}{3} \cdot \tfrac{3}{2}(x^2 + 1)^{1/2} \cdot 2x = 2x\sqrt{x^2 + 1}$$

In general, this method works whenever we have an integral that we can write in the form $\int f(g(x))\, g'(x)\, dx$. Observe that if $F' = f$, then

$$\boxed{3} \qquad \int F'(g(x))\, g'(x)\, dx = F(g(x)) + C$$

because, by the Chain Rule,

$$\frac{d}{dx}\,[F(g(x))] = F'(g(x))\,g'(x)$$

If we make the "change of variable" or "substitution" $u = g(x)$, then from Equation 3 we have

$$\int F'(g(x))\,g'(x)\,dx = F(g(x)) + C = F(u) + C = \int F'(u)\,du$$

or, writing $F' = f$, we get

$$\int f(g(x))\,g'(x)\,dx = \int f(u)\,du$$

Thus we have proved the following rule.

> **4   The Substitution Rule**   If $u = g(x)$ is a differentiable function whose range is an interval $I$ and $f$ is continuous on $I$, then
>
> $$\int f(g(x))\,g'(x)\,dx = \int f(u)\,du$$

Notice that the Substitution Rule for integration was proved using the Chain Rule for differentiation. Notice also that if $u = g(x)$, then $du = g'(x)\,dx$, so a way to remember the Substitution Rule is to think of $dx$ and $du$ in $\boxed{4}$ as differentials.

Thus the Substitution Rule says: **It is permissible to operate with $dx$ and $du$ after integral signs as if they were differentials.**

**EXAMPLE 1**   Find $\displaystyle\int x^3 \cos(x^4 + 2)\,dx$.

SOLUTION   We make the substitution $u = x^4 + 2$ because its differential is $du = 4x^3\,dx$, which, apart from the constant factor 4, occurs in the integral. Thus, using $x^3\,dx = \tfrac{1}{4}\,du$ and the Substitution Rule, we have

$$\int x^3 \cos(x^4 + 2)\,dx = \int \cos u \cdot \tfrac{1}{4}\,du = \tfrac{1}{4}\int \cos u\,du$$

$$= \tfrac{1}{4}\sin u + C$$

$$= \tfrac{1}{4}\sin(x^4 + 2) + C$$

Check the answer by differentiating it.

Notice that at the final stage we had to return to the original variable $x$.

The idea behind the Substitution Rule is to replace a relatively complicated integral by a simpler integral. This is accomplished by changing from the original variable $x$ to a new variable $u$ that is a function of $x$. Thus in Example 1 we replaced the integral $\int x^3 \cos(x^4 + 2)\,dx$ by the simpler integral $\tfrac{1}{4}\int \cos u\,du$.

The main challenge in using the Substitution Rule is to think of an appropriate substitution. You should try to choose $u$ to be some function in the integrand whose differential also occurs (except for a constant factor). This was the case in Example 1. If that is not possible, try choosing $u$ to be some complicated part of the integrand (perhaps the inner function in a composite function). Finding the right substitution is a bit of an art. It's not unusual to guess wrong; if your first guess doesn't work, try another substitution.

**EXAMPLE 2**  Evaluate $\int \sqrt{2x + 1}\, dx$.

SOLUTION 1  Let $u = 2x + 1$. Then $du = 2\, dx$, so $dx = \frac{1}{2}\, du$. Thus the Substitution Rule gives

$$\int \sqrt{2x + 1}\, dx = \int \sqrt{u}\, \cdot \tfrac{1}{2}\, du = \tfrac{1}{2} \int u^{1/2}\, du$$

$$= \frac{1}{2} \cdot \frac{u^{3/2}}{3/2} + C = \tfrac{1}{3} u^{3/2} + C$$

$$= \tfrac{1}{3}(2x + 1)^{3/2} + C$$

SOLUTION 2  Another possible substitution is $u = \sqrt{2x + 1}$. Then

$$du = \frac{dx}{\sqrt{2x + 1}} \qquad \text{so} \qquad dx = \sqrt{2x + 1}\, du = u\, du$$

(Or observe that $u^2 = 2x + 1$, so $2u\, du = 2\, dx$.) Therefore

$$\int \sqrt{2x + 1}\, dx = \int u \cdot u\, du = \int u^2\, du$$

$$= \frac{u^3}{3} + C = \tfrac{1}{3}(2x + 1)^{3/2} + C$$

**V  EXAMPLE 3**  Find $\int \dfrac{x}{\sqrt{1 - 4x^2}}\, dx$.

SOLUTION  Let $u = 1 - 4x^2$. Then $du = -8x\, dx$, so $x\, dx = -\frac{1}{8}\, du$ and

$$\int \frac{x}{\sqrt{1 - 4x^2}}\, dx = -\tfrac{1}{8} \int \frac{1}{\sqrt{u}}\, du = -\tfrac{1}{8} \int u^{-1/2}\, du$$

$$= -\tfrac{1}{8}(2\sqrt{u}) + C = -\tfrac{1}{4}\sqrt{1 - 4x^2} + C$$

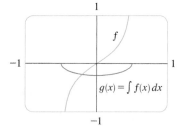

**FIGURE 1**

$$f(x) = \frac{x}{\sqrt{1 - 4x^2}}$$

$$g(x) = \int f(x)\, dx = -\tfrac{1}{4}\sqrt{1 - 4x^2}$$

The answer to Example 3 could be checked by differentiation, but instead let's check it with a graph. In Figure 1 we have used a computer to graph both the integrand $f(x) = x/\sqrt{1 - 4x^2}$ and its indefinite integral $g(x) = -\frac{1}{4}\sqrt{1 - 4x^2}$ (we take the case $C = 0$). Notice that $g(x)$ decreases when $f(x)$ is negative, increases when $f(x)$ is positive, and has its minimum value when $f(x) = 0$. So it seems reasonable, from the graphical evidence, that $g$ is an antiderivative of $f$.

**EXAMPLE 4**  Calculate $\int \cos 5x\, dx$.

SOLUTION  If we let $u = 5x$, then $du = 5\, dx$, so $dx = \frac{1}{5}\, du$. Therefore

$$\int \cos 5x\, dx = \tfrac{1}{5} \int \cos u\, du = \tfrac{1}{5} \sin u + C = \tfrac{1}{5} \sin 5x + C$$

**NOTE**  With some experience, you might be able to evaluate integrals like those in Examples 1–4 without going to the trouble of making an explicit substitution. By recognizing the pattern in Equation 3, where the integrand on the left side is the product of the derivative of an outer function and the derivative of the inner function, we could work

Example 1 as follows:

$$\int x^3 \cos(x^4 + 2)\, dx = \int \cos(x^4 + 2) \cdot x^3\, dx = \tfrac{1}{4} \int \cos(x^4 + 2) \cdot (4x^3)\, dx$$

$$= \tfrac{1}{4} \int \cos(x^4 + 2) \cdot \frac{d}{dx}(x^4 + 2)\, dx = \tfrac{1}{4} \sin(x^4 + 2) + C$$

Similarly, the solution to Example 4 could be written like this:

$$\int \cos 5x\, dx = \tfrac{1}{5} \int 5 \cos 5x\, dx = \tfrac{1}{5} \int \frac{d}{dx}(\sin 5x)\, dx = \tfrac{1}{5} \sin 5x + C$$

The following example, however, is more complicated and so an explicit substitution is advisable.

**EXAMPLE 5** Find $\displaystyle \int \sqrt{1 + x^2}\, x^5\, dx$.

SOLUTION An appropriate substitution becomes more obvious if we factor $x^5$ as $x^4 \cdot x$. Let $u = 1 + x^2$. Then $du = 2x\, dx$, so $x\, dx = \tfrac{1}{2} du$. Also $x^2 = u - 1$, so $x^4 = (u - 1)^2$:

$$\int \sqrt{1 + x^2}\, x^5\, dx = \int \sqrt{1 + x^2}\, x^4 \cdot x\, dx$$

$$= \int \sqrt{u}\,(u - 1)^2 \cdot \tfrac{1}{2}\, du = \tfrac{1}{2} \int \sqrt{u}\,(u^2 - 2u + 1)\, du$$

$$= \tfrac{1}{2} \int (u^{5/2} - 2u^{3/2} + u^{1/2})\, du$$

$$= \tfrac{1}{2}\left(\tfrac{2}{7}u^{7/2} - 2 \cdot \tfrac{2}{5}u^{5/2} + \tfrac{2}{3}u^{3/2}\right) + C$$

$$= \tfrac{1}{7}(1 + x^2)^{7/2} - \tfrac{2}{5}(1 + x^2)^{5/2} + \tfrac{1}{3}(1 + x^2)^{3/2} + C$$

## Definite Integrals

When evaluating a *definite* integral by substitution, two methods are possible. One method is to evaluate the indefinite integral first and then use the Fundamental Theorem. For instance, using the result of Example 2, we have

$$\int_0^4 \sqrt{2x + 1}\, dx = \int \sqrt{2x + 1}\, dx\Big]_0^4$$

$$= \tfrac{1}{3}(2x + 1)^{3/2}\Big]_0^4 = \tfrac{1}{3}(9)^{3/2} - \tfrac{1}{3}(1)^{3/2}$$

$$= \tfrac{1}{3}(27 - 1) = \tfrac{26}{3}$$

Another method, which is usually preferable, is to change the limits of integration when the variable is changed.

This rule says that when using a substitution in a definite integral, we must put everything in terms of the new variable $u$, not only $x$ and $dx$ but also the limits of integration. The new limits of integration are the values of $u$ that correspond to $x = a$ and $x = b$.

**5   The Substitution Rule for Definite Integrals**   If $g'$ is continuous on $[a, b]$ and $f$ is continuous on the range of $u = g(x)$, then

$$\int_a^b f(g(x))\, g'(x)\, dx = \int_{g(a)}^{g(b)} f(u)\, du$$

PROOF Let $F$ be an antiderivative of $f$. Then, by $\boxed{3}$, $F(g(x))$ is an antiderivative of $f(g(x))\,g'(x)$, so by Part 2 of the Fundamental Theorem, we have

$$\int_a^b f(g(x))\,g'(x)\,dx = F(g(x))\Big]_a^b = F(g(b)) - F(g(a))$$

But, applying FTC2 a second time, we also have

$$\int_{g(a)}^{g(b)} f(u)\,du = F(u)\Big]_{g(a)}^{g(b)} = F(g(b)) - F(g(a))$$

$\boxed{V}$ **EXAMPLE 6** Evaluate $\int_0^4 \sqrt{2x+1}\,dx$ using $\boxed{5}$.

SOLUTION Using the substitution from Solution 1 of Example 2, we have $u = 2x + 1$ and $dx = \frac{1}{2}\,du$. To find the new limits of integration we note that

$$\text{when } x = 0, \quad u = 2(0) + 1 = 1 \qquad \text{and} \qquad \text{when } x = 4, \quad u = 2(4) + 1 = 9$$

Therefore

$$\int_0^4 \sqrt{2x+1}\,dx = \int_1^9 \tfrac{1}{2}\sqrt{u}\,du$$

$$= \tfrac{1}{2} \cdot \tfrac{2}{3} u^{3/2}\Big]_1^9$$

$$= \tfrac{1}{3}(9^{3/2} - 1^{3/2}) = \tfrac{26}{3}$$

Observe that when using $\boxed{5}$ we do *not* return to the variable $x$ after integrating. We simply evaluate the expression in $u$ between the appropriate values of $u$.

**EXAMPLE 7** Evaluate $\displaystyle\int_1^2 \frac{dx}{(3-5x)^2}$.

The integral given in Example 7 is an abbreviation for

$$\int_1^2 \frac{1}{(3-5x)^2}\,dx$$

SOLUTION Let $u = 3 - 5x$. Then $du = -5\,dx$, so $dx = -\frac{1}{5}\,du$. When $x = 1$, $u = -2$ and when $x = 2$, $u = -7$. Thus

$$\int_1^2 \frac{dx}{(3-5x)^2} = -\frac{1}{5}\int_{-2}^{-7} \frac{du}{u^2}$$

$$= -\frac{1}{5}\left[-\frac{1}{u}\right]_{-2}^{-7} = \frac{1}{5u}\Big]_{-2}^{-7}$$

$$= \frac{1}{5}\left(-\frac{1}{7} + \frac{1}{2}\right) = \frac{1}{14}$$

## Symmetry

The next theorem uses the Substitution Rule for Definite Integrals $\boxed{5}$ to simplify the calculation of integrals of functions that possess symmetry properties.

---

$\boxed{6}$ **Integrals of Symmetric Functions** Suppose $f$ is continuous on $[-a, a]$.

(a) If $f$ is even $[f(-x) = f(x)]$, then $\int_{-a}^a f(x)\,dx = 2\int_0^a f(x)\,dx$.

(b) If $f$ is odd $[f(-x) = -f(x)]$, then $\int_{-a}^a f(x)\,dx = 0$.

---

PROOF   We split the integral in two:

$$\boxed{7} \qquad \int_{-a}^{a} f(x)\, dx = \int_{-a}^{0} f(x)\, dx + \int_{0}^{a} f(x)\, dx = -\int_{0}^{-a} f(x)\, dx + \int_{0}^{a} f(x)\, dx$$

In the first integral on the far right side we make the substitution $u = -x$. Then $du = -dx$ and when $x = -a$, $u = a$. Therefore

$$-\int_{0}^{-a} f(x)\, dx = -\int_{0}^{a} f(-u)\,(-du) = \int_{0}^{a} f(-u)\, du$$

and so Equation 7 becomes

$$\boxed{8} \qquad\qquad \int_{-a}^{a} f(x)\, dx = \int_{0}^{a} f(-u)\, du + \int_{0}^{a} f(x)\, dx$$

(a) If $f$ is even, then $f(-u) = f(u)$ so Equation 8 gives

$$\int_{-a}^{a} f(x)\, dx = \int_{0}^{a} f(u)\, du + \int_{0}^{a} f(x)\, dx = 2\int_{0}^{a} f(x)\, dx$$

(b) If $f$ is odd, then $f(-u) = -f(u)$ and so Equation 8 gives

$$\int_{-a}^{a} f(x)\, dx = -\int_{0}^{a} f(u)\, du + \int_{0}^{a} f(x)\, dx = 0$$

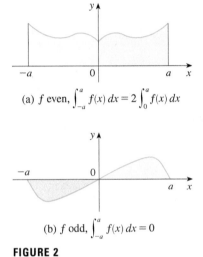

(a) $f$ even, $\int_{-a}^{a} f(x)\, dx = 2 \int_{0}^{a} f(x)\, dx$

(b) $f$ odd, $\int_{-a}^{a} f(x)\, dx = 0$

**FIGURE 2**

Theorem 6 is illustrated by Figure 2. For the case where $f$ is positive and even, part (a) says that the area under $y = f(x)$ from $-a$ to $a$ is twice the area from 0 to $a$ because of symmetry. Recall that an integral $\int_{a}^{b} f(x)\, dx$ can be expressed as the area above the $x$-axis and below $y = f(x)$ minus the area below the axis and above the curve. Thus part (b) says the integral is 0 because the areas cancel.

**V  EXAMPLE 8**   Since $f(x) = x^6 + 1$ satisfies $f(-x) = f(x)$, it is even and so

$$\int_{-2}^{2} (x^6 + 1)\, dx = 2\int_{0}^{2} (x^6 + 1)\, dx$$

$$= 2\left[\tfrac{1}{7}x^7 + x\right]_{0}^{2} = 2\left(\tfrac{128}{7} + 2\right) = \tfrac{284}{7}$$

**EXAMPLE 9**   Since $f(x) = (\tan x)/(1 + x^2 + x^4)$ satisfies $f(-x) = -f(x)$, it is odd and so

$$\int_{-1}^{1} \frac{\tan x}{1 + x^2 + x^4}\, dx = 0$$

## 4.5   Exercises

1–6 Evaluate the integral by making the given substitution.

**1.** $\displaystyle\int \sin \pi x\, dx, \quad u = \pi x$

**2.** $\displaystyle\int x^3 (2 + x^4)^5\, dx, \quad u = 2 + x^4$

**3.** $\displaystyle\int x^2 \sqrt{x^3 + 1}\, dx, \quad u = x^3 + 1$

**4.** $\displaystyle\int \frac{dt}{(1 - 6t)^4}, \quad u = 1 - 6t$

**5.** $\int \cos^3\theta \sin\theta \, d\theta, \quad u = \cos\theta$

**6.** $\int \dfrac{\sec^2(1/x)}{x^2} \, dx, \quad u = 1/x$

**7–30** Evaluate the indefinite integral.

**7.** $\int x \sin(x^2) \, dx$

**8.** $\int x^2 \cos(x^3) \, dx$

**9.** $\int (1 - 2x)^9 \, dx$

**10.** $\int (3t + 2)^{2.4} \, dt$

**11.** $\int (x + 1)\sqrt{2x + x^2} \, dx$

**12.** $\int \sec^2 2\theta \, d\theta$

**13.** $\int \sec 3t \tan 3t \, dt$

**14.** $\int u\sqrt{1 - u^2} \, du$

**15.** $\int \dfrac{a + bx^2}{\sqrt{3ax + bx^3}} \, dx$

**16.** $\int \dfrac{\sin\sqrt{x}}{\sqrt{x}} \, dx$

**17.** $\int \sec^2\theta \tan^3\theta \, d\theta$

**18.** $\int \cos^4\theta \sin\theta \, d\theta$

**19.** $\int (x^2 + 1)(x^3 + 3x)^4 \, dx$

**20.** $\int \sqrt{x}\,\sin(1 + x^{3/2}) \, dx$

**21.** $\int \dfrac{\cos x}{\sin^2 x} \, dx$

**22.** $\int \dfrac{\cos(\pi/x)}{x^2} \, dx$

**23.** $\int \dfrac{z^2}{\sqrt[3]{1 + z^3}} \, dz$

**24.** $\int \dfrac{dt}{\cos^2 t\sqrt{1 + \tan t}}$

**25.** $\int \sqrt{\cot x}\,\csc^2 x \, dx$

**26.** $\int \sin t \sec^2(\cos t) \, dt$

**27.** $\int \sec^3 x \tan x \, dx$

**28.** $\int x^2\sqrt{2 + x} \, dx$

**29.** $\int x(2x + 5)^8 \, dx$

**30.** $\int x^3\sqrt{x^2 + 1} \, dx$

 **31–34** Evaluate the indefinite integral. Illustrate and check that your answer is reasonable by graphing both the function and its antiderivative (take $C = 0$).

**31.** $\int x(x^2 - 1)^3 \, dx$

**32.** $\int \tan^2\theta \sec^2\theta \, d\theta$

**33.** $\int \sin^3 x \cos x \, dx$

**34.** $\int \sin x \cos^4 x \, dx$

**35–51** Evaluate the definite integral.

**35.** $\int_0^1 \cos(\pi t/2) \, dt$

**36.** $\int_0^1 (3t - 1)^{50} \, dt$

**37.** $\int_0^1 \sqrt[3]{1 + 7x} \, dx$

**38.** $\int_0^{\sqrt{\pi}} x \cos(x^2) \, dx$

**39.** $\int_0^\pi \sec^2(t/4) \, dt$

**40.** $\int_{1/6}^{1/2} \csc \pi t \cot \pi t \, dt$

**41.** $\int_{-\pi/4}^{\pi/4} (x^3 + x^4 \tan x) \, dx$

**42.** $\int_0^{\pi/2} \cos x \sin(\sin x) \, dx$

**43.** $\int_0^{13} \dfrac{dx}{\sqrt[3]{(1 + 2x)^2}}$

**44.** $\int_0^a x\sqrt{a^2 - x^2} \, dx$

**45.** $\int_0^a x\sqrt{x^2 + a^2} \, dx \quad (a > 0)$

**46.** $\int_{-\pi/3}^{\pi/3} x^4 \sin x \, dx$

**47.** $\int_1^2 x\sqrt{x - 1} \, dx$

**48.** $\int_0^4 \dfrac{x}{\sqrt{1 + 2x}} \, dx$

**49.** $\int_{1/2}^1 \dfrac{\cos(x^{-2})}{x^3} \, dx$

**50.** $\int_0^{T/2} \sin(2\pi t/T - \alpha) \, dt$

**51.** $\int_0^1 \dfrac{dx}{(1 + \sqrt{x})^4}$

**52.** Verify that $f(x) = \sin\sqrt[3]{x}$ is an odd function and use that fact to show that
$$0 \le \int_{-2}^3 \sin\sqrt[3]{x} \, dx \le 1$$

**53–54** Use a graph to give a rough estimate of the area of the region that lies under the given curve. Then find the exact area.

**53.** $y = \sqrt{2x + 1}, \quad 0 \le x \le 1$

**54.** $y = 2\sin x - \sin 2x, \quad 0 \le x \le \pi$

**55.** Evaluate $\int_{-2}^2 (x + 3)\sqrt{4 - x^2} \, dx$ by writing it as a sum of two integrals and interpreting one of those integrals in terms of an area.

**56.** Evaluate $\int_0^1 x\sqrt{1 - x^4} \, dx$ by making a substitution and interpreting the resulting integral in terms of an area.

**57.** Breathing is cyclic and a full respiratory cycle from the beginning of inhalation to the end of exhalation takes about 5 s. The maximum rate of air flow into the lungs is about 0.5 L/s. This explains, in part, why the function $f(t) = \frac{1}{2}\sin(2\pi t/5)$ has often been used to model the rate of air flow into the lungs. Use this model to find the volume of inhaled air in the lungs at time $t$.

**58.** A model for the basal metabolism rate, in kcal/h, of a young man is $R(t) = 85 - 0.18\cos(\pi t/12)$, where $t$ is the time in hours measured from 5:00 AM. What is the total basal metabolism of this man, $\int_0^{24} R(t) \, dt$, over a 24-hour time period?

**59.** If $f$ is continuous and $\int_0^4 f(x) \, dx = 10$, find $\int_0^2 f(2x) \, dx$.

**60.** If $f$ is continuous and $\int_0^9 f(x) \, dx = 4$, find $\int_0^3 xf(x^2) \, dx$.

**61.** If $f$ is continuous on $\mathbb{R}$, prove that

$$\int_a^b f(-x)\, dx = \int_{-b}^{-a} f(x)\, dx$$

For the case where $f(x) \geq 0$ and $0 < a < b$, draw a diagram to interpret this equation geometrically as an equality of areas.

**62.** If $f$ is continuous on $\mathbb{R}$, prove that

$$\int_a^b f(x + c)\, dx = \int_{a+c}^{b+c} f(x)\, dx$$

For the case where $f(x) \geq 0$, draw a diagram to interpret this equation geometrically as an equality of areas.

**63.** If $a$ and $b$ are positive numbers, show that

$$\int_0^1 x^a (1 - x)^b\, dx = \int_0^1 x^b (1 - x)^a\, dx$$

**64.** If $f$ is continuous on $[0, \pi]$, use the substitution $u = \pi - x$ to show that

$$\int_0^\pi x f(\sin x)\, dx = \frac{\pi}{2} \int_0^\pi f(\sin x)\, dx$$

**65.** If $f$ is continuous, prove that

$$\int_0^{\pi/2} f(\cos x)\, dx = \int_0^{\pi/2} f(\sin x)\, dx$$

**66.** Use Exercise 65 to evaluate $\int_0^{\pi/2} \cos^2 x\, dx$ and $\int_0^{\pi/2} \sin^2 x\, dx$.

---

The following exercises are intended only for those who have already covered Chapter 6.

**67–84** Evaluate the integral.

**67.** $\displaystyle\int \frac{dx}{5 - 3x}$

**68.** $\displaystyle\int e^x \sin(e^x)\, dx$

**69.** $\displaystyle\int \frac{(\ln x)^2}{x}\, dx$

**70.** $\displaystyle\int \frac{dx}{ax + b}$  $(a \neq 0)$

**71.** $\displaystyle\int e^x \sqrt{1 + e^x}\, dx$

**72.** $\displaystyle\int e^{\cos t} \sin t\, dt$

**73.** $\displaystyle\int e^{\tan x} \sec^2 x\, dx$

**74.** $\displaystyle\int \frac{\tan^{-1} x}{1 + x^2}\, dx$

**75.** $\displaystyle\int \frac{1 + x}{1 + x^2}\, dx$

**76.** $\displaystyle\int \frac{\sin(\ln x)}{x}\, dx$

**77.** $\displaystyle\int \frac{\sin 2x}{1 + \cos^2 x}\, dx$

**78.** $\displaystyle\int \frac{\sin x}{1 + \cos^2 x}\, dx$

**79.** $\displaystyle\int \cot x\, dx$

**80.** $\displaystyle\int \frac{x}{1 + x^4}\, dx$

**81.** $\displaystyle\int_e^{e^4} \frac{dx}{x\sqrt{\ln x}}$

**82.** $\displaystyle\int_0^1 x e^{-x^2}\, dx$

**83.** $\displaystyle\int_0^1 \frac{e^z + 1}{e^z + z}\, dz$

**84.** $\displaystyle\int_0^{1/2} \frac{\sin^{-1} x}{\sqrt{1 - x^2}}\, dx$

**85.** Use Exercise 64 to evaluate the integral

$$\int_0^\pi \frac{x \sin x}{1 + \cos^2 x}\, dx$$

---

## 4   Review

### Concept Check

**1.** (a) Write an expression for a Riemann sum of a function $f$. Explain the meaning of the notation that you use.

(b) If $f(x) \geq 0$, what is the geometric interpretation of a Riemann sum? Illustrate with a diagram.

(c) If $f(x)$ takes on both positive and negative values, what is the geometric interpretation of a Riemann sum? Illustrate with a diagram.

**2.** (a) Write the definition of the definite integral of a continuous function from $a$ to $b$.

(b) What is the geometric interpretation of $\int_a^b f(x)\, dx$ if $f(x) \geq 0$?

(c) What is the geometric interpretation of $\int_a^b f(x)\, dx$ if $f(x)$ takes on both positive and negative values? Illustrate with a diagram.

**3.** State both parts of the Fundamental Theorem of Calculus.

**4.** (a) State the Net Change Theorem.

(b) If $r(t)$ is the rate at which water flows into a reservoir, what does $\int_{t_1}^{t_2} r(t)\, dt$ represent?

**5.** Suppose a particle moves back and forth along a straight line with velocity $v(t)$, measured in feet per second, and acceleration $a(t)$.

(a) What is the meaning of $\int_{60}^{120} v(t)\, dt$?

(b) What is the meaning of $\int_{60}^{120} |v(t)|\, dt$?

(c) What is the meaning of $\int_{60}^{120} a(t)\, dt$?

**6.** (a) Explain the meaning of the indefinite integral $\int f(x)\, dx$.

(b) What is the connection between the definite integral $\int_a^b f(x)\, dx$ and the indefinite integral $\int f(x)\, dx$?

**7.** Explain exactly what is meant by the statement that "differentiation and integration are inverse processes."

**8.** State the Substitution Rule. In practice, how do you use it?

## True-False Quiz

Determine whether the statement is true or false. If it is true, explain why. If it is false, explain why or give an example that disproves the statement.

**1.** If $f$ and $g$ are continuous on $[a, b]$, then

$$\int_a^b [f(x) + g(x)]\, dx = \int_a^b f(x)\, dx + \int_a^b g(x)\, dx$$

**2.** If $f$ and $g$ are continuous on $[a, b]$, then

$$\int_a^b [f(x)g(x)]\, dx = \left( \int_a^b f(x)\, dx \right)\left( \int_a^b g(x)\, dx \right)$$

**3.** If $f$ is continuous on $[a, b]$, then

$$\int_a^b 5f(x)\, dx = 5\int_a^b f(x)\, dx$$

**4.** If $f$ is continuous on $[a, b]$, then

$$\int_a^b xf(x)\, dx = x\int_a^b f(x)\, dx$$

**5.** If $f$ is continuous on $[a, b]$ and $f(x) \geq 0$, then

$$\int_a^b \sqrt{f(x)}\, dx = \sqrt{\int_a^b f(x)\, dx}$$

**6.** If $f'$ is continuous on $[1, 3]$, then $\int_1^3 f'(v)\, dv = f(3) - f(1)$.

**7.** If $f$ and $g$ are continuous and $f(x) \geq g(x)$ for $a \leq x \leq b$, then

$$\int_a^b f(x)\, dx \geq \int_a^b g(x)\, dx$$

**8.** If $f$ and $g$ are differentiable and $f(x) \geq g(x)$ for $a < x < b$, then $f'(x) \geq g'(x)$ for $a < x < b$.

**9.** $\displaystyle\int_{-1}^1 \left( x^5 - 6x^9 + \frac{\sin x}{(1 + x^4)^2} \right) dx = 0$

**10.** $\displaystyle\int_{-5}^5 (ax^2 + bx + c)\, dx = 2\int_0^5 (ax^2 + c)\, dx$

**11.** All continuous functions have derivatives.

**12.** All continuous functions have antiderivatives.

**13.** $\displaystyle\int_\pi^{2\pi} \frac{\sin x}{x}\, dx = \int_\pi^{3\pi} \frac{\sin x}{x}\, dx + \int_{3\pi}^{2\pi} \frac{\sin x}{x}\, dx$

**14.** If $\int_0^1 f(x)\, dx = 0$, then $f(x) = 0$ for $0 \leq x \leq 1$.

**15.** If $f$ is continuous on $[a, b]$, then

$$\frac{d}{dx}\left( \int_a^b f(x)\, dx \right) = f(x)$$

**16.** $\int_0^2 (x - x^3)\, dx$ represents the area under the curve $y = x - x^3$ from 0 to 2.

**17.** $\displaystyle\int_{-2}^1 \frac{1}{x^4}\, dx = -\frac{3}{8}$

**18.** If $f$ has a discontinuity at 0, then $\int_{-1}^1 f(x)\, dx$ does not exist.

## Exercises

**1.** Use the given graph of $f$ to find the Riemann sum with six subintervals. Take the sample points to be (a) left endpoints and (b) midpoints. In each case draw a diagram and explain what the Riemann sum represents.

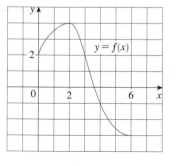

**2.** (a) Evaluate the Riemann sum for

$$f(x) = x^2 - x \qquad 0 \leq x \leq 2$$

with four subintervals, taking the sample points to be right endpoints. Explain, with the aid of a diagram, what the Riemann sum represents.

(b) Use the definition of a definite integral (with right endpoints) to calculate the value of the integral

$$\int_0^2 (x^2 - x)\, dx$$

(c) Use the Fundamental Theorem to check your answer to part (b).

(d) Draw a diagram to explain the geometric meaning of the integral in part (b).

**3.** Evaluate

$$\int_0^1 \left(x + \sqrt{1 - x^2}\right) dx$$

by interpreting it in terms of areas.

**4.** Express

$$\lim_{n \to \infty} \sum_{i=1}^n \sin x_i \, \Delta x$$

as a definite integral on the interval $[0, \pi]$ and then evaluate the integral.

**5.** If $\int_0^6 f(x)\, dx = 10$ and $\int_0^4 f(x)\, dx = 7$, find $\int_4^6 f(x)\, dx$.

CAS **6.** (a) Write $\int_1^5 (x + 2x^5)\, dx$ as a limit of Riemann sums, taking the sample points to be right endpoints. Use a computer algebra system to evaluate the sum and to compute the limit.
(b) Use the Fundamental Theorem to check your answer to part (a).

**7.** The following figure shows the graphs of $f$, $f'$, and $\int_0^x f(t)\, dt$. Identify each graph, and explain your choices.

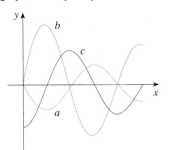

**8.** Evaluate:

(a) $\displaystyle\int_0^{\pi/2} \frac{d}{dx}\left(\sin\frac{x}{2}\cos\frac{x}{3}\right) dx$

(b) $\displaystyle\frac{d}{dx}\int_0^{\pi/2} \sin\frac{x}{2}\cos\frac{x}{3}\, dx$

(c) $\displaystyle\frac{d}{dx}\int_x^{\pi/2} \sin\frac{t}{2}\cos\frac{t}{3}\, dt$

**9–28** Evaluate the integral.

**9.** $\displaystyle\int_1^2 (8x^3 + 3x^2)\, dx$

**10.** $\displaystyle\int_0^T (x^4 - 8x + 7)\, dx$

**11.** $\displaystyle\int_0^1 (1 - x^9)\, dx$

**12.** $\displaystyle\int_0^1 (1 - x)^9\, dx$

**13.** $\displaystyle\int_1^9 \frac{\sqrt{u} - 2u^2}{u}\, du$

**14.** $\displaystyle\int_0^1 (\sqrt[4]{u} + 1)^2\, du$

**15.** $\displaystyle\int_0^1 y(y^2 + 1)^5\, dy$

**16.** $\displaystyle\int_0^2 y^2\sqrt{1 + y^3}\, dy$

**17.** $\displaystyle\int_1^5 \frac{dt}{(t - 4)^2}$

**18.** $\displaystyle\int_0^1 \sin(3\pi t)\, dt$

**19.** $\displaystyle\int_0^1 v^2 \cos(v^3)\, dv$

**20.** $\displaystyle\int_{-1}^1 \frac{\sin x}{1 + x^2}\, dx$

**21.** $\displaystyle\int_{-\pi/4}^{\pi/4} \frac{t^4 \tan t}{2 + \cos t}\, dt$

**22.** $\displaystyle\int \frac{x + 2}{\sqrt{x^2 + 4x}}\, dx$

**23.** $\displaystyle\int \sin \pi t \cos \pi t\, dt$

**24.** $\displaystyle\int \sin x \cos(\cos x)\, dx$

**25.** $\displaystyle\int_0^{\pi/8} \sec 2\theta \tan 2\theta\, d\theta$

**26.** $\displaystyle\int_0^{\pi/4} (1 + \tan t)^3 \sec^2 t\, dt$

**27.** $\displaystyle\int_0^3 |x^2 - 4|\, dx$

**28.** $\displaystyle\int_0^4 |\sqrt{x} - 1|\, dx$

**29–30** Evaluate the indefinite integral. Illustrate and check that your answer is reasonable by graphing both the function and its antiderivative (take $C = 0$).

**29.** $\displaystyle\int \frac{\cos x}{\sqrt{1 + \sin x}}\, dx$

**30.** $\displaystyle\int \frac{x^3}{\sqrt{x^2 + 1}}\, dx$

**31.** Use a graph to give a rough estimate of the area of the region that lies under the curve $y = x\sqrt{x}$, $0 \le x \le 4$. Then find the exact area.

**32.** Graph the function $f(x) = \cos^2 x \sin x$ and use the graph to guess the value of the integral $\int_0^{2\pi} f(x)\, dx$. Then evaluate the integral to confirm your guess.

**33–38** Find the derivative of the function.

**33.** $\displaystyle F(x) = \int_0^x \frac{t^2}{1 + t^3}\, dt$

**34.** $\displaystyle F(x) = \int_x^1 \sqrt{t + \sin t}\, dt$

**35.** $\displaystyle g(x) = \int_0^{x^4} \cos(t^2)\, dt$

**36.** $\displaystyle g(x) = \int_1^{\sin x} \frac{1 - t^2}{1 + t^4}\, dt$

**37.** $\displaystyle y = \int_{\sqrt{x}}^x \frac{\cos \theta}{\theta}\, d\theta$

**38.** $\displaystyle y = \int_{2x}^{3x+1} \sin(t^4)\, dt$

**39–40** Use Property 8 of integrals to estimate the value of the integral.

**39.** $\displaystyle\int_1^3 \sqrt{x^2 + 3}\, dx$

**40.** $\displaystyle\int_3^5 \frac{1}{x + 1}\, dx$

**41–42** Use the properties of integrals to verify the inequality.

**41.** $\displaystyle\int_0^1 x^2 \cos x\, dx \le \frac{1}{3}$

**42.** $\displaystyle\int_{\pi/4}^{\pi/2} \frac{\sin x}{x}\, dx \le \frac{\sqrt{2}}{2}$

**43.** Use the Midpoint Rule with $n = 6$ to approximate $\int_0^3 \sin(x^3)\, dx$.

**44.** A particle moves along a line with velocity function $v(t) = t^2 - t$, where $v$ is measured in meters per second. Find (a) the displacement and (b) the distance traveled by the particle during the time interval $[0, 5]$.

**45.** Let $r(t)$ be the rate at which the world's oil is consumed, where $t$ is measured in years starting at $t = 0$ on January 1, 2000, and $r(t)$ is measured in barrels per year. What does $\int_0^8 r(t)\, dt$ represent?

**46.** A radar gun was used to record the speed of a runner at the times given in the table. Use the Midpoint Rule to estimate the distance the runner covered during those 5 seconds.

| $t$ (s) | $v$ (m/s) | $t$ (s) | $v$ (m/s) |
|---------|-----------|---------|-----------|
| 0 | 0 | 3.0 | 10.51 |
| 0.5 | 4.67 | 3.5 | 10.67 |
| 1.0 | 7.34 | 4.0 | 10.76 |
| 1.5 | 8.86 | 4.5 | 10.81 |
| 2.0 | 9.73 | 5.0 | 10.81 |
| 2.5 | 10.22 | | |

**47.** A population of honeybees increased at a rate of $r(t)$ bees per week, where the graph of $r$ is as shown. Use the Midpoint Rule with six subintervals to estimate the increase in the bee population during the first 24 weeks.

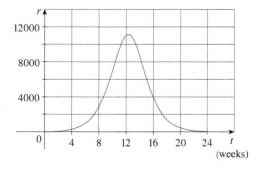

**48.** Let
$$f(x) = \begin{cases} -x - 1 & \text{if } -3 \le x \le 0 \\ -\sqrt{1 - x^2} & \text{if } 0 \le x \le 1 \end{cases}$$

Evaluate $\int_{-3}^1 f(x)\, dx$ by interpreting the integral as a difference of areas.

**49.** If $f$ is continuous and $\int_0^2 f(x)\, dx = 6$, evaluate $\int_0^{\pi/2} f(2 \sin \theta) \cos \theta\, d\theta$.

**50.** The Fresnel function $S(x) = \int_0^x \sin\left(\frac{1}{2} \pi t^2\right) dt$ was introduced in Section 4.3. Fresnel also used the function
$$C(x) = \int_0^x \cos\left(\frac{1}{2} \pi t^2\right) dt$$

in his theory of the diffraction of light waves.
(a) On what intervals is $C$ increasing?
(b) On what intervals is $C$ concave upward?
(c) Use a graph to solve the following equation correct to two decimal places:
$$\int_0^x \cos\left(\frac{1}{2} \pi t^2\right) dt = 0.7$$

(d) Plot the graphs of $C$ and $S$ on the same screen. How are these graphs related?

**51.** If $f$ is a continuous function such that
$$\int_0^x f(t)\, dt = x \sin x + \int_0^x \frac{f(t)}{1 + t^2}\, dt$$

for all $x$, find an explicit formula for $f(x)$.

**52.** Find a function $f$ and a value of the constant $a$ such that
$$2 \int_a^x f(t)\, dt = 2 \sin x - 1$$

**53.** If $f'$ is continuous on $[a, b]$, show that
$$2 \int_a^b f(x) f'(x)\, dx = [f(b)]^2 - [f(a)]^2$$

**54.** Find $\displaystyle \lim_{h \to 0} \frac{1}{h} \int_2^{2+h} \sqrt{1 + t^3}\, dt$.

**55.** If $f$ is continuous on $[0, 1]$, prove that
$$\int_0^1 f(x)\, dx = \int_0^1 f(1 - x)\, dx$$

**56.** Evaluate
$$\lim_{n \to \infty} \frac{1}{n} \left[ \left(\frac{1}{n}\right)^9 + \left(\frac{2}{n}\right)^9 + \left(\frac{3}{n}\right)^9 + \cdots + \left(\frac{n}{n}\right)^9 \right]$$

# Problems Plus

Before you look at the solution of the following example, cover it up and first try to solve the problem yourself.

**EXAMPLE 1** Evaluate $\displaystyle\lim_{x \to 3} \left( \frac{x}{x-3} \int_3^x \frac{\sin t}{t}\, dt \right)$.

SOLUTION Let's start by having a preliminary look at the ingredients of the function. What happens to the first factor, $x/(x-3)$, when $x$ approaches 3? The numerator approaches 3 and the denominator approaches 0, so we have

$$\frac{x}{x-3} \to \infty \quad \text{as} \quad x \to 3^+ \qquad \text{and} \qquad \frac{x}{x-3} \to -\infty \quad \text{as} \quad x \to 3^-$$

The second factor approaches $\int_3^3 (\sin t)/t\, dt$, which is 0. It's not clear what happens to the function as a whole. (One factor is becoming large while the other is becoming small.) So how do we proceed?

One of the principles of problem solving is *recognizing something familiar.* Is there a part of the function that reminds us of something we've seen before? Well, the integral

$$\int_3^x \frac{\sin t}{t}\, dt$$

has $x$ as its upper limit of integration and that type of integral occurs in Part 1 of the Fundamental Theorem of Calculus:

$$\frac{d}{dx} \int_a^x f(t)\, dt = f(x)$$

This suggests that differentiation might be involved.

Once we start thinking about differentiation, the denominator $(x-3)$ reminds us of something else that should be familiar: One of the forms of the definition of the derivative in Chapter 2 is

$$F'(a) = \lim_{x \to a} \frac{F(x) - F(a)}{x - a}$$

and with $a = 3$ this becomes

$$F'(3) = \lim_{x \to 3} \frac{F(x) - F(3)}{x - 3}$$

So what is the function $F$ in our situation? Notice that if we define

$$F(x) = \int_3^x \frac{\sin t}{t}\, dt$$

then $F(3) = 0$. What about the factor $x$ in the numerator? That's just a red herring, so let's factor it out and put together the calculation:

$$\lim_{x \to 3} \left( \frac{x}{x-3} \int_3^x \frac{\sin t}{t}\, dt \right) = \lim_{x \to 3} x \cdot \lim_{x \to 3} \frac{\displaystyle\int_3^x \frac{\sin t}{t}\, dt}{x - 3}$$

$$= 3 \lim_{x \to 3} \frac{F(x) - F(3)}{x - 3}$$

$$= 3 F'(3) = 3 \,\frac{\sin 3}{3} \qquad \text{(FTC1)}$$

$$= \sin 3$$

PS The principles of problem solving are discussed on page 97.

1. If $x \sin \pi x = \int_0^{x^2} f(t)\, dt$, where $f$ is a continuous function, find $f(4)$.

2. Find the maximum value of the area of the region under the curve $y = 4x - x^3$ from $x = a$ to $x = a + 1$, for all $a > 0$.

3. If $f$ is a differentiable function such that $f(x)$ is never 0 and $\int_0^x f(t)\, dt = [f(x)]^2$ for all $x$, find $f$.

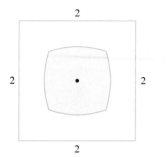 4. (a) Graph several members of the family of functions $f(x) = (2cx - x^2)/c^3$ for $c > 0$ and look at the regions enclosed by these curves and the $x$-axis. Make a conjecture about how the areas of these regions are related.
   (b) Prove your conjecture in part (a).
   (c) Take another look at the graphs in part (a) and use them to sketch the curve traced out by the vertices (highest points) of the family of functions. Can you guess what kind of curve this is?
   (d) Find an equation of the curve you sketched in part (c).

5. If $f(x) = \int_0^{g(x)} \dfrac{1}{\sqrt{1 + t^3}}\, dt$, where $g(x) = \int_0^{\cos x} [1 + \sin(t^2)]\, dt$, find $f'(\pi/2)$.

6. If $f(x) = \int_0^x x^2 \sin(t^2)\, dt$, find $f'(x)$.

7. Find the interval $[a, b]$ for which the value of the integral $\int_a^b (2 + x - x^2)\, dx$ is a maximum.

8. Use an integral to estimate the sum $\displaystyle\sum_{i=1}^{10000} \sqrt{i}$.

9. (a) Evaluate $\int_0^n [\![x]\!]\, dx$, where $n$ is a positive integer.
   (b) Evaluate $\int_a^b [\![x]\!]\, dx$, where $a$ and $b$ are real numbers with $0 \leqslant a < b$.

10. Find $\dfrac{d^2}{dx^2} \displaystyle\int_0^x \left( \int_1^{\sin t} \sqrt{1 + u^4}\, du \right) dt$.

11. Suppose the coefficients of the cubic polynomial $P(x) = a + bx + cx^2 + dx^3$ satisfy the equation

$$ a + \frac{b}{2} + \frac{c}{3} + \frac{d}{4} = 0 $$

Show that the equation $P(x) = 0$ has a root between 0 and 1. Can you generalize this result for an $n$th-degree polynomial?

12. A circular disk of radius $r$ is used in an evaporator and is rotated in a vertical plane. If it is to be partially submerged in the liquid so as to maximize the exposed wetted area of the disk, show that the center of the disk should be positioned at a height $r/\sqrt{1 + \pi^2}$ above the surface of the liquid.

13. Prove that if $f$ is continuous, then $\displaystyle\int_0^x f(u)(x - u)\, du = \int_0^x \left( \int_0^u f(t)\, dt \right) du$.

14. The figure shows a region consisting of all points inside a square that are closer to the center than to the sides of the square. Find the area of the region.

15. Evaluate $\displaystyle\lim_{n \to \infty} \left( \frac{1}{\sqrt{n}\,\sqrt{n+1}} + \frac{1}{\sqrt{n}\,\sqrt{n+2}} + \cdots + \frac{1}{\sqrt{n}\,\sqrt{n+n}} \right)$.

16. For any number $c$, we let $f_c(x)$ be the smaller of the two numbers $(x - c)^2$ and $(x - c - 2)^2$. Then we define $g(c) = \int_0^1 f_c(x)\, dx$. Find the maximum and minimum values of $g(c)$ if $-2 \leqslant c \leqslant 2$.

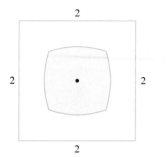

**FIGURE FOR PROBLEM 14**

---

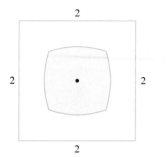 Graphing calculator or computer required

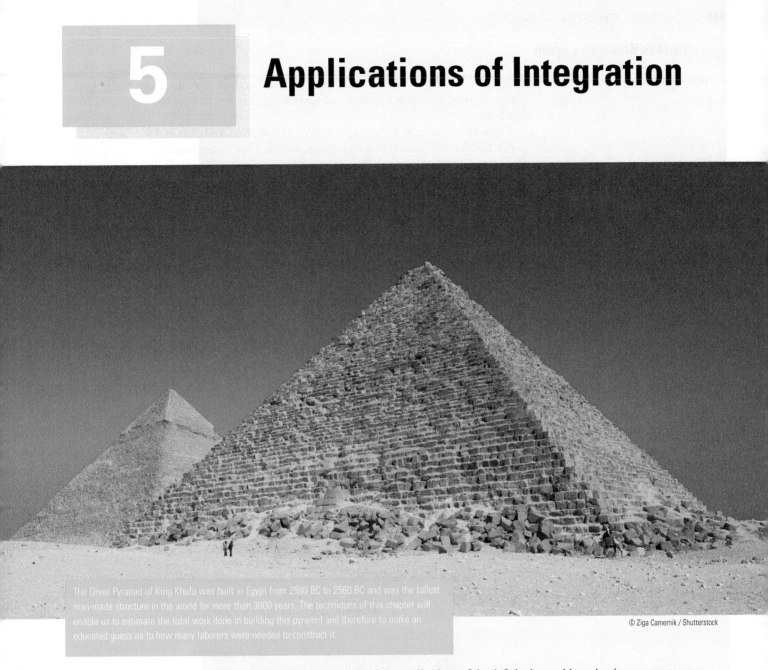

# 5 Applications of Integration

The Great Pyramid of King Khufu was built in Egypt from 2580 BC to 2560 BC and was the tallest man-made structure in the world for more than 3800 years. The techniques of this chapter will enable us to estimate the total work done in building this pyramid and therefore to make an educated guess as to how many laborers were needed to construct it.

© Ziga Camernik / Shutterstock

In this chapter we explore some of the applications of the definite integral by using it to compute areas between curves, volumes of solids, and the work done by a varying force. The common theme is the following general method, which is similar to the one we used to find areas under curves: We break up a quantity $Q$ into a large number of small parts. We next approximate each small part by a quantity of the form $f(x_i^*)\,\Delta x$ and thus approximate $Q$ by a Riemann sum. Then we take the limit and express $Q$ as an integral. Finally we evaluate the integral using the Fundamental Theorem of Calculus or the Midpoint Rule.

**Areas Between Curves**

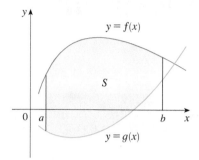

**FIGURE 1**
$S = \{(x, y) \mid a \leqslant x \leqslant b, g(x) \leqslant y \leqslant f(x)\}$

In Chapter 4 we defined and calculated areas of regions that lie under the graphs of functions. Here we use integrals to find areas of regions that lie between the graphs of two functions.

Consider the region $S$ that lies between two curves $y = f(x)$ and $y = g(x)$ and between the vertical lines $x = a$ and $x = b$, where $f$ and $g$ are continuous functions and $f(x) \geqslant g(x)$ for all $x$ in $[a, b]$. (See Figure 1.)

Just as we did for areas under curves in Section 4.1, we divide $S$ into $n$ strips of equal width and then we approximate the $i$th strip by a rectangle with base $\Delta x$ and height $f(x_i^*) - g(x_i^*)$. (See Figure 2. If we liked, we could take all of the sample points to be right endpoints, in which case $x_i^* = x_i$.) The Riemann sum

$$\sum_{i=1}^{n} [f(x_i^*) - g(x_i^*)] \Delta x$$

is therefore an approximation to what we intuitively think of as the area of $S$.

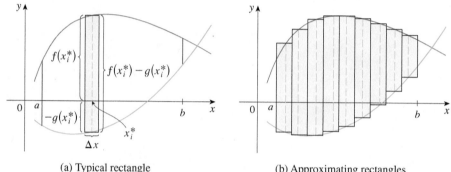

**FIGURE 2**      (a) Typical rectangle                  (b) Approximating rectangles

This approximation appears to become better and better as $n \to \infty$. Therefore we define the **area** $A$ of the region $S$ as the limiting value of the sum of the areas of these approximating rectangles.

$$\boxed{1} \qquad A = \lim_{n \to \infty} \sum_{i=1}^{n} [f(x_i^*) - g(x_i^*)] \Delta x$$

We recognize the limit in $\boxed{1}$ as the definite integral of $f - g$. Therefore we have the following formula for area.

> $\boxed{2}$    The area $A$ of the region bounded by the curves $y = f(x)$, $y = g(x)$, and the lines $x = a$, $x = b$, where $f$ and $g$ are continuous and $f(x) \geqslant g(x)$ for all $x$ in $[a, b]$, is
>
> $$A = \int_a^b [f(x) - g(x)]\,dx$$

Notice that in the special case where $g(x) = 0$, $S$ is the region under the graph of $f$ and our general definition of area $\boxed{1}$ reduces to our previous definition (Definition 2 in Section 4.1).

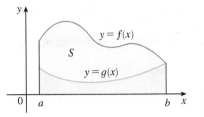

**FIGURE 3**
$$A = \int_a^b f(x)\, dx - \int_a^b g(x)\, dx$$

**FIGURE 4**

**FIGURE 5**

**FIGURE 6**

In the case where both $f$ and $g$ are positive, you can see from Figure 3 why $\boxed{2}$ is true:

$$A = [\text{area under } y = f(x)] - [\text{area under } y = g(x)]$$

$$= \int_a^b f(x)\, dx - \int_a^b g(x)\, dx = \int_a^b [f(x) - g(x)]\, dx$$

 **EXAMPLE 1** Find the area of the region bounded above by $y = x^2 + 1$, bounded below by $y = x$, and bounded on the sides by $x = 0$ and $x = 1$.

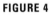 **SOLUTION** The region is shown in Figure 4. The upper boundary curve is $y = x^2 + 1$ and the lower boundary curve is $y = x$. So we use the area formula $\boxed{2}$ with $f(x) = x^2 + 1$, $g(x) = x$, $a = 0$, and $b = 1$:

$$A = \int_0^1 [(x^2 + 1) - x]\, dx = \int_0^1 (x^2 - x + 1)\, dx$$

$$= \frac{x^3}{3} - \frac{x^2}{2} + x \Big]_0^1 = \frac{1}{3} - \frac{1}{2} + 1 = \frac{5}{6}$$

In Figure 4 we drew a typical approximating rectangle with width $\Delta x$ as a reminder of the procedure by which the area is defined in $\boxed{1}$. In general, when we set up an integral for an area, it's helpful to sketch the region to identify the top curve $y_T$, the bottom curve $y_B$, and a typical approximating rectangle as in Figure 5. Then the area of a typical rectangle is $(y_T - y_B)\, \Delta x$ and the equation

$$A = \lim_{n \to \infty} \sum_{i=1}^n (y_T - y_B)\, \Delta x = \int_a^b (y_T - y_B)\, dx$$

summarizes the procedure of adding (in a limiting sense) the areas of all the typical rectangles.

Notice that in Figure 5 the left-hand boundary reduces to a point, whereas in Figure 3 the right-hand boundary reduces to a point. In the next example both of the side boundaries reduce to a point, so the first step is to find $a$ and $b$.

**V EXAMPLE 2** Find the area of the region enclosed by the parabolas $y = x^2$ and $y = 2x - x^2$.

 **SOLUTION** We first find the points of intersection of the parabolas by solving their equations simultaneously. This gives $x^2 = 2x - x^2$, or $2x^2 - 2x = 0$. Thus $2x(x - 1) = 0$, so $x = 0$ or $1$. The points of intersection are $(0, 0)$ and $(1, 1)$.

We see from Figure 6 that the top and bottom boundaries are

$$y_T = 2x - x^2 \qquad \text{and} \qquad y_B = x^2$$

The area of a typical rectangle is

$$(y_T - y_B)\, \Delta x = (2x - x^2 - x^2)\, \Delta x$$

and the region lies between $x = 0$ and $x = 1$. So the total area is

$$A = \int_0^1 (2x - 2x^2)\, dx = 2 \int_0^1 (x - x^2)\, dx$$

$$= 2 \left[ \frac{x^2}{2} - \frac{x^3}{3} \right]_0^1 = 2 \left( \frac{1}{2} - \frac{1}{3} \right) = \frac{1}{3}$$

Sometimes it's difficult, or even impossible, to find the points of intersection of two curves exactly. As shown in the following example, we can use a graphing calculator or computer to find approximate values for the intersection points and then proceed as before.

**EXAMPLE 3**    Find the approximate area of the region bounded by the curves $y = x/\sqrt{x^2 + 1}$ and $y = x^4 - x$.

SOLUTION    If we were to try to find the exact intersection points, we would have to solve the equation

$$\frac{x}{\sqrt{x^2 + 1}} = x^4 - x$$

This looks like a very difficult equation to solve exactly (in fact, it's impossible), so instead we use a graphing device to draw the graphs of the two curves in Figure 7. One intersection point is the origin. We zoom in toward the other point of intersection and find that $x \approx 1.18$. (If greater accuracy is required, we could use Newton's method or a rootfinder, if available on our graphing device.) Thus an approximation to the area between the curves is

$$A \approx \int_0^{1.18} \left[ \frac{x}{\sqrt{x^2 + 1}} - (x^4 - x) \right] dx$$

**FIGURE 7**

To integrate the first term we use the substitution $u = x^2 + 1$. Then $du = 2x\,dx$, and when $x = 1.18$, we have $u \approx 2.39$. So

$$A \approx \tfrac{1}{2} \int_1^{2.39} \frac{du}{\sqrt{u}} - \int_0^{1.18} (x^4 - x)\,dx$$

$$= \sqrt{u}\,\Big]_1^{2.39} - \left[ \frac{x^5}{5} - \frac{x^2}{2} \right]_0^{1.18}$$

$$= \sqrt{2.39} - 1 - \frac{(1.18)^5}{5} + \frac{(1.18)^2}{2}$$

$$\approx 0.785$$

**FIGURE 8**

**EXAMPLE 4**    Figure 8 shows velocity curves for two cars, A and B, that start side by side and move along the same road. What does the area between the curves represent? Use the Midpoint Rule to estimate it.

SOLUTION    We know from Section 4.4 that the area under the velocity curve A represents the distance traveled by car A during the first 16 seconds. Similarly, the area under curve B is the distance traveled by car B during that time period. So the area between these curves, which is the difference of the areas under the curves, is the distance between the cars after 16 seconds. We read the velocities from the graph and convert them to feet per second (1 mi/h = $\frac{5280}{3600}$ ft/s).

| $t$ | 0 | 2 | 4 | 6 | 8 | 10 | 12 | 14 | 16 |
|---|---|---|---|---|---|---|---|---|---|
| $v_A$ | 0 | 34 | 54 | 67 | 76 | 84 | 89 | 92 | 95 |
| $v_B$ | 0 | 21 | 34 | 44 | 51 | 56 | 60 | 63 | 65 |
| $v_A - v_B$ | 0 | 13 | 20 | 23 | 25 | 28 | 29 | 29 | 30 |

We use the Midpoint Rule with $n = 4$ intervals, so that $\Delta t = 4$. The midpoints of the intervals are $\bar{t}_1 = 2$, $\bar{t}_2 = 6$, $\bar{t}_3 = 10$, and $\bar{t}_4 = 14$. We estimate the distance between the cars after 16 seconds as follows:

$$\int_0^{16} (v_A - v_B)\, dt \approx \Delta t\, [13 + 23 + 28 + 29]$$

$$= 4(93) = 372 \text{ ft}$$

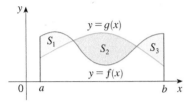

**FIGURE 9**

If we are asked to find the area between the curves $y = f(x)$ and $y = g(x)$ where $f(x) \geqslant g(x)$ for some values of $x$ but $g(x) \geqslant f(x)$ for other values of $x$, then we split the given region $S$ into several regions $S_1$, $S_2$, ... with areas $A_1$, $A_2$, ... as shown in Figure 9. We then define the area of the region $S$ to be the sum of the areas of the smaller regions $S_1$, $S_2$, ..., that is, $A = A_1 + A_2 + \cdots$. Since

$$|f(x) - g(x)| = \begin{cases} f(x) - g(x) & \text{when } f(x) \geqslant g(x) \\ g(x) - f(x) & \text{when } g(x) \geqslant f(x) \end{cases}$$

we have the following expression for $A$.

---

**3** The area between the curves $y = f(x)$ and $y = g(x)$ and between $x = a$ and $x = b$ is

$$A = \int_a^b |f(x) - g(x)|\, dx$$

---

When evaluating the integral in **3**, however, we must still split it into integrals corresponding to $A_1$, $A_2$, ....

**V EXAMPLE 5** Find the area of the region bounded by the curves $y = \sin x$, $y = \cos x$, $x = 0$, and $x = \pi/2$.

SOLUTION The points of intersection occur when $\sin x = \cos x$, that is, when $x = \pi/4$ (since $0 \leqslant x \leqslant \pi/2$). The region is sketched in Figure 10. Observe that $\cos x \geqslant \sin x$ when $0 \leqslant x \leqslant \pi/4$ but $\sin x \geqslant \cos x$ when $\pi/4 \leqslant x \leqslant \pi/2$. Therefore the required area is

$$A = \int_0^{\pi/2} |\cos x - \sin x|\, dx = A_1 + A_2$$

$$= \int_0^{\pi/4} (\cos x - \sin x)\, dx + \int_{\pi/4}^{\pi/2} (\sin x - \cos x)\, dx$$

$$= \Big[\sin x + \cos x\Big]_0^{\pi/4} + \Big[-\cos x - \sin x\Big]_{\pi/4}^{\pi/2}$$

$$= \left(\frac{1}{\sqrt{2}} + \frac{1}{\sqrt{2}} - 0 - 1\right) + \left(-0 - 1 + \frac{1}{\sqrt{2}} + \frac{1}{\sqrt{2}}\right)$$

$$= 2\sqrt{2} - 2$$

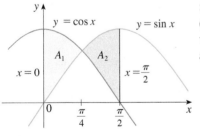

**FIGURE 10**

In this particular example we could have saved some work by noticing that the region is symmetric about $x = \pi/4$ and so

$$A = 2A_1 = 2\int_0^{\pi/4} (\cos x - \sin x)\, dx$$

Some regions are best treated by regarding $x$ as a function of $y$. If a region is bounded by curves with equations $x = f(y)$, $x = g(y)$, $y = c$, and $y = d$, where $f$ and $g$ are continuous and $f(y) \geq g(y)$ for $c \leq y \leq d$ (see Figure 11), then its area is

$$A = \int_c^d [f(y) - g(y)]\, dy$$

**FIGURE 11**          **FIGURE 12**

If we write $x_R$ for the right boundary and $x_L$ for the left boundary, then, as Figure 12 illustrates, we have

$$A = \int_c^d (x_R - x_L)\, dy$$

Here a typical approximating rectangle has dimensions $x_R - x_L$ and $\Delta y$.

**FIGURE 13**

**FIGURE 14**

**V** **EXAMPLE 6**  Find the area enclosed by the line $y = x - 1$ and the parabola $y^2 = 2x + 6$.

SOLUTION  By solving the two equations we find that the points of intersection are $(-1, -2)$ and $(5, 4)$. We solve the equation of the parabola for $x$ and notice from Figure 13 that the left and right boundary curves are

$$x_L = \tfrac{1}{2}y^2 - 3 \qquad \text{and} \qquad x_R = y + 1$$

We must integrate between the appropriate $y$-values, $y = -2$ and $y = 4$. Thus

$$A = \int_{-2}^{4} (x_R - x_L)\, dy = \int_{-2}^{4} \left[(y + 1) - \left(\tfrac{1}{2}y^2 - 3\right)\right] dy$$

$$= \int_{-2}^{4} \left(-\tfrac{1}{2}y^2 + y + 4\right) dy$$

$$= -\frac{1}{2}\left(\frac{y^3}{3}\right) + \frac{y^2}{2} + 4y \Bigg]_{-2}^{4}$$

$$= -\tfrac{1}{6}(64) + 8 + 16 - \left(\tfrac{4}{3} + 2 - 8\right) = 18$$

**NOTE**  We could have found the area in Example 6 by integrating with respect to $x$ instead of $y$, but the calculation is much more involved. It would have meant splitting the region in two and computing the areas labeled $A_1$ and $A_2$ in Figure 14. The method we used in Example 6 is *much* easier.

## 5.1 Exercises

**1–4** Find the area of the shaded region.

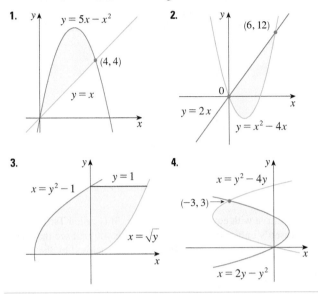

**1.**  $y = 5x - x^2$,  $(4, 4)$,  $y = x$

**2.**  $(6, 12)$,  $y = 2x$,  $y = x^2 - 4x$

**3.**  $x = y^2 - 1$,  $y = 1$,  $x = \sqrt{y}$

**4.**  $x = y^2 - 4y$,  $(-3, 3)$,  $x = 2y - y^2$

**5–12** Sketch the region enclosed by the given curves. Decide whether to integrate with respect to $x$ or $y$. Draw a typical approximating rectangle and label its height and width. Then find the area of the region.

**5.**  $y = x + 1$,  $y = 9 - x^2$,  $x = -1$,  $x = 2$

**6.**  $y = \sin x$,  $y = x$,  $x = \pi/2$,  $x = \pi$

**7.**  $y = (x - 2)^2$,  $y = x$

**8.**  $y = x^2 - 2x$,  $y = x + 4$

**9.**  $y = \sqrt{x + 3}$,  $y = (x + 3)/2$

**10.**  $y = \sin x$,  $y = 2x/\pi$,  $x \geqslant 0$

**11.**  $x = 1 - y^2$,  $x = y^2 - 1$

**12.**  $4x + y^2 = 12$,  $x = y$

**13–28** Sketch the region enclosed by the given curves and find its area.

**13.**  $y = 12 - x^2$,  $y = x^2 - 6$

**14.**  $y = x^2$,  $y = 4x - x^2$

**15.**  $y = \sec^2 x$,  $y = 8 \cos x$,  $-\pi/3 \leqslant x \leqslant \pi/3$

**16.**  $y = \cos x$,  $y = 2 - \cos x$,  $0 \leqslant x \leqslant 2\pi$

**17.**  $x = 2y^2$,  $x = 4 + y^2$

**18.**  $y = \sqrt{x - 1}$,  $x - y = 1$

**19.**  $y = \cos \pi x$,  $y = 4x^2 - 1$

**20.**  $x = y^4$,  $y = \sqrt{2 - x}$,  $y = 0$

**21.**  $y = \cos x$,  $y = 1 - 2x/\pi$

**22.**  $y = x^3$,  $y = x$

**23.**  $y = \cos x$,  $y = \sin 2x$,  $x = 0$,  $x = \pi/2$

**24.**  $y = \cos x$,  $y = 1 - \cos x$,  $0 \leqslant x \leqslant \pi$

**25.**  $y = \sqrt{x}$,  $y = \frac{1}{2}x$,  $x = 9$

**26.**  $y = |x|$,  $y = x^2 - 2$

**27.**  $y = 1/x^2$,  $y = x$,  $y = \frac{1}{8}x$

**28.**  $y = \frac{1}{4}x^2$,  $y = 2x^2$,  $x + y = 3$,  $x \geqslant 0$

**29–30** Use calculus to find the area of the triangle with the given vertices.

**29.**  $(0, 0)$,  $(3, 1)$,  $(1, 2)$          **30.**  $(2, 0)$,  $(0, 2)$,  $(-1, 1)$

**31–32** Evaluate the integral and interpret it as the area of a region. Sketch the region.

**31.**  $\displaystyle\int_0^{\pi/2} | \sin x - \cos 2x | \, dx$       **32.**  $\displaystyle\int_0^4 | \sqrt{x + 2} - x | \, dx$

**33–36** Use a graph to find approximate $x$-coordinates of the points of intersection of the given curves. Then find (approximately) the area of the region bounded by the curves.

**33.**  $y = x \sin(x^2)$,  $y = x^4$,  $x \geqslant 0$

**34.**  $y = \dfrac{x}{(x^2 + 1)^2}$,  $y = x^5 - x$,  $x \geqslant 0$

**35.**  $y = 3x^2 - 2x$,  $y = x^3 - 3x + 4$

**36.**  $y = x^2 \cos(x^3)$,  $y = x^{10}$

**37–40** Graph the region between the curves and use your calculator to compute the area correct to five decimal places.

**37.**  $y = \dfrac{2}{1 + x^4}$,  $y = x^2$       **38.**  $y = x^6$,  $y = \sqrt{2 - x^4}$

**39.**  $y = \tan^2 x$,  $y = \sqrt{x}$       **40.**  $y = \cos x$,  $y = x + 2\sin^4 x$

**41.** Use a computer algebra system to find the exact area enclosed by the curves $y = x^5 - 6x^3 + 4x$ and $y = x$.

**42.** Sketch the region in the $xy$-plane defined by the inequalities $x - 2y^2 \geqslant 0$, $1 - x - |y| \geqslant 0$ and find its area.

**43.** Racing cars driven by Chris and Kelly are side by side at the start of a race. The table shows the velocities of each car (in miles per hour) during the first ten seconds of the race. Use the

Midpoint Rule to estimate how much farther Kelly travels than Chris does during the first ten seconds.

| $t$ | $v_C$ | $v_K$ | $t$ | $v_C$ | $v_K$ |
|---|---|---|---|---|---|
| 0 | 0 | 0 | 6 | 69 | 80 |
| 1 | 20 | 22 | 7 | 75 | 86 |
| 2 | 32 | 37 | 8 | 81 | 93 |
| 3 | 46 | 52 | 9 | 86 | 98 |
| 4 | 54 | 61 | 10 | 90 | 102 |
| 5 | 62 | 71 | | | |

**44.** The widths (in meters) of a kidney-shaped swimming pool were measured at 2-meter intervals as indicated in the figure. Use the Midpoint Rule to estimate the area of the pool.

**45.** A cross-section of an airplane wing is shown. Measurements of the thickness of the wing, in centimeters, at 20-centimeter intervals are 5.8, 20.3, 26.7, 29.0, 27.6, 27.3, 23.8, 20.5, 15.1, 8.7, and 2.8. Use the Midpoint Rule to estimate the area of the wing's cross-section.

**46.** If the birth rate of a population is $b(t) = 2200 + 52.3t + 0.74t^2$ people per year and the death rate is $d(t) = 1460 + 28.8t$ people per year, find the area between these curves for $0 \le t \le 10$. What does this area represent?

**47.** Two cars, A and B, start side by side and accelerate from rest. The figure shows the graphs of their velocity functions.
(a) Which car is ahead after one minute? Explain.
(b) What is the meaning of the area of the shaded region?
(c) Which car is ahead after two minutes? Explain.
(d) Estimate the time at which the cars are again side by side.

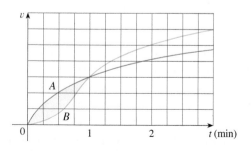

**48.** The figure shows graphs of the marginal revenue function $R'$ and the marginal cost function $C'$ for a manufacturer. [Recall from Section 3.7 that $R(x)$ and $C(x)$ represent the revenue and cost when $x$ units are manufactured. Assume that $R$ and $C$ are measured in thousands of dollars.] What is the meaning of the area of the shaded region? Use the Midpoint Rule to estimate the value of this quantity.

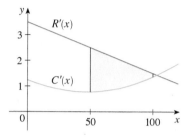

**49.** The curve with equation $y^2 = x^2(x + 3)$ is called **Tschirnhausen's cubic**. If you graph this curve you will see that part of the curve forms a loop. Find the area enclosed by the loop.

**50.** Find the area of the region bounded by the parabola $y = x^2$, the tangent line to this parabola at $(1, 1)$, and the $x$-axis.

**51.** Find the number $b$ such that the line $y = b$ divides the region bounded by the curves $y = x^2$ and $y = 4$ into two regions with equal area.

**52.** (a) Find the number $a$ such that the line $x = a$ bisects the area under the curve $y = 1/x^2$, $1 \le x \le 4$.
(b) Find the number $b$ such that the line $y = b$ bisects the area in part (a).

**53.** Find the values of $c$ such that the area of the region bounded by the parabolas $y = x^2 - c^2$ and $y = c^2 - x^2$ is 576.

**54.** Suppose that $0 < c < \pi/2$. For what value of $c$ is the area of the region enclosed by the curves $y = \cos x$, $y = \cos(x - c)$, and $x = 0$ equal to the area of the region enclosed by the curves $y = \cos(x - c)$, $x = \pi$, and $y = 0$?

The following exercises are intended only for those who have already covered Chapter 6.

**55–57** Sketch the region bounded by the given curves and find the area of the region.

**55.** $y = 1/x$,   $y = 1/x^2$,   $x = 2$

**56.** $y = \sin x$,   $y = e^x$,   $x = 0$,   $x = \pi/2$

**57.** $y = \tan x$,   $y = 2 \sin x$,   $-\pi/3 \le x \le \pi/3$

**58.** For what values of $m$ do the line $y = mx$ and the curve $y = x/(x^2 + 1)$ enclose a region? Find the area of the region.

**THE GINI INDEX**

How is it possible to measure the distribution of income among the inhabitants of a given country? One such measure is the *Gini index*, named after the Italian economist Corrado Gini who first devised it in 1912.

We first rank all households in a country by income and then we compute the percentage of households whose income is at most a given percentage of the country's total income. We define a **Lorenz curve** $y = L(x)$ on the interval $[0, 1]$ by plotting the point $(a/100, b/100)$ on the curve if the bottom $a\%$ of households receive at most $b\%$ of the total income. For instance, in Figure 1 the point $(0.4, 0.12)$ is on the Lorenz curve for the United States in 2008 because the poorest 40% of the population received just 12% of the total income. Likewise, the bottom 80% of the population received 50% of the total income, so the point $(0.8, 0.5)$ lies on the Lorenz curve. (The Lorenz curve is named after the American economist Max Lorenz.)

Figure 2 shows some typical Lorenz curves. They all pass through the points $(0, 0)$ and $(1, 1)$ and are concave upward. In the extreme case $L(x) = x$, society is perfectly egalitarian: The poorest $a\%$ of the population receives $a\%$ of the total income and so everybody receives the same income. The area between a Lorenz curve $y = L(x)$ and the line $y = x$ measures how much the income distribution differs from absolute equality. The **Gini index** (sometimes called the **Gini coefficient** or the **coefficient of inequality**) is the area between the Lorenz curve and the line $y = x$ (shaded in Figure 3) divided by the area under $y = x$.

1. (a) Show that the Gini index $G$ is twice the area between the Lorenz curve and the line $y = x$, that is,
$$G = 2 \int_0^1 [x - L(x)]\, dx$$

   (b) What is the value of $G$ for a perfectly egalitarian society (everybody has the same income)? What is the value of $G$ for a perfectly totalitarian society (a single person receives all the income?)

2. The following table (derived from data supplied by the US Census Bureau) shows values of the Lorenz function for income distribution in the United States for the year 2008.

| $x$ | 0.0 | 0.2 | 0.4 | 0.6 | 0.8 | 1.0 |
|------|-------|-------|-------|-------|-------|-------|
| $L(x)$ | 0.000 | 0.034 | 0.120 | 0.267 | 0.500 | 1.000 |

   (a) What percentage of the total US income was received by the richest 20% of the population in 2008?

   (b) Use a calculator or computer to fit a quadratic function to the data in the table. Graph the data points and the quadratic function. Is the quadratic model a reasonable fit?

   (c) Use the quadratic model for the Lorenz function to estimate the Gini index for the United States in 2008.

3. The following table gives values for the Lorenz function in the years 1970, 1980, 1990, and 2000. Use the method of Problem 2 to estimate the Gini index for the United States for those years and compare with your answer to Problem 2(c). Do you notice a trend?

| $x$ | 0.0 | 0.2 | 0.4 | 0.6 | 0.8 | 1.0 |
|------|-------|-------|-------|-------|-------|-------|
| 1970 | 0.000 | 0.041 | 0.149 | 0.323 | 0.568 | 1.000 |
| 1980 | 0.000 | 0.042 | 0.144 | 0.312 | 0.559 | 1.000 |
| 1990 | 0.000 | 0.038 | 0.134 | 0.293 | 0.530 | 1.000 |
| 2000 | 0.000 | 0.036 | 0.125 | 0.273 | 0.503 | 1.000 |

CAS 4. A power model often provides a more accurate fit than a quadratic model for a Lorenz function. If you have a computer with Maple or Mathematica, fit a power function $(y = ax^k)$ to the data in Problem 2 and use it to estimate the Gini index for the United States in 2008. Compare with your answer to parts (b) and (c) of Problem 2.

**FIGURE 1**
Lorenz curve for the US in 2008

**FIGURE 2**

**FIGURE 3**

## 5.2    Volumes

In trying to find the volume of a solid we face the same type of problem as in finding areas. We have an intuitive idea of what volume means, but we must make this idea precise by using calculus to give an exact definition of volume.

We start with a simple type of solid called a **cylinder** (or, more precisely, a *right cylinder*). As illustrated in Figure 1(a), a cylinder is bounded by a plane region $B_1$, called the **base**, and a congruent region $B_2$ in a parallel plane. The cylinder consists of all points on line segments that are perpendicular to the base and join $B_1$ to $B_2$. If the area of the base is $A$ and the height of the cylinder (the distance from $B_1$ to $B_2$) is $h$, then the volume $V$ of the cylinder is defined as

$$V = Ah$$

In particular, if the base is a circle with radius $r$, then the cylinder is a circular cylinder with volume $V = \pi r^2 h$ [see Figure 1(b)], and if the base is a rectangle with length $l$ and width $w$, then the cylinder is a rectangular box (also called a *rectangular parallelepiped*) with volume $V = lwh$ [see Figure 1(c)].

**FIGURE 1**    (a) Cylinder $V = Ah$    (b) Circular cylinder $V = \pi r^2 h$    (c) Rectangular box $V = lwh$

For a solid $S$ that isn't a cylinder we first "cut" $S$ into pieces and approximate each piece by a cylinder. We estimate the volume of $S$ by adding the volumes of the cylinders. We arrive at the exact volume of $S$ through a limiting process in which the number of pieces becomes large.

We start by intersecting $S$ with a plane and obtaining a plane region that is called a **cross-section** of $S$. Let $A(x)$ be the area of the cross-section of $S$ in a plane $P_x$ perpendicular to the $x$-axis and passing through the point $x$, where $a \leqslant x \leqslant b$. (See Figure 2. Think of slicing $S$ with a knife through $x$ and computing the area of this slice.) The cross-sectional area $A(x)$ will vary as $x$ increases from $a$ to $b$.

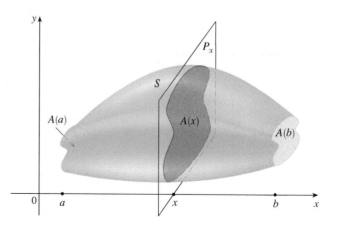

**FIGURE 2**

Let's divide $S$ into $n$ "slabs" of equal width $\Delta x$ by using the planes $P_{x_1}$, $P_{x_2}$, ... to slice the solid. (Think of slicing a loaf of bread.) If we choose sample points $x_i^*$ in $[x_{i-1}, x_i]$, we can approximate the $i$th slab $S_i$ (the part of $S$ that lies between the planes $P_{x_{i-1}}$ and $P_{x_i}$) by a cylinder with base area $A(x_i^*)$ and "height" $\Delta x$. (See Figure 3.)

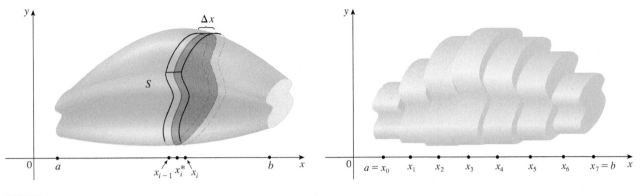

**FIGURE 3**

The volume of this cylinder is $A(x_i^*)\,\Delta x$, so an approximation to our intuitive conception of the volume of the $i$th slab $S_i$ is

$$V(S_i) \approx A(x_i^*)\,\Delta x$$

Adding the volumes of these slabs, we get an approximation to the total volume (that is, what we think of intuitively as the volume):

$$V \approx \sum_{i=1}^{n} A(x_i^*)\,\Delta x$$

This approximation appears to become better and better as $n \to \infty$. (Think of the slices as becoming thinner and thinner.) Therefore we *define* the volume as the limit of these sums as $n \to \infty$. But we recognize the limit of Riemann sums as a definite integral and so we have the following definition.

It can be proved that this definition is independent of how $S$ is situated with respect to the $x$-axis. In other words, no matter how we slice $S$ with parallel planes, we always get the same answer for $V$.

**Definition of Volume** Let $S$ be a solid that lies between $x = a$ and $x = b$. If the cross-sectional area of $S$ in the plane $P_x$, through $x$ and perpendicular to the $x$-axis, is $A(x)$, where $A$ is a continuous function, then the **volume** of $S$ is

$$V = \lim_{n \to \infty} \sum_{i=1}^{n} A(x_i^*)\,\Delta x = \int_a^b A(x)\,dx$$

When we use the volume formula $V = \int_a^b A(x)\,dx$, it is important to remember that $A(x)$ is the area of a moving cross-section obtained by slicing through $x$ perpendicular to the $x$-axis.

Notice that, for a cylinder, the cross-sectional area is constant: $A(x) = A$ for all $x$. So our definition of volume gives $V = \int_a^b A\,dx = A(b - a)$; this agrees with the formula $V = Ah$.

**EXAMPLE 1** Show that the volume of a sphere of radius $r$ is $V = \frac{4}{3}\pi r^3$.

SOLUTION If we place the sphere so that its center is at the origin (see Figure 4), then the plane $P_x$ intersects the sphere in a circle whose radius (from the Pythagorean Theorem)

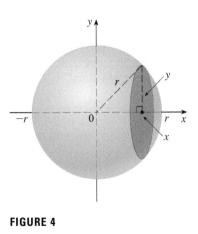

**FIGURE 4**

is $y = \sqrt{r^2 - x^2}$. So the cross-sectional area is

$$A(x) = \pi y^2 = \pi(r^2 - x^2)$$

Using the definition of volume with $a = -r$ and $b = r$, we have

$$V = \int_{-r}^{r} A(x)\,dx = \int_{-r}^{r} \pi(r^2 - x^2)\,dx$$

$$= 2\pi \int_{0}^{r} (r^2 - x^2)\,dx \qquad \text{(The integrand is even.)}$$

$$= 2\pi \left[ r^2 x - \frac{x^3}{3} \right]_0^r = 2\pi \left( r^3 - \frac{r^3}{3} \right)$$

$$= \tfrac{4}{3}\pi r^3$$

Figure 5 illustrates the definition of volume when the solid is a sphere with radius $r = 1$. From the result of Example 1, we know that the volume of the sphere is $\frac{4}{3}\pi$, which is approximately 4.18879. Here the slabs are circular cylinders, or *disks*, and the three parts of Figure 5 show the geometric interpretations of the Riemann sums

$$\sum_{i=1}^{n} A(\bar{x}_i)\,\Delta x = \sum_{i=1}^{n} \pi(1^2 - \bar{x}_i^2)\,\Delta x$$

when $n = 5$, 10, and 20 if we choose the sample points $x_i^*$ to be the midpoints $\bar{x}_i$. Notice that as we increase the number of approximating cylinders, the corresponding Riemann sums become closer to the true volume.

**TEC** Visual 5.2A shows an animation of Figure 5.

(a) Using 5 disks, $V \approx 4.2726$　　　(b) Using 10 disks, $V \approx 4.2097$　　　(c) Using 20 disks, $V \approx 4.1940$

**FIGURE 5** Approximating the volume of a sphere with radius 1

**V EXAMPLE 2** Find the volume of the solid obtained by rotating about the *x*-axis the region under the curve $y = \sqrt{x}$ from 0 to 1. Illustrate the definition of volume by sketching a typical approximating cylinder.

SOLUTION The region is shown in Figure 6(a). If we rotate about the *x*-axis, we get the solid shown in Figure 6(b). When we slice through the point *x*, we get a disk with radius $\sqrt{x}$. The area of this cross-section is

$$A(x) = \pi\left(\sqrt{x}\right)^2 = \pi x$$

and the volume of the approximating cylinder (a disk with thickness $\Delta x$) is

$$A(x)\,\Delta x = \pi x\,\Delta x$$

The solid lies between $x = 0$ and $x = 1$, so its volume is

$$V = \int_0^1 A(x)\,dx = \int_0^1 \pi x\,dx = \pi\frac{x^2}{2}\bigg]_0^1 = \frac{\pi}{2}$$

Did we get a reasonable answer in Example 2? As a check on our work, let's replace the given region by a square with base $[0, 1]$ and height 1. If we rotate this square, we get a cylinder with radius 1, height 1, and volume $\pi \cdot 1^2 \cdot 1 = \pi$. We computed that the given solid has half this volume. That seems about right.

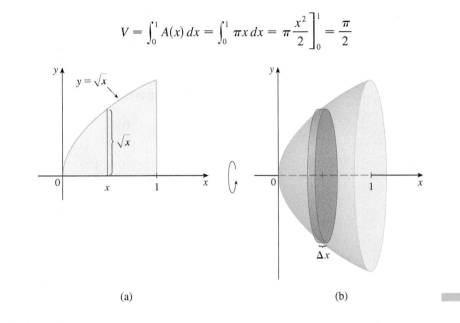

**FIGURE 6**

(a)    (b)

**EXAMPLE 3** Find the volume of the solid obtained by rotating the region bounded by $y = x^3$, $y = 8$, and $x = 0$ about the $y$-axis.

SOLUTION The region is shown in Figure 7(a) and the resulting solid is shown in Figure 7(b). Because the region is rotated about the $y$-axis, it makes sense to slice the solid perpendicular to the $y$-axis and therefore to integrate with respect to $y$. If we slice at height $y$, we get a circular disk with radius $x$, where $x = \sqrt[3]{y}$. So the area of a cross-section through $y$ is

$$A(y) = \pi x^2 = \pi(\sqrt[3]{y})^2 = \pi y^{2/3}$$

and the volume of the approximating cylinder pictured in Figure 7(b) is

$$A(y)\,\Delta y = \pi y^{2/3}\,\Delta y$$

Since the solid lies between $y = 0$ and $y = 8$, its volume is

$$V = \int_0^8 A(y)\,dy = \int_0^8 \pi y^{2/3}\,dy = \pi\left[\tfrac{3}{5}y^{5/3}\right]_0^8 = \frac{96\pi}{5}$$

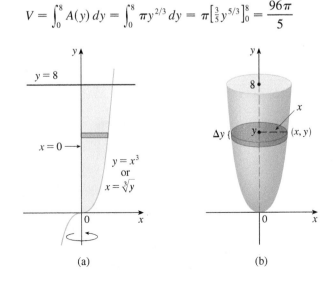

**FIGURE 7**

(a)    (b)

**EXAMPLE 4** The region $\mathcal{R}$ enclosed by the curves $y = x$ and $y = x^2$ is rotated about the $x$-axis. Find the volume of the resulting solid.

**SOLUTION** The curves $y = x$ and $y = x^2$ intersect at the points $(0, 0)$ and $(1, 1)$. The region between them, the solid of rotation, and a cross-section perpendicular to the $x$-axis are shown in Figure 8. A cross-section in the plane $P_x$ has the shape of a *washer* (an annular ring) with inner radius $x^2$ and outer radius $x$, so we find the cross-sectional area by subtracting the area of the inner circle from the area of the outer circle:

$$A(x) = \pi x^2 - \pi(x^2)^2 = \pi(x^2 - x^4)$$

Therefore we have

$$V = \int_0^1 A(x)\, dx = \int_0^1 \pi(x^2 - x^4)\, dx$$

$$= \pi \left[ \frac{x^3}{3} - \frac{x^5}{5} \right]_0^1 = \frac{2\pi}{15}$$

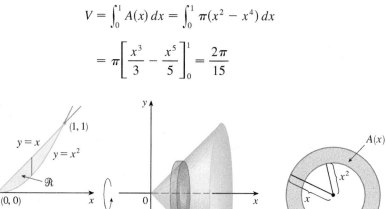

**FIGURE 8**            (a)                              (b)                              (c)

**EXAMPLE 5** Find the volume of the solid obtained by rotating the region in Example 4 about the line $y = 2$.

**SOLUTION** The solid and a cross-section are shown in Figure 9. Again the cross-section is a washer, but this time the inner radius is $2 - x$ and the outer radius is $2 - x^2$.

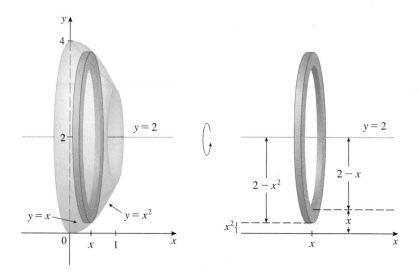

**FIGURE 9**

The cross-sectional area is

$$A(x) = \pi(2 - x^2)^2 - \pi(2 - x)^2$$

and so the volume of $S$ is

$$V = \int_0^1 A(x)\, dx$$

$$= \pi \int_0^1 [(2 - x^2)^2 - (2 - x)^2]\, dx$$

$$= \pi \int_0^1 (x^4 - 5x^2 + 4x)\, dx$$

$$= \pi \left[ \frac{x^5}{5} - 5\frac{x^3}{3} + 4\frac{x^2}{2} \right]_0^1$$

$$= \frac{8\pi}{15}$$

The solids in Examples 1–5 are all called **solids of revolution** because they are obtained by revolving a region about a line. In general, we calculate the volume of a solid of revolution by using the basic defining formula

$$V = \int_a^b A(x)\, dx \qquad \text{or} \qquad V = \int_c^d A(y)\, dy$$

and we find the cross-sectional area $A(x)$ or $A(y)$ in one of the following ways:

- If the cross-section is a disk (as in Examples 1–3), we find the radius of the disk (in terms of $x$ or $y$) and use

$$A = \pi(\text{radius})^2$$

- If the cross-section is a washer (as in Examples 4 and 5), we find the inner radius $r_{\text{in}}$ and outer radius $r_{\text{out}}$ from a sketch (as in Figures 8, 9, and 10) and compute the area of the washer by subtracting the area of the inner disk from the area of the outer disk:

$$A = \pi(\text{outer radius})^2 - \pi(\text{inner radius})^2$$

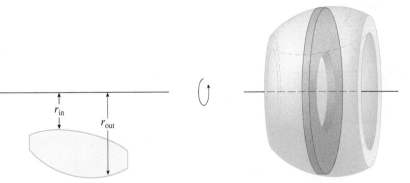

**FIGURE 10**

The next example gives a further illustration of the procedure.

**EXAMPLE 6**   Find the volume of the solid obtained by rotating the region in Example 4 about the line $x = -1$.

SOLUTION   Figure 11 shows a horizontal cross-section. It is a washer with inner radius $1 + y$ and outer radius $1 + \sqrt{y}$, so the cross-sectional area is

$$A(y) = \pi(\text{outer radius})^2 - \pi(\text{inner radius})^2$$

$$= \pi\left(1 + \sqrt{y}\right)^2 - \pi(1 + y)^2$$

The volume is

$$V = \int_0^1 A(y)\, dy = \pi \int_0^1 \left[\left(1 + \sqrt{y}\right)^2 - (1 + y)^2\right] dy$$

$$= \pi \int_0^1 \left(2\sqrt{y} - y - y^2\right) dy = \pi \left[\frac{4y^{3/2}}{3} - \frac{y^2}{2} - \frac{y^3}{3}\right]_0^1 = \frac{\pi}{2}$$

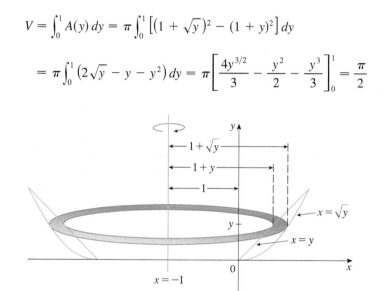

**FIGURE 11**

We now find the volumes of two solids that are *not* solids of revolution.

**EXAMPLE 7**   Figure 12 shows a solid with a circular base of radius 1. Parallel cross-sections perpendicular to the base are equilateral triangles. Find the volume of the solid.

SOLUTION   Let's take the circle to be $x^2 + y^2 = 1$. The solid, its base, and a typical cross-section at a distance $x$ from the origin are shown in Figure 13.

**FIGURE 12**
Computer-generated picture
of the solid in Example 7

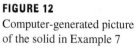

(a) The solid          (b) Its base          (c) A cross-section

**FIGURE 13**

Since $B$ lies on the circle, we have $y = \sqrt{1 - x^2}$ and so the base of the triangle $ABC$ is $|AB| = 2\sqrt{1 - x^2}$. Since the triangle is equilateral, we see from Figure 13(c) that its height is $\sqrt{3}\,y = \sqrt{3}\,\sqrt{1-x^2}$. The cross-sectional area is therefore

$$A(x) = \tfrac{1}{2} \cdot 2\sqrt{1-x^2} \cdot \sqrt{3}\,\sqrt{1-x^2} = \sqrt{3}\,(1-x^2)$$

and the volume of the solid is

$$V = \int_{-1}^{1} A(x)\,dx = \int_{-1}^{1} \sqrt{3}\,(1-x^2)\,dx$$

$$= 2\int_{0}^{1} \sqrt{3}\,(1-x^2)\,dx = 2\sqrt{3}\left[x - \frac{x^3}{3}\right]_{0}^{1} = \frac{4\sqrt{3}}{3}$$

**V** **EXAMPLE 8** Find the volume of a pyramid whose base is a square with side $L$ and whose height is $h$.

SOLUTION We place the origin $O$ at the vertex of the pyramid and the $x$-axis along its central axis as in Figure 14. Any plane $P_x$ that passes through $x$ and is perpendicular to the $x$-axis intersects the pyramid in a square with side of length $s$, say. We can express $s$ in terms of $x$ by observing from the similar triangles in Figure 15 that

$$\frac{x}{h} = \frac{s/2}{L/2} = \frac{s}{L}$$

and so $s = Lx/h$. [Another method is to observe that the line $OP$ has slope $L/(2h)$ and so its equation is $y = Lx/(2h)$.] Thus the cross-sectional area is

$$A(x) = s^2 = \frac{L^2}{h^2} x^2$$

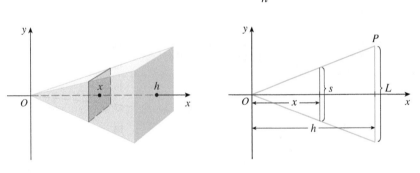

**FIGURE 14**          **FIGURE 15**

The pyramid lies between $x = 0$ and $x = h$, so its volume is

$$V = \int_{0}^{h} A(x)\,dx = \int_{0}^{h} \frac{L^2}{h^2} x^2\,dx = \frac{L^2}{h^2}\frac{x^3}{3}\bigg]_{0}^{h} = \frac{L^2 h}{3}$$

**NOTE** We didn't need to place the vertex of the pyramid at the origin in Example 8. We did so merely to make the equations simple. If, instead, we had placed the center of the base at the origin and the vertex on the positive $y$-axis, as in Figure 16, you can verify that we would have obtained the integral

$$V = \int_{0}^{h} \frac{L^2}{h^2} (h-y)^2\,dy = \frac{L^2 h}{3}$$

**FIGURE 16**

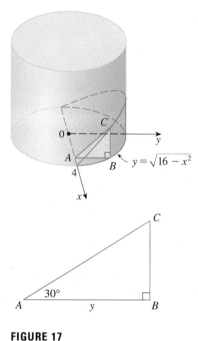

**EXAMPLE 9** A wedge is cut out of a circular cylinder of radius 4 by two planes. One plane is perpendicular to the axis of the cylinder. The other intersects the first at an angle of 30° along a diameter of the cylinder. Find the volume of the wedge.

SOLUTION If we place the $x$-axis along the diameter where the planes meet, then the base of the solid is a semicircle with equation $y = \sqrt{16 - x^2}$, $-4 \le x \le 4$. A cross-section perpendicular to the $x$-axis at a distance $x$ from the origin is a triangle $ABC$, as shown in Figure 17, whose base is $y = \sqrt{16 - x^2}$ and whose height is $|BC| = y \tan 30° = \sqrt{16 - x^2}/\sqrt{3}$. Thus the cross-sectional area is

$$A(x) = \tfrac{1}{2}\sqrt{16 - x^2} \cdot \frac{1}{\sqrt{3}}\sqrt{16 - x^2} = \frac{16 - x^2}{2\sqrt{3}}$$

and the volume is

$$V = \int_{-4}^{4} A(x)\, dx = \int_{-4}^{4} \frac{16 - x^2}{2\sqrt{3}}\, dx$$

$$= \frac{1}{\sqrt{3}} \int_{0}^{4} (16 - x^2)\, dx = \frac{1}{\sqrt{3}}\left[ 16x - \frac{x^3}{3} \right]_{0}^{4}$$

$$= \frac{128}{3\sqrt{3}}$$

**FIGURE 17**

For another method see Exercise 62.

---

## 5.2 Exercises

**1–18** Find the volume of the solid obtained by rotating the region bounded by the given curves about the specified line. Sketch the region, the solid, and a typical disk or washer.

**1.** $y = 2 - \tfrac{1}{2}x$, $y = 0$, $x = 1$, $x = 2$;   about the $x$-axis

**2.** $y = 1 - x^2$, $y = 0$;   about the $x$-axis

**3.** $y = \sqrt{x - 1}$, $y = 0$, $x = 5$;   about the $x$-axis

**4.** $y = \sqrt{25 - x^2}$, $y = 0$, $x = 2$, $x = 4$;   about the $x$-axis

**5.** $x = 2\sqrt{y}$, $x = 0$, $y = 9$;   about the $y$-axis

**6.** $x = y - y^2$, $x = 0$;   about the $y$-axis

**7.** $y = x^3$, $y = x$, $x \ge 0$;   about the $x$-axis

**8.** $y = \tfrac{1}{4}x^2$, $y = 5 - x^2$;   about the $x$-axis

**9.** $y^2 = x$, $x = 2y$;   about the $y$-axis

**10.** $y = \tfrac{1}{4}x^2$, $x = 2$, $y = 0$;   about the $y$-axis

**11.** $y = x^2$, $x = y^2$;   about $y = 1$

**12.** $y = x^2$, $y = 4$;   about $y = 4$

**13.** $y = 1 + \sec x$, $y = 3$;   about $y = 1$

**14.** $y = \sin x$, $y = \cos x$, $0 \le x \le \pi/4$;   about $y = -1$

**15.** $y = x^3$, $y = 0$, $x = 1$;   about $x = 2$

**16.** $y = x^2$, $x = y^2$;   about $x = -1$

**17.** $x = y^2$, $x = 1 - y^2$;   about $x = 3$

**18.** $y = x$, $y = 0$, $x = 2$, $x = 4$;   about $x = 1$

---

**19–30** Refer to the figure and find the volume generated by rotating the given region about the specified line.

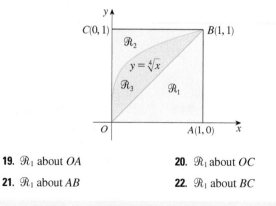

**19.** $\mathcal{R}_1$ about $OA$

**20.** $\mathcal{R}_1$ about $OC$

**21.** $\mathcal{R}_1$ about $AB$

**22.** $\mathcal{R}_1$ about $BC$

---

⊞ Graphing calculator or computer required    CAS Computer algebra system required    **1.** Homework Hints available at stewartcalculus.com

**23.** $\mathcal{R}_2$ about $OA$

**24.** $\mathcal{R}_2$ about $OC$

**25.** $\mathcal{R}_2$ about $AB$

**26.** $\mathcal{R}_2$ about $BC$

**27.** $\mathcal{R}_3$ about $OA$

**28.** $\mathcal{R}_3$ about $OC$

**29.** $\mathcal{R}_3$ about $AB$

**30.** $\mathcal{R}_3$ about $BC$

**31–34** Set up an integral for the volume of the solid obtained by rotating the region bounded by the given curves about the specified line. Then use your calculator to evaluate the integral correct to five decimal places.

**31.** $y = \tan x$, $y = 0$, $x = \pi/4$
   (a) About the $x$-axis       (b) About $y = -1$

**32.** $y = 0$, $y = \cos^2 x$, $-\pi/2 \leqslant x \leqslant \pi/2$
   (a) About the $x$-axis       (b) About $y = 1$

**33.** $x^2 + 4y^2 = 4$
   (a) About $y = 2$          (b) About $x = 2$

**34.** $y = x^2$, $x^2 + y^2 = 1$, $y \geqslant 0$
   (a) About the $x$-axis       (b) About the $y$-axis

**35–36** Use a graph to find approximate $x$-coordinates of the points of intersection of the given curves. Then use your calculator to find (approximately) the volume of the solid obtained by rotating about the $x$-axis the region bounded by these curves.

**35.** $y = 2 + x^2 \cos x$,   $y = x^4 + x + 1$

**36.** $y = x^4$,   $y = 3x - x^3$

**CAS** **37–38** Use a computer algebra system to find the exact volume of the solid obtained by rotating the region bounded by the given curves about the specified line.

**37.** $y = \sin^2 x$, $y = 0$, $0 \leqslant x \leqslant \pi$;   about $y = -1$

**38.** $y = x^2 - 2x$, $y = x \cos(\pi x/4)$;   about $y = 2$

**39–42** Each integral represents the volume of a solid. Describe the solid.

**39.** $\pi \displaystyle\int_0^{\pi} \sin x \, dx$         **40.** $\pi \displaystyle\int_{-1}^{1} (1 - y^2)^2 \, dy$

**41.** $\pi \displaystyle\int_0^{1} (y^4 - y^8) \, dy$

**42.** $\pi \displaystyle\int_0^{\pi/2} [(1 + \cos x)^2 - 1^2] \, dx$

**43.** A CAT scan produces equally spaced cross-sectional views of a human organ that provide information about the organ otherwise obtained only by surgery. Suppose that a CAT scan of a human liver shows cross-sections spaced 1.5 cm apart. The liver is 15 cm long and the cross-sectional areas, in square centimeters, are 0, 18, 58, 79, 94, 106, 117, 128, 63, 39, and 0. Use the Midpoint Rule to estimate the volume of the liver.

**44.** A log 10 m long is cut at 1-meter intervals and its cross-sectional areas $A$ (at a distance $x$ from the end of the log) are listed in the table. Use the Midpoint Rule with $n = 5$ to estimate the volume of the log.

| $x$ (m) | $A$ (m²) | $x$ (m) | $A$ (m²) |
|---------|----------|---------|----------|
| 0 | 0.68 | 6 | 0.53 |
| 1 | 0.65 | 7 | 0.55 |
| 2 | 0.64 | 8 | 0.52 |
| 3 | 0.61 | 9 | 0.50 |
| 4 | 0.58 | 10 | 0.48 |
| 5 | 0.59 | | |

**45.** (a) If the region shown in the figure is rotated about the $x$-axis to form a solid, use the Midpoint Rule with $n = 4$ to estimate the volume of the solid.

   (b) Estimate the volume if the region is rotated about the $y$-axis. Again use the Midpoint Rule with $n = 4$.

**CAS** **46.** (a) A model for the shape of a bird's egg is obtained by rotating about the $x$-axis the region under the graph of
$$f(x) = (ax^3 + bx^2 + cx + d)\sqrt{1 - x^2}$$
   Use a CAS to find the volume of such an egg.

   (b) For a Red-throated Loon, $a = -0.06$, $b = 0.04$, $c = 0.1$, and $d = 0.54$. Graph $f$ and find the volume of an egg of this species.

**47–59** Find the volume of the described solid $S$.

**47.** A right circular cone with height $h$ and base radius $r$

**48.** A frustum of a right circular cone with height $h$, lower base radius $R$, and top radius $r$

**49.** A cap of a sphere with radius $r$ and cap height $h$

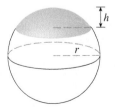

**50.** A frustum of a pyramid with square base of side $b$, square top of side $a$, and height $h$

What happens if $a = b$? What happens if $a = 0$?

**51.** A pyramid with height $h$ and rectangular base with dimensions $b$ and $2b$

**52.** A pyramid with height $h$ and base an equilateral triangle with side $a$ (a tetrahedron)

**53.** A tetrahedron with three mutually perpendicular faces and three mutually perpendicular edges with lengths 3 cm, 4 cm, and 5 cm

**54.** The base of $S$ is a circular disk with radius $r$. Parallel cross-sections perpendicular to the base are squares.

**55.** The base of $S$ is an elliptical region with boundary curve $9x^2 + 4y^2 = 36$. Cross-sections perpendicular to the $x$-axis are isosceles right triangles with hypotenuse in the base.

**56.** The base of $S$ is the triangular region with vertices $(0, 0)$, $(1, 0)$, and $(0, 1)$. Cross-sections perpendicular to the $y$-axis are equilateral triangles.

**57.** The base of $S$ is the same base as in Exercise 56, but cross-sections perpendicular to the $x$-axis are squares.

**58.** The base of $S$ is the region enclosed by the parabola $y = 1 - x^2$ and the $x$-axis. Cross-sections perpendicular to the $y$-axis are squares.

**59.** The base of $S$ is the same base as in Exercise 58, but cross-sections perpendicular to the $x$-axis are isosceles triangles with height equal to the base.

**60.** The base of $S$ is a circular disk with radius $r$. Parallel cross-sections perpendicular to the base are isosceles triangles with height $h$ and unequal side in the base.
(a) Set up an integral for the volume of $S$.
(b) By interpreting the integral as an area, find the volume of $S$.

**61.** (a) Set up an integral for the volume of a solid *torus* (the donut-shaped solid shown in the figure) with radii $r$ and $R$.
(b) By interpreting the integral as an area, find the volume of the torus.

**62.** Solve Example 9 taking cross-sections to be parallel to the line of intersection of the two planes.

**63.** (a) Cavalieri's Principle states that if a family of parallel planes gives equal cross-sectional areas for two solids $S_1$ and $S_2$, then the volumes of $S_1$ and $S_2$ are equal. Prove this principle.
(b) Use Cavalieri's Principle to find the volume of the oblique cylinder shown in the figure.

**64.** Find the volume common to two circular cylinders, each with radius $r$, if the axes of the cylinders intersect at right angles.

**65.** Find the volume common to two spheres, each with radius $r$, if the center of each sphere lies on the surface of the other sphere.

**66.** A bowl is shaped like a hemisphere with diameter 30 cm. A heavy ball with diameter 10 cm is placed in the bowl and water is poured into the bowl to a depth of $h$ centimeters. Find the volume of water in the bowl.

**67.** A hole of radius $r$ is bored through the middle of a cylinder of radius $R > r$ at right angles to the axis of the cylinder. Set up, but do not evaluate, an integral for the volume cut out.

**68.** A hole of radius $r$ is bored through the center of a sphere of radius $R > r$. Find the volume of the remaining portion of the sphere.

**69.** Some of the pioneers of calculus, such as Kepler and Newton, were inspired by the problem of finding the volumes of wine barrels. (In fact Kepler published a book *Stereometria doliorum* in 1615 devoted to methods for finding the volumes of barrels.) They often approximated the shape of the sides by parabolas.

(a) A barrel with height $h$ and maximum radius $R$ is constructed by rotating about the $x$-axis the parabola $y = R - cx^2$, $-h/2 \leqslant x \leqslant h/2$, where $c$ is a positive constant. Show that the radius of each end of the barrel is $r = R - d$, where $d = ch^2/4$.

(b) Show that the volume enclosed by the barrel is
$$V = \tfrac{1}{3}\pi h\left(2R^2 + r^2 - \tfrac{2}{5}d^2\right)$$

**70.** Suppose that a region $\mathcal{R}$ has area $A$ and lies above the $x$-axis. When $\mathcal{R}$ is rotated about the $x$-axis, it sweeps out a solid with volume $V_1$. When $\mathcal{R}$ is rotated about the line $y = -k$ (where $k$ is a positive number), it sweeps out a solid with volume $V_2$. Express $V_2$ in terms of $V_1$, $k$, and $A$.

## 5.3  Volumes by Cylindrical Shells

$y = 2x^2 - x^3$

$x_L = ?$     $x_R = ?$

**FIGURE 1**

**FIGURE 2**

Some volume problems are very difficult to handle by the methods of the preceding section. For instance, let's consider the problem of finding the volume of the solid obtained by rotating about the $y$-axis the region bounded by $y = 2x^2 - x^3$ and $y = 0$. (See Figure 1.) If we slice perpendicular to the $y$-axis, we get a washer. But to compute the inner radius and the outer radius of the washer, we'd have to solve the cubic equation $y = 2x^2 - x^3$ for $x$ in terms of $y$; that's not easy.

Fortunately, there is a method, called the **method of cylindrical shells**, that is easier to use in such a case. Figure 2 shows a cylindrical shell with inner radius $r_1$, outer radius $r_2$, and height $h$. Its volume $V$ is calculated by subtracting the volume $V_1$ of the inner cylinder from the volume $V_2$ of the outer cylinder:

$$V = V_2 - V_1$$
$$= \pi r_2^2 h - \pi r_1^2 h = \pi(r_2^2 - r_1^2)h$$
$$= \pi(r_2 + r_1)(r_2 - r_1)h$$
$$= 2\pi \frac{r_2 + r_1}{2} h(r_2 - r_1)$$

If we let $\Delta r = r_2 - r_1$ (the thickness of the shell) and $r = \tfrac{1}{2}(r_2 + r_1)$ (the average radius of the shell), then this formula for the volume of a cylindrical shell becomes

**1**
$$\boxed{V = 2\pi r h \, \Delta r}$$

and it can be remembered as

$$V = [\text{circumference}][\text{height}][\text{thickness}]$$

Now let $S$ be the solid obtained by rotating about the $y$-axis the region bounded by $y = f(x)$ [where $f(x) \geqslant 0$], $y = 0$, $x = a$, and $x = b$, where $b > a \geqslant 0$. (See Figure 3.)

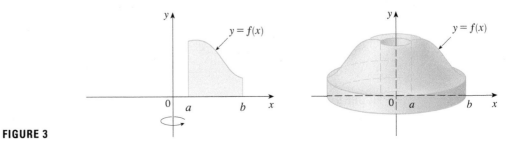

**FIGURE 3**

We divide the interval $[a, b]$ into $n$ subintervals $[x_{i-1}, x_i]$ of equal width $\Delta x$ and let $\bar{x}_i$ be the midpoint of the $i$th subinterval. If the rectangle with base $[x_{i-1}, x_i]$ and height $f(\bar{x}_i)$ is rotated about the $y$-axis, then the result is a cylindrical shell with average radius $\bar{x}_i$, height $f(\bar{x}_i)$, and thickness $\Delta x$ (see Figure 4), so by Formula 1 its volume is

$$V_i = (2\pi\bar{x}_i)[f(\bar{x}_i)]\,\Delta x$$

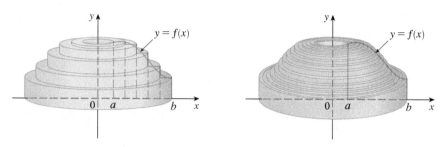

**FIGURE 4**

Therefore an approximation to the volume $V$ of $S$ is given by the sum of the volumes of these shells:

$$V \approx \sum_{i=1}^{n} V_i = \sum_{i=1}^{n} 2\pi\bar{x}_i f(\bar{x}_i)\,\Delta x$$

This approximation appears to become better as $n \to \infty$. But, from the definition of an integral, we know that

$$\lim_{n\to\infty} \sum_{i=1}^{n} 2\pi\bar{x}_i f(\bar{x}_i)\,\Delta x = \int_a^b 2\pi x f(x)\,dx$$

Thus the following appears plausible:

---

**2**    The volume of the solid in Figure 3, obtained by rotating about the $y$-axis the region under the curve $y = f(x)$ from $a$ to $b$, is

$$V = \int_a^b 2\pi x f(x)\,dx \qquad \text{where } 0 \leq a < b$$

---

The argument using cylindrical shells makes Formula 2 seem reasonable, but later we will be able to prove it (see Exercise 71 in Section 7.1).

The best way to remember Formula 2 is to think of a typical shell, cut and flattened as in Figure 5, with radius $x$, circumference $2\pi x$, height $f(x)$, and thickness $\Delta x$ or $dx$:

$$\int_a^b \underbrace{(2\pi x)}_{\text{circumference}} \underbrace{[f(x)]}_{\text{height}} \underbrace{dx}_{\text{thickness}}$$

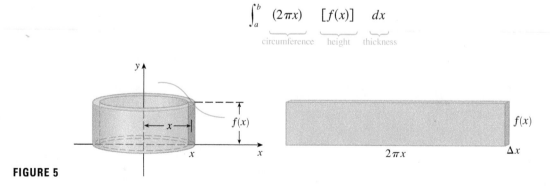

**FIGURE 5**

This type of reasoning will be helpful in other situations, such as when we rotate about lines other than the y-axis.

**EXAMPLE 1** Find the volume of the solid obtained by rotating about the y-axis the region bounded by $y = 2x^2 - x^3$ and $y = 0$.

**SOLUTION** From the sketch in Figure 6 we see that a typical shell has radius $x$, circumference $2\pi x$, and height $f(x) = 2x^2 - x^3$. So, by the shell method, the volume is

$$V = \int_0^2 (2\pi x)(2x^2 - x^3)\, dx = 2\pi \int_0^2 (2x^3 - x^4)\, dx$$

$$= 2\pi\left[\tfrac{1}{2}x^4 - \tfrac{1}{5}x^5\right]_0^2 = 2\pi\left(8 - \tfrac{32}{5}\right) = \tfrac{16}{5}\pi$$

It can be verified that the shell method gives the same answer as slicing.

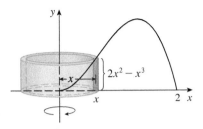

**FIGURE 6**

Figure 7 shows a computer-generated picture of the solid whose volume we computed in Example 1.

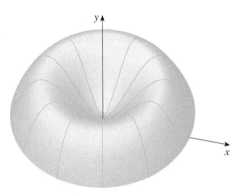

**FIGURE 7**

**NOTE** Comparing the solution of Example 1 with the remarks at the beginning of this section, we see that the method of cylindrical shells is much easier than the washer method for this problem. We did not have to find the coordinates of the local maximum and we did not have to solve the equation of the curve for $x$ in terms of $y$. However, in other examples the methods of the preceding section may be easier.

**EXAMPLE 2** Find the volume of the solid obtained by rotating about the y-axis the region between $y = x$ and $y = x^2$.

**SOLUTION** The region and a typical shell are shown in Figure 8. We see that the shell has radius $x$, circumference $2\pi x$, and height $x - x^2$. So the volume is

$$V = \int_0^1 (2\pi x)(x - x^2)\, dx = 2\pi \int_0^1 (x^2 - x^3)\, dx$$

$$= 2\pi\left[\frac{x^3}{3} - \frac{x^4}{4}\right]_0^1 = \frac{\pi}{6}$$

**FIGURE 8**

As the following example shows, the shell method works just as well if we rotate about the x-axis. We simply have to draw a diagram to identify the radius and height of a shell.

**EXAMPLE 3** Use cylindrical shells to find the volume of the solid obtained by rotating about the x-axis the region under the curve $y = \sqrt{x}$ from 0 to 1.

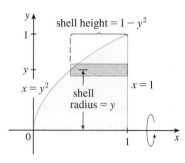

**FIGURE 9**

SOLUTION  This problem was solved using disks in Example 2 in Section 5.2. To use shells we relabel the curve $y = \sqrt{x}$ (in the figure in that example) as $x = y^2$ in Figure 9. For rotation about the $x$-axis we see that a typical shell has radius $y$, circumference $2\pi y$, and height $1 - y^2$. So the volume is

$$V = \int_0^1 (2\pi y)(1 - y^2)\,dy = 2\pi \int_0^1 (y - y^3)\,dy$$

$$= 2\pi \left[ \frac{y^2}{2} - \frac{y^4}{4} \right]_0^1 = \frac{\pi}{2}$$

In this problem the disk method was simpler.

**V  EXAMPLE 4**  Find the volume of the solid obtained by rotating the region bounded by $y = x - x^2$ and $y = 0$ about the line $x = 2$.

SOLUTION  Figure 10 shows the region and a cylindrical shell formed by rotation about the line $x = 2$. It has radius $2 - x$, circumference $2\pi(2 - x)$, and height $x - x^2$.

**FIGURE 10**

The volume of the given solid is

$$V = \int_0^1 2\pi(2 - x)(x - x^2)\,dx$$

$$= 2\pi \int_0^1 (x^3 - 3x^2 + 2x)\,dx$$

$$= 2\pi \left[ \frac{x^4}{4} - x^3 + x^2 \right]_0^1 = \frac{\pi}{2}$$

**5.3    Exercises**

**1.** Let $S$ be the solid obtained by rotating the region shown in the figure about the $y$-axis. Explain why it is awkward to use slicing to find the volume $V$ of $S$. Sketch a typical approximating shell. What are its circumference and height? Use shells to find $V$.

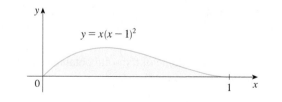

**2.** Let $S$ be the solid obtained by rotating the region shown in the figure about the $y$-axis. Sketch a typical cylindrical shell and find its circumference and height. Use shells to find the volume of $S$. Do you think this method is preferable to slicing? Explain.

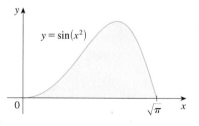

**3–7** Use the method of cylindrical shells to find the volume generated by rotating the region bounded by the given curves about the $y$-axis.

**3.** $y = \sqrt[3]{x}$, $\quad y = 0$, $\quad x = 1$

**4.** $y = x^3$, $\quad y = 0$, $\quad x = 1$, $\quad x = 2$

**5.** $y = x^2$, $\quad 0 \le x \le 2$, $\quad y = 4$, $\quad x = 0$

**6.** $y = 4x - x^2$, $\quad y = x$

**7.** $y = x^2$, $\quad y = 6x - 2x^2$

**8.** Let $V$ be the volume of the solid obtained by rotating about the $y$-axis the region bounded by $y = \sqrt{x}$ and $y = x^2$. Find $V$ both by slicing and by cylindrical shells. In both cases draw a diagram to explain your method.

**9–14** Use the method of cylindrical shells to find the volume of the solid obtained by rotating the region bounded by the given curves about the $x$-axis.

**9.** $xy = 1$, $\quad x = 0$, $\quad y = 1$, $\quad y = 3$

**10.** $y = \sqrt{x}$, $\quad x = 0$, $\quad y = 2$

**11.** $y = x^3$, $\quad y = 8$, $\quad x = 0$

**12.** $x = 4y^2 - y^3$, $\quad x = 0$

**13.** $x = 1 + (y - 2)^2$, $\quad x = 2$

**14.** $x + y = 3$, $\quad x = 4 - (y - 1)^2$

**15–20** Use the method of cylindrical shells to find the volume generated by rotating the region bounded by the given curves about the specified axis.

**15.** $y = x^4$, $y = 0$, $x = 1$; $\quad$ about $x = 2$

**16.** $y = \sqrt{x}$, $y = 0$, $x = 1$; $\quad$ about $x = -1$

**17.** $y = 4x - x^2$, $y = 3$; $\quad$ about $x = 1$

**18.** $y = x^2$, $y = 2 - x^2$; $\quad$ about $x = 1$

**19.** $y = x^3$, $y = 0$, $x = 1$; $\quad$ about $y = 1$

**20.** $x = y^2 + 1$, $x = 2$; $\quad$ about $y = -2$

**21–26**

(a) Set up an integral for the volume of the solid obtained by rotating the region bounded by the given curve about the specified axis.

(b) Use your calculator to evaluate the integral correct to five decimal places.

**21.** $y = \sin x$, $y = 0$, $x = 2\pi$, $x = 3\pi$; $\quad$ about the $y$-axis

**22.** $y = \tan x$, $y = 0$, $x = \pi/4$; $\quad$ about $x = \pi/2$

**23.** $y = \cos^4 x$, $y = -\cos^4 x$, $-\pi/2 \le x \le \pi/2$; $\quad$ about $x = \pi$

**24.** $y = x$, $y = 2x/(1 + x^3)$; $\quad$ about $x = -1$

**25.** $x = \sqrt{\sin y}$, $0 \le y \le \pi$, $x = 0$; $\quad$ about $y = 4$

**26.** $x^2 - y^2 = 7$, $x = 4$; $\quad$ about $y = 5$

**27.** Use the Midpoint Rule with $n = 5$ to estimate the volume obtained by rotating about the $y$-axis the region under the curve $y = \sqrt{1 + x^3}$, $0 \le x \le 1$.

**28.** If the region shown in the figure is rotated about the $y$-axis to form a solid, use the Midpoint Rule with $n = 5$ to estimate the volume of the solid.

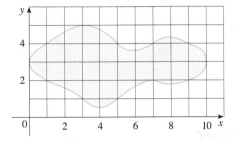

**29–32** Each integral represents the volume of a solid. Describe the solid.

**29.** $\displaystyle\int_0^3 2\pi x^5 \, dx$

**30.** $\displaystyle 2\pi \int_0^2 \frac{y}{1 + y^2} \, dy$

**31.** $\displaystyle\int_0^1 2\pi(3 - y)(1 - y^2) \, dy$

**32.** $\displaystyle\int_0^{\pi/4} 2\pi(\pi - x)(\cos x - \sin x) \, dx$

**33–34** Use a graph to estimate the $x$-coordinates of the points of intersection of the given curves. Then use this information and your calculator to estimate the volume of the solid obtained by rotating about the $y$-axis the region enclosed by these curves.

**33.** $y = 0$, $\quad y = x + x^2 - x^4$

**34.** $y = x^3 - x + 1$, $\quad y = -x^4 + 4x - 1$

**35–36** Use a computer algebra system to find the exact volume of the solid obtained by rotating the region bounded by the given curves about the specified line.

**35.** $y = \sin^2 x$, $y = \sin^4 x$, $0 \le x \le \pi$; $\quad$ about $x = \pi/2$

**36.** $y = x^3 \sin x$, $y = 0$, $0 \le x \le \pi$; $\quad$ about $x = -1$

**37–43** The region bounded by the given curves is rotated about the specified axis. Find the volume of the resulting solid by any method.

**37.** $y = -x^2 + 6x - 8$, $y = 0$; $\quad$ about the $y$-axis

**38.** $y = -x^2 + 6x - 8$, $y = 0$; $\quad$ about the $x$-axis

**39.** $y^2 - x^2 = 1$, $y = 2$; $\quad$ about the $x$-axis

**40.** $y^2 - x^2 = 1$, $y = 2$; $\quad$ about the $y$-axis

**41.** $x^2 + (y - 1)^2 = 1$;  about the $y$-axis

**42.** $x = (y - 3)^2$, $x = 4$;  about $y = 1$

**43.** $x = (y - 1)^2$, $x - y = 1$;  about $x = -1$

**44.** Let $T$ be the triangular region with vertices $(0, 0)$, $(1, 0)$, and $(1, 2)$, and let $V$ be the volume of the solid generated when $T$ is rotated about the line $x = a$, where $a > 1$. Express $a$ in terms of $V$.

45–47 Use cylindrical shells to find the volume of the solid.

**45.** A sphere of radius $r$

**46.** The solid torus of Exercise 61 in Section 5.2

**47.** A right circular cone with height $h$ and base radius $r$

**48.** Suppose you make napkin rings by drilling holes with different diameters through two wooden balls (which also have different diameters). You discover that both napkin rings have the same height $h$, as shown in the figure.
(a) Guess which ring has more wood in it.
(b) Check your guess: Use cylindrical shells to compute the volume of a napkin ring created by drilling a hole with radius $r$ through the center of a sphere of radius $R$ and express the answer in terms of $h$.

# 5.4 Work

The term *work* is used in everyday language to mean the total amount of effort required to perform a task. In physics it has a technical meaning that depends on the idea of a *force*. Intuitively, you can think of a force as describing a push or pull on an object—for example, a horizontal push of a book across a table or the downward pull of the earth's gravity on a ball. In general, if an object moves along a straight line with position function $s(t)$, then the **force** $F$ on the object (in the same direction) is given by Newton's Second Law of Motion as the product of its mass $m$ and its acceleration:

$$\boxed{1} \qquad\qquad\qquad F = m\frac{d^2 s}{dt^2}$$

In the SI metric system, the mass is measured in kilograms (kg), the displacement in meters (m), the time in seconds (s), and the force in newtons (N $=$ kg·m/s²). Thus a force of 1 N acting on a mass of 1 kg produces an acceleration of 1 m/s². In the US Customary system the fundamental unit is chosen to be the unit of force, which is the pound.

In the case of constant acceleration, the force $F$ is also constant and the work done is defined to be the product of the force $F$ and the distance $d$ that the object moves:

$$\boxed{2} \qquad\qquad W = Fd \qquad \text{work} = \text{force} \times \text{distance}$$

If $F$ is measured in newtons and $d$ in meters, then the unit for $W$ is a newton-meter, which is called a joule (J). If $F$ is measured in pounds and $d$ in feet, then the unit for $W$ is a foot-pound (ft-lb), which is about 1.36 J.

**V EXAMPLE 1**
(a) How much work is done in lifting a 1.2-kg book off the floor to put it on a desk that is 0.7 m high? Use the fact that the acceleration due to gravity is $g = 9.8$ m/s².
(b) How much work is done in lifting a 20-lb weight 6 ft off the ground?

SOLUTION
(a) The force exerted is equal and opposite to that exerted by gravity, so Equation 1 gives

$$F = mg = (1.2)(9.8) = 11.76 \text{ N}$$

and then Equation 2 gives the work done as

$$W = Fd = (11.76)(0.7) \approx 8.2 \text{ J}$$

(b)  Here the force is given as $F = 20$ lb, so the work done is

$$W = Fd = 20 \cdot 6 = 120 \text{ ft-lb}$$

Notice that in part (b), unlike part (a), we did not have to multiply by $g$ because we were given the *weight* (which is a force) and not the mass of the object.

Equation 2 defines work as long as the force is constant, but what happens if the force is variable? Let's suppose that the object moves along the $x$-axis in the positive direction, from $x = a$ to $x = b$, and at each point $x$ between $a$ and $b$ a force $f(x)$ acts on the object, where $f$ is a continuous function. We divide the interval $[a, b]$ into $n$ subintervals with endpoints $x_0, x_1, \ldots, x_n$ and equal width $\Delta x$. We choose a sample point $x_i^*$ in the $i$th subinterval $[x_{i-1}, x_i]$. Then the force at that point is $f(x_i^*)$. If $n$ is large, then $\Delta x$ is small, and since $f$ is continuous, the values of $f$ don't change very much over the interval $[x_{i-1}, x_i]$. In other words, $f$ is almost constant on the interval and so the work $W_i$ that is done in moving the particle from $x_{i-1}$ to $x_i$ is approximately given by Equation 2:

$$W_i \approx f(x_i^*)\, \Delta x$$

Thus we can approximate the total work by

$$\boxed{3} \qquad W \approx \sum_{i=1}^{n} f(x_i^*)\, \Delta x$$

It seems that this approximation becomes better as we make $n$ larger. Therefore we define the **work done in moving the object from $a$ to $b$** as the limit of this quantity as $n \to \infty$. Since the right side of $\boxed{3}$ is a Riemann sum, we recognize its limit as being a definite integral and so

$$\boxed{4} \qquad W = \lim_{n \to \infty} \sum_{i=1}^{n} f(x_i^*)\, \Delta x = \int_a^b f(x)\, dx$$

frictionless   0                 $x$
surface

(a) Natural position of spring

$f(x) = kx$

0        $x$       $x$

(b) Stretched position of spring

**FIGURE 1**
Hooke's Law

**EXAMPLE 2**   When a particle is located a distance $x$ feet from the origin, a force of $x^2 + 2x$ pounds acts on it. How much work is done in moving it from $x = 1$ to $x = 3$?

SOLUTION
$$W = \int_1^3 (x^2 + 2x)\, dx = \frac{x^3}{3} + x^2 \Big]_1^3 = \frac{50}{3}$$

The work done is $16\frac{2}{3}$ ft-lb.

In the next example we use a law from physics: **Hooke's Law** states that the force required to maintain a spring stretched $x$ units beyond its natural length is proportional to $x$:

$$f(x) = kx$$

where $k$ is a positive constant (called the **spring constant**). Hooke's Law holds provided that $x$ is not too large (see Figure 1).

**◢ EXAMPLE 3**  A force of 40 N is required to hold a spring that has been stretched from its natural length of 10 cm to a length of 15 cm. How much work is done in stretching the spring from 15 cm to 18 cm?

SOLUTION  According to Hooke's Law, the force required to hold the spring stretched $x$ meters beyond its natural length is $f(x) = kx$. When the spring is stretched from 10 cm to 15 cm, the amount stretched is 5 cm = 0.05 m. This means that $f(0.05) = 40$, so

$$0.05k = 40 \qquad k = \frac{40}{0.05} = 800$$

Thus $f(x) = 800x$ and the work done in stretching the spring from 15 cm to 18 cm is

$$W = \int_{0.05}^{0.08} 800x \, dx = 800 \left. \frac{x^2}{2} \right]_{0.05}^{0.08}$$

$$= 400[(0.08)^2 - (0.05)^2] = 1.56 \text{ J}$$

**◢ EXAMPLE 4**  A 200-lb cable is 100 ft long and hangs vertically from the top of a tall building. How much work is required to lift the cable to the top of the building?

SOLUTION  Here we don't have a formula for the force function, but we can use an argument similar to the one that led to Definition 4.

Let's place the origin at the top of the building and the $x$-axis pointing downward as in Figure 2. We divide the cable into small parts with length $\Delta x$. If $x_i^*$ is a point in the $i$th such interval, then all points in the interval are lifted by approximately the same amount, namely $x_i^*$. The cable weighs 2 pounds per foot, so the weight of the $i$th part is $2\Delta x$. Thus the work done on the $i$th part, in foot-pounds, is

$$\underbrace{(2\,\Delta x)}_{\text{force}} \cdot \underbrace{x_i^*}_{\text{distance}} = 2x_i^* \, \Delta x$$

We get the total work done by adding all these approximations and letting the number of parts become large (so $\Delta x \to 0$):

$$W = \lim_{n \to \infty} \sum_{i=1}^{n} 2x_i^* \, \Delta x = \int_0^{100} 2x \, dx$$

$$= x^2 \Big]_0^{100} = 10{,}000 \text{ ft-lb}$$

**EXAMPLE 5**  A tank has the shape of an inverted circular cone with height 10 m and base radius 4 m. It is filled with water to a height of 8 m. Find the work required to empty the tank by pumping all of the water to the top of the tank. (The density of water is 1000 kg/m³.)

SOLUTION  Let's measure depths from the top of the tank by introducing a vertical coordinate line as in Figure 3. The water extends from a depth of 2 m to a depth of 10 m and so we divide the interval $[2, 10]$ into $n$ subintervals with endpoints $x_0, x_1, \ldots, x_n$ and choose $x_i^*$ in the $i$th subinterval. This divides the water into $n$ layers. The $i$th layer is approximated by a circular cylinder with radius $r_i$ and height $\Delta x$. We can compute $r_i$ from similar triangles, using Figure 4, as follows:

$$\frac{r_i}{10 - x_i^*} = \frac{4}{10} \qquad r_i = \tfrac{2}{5}(10 - x_i^*)$$

**FIGURE 2**

If we had placed the origin at the bottom of the cable and the $x$-axis upward, we would have gotten

$$W = \int_0^{100} 2(100 - x) \, dx$$

which gives the same answer.

**FIGURE 3**

**FIGURE 4**

Thus an approximation to the volume of the $i$th layer of water is

$$V_i \approx \pi r_i^2 \, \Delta x = \frac{4\pi}{25}(10 - x_i^*)^2 \, \Delta x$$

and so its mass is

$$m_i = \text{density} \times \text{volume}$$

$$\approx 1000 \cdot \frac{4\pi}{25}(10 - x_i^*)^2 \, \Delta x = 160\pi(10 - x_i^*)^2 \, \Delta x$$

The force required to raise this layer must overcome the force of gravity and so

$$F_i = m_i g \approx (9.8)160\pi(10 - x_i^*)^2 \, \Delta x$$

$$= 1568\pi(10 - x_i^*)^2 \, \Delta x$$

Each particle in the layer must travel a distance of approximately $x_i^*$. The work $W_i$ done to raise this layer to the top is approximately the product of the force $F_i$ and the distance $x_i^*$:

$$W_i \approx F_i x_i^* \approx 1568\pi x_i^*(10 - x_i^*)^2 \, \Delta x$$

To find the total work done in emptying the entire tank, we add the contributions of each of the $n$ layers and then take the limit as $n \to \infty$:

$$W = \lim_{n \to \infty} \sum_{i=1}^{n} 1568\pi x_i^*(10 - x_i^*)^2 \, \Delta x = \int_2^{10} 1568\pi x(10 - x)^2 \, dx$$

$$= 1568\pi \int_2^{10} (100x - 20x^2 + x^3) \, dx = 1568\pi \left[ 50x^2 - \frac{20x^3}{3} + \frac{x^4}{4} \right]_2^{10}$$

$$= 1568\pi \left( \tfrac{2048}{3} \right) \approx 3.4 \times 10^6 \text{ J}$$

## 5.4   Exercises

**1.** A 360-lb gorilla climbs a tree to a height of 20 ft. Find the work done if the gorilla reaches that height in
(a) 10 seconds         (b) 5 seconds

**2.** How much work is done when a hoist lifts a 200-kg rock to a height of 3 m?

**3.** A variable force of $5x^{-2}$ pounds moves an object along a straight line when it is $x$ feet from the origin. Calculate the work done in moving the object from $x = 1$ ft to $x = 10$ ft.

**4.** When a particle is located a distance $x$ meters from the origin, a force of $\cos(\pi x/3)$ newtons acts on it. How much work is done in moving the particle from $x = 1$ to $x = 2$? Interpret your answer by considering the work done from $x = 1$ to $x = 1.5$ and from $x = 1.5$ to $x = 2$.

**5.** Shown is the graph of a force function (in newtons) that increases to its maximum value and then remains constant.

How much work is done by the force in moving an object a distance of 8 m?

**6.** The table shows values of a force function $f(x)$, where $x$ is measured in meters and $f(x)$ in newtons. Use the Midpoint Rule to estimate the work done by the force in moving an object from $x = 4$ to $x = 20$.

| $x$ | 4 | 6 | 8 | 10 | 12 | 14 | 16 | 18 | 20 |
|-----|---|-----|-----|-----|-----|-----|-----|-----|-----|
| $f(x)$ | 5 | 5.8 | 7.0 | 8.8 | 9.6 | 8.2 | 6.7 | 5.2 | 4.1 |

Graphing calculator or computer required        1. Homework Hints available at stewartcalculus.com

**7.** A force of 10 lb is required to hold a spring stretched 4 in. beyond its natural length. How much work is done in stretching it from its natural length to 6 in. beyond its natural length?

**8.** A spring has a natural length of 20 cm. If a 25-N force is required to keep it stretched to a length of 30 cm, how much work is required to stretch it from 20 cm to 25 cm?

**9.** Suppose that 2 J of work is needed to stretch a spring from its natural length of 30 cm to a length of 42 cm.
 (a) How much work is needed to stretch the spring from 35 cm to 40 cm?
 (b) How far beyond its natural length will a force of 30 N keep the spring stretched?

**10.** If the work required to stretch a spring 1 ft beyond its natural length is 12 ft-lb, how much work is needed to stretch it 9 in. beyond its natural length?

**11.** A spring has natural length 20 cm. Compare the work $W_1$ done in stretching the spring from 20 cm to 30 cm with the work $W_2$ done in stretching it from 30 cm to 40 cm. How are $W_2$ and $W_1$ related?

**12.** If 6 J of work is needed to stretch a spring from 10 cm to 12 cm and another 10 J is needed to stretch it from 12 cm to 14 cm, what is the natural length of the spring?

**13–20** Show how to approximate the required work by a Riemann sum. Then express the work as an integral and evaluate it.

**13.** A heavy rope, 50 ft long, weighs 0.5 lb/ft and hangs over the edge of a building 120 ft high.
 (a) How much work is done in pulling the rope to the top of the building?
 (b) How much work is done in pulling half the rope to the top of the building?

**14.** A chain lying on the ground is 10 m long and its mass is 80 kg. How much work is required to raise one end of the chain to a height of 6 m?

**15.** A cable that weighs 2 lb/ft is used to lift 800 lb of coal up a mine shaft 500 ft deep. Find the work done.

**16.** A bucket that weighs 4 lb and a rope of negligible weight are used to draw water from a well that is 80 ft deep. The bucket is filled with 40 lb of water and is pulled up at a rate of 2 ft/s, but water leaks out of a hole in the bucket at a rate of 0.2 lb/s. Find the work done in pulling the bucket to the top of the well.

**17.** A leaky 10-kg bucket is lifted from the ground to a height of 12 m at a constant speed with a rope that weighs 0.8 kg/m. Initially the bucket contains 36 kg of water, but the water leaks at a constant rate and finishes draining just as the bucket reaches the 12-m level. How much work is done?

**18.** A 10-ft chain weighs 25 lb and hangs from a ceiling. Find the work done in lifting the lower end of the chain to the ceiling so that it's level with the upper end.

**19.** An aquarium 2 m long, 1 m wide, and 1 m deep is full of water. Find the work needed to pump half of the water out of the aquarium. (Use the fact that the density of water is 1000 kg/m³.)

**20.** A circular swimming pool has a diameter of 24 ft, the sides are 5 ft high, and the depth of the water is 4 ft. How much work is required to pump all of the water out over the side? (Use the fact that water weighs 62.5 lb/ft³.)

**21–24** A tank is full of water. Find the work required to pump the water out of the spout. In Exercises 23 and 24 use the fact that water weighs 62.5 lb/ft³.

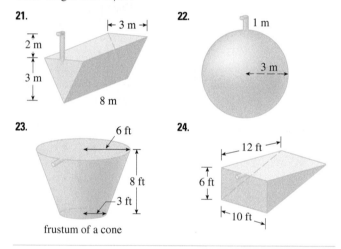

**21.** **22.**

**23.** 6 ft **24.**

frustum of a cone

**25.** Suppose that for the tank in Exercise 21 the pump breaks down after $4.7 \times 10^5$ J of work has been done. What is the depth of the water remaining in the tank?

**26.** Solve Exercise 22 if the tank is half full of oil that has a density of 900 kg/m³.

**27.** When gas expands in a cylinder with radius $r$, the pressure at any given time is a function of the volume: $P = P(V)$. The force exerted by the gas on the piston (see the figure) is the product of the pressure and the area: $F = \pi r^2 P$. Show that the work done by the gas when the volume expands from volume $V_1$ to volume $V_2$ is

$$W = \int_{V_1}^{V_2} P \, dV$$

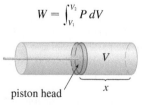

piston head

**28.** In a steam engine the pressure $P$ and volume $V$ of steam satisfy the equation $PV^{1.4} = k$, where $k$ is a constant. (This is true for adiabatic expansion, that is, expansion in which there is no heat transfer between the cylinder and its surroundings.) Use Exercise 27 to calculate the work done by the engine during a cycle when the steam starts at a pressure of 160 lb/in² and a volume of 100 in³ and expands to a volume of 800 in³.

**29.** (a) Newton's Law of Gravitation states that two bodies with masses $m_1$ and $m_2$ attract each other with a force

$$F = G\frac{m_1 m_2}{r^2}$$

where $r$ is the distance between the bodies and $G$ is the gravitational constant. If one of the bodies is fixed, find the work needed to move the other from $r = a$ to $r = b$.

(b) Compute the work required to launch a 1000-kg satellite vertically to a height of 1000 km. You may assume that the earth's mass is $5.98 \times 10^{24}$ kg and is concentrated at its center. Take the radius of the earth to be $6.37 \times 10^6$ m and $G = 6.67 \times 10^{-11}$ N·m²/kg².

**30.** The Great Pyramid of King Khufu was built of limestone in Egypt over a 20-year time period from 2580 BC to 2560 BC. Its base is a square with side length 756 ft and its height when built was 481 ft. (It was the tallest man-made structure in the world for more than 3800 years.) The density of the limestone is about 150 lb/ft³.

(a) Estimate the total work done in building the pyramid.

(b) If each laborer worked 10 hours a day for 20 years, for 340 days a year, and did 200 ft-lb/h of work in lifting the limestone blocks into place, about how many laborers were needed to construct the pyramid?

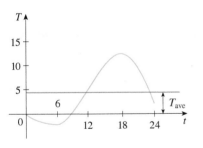

© Vladimir Korostyshevskiy / Shutterstock

## 5.5 Average Value of a Function

It is easy to calculate the average value of finitely many numbers $y_1, y_2, \ldots, y_n$:

$$y_{\text{ave}} = \frac{y_1 + y_2 + \cdots + y_n}{n}$$

But how do we compute the average temperature during a day if infinitely many temperature readings are possible? Figure 1 shows the graph of a temperature function $T(t)$, where $t$ is measured in hours and $T$ in °C, and a guess at the average temperature, $T_{\text{ave}}$.

In general, let's try to compute the average value of a function $y = f(x)$, $a \le x \le b$. We start by dividing the interval $[a, b]$ into $n$ equal subintervals, each with length $\Delta x = (b - a)/n$. Then we choose points $x_1^*, \ldots, x_n^*$ in successive subintervals and calculate the average of the numbers $f(x_1^*), \ldots, f(x_n^*)$:

$$\frac{f(x_1^*) + \cdots + f(x_n^*)}{n}$$

**FIGURE 1**

(For example, if $f$ represents a temperature function and $n = 24$, this means that we take temperature readings every hour and then average them.) Since $\Delta x = (b - a)/n$, we can write $n = (b - a)/\Delta x$ and the average value becomes

$$\frac{f(x_1^*) + \cdots + f(x_n^*)}{\dfrac{b - a}{\Delta x}} = \frac{1}{b - a}\left[f(x_1^*)\,\Delta x + \cdots + f(x_n^*)\,\Delta x\right]$$

$$= \frac{1}{b - a}\sum_{i=1}^{n} f(x_i^*)\,\Delta x$$

If we let $n$ increase, we would be computing the average value of a large number of closely spaced values. (For example, we would be averaging temperature readings taken every minute or even every second.) The limiting value is

$$\lim_{n \to \infty} \frac{1}{b - a} \sum_{i=1}^{n} f(x_i^*) \, \Delta x = \frac{1}{b - a} \int_a^b f(x) \, dx$$

by the definition of a definite integral.

Therefore we define the **average value of $f$** on the interval $[a, b]$ as

$$f_{\text{ave}} = \frac{1}{b - a} \int_a^b f(x) \, dx$$

For a positive function, we can think of this definition as saying

$$\frac{\text{area}}{\text{width}} = \text{average height}$$

**V EXAMPLE 1** Find the average value of the function $f(x) = 1 + x^2$ on the interval $[-1, 2]$.

SOLUTION With $a = -1$ and $b = 2$ we have

$$f_{\text{ave}} = \frac{1}{b - a} \int_a^b f(x) \, dx = \frac{1}{2 - (-1)} \int_{-1}^{2} (1 + x^2) \, dx$$

$$= \frac{1}{3} \left[ x + \frac{x^3}{3} \right]_{-1}^{2} = 2$$

If $T(t)$ is the temperature at time $t$, we might wonder if there is a specific time when the temperature is the same as the average temperature. For the temperature function graphed in Figure 1, we see that there are two such times—just before noon and just before midnight. In general, is there a number $c$ at which the value of a function $f$ is exactly equal to the average value of the function, that is, $f(c) = f_{\text{ave}}$? The following theorem says that this is true for continuous functions.

---

**The Mean Value Theorem for Integrals** If $f$ is continuous on $[a, b]$, then there exists a number $c$ in $[a, b]$ such that

$$f(c) = f_{\text{ave}} = \frac{1}{b - a} \int_a^b f(x) \, dx$$

that is,

$$\int_a^b f(x) \, dx = f(c)(b - a)$$

---

The Mean Value Theorem for Integrals is a consequence of the Mean Value Theorem for derivatives and the Fundamental Theorem of Calculus. The proof is outlined in Exercise 23.

The geometric interpretation of the Mean Value Theorem for Integrals is that, for *positive* functions $f$, there is a number $c$ such that the rectangle with base $[a, b]$ and height $f(c)$ has the same area as the region under the graph of $f$ from $a$ to $b$. (See Figure 2 and the more picturesque interpretation in the margin note.)

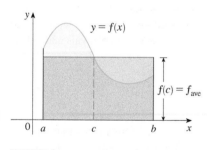

**FIGURE 2**

You can always chop off the top of a (two-dimensional) mountain at a certain height and use it to fill in the valleys so that the mountain becomes completely flat.

**V EXAMPLE 2** Since $f(x) = 1 + x^2$ is continuous on the interval $[-1, 2]$, the Mean Value Theorem for Integrals says there is a number $c$ in $[-1, 2]$ such that

$$\int_{-1}^{2} (1 + x^2) \, dx = f(c)[2 - (-1)]$$

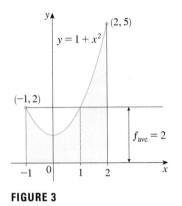

**FIGURE 3**

In this particular case we can find $c$ explicitly. From Example 1 we know that $f_{ave} = 2$, so the value of $c$ satisfies

$$f(c) = f_{ave} = 2$$

Therefore                    $1 + c^2 = 2$    so    $c^2 = 1$

So in this case there happen to be two numbers $c = \pm 1$ in the interval $[-1, 2]$ that work in the Mean Value Theorem for Integrals.

Examples 1 and 2 are illustrated by Figure 3.

**V  EXAMPLE 3**  Show that the average velocity of a car over a time interval $[t_1, t_2]$ is the same as the average of its velocities during the trip.

SOLUTION  If $s(t)$ is the displacement of the car at time $t$, then, by definition, the average velocity of the car over the interval is

$$\frac{\Delta s}{\Delta t} = \frac{s(t_2) - s(t_1)}{t_2 - t_1}$$

On the other hand, the average value of the velocity function on the interval is

$$v_{ave} = \frac{1}{t_2 - t_1} \int_{t_1}^{t_2} v(t)\, dt = \frac{1}{t_2 - t_1} \int_{t_1}^{t_2} s'(t)\, dt$$

$$= \frac{1}{t_2 - t_1} [s(t_2) - s(t_1)] \qquad \text{(by the Net Change Theorem)}$$

$$= \frac{s(t_2) - s(t_1)}{t_2 - t_1} = \text{average velocity}$$

## 5.5  Exercises

1–8  Find the average value of the function on the given interval.

**1.** $f(x) = 4x - x^2$,  $[0, 4]$

**2.** $f(x) = \sin 4x$,  $[-\pi, \pi]$

**3.** $g(x) = \sqrt[3]{x}$,  $[1, 8]$

**4.** $g(t) = \dfrac{t}{\sqrt{3 + t^2}}$,  $[1, 3]$

**5.** $f(t) = t^2(1 + t^3)^4$,  $[0, 2]$

**6.** $f(\theta) = \sec^2(\theta/2)$,  $[0, \pi/2]$

**7.** $h(x) = \cos^4 x \sin x$,  $[0, \pi]$

**8.** $h(r) = 3/(1 + r)^2$,  $[1, 6]$

9–12
(a) Find the average value of $f$ on the given interval.
(b) Find $c$ such that $f_{ave} = f(c)$.
(c) Sketch the graph of $f$ and a rectangle whose area is the same as the area under the graph of $f$.

**9.** $f(x) = (x - 3)^2$,  $[2, 5]$

**10.** $f(x) = \sqrt{x}$,  $[0, 4]$

**11.** $f(x) = 2 \sin x - \sin 2x$,  $[0, \pi]$

**12.** $f(x) = 2x/(1 + x^2)^2$,  $[0, 2]$

**13.** If $f$ is continuous and $\int_1^3 f(x)\, dx = 8$, show that $f$ takes on the value 4 at least once on the interval $[1, 3]$.

**14.** Find the numbers $b$ such that the average value of $f(x) = 2 + 6x - 3x^2$ on the interval $[0, b]$ is equal to 3.

**15.** Find the average value of $f$ on $[0, 8]$.

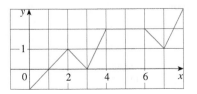

**16.** The velocity graph of an accelerating car is shown.

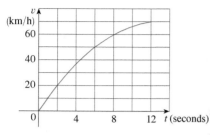

(a) Use the Midpoint rule to estimate the average velocity of the car during the first 12 seconds.
(b) At what time was the instantaneous velocity equal to the average velocity?

**17.** In a certain city the temperature (in °F) $t$ hours after 9 AM was modeled by the function

$$T(t) = 50 + 14 \sin \frac{\pi t}{12}$$

Find the average temperature during the period from 9 AM to 9 PM.

**18.** The velocity $v$ of blood that flows in a blood vessel with radius $R$ and length $l$ at a distance $r$ from the central axis is

$$v(r) = \frac{P}{4\eta l}(R^2 - r^2)$$

where $P$ is the pressure difference between the ends of the vessel and $\eta$ is the viscosity of the blood (see Example 7 in Section 2.7). Find the average velocity (with respect to $r$) over the interval $0 \le r \le R$. Compare the average velocity with the maximum velocity.

**19.** The linear density in a rod 8 m long is $12/\sqrt{x + 1}$ kg/m, where $x$ is measured in meters from one end of the rod. Find the average density of the rod.

**20.** If a freely falling body starts from rest, then its displacement is given by $s = \frac{1}{2}gt^2$. Let the velocity after a time $T$ be $v_T$. Show that if we compute the average of the velocities with respect to $t$ we get $v_{ave} = \frac{1}{2}v_T$, but if we compute the average of the velocities with respect to $s$ we get $v_{ave} = \frac{2}{3}v_T$.

**21.** Use the result of Exercise 57 in Section 4.5 to compute the average volume of inhaled air in the lungs in one respiratory cycle.

**22.** Use the diagram to show that if $f$ is concave upward on $[a, b]$, then

$$f_{ave} > f\left(\frac{a + b}{2}\right)$$

**23.** Prove the Mean Value Theorem for Integrals by applying the Mean Value Theorem for derivatives (see Section 3.2) to the function $F(x) = \int_a^x f(t)\, dt$.

**24.** If $f_{ave}[a, b]$ denotes the average value of $f$ on the interval $[a, b]$ and $a < c < b$, show that

$$f_{ave}[a, b] = \frac{c - a}{b - a}\, f_{ave}[a, c] + \frac{b - c}{b - a}\, f_{ave}[c, b]$$

---

**APPLIED PROJECT**     **CALCULUS AND BASEBALL**

An overhead view of the position of a baseball bat, shown every fiftieth of a second during a typical swing.
(Adapted from *The Physics of Baseball*)

In this project we explore two of the many applications of calculus to baseball. The physical interactions of the game, especially the collision of ball and bat, are quite complex and their models are discussed in detail in a book by Robert Adair, *The Physics of Baseball*, 3d ed. (New York, 2002).

**1.** It may surprise you to learn that the collision of baseball and bat lasts only about a thousandth of a second. Here we calculate the average force on the bat during this collision by first computing the change in the ball's momentum.

The *momentum p* of an object is the product of its mass $m$ and its velocity $v$, that is, $p = mv$. Suppose an object, moving along a straight line, is acted on by a force $F = F(t)$ that is a continuous function of time.

(a) Show that the change in momentum over a time interval $[t_0, t_1]$ is equal to the integral of $F$ from $t_0$ to $t_1$; that is, show that

$$p(t_1) - p(t_0) = \int_{t_0}^{t_1} F(t)\, dt$$

This integral is called the *impulse* of the force over the time interval.

(b) A pitcher throws a 90-mi/h fastball to a batter, who hits a line drive directly back to the pitcher. The ball is in contact with the bat for 0.001 s and leaves the bat with velocity 110 mi/h. A baseball weighs 5 oz and, in US Customary units, its mass is measured in slugs: $m = w/g$ where $g = 32$ ft/s$^2$.

(i) Find the change in the ball's momentum.

(ii) Find the average force on the bat.

2. In this problem we calculate the work required for a pitcher to throw a 90-mi/h fastball by first considering kinetic energy.

The *kinetic energy* K of an object of mass $m$ and velocity $v$ is given by $K = \frac{1}{2}mv^2$. Suppose an object of mass $m$, moving in a straight line, is acted on by a force $F = F(s)$ that depends on its position $s$. According to Newton's Second Law

$$F(s) = ma = m\frac{dv}{dt}$$

where $a$ and $v$ denote the acceleration and velocity of the object.

(a) Show that the work done in moving the object from a position $s_0$ to a position $s_1$ is equal to the change in the object's kinetic energy; that is, show that

$$W = \int_{s_0}^{s_1} F(s)\, ds = \frac{1}{2}mv_1^2 - \frac{1}{2}mv_0^2$$

where $v_0 = v(s_0)$ and $v_1 = v(s_1)$ are the velocities of the object at the positions $s_0$ and $s_1$. *Hint:* By the Chain Rule,

$$m\frac{dv}{dt} = m\frac{dv}{ds}\frac{ds}{dt} = mv\frac{dv}{ds}$$

(b) How many foot-pounds of work does it take to throw a baseball at a speed of 90 mi/h?

NOTE: Another application of calculus to baseball can be found in Problem 16 on page 658.

## 5   Review

### Concept Check

1. (a) Draw two typical curves $y = f(x)$ and $y = g(x)$, where $f(x) \geqslant g(x)$ for $a \leqslant x \leqslant b$. Show how to approximate the area between these curves by a Riemann sum and sketch the corresponding approximating rectangles. Then write an expression for the exact area.

   (b) Explain how the situation changes if the curves have equations $x = f(y)$ and $x = g(y)$, where $f(y) \geqslant g(y)$ for $c \leqslant y \leqslant d$.

2. Suppose that Sue runs faster than Kathy throughout a 1500-meter race. What is the physical meaning of the area between their velocity curves for the first minute of the race?

3. (a) Suppose $S$ is a solid with known cross-sectional areas. Explain how to approximate the volume of $S$ by a Riemann sum. Then write an expression for the exact volume.

   (b) If $S$ is a solid of revolution, how do you find the cross-sectional areas?

4. (a) What is the volume of a cylindrical shell?

   (b) Explain how to use cylindrical shells to find the volume of a solid of revolution.

   (c) Why might you want to use the shell method instead of slicing?

5. Suppose that you push a book across a 6-meter-long table by exerting a force $f(x)$ at each point from $x = 0$ to $x = 6$. What does $\int_0^6 f(x)\, dx$ represent? If $f(x)$ is measured in newtons, what are the units for the integral?

6. (a) What is the average value of a function $f$ on an interval $[a, b]$?

   (b) What does the Mean Value Theorem for Integrals say? What is its geometric interpretation?

## Exercises

**1–6** Find the area of the region bounded by the given curves.

**1.** $y = x^2$,   $y = 4x - x^2$

**2.** $y = 20 - x^2$,   $y = x^2 - 12$

**3.** $y = 1 - 2x^2$,   $y = |x|$

**4.** $x + y = 0$,   $x = y^2 + 3y$

**5.** $y = \sin(\pi x/2)$,   $y = x^2 - 2x$

**6.** $y = \sqrt{x}$,   $y = x^2$,   $x = 2$

**7–11** Find the volume of the solid obtained by rotating the region bounded by the given curves about the specified axis.

**7.** $y = 2x$, $y = x^2$;   about the $x$-axis

**8.** $x = 1 + y^2$, $y = x - 3$;   about the $y$-axis

**9.** $x = 0$, $x = 9 - y^2$;   about $x = -1$

**10.** $y = x^2 + 1$, $y = 9 - x^2$;   about $y = -1$

**11.** $x^2 - y^2 = a^2$, $x = a + h$ (where $a > 0, h > 0$); about the $y$-axis

**12–14** Set up, but do not evaluate, an integral for the volume of the solid obtained by rotating the region bounded by the given curves about the specified axis.

**12.** $y = \tan x$, $y = x$, $x = \pi/3$;   about the $y$-axis

**13.** $y = \cos^2 x$, $|x| \leq \pi/2$, $y = \frac{1}{4}$;   about $x = \pi/2$

**14.** $y = \sqrt{x}$, $y = x^2$;   about $y = 2$

**15.** Find the volumes of the solids obtained by rotating the region bounded by the curves $y = x$ and $y = x^2$ about the following lines.
(a) The $x$-axis      (b) The $y$-axis      (c) $y = 2$

**16.** Let $\mathcal{R}$ be the region in the first quadrant bounded by the curves $y = x^3$ and $y = 2x - x^2$. Calculate the following quantities.
(a) The area of $\mathcal{R}$
(b) The volume obtained by rotating $\mathcal{R}$ about the $x$-axis
(c) The volume obtained by rotating $\mathcal{R}$ about the $y$-axis

**17.** Let $\mathcal{R}$ be the region bounded by the curves $y = \tan(x^2)$, $x = 1$, and $y = 0$. Use the Midpoint Rule with $n = 4$ to estimate the following quantities.
(a) The area of $\mathcal{R}$
(b) The volume obtained by rotating $\mathcal{R}$ about the $x$-axis

**18.** Let $\mathcal{R}$ be the region bounded by the curves $y = 1 - x^2$ and $y = x^6 - x + 1$. Estimate the following quantities.
(a) The $x$-coordinates of the points of intersection of the curves
(b) The area of $\mathcal{R}$
(c) The volume generated when $\mathcal{R}$ is rotated about the $x$-axis
(d) The volume generated when $\mathcal{R}$ is rotated about the $y$-axis

**19–22** Each integral represents the volume of a solid. Describe the solid.

**19.** $\displaystyle\int_0^{\pi/2} 2\pi x \cos x \, dx$

**20.** $\displaystyle\int_0^{\pi/2} 2\pi \cos^2 x \, dx$

**21.** $\displaystyle\int_0^{\pi} \pi(2 - \sin x)^2 \, dx$

**22.** $\displaystyle\int_0^4 2\pi(6 - y)(4y - y^2) \, dy$

**23.** The base of a solid is a circular disk with radius 3. Find the volume of the solid if parallel cross-sections perpendicular to the base are isosceles right triangles with hypotenuse lying along the base.

**24.** The base of a solid is the region bounded by the parabolas $y = x^2$ and $y = 2 - x^2$. Find the volume of the solid if the cross-sections perpendicular to the $x$-axis are squares with one side lying along the base.

**25.** The height of a monument is 20 m. A horizontal cross-section at a distance $x$ meters from the top is an equilateral triangle with side $\frac{1}{4}x$ meters. Find the volume of the monument.

**26.** (a) The base of a solid is a square with vertices located at $(1, 0)$, $(0, 1)$, $(-1, 0)$, and $(0, -1)$. Each cross-section perpendicular to the $x$-axis is a semicircle. Find the volume of the solid.
(b) Show that by cutting the solid of part (a), we can rearrange it to form a cone. Thus compute its volume more simply.

**27.** A force of 30 N is required to maintain a spring stretched from its natural length of 12 cm to a length of 15 cm. How much work is done in stretching the spring from 12 cm to 20 cm?

**28.** A 1600-lb elevator is suspended by a 200-ft cable that weighs 10 lb/ft. How much work is required to raise the elevator from the basement to the third floor, a distance of 30 ft?

**29.** A tank full of water has the shape of a paraboloid of revolution as shown in the figure; that is, its shape is obtained by rotating a parabola about a vertical axis.
(a) If its height is 4 ft and the radius at the top is 4 ft, find the work required to pump the water out of the tank.

⌂ Graphing calculator or computer required

(b) After 4000 ft-lb of work has been done, what is the depth of the water remaining in the tank?

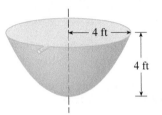

**30.** Find the average value of the function $f(t) = t \sin(t^2)$ on the interval $[0, 10]$.

**31.** If $f$ is a continuous function, what is the limit as $h \to 0$ of the average value of $f$ on the interval $[x, x + h]$?

**32.** Let $\mathcal{R}_1$ be the region bounded by $y = x^2$, $y = 0$, and $x = b$, where $b > 0$. Let $\mathcal{R}_2$ be the region bounded by $y = x^2$, $x = 0$, and $y = b^2$.
(a) Is there a value of $b$ such that $\mathcal{R}_1$ and $\mathcal{R}_2$ have the same area?
(b) Is there a value of $b$ such that $\mathcal{R}_1$ sweeps out the same volume when rotated about the $x$-axis and the $y$-axis?
(c) Is there a value of $b$ such that $\mathcal{R}_1$ and $\mathcal{R}_2$ sweep out the same volume when rotated about the $x$-axis?
(d) Is there a value of $b$ such that $\mathcal{R}_1$ and $\mathcal{R}_2$ sweep out the same volume when rotated about the $y$-axis?

# Problems Plus

1. (a) Find a positive continuous function such that the area under the graph of from 0 to is for all .
   (b) A solid is generated by rotating about the -axis the region under the curve , where is a positive function and . The volume generated by the part of the curve from to $x = b$ is $b^2$ for all $b > 0$. Find the function $f$.

2. There is a line through the origin that divides the region bounded by the parabola $y = x - x^2$ and the $x$-axis into two regions with equal area. What is the slope of that line?

3. The figure shows a horizontal line $y = c$ intersecting the curve $y = 8x - 27x^3$. Find the number $c$ such that the areas of the shaded regions are equal.

$y = 8x - 27x^3$

$y = c$

**FIGURE FOR PROBLEM 3**

4. A cylindrical glass of radius $r$ and height $L$ is filled with water and then tilted until the water remaining in the glass exactly covers its base.
   (a) Determine a way to "slice" the water into parallel rectangular cross-sections and then *set up* a definite integral for the volume of the water in the glass.
   (b) Determine a way to "slice" the water into parallel cross-sections that are trapezoids and then *set up* a definite integral for the volume of the water.
   (c) Find the volume of water in the glass by evaluating one of the integrals in part (a) or part (b).
   (d) Find the volume of the water in the glass from purely geometric considerations.
   (e) Suppose the glass is tilted until the water exactly covers half the base. In what direction can you "slice" the water into triangular cross-sections? Rectangular cross-sections? Cross-sections that are segments of circles? Find the volume of water in the glass.

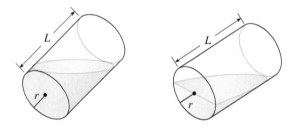

5. (a) Show that the volume of a segment of height $h$ of a sphere of radius $r$ is

$$V = \tfrac{1}{3}\pi h^2(3r - h)$$

   (See the figure.)
   (b) Show that if a sphere of radius 1 is sliced by a plane at a distance $x$ from the center in such a way that the volume of one segment is twice the volume of the other, then $x$ is a solution of the equation

$$3x^3 - 9x + 2 = 0$$

   where $0 < x < 1$. Use Newton's method to find $x$ accurate to four decimal places.
   (c) Using the formula for the volume of a segment of a sphere, it can be shown that the depth $x$ to which a floating sphere of radius $r$ sinks in water is a root of the equation

$$x^3 - 3rx^2 + 4r^3s = 0$$

   where $s$ is the specific gravity of the sphere. Suppose a wooden sphere of radius 0.5 m has specific gravity 0.75. Calculate, to four-decimal-place accuracy, the depth to which the sphere will sink.

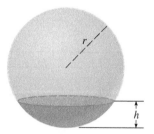

$r$

$h$

**FIGURE FOR PROBLEM 5**

(d) A hemispherical bowl has radius 5 inches and water is running into the bowl at the rate of 0.2 in³/s.
  (i) How fast is the water level in the bowl rising at the instant the water is 3 inches deep?
  (ii) At a certain instant, the water is 4 inches deep. How long will it take to fill the bowl?

$y = L - h$

$y = 0$

$L$

$h$

$y = -h$

**FIGURE FOR PROBLEM 6**

6. Archimedes' Principle states that the buoyant force on an object partially or fully submerged in a fluid is equal to the weight of the fluid that the object displaces. Thus, for an object of density $\rho_0$ floating partly submerged in a fluid of density $\rho_f$, the buoyant force is given by $F = \rho_f g \int_{-h}^{0} A(y)\, dy$, where $g$ is the acceleration due to gravity and $A(y)$ is the area of a typical cross-section of the object (see the figure). The weight of the object is given by

$$W = \rho_0 g \int_{-h}^{L-h} A(y)\, dy$$

(a) Show that the percentage of the volume of the object above the surface of the liquid is

$$100 \frac{\rho_f - \rho_0}{\rho_f}$$

(b) The density of ice is 917 kg/m³ and the density of seawater is 1030 kg/m³. What percentage of the volume of an iceberg is above water?
(c) An ice cube floats in a glass filled to the brim with water. Does the water overflow when the ice melts?
(d) A sphere of radius 0.4 m and having negligible weight is floating in a large freshwater lake. How much work is required to completely submerge the sphere? The density of the water is 1000 kg/m³.

7. Water in an open bowl evaporates at a rate proportional to the area of the surface of the water. (This means that the rate of decrease of the volume is proportional to the area of the surface.) Show that the depth of the water decreases at a constant rate, regardless of the shape of the bowl.

8. A sphere of radius 1 overlaps a smaller sphere of radius $r$ in such a way that their intersection is a circle of radius $r$. (In other words, they intersect in a great circle of the small sphere.) Find $r$ so that the volume inside the small sphere and outside the large sphere is as large as possible.

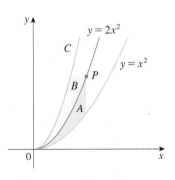

$y$

$y = 2x^2$

$C$

$P$

$y = x^2$

$B$

$A$

$0$

$x$

**FIGURE FOR PROBLEM 9**

9. The figure shows a curve $C$ with the property that, for every point $P$ on the middle curve $y = 2x^2$, the areas $A$ and $B$ are equal. Find an equation for $C$.

10. A paper drinking cup filled with water has the shape of a cone with height $h$ and semivertical angle $\theta$. (See the figure.) A ball is placed carefully in the cup, thereby displacing some of the water and making it overflow. What is the radius of the ball that causes the greatest volume of water to spill out of the cup?

$h$

$\theta$

**11.** A *clepsydra*, or water clock, is a glass container with a small hole in the bottom through which water can flow. The "clock" is calibrated for measuring time by placing markings on the container corresponding to water levels at equally spaced times. Let $x = f(y)$ be continuous on the interval $[0, b]$ and assume that the container is formed by rotating the graph of $f$ about the $y$-axis. Let $V$ denote the volume of water and $h$ the height of the water level at time $t$.

(a) Determine $V$ as a function of $h$.

(b) Show that

$$\frac{dV}{dt} = \pi[f(h)]^2 \frac{dh}{dt}$$

(c) Suppose that $A$ is the area of the hole in the bottom of the container. It follows from Torricelli's Law that the rate of change of the volume of the water is given by

$$\frac{dV}{dt} = kA\sqrt{h}$$

where $k$ is a negative constant. Determine a formula for the function $f$ such that $dh/dt$ is a constant $C$. What is the advantage in having $dh/dt = C$?

**12.** A cylindrical container of radius $r$ and height $L$ is partially filled with a liquid whose volume is $V$. If the container is rotated about its axis of symmetry with constant angular speed $\omega$, then the container will induce a rotational motion in the liquid around the same axis. Eventually, the liquid will be rotating at the same angular speed as the container. The surface of the liquid will be convex, as indicated in the figure, because the centrifugal force on the liquid particles increases with the distance from the axis of the container. It can be shown that the surface of the liquid is a paraboloid of revolution generated by rotating the parabola

$$y = h + \frac{\omega^2 x^2}{2g}$$

**FIGURE FOR PROBLEM 12**

about the $y$-axis, where $g$ is the acceleration due to gravity.

(a) Determine $h$ as a function of $\omega$.

(b) At what angular speed will the surface of the liquid touch the bottom? At what speed will it spill over the top?

(c) Suppose the radius of the container is 2 ft, the height is 7 ft, and the container and liquid are rotating at the same constant angular speed. The surface of the liquid is 5 ft below the top of the tank at the central axis and 4 ft below the top of the tank 1 ft out from the central axis.
   (i) Determine the angular speed of the container and the volume of the fluid.
   (ii) How far below the top of the tank is the liquid at the wall of the container?

**13.** Suppose the graph of a cubic polynomial intersects the parabola $y = x^2$ when $x = 0$, $x = a$, and $x = b$, where $0 < a < b$. If the two regions between the curves have the same area, how is $b$ related to $a$?

# 9 Differential Equations

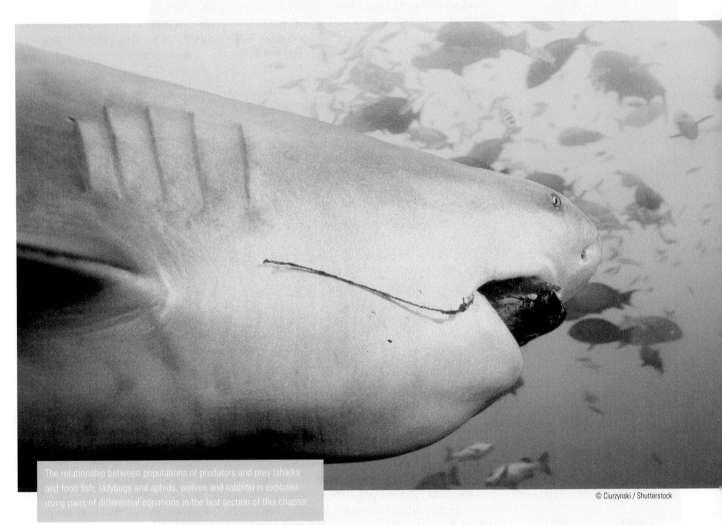

The relationship between populations of predators and prey (sharks and food fish, ladybugs and aphids, wolves and rabbits) is explored using pairs of differential equations in the last section of this chapter.

© Ciurzynski / Shutterstock

Perhaps the most important of all the applications of calculus is to differential equations. When physical scientists or social scientists use calculus, more often than not it is to analyze a differential equation that has arisen in the process of modeling some phenomenon that they are studying. Although it is often impossible to find an explicit formula for the solution of a differential equation, we will see that graphical and numerical approaches provide the needed information.

## 9.1 Modeling with Differential Equations

Now is a good time to read (or reread) the discussion of mathematical modeling on page 23.

In describing the process of modeling in Section 1.2, we talked about formulating a mathematical model of a real-world problem either through intuitive reasoning about the phenomenon or from a physical law based on evidence from experiments. The mathematical model often takes the form of a *differential equation,* that is, an equation that contains an unknown function and some of its derivatives. This is not surprising because in a real-world problem we often notice that changes occur and we want to predict future behavior on the basis of how current values change. Let's begin by examining several examples of how differential equations arise when we model physical phenomena.

### Models of Population Growth

One model for the growth of a population is based on the assumption that the population grows at a rate proportional to the size of the population. That is a reasonable assumption for a population of bacteria or animals under ideal conditions (unlimited environment, adequate nutrition, absence of predators, immunity from disease).

Let's identify and name the variables in this model:

$$t = \text{time} \quad \text{(the independent variable)}$$

$$P = \text{the number of individuals in the population} \quad \text{(the dependent variable)}$$

The rate of growth of the population is the derivative $dP/dt$. So our assumption that the rate of growth of the population is proportional to the population size is written as the equation

$$\boxed{1} \qquad\qquad \frac{dP}{dt} = kP$$

where $k$ is the proportionality constant. Equation 1 is our first model for population growth; it is a differential equation because it contains an unknown function $P$ and its derivative $dP/dt$.

Having formulated a model, let's look at its consequences. If we rule out a population of 0, then $P(t) > 0$ for all $t$. So, if $k > 0$, then Equation 1 shows that $P'(t) > 0$ for all $t$. This means that the population is always increasing. In fact, as $P(t)$ increases, Equation 1 shows that $dP/dt$ becomes larger. In other words, the growth rate increases as the population increases.

Let's try to think of a solution of Equation 1. This equation asks us to find a function whose derivative is a constant multiple of itself. We know from Chapter 6 that exponential functions have that property. In fact, if we let $P(t) = Ce^{kt}$, then

$$P'(t) = C(ke^{kt}) = k(Ce^{kt}) = kP(t)$$

Thus any exponential function of the form $P(t) = Ce^{kt}$ is a solution of Equation 1. In Section 9.4, we will see that there is no other solution.

Allowing $C$ to vary through all the real numbers, we get the *family* of solutions $P(t) = Ce^{kt}$ whose graphs are shown in Figure 1. But populations have only positive values and so we are interested only in the solutions with $C > 0$. And we are probably con-

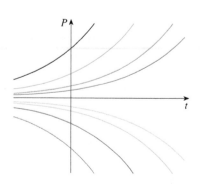

**FIGURE 1**
The family of solutions of $dP/dt = kP$

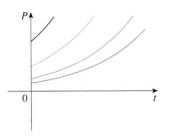

**FIGURE 2**
The family of solutions $P(t) = Ce^{kt}$
with $C > 0$ and $t \geq 0$

cerned only with values of $t$ greater than the initial time $t = 0$. Figure 2 shows the physically meaningful solutions. Putting $t = 0$, we get $P(0) = Ce^{k(0)} = C$, so the constant $C$ turns out to be the initial population, $P(0)$.

Equation 1 is appropriate for modeling population growth under ideal conditions, but we have to recognize that a more realistic model must reflect the fact that a given environment has limited resources. Many populations start by increasing in an exponential manner, but the population levels off when it approaches its *carrying capacity M* (or decreases toward $M$ if it ever exceeds $M$). For a model to take into account both trends, we make two assumptions:

- $\dfrac{dP}{dt} \approx kP$ if $P$ is small   (Initially, the growth rate is proportional to $P$.)

- $\dfrac{dP}{dt} < 0$ if $P > M$   ($P$ decreases if it ever exceeds $M$.)

A simple expression that incorporates both assumptions is given by the equation

$$\boxed{2} \qquad \frac{dP}{dt} = kP\left(1 - \frac{P}{M}\right)$$

Notice that if $P$ is small compared with $M$, then $P/M$ is close to 0 and so $dP/dt \approx kP$. If $P > M$, then $1 - P/M$ is negative and so $dP/dt < 0$.

Equation 2 is called the *logistic differential equation* and was proposed by the Dutch mathematical biologist Pierre-François Verhulst in the 1840s as a model for world population growth. We will develop techniques that enable us to find explicit solutions of the logistic equation in Section 9.4, but for now we can deduce qualitative characteristics of the solutions directly from Equation 2. We first observe that the constant functions $P(t) = 0$ and $P(t) = M$ are solutions because, in either case, one of the factors on the right side of Equation 2 is zero. (This certainly makes physical sense: If the population is ever either 0 or at the carrying capacity, it stays that way.) These two constant solutions are called *equilibrium solutions*.

If the initial population $P(0)$ lies between 0 and $M$, then the right side of Equation 2 is positive, so $dP/dt > 0$ and the population increases. But if the population exceeds the carrying capacity ($P > M$), then $1 - P/M$ is negative, so $dP/dt < 0$ and the population decreases. Notice that, in either case, if the population approaches the carrying capacity ($P \to M$), then $dP/dt \to 0$, which means the population levels off. So we expect that the solutions of the logistic differential equation have graphs that look something like the ones in Figure 3. Notice that the graphs move away from the equilibrium solution $P = 0$ and move toward the equilibrium solution $P = M$.

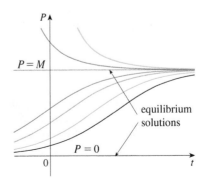

**FIGURE 3**
Solutions of the logistic equation

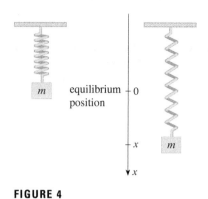

**FIGURE 4**

## A Model for the Motion of a Spring

Let's now look at an example of a model from the physical sciences. We consider the motion of an object with mass $m$ at the end of a vertical spring (as in Figure 4). In Section 5.4 we discussed Hooke's Law, which says that if the spring is stretched (or compressed) $x$ units from its natural length, then it exerts a force that is proportional to $x$:

$$\text{restoring force} = -kx$$

where $k$ is a positive constant (called the *spring constant*). If we ignore any external resisting forces (due to air resistance or friction) then, by Newton's Second Law (force equals mass times acceleration), we have

$$\boxed{3} \qquad m\frac{d^2x}{dt^2} = -kx$$

This is an example of what is called a *second-order differential equation* because it involves second derivatives. Let's see what we can guess about the form of the solution directly from the equation. We can rewrite Equation 3 in the form

$$\frac{d^2x}{dt^2} = -\frac{k}{m}x$$

which says that the second derivative of $x$ is proportional to $x$ but has the opposite sign. We know two functions with this property, the sine and cosine functions. In fact, it turns out that all solutions of Equation 3 can be written as combinations of certain sine and cosine functions (see Exercise 4). This is not surprising; we expect the spring to oscillate about its equilibrium position and so it is natural to think that trigonometric functions are involved.

## General Differential Equations

In general, a **differential equation** is an equation that contains an unknown function and one or more of its derivatives. The **order** of a differential equation is the order of the highest derivative that occurs in the equation. Thus Equations 1 and 2 are first-order equations and Equation 3 is a second-order equation. In all three of those equations the independent variable is called $t$ and represents time, but in general the independent variable doesn't have to represent time. For example, when we consider the differential equation

$$\boxed{4} \qquad\qquad y' = xy$$

it is understood that $y$ is an unknown function of $x$.

A function $f$ is called a **solution** of a differential equation if the equation is satisfied when $y = f(x)$ and its derivatives are substituted into the equation. Thus $f$ is a solution of Equation 4 if

$$f'(x) = xf(x)$$

for all values of $x$ in some interval.

When we are asked to *solve* a differential equation we are expected to find all possible solutions of the equation. We have already solved some particularly simple differential equations, namely, those of the form

$$y' = f(x)$$

For instance, we know that the general solution of the differential equation

$$y' = x^3$$

is given by

$$y = \frac{x^4}{4} + C$$

where $C$ is an arbitrary constant.

But, in general, solving a differential equation is not an easy matter. There is no systematic technique that enables us to solve all differential equations. In Section 9.2, however, we will see how to draw rough graphs of solutions even when we have no explicit formula. We will also learn how to find numerical approximations to solutions.

**V EXAMPLE 1** Show that every member of the family of functions

$$y = \frac{1 + ce^t}{1 - ce^t}$$

is a solution of the differential equation $y' = \frac{1}{2}(y^2 - 1)$.

SOLUTION We use the Quotient Rule to differentiate the expression for $y$:

$$y' = \frac{(1 - ce^t)(ce^t) - (1 + ce^t)(-ce^t)}{(1 - ce^t)^2}$$

$$= \frac{ce^t - c^2e^{2t} + ce^t + c^2e^{2t}}{(1 - ce^t)^2} = \frac{2ce^t}{(1 - ce^t)^2}$$

The right side of the differential equation becomes

$$\frac{1}{2}(y^2 - 1) = \frac{1}{2}\left[\left(\frac{1 + ce^t}{1 - ce^t}\right)^2 - 1\right]$$

$$= \frac{1}{2}\left[\frac{(1 + ce^t)^2 - (1 - ce^t)^2}{(1 - ce^t)^2}\right]$$

$$= \frac{1}{2}\frac{4ce^t}{(1 - ce^t)^2} = \frac{2ce^t}{(1 - ce^t)^2}$$

Therefore, for every value of $c$, the given function is a solution of the differential equation.

Figure 5 shows graphs of seven members of the family in Example 1. The differential equation shows that if $y \approx \pm 1$, then $y' \approx 0$. That is borne out by the flatness of the graphs near $y = 1$ and $y = -1$.

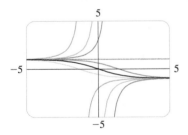

**FIGURE 5**

When applying differential equations, we are usually not as interested in finding a family of solutions (the *general solution*) as we are in finding a solution that satisfies some additional requirement. In many physical problems we need to find the particular solution that satisfies a condition of the form $y(t_0) = y_0$. This is called an **initial condition**, and the problem of finding a solution of the differential equation that satisfies the initial condition is called an **initial-value problem**.

Geometrically, when we impose an initial condition, we look at the family of solution curves and pick the one that passes through the point $(t_0, y_0)$. Physically, this corresponds to measuring the state of a system at time $t_0$ and using the solution of the initial-value problem to predict the future behavior of the system.

**V** **EXAMPLE 2** Find a solution of the differential equation $y' = \frac{1}{2}(y^2 - 1)$ that satisfies the initial condition $y(0) = 2$.

SOLUTION Substituting the values $t = 0$ and $y = 2$ into the formula

$$y = \frac{1 + ce^t}{1 - ce^t}$$

from Example 1, we get

$$2 = \frac{1 + ce^0}{1 - ce^0} = \frac{1 + c}{1 - c}$$

Solving this equation for $c$, we get $2 - 2c = 1 + c$, which gives $c = \frac{1}{3}$. So the solution of the initial-value problem is

$$y = \frac{1 + \frac{1}{3}e^t}{1 - \frac{1}{3}e^t} = \frac{3 + e^t}{3 - e^t}$$

---

## 9.1 Exercises

**1.** Show that $y = \frac{2}{3}e^x + e^{-2x}$ is a solution of the differential equation $y' + 2y = 2e^x$.

**2.** Verify that $y = -t \cos t - t$ is a solution of the initial-value problem

$$t \frac{dy}{dt} = y + t^2 \sin t \qquad y(\pi) = 0$$

**3.** (a) For what values of $r$ does the function $y = e^{rx}$ satisfy the differential equation $2y'' + y' - y = 0$?
   (b) If $r_1$ and $r_2$ are the values of $r$ that you found in part (a), show that every member of the family of functions $y = ae^{r_1x} + be^{r_2x}$ is also a solution.

**4.** (a) For what values of $k$ does the function $y = \cos kt$ satisfy the differential equation $4y'' = -25y$?
   (b) For those values of $k$, verify that every member of the family of functions $y = A \sin kt + B \cos kt$ is also a solution.

**5.** Which of the following functions are solutions of the differential equation $y'' + y = \sin x$?
   (a) $y = \sin x$               (b) $y = \cos x$
   (c) $y = \frac{1}{2}x \sin x$        (d) $y = -\frac{1}{2}x \cos x$

**6.** (a) Show that every member of the family of functions $y = (\ln x + C)/x$ is a solution of the differential equation $x^2 y' + xy = 1$.
   (b) Illustrate part (a) by graphing several members of the family of solutions on a common screen.
   (c) Find a solution of the differential equation that satisfies the initial condition $y(1) = 2$.
   (d) Find a solution of the differential equation that satisfies the initial condition $y(2) = 1$.

**7.** (a) What can you say about a solution of the equation $y' = -y^2$ just by looking at the differential equation?
   (b) Verify that all members of the family $y = 1/(x + C)$ are solutions of the equation in part (a).
   (c) Can you think of a solution of the differential equation $y' = -y^2$ that is not a member of the family in part (b)?
   (d) Find a solution of the initial-value problem

$$y' = -y^2 \qquad y(0) = 0.5$$

**8.** (a) What can you say about the graph of a solution of the equation $y' = xy^3$ when $x$ is close to 0? What if $x$ is large?
   (b) Verify that all members of the family $y = (c - x^2)^{-1/2}$ are solutions of the differential equation $y' = xy^3$.
   (c) Graph several members of the family of solutions on a common screen. Do the graphs confirm what you predicted in part (a)?
   (d) Find a solution of the initial-value problem

$$y' = xy^3 \qquad y(0) = 2$$

**9.** A population is modeled by the differential equation

$$\frac{dP}{dt} = 1.2P\left(1 - \frac{P}{4200}\right)$$

   (a) For what values of $P$ is the population increasing?
   (b) For what values of $P$ is the population decreasing?
   (c) What are the equilibrium solutions?

**10.** A function $y(t)$ satisfies the differential equation

$$\frac{dy}{dt} = y^4 - 6y^3 + 5y^2$$

   (a) What are the constant solutions of the equation?

---

(b) For what values of $y$ is $y$ increasing?

(c) For what values of $y$ is $y$ decreasing?

**11.** Explain why the functions with the given graphs *can't* be solutions of the differential equation

$$\frac{dy}{dt} = e^t(y - 1)^2$$

(a)

(b)

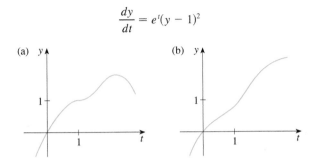

**12.** The function with the given graph is a solution of one of the following differential equations. Decide which is the correct equation and justify your answer.

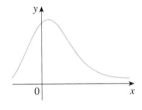

**A.** $y' = 1 + xy$    **B.** $y' = -2xy$    **C.** $y' = 1 - 2xy$

**13.** Match the differential equations with the solution graphs labeled I–IV. Give reasons for your choices.

(a) $y' = 1 + x^2 + y^2$    (b) $y' = xe^{-x^2 - y^2}$

(c) $y' = \dfrac{1}{1 + e^{x^2 + y^2}}$    (d) $y' = \sin(xy)\cos(xy)$

I

II

III

IV

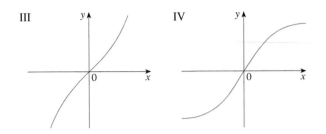

**14.** Suppose you have just poured a cup of freshly brewed coffee with temperature 95°C in a room where the temperature is 20°C.

(a) When do you think the coffee cools most quickly? What happens to the rate of cooling as time goes by? Explain.

(b) Newton's Law of Cooling states that the rate of cooling of an object is proportional to the temperature difference between the object and its surroundings, provided that this difference is not too large. Write a differential equation that expresses Newton's Law of Cooling for this particular situation. What is the initial condition? In view of your answer to part (a), do you think this differential equation is an appropriate model for cooling?

(c) Make a rough sketch of the graph of the solution of the initial-value problem in part (b).

**15.** Psychologists interested in learning theory study **learning curves**. A learning curve is the graph of a function $P(t)$, the performance of someone learning a skill as a function of the training time $t$. The derivative $dP/dt$ represents the rate at which performance improves.

(a) When do you think $P$ increases most rapidly? What happens to $dP/dt$ as $t$ increases? Explain.

(b) If $M$ is the maximum level of performance of which the learner is capable, explain why the differential equation

$$\frac{dP}{dt} = k(M - P) \qquad k \text{ a positive constant}$$

is a reasonable model for learning.

(c) Make a rough sketch of a possible solution of this differential equation.

# Appendixes

Calculus is based on the real number system. We start with the **integers**:

$$\ldots, \quad -3, \quad -2, \quad -1, \quad 0, \quad 1, \quad 2, \quad 3, \quad 4, \quad \ldots$$

Then we construct the **rational numbers**, which are ratios of integers. Thus any rational number $r$ can be expressed as

$$r = \frac{m}{n} \qquad \text{where } m \text{ and } n \text{ are integers and } n \neq 0$$

Examples are

$$\tfrac{1}{2} \qquad -\tfrac{3}{7} \qquad 46 = \tfrac{46}{1} \qquad 0.17 = \tfrac{17}{100}$$

(Recall that division by 0 is always ruled out, so expressions like $\frac{3}{0}$ and $\frac{0}{0}$ are undefined.) Some real numbers, such as $\sqrt{2}$, can't be expressed as a ratio of integers and are therefore called **irrational numbers**. It can be shown, with varying degrees of difficulty, that the following are also irrational numbers:

$$\sqrt{3} \qquad \sqrt{5} \qquad \sqrt[3]{2} \qquad \pi \qquad \sin 1° \qquad \log_{10} 2$$

The set of all real numbers is usually denoted by the symbol $\mathbb{R}$. When we use the word *number* without qualification, we mean "real number."

Every number has a decimal representation. If the number is rational, then the corresponding decimal is repeating. For example,

$$\tfrac{1}{2} = 0.5000\ldots = 0.5\overline{0} \qquad\qquad \tfrac{2}{3} = 0.66666\ldots = 0.\overline{6}$$

$$\tfrac{157}{495} = 0.317171717\ldots = 0.3\overline{17} \qquad \tfrac{9}{7} = 1.285714285714\ldots = 1.\overline{285714}$$

(The bar indicates that the sequence of digits repeats forever.) On the other hand, if the number is irrational, the decimal is nonrepeating:

$$\sqrt{2} = 1.414213562373095\ldots \qquad \pi = 3.141592653589793\ldots$$

If we stop the decimal expansion of any number at a certain place, we get an approximation to the number. For instance, we can write

$$\pi \approx 3.14159265$$

where the symbol $\approx$ is read "is approximately equal to." The more decimal places we retain, the better the approximation we get.

The real numbers can be represented by points on a line as in Figure 1. The positive direction (to the right) is indicated by an arrow. We choose an arbitrary reference point $O$, called the **origin**, which corresponds to the real number 0. Given any convenient unit of measurement, each positive number $x$ is represented by the point on the line a distance of $x$ units to the right of the origin, and each negative number $-x$ is represented by the point $x$ units to the left of the origin. Thus every real number is represented by a point on the line, and every point $P$ on the line corresponds to exactly one real number. The number associated with the point $P$ is called the **coordinate** of $P$ and the line is then called a **coordinate**

**line**, or a **real number line**, or simply a **real line**. Often we identify the point with its coordinate and think of a number as being a point on the real line.

**FIGURE 1**

The real numbers are ordered. We say *a is less than b* and write $a < b$ if $b - a$ is a positive number. Geometrically this means that *a* lies to the left of *b* on the number line. (Equivalently, we say *b is greater than a* and write $b > a$.) The symbol $a \leq b$ (or $b \geq a$) means that either $a < b$ or $a = b$ and is read "*a* is less than or equal to *b*." For instance, the following are true inequalities:

$$7 < 7.4 < 7.5 \qquad -3 > -\pi \qquad \sqrt{2} < 2 \qquad \sqrt{2} \leq 2 \qquad 2 \leq 2$$

In what follows we need to use *set notation*. A **set** is a collection of objects, and these objects are called the **elements** of the set. If *S* is a set, the notation $a \in S$ means that *a* is an element of *S*, and $a \notin S$ means that *a* is not an element of *S*. For example, if *Z* represents the set of integers, then $-3 \in Z$ but $\pi \notin Z$. If *S* and *T* are sets, then their **union** $S \cup T$ is the set consisting of all elements that are in *S* or *T* (or in both *S* and *T*). The **intersection** of *S* and *T* is the set $S \cap T$ consisting of all elements that are in both *S* and *T*. In other words, $S \cap T$ is the common part of *S* and *T*. The empty set, denoted by $\varnothing$, is the set that contains no element.

Some sets can be described by listing their elements between braces. For instance, the set *A* consisting of all positive integers less than 7 can be written as

$$A = \{1, 2, 3, 4, 5, 6\}$$

We could also write *A* in *set-builder notation* as

$$A = \{x \mid x \text{ is an integer and } 0 < x < 7\}$$

which is read "*A* is the set of *x* such that *x* is an integer and $0 < x < 7$."

## Intervals

Certain sets of real numbers, called **intervals**, occur frequently in calculus and correspond geometrically to line segments. For example, if $a < b$, the **open interval** from *a* to *b* consists of all numbers between *a* and *b* and is denoted by the symbol $(a, b)$. Using set-builder notation, we can write

$$(a, b) = \{x \mid a < x < b\}$$

Notice that the endpoints of the interval—namely, *a* and *b*—are excluded. This is indicated by the round brackets ( ) and by the open dots in Figure 2. The **closed interval** from *a* to *b* is the set

$$[a, b] = \{x \mid a \leq x \leq b\}$$

Here the endpoints of the interval are included. This is indicated by the square brackets [ ] and by the solid dots in Figure 3. It is also possible to include only one endpoint in an interval, as shown in Table 1.

**FIGURE 2**
Open interval $(a, b)$

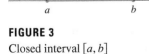

**FIGURE 3**
Closed interval $[a, b]$

1  **Table of Intervals**

Table 1 lists the nine possible types of intervals. When these intervals are discussed, it is always assumed that $a < b$.

| Notation | Set description | Picture |
|---|---|---|
| $(a, b)$ | $\{x \mid a < x < b\}$ | |
| $[a, b]$ | $\{x \mid a \leq x \leq b\}$ | |
| $[a, b)$ | $\{x \mid a \leq x < b\}$ | |
| $(a, b]$ | $\{x \mid a < x \leq b\}$ | |
| $(a, \infty)$ | $\{x \mid x > a\}$ | |
| $[a, \infty)$ | $\{x \mid x \geq a\}$ | |
| $(-\infty, b)$ | $\{x \mid x < b\}$ | |
| $(-\infty, b]$ | $\{x \mid x \leq b\}$ | |
| $(-\infty, \infty)$ | $\mathbb{R}$ (set of all real numbers) | |

We also need to consider infinite intervals such as

$$(a, \infty) = \{x \mid x > a\}$$

This does not mean that $\infty$ ("infinity") is a number. The notation $(a, \infty)$ stands for the set of all numbers that are greater than $a$, so the symbol $\infty$ simply indicates that the interval extends indefinitely far in the positive direction.

## Inequalities

When working with inequalities, note the following rules.

---

2  **Rules for Inequalities**

**1.** If $a < b$, then $a + c < b + c$.

**2.** If $a < b$ and $c < d$, then $a + c < b + d$.

**3.** If $a < b$ and $c > 0$, then $ac < bc$.

**4.** If $a < b$ and $c < 0$, then $ac > bc$.

**5.** If $0 < a < b$, then $1/a > 1/b$.

---

Rule 1 says that we can add any number to both sides of an inequality, and Rule 2 says that two inequalities can be added. However, we have to be careful with multiplication. Rule 3 says that we can multiply both sides of an inequality by a *positive* number, but Rule 4 says that if we multiply both sides of an inequality by a negative number, then we reverse the direction of the inequality. For example, if we take the inequality $3 < 5$ and multiply by 2, we get $6 < 10$, but if we multiply by $-2$, we get $-6 > -10$. Finally, Rule 5 says that if we take reciprocals, then we reverse the direction of an inequality (provided the numbers are positive).

**EXAMPLE 1** Solve the inequality $1 + x < 7x + 5$.

SOLUTION  The given inequality is satisfied by some values of $x$ but not by others. To *solve* an inequality means to determine the set of numbers $x$ for which the inequality is true. This is called the *solution set*.

First we subtract 1 from each side of the inequality (using Rule 1 with $c = -1$):

$$x < 7x + 4$$

Then we subtract $7x$ from both sides (Rule 1 with $c = -7x$):

$$-6x < 4$$

Now we divide both sides by $-6$ $\left(\text{Rule 4 with } c = -\frac{1}{6}\right)$:

$$x > -\tfrac{4}{6} = -\tfrac{2}{3}$$

These steps can all be reversed, so the solution set consists of all numbers greater than $-\frac{2}{3}$. In other words, the solution of the inequality is the interval $\left(-\frac{2}{3}, \infty\right)$.

**EXAMPLE 2** Solve the inequalities $4 \leqslant 3x - 2 < 13$.

SOLUTION Here the solution set consists of all values of $x$ that satisfy both inequalities. Using the rules given in $\boxed{2}$, we see that the following inequalities are equivalent:

$$4 \leqslant 3x - 2 < 13$$

$$6 \leqslant 3x < 15 \qquad \text{(add 2)}$$

$$2 \leqslant x < 5 \qquad \text{(divide by 3)}$$

Therefore the solution set is $[2, 5)$.

**EXAMPLE 3** Solve the inequality $x^2 - 5x + 6 \leqslant 0$.

SOLUTION First we factor the left side:

$$(x - 2)(x - 3) \leqslant 0$$

We know that the corresponding equation $(x - 2)(x - 3) = 0$ has the solutions 2 and 3. The numbers 2 and 3 divide the real line into three intervals:

$$(-\infty, 2) \qquad (2, 3) \qquad (3, \infty)$$

On each of these intervals we determine the signs of the factors. For instance,

$$x \in (-\infty, 2) \quad \Rightarrow \quad x < 2 \quad \Rightarrow \quad x - 2 < 0$$

Then we record these signs in the following chart:

| Interval | $x - 2$ | $x - 3$ | $(x - 2)(x - 3)$ |
|---|---|---|---|
| $x < 2$ | $-$ | $-$ | $+$ |
| $2 < x < 3$ | $+$ | $-$ | $-$ |
| $x > 3$ | $+$ | $+$ | $+$ |

Another method for obtaining the information in the chart is to use *test values*. For instance, if we use the test value $x = 1$ for the interval $(-\infty, 2)$, then substitution in $x^2 - 5x + 6$ gives

$$1^2 - 5(1) + 6 = 2$$

A visual method for solving Example 3 is to use a graphing device to graph the parabola $y = x^2 - 5x + 6$ (as in Figure 4) and observe that the curve lies on or below the $x$-axis when $2 \leqslant x \leqslant 3$.

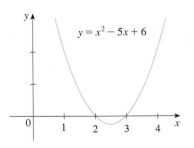

**FIGURE 4**

The polynomial $x^2 - 5x + 6$ doesn't change sign inside any of the three intervals, so we conclude that it is positive on $(-\infty, 2)$.

Then we read from the chart that $(x - 2)(x - 3)$ is negative when $2 < x < 3$. Thus the solution of the inequality $(x - 2)(x - 3) \le 0$ is

$$\{x \mid 2 \le x \le 3\} = [2, 3]$$

Notice that we have included the endpoints 2 and 3 because we are looking for values of $x$ such that the product is either negative or zero. The solution is illustrated in Figure 5.

**FIGURE 5**

**EXAMPLE 4**  Solve $x^3 + 3x^2 > 4x$.

SOLUTION  First we take all nonzero terms to one side of the inequality sign and factor the resulting expression:

$$x^3 + 3x^2 - 4x > 0 \qquad \text{or} \qquad x(x - 1)(x + 4) > 0$$

As in Example 3 we solve the corresponding equation $x(x - 1)(x + 4) = 0$ and use the solutions $x = -4$, $x = 0$, and $x = 1$ to divide the real line into four intervals $(-\infty, -4)$, $(-4, 0)$, $(0, 1)$, and $(1, \infty)$. On each interval the product keeps a constant sign as shown in the following chart:

| Interval | $x$ | $x - 1$ | $x + 4$ | $x(x - 1)(x + 4)$ |
|---|---|---|---|---|
| $x < -4$ | $-$ | $-$ | $-$ | $-$ |
| $-4 < x < 0$ | $-$ | $-$ | $+$ | $+$ |
| $0 < x < 1$ | $+$ | $-$ | $+$ | $-$ |
| $x > 1$ | $+$ | $+$ | $+$ | $+$ |

Then we read from the chart that the solution set is

$$\{x \mid -4 < x < 0 \text{ or } x > 1\} = (-4, 0) \cup (1, \infty)$$

**FIGURE 6**

The solution is illustrated in Figure 6.

## Absolute Value

The **absolute value** of a number $a$, denoted by $|a|$, is the distance from $a$ to 0 on the real number line. Distances are always positive or 0, so we have

$$|a| \ge 0 \qquad \text{for every number } a$$

For example,

$$|3| = 3 \qquad |-3| = 3 \qquad |0| = 0 \qquad |\sqrt{2} - 1| = \sqrt{2} - 1 \qquad |3 - \pi| = \pi - 3$$

In general, we have

Remember that if $a$ is negative, then $-a$ is positive.

$$\boxed{3} \qquad \begin{aligned} |a| &= a \qquad \text{if } a \ge 0 \\ |a| &= -a \qquad \text{if } a < 0 \end{aligned}$$

**EXAMPLE 5** Express $|3x - 2|$ without using the absolute-value symbol.

SOLUTION

$$|3x - 2| = \begin{cases} 3x - 2 & \text{if } 3x - 2 \geq 0 \\ -(3x - 2) & \text{if } 3x - 2 < 0 \end{cases}$$

$$= \begin{cases} 3x - 2 & \text{if } x \geq \frac{2}{3} \\ 2 - 3x & \text{if } x < \frac{2}{3} \end{cases}$$

Recall that the symbol $\sqrt{\ }$ means "the positive square root of." Thus $\sqrt{r} = s$ means $s^2 = r$ and $s \geq 0$. Therefore the equation $\sqrt{a^2} = a$ is not always true. It is true only when $a \geq 0$. If $a < 0$, then $-a > 0$, so we have $\sqrt{a^2} = -a$. In view of $\boxed{3}$, we then have the equation

$$\boxed{4} \qquad \boxed{\sqrt{a^2} = |a|}$$

which is true for all values of $a$.

Hints for the proofs of the following properties are given in the exercises.

---

$\boxed{5}$ **Properties of Absolute Values** Suppose $a$ and $b$ are any real numbers and $n$ is an integer. Then

**1.** $|ab| = |a||b|$      **2.** $\left|\dfrac{a}{b}\right| = \dfrac{|a|}{|b|}$    $(b \neq 0)$      **3.** $|a^n| = |a|^n$

---

For solving equations or inequalities involving absolute values, it's often very helpful to use the following statements.

---

$\boxed{6}$ Suppose $a > 0$. Then

**4.** $|x| = a$   if and only if   $x = \pm a$

**5.** $|x| < a$   if and only if   $-a < x < a$

**6.** $|x| > a$   if and only if   $x > a$ or $x < -a$

---

**FIGURE 7**

For instance, the inequality $|x| < a$ says that the distance from $x$ to the origin is less than $a$, and you can see from Figure 7 that this is true if and only if $x$ lies between $-a$ and $a$.

If $a$ and $b$ are any real numbers, then the distance between $a$ and $b$ is the absolute value of the difference, namely, $|a - b|$, which is also equal to $|b - a|$. (See Figure 8.)

**FIGURE 8**
Length of a line segment $= |a - b|$

**EXAMPLE 6** Solve $|2x - 5| = 3$.

SOLUTION By Property 4 of $\boxed{6}$, $|2x - 5| = 3$ is equivalent to

$$2x - 5 = 3 \qquad \text{or} \qquad 2x - 5 = -3$$

So $2x = 8$ or $2x = 2$. Thus $x = 4$ or $x = 1$.

**EXAMPLE 7**   Solve $|x - 5| < 2$.

SOLUTION 1   By Property 5 of $\boxed{6}$, $|x - 5| < 2$ is equivalent to

$$-2 < x - 5 < 2$$

Therefore, adding 5 to each side, we have

$$3 < x < 7$$

and the solution set is the open interval $(3, 7)$.

SOLUTION 2   Geometrically the solution set consists of all numbers $x$ whose distance from 5 is less than 2. From Figure 9 we see that this is the interval $(3, 7)$.

**FIGURE 9**

**EXAMPLE 8**   Solve $|3x + 2| \geqslant 4$.

SOLUTION   By Properties 4 and 6 of $\boxed{6}$, $|3x + 2| \geqslant 4$ is equivalent to

$$3x + 2 \geqslant 4 \qquad \text{or} \qquad 3x + 2 \leqslant -4$$

In the first case $3x \geqslant 2$, which gives $x \geqslant \frac{2}{3}$. In the second case $3x \leqslant -6$, which gives $x \leqslant -2$. So the solution set is

$$\left\{ x \mid x \leqslant -2 \ \text{or} \ x \geqslant \tfrac{2}{3} \right\} = (-\infty, -2] \cup \left[ \tfrac{2}{3}, \infty \right)$$

Another important property of absolute value, called the Triangle Inequality, is used frequently not only in calculus but throughout mathematics in general.

---

$\boxed{7}$   **The Triangle Inequality**   If $a$ and $b$ are any real numbers, then

$$|a + b| \leqslant |a| + |b|$$

---

Observe that if the numbers $a$ and $b$ are both positive or both negative, then the two sides in the Triangle Inequality are actually equal. But if $a$ and $b$ have opposite signs, the left side involves a subtraction and the right side does not. This makes the Triangle Inequality seem reasonable, but we can prove it as follows.

Notice that

$$-|a| \leqslant a \leqslant |a|$$

is always true because $a$ equals either $|a|$ or $-|a|$. The corresponding statement for $b$ is

$$-|b| \leqslant b \leqslant |b|$$

Adding these inequalities, we get

$$-(|a| + |b|) \leqslant a + b \leqslant |a| + |b|$$

If we now apply Properties 4 and 5 (with $x$ replaced by $a + b$ and $a$ by $|a| + |b|$), we obtain

$$|a + b| \leqslant |a| + |b|$$

which is what we wanted to show.

**EXAMPLE 9** If $|x - 4| < 0.1$ and $|y - 7| < 0.2$, use the Triangle Inequality to estimate $|(x + y) - 11|$.

**SOLUTION** In order to use the given information, we use the Triangle Inequality with $a = x - 4$ and $b = y - 7$:

$$
\begin{aligned}
|(x + y) - 11| &= |(x - 4) + (y - 7)| \\
&\leq |x - 4| + |y - 7| \\
&< 0.1 + 0.2 = 0.3
\end{aligned}
$$

Thus $$|(x + y) - 11| < 0.3$$

---

## A Exercises

**1–12** Rewrite the expression without using the absolute value symbol.

**1.** $|5 - 23|$

**2.** $|5| - |-23|$

**3.** $|-\pi|$

**4.** $|\pi - 2|$

**5.** $|\sqrt{5} - 5|$

**6.** $||-2| - |-3||$

**7.** $|x - 2|$  if $x < 2$

**8.** $|x - 2|$  if $x > 2$

**9.** $|x + 1|$

**10.** $|2x - 1|$

**11.** $|x^2 + 1|$

**12.** $|1 - 2x^2|$

---

**13–38** Solve the inequality in terms of intervals and illustrate the solution set on the real number line.

**13.** $2x + 7 > 3$

**14.** $3x - 11 < 4$

**15.** $1 - x \leq 2$

**16.** $4 - 3x \geq 6$

**17.** $2x + 1 < 5x - 8$

**18.** $1 + 5x > 5 - 3x$

**19.** $-1 < 2x - 5 < 7$

**20.** $1 < 3x + 4 \leq 16$

**21.** $0 \leq 1 - x < 1$

**22.** $-5 \leq 3 - 2x \leq 9$

**23.** $4x < 2x + 1 \leq 3x + 2$

**24.** $2x - 3 < x + 4 < 3x - 2$

**25.** $(x - 1)(x - 2) > 0$

**26.** $(2x + 3)(x - 1) \geq 0$

**27.** $2x^2 + x \leq 1$

**28.** $x^2 < 2x + 8$

**29.** $x^2 + x + 1 > 0$

**30.** $x^2 + x > 1$

**31.** $x^2 < 3$

**32.** $x^2 \geq 5$

**33.** $x^3 - x^2 \leq 0$

**34.** $(x + 1)(x - 2)(x + 3) \geq 0$

**35.** $x^3 > x$

**36.** $x^3 + 3x < 4x^2$

**37.** $\dfrac{1}{x} < 4$

**38.** $-3 < \dfrac{1}{x} \leq 1$

---

**39.** The relationship between the Celsius and Fahrenheit temperature scales is given by $C = \frac{5}{9}(F - 32)$, where $C$ is the temper-ature in degrees Celsius and $F$ is the temperature in degrees Fahrenheit. What interval on the Celsius scale corresponds to the temperature range $50 \leq F \leq 95$?

**40.** Use the relationship between $C$ and $F$ given in Exercise 39 to find the interval on the Fahrenheit scale corresponding to the temperature range $20 \leq C \leq 30$.

**41.** As dry air moves upward, it expands and in so doing cools at a rate of about 1°C for each 100-m rise, up to about 12 km.
(a) If the ground temperature is 20°C, write a formula for the temperature at height $h$.
(b) What range of temperature can be expected if a plane takes off and reaches a maximum height of 5 km?

**42.** If a ball is thrown upward from the top of a building 128 ft high with an initial velocity of 16 ft/s, then the height $h$ above the ground $t$ seconds later will be

$$h = 128 + 16t - 16t^2$$

During what time interval will the ball be at least 32 ft above the ground?

---

**43–46** Solve the equation for $x$.

**43.** $|2x| = 3$

**44.** $|3x + 5| = 1$

**45.** $|x + 3| = |2x + 1|$

**46.** $\left|\dfrac{2x - 1}{x + 1}\right| = 3$

---

**47–56** Solve the inequality.

**47.** $|x| < 3$

**48.** $|x| \geq 3$

**49.** $|x - 4| < 1$

**50.** $|x - 6| < 0.1$

**51.** $|x + 5| \geq 2$

**52.** $|x + 1| \geq 3$

**53.** $|2x - 3| \leq 0.4$

**54.** $|5x - 2| < 6$

**55.** $1 \leq |x| \leq 4$

**56.** $0 < |x - 5| < \frac{1}{2}$

57–58 Solve for $x$, assuming $a$, $b$, and $c$ are positive constants.

**57.** $a(bx - c) \geqslant bc$

**58.** $a \leqslant bx + c < 2a$

59–60 Solve for $x$, assuming $a$, $b$, and $c$ are negative constants.

**59.** $ax + b < c$

**60.** $\dfrac{ax + b}{c} \leqslant b$

**61.** Suppose that $|x - 2| < 0.01$ and $|y - 3| < 0.04$. Use the Triangle Inequality to show that $|(x + y) - 5| < 0.05$.

**62.** Show that if $|x + 3| < \frac{1}{2}$, then $|4x + 13| < 3$.

**63.** Show that if $a < b$, then $a < \dfrac{a + b}{2} < b$.

**64.** Use Rule 3 to prove Rule 5 of $\boxed{2}$.

**65.** Prove that $|ab| = |a||b|$. [*Hint:* Use Equation 4.]

**66.** Prove that $\left|\dfrac{a}{b}\right| = \dfrac{|a|}{|b|}$.

**67.** Show that if $0 < a < b$, then $a^2 < b^2$.

**68.** Prove that $|x - y| \geqslant |x| - |y|$. [*Hint:* Use the Triangle Inequality with $a = x - y$ and $b = y$.]

**69.** Show that the sum, difference, and product of rational numbers are rational numbers.

**70.** (a) Is the sum of two irrational numbers always an irrational number?
(b) Is the product of two irrational numbers always an irrational number?

<br>

| B | **Coordinate Geometry and Lines** |
|---|---|

Just as the points on a line can be identified with real numbers by assigning them coordinates, as described in Appendix A, so the points in a plane can be identified with ordered pairs of real numbers. We start by drawing two perpendicular coordinate lines that intersect at the origin $O$ on each line. Usually one line is horizontal with positive direction to the right and is called the *x*-axis; the other line is vertical with positive direction upward and is called the *y*-axis.

Any point $P$ in the plane can be located by a unique ordered pair of numbers as follows. Draw lines through $P$ perpendicular to the *x*- and *y*-axes. These lines intersect the axes in points with coordinates $a$ and $b$ as shown in Figure 1. Then the point $P$ is assigned the ordered pair $(a, b)$. The first number $a$ is called the **x-coordinate** of $P$; the second number $b$ is called the **y-coordinate** of $P$. We say that $P$ is the point with coordinates $(a, b)$, and we denote the point by the symbol $P(a, b)$. Several points are labeled with their coordinates in Figure 2.

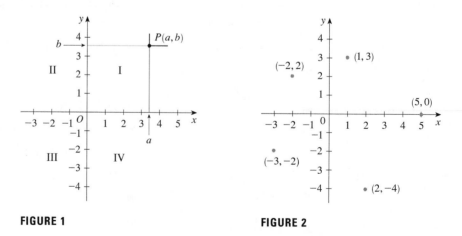

**FIGURE 1**                    **FIGURE 2**

By reversing the preceding process we can start with an ordered pair $(a, b)$ and arrive at the corresponding point $P$. Often we identify the point $P$ with the ordered pair $(a, b)$ and refer to "the point $(a, b)$." [Although the notation used for an open interval $(a, b)$ is the same as the notation used for a point $(a, b)$, you will be able to tell from the context which meaning is intended.]

This coordinate system is called the **rectangular coordinate system** or the **Cartesian coordinate system** in honor of the French mathematician René Descartes (1596–1650), even though another Frenchman, Pierre Fermat (1601–1665), invented the principles of analytic geometry at about the same time as Descartes. The plane supplied with this coordinate system is called the **coordinate plane** or the **Cartesian plane** and is denoted by $\mathbb{R}^2$.

The $x$- and $y$-axes are called the **coordinate axes** and divide the Cartesian plane into four quadrants, which are labeled I, II, III, and IV in Figure 1. Notice that the first quadrant consists of those points whose $x$- and $y$-coordinates are both positive.

**EXAMPLE 1**  Describe and sketch the regions given by the following sets.

(a) $\{(x, y) \mid x \geqslant 0\}$    (b) $\{(x, y) \mid y = 1\}$    (c) $\{(x, y) \mid |y| < 1\}$

SOLUTION

(a) The points whose $x$-coordinates are 0 or positive lie on the $y$-axis or to the right of it as indicated by the shaded region in Figure 3(a).

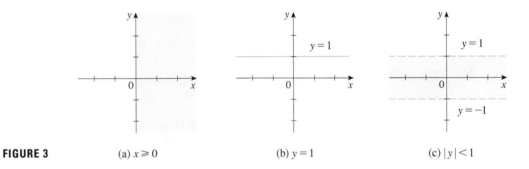

**FIGURE 3**        (a) $x \geqslant 0$            (b) $y = 1$            (c) $|y| < 1$

(b) The set of all points with $y$-coordinate 1 is a horizontal line one unit above the $x$-axis [see Figure 3(b)].

(c) Recall from Appendix A that

$$|y| < 1 \qquad \text{if and only if} \qquad -1 < y < 1$$

The given region consists of those points in the plane whose $y$-coordinates lie between $-1$ and 1. Thus the region consists of all points that lie between (but not on) the horizontal lines $y = 1$ and $y = -1$. [These lines are shown as dashed lines in Figure 3(c) to indicate that the points on these lines don't lie in the set.]

Recall from Appendix A that the distance between points $a$ and $b$ on a number line is $|a - b| = |b - a|$. Thus the distance between points $P_1(x_1, y_1)$ and $P_3(x_2, y_1)$ on a horizontal line must be $|x_2 - x_1|$ and the distance between $P_2(x_2, y_2)$ and $P_3(x_2, y_1)$ on a vertical line must be $|y_2 - y_1|$. (See Figure 4.)

To find the distance $|P_1P_2|$ between any two points $P_1(x_1, y_1)$ and $P_2(x_2, y_2)$, we note that triangle $P_1P_2P_3$ in Figure 4 is a right triangle, and so by the Pythagorean Theorem we have

$$|P_1P_2| = \sqrt{|P_1P_3|^2 + |P_2P_3|^2} = \sqrt{|x_2 - x_1|^2 + |y_2 - y_1|^2}$$

$$= \sqrt{(x_2 - x_1)^2 + (y_2 - y_1)^2}$$

**FIGURE 4**

**1 Distance Formula** The distance between the points $P_1(x_1, y_1)$ and $P_2(x_2, y_2)$ is

$$|P_1P_2| = \sqrt{(x_2 - x_1)^2 + (y_2 - y_1)^2}$$

**EXAMPLE 2** The distance between $(1, -2)$ and $(5, 3)$ is

$$\sqrt{(5 - 1)^2 + [3 - (-2)]^2} = \sqrt{4^2 + 5^2} = \sqrt{41}$$

## Lines

We want to find an equation of a given line $L$; such an equation is satisfied by the coordinates of the points on $L$ and by no other point. To find the equation of $L$ we use its *slope*, which is a measure of the steepness of the line.

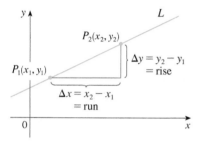

**FIGURE 5**

**2 Definition** The **slope** of a nonvertical line that passes through the points $P_1(x_1, y_1)$ and $P_2(x_2, y_2)$ is

$$m = \frac{\Delta y}{\Delta x} = \frac{y_2 - y_1}{x_2 - x_1}$$

The slope of a vertical line is not defined.

Thus the slope of a line is the ratio of the change in $y$, $\Delta y$, to the change in $x$, $\Delta x$. (See Figure 5.) The slope is therefore the rate of change of $y$ with respect to $x$. The fact that the line is straight means that the rate of change is constant.

Figure 6 shows several lines labeled with their slopes. Notice that lines with positive slope slant upward to the right, whereas lines with negative slope slant downward to the right. Notice also that the steepest lines are the ones for which the absolute value of the slope is largest, and a horizontal line has slope 0.

Now let's find an equation of the line that passes through a given point $P_1(x_1, y_1)$ and has slope $m$. A point $P(x, y)$ with $x \neq x_1$ lies on this line if and only if the slope of the line through $P_1$ and $P$ is equal to $m$; that is,

$$\frac{y - y_1}{x - x_1} = m$$

This equation can be rewritten in the form

$$y - y_1 = m(x - x_1)$$

and we observe that this equation is also satisfied when $x = x_1$ and $y = y_1$. Therefore it is an equation of the given line.

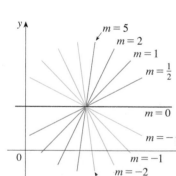

**FIGURE 6**

**3 Point-Slope Form of the Equation of a Line** An equation of the line passing through the point $P_1(x_1, y_1)$ and having slope $m$ is

$$y - y_1 = m(x - x_1)$$

**EXAMPLE 3** Find an equation of the line through $(1, -7)$ with slope $-\frac{1}{2}$.

SOLUTION Using $\boxed{3}$ with $m = -\frac{1}{2}$, $x_1 = 1$, and $y_1 = -7$, we obtain an equation of the line as

$$y + 7 = -\tfrac{1}{2}(x - 1)$$

which we can rewrite as

$$2y + 14 = -x + 1 \qquad \text{or} \qquad x + 2y + 13 = 0$$

**EXAMPLE 4** Find an equation of the line through the points $(-1, 2)$ and $(3, -4)$.

SOLUTION By Definition 2 the slope of the line is

$$m = \frac{-4 - 2}{3 - (-1)} = -\frac{3}{2}$$

Using the point-slope form with $x_1 = -1$ and $y_1 = 2$, we obtain

$$y - 2 = -\tfrac{3}{2}(x + 1)$$

which simplifies to $\qquad\qquad 3x + 2y = 1$

Suppose a nonvertical line has slope $m$ and $y$-intercept $b$. (See Figure 7.) This means it intersects the $y$-axis at the point $(0, b)$, so the point-slope form of the equation of the line, with $x_1 = 0$ and $y_1 = b$, becomes

$$y - b = m(x - 0)$$

This simplifies as follows.

**FIGURE 7**

---

$\boxed{4}$  **Slope-Intercept Form of the Equation of a Line**  An equation of the line with slope $m$ and $y$-intercept $b$ is

$$y = mx + b$$

---

In particular, if a line is horizontal, its slope is $m = 0$, so its equation is $y = b$, where $b$ is the $y$-intercept (see Figure 8). A vertical line does not have a slope, but we can write its equation as $x = a$, where $a$ is the $x$-intercept, because the $x$-coordinate of every point on the line is $a$.

Observe that the equation of every line can be written in the form

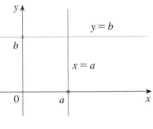

**FIGURE 8**

$\boxed{5}$
$$Ax + By + C = 0$$

because a vertical line has the equation $x = a$ or $x - a = 0$ ($A = 1, B = 0, C = -a$) and a nonvertical line has the equation $y = mx + b$ or $-mx + y - b = 0$ ($A = -m, B = 1$, $C = -b$). Conversely, if we start with a general first-degree equation, that is, an equation of the form $\boxed{5}$, where $A$, $B$, and $C$ are constants and $A$ and $B$ are not both 0, then we can show that it is the equation of a line. If $B = 0$, the equation becomes $Ax + C = 0$ or $x = -C/A$, which represents a vertical line with $x$-intercept $-C/A$. If $B \neq 0$, the equation can be rewritten by solving for $y$:

$$y = -\frac{A}{B}x - \frac{C}{B}$$

and we recognize this as being the slope-intercept form of the equation of a line ($m = -A/B$, $b = -C/B$). Therefore an equation of the form $\boxed{5}$ is called a **linear equation** or the **general equation of a line**. For brevity, we often refer to "the line $Ax + By + C = 0$" instead of "the line whose equation is $Ax + By + C = 0$."

**FIGURE 9**

**EXAMPLE 5** Sketch the graph of the equation $3x - 5y = 15$.

SOLUTION Since the equation is linear, its graph is a line. To draw the graph, we can simply find two points on the line. It's easiest to find the intercepts. Substituting $y = 0$ (the equation of the $x$-axis) in the given equation, we get $3x = 15$, so $x = 5$ is the $x$-intercept. Substituting $x = 0$ in the equation, we see that the $y$-intercept is $-3$. This allows us to sketch the graph as in Figure 9.

**EXAMPLE 6** Graph the inequality $x + 2y > 5$.

SOLUTION We are asked to sketch the graph of the set $\{(x, y) \mid x + 2y > 5\}$ and we begin by solving the inequality for $y$:

$$x + 2y > 5$$

$$2y > -x + 5$$

$$y > -\tfrac{1}{2}x + \tfrac{5}{2}$$

**FIGURE 10**

Compare this inequality with the equation $y = -\tfrac{1}{2}x + \tfrac{5}{2}$, which represents a line with slope $-\tfrac{1}{2}$ and $y$-intercept $\tfrac{5}{2}$. We see that the given graph consists of points whose $y$-coordinates are *larger* than those on the line $y = -\tfrac{1}{2}x + \tfrac{5}{2}$. Thus the graph is the region that lies *above* the line, as illustrated in Figure 10.

### Parallel and Perpendicular Lines

Slopes can be used to show that lines are parallel or perpendicular. The following facts are proved, for instance, in *Precalculus: Mathematics for Calculus, Sixth Edition* by Stewart, Redlin, and Watson (Belmont, CA, 2012).

---

$\boxed{6}$  **Parallel and Perpendicular Lines**

**1.** Two nonvertical lines are parallel if and only if they have the same slope.

**2.** Two lines with slopes $m_1$ and $m_2$ are perpendicular if and only if $m_1m_2 = -1$; that is, their slopes are negative reciprocals:

$$m_2 = -\frac{1}{m_1}$$

---

**EXAMPLE 7** Find an equation of the line through the point $(5, 2)$ that is parallel to the line $4x + 6y + 5 = 0$.

SOLUTION The given line can be written in the form

$$y = -\tfrac{2}{3}x - \tfrac{5}{6}$$

which is in slope-intercept form with $m = -\frac{2}{3}$. Parallel lines have the same slope, so the required line has slope $-\frac{2}{3}$ and its equation in point-slope form is

$$y - 2 = -\frac{2}{3}(x - 5)$$

We can write this equation as $2x + 3y = 16$.

**EXAMPLE 8** Show that the lines $2x + 3y = 1$ and $6x - 4y - 1 = 0$ are perpendicular.

SOLUTION The equations can be written as

$$y = -\frac{2}{3}x + \frac{1}{3} \quad \text{and} \quad y = \frac{3}{2}x - \frac{1}{4}$$

from which we see that the slopes are

$$m_1 = -\frac{2}{3} \quad \text{and} \quad m_2 = \frac{3}{2}$$

Since $m_1 m_2 = -1$, the lines are perpendicular.

## B  Exercises

1–6 Find the distance between the points.

**1.** $(1, 1)$,  $(4, 5)$

**2.** $(1, -3)$,  $(5, 7)$

**3.** $(6, -2)$,  $(-1, 3)$

**4.** $(1, -6)$,  $(-1, -3)$

**5.** $(2, 5)$,  $(4, -7)$

**6.** $(a, b)$,  $(b, a)$

7–10 Find the slope of the line through $P$ and $Q$.

**7.** $P(1, 5)$,  $Q(4, 11)$

**8.** $P(-1, 6)$,  $Q(4, -3)$

**9.** $P(-3, 3)$,  $Q(-1, -6)$

**10.** $P(-1, -4)$,  $Q(6, 0)$

**11.** Show that the triangle with vertices $A(0, 2)$, $B(-3, -1)$, and $C(-4, 3)$ is isosceles.

**12.** (a) Show that the triangle with vertices $A(6, -7)$, $B(11, -3)$, and $C(2, -2)$ is a right triangle using the converse of the Pythagorean Theorem.
(b) Use slopes to show that $ABC$ is a right triangle.
(c) Find the area of the triangle.

**13.** Show that the points $(-2, 9)$, $(4, 6)$, $(1, 0)$, and $(-5, 3)$ are the vertices of a square.

**14.** (a) Show that the points $A(-1, 3)$, $B(3, 11)$, and $C(5, 15)$ are collinear (lie on the same line) by showing that $|AB| + |BC| = |AC|$.
(b) Use slopes to show that $A$, $B$, and $C$ are collinear.

**15.** Show that $A(1, 1)$, $B(7, 4)$, $C(5, 10)$, and $D(-1, 7)$ are vertices of a parallelogram.

**16.** Show that $A(1, 1)$, $B(11, 3)$, $C(10, 8)$, and $D(0, 6)$ are vertices of a rectangle.

17–20 Sketch the graph of the equation.

**17.** $x = 3$

**18.** $y = -2$

**19.** $xy = 0$

**20.** $|y| = 1$

21–36 Find an equation of the line that satisfies the given conditions.

**21.** Through $(2, -3)$,  slope 6

**22.** Through $(-1, 4)$,  slope $-3$

**23.** Through $(1, 7)$,  slope $\frac{2}{3}$

**24.** Through $(-3, -5)$,  slope $-\frac{7}{2}$

**25.** Through $(2, 1)$ and $(1, 6)$

**26.** Through $(-1, -2)$ and $(4, 3)$

**27.** Slope 3,  $y$-intercept $-2$

**28.** Slope $\frac{2}{5}$,  $y$-intercept 4

**29.** $x$-intercept 1,  $y$-intercept $-3$

**30.** $x$-intercept $-8$,  $y$-intercept 6

**31.** Through $(4, 5)$,  parallel to the $x$-axis

**32.** Through $(4, 5)$,  parallel to the $y$-axis

**33.** Through $(1, -6)$,  parallel to the line $x + 2y = 6$

**34.** $y$-intercept 6,  parallel to the line $2x + 3y + 4 = 0$

**35.** Through $(-1, -2)$,  perpendicular to the line $2x + 5y + 8 = 0$

**36.** Through $\left(\frac{1}{2}, -\frac{2}{3}\right)$,  perpendicular to the line $4x - 8y = 1$

37–42 Find the slope and $y$-intercept of the line and draw its graph.

**37.** $x + 3y = 0$

**38.** $2x - 5y = 0$

**39.** $y = -2$

**40.** $2x - 3y + 6 = 0$

**41.** $3x - 4y = 12$

**42.** $4x + 5y = 10$

**43–52** Sketch the region in the $xy$-plane.

**43.** $\{(x, y) \mid x < 0\}$

**44.** $\{(x, y) \mid y > 0\}$

**45.** $\{(x, y) \mid xy < 0\}$

**46.** $\{(x, y) \mid x \geq 1 \text{ and } y < 3\}$

**47.** $\{(x, y) \mid |x| \leq 2\}$

**48.** $\{(x, y) \mid |x| < 3 \text{ and } |y| < 2\}$

**49.** $\{(x, y) \mid 0 \leq y \leq 4 \text{ and } x \leq 2\}$

**50.** $\{(x, y) \mid y > 2x - 1\}$

**51.** $\{(x, y) \mid 1 + x \leq y \leq 1 - 2x\}$

**52.** $\{(x, y) \mid -x \leq y < \frac{1}{2}(x + 3)\}$

**53.** Find a point on the $y$-axis that is equidistant from $(5, -5)$ and $(1, 1)$.

**54.** Show that the midpoint of the line segment from $P_1(x_1, y_1)$ to $P_2(x_2, y_2)$ is
$$\left( \frac{x_1 + x_2}{2}, \frac{y_1 + y_2}{2} \right)$$

**55.** Find the midpoint of the line segment joining the given points.
(a) $(1, 3)$ and $(7, 15)$
(b) $(-1, 6)$ and $(8, -12)$

**56.** Find the lengths of the medians of the triangle with vertices $A(1, 0)$, $B(3, 6)$, and $C(8, 2)$. (A median is a line segment from a vertex to the midpoint of the opposite side.)

**57.** Show that the lines $2x - y = 4$ and $6x - 2y = 10$ are not parallel and find their point of intersection.

**58.** Show that the lines $3x - 5y + 19 = 0$ and $10x + 6y - 50 = 0$ are perpendicular and find their point of intersection.

**59.** Find an equation of the perpendicular bisector of the line segment joining the points $A(1, 4)$ and $B(7, -2)$.

**60.** (a) Find equations for the sides of the triangle with vertices $P(1, 0)$, $Q(3, 4)$, and $R(-1, 6)$.
(b) Find equations for the medians of this triangle. Where do they intersect?

**61.** (a) Show that if the $x$- and $y$-intercepts of a line are nonzero numbers $a$ and $b$, then the equation of the line can be put in the form
$$\frac{x}{a} + \frac{y}{b} = 1$$
This equation is called the **two-intercept form** of an equation of a line.
(b) Use part (a) to find an equation of the line whose $x$-intercept is 6 and whose $y$-intercept is $-8$.

**62.** A car leaves Detroit at 2:00 PM, traveling at a constant speed west along I-96. It passes Ann Arbor, 40 mi from Detroit, at 2:50 PM.
(a) Express the distance traveled in terms of the time elapsed.
(b) Draw the graph of the equation in part (a).
(c) What is the slope of this line? What does it represent?

## C Graphs of Second-Degree Equations

In Appendix B we saw that a first-degree, or linear, equation $Ax + By + C = 0$ represents a line. In this section we discuss second-degree equations such as

$$x^2 + y^2 = 1 \qquad y = x^2 + 1 \qquad \frac{x^2}{9} + \frac{y^2}{4} = 1 \qquad x^2 - y^2 = 1$$

which represent a circle, a parabola, an ellipse, and a hyperbola, respectively.

The graph of such an equation in $x$ and $y$ is the set of all points $(x, y)$ that satisfy the equation; it gives a visual representation of the equation. Conversely, given a curve in the $xy$-plane, we may have to find an equation that represents it, that is, an equation satisfied by the coordinates of the points on the curve and by no other point. This is the other half of the basic principle of analytic geometry as formulated by Descartes and Fermat. The idea is that if a geometric curve can be represented by an algebraic equation, then the rules of algebra can be used to analyze the geometric problem.

### Circles

As an example of this type of problem, let's find an equation of the circle with radius $r$ and center $(h, k)$. By definition, the circle is the set of all points $P(x, y)$ whose distance from the center $C(h, k)$ is $r$. (See Figure 1.) Thus $P$ is on the circle if and only if $|PC| = r$. From the distance formula, we have

$$\sqrt{(x - h)^2 + (y - k)^2} = r$$

or equivalently, squaring both sides, we get

$$(x - h)^2 + (y - k)^2 = r^2$$

This is the desired equation.

---

> **1  Equation of a Circle**  An equation of the circle with center $(h, k)$ and radius $r$ is
>
> $$(x - h)^2 + (y - k)^2 = r^2$$
>
> In particular, if the center is the origin $(0, 0)$, the equation is
>
> $$x^2 + y^2 = r^2$$

---

**EXAMPLE 1**  Find an equation of the circle with radius 3 and center $(2, -5)$.

SOLUTION  From Equation 1 with $r = 3$, $h = 2$, and $k = -5$, we obtain

$$(x - 2)^2 + (y + 5)^2 = 9$$

**EXAMPLE 2**  Sketch the graph of the equation $x^2 + y^2 + 2x - 6y + 7 = 0$ by first showing that it represents a circle and then finding its center and radius.

SOLUTION  We first group the $x$-terms and $y$-terms as follows:

$$(x^2 + 2x) + (y^2 - 6y) = -7$$

Then we complete the square within each grouping, adding the appropriate constants (the squares of half the coefficients of $x$ and $y$) to both sides of the equation:

$$(x^2 + 2x + 1) + (y^2 - 6y + 9) = -7 + 1 + 9$$

or                   $(x + 1)^2 + (y - 3)^2 = 3$

Comparing this equation with the standard equation of a circle $\boxed{1}$, we see that $h = -1$, $k = 3$, and $r = \sqrt{3}$, so the given equation represents a circle with center $(-1, 3)$ and radius $\sqrt{3}$. It is sketched in Figure 2.

**FIGURE 1**

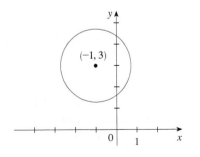

**FIGURE 2**

$x^2 + y^2 + 2x - 6y + 7 = 0$

### Parabolas

The geometric properties of parabolas are reviewed in Section 10.5. Here we regard a parabola as a graph of an equation of the form $y = ax^2 + bx + c$.

**EXAMPLE 3** Draw the graph of the parabola $y = x^2$.

**SOLUTION** We set up a table of values, plot points, and join them by a smooth curve to obtain the graph in Figure 3.

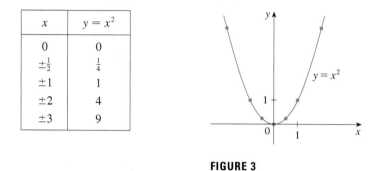

| $x$ | $y = x^2$ |
|:---:|:---:|
| 0 | 0 |
| $\pm\frac{1}{2}$ | $\frac{1}{4}$ |
| $\pm 1$ | 1 |
| $\pm 2$ | 4 |
| $\pm 3$ | 9 |

**FIGURE 3**

Figure 4 shows the graphs of several parabolas with equations of the form $y = ax^2$ for various values of the number $a$. In each case the *vertex*, the point where the parabola changes direction, is the origin. We see that the parabola $y = ax^2$ opens upward if $a > 0$ and downward if $a < 0$ (as in Figure 5).

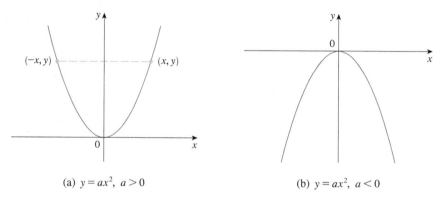

(a) $y = ax^2$, $a > 0$

(b) $y = ax^2$, $a < 0$

**FIGURE 4**

**FIGURE 5**

Notice that if $(x, y)$ satisfies $y = ax^2$, then so does $(-x, y)$. This corresponds to the geometric fact that if the right half of the graph is reflected about the $y$-axis, then the left half of the graph is obtained. We say that the graph is **symmetric with respect to the $y$-axis**.

> The graph of an equation is symmetric with respect to the $y$-axis if the equation is unchanged when $x$ is replaced by $-x$.

If we interchange $x$ and $y$ in the equation $y = ax^2$, the result is $x = ay^2$, which also represents a parabola. (Interchanging $x$ and $y$ amounts to reflecting about the diagonal line $y = x$.) The parabola $x = ay^2$ opens to the right if $a > 0$ and to the left if $a < 0$. (See

Figure 6.) This time the parabola is symmetric with respect to the $x$-axis because if $(x, y)$ satisfies $x = ay^2$, then so does $(x, -y)$.

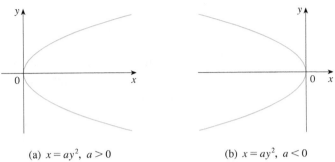

**FIGURE 6**

(a) $x = ay^2$, $a > 0$    (b) $x = ay^2$, $a < 0$

> The graph of an equation is symmetric with respect to the $x$-axis if the equation is unchanged when $y$ is replaced by $-y$.

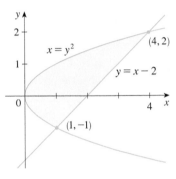

**FIGURE 7**

**EXAMPLE 4**  Sketch the region bounded by the parabola $x = y^2$ and the line $y = x - 2$.

SOLUTION  First we find the points of intersection by solving the two equations. Substituting $x = y + 2$ into the equation $x = y^2$, we get $y + 2 = y^2$, which gives

$$0 = y^2 - y - 2 = (y - 2)(y + 1)$$

so $y = 2$ or $-1$. Thus the points of intersection are $(4, 2)$ and $(1, -1)$, and we draw the line $y = x - 2$ passing through these points. We then sketch the parabola $x = y^2$ by referring to Figure 6(a) and having the parabola pass through $(4, 2)$ and $(1, -1)$. The region bounded by $x = y^2$ and $y = x - 2$ means the finite region whose boundaries are these curves. It is sketched in Figure 7.

### Ellipses

The curve with equation

 $$\frac{x^2}{a^2} + \frac{y^2}{b^2} = 1$$

where $a$ and $b$ are positive numbers, is called an **ellipse** in standard position. (Geometric properties of ellipses are discussed in Section 10.5.) Observe that Equation 2 is unchanged if $x$ is replaced by $-x$ or $y$ is replaced by $-y$, so the ellipse is symmetric with respect to both axes. As a further aid to sketching the ellipse, we find its intercepts.

> The **$x$-intercepts** of a graph are the $x$-coordinates of the points where the graph intersects the $x$-axis. They are found by setting $y = 0$ in the equation of the graph.
>
> The **$y$-intercepts** are the $y$-coordinates of the points where the graph intersects the $y$-axis. They are found by setting $x = 0$ in its equation.

**FIGURE 8**

$$\frac{x^2}{a^2} + \frac{y^2}{b^2} = 1$$

If we set $y = 0$ in Equation 2, we get $x^2 = a^2$ and so the $x$-intercepts are $\pm a$. Setting $x = 0$, we get $y^2 = b^2$, so the $y$-intercepts are $\pm b$. Using this information, together with symmetry, we sketch the ellipse in Figure 8. If $a = b$, the ellipse is a circle with radius $a$.

**EXAMPLE 5** Sketch the graph of $9x^2 + 16y^2 = 144$.

SOLUTION We divide both sides of the equation by 144:

$$\frac{x^2}{16} + \frac{y^2}{9} = 1$$

The equation is now in the standard form for an ellipse $\boxed{2}$, so we have $a^2 = 16$, $b^2 = 9$, $a = 4$, and $b = 3$. The $x$-intercepts are $\pm 4$; the $y$-intercepts are $\pm 3$. The graph is sketched in Figure 9.

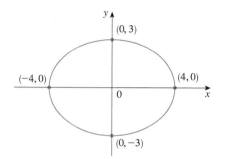

**FIGURE 9**
$9x^2 + 16y^2 = 144$

### Hyperbolas

The curve with equation

$\boxed{3}$

$$\frac{x^2}{a^2} - \frac{y^2}{b^2} = 1$$

**FIGURE 10**

The hyperbola $\dfrac{x^2}{a^2} - \dfrac{y^2}{b^2} = 1$

is called a **hyperbola** in standard position. Again, Equation 3 is unchanged when $x$ is replaced by $-x$ or $y$ is replaced by $-y$, so the hyperbola is symmetric with respect to both axes. To find the $x$-intercepts we set $y = 0$ and obtain $x^2 = a^2$ and $x = \pm a$. However, if we put $x = 0$ in Equation 3, we get $y^2 = -b^2$, which is impossible, so there is no $y$-intercept. In fact, from Equation 3 we obtain

$$\frac{x^2}{a^2} = 1 + \frac{y^2}{b^2} \geqslant 1$$

which shows that $x^2 \geqslant a^2$ and so $|x| = \sqrt{x^2} \geqslant a$. Therefore we have $x \geqslant a$ or $x \leqslant -a$. This means that the hyperbola consists of two parts, called its *branches*. It is sketched in Figure 10.

In drawing a hyperbola it is useful to draw first its *asymptotes*, which are the lines $y = (b/a)x$ and $y = -(b/a)x$ shown in Figure 10. Both branches of the hyperbola approach the asymptotes; that is, they come arbitrarily close to the asymptotes. This involves the idea of a limit, which is discussed in Chapter 1. (See also Exercise 57 in Section 3.5.)

By interchanging the roles of $x$ and $y$ we get an equation of the form

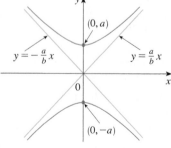

**FIGURE 11**

The hyperbola $\dfrac{y^2}{a^2} - \dfrac{x^2}{b^2} = 1$

$$\frac{y^2}{a^2} - \frac{x^2}{b^2} = 1$$

which also represents a hyperbola and is sketched in Figure 11.

EXAMPLE 6  Sketch the curve $9x^2 - 4y^2 = 36$.

SOLUTION  Dividing both sides by 36, we obtain

$$\frac{x^2}{4} - \frac{y^2}{9} = 1$$

which is the standard form of the equation of a hyperbola (Equation 3). Since $a^2 = 4$, the $x$-intercepts are $\pm 2$. Since $b^2 = 9$, we have $b = 3$ and the asymptotes are $y = \pm(\frac{3}{2})x$. The hyperbola is sketched in Figure 12.

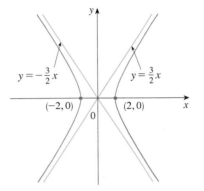

**FIGURE 12**
The hyperbola $9x^2 - 4y^2 = 36$

If $b = a$, a hyperbola has the equation $x^2 - y^2 = a^2$ (or $y^2 - x^2 = a^2$) and is called an *equilateral hyperbola* [see Figure 13(a)]. Its asymptotes are $y = \pm x$, which are perpendicular. If an equilateral hyperbola is rotated by 45°, the asymptotes become the $x$- and $y$-axes, and it can be shown that the new equation of the hyperbola is $xy = k$, where $k$ is a constant [see Figure 13(b)].

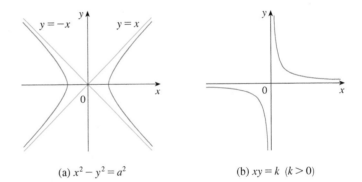

**FIGURE 13**
Equilateral hyperbolas

(a) $x^2 - y^2 = a^2$

(b) $xy = k$  $(k > 0)$

## Shifted Conics

Recall that an equation of the circle with center the origin and radius $r$ is $x^2 + y^2 = r^2$, but if the center is the point $(h, k)$, then the equation of the circle becomes

$$(x - h)^2 + (y - k)^2 = r^2$$

Similarly, if we take the ellipse with equation

**4**

$$\frac{x^2}{a^2} + \frac{y^2}{b^2} = 1$$

and translate it (shift it) so that its center is the point $(h, k)$, then its equation becomes

$$\boxed{5} \qquad \boxed{\frac{(x - h)^2}{a^2} + \frac{(y - k)^2}{b^2} = 1}$$

(See Figure 14.)

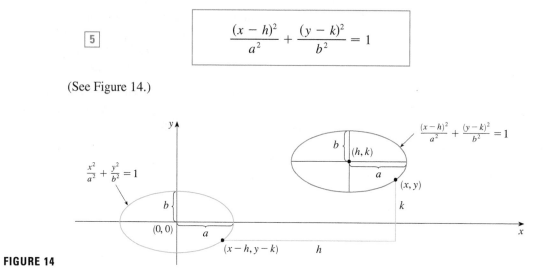

**FIGURE 14**

Notice that in shifting the ellipse, we replaced $x$ by $x - h$ and $y$ by $y - k$ in Equation 4 to obtain Equation 5. We use the same procedure to shift the parabola $y = ax^2$ so that its vertex (the origin) becomes the point $(h, k)$ as in Figure 15. Replacing $x$ by $x - h$ and $y$ by $y - k$, we see that the new equation is

$$y - k = a(x - h)^2 \qquad \text{or} \qquad y = a(x - h)^2 + k$$

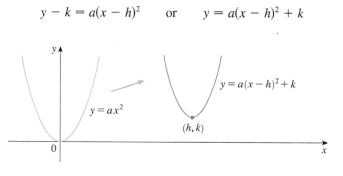

**FIGURE 15**

**EXAMPLE 7**  Sketch the graph of the equation $y = 2x^2 - 4x + 1$.

**SOLUTION**  First we complete the square:

$$y = 2(x^2 - 2x) + 1 = 2(x - 1)^2 - 1$$

In this form we see that the equation represents the parabola obtained by shifting $y = 2x^2$ so that its vertex is at the point $(1, -1)$. The graph is sketched in Figure 16.

**FIGURE 16**
$y = 2x^2 - 4x + 1$

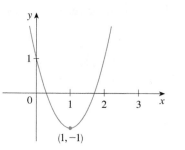

**EXAMPLE 8** Sketch the curve $x = 1 - y^2$.

SOLUTION This time we start with the parabola $x = -y^2$ (as in Figure 6 with $a = -1$) and shift one unit to the right to get the graph of $x = 1 - y^2$. (See Figure 17.)

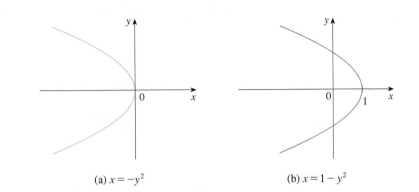

**FIGURE 17**

(a) $x = -y^2$

(b) $x = 1 - y^2$

## C Exercises

1–4 Find an equation of a circle that satisfies the given conditions.

**1.** Center $(3, -1)$, radius 5

**2.** Center $(-2, -8)$, radius 10

**3.** Center at the origin, passes through $(4, 7)$

**4.** Center $(-1, 5)$, passes through $(-4, -6)$

5–9 Show that the equation represents a circle and find the center and radius.

**5.** $x^2 + y^2 - 4x + 10y + 13 = 0$

**6.** $x^2 + y^2 + 6y + 2 = 0$

**7.** $x^2 + y^2 + x = 0$

**8.** $16x^2 + 16y^2 + 8x + 32y + 1 = 0$

**9.** $2x^2 + 2y^2 - x + y = 1$

**10.** Under what condition on the coefficients $a$, $b$, and $c$ does the equation $x^2 + y^2 + ax + by + c = 0$ represent a circle? When that condition is satisfied, find the center and radius of the circle.

11–32 Identify the type of curve and sketch the graph. Do not plot points. Just use the standard graphs given in Figures 5, 6, 8, 10, and 11 and shift if necessary.

**11.** $y = -x^2$

**12.** $y^2 - x^2 = 1$

**13.** $x^2 + 4y^2 = 16$

**14.** $x = -2y^2$

**15.** $16x^2 - 25y^2 = 400$

**16.** $25x^2 + 4y^2 = 100$

**17.** $4x^2 + y^2 = 1$

**18.** $y = x^2 + 2$

**19.** $x = y^2 - 1$

**20.** $9x^2 - 25y^2 = 225$

**21.** $9y^2 - x^2 = 9$

**22.** $2x^2 + 5y^2 = 10$

**23.** $xy = 4$

**24.** $y = x^2 + 2x$

**25.** $9(x - 1)^2 + 4(y - 2)^2 = 36$

**26.** $16x^2 + 9y^2 - 36y = 108$

**27.** $y = x^2 - 6x + 13$

**28.** $x^2 - y^2 - 4x + 3 = 0$

**29.** $x = 4 - y^2$

**30.** $y^2 - 2x + 6y + 5 = 0$

**31.** $x^2 + 4y^2 - 6x + 5 = 0$

**32.** $4x^2 + 9y^2 - 16x + 54y + 61 = 0$

33–34 Sketch the region bounded by the curves.

**33.** $y = 3x$, $y = x^2$

**34.** $y = 4 - x^2$, $x - 2y = 2$

**35.** Find an equation of the parabola with vertex $(1, -1)$ that passes through the points $(-1, 3)$ and $(3, 3)$.

**36.** Find an equation of the ellipse with center at the origin that passes through the points $\left(1, -10\sqrt{2}/3\right)$ and $\left(-2, 5\sqrt{5}/3\right)$.

37–40 Sketch the graph of the set.

**37.** $\{(x, y) \mid x^2 + y^2 \le 1\}$

**38.** $\{(x, y) \mid x^2 + y^2 > 4\}$

**39.** $\{(x, y) \mid y \ge x^2 - 1\}$

**40.** $\{(x, y) \mid x^2 + 4y^2 \le 4\}$

## D  Trigonometry

### Angles

Angles can be measured in degrees or in radians (abbreviated as rad). The angle given by a complete revolution contains 360°, which is the same as $2\pi$ rad. Therefore

$$\boxed{1} \qquad \pi \text{ rad} = 180°$$

and

$$\boxed{2} \qquad 1 \text{ rad} = \left(\frac{180}{\pi}\right)° \approx 57.3° \qquad 1° = \frac{\pi}{180} \text{ rad} \approx 0.017 \text{ rad}$$

**EXAMPLE 1**

(a) Find the radian measure of 60°.  (b) Express $5\pi/4$ rad in degrees.

SOLUTION

(a) From Equation 1 or 2 we see that to convert from degrees to radians we multiply by $\pi/180$. Therefore

$$60° = 60\left(\frac{\pi}{180}\right) = \frac{\pi}{3} \text{ rad}$$

(b) To convert from radians to degrees we multiply by $180/\pi$. Thus

$$\frac{5\pi}{4} \text{ rad} = \frac{5\pi}{4}\left(\frac{180}{\pi}\right) = 225°$$

In calculus we use radians to measure angles except when otherwise indicated. The following table gives the correspondence between degree and radian measures of some common angles.

| Degrees | 0° | 30° | 45° | 60° | 90° | 120° | 135° | 150° | 180° | 270° | 360° |
|---|---|---|---|---|---|---|---|---|---|---|---|
| Radians | 0 | $\frac{\pi}{6}$ | $\frac{\pi}{4}$ | $\frac{\pi}{3}$ | $\frac{\pi}{2}$ | $\frac{2\pi}{3}$ | $\frac{3\pi}{4}$ | $\frac{5\pi}{6}$ | $\pi$ | $\frac{3\pi}{2}$ | $2\pi$ |

Figure 1 shows a sector of a circle with central angle $\theta$ and radius $r$ subtending an arc with length $a$. Since the length of the arc is proportional to the size of the angle, and since the entire circle has circumference $2\pi r$ and central angle $2\pi$, we have

$$\frac{\theta}{2\pi} = \frac{a}{2\pi r}$$

Solving this equation for $\theta$ and for $a$, we obtain

$$\boxed{3} \qquad \theta = \frac{a}{r} \qquad\qquad a = r\theta$$

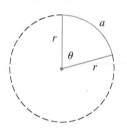

**FIGURE 1**

Remember that Equations 3 are valid only when $\theta$ is measured in radians.

**FIGURE 2**

In particular, putting $a = r$ in Equation 3, we see that an angle of 1 rad is the angle subtended at the center of a circle by an arc equal in length to the radius of the circle (see Figure 2).

**EXAMPLE 2**

(a) If the radius of a circle is 5 cm, what angle is subtended by an arc of 6 cm?
(b) If a circle has radius 3 cm, what is the length of an arc subtended by a central angle of $3\pi/8$ rad?

SOLUTION

(a) Using Equation 3 with $a = 6$ and $r = 5$, we see that the angle is

$$\theta = \tfrac{6}{5} = 1.2 \text{ rad}$$

(b) With $r = 3$ cm and $\theta = 3\pi/8$ rad, the arc length is

$$a = r\theta = 3\left(\frac{3\pi}{8}\right) = \frac{9\pi}{8} \text{ cm}$$

The **standard position** of an angle occurs when we place its vertex at the origin of a coordinate system and its initial side on the positive $x$-axis as in Figure 3. A **positive** angle is obtained by rotating the initial side counterclockwise until it coincides with the terminal side. Likewise, **negative** angles are obtained by clockwise rotation as in Figure 4.

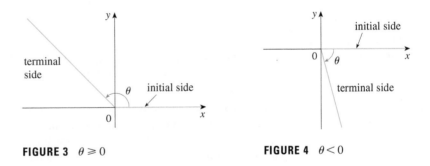

**FIGURE 3**  $\theta \geq 0$          **FIGURE 4**  $\theta < 0$

Figure 5 shows several examples of angles in standard position. Notice that different angles can have the same terminal side. For instance, the angles $3\pi/4$, $-5\pi/4$, and $11\pi/4$ have the same initial and terminal sides because

$$\frac{3\pi}{4} - 2\pi = -\frac{5\pi}{4} \qquad \frac{3\pi}{4} + 2\pi = \frac{11\pi}{4}$$

and $2\pi$ rad represents a complete revolution.

**FIGURE 5**
Angles in standard position

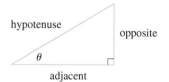

**FIGURE 6**

## The Trigonometric Functions

For an acute angle $\theta$ the six trigonometric functions are defined as ratios of lengths of sides of a right triangle as follows (see Figure 6).

$$\boxed{4}$$

$$\sin \theta = \frac{\text{opp}}{\text{hyp}} \qquad \csc \theta = \frac{\text{hyp}}{\text{opp}}$$

$$\cos \theta = \frac{\text{adj}}{\text{hyp}} \qquad \sec \theta = \frac{\text{hyp}}{\text{adj}}$$

$$\tan \theta = \frac{\text{opp}}{\text{adj}} \qquad \cot \theta = \frac{\text{adj}}{\text{opp}}$$

**FIGURE 7**

This definition doesn't apply to obtuse or negative angles, so for a general angle $\theta$ in standard position we let $P(x, y)$ be any point on the terminal side of $\theta$ and we let $r$ be the distance $|OP|$ as in Figure 7. Then we define

$$\boxed{5}$$

$$\sin \theta = \frac{y}{r} \qquad \csc \theta = \frac{r}{y}$$

$$\cos \theta = \frac{x}{r} \qquad \sec \theta = \frac{r}{x}$$

$$\tan \theta = \frac{y}{x} \qquad \cot \theta = \frac{x}{y}$$

If we put $r = 1$ in Definition 5 and draw a unit circle with center the origin and label $\theta$ as in Figure 8, then the coordinates of $P$ are $(\cos \theta, \sin \theta)$.

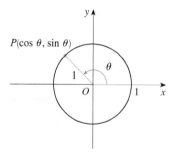

**FIGURE 8**

Since division by 0 is not defined, $\tan \theta$ and $\sec \theta$ are undefined when $x = 0$ and $\csc \theta$ and $\cot \theta$ are undefined when $y = 0$. Notice that the definitions in $\boxed{4}$ and $\boxed{5}$ are consistent when $\theta$ is an acute angle.

If $\theta$ is a number, the convention is that $\sin \theta$ means the sine of the angle whose *radian* measure is $\theta$. For example, the expression $\sin 3$ implies that we are dealing with an angle of 3 rad. When finding a calculator approximation to this number, we must remember to set our calculator in radian mode, and then we obtain

$$\sin 3 \approx 0.14112$$

If we want to know the sine of the angle 3° we would write $\sin 3°$ and, with our calculator in degree mode, we find that

$$\sin 3° \approx 0.05234$$

The exact trigonometric ratios for certain angles can be read from the triangles in Figure 9. For instance,

$$\sin \frac{\pi}{4} = \frac{1}{\sqrt{2}} \qquad \sin \frac{\pi}{6} = \frac{1}{2} \qquad \sin \frac{\pi}{3} = \frac{\sqrt{3}}{2}$$

$$\cos \frac{\pi}{4} = \frac{1}{\sqrt{2}} \qquad \cos \frac{\pi}{6} = \frac{\sqrt{3}}{2} \qquad \cos \frac{\pi}{3} = \frac{1}{2}$$

$$\tan \frac{\pi}{4} = 1 \qquad \tan \frac{\pi}{6} = \frac{1}{\sqrt{3}} \qquad \tan \frac{\pi}{3} = \sqrt{3}$$

**FIGURE 9**

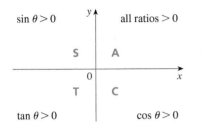

**FIGURE 10**

The signs of the trigonometric functions for angles in each of the four quadrants can be remembered by means of the rule "All Students Take Calculus" shown in Figure 10.

**EXAMPLE 3** Find the exact trigonometric ratios for $\theta = 2\pi/3$.

**SOLUTION** From Figure 11 we see that a point on the terminal line for $\theta = 2\pi/3$ is $P\left(-1, \sqrt{3}\,\right)$. Therefore, taking

$$x = -1 \qquad y = \sqrt{3} \qquad r = 2$$

in the definitions of the trigonometric ratios, we have

$$\sin \frac{2\pi}{3} = \frac{\sqrt{3}}{2} \qquad \cos \frac{2\pi}{3} = -\frac{1}{2} \qquad \tan \frac{2\pi}{3} = -\sqrt{3}$$

$$\csc \frac{2\pi}{3} = \frac{2}{\sqrt{3}} \qquad \sec \frac{2\pi}{3} = -2 \qquad \cot \frac{2\pi}{3} = -\frac{1}{\sqrt{3}}$$

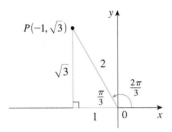

**FIGURE 11**

The following table gives some values of $\sin \theta$ and $\cos \theta$ found by the method of Example 3.

| $\theta$ | $0$ | $\dfrac{\pi}{6}$ | $\dfrac{\pi}{4}$ | $\dfrac{\pi}{3}$ | $\dfrac{\pi}{2}$ | $\dfrac{2\pi}{3}$ | $\dfrac{3\pi}{4}$ | $\dfrac{5\pi}{6}$ | $\pi$ | $\dfrac{3\pi}{2}$ | $2\pi$ |
|---|---|---|---|---|---|---|---|---|---|---|---|
| $\sin \theta$ | $0$ | $\dfrac{1}{2}$ | $\dfrac{1}{\sqrt{2}}$ | $\dfrac{\sqrt{3}}{2}$ | $1$ | $\dfrac{\sqrt{3}}{2}$ | $\dfrac{1}{\sqrt{2}}$ | $\dfrac{1}{2}$ | $0$ | $-1$ | $0$ |
| $\cos \theta$ | $1$ | $\dfrac{\sqrt{3}}{2}$ | $\dfrac{1}{\sqrt{2}}$ | $\dfrac{1}{2}$ | $0$ | $-\dfrac{1}{2}$ | $-\dfrac{1}{\sqrt{2}}$ | $-\dfrac{\sqrt{3}}{2}$ | $-1$ | $0$ | $1$ |

**EXAMPLE 4** If $\cos \theta = \frac{2}{5}$ and $0 < \theta < \pi/2$, find the other five trigonometric functions of $\theta$.

**SOLUTION** Since $\cos \theta = \frac{2}{5}$, we can label the hypotenuse as having length 5 and the adjacent side as having length 2 in Figure 12. If the opposite side has length $x$, then the Pythagorean Theorem gives $x^2 + 4 = 25$ and so $x^2 = 21$, $x = \sqrt{21}$. We can now use the diagram to write the other five trigonometric functions:

$$\sin \theta = \frac{\sqrt{21}}{5} \qquad \tan \theta = \frac{\sqrt{21}}{2}$$

$$\csc \theta = \frac{5}{\sqrt{21}} \qquad \sec \theta = \frac{5}{2} \qquad \cot \theta = \frac{2}{\sqrt{21}}$$

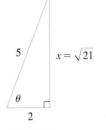

**FIGURE 12**

**EXAMPLE 5** Use a calculator to approximate the value of $x$ in Figure 13.

**SOLUTION** From the diagram we see that

$$\tan 40° = \frac{16}{x}$$

Therefore

$$x = \frac{16}{\tan 40°} \approx 19.07$$

**FIGURE 13**

### Trigonometric Identities

A trigonometric identity is a relationship among the trigonometric functions. The most elementary are the following, which are immediate consequences of the definitions of the trigonometric functions.

**6**
$$\csc\theta = \frac{1}{\sin\theta} \qquad \sec\theta = \frac{1}{\cos\theta} \qquad \cot\theta = \frac{1}{\tan\theta}$$

$$\tan\theta = \frac{\sin\theta}{\cos\theta} \qquad \cot\theta = \frac{\cos\theta}{\sin\theta}$$

For the next identity we refer back to Figure 7. The distance formula (or, equivalently, the Pythagorean Theorem) tells us that $x^2 + y^2 = r^2$. Therefore

$$\sin^2\theta + \cos^2\theta = \frac{y^2}{r^2} + \frac{x^2}{r^2} = \frac{x^2 + y^2}{r^2} = \frac{r^2}{r^2} = 1$$

We have therefore proved one of the most useful of all trigonometric identities:

**7**
$$\sin^2\theta + \cos^2\theta = 1$$

If we now divide both sides of Equation 7 by $\cos^2\theta$ and use Equations 6, we get

**8**
$$\tan^2\theta + 1 = \sec^2\theta$$

Similarly, if we divide both sides of Equation 7 by $\sin^2\theta$, we get

**9**
$$1 + \cot^2\theta = \csc^2\theta$$

The identities

**10a**
$$\sin(-\theta) = -\sin\theta$$

**10b**
$$\cos(-\theta) = \cos\theta$$

Odd functions and even functions are discussed in Section 1.1.

show that sine is an odd function and cosine is an even function. They are easily proved by drawing a diagram showing $\theta$ and $-\theta$ in standard position (see Exercise 39).

Since the angles $\theta$ and $\theta + 2\pi$ have the same terminal side, we have

**11**
$$\sin(\theta + 2\pi) = \sin\theta \qquad \cos(\theta + 2\pi) = \cos\theta$$

These identities show that the sine and cosine functions are periodic with period $2\pi$.

The remaining trigonometric identities are all consequences of two basic identities called the **addition formulas**:

12a
$$\sin(x + y) = \sin x \cos y + \cos x \sin y$$

12b
$$\cos(x + y) = \cos x \cos y - \sin x \sin y$$

The proofs of these addition formulas are outlined in Exercises 85, 86, and 87.

By substituting $-y$ for $y$ in Equations 12a and 12b and using Equations 10a and 10b, we obtain the following **subtraction formulas**:

13a
$$\sin(x - y) = \sin x \cos y - \cos x \sin y$$

13b
$$\cos(x - y) = \cos x \cos y + \sin x \sin y$$

Then, by dividing the formulas in Equations 12 or Equations 13, we obtain the corresponding formulas for $\tan(x \pm y)$:

14a
$$\tan(x + y) = \frac{\tan x + \tan y}{1 - \tan x \tan y}$$

14b
$$\tan(x - y) = \frac{\tan x - \tan y}{1 + \tan x \tan y}$$

If we put $y = x$ in the addition formulas $\boxed{12}$, we get the **double-angle formulas**:

15a
$$\sin 2x = 2 \sin x \cos x$$

15b
$$\cos 2x = \cos^2 x - \sin^2 x$$

Then, by using the identity $\sin^2 x + \cos^2 x = 1$, we obtain the following alternate forms of the double-angle formulas for $\cos 2x$:

16a
$$\cos 2x = 2 \cos^2 x - 1$$

16b
$$\cos 2x = 1 - 2 \sin^2 x$$

If we now solve these equations for $\cos^2 x$ and $\sin^2 x$, we get the following **half-angle formulas**, which are useful in integral calculus:

17a
$$\cos^2 x = \frac{1 + \cos 2x}{2}$$

17b
$$\sin^2 x = \frac{1 - \cos 2x}{2}$$

Finally, we state the **product formulas**, which can be deduced from Equations 12 and 13:

18a
$$\sin x \cos y = \tfrac{1}{2}[\sin(x + y) + \sin(x - y)]$$

18b
$$\cos x \cos y = \tfrac{1}{2}[\cos(x + y) + \cos(x - y)]$$

18c
$$\sin x \sin y = \tfrac{1}{2}[\cos(x - y) - \cos(x + y)]$$

There are many other trigonometric identities, but those we have stated are the ones used most often in calculus. If you forget any of the identities 13–18, remember that they can all be deduced from Equations 12a and 12b.

**EXAMPLE 6** Find all values of $x$ in the interval $[0, 2\pi]$ such that $\sin x = \sin 2x$.

**SOLUTION** Using the double-angle formula (15a), we rewrite the given equation as

$$\sin x = 2 \sin x \cos x \qquad \text{or} \qquad \sin x(1 - 2 \cos x) = 0$$

Therefore there are two possibilities:

$$\sin x = 0 \qquad \text{or} \qquad 1 - 2 \cos x = 0$$

$$x = 0, \pi, 2\pi \qquad\qquad\qquad \cos x = \tfrac{1}{2}$$

$$x = \frac{\pi}{3}, \frac{5\pi}{3}$$

The given equation has five solutions: $0$, $\pi/3$, $\pi$, $5\pi/3$, and $2\pi$.

### Graphs of the Trigonometric Functions

The graph of the function $f(x) = \sin x$, shown in Figure 14(a), is obtained by plotting points for $0 \le x \le 2\pi$ and then using the periodic nature of the function (from Equation 11) to complete the graph. Notice that the zeros of the sine function occur at the

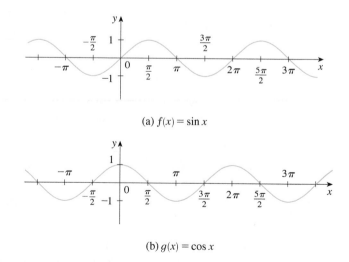

(a) $f(x) = \sin x$

(b) $g(x) = \cos x$

**FIGURE 14**

integer multiples of $\pi$, that is,

$$\sin x = 0 \qquad \text{whenever } x = n\pi, \quad n \text{ an integer}$$

Because of the identity

$$\cos x = \sin\left(x + \frac{\pi}{2}\right)$$

(which can be verified using Equation 12a), the graph of cosine is obtained by shifting the graph of sine by an amount $\pi/2$ to the left [see Figure 14(b)]. Note that for both the sine and cosine functions the domain is $(-\infty, \infty)$ and the range is the closed interval $[-1, 1]$. Thus, for all values of $x$, we have

$$-1 \leqslant \sin x \leqslant 1 \qquad -1 \leqslant \cos x \leqslant 1$$

The graphs of the remaining four trigonometric functions are shown in Figure 15 and their domains are indicated there. Notice that tangent and cotangent have range $(-\infty, \infty)$, whereas cosecant and secant have range $(-\infty, -1] \cup [1, \infty)$. All four functions are periodic: tangent and cotangent have period $\pi$, whereas cosecant and secant have period $2\pi$.

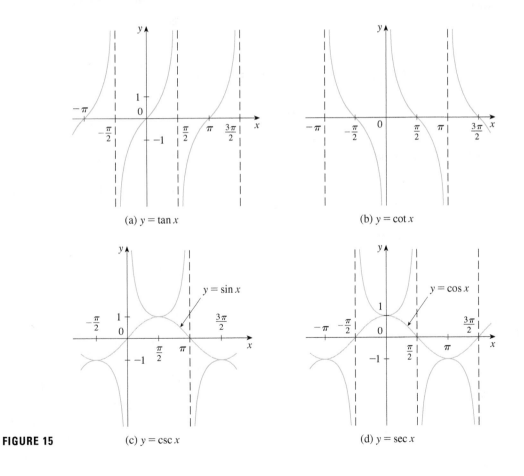

(a) $y = \tan x$

(b) $y = \cot x$

(c) $y = \csc x$

(d) $y = \sec x$

**FIGURE 15**

## D    Exercises

**1–6** Convert from degrees to radians.

**1.** $210°$

**2.** $300°$

**3.** $9°$

**4.** $-315°$

**5.** $900°$

**6.** $36°$

**7–12** Convert from radians to degrees.

**7.** $4\pi$

**8.** $-\dfrac{7\pi}{2}$

**9.** $\dfrac{5\pi}{12}$

**10.** $\dfrac{8\pi}{3}$

**11.** $-\dfrac{3\pi}{8}$

**12.** $5$

**13.** Find the length of a circular arc subtended by an angle of $\pi/12$ rad if the radius of the circle is 36 cm.

**14.** If a circle has radius 10 cm, find the length of the arc subtended by a central angle of $72°$.

**15.** A circle has radius 1.5 m. What angle is subtended at the center of the circle by an arc 1 m long?

**16.** Find the radius of a circular sector with angle $3\pi/4$ and arc length 6 cm.

**17–22** Draw, in standard position, the angle whose measure is given.

**17.** $315°$

**18.** $-150°$

**19.** $-\dfrac{3\pi}{4}$ rad

**20.** $\dfrac{7\pi}{3}$ rad

**21.** $2$ rad

**22.** $-3$ rad

**23–28** Find the exact trigonometric ratios for the angle whose radian measure is given.

**23.** $\dfrac{3\pi}{4}$

**24.** $\dfrac{4\pi}{3}$

**25.** $\dfrac{9\pi}{2}$

**26.** $-5\pi$

**27.** $\dfrac{5\pi}{6}$

**28.** $\dfrac{11\pi}{4}$

**29–34** Find the remaining trigonometric ratios.

**29.** $\sin\theta = \dfrac{3}{5}, \quad 0 < \theta < \dfrac{\pi}{2}$

**30.** $\tan\alpha = 2, \quad 0 < \alpha < \dfrac{\pi}{2}$

**31.** $\sec\phi = -1.5, \quad \dfrac{\pi}{2} < \phi < \pi$

**32.** $\cos x = -\dfrac{1}{3}, \quad \pi < x < \dfrac{3\pi}{2}$

**33.** $\cot\beta = 3, \quad \pi < \beta < 2\pi$

**34.** $\csc\theta = -\dfrac{4}{3}, \quad \dfrac{3\pi}{2} < \theta < 2\pi$

**35–38** Find, correct to five decimal places, the length of the side labeled $x$.

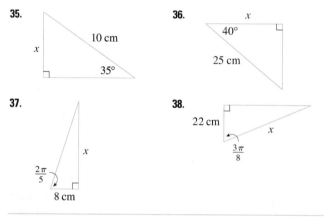

**39–41** Prove each equation.

**39.** (a) Equation 10a          (b) Equation 10b

**40.** (a) Equation 14a          (b) Equation 14b

**41.** (a) Equation 18a          (b) Equation 18b
      (c) Equation 18c

**42–58** Prove the identity.

**42.** $\cos\left(\dfrac{\pi}{2} - x\right) = \sin x$

**43.** $\sin\left(\dfrac{\pi}{2} + x\right) = \cos x$          **44.** $\sin(\pi - x) = \sin x$

**45.** $\sin\theta\cot\theta = \cos\theta$          **46.** $(\sin x + \cos x)^2 = 1 + \sin 2x$

**47.** $\sec y - \cos y = \tan y\sin y$

**48.** $\tan^2\alpha - \sin^2\alpha = \tan^2\alpha\sin^2\alpha$

**49.** $\cot^2\theta + \sec^2\theta = \tan^2\theta + \csc^2\theta$

**50.** $2\csc 2t = \sec t\csc t$

**51.** $\tan 2\theta = \dfrac{2\tan\theta}{1 - \tan^2\theta}$

**52.** $\dfrac{1}{1 - \sin\theta} + \dfrac{1}{1 + \sin\theta} = 2\sec^2\theta$

**53.** $\sin x\sin 2x + \cos x\cos 2x = \cos x$

**54.** $\sin^2 x - \sin^2 y = \sin(x + y)\sin(x - y)$

**55.** $\dfrac{\sin\phi}{1 - \cos\phi} = \csc\phi + \cot\phi$

**56.** $\tan x + \tan y = \dfrac{\sin(x + y)}{\cos x \cos y}$

**57.** $\sin 3\theta + \sin \theta = 2 \sin 2\theta \cos \theta$

**58.** $\cos 3\theta = 4 \cos^3\theta - 3 \cos \theta$

**59–64** If $\sin x = \frac{1}{3}$ and $\sec y = \frac{5}{4}$, where $x$ and $y$ lie between 0 and $\pi/2$, evaluate the expression.

**59.** $\sin(x + y)$             **60.** $\cos(x + y)$

**61.** $\cos(x - y)$             **62.** $\sin(x - y)$

**63.** $\sin 2y$                  **64.** $\cos 2y$

**65–72** Find all values of $x$ in the interval $[0, 2\pi]$ that satisfy the equation.

**65.** $2 \cos x - 1 = 0$       **66.** $3 \cot^2 x = 1$

**67.** $2 \sin^2 x = 1$           **68.** $|\tan x| = 1$

**69.** $\sin 2x = \cos x$        **70.** $2 \cos x + \sin 2x = 0$

**71.** $\sin x = \tan x$         **72.** $2 + \cos 2x = 3 \cos x$

**73–76** Find all values of $x$ in the interval $[0, 2\pi]$ that satisfy the inequality.

**73.** $\sin x \le \frac{1}{2}$           **74.** $2 \cos x + 1 > 0$

**75.** $-1 < \tan x < 1$      **76.** $\sin x > \cos x$

**77–82** Graph the function by starting with the graphs in Figures 14 and 15 and applying the transformations of Section 1.3 where appropriate.

**77.** $y = \cos\left(x - \dfrac{\pi}{3}\right)$       **78.** $y = \tan 2x$

**79.** $y = \dfrac{1}{3} \tan\left(x - \dfrac{\pi}{2}\right)$      **80.** $y = 1 + \sec x$

**81.** $y = |\sin x|$              **82.** $y = 2 + \sin\left(x + \dfrac{\pi}{4}\right)$

**83.** Prove the **Law of Cosines**: If a triangle has sides with lengths $a$, $b$, and $c$, and $\theta$ is the angle between the sides with lengths $a$ and $b$, then
$$c^2 = a^2 + b^2 - 2ab \cos \theta$$

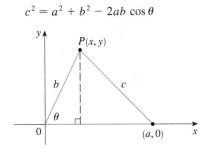

[*Hint:* Introduce a coordinate system so that $\theta$ is in standard

position, as in the figure. Express $x$ and $y$ in terms of $\theta$ and then use the distance formula to compute $c$.]

**84.** In order to find the distance $|AB|$ across a small inlet, a point $C$ was located as in the figure and the following measurements were recorded:
$$\angle C = 103° \qquad |AC| = 820 \text{ m} \qquad |BC| = 910 \text{ m}$$
Use the Law of Cosines from Exercise 83 to find the required distance.

**85.** Use the figure to prove the subtraction formula
$$\cos(\alpha - \beta) = \cos \alpha \cos \beta + \sin \alpha \sin \beta$$

[*Hint:* Compute $c^2$ in two ways (using the Law of Cosines from Exercise 83 and also using the distance formula) and compare the two expressions.]

**86.** Use the formula in Exercise 85 to prove the addition formula for cosine (12b).

**87.** Use the addition formula for cosine and the identities
$$\cos\left(\dfrac{\pi}{2} - \theta\right) = \sin \theta \qquad \sin\left(\dfrac{\pi}{2} - \theta\right) = \cos \theta$$
to prove the subtraction formula (13a) for the sine function.

**88.** Show that the area of a triangle with sides of lengths $a$ and $b$ and with included angle $\theta$ is
$$A = \tfrac{1}{2}ab \sin \theta$$

**89.** Find the area of triangle $ABC$, correct to five decimal places, if
$$|AB| = 10 \text{ cm} \qquad |BC| = 3 \text{ cm} \qquad \angle ABC = 107°$$

## E    Sigma Notation

A convenient way of writing sums uses the Greek letter $\Sigma$ (capital sigma, corresponding to our letter S) and is called **sigma notation**.

This tells us to end with $i = n$.

This tells us to add.

This tells us to start with $i = m$.

$$\sum_{i=m}^{n} a_i$$

---

**1**  **Definition**   If $a_m, a_{m+1}, \ldots, a_n$ are real numbers and $m$ and $n$ are integers such that $m \leq n,$ then

$$\sum_{i=m}^{n} a_i = a_m + a_{m+1} + a_{m+2} + \cdots + a_{n-1} + a_n$$

---

With function notation, Definition 1 can be written as

$$\sum_{i=m}^{n} f(i) = f(m) + f(m + 1) + f(m + 2) + \cdots + f(n - 1) + f(n)$$

Thus the symbol $\sum_{i=m}^{n}$ indicates a summation in which the letter $i$ (called the **index of summation**) takes on consecutive integer values beginning with $m$ and ending with $n$, that is, $m, m + 1, \ldots, n$. Other letters can also be used as the index of summation.

**EXAMPLE 1**

(a) $\displaystyle\sum_{i=1}^{4} i^2 = 1^2 + 2^2 + 3^2 + 4^2 = 30$

(b) $\displaystyle\sum_{i=3}^{n} i = 3 + 4 + 5 + \cdots + (n - 1) + n$

(c) $\displaystyle\sum_{j=0}^{5} 2^j = 2^0 + 2^1 + 2^2 + 2^3 + 2^4 + 2^5 = 63$

(d) $\displaystyle\sum_{k=1}^{n} \frac{1}{k} = 1 + \frac{1}{2} + \frac{1}{3} + \cdots + \frac{1}{n}$

(e) $\displaystyle\sum_{i=1}^{3} \frac{i - 1}{i^2 + 3} = \frac{1 - 1}{1^2 + 3} + \frac{2 - 1}{2^2 + 3} + \frac{3 - 1}{3^2 + 3} = 0 + \frac{1}{7} + \frac{1}{6} = \frac{13}{42}$

(f) $\displaystyle\sum_{i=1}^{4} 2 = 2 + 2 + 2 + 2 = 8$

**EXAMPLE 2**   Write the sum $2^3 + 3^3 + \cdots + n^3$ in sigma notation.

SOLUTION   There is no unique way of writing a sum in sigma notation. We could write

$$2^3 + 3^3 + \cdots + n^3 = \sum_{i=2}^{n} i^3$$

or

$$2^3 + 3^3 + \cdots + n^3 = \sum_{j=1}^{n-1} (j + 1)^3$$

or

$$2^3 + 3^3 + \cdots + n^3 = \sum_{k=0}^{n-2} (k + 2)^3$$

The following theorem gives three simple rules for working with sigma notation.

2 **Theorem**  If $c$ is any constant (that is, it does not depend on $i$), then

(a) $\displaystyle\sum_{i=m}^{n} ca_i = c \sum_{i=m}^{n} a_i$

(b) $\displaystyle\sum_{i=m}^{n} (a_i + b_i) = \sum_{i=m}^{n} a_i + \sum_{i=m}^{n} b_i$

(c) $\displaystyle\sum_{i=m}^{n} (a_i - b_i) = \sum_{i=m}^{n} a_i - \sum_{i=m}^{n} b_i$

PROOF  To see why these rules are true, all we have to do is write both sides in expanded form. Rule (a) is just the distributive property of real numbers:

$$ca_m + ca_{m+1} + \cdots + ca_n = c(a_m + a_{m+1} + \cdots + a_n)$$

Rule (b) follows from the associative and commutative properties:

$$(a_m + b_m) + (a_{m+1} + b_{m+1}) + \cdots + (a_n + b_n)$$
$$= (a_m + a_{m+1} + \cdots + a_n) + (b_m + b_{m+1} + \cdots + b_n)$$

Rule (c) is proved similarly.

**EXAMPLE 3**  Find $\displaystyle\sum_{i=1}^{n} 1$.

SOLUTION
$$\sum_{i=1}^{n} 1 = \underbrace{1 + 1 + \cdots + 1}_{n \text{ terms}} = n$$

**EXAMPLE 4**  Prove the formula for the sum of the first $n$ positive integers:

$$\sum_{i=1}^{n} i = 1 + 2 + 3 + \cdots + n = \frac{n(n + 1)}{2}$$

SOLUTION  This formula can be proved by mathematical induction (see page 98) or by the following method used by the German mathematician Karl Friedrich Gauss (1777–1855) when he was ten years old.

Write the sum $S$ twice, once in the usual order and once in reverse order:

$$S = 1 + \quad 2 \quad + \quad 3 \quad + \cdots + (n - 1) + n$$
$$S = n + (n - 1) + (n - 2) + \cdots + \quad 2 \quad + 1$$

Adding all columns vertically, we get

$$2S = (n + 1) + (n + 1) + (n + 1) + \cdots + (n + 1) + (n + 1)$$

On the right side there are $n$ terms, each of which is $n + 1$, so

$$2S = n(n + 1) \quad \text{or} \quad S = \frac{n(n + 1)}{2}$$

**EXAMPLE 5**  Prove the formula for the sum of the squares of the first $n$ positive integers:

$$\sum_{i=1}^{n} i^2 = 1^2 + 2^2 + 3^2 + \cdots + n^2 = \frac{n(n + 1)(2n + 1)}{6}$$

SOLUTION 1  Let $S$ be the desired sum. We start with the *telescoping sum* (or collapsing sum):

Most terms cancel in pairs.

$$\sum_{i=1}^{n} [(1+i)^3 - i^3] = (2^3 - 1^3) + (3^3 - 2^3) + (4^3 - 3^3) + \cdots + [(n+1)^3 - n^3]$$

$$= (n+1)^3 - 1^3 = n^3 + 3n^2 + 3n$$

On the other hand, using Theorem 2 and Examples 3 and 4, we have

$$\sum_{i=1}^{n} [(1+i)^3 - i^3] = \sum_{i=1}^{n} [3i^2 + 3i + 1] = 3\sum_{i=1}^{n} i^2 + 3\sum_{i=1}^{n} i + \sum_{i=1}^{n} 1$$

$$= 3S + 3\frac{n(n+1)}{2} + n = 3S + \tfrac{3}{2}n^2 + \tfrac{5}{2}n$$

Thus we have

$$n^3 + 3n^2 + 3n = 3S + \tfrac{3}{2}n^2 + \tfrac{5}{2}n$$

Solving this equation for $S$, we obtain

$$3S = n^3 + \tfrac{3}{2}n^2 + \tfrac{1}{2}n$$

or

$$S = \frac{2n^3 + 3n^2 + n}{6} = \frac{n(n+1)(2n+1)}{6}$$

**Principle of Mathematical Induction**
Let $S_n$ be a statement involving the positive integer $n$. Suppose that
 1. $S_1$ is true.
 2. If $S_k$ is true, then $S_{k+1}$ is true.
Then $S_n$ is true for all positive integers $n$.

SOLUTION 2  Let $S_n$ be the given formula.

**1.** $S_1$ is true because

$$1^2 = \frac{1(1+1)(2 \cdot 1 + 1)}{6}$$

**2.** Assume that $S_k$ is true; that is,

$$1^2 + 2^2 + 3^2 + \cdots + k^2 = \frac{k(k+1)(2k+1)}{6}$$

See pages 98 and 100 for a more thorough discussion of mathematical induction.

Then

$$1^2 + 2^2 + 3^2 + \cdots + (k+1)^2 = (1^2 + 2^2 + 3^2 + \cdots + k^2) + (k+1)^2$$

$$= \frac{k(k+1)(2k+1)}{6} + (k+1)^2$$

$$= (k+1)\frac{k(2k+1) + 6(k+1)}{6}$$

$$= (k+1)\frac{2k^2 + 7k + 6}{6}$$

$$= \frac{(k+1)(k+2)(2k+3)}{6}$$

$$= \frac{(k+1)[(k+1)+1][2(k+1)+1]}{6}$$

So $S_{k+1}$ is true.

By the Principle of Mathematical Induction, $S_n$ is true for all $n$.

We list the results of Examples 3, 4, and 5 together with a similar result for cubes (see Exercises 37–40) as Theorem 3. These formulas are needed for finding areas and evaluating integrals in Chapter 4.

---

**3** **Theorem** Let $c$ be a constant and $n$ a positive integer. Then

(a) $\displaystyle\sum_{i=1}^{n} 1 = n$ 　　　　　　　　(b) $\displaystyle\sum_{i=1}^{n} c = nc$

(c) $\displaystyle\sum_{i=1}^{n} i = \frac{n(n+1)}{2}$ 　　　　(d) $\displaystyle\sum_{i=1}^{n} i^2 = \frac{n(n+1)(2n+1)}{6}$

(e) $\displaystyle\sum_{i=1}^{n} i^3 = \left[\frac{n(n+1)}{2}\right]^2$

---

**EXAMPLE 6** Evaluate $\displaystyle\sum_{i=1}^{n} i(4i^2 - 3)$.

**SOLUTION** Using Theorems 2 and 3, we have

$$\sum_{i=1}^{n} i(4i^2 - 3) = \sum_{i=1}^{n} (4i^3 - 3i) = 4\sum_{i=1}^{n} i^3 - 3\sum_{i=1}^{n} i$$

$$= 4\left[\frac{n(n+1)}{2}\right]^2 - 3\frac{n(n+1)}{2}$$

$$= \frac{n(n+1)[2n(n+1) - 3]}{2}$$

$$= \frac{n(n+1)(2n^2 + 2n - 3)}{2}$$

**EXAMPLE 7** Find $\displaystyle\lim_{n\to\infty} \sum_{i=1}^{n} \frac{3}{n}\left[\left(\frac{i}{n}\right)^2 + 1\right]$.

The type of calculation in Example 7 arises in Chapter 4 when we compute areas.

**SOLUTION**

$$\lim_{n\to\infty} \sum_{i=1}^{n} \frac{3}{n}\left[\left(\frac{i}{n}\right)^2 + 1\right] = \lim_{n\to\infty} \sum_{i=1}^{n} \left[\frac{3}{n^3} i^2 + \frac{3}{n}\right]$$

$$= \lim_{n\to\infty} \left[\frac{3}{n^3} \sum_{i=1}^{n} i^2 + \frac{3}{n} \sum_{i=1}^{n} 1\right]$$

$$= \lim_{n\to\infty} \left[\frac{3}{n^3} \frac{n(n+1)(2n+1)}{6} + \frac{3}{n} \cdot n\right]$$

$$= \lim_{n\to\infty} \left[\frac{1}{2} \cdot \frac{n}{n} \cdot \left(\frac{n+1}{n}\right)\left(\frac{2n+1}{n}\right) + 3\right]$$

$$= \lim_{n\to\infty} \left[\frac{1}{2} \cdot 1\left(1 + \frac{1}{n}\right)\left(2 + \frac{1}{n}\right) + 3\right]$$

$$= \tfrac{1}{2} \cdot 1 \cdot 1 \cdot 2 + 3 = 4$$

**Exercises**

**1–10** Write the sum in expanded form.

**1.** $\displaystyle\sum_{i=1}^{5} \sqrt{i}$

**2.** $\displaystyle\sum_{i=1}^{6} \frac{1}{i+1}$

**3.** $\displaystyle\sum_{i=4}^{6} 3^i$

**4.** $\displaystyle\sum_{i=4}^{6} i^3$

**5.** $\displaystyle\sum_{k=0}^{4} \frac{2k-1}{2k+1}$

**6.** $\displaystyle\sum_{k=5}^{8} x^k$

**7.** $\displaystyle\sum_{i=1}^{n} i^{10}$

**8.** $\displaystyle\sum_{j=n}^{n+3} j^2$

**9.** $\displaystyle\sum_{j=0}^{n-1} (-1)^j$

**10.** $\displaystyle\sum_{i=1}^{n} f(x_i)\,\Delta x_i$

**11–20** Write the sum in sigma notation.

**11.** $1 + 2 + 3 + 4 + \cdots + 10$

**12.** $\sqrt{3} + \sqrt{4} + \sqrt{5} + \sqrt{6} + \sqrt{7}$

**13.** $\frac{1}{2} + \frac{2}{3} + \frac{3}{4} + \frac{4}{5} + \cdots + \frac{19}{20}$

**14.** $\frac{3}{7} + \frac{4}{8} + \frac{5}{9} + \frac{6}{10} + \cdots + \frac{23}{27}$

**15.** $2 + 4 + 6 + 8 + \cdots + 2n$

**16.** $1 + 3 + 5 + 7 + \cdots + (2n-1)$

**17.** $1 + 2 + 4 + 8 + 16 + 32$

**18.** $\frac{1}{1} + \frac{1}{4} + \frac{1}{9} + \frac{1}{16} + \frac{1}{25} + \frac{1}{36}$

**19.** $x + x^2 + x^3 + \cdots + x^n$

**20.** $1 - x + x^2 - x^3 + \cdots + (-1)^n x^n$

**21–35** Find the value of the sum.

**21.** $\displaystyle\sum_{i=4}^{8} (3i-2)$

**22.** $\displaystyle\sum_{i=3}^{6} i(i+2)$

**23.** $\displaystyle\sum_{j=1}^{6} 3^{j+1}$

**24.** $\displaystyle\sum_{k=0}^{8} \cos k\pi$

**25.** $\displaystyle\sum_{n=1}^{20} (-1)^n$

**26.** $\displaystyle\sum_{i=1}^{100} 4$

**27.** $\displaystyle\sum_{i=0}^{4} (2^i + i^2)$

**28.** $\displaystyle\sum_{i=-2}^{4} 2^{3-i}$

**29.** $\displaystyle\sum_{i=1}^{n} 2i$

**30.** $\displaystyle\sum_{i=1}^{n} (2-5i)$

**31.** $\displaystyle\sum_{i=1}^{n} (i^2 + 3i + 4)$

**32.** $\displaystyle\sum_{i=1}^{n} (3+2i)^2$

**33.** $\displaystyle\sum_{i=1}^{n} (i+1)(i+2)$

**34.** $\displaystyle\sum_{i=1}^{n} i(i+1)(i+2)$

**35.** $\displaystyle\sum_{i=1}^{n} (i^3 - i - 2)$

**36.** Find the number $n$ such that $\displaystyle\sum_{i=1}^{n} i = 78$.

**37.** Prove formula (b) of Theorem 3.

**38.** Prove formula (e) of Theorem 3 using mathematical induction.

**39.** Prove formula (e) of Theorem 3 using a method similar to that of Example 5, Solution 1 [start with $(1+i)^4 - i^4$].

**40.** Prove formula (e) of Theorem 3 using the following method published by Abu Bekr Mohammed ibn Alhusain Alkarchi in about AD 1010. The figure shows a square $ABCD$ in which sides $AB$ and $AD$ have been divided into segments of lengths 1, 2, 3, ..., $n$. Thus the side of the square has length $n(n+1)/2$ so the area is $[n(n+1)/2]^2$. But the area is also the sum of the areas of the $n$ "gnomons" $G_1, G_2, \ldots, G_n$ shown in the figure. Show that the area of $G_i$ is $i^3$ and conclude that formula (e) is true.

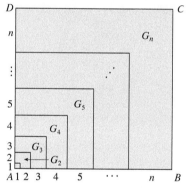

**41.** Evaluate each telescoping sum.

(a) $\displaystyle\sum_{i=1}^{n} [i^4 - (i-1)^4]$

(b) $\displaystyle\sum_{i=1}^{100} (5^i - 5^{i-1})$

(c) $\displaystyle\sum_{i=3}^{99} \left(\frac{1}{i} - \frac{1}{i+1}\right)$

(d) $\displaystyle\sum_{i=1}^{n} (a_i - a_{i-1})$

**42.** Prove the generalized triangle inequality:

$$\left| \sum_{i=1}^{n} a_i \right| \le \sum_{i=1}^{n} |a_i|$$

**43–46** Find the limit.

**43.** $\displaystyle\lim_{n\to\infty} \sum_{i=1}^{n} \frac{1}{n}\left(\frac{i}{n}\right)^2$

**44.** $\displaystyle\lim_{n\to\infty} \sum_{i=1}^{n} \frac{1}{n}\left[\left(\frac{i}{n}\right)^3 + 1\right]$

**45.** $\displaystyle\lim_{n\to\infty} \sum_{i=1}^{n} \frac{2}{n}\left[\left(\frac{2i}{n}\right)^3 + 5\left(\frac{2i}{n}\right)\right]$

**46.** $\lim\limits_{n\to\infty} \sum\limits_{i=1}^{n} \dfrac{3}{n}\left[\left(1 + \dfrac{3i}{n}\right)^3 - 2\left(1 + \dfrac{3i}{n}\right)\right]$

**48.** Evaluate $\sum\limits_{i=1}^{n} \dfrac{3}{2^{i-1}}$.

**49.** Evaluate $\sum\limits_{i=1}^{n} (2i + 2^i)$.

**47.** Prove the formula for the sum of a finite geometric series with first term $a$ and common ratio $r \neq 1$:

$$\sum_{i=1}^{n} ar^{i-1} = a + ar + ar^2 + \cdots + ar^{n-1} = \frac{a(r^n - 1)}{r - 1}$$

**50.** Evaluate $\sum\limits_{i=1}^{m}\left[\sum\limits_{j=1}^{n}(i + j)\right]$.

---

| F | **Proofs of Theorems** |
|---|---|

In this appendix we present proofs of several theorems that are stated in the main body of the text. The sections in which they occur are indicated in the margin.

Section 1.6

**Limit Laws**  Suppose that $c$ is a constant and the limits

$$\lim_{x\to a} f(x) = L \qquad \text{and} \qquad \lim_{x\to a} g(x) = M$$

exist. Then

**1.** $\lim\limits_{x\to a} [f(x) + g(x)] = L + M$ 　　 **2.** $\lim\limits_{x\to a} [f(x) - g(x)] = L - M$

**3.** $\lim\limits_{x\to a} [cf(x)] = cL$ 　　 **4.** $\lim\limits_{x\to a} [f(x)g(x)] = LM$

**5.** $\lim\limits_{x\to a} \dfrac{f(x)}{g(x)} = \dfrac{L}{M}$  if $M \neq 0$

PROOF OF LAW 4  Let $\varepsilon > 0$ be given. We want to find $\delta > 0$ such that

$$\text{if} \qquad 0 < |x - a| < \delta \qquad \text{then} \qquad |f(x)g(x) - LM| < \varepsilon$$

In order to get terms that contain $|f(x) - L|$ and $|g(x) - M|$, we add and subtract $Lg(x)$ as follows:

$$\begin{aligned}
|f(x)g(x) - LM| &= |f(x)g(x) - Lg(x) + Lg(x) - LM| \\
&= |[f(x) - L]g(x) + L[g(x) - M]| \\
&\leq |[f(x) - L]g(x)| + |L[g(x) - M]| \qquad \text{(Triangle Inequality)} \\
&= |f(x) - L||g(x)| + |L||g(x) - M|
\end{aligned}$$

We want to make each of these terms less than $\varepsilon/2$.

Since $\lim_{x\to a} g(x) = M$, there is a number $\delta_1 > 0$ such that

$$\text{if} \qquad 0 < |x - a| < \delta_1 \qquad \text{then} \qquad |g(x) - M| < \frac{\varepsilon}{2(1 + |L|)}$$

Also, there is a number $\delta_2 > 0$ such that if $0 < |x - a| < \delta_2$, then

$$|g(x) - M| < 1$$

and therefore

$$|g(x)| = |g(x) - M + M| \leq |g(x) - M| + |M| < 1 + |M|$$

Since $\lim_{x \to a} f(x) = L$, there is a number $\delta_3 > 0$ such that

$$\text{if} \quad 0 < |x - a| < \delta_3 \quad \text{then} \quad |f(x) - L| < \frac{\varepsilon}{2(1 + |M|)}$$

Let $\delta = \min\{\delta_1, \delta_2, \delta_3\}$. If $0 < |x - a| < \delta$, then we have $0 < |x - a| < \delta_1$, $0 < |x - a| < \delta_2$, and $0 < |x - a| < \delta_3$, so we can combine the inequalities to obtain

$$|f(x)g(x) - LM| \leq |f(x) - L||g(x)| + |L||g(x) - M|$$

$$< \frac{\varepsilon}{2(1 + |M|)}(1 + |M|) + |L|\frac{\varepsilon}{2(1 + |L|)}$$

$$< \frac{\varepsilon}{2} + \frac{\varepsilon}{2} = \varepsilon$$

This shows that $\lim_{x \to a}[f(x)g(x)] = LM$. ▪

**PROOF OF LAW 3** If we take $g(x) = c$ in Law 4, we get

$$\lim_{x \to a}[cf(x)] = \lim_{x \to a}[g(x)f(x)] = \lim_{x \to a} g(x) \cdot \lim_{x \to a} f(x)$$

$$= \lim_{x \to a} c \cdot \lim_{x \to a} f(x)$$

$$= c \lim_{x \to a} f(x) \qquad \text{(by Law 7)} \qquad ▪$$

**PROOF OF LAW 2** Using Law 1 and Law 3 with $c = -1$, we have

$$\lim_{x \to a}[f(x) - g(x)] = \lim_{x \to a}[f(x) + (-1)g(x)] = \lim_{x \to a} f(x) + \lim_{x \to a}(-1)g(x)$$

$$= \lim_{x \to a} f(x) + (-1)\lim_{x \to a} g(x) = \lim_{x \to a} f(x) - \lim_{x \to a} g(x) \quad ▪$$

**PROOF OF LAW 5** First let us show that

$$\lim_{x \to a}\frac{1}{g(x)} = \frac{1}{M}$$

To do this we must show that, given $\varepsilon > 0$, there exists $\delta > 0$ such that

$$\text{if} \quad 0 < |x - a| < \delta \quad \text{then} \quad \left|\frac{1}{g(x)} - \frac{1}{M}\right| < \varepsilon$$

Observe that $\qquad \left|\frac{1}{g(x)} - \frac{1}{M}\right| = \frac{|M - g(x)|}{|Mg(x)|}$

We know that we can make the numerator small. But we also need to know that the denominator is not small when $x$ is near $a$. Since $\lim_{x \to a} g(x) = M$, there is a number $\delta_1 > 0$ such that, whenever $0 < |x - a| < \delta_1$, we have

$$|g(x) - M| < \frac{|M|}{2}$$

and therefore $\qquad |M| = |M - g(x) + g(x)| \leq |M - g(x)| + |g(x)|$

$$< \frac{|M|}{2} + |g(x)|$$

This shows that

$$\text{if} \quad 0 < |x - a| < \delta_1 \quad \text{then} \quad |g(x)| > \frac{|M|}{2}$$

and so, for these values of $x$,

$$\frac{1}{|Mg(x)|} = \frac{1}{|M||g(x)|} < \frac{1}{|M|} \cdot \frac{2}{|M|} = \frac{2}{M^2}$$

Also, there exists $\delta_2 > 0$ such that

$$\text{if} \quad 0 < |x - a| < \delta_2 \quad \text{then} \quad |g(x) - M| < \frac{M^2}{2} \varepsilon$$

Let $\delta = \min\{\delta_1, \delta_2\}$. Then, for $0 < |x - a| < \delta$, we have

$$\left| \frac{1}{g(x)} - \frac{1}{M} \right| = \frac{|M - g(x)|}{|Mg(x)|} < \frac{2}{M^2} \frac{M^2}{2} \varepsilon = \varepsilon$$

It follows that $\lim_{x \to a} 1/g(x) = 1/M$. Finally, using Law 4, we obtain

$$\lim_{x \to a} \frac{f(x)}{g(x)} = \lim_{x \to a} f(x) \left( \frac{1}{g(x)} \right) = \lim_{x \to a} f(x) \lim_{x \to a} \frac{1}{g(x)} = L \cdot \frac{1}{M} = \frac{L}{M} \qquad \blacksquare$$

---

**2**   **Theorem**   If $f(x) \leqslant g(x)$ for all $x$ in an open interval that contains $a$ (except possibly at $a$) and

$$\lim_{x \to a} f(x) = L \quad \text{and} \quad \lim_{x \to a} g(x) = M$$

then $L \leqslant M$.

---

PROOF   We use the method of proof by contradiction. Suppose, if possible, that $L > M$. Law 2 of limits says that

$$\lim_{x \to a} [g(x) - f(x)] = M - L$$

Therefore, for any $\varepsilon > 0$, there exists $\delta > 0$ such that

$$\text{if} \quad 0 < |x - a| < \delta \quad \text{then} \quad |[g(x) - f(x)] - (M - L)| < \varepsilon$$

In particular, taking $\varepsilon = L - M$ (noting that $L - M > 0$ by hypothesis), we have a number $\delta > 0$ such that

$$\text{if} \quad 0 < |x - a| < \delta \quad \text{then} \quad |[g(x) - f(x)] - (M - L)| < L - M$$

Since $a \leqslant |a|$ for any number $a$, we have

$$\text{if} \quad 0 < |x - a| < \delta \quad \text{then} \quad [g(x) - f(x)] - (M - L) < L - M$$

which simplifies to

$$\text{if} \quad 0 < |x - a| < \delta \quad \text{then} \quad g(x) < f(x)$$

But this contradicts $f(x) \leqslant g(x)$. Thus the inequality $L > M$ must be false. Therefore $L \leqslant M$.   $\blacksquare$

---

**3**  **The Squeeze Theorem**  If $f(x) \leqslant g(x) \leqslant h(x)$ for all $x$ in an open interval that contains $a$ (except possibly at $a$) and

$$\lim_{x \to a} f(x) = \lim_{x \to a} h(x) = L$$

then

$$\lim_{x \to a} g(x) = L$$

---

PROOF  Let $\varepsilon > 0$ be given. Since $\lim_{x \to a} f(x) = L$, there is a number $\delta_1 > 0$ such that

$$\text{if} \qquad 0 < |x - a| < \delta_1 \qquad \text{then} \qquad |f(x) - L| < \varepsilon$$

that is,

$$\text{if} \qquad 0 < |x - a| < \delta_1 \qquad \text{then} \qquad L - \varepsilon < f(x) < L + \varepsilon$$

Since $\lim_{x \to a} h(x) = L$, there is a number $\delta_2 > 0$ such that

$$\text{if} \qquad 0 < |x - a| < \delta_2 \qquad \text{then} \qquad |h(x) - L| < \varepsilon$$

that is,

$$\text{if} \qquad 0 < |x - a| < \delta_2 \qquad \text{then} \qquad L - \varepsilon < h(x) < L + \varepsilon$$

Let $\delta = \min\{\delta_1, \delta_2\}$. If $0 < |x - a| < \delta$, then $0 < |x - a| < \delta_1$ and $0 < |x - a| < \delta_2$, so

$$L - \varepsilon < f(x) \leqslant g(x) \leqslant h(x) < L + \varepsilon$$

In particular,   $\qquad\qquad\qquad L - \varepsilon < g(x) < L + \varepsilon$

and so $|g(x) - L| < \varepsilon$. Therefore $\lim_{x \to a} g(x) = L$.   ▬

Section 1.8

---

**8**  **Theorem**  If $f$ is continuous at $b$ and $\lim_{x \to a} g(x) = b$, then

$$\lim_{x \to a} f(g(x)) = f(b)$$

---

PROOF  Let $\varepsilon > 0$ be given. We want to find a number $\delta > 0$ such that

$$\text{if} \qquad 0 < |x - a| < \delta \qquad \text{then} \qquad |f(g(x)) - f(b)| < \varepsilon$$

Since $f$ is continuous at $b$, we have

$$\lim_{y \to b} f(y) = f(b)$$

and so there exists $\delta_1 > 0$ such that

$$\text{if} \qquad 0 < |y - b| < \delta_1 \qquad \text{then} \qquad |f(y) - f(b)| < \varepsilon$$

Since $\lim_{x \to a} g(x) = b$, there exists $\delta > 0$ such that

$$\text{if} \qquad 0 < |x - a| < \delta \qquad \text{then} \qquad |g(x) - b| < \delta_1$$

Combining these two statements, we see that whenever $0 < |x - a| < \delta$ we have $|g(x) - b| < \delta_1$, which implies that $|f(g(x)) - f(b)| < \varepsilon$. Therefore we have proved that $\lim_{x \to a} f(g(x)) = f(b)$.

Section 2.4

The proof of the following result was promised when we proved that $\lim_{\theta \to 0} \dfrac{\sin \theta}{\theta} = 1$.

---

**Theorem**   If $0 < \theta < \pi/2$, then $\theta \leq \tan \theta$.

---

PROOF   Figure 1 shows a sector of a circle with center $O$, central angle $\theta$, and radius 1. Then

$$|AD| = |OA| \tan \theta = \tan \theta$$

We approximate the arc $AB$ by an inscribed polygon consisting of $n$ equal line segments and we look at a typical segment $PQ$. We extend the lines $OP$ and $OQ$ to meet $AD$ in the points $R$ and $S$. Then we draw $RT \parallel PQ$ as in Figure 2. Observe that

$$\angle RTO = \angle PQO < 90°$$

and so $\angle RTS > 90°$. Therefore we have

$$|PQ| < |RT| < |RS|$$

If we add $n$ such inequalities, we get

$$L_n < |AD| = \tan \theta$$

where $L_n$ is the length of the inscribed polygon. Thus, by Theorem 1.6.2, we have

$$\lim_{n \to \infty} L_n \leq \tan \theta$$

But the arc length is defined in Equation 8.1.1 as the limit of the lengths of inscribed polygons, so

$$\theta = \lim_{n \to \infty} L_n \leq \tan \theta$$

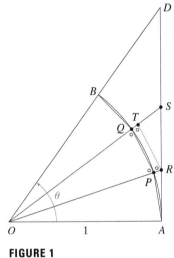

**FIGURE 1**

---

Section 3.3

**Concavity Test**

(a) If $f''(x) > 0$ for all $x$ in $I$, then the graph of $f$ is concave upward on $I$.

(b) If $f''(x) < 0$ for all $x$ in $I$, then the graph of $f$ is concave downward on $I$.

---

PROOF OF (a)   Let $a$ be any number in $I$. We must show that the curve $y = f(x)$ lies above the tangent line at the point $(a, f(a))$. The equation of this tangent is

$$y = f(a) + f'(a)(x - a)$$

So we must show that

$$f(x) > f(a) + f'(a)(x - a)$$

whenever $x \in I$ ($x \neq a$). (See Figure 2.)

**FIGURE 2**

First let us take the case where $x > a$. Applying the Mean Value Theorem to $f$ on the interval $[a, x]$, we get a number $c$, with $a < c < x$, such that

$$\boxed{1} \qquad\qquad f(x) - f(a) = f'(c)(x - a)$$

Since $f'' > 0$ on $I$, we know from the Increasing/Decreasing Test that $f'$ is increasing on $I$. Thus, since $a < c$, we have

$$f'(a) < f'(c)$$

and so, multiplying this inequality by the positive number $x - a$, we get

$$\boxed{2} \qquad\qquad f'(a)(x - a) < f'(c)(x - a)$$

Now we add $f(a)$ to both sides of this inequality:

$$f(a) + f'(a)(x - a) < f(a) + f'(c)(x - a)$$

But from Equation 1 we have $f(x) = f(a) + f'(c)(x - a)$. So this inequality becomes

$$\boxed{3} \qquad\qquad f(x) > f(a) + f'(a)(x - a)$$

which is what we wanted to prove.

For the case where $x < a$ we have $f'(c) < f'(a)$, but multiplication by the negative number $x - a$ reverses the inequality, so we get $\boxed{2}$ and $\boxed{3}$ as before.   ▬

Section 6.1

---

**Theorem**  If $f$ is a one-to-one continuous function defined on an interval $(a, b)$, then its inverse function $f^{-1}$ is also continuous.

---

PROOF  First we show that if $f$ is both one-to-one and continuous on $(a, b)$, then it must be either increasing or decreasing on $(a, b)$. If it were neither increasing nor decreasing, then there would exist numbers $x_1$, $x_2$, and $x_3$ in $(a, b)$ with $x_1 < x_2 < x_3$ such that $f(x_2)$ does not lie between $f(x_1)$ and $f(x_3)$. There are two possibilities: either (1) $f(x_3)$ lies between $f(x_1)$ and $f(x_2)$ or (2) $f(x_1)$ lies between $f(x_2)$ and $f(x_3)$. (Draw a picture.) In case (1) we apply the Intermediate Value Theorem to the continuous function $f$ to get a number $c$ between $x_1$ and $x_2$ such that $f(c) = f(x_3)$. In case (2) the Intermediate Value Theorem gives a number $c$ between $x_2$ and $x_3$ such that $f(c) = f(x_1)$. In either case we have contradicted the fact that $f$ is one-to-one.

Let us assume, for the sake of definiteness, that $f$ is increasing on $(a, b)$. We take any number $y_0$ in the domain of $f^{-1}$ and we let $f^{-1}(y_0) = x_0$; that is, $x_0$ is the number in $(a, b)$ such that $f(x_0) = y_0$. To show that $f^{-1}$ is continuous at $y_0$ we take any $\varepsilon > 0$ such that the interval $(x_0 - \varepsilon, x_0 + \varepsilon)$ is contained in the interval $(a, b)$. Since $f$ is increasing, it maps the numbers in the interval $(x_0 - \varepsilon, x_0 + \varepsilon)$ onto the numbers in the interval $(f(x_0 - \varepsilon), f(x_0 + \varepsilon))$ and $f^{-1}$ reverses the correspondence. If we let $\delta$ denote the smaller of the numbers $\delta_1 = y_0 - f(x_0 - \varepsilon)$ and $\delta_2 = f(x_0 + \varepsilon) - y_0$, then the interval $(y_0 - \delta, y_0 + \delta)$ is contained in the interval $(f(x_0 - \varepsilon), f(x_0 + \varepsilon))$ and so is mapped into the interval $(x_0 - \varepsilon, x_0 + \varepsilon)$ by $f^{-1}$. (See the arrow diagram in Figure 3.) We have therefore found a number $\delta > 0$ such that

$$\text{if} \qquad |y - y_0| < \delta \qquad \text{then} \qquad |f^{-1}(y) - f^{-1}(y_0)| < \varepsilon$$

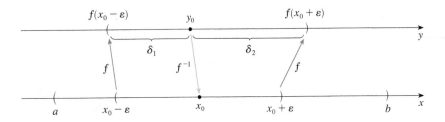

**FIGURE 3**

This shows that $\lim_{y \to y_0} f^{-1}(y) = f^{-1}(y_0)$ and so $f^{-1}$ is continuous at any number $y_0$ in its domain.

Section 11.8

In order to prove Theorem 11.8.3, we first need the following results.

---

**Theorem**

**1.** If a power series $\sum c_n x^n$ converges when $x = b$ (where $b \neq 0$), then it converges whenever $|x| < |b|$.

**2.** If a power series $\sum c_n x^n$ diverges when $x = d$ (where $d \neq 0$), then it diverges whenever $|x| > |d|$.

---

PROOF OF 1  Suppose that $\sum c_n b^n$ converges. Then, by Theorem 11.2.6, we have $\lim_{n \to \infty} c_n b^n = 0$. According to Definition 11.1.2 with $\varepsilon = 1$, there is a positive integer $N$ such that $|c_n b^n| < 1$ whenever $n \geq N$. Thus, for $n \geq N$, we have

$$|c_n x^n| = \left| \frac{c_n b^n x^n}{b^n} \right| = |c_n b^n| \left| \frac{x}{b} \right|^n < \left| \frac{x}{b} \right|^n$$

If $|x| < |b|$, then $|x/b| < 1$, so $\sum |x/b|^n$ is a convergent geometric series. Therefore, by the Comparison Test, the series $\sum_{n=N}^{\infty} |c_n x^n|$ is convergent. Thus the series $\sum c_n x^n$ is absolutely convergent and therefore convergent.

PROOF OF 2  Suppose that $\sum c_n d^n$ diverges. If $x$ is any number such that $|x| > |d|$, then $\sum c_n x^n$ cannot converge because, by part 1, the convergence of $\sum c_n x^n$ would imply the convergence of $\sum c_n d^n$. Therefore $\sum c_n x^n$ diverges whenever $|x| > |d|$.

---

**Theorem**  For a power series $\sum c_n x^n$ there are only three possibilities:

**1.** The series converges only when $x = 0$.

**2.** The series converges for all $x$.

**3.** There is a positive number $R$ such that the series converges if $|x| < R$ and diverges if $|x| > R$.

---

PROOF  Suppose that neither case 1 nor case 2 is true. Then there are nonzero numbers $b$ and $d$ such that $\sum c_n x^n$ converges for $x = b$ and diverges for $x = d$. Therefore the set $S = \{x \mid \sum c_n x^n$ converges$\}$ is not empty. By the preceding theorem, the series diverges if $|x| > |d|$, so $|x| \leq |d|$ for all $x \in S$. This says that $|d|$ is an upper bound for the set $S$. Thus, by the Completeness Axiom (see Section 11.1), $S$ has a least upper bound $R$. If $|x| > R$, then $x \notin S$, so $\sum c_n x^n$ diverges. If $|x| < R$, then $|x|$ is not an upper bound for $S$ and so there exists $b \in S$ such that $b > |x|$. Since $b \in S$, $\sum c_n b^n$ converges, so by the preceding theorem $\sum c_n x^n$ converges.

> **3** | **Theorem**   For a power series $\sum c_n(x - a)^n$ there are only three possibilities:
>
> **1.** The series converges only when $x = a$.
>
> **2.** The series converges for all $x$.
>
> **3.** There is a positive number $R$ such that the series converges if $|x - a| < R$ and diverges if $|x - a| > R$.

PROOF   If we make the change of variable $u = x - a$, then the power series becomes $\sum c_n u^n$ and we can apply the preceding theorem to this series. In case 3 we have convergence for $|u| < R$ and divergence for $|u| > R$. Thus we have convergence for $|x - a| < R$ and divergence for $|x - a| > R$. ∎

Section 14.3

> **Clairaut's Theorem**   Suppose $f$ is defined on a disk $D$ that contains the point $(a, b)$. If the functions $f_{xy}$ and $f_{yx}$ are both continuous on $D$, then $f_{xy}(a, b) = f_{yx}(a, b)$.

PROOF   For small values of $h$, $h \neq 0$, consider the difference

$$\Delta(h) = [f(a + h, b + h) - f(a + h, b)] - [f(a, b + h) - f(a, b)]$$

Notice that if we let $g(x) = f(x, b + h) - f(x, b)$, then

$$\Delta(h) = g(a + h) - g(a)$$

By the Mean Value Theorem, there is a number $c$ between $a$ and $a + h$ such that

$$g(a + h) - g(a) = g'(c)h = h[f_x(c, b + h) - f_x(c, b)]$$

Applying the Mean Value Theorem again, this time to $f_x$, we get a number $d$ between $b$ and $b + h$ such that

$$f_x(c, b + h) - f_x(c, b) = f_{xy}(c, d)h$$

Combining these equations, we obtain

$$\Delta(h) = h^2 f_{xy}(c, d)$$

If $h \rightarrow 0$, then $(c, d) \rightarrow (a, b)$, so the continuity of $f_{xy}$ at $(a, b)$ gives

$$\lim_{h \to 0} \frac{\Delta(h)}{h^2} = \lim_{(c,\, d) \to (a,\, b)} f_{xy}(c, d) = f_{xy}(a, b)$$

Similarly, by writing

$$\Delta(h) = [f(a + h, b + h) - f(a, b + h)] - [f(a + h, b) - f(a, b)]$$

and using the Mean Value Theorem twice and the continuity of $f_{yx}$ at $(a, b)$, we obtain

$$\lim_{h \to 0} \frac{\Delta(h)}{h^2} = f_{yx}(a, b)$$

It follows that $f_{xy}(a, b) = f_{yx}(a, b)$. ∎

> **8**  **Theorem**  If the partial derivatives $f_x$ and $f_y$ exist near $(a, b)$ and are continuous at $(a, b)$, then $f$ is differentiable at $(a, b)$.

PROOF  Let

$$\Delta z = f(a + \Delta x, b + \Delta y) - f(a, b)$$

According to (14.4.7), to prove that $f$ is differentiable at $(a, b)$ we have to show that we can write $\Delta z$ in the form

$$\Delta z = f_x(a, b)\, \Delta x + f_y(a, b)\, \Delta y + \varepsilon_1\, \Delta x + \varepsilon_2\, \Delta y$$

where $\varepsilon_1$ and $\varepsilon_2 \to 0$ as $(\Delta x, \Delta y) \to (0, 0)$.

Referring to Figure 4, we write

$$\boxed{1} \quad \Delta z = [f(a + \Delta x, b + \Delta y) - f(a, b + \Delta y)] + [f(a, b + \Delta y) - f(a, b)]$$

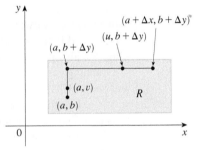

**FIGURE 4**

Observe that the function of a single variable

$$g(x) = f(x, b + \Delta y)$$

is defined on the interval $[a, a + \Delta x]$ and $g'(x) = f_x(x, b + \Delta y)$. If we apply the Mean Value Theorem to $g$, we get

$$g(a + \Delta x) - g(a) = g'(u)\, \Delta x$$

where $u$ is some number between $a$ and $a + \Delta x$. In terms of $f$, this equation becomes

$$f(a + \Delta x, b + \Delta y) - f(a, b + \Delta y) = f_x(u, b + \Delta y)\, \Delta x$$

This gives us an expression for the first part of the right side of Equation 1. For the second part we let $h(y) = f(a, y)$. Then $h$ is a function of a single variable defined on the interval $[b, b + \Delta y]$ and $h'(y) = f_y(a, y)$. A second application of the Mean Value Theorem then gives

$$h(b + \Delta y) - h(b) = h'(v)\, \Delta y$$

where $v$ is some number between $b$ and $b + \Delta y$. In terms of $f$, this becomes

$$f(a, b + \Delta y) - f(a, b) = f_y(a, v)\, \Delta y$$

We now substitute these expressions into Equation 1 and obtain

$$\Delta z = f_x(u, b + \Delta y)\, \Delta x + f_y(a, v)\, \Delta y$$

$$= f_x(a, b)\, \Delta x + [f_x(u, b + \Delta y) - f_x(a, b)]\, \Delta x + f_y(a, b)\, \Delta y$$

$$+ [f_y(a, v) - f_y(a, b)]\, \Delta y$$

$$= f_x(a, b)\, \Delta x + f_y(a, b)\, \Delta y + \varepsilon_1\, \Delta x + \varepsilon_2\, \Delta y$$

where

$$\varepsilon_1 = f_x(u, b + \Delta y) - f_x(a, b)$$

$$\varepsilon_2 = f_y(a, v) - f_y(a, b)$$

Since $(u, b + \Delta y) \to (a, b)$ and $(a, v) \to (a, b)$ as $(\Delta x, \Delta y) \to (0, 0)$ and since $f_x$ and $f_y$ are continuous at $(a, b)$, we see that $\varepsilon_1 \to 0$ and $\varepsilon_2 \to 0$ as $(\Delta x, \Delta y) \to (0, 0)$.

Therefore $f$ is differentiable at $(a, b)$. ∎

# G   Graphing Calculators and Computers

In this appendix we assume that you have access to a graphing calculator or a computer with graphing software. We see that the use of such a device enables us to graph more complicated functions and to solve more complex problems than would otherwise be possible. We also point out some of the pitfalls that can occur with these machines.

Graphing calculators and computers can give very accurate graphs of functions. But we will see in Chapter 3 that only through the use of calculus can we be sure that we have uncovered all the interesting aspects of a graph.

A graphing calculator or computer displays a rectangular portion of the graph of a function in a **display window** or **viewing screen**, which we refer to as a **viewing rectangle**. The default screen often gives an incomplete or misleading picture, so it is important to choose the viewing rectangle with care. If we choose the $x$-values to range from a minimum value of $Xmin = a$ to a maximum value of $Xmax = b$ and the $y$-values to range from a minimum of $Ymin = c$ to a maximum of $Ymax = d$, then the visible portion of the graph lies in the rectangle

$$[a, b] \times [c, d] = \{(x, y) \mid a \leqslant x \leqslant b, c \leqslant y \leqslant d\}$$

shown in Figure 1. We refer to this rectangle as the $[a, b]$ *by* $[c, d]$ *viewing rectangle*.

**FIGURE 1**
The viewing rectangle $[a, b]$ by $[c, d]$

The machine draws the graph of a function $f$ much as you would. It plots points of the form $(x, f(x))$ for a certain number of equally spaced values of $x$ between $a$ and $b$. If an $x$-value is not in the domain of $f$, or if $f(x)$ lies outside the viewing rectangle, it moves on to the next $x$-value. The machine connects each point to the preceding plotted point to form a representation of the graph of $f$.

**EXAMPLE 1** Draw the graph of the function $f(x) = x^2 + 3$ in each of the following viewing rectangles.

(a) $[-2, 2]$ by $[-2, 2]$       (b) $[-4, 4]$ by $[-4, 4]$

(c) $[-10, 10]$ by $[-5, 30]$       (d) $[-50, 50]$ by $[-100, 1000]$

SOLUTION For part (a) we select the range by setting $Xmin = -2$, $Xmax = 2$, $Ymin = -2$, and $Ymax = 2$. The resulting graph is shown in Figure 2(a). The display window is blank! A moment's thought provides the explanation: Notice that $x^2 \geqslant 0$ for all $x$, so $x^2 + 3 \geqslant 3$ for all $x$. Thus the range of the function $f(x) = x^2 + 3$ is $[3, \infty)$. This means that the graph of $f$ lies entirely outside the viewing rectangle $[-2, 2]$ by $[-2, 2]$.

The graphs for the viewing rectangles in parts (b), (c), and (d) are also shown in Figure 2. Observe that we get a more complete picture in parts (c) and (d), but in part (d) it is not clear that the $y$-intercept is 3.

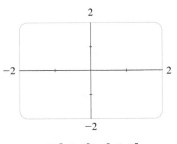

(a) $[-2, 2]$ by $[-2, 2]$

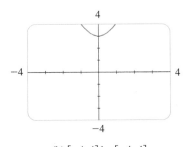

(b) $[-4, 4]$ by $[-4, 4]$

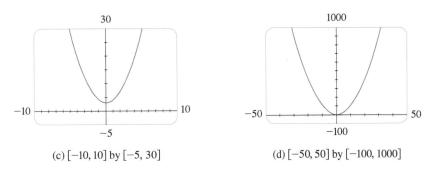

(c) $[-10, 10]$ by $[-5, 30]$           (d) $[-50, 50]$ by $[-100, 1000]$

**FIGURE 2** Graphs of $f(x) = x^2 + 3$

We see from Example 1 that the choice of a viewing rectangle can make a big difference in the appearance of a graph. Often it's necessary to change to a larger viewing rectangle to obtain a more complete picture, a more global view, of the graph. In the next example we see that knowledge of the domain and range of a function sometimes provides us with enough information to select a good viewing rectangle.

**EXAMPLE 2** Determine an appropriate viewing rectangle for the function $f(x) = \sqrt{8 - 2x^2}$ and use it to graph $f$.

SOLUTION The expression for $f(x)$ is defined when

$$8 - 2x^2 \geqslant 0 \iff 2x^2 \leqslant 8 \iff x^2 \leqslant 4$$

$$\iff |x| \leqslant 2 \iff -2 \leqslant x \leqslant 2$$

Therefore the domain of $f$ is the interval $[-2, 2]$. Also,

$$0 \leqslant \sqrt{8 - 2x^2} \leqslant \sqrt{8} = 2\sqrt{2} \approx 2.83$$

so the range of $f$ is the interval $[0, 2\sqrt{2}]$.

We choose the viewing rectangle so that the $x$-interval is somewhat larger than the domain and the $y$-interval is larger than the range. Taking the viewing rectangle to be $[-3, 3]$ by $[-1, 4]$, we get the graph shown in Figure 3.

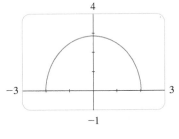

**FIGURE 3**
$f(x) = \sqrt{8 - 2x^2}$

**EXAMPLE 3** Graph the function $y = x^3 - 150x$.

SOLUTION Here the domain is $\mathbb{R}$, the set of all real numbers. That doesn't help us choose a viewing rectangle. Let's experiment. If we start with the viewing rectangle $[-5, 5]$ by

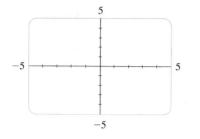

**FIGURE 4**

$[-5, 5]$, we get the graph in Figure 4. It appears blank, but actually the graph is so nearly vertical that it blends in with the $y$-axis.

If we change the viewing rectangle to $[-20, 20]$ by $[-20, 20]$, we get the picture shown in Figure 5(a). The graph appears to consist of vertical lines, but we know that can't be correct. If we look carefully while the graph is being drawn, we see that the graph leaves the screen and reappears during the graphing process. This indicates that we need to see more in the vertical direction, so we change the viewing rectangle to $[-20, 20]$ by $[-500, 500]$. The resulting graph is shown in Figure 5(b). It still doesn't quite reveal all the main features of the function, so we try $[-20, 20]$ by $[-1000, 1000]$ in Figure 5(c). Now we are more confident that we have arrived at an appropriate viewing rectangle. In Chapter 4 we will be able to see that the graph shown in Figure 5(c) does indeed reveal all the main features of the function.

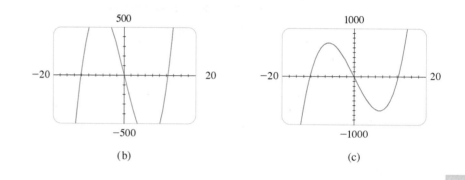

| (a) | (b) | (c) |

**FIGURE 5** Graphs of $y = x^3 - 150x$

**V** **EXAMPLE 4** Graph the function $f(x) = \sin 50x$ in an appropriate viewing rectangle.

SOLUTION Figure 6(a) shows the graph of $f$ produced by a graphing calculator using the viewing rectangle $[-12, 12]$ by $[-1.5, 1.5]$. At first glance the graph appears to be reasonable. But if we change the viewing rectangle to the ones shown in the following parts of Figure 6, the graphs look very different. Something strange is happening.

The appearance of the graphs in Figure 6 depends on the machine used. The graphs you get with your own graphing device might not look like these figures, but they will also be quite inaccurate.

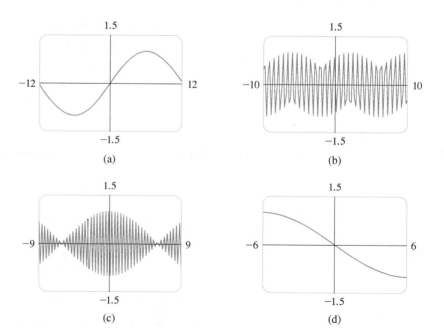

| (a) | (b) |
| (c) | (d) |

**FIGURE 6**
Graphs of $f(x) = \sin 50x$
in four viewing rectangles

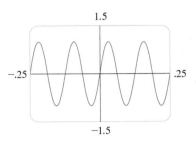

**FIGURE 7**

$f(x) = \sin 50x$

In order to explain the big differences in appearance of these graphs and to find an appropriate viewing rectangle, we need to find the period of the function $y = \sin 50x$. We know that the function $y = \sin x$ has period $2\pi$ and the graph of $y = \sin 50x$ is shrunk horizontally by a factor of 50, so the period of $y = \sin 50x$ is

$$\frac{2\pi}{50} = \frac{\pi}{25} \approx 0.126$$

This suggests that we should deal only with small values of $x$ in order to show just a few oscillations of the graph. If we choose the viewing rectangle $[-0.25, 0.25]$ by $[-1.5, 1.5]$, we get the graph shown in Figure 7.

Now we see what went wrong in Figure 6. The oscillations of $y = \sin 50x$ are so rapid that when the calculator plots points and joins them, it misses most of the maximum and minimum points and therefore gives a very misleading impression of the graph.

We have seen that the use of an inappropriate viewing rectangle can give a misleading impression of the graph of a function. In Examples 1 and 3 we solved the problem by changing to a larger viewing rectangle. In Example 4 we had to make the viewing rectangle smaller. In the next example we look at a function for which there is no single viewing rectangle that reveals the true shape of the graph.

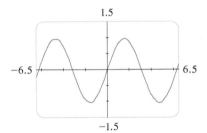

**FIGURE 8**

**V EXAMPLE 5** Graph the function $f(x) = \sin x + \frac{1}{100}\cos 100x$.

SOLUTION Figure 8 shows the graph of $f$ produced by a graphing calculator with viewing rectangle $[-6.5, 6.5]$ by $[-1.5, 1.5]$. It looks much like the graph of $y = \sin x$, but perhaps with some bumps attached. If we zoom in to the viewing rectangle $[-0.1, 0.1]$ by $[-0.1, 0.1]$, we can see much more clearly the shape of these bumps in Figure 9. The reason for this behavior is that the second term, $\frac{1}{100}\cos 100x$, is very small in comparison with the first term, $\sin x$. Thus we really need two graphs to see the true nature of this function.

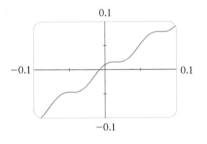

**FIGURE 9**

**EXAMPLE 6** Draw the graph of the function $y = \dfrac{1}{1 - x}$.

SOLUTION Figure 10(a) shows the graph produced by a graphing calculator with viewing rectangle $[-9, 9]$ by $[-9, 9]$. In connecting successive points on the graph, the calculator produced a steep line segment from the top to the bottom of the screen. That line segment is not truly part of the graph. Notice that the domain of the function $y = 1/(1 - x)$ is $\{x \mid x \neq 1\}$. We can eliminate the extraneous near-vertical line by experimenting with a change of scale. When we change to the smaller viewing rectangle $[-4.7, 4.7]$ by $[-4.7, 4.7]$ on this particular calculator, we obtain the much better graph in Figure 10(b).

Another way to avoid the extraneous line is to change the graphing mode on the calculator so that the dots are not connected.

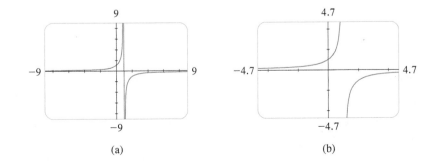

**FIGURE 10**

(a)

(b)

**EXAMPLE 7** Graph the function $y = \sqrt[3]{x}$.

**SOLUTION** Some graphing devices display the graph shown in Figure 11, whereas others produce a graph like that in Figure 12. We know from Section 1.2 (Figure 13) that the graph in Figure 12 is correct, so what happened in Figure 11? The explanation is that some machines compute the cube root of $x$ using a logarithm, which is not defined if $x$ is negative, so only the right half of the graph is produced.

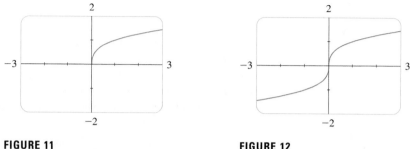

**FIGURE 11**                                              **FIGURE 12**

You can get the correct graph with Maple if you first type

    with(RealDomain);

You should experiment with your own machine to see which of these two graphs is produced. If you get the graph in Figure 11, you can obtain the correct picture by graphing the function

$$f(x) = \frac{x}{|x|} \cdot |x|^{1/3}$$

Notice that this function is equal to $\sqrt[3]{x}$ (except when $x = 0$).

To understand how the expression for a function relates to its graph, it's helpful to graph a **family of functions**, that is, a collection of functions whose equations are related. In the next example we graph members of a family of cubic polynomials.

**V EXAMPLE 8** Graph the function $y = x^3 + cx$ for various values of the number $c$. How does the graph change when $c$ is changed?

**SOLUTION** Figure 13 shows the graphs of $y = x^3 + cx$ for $c = 2, 1, 0, -1$, and $-2$. We see that, for positive values of $c$, the graph increases from left to right with no maximum or minimum points (peaks or valleys). When $c = 0$, the curve is flat at the origin. When $c$ is negative, the curve has a maximum point and a minimum point. As $c$ decreases, the maximum point becomes higher and the minimum point lower.

**TEC** In Visual G you can see an animation of Figure 13.

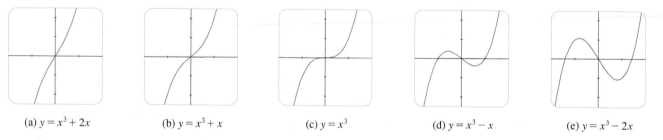

(a) $y = x^3 + 2x$    (b) $y = x^3 + x$    (c) $y = x^3$    (d) $y = x^3 - x$    (e) $y = x^3 - 2x$

**FIGURE 13**

Several members of the family of functions $y = x^3 + cx$, all graphed in the viewing rectangle $[-2, 2]$ by $[-2.5, 2.5]$

**EXAMPLE 9** Find the solution of the equation $\cos x = x$ correct to two decimal places.

**SOLUTION** The solutions of the equation $\cos x = x$ are the $x$-coordinates of the points of intersection of the curves $y = \cos x$ and $y = x$. From Figure 14(a) we see that there is only one solution and it lies between 0 and 1. Zooming in to the viewing rectangle $[0, 1]$ by $[0, 1]$, we see from Figure 14(b) that the root lies between 0.7 and 0.8. So we zoom in further to the viewing rectangle $[0.7, 0.8]$ by $[0.7, 0.8]$ in Figure 14(c). By moving the cursor to the intersection point of the two curves, or by inspection and the fact that the $x$-scale is 0.01, we see that the solution of the equation is about 0.74. (Many calculators have a built-in intersection feature.)

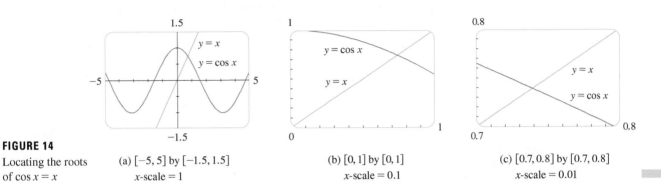

**FIGURE 14**

Locating the roots
of $\cos x = x$

(a) $[-5, 5]$ by $[-1.5, 1.5]$
$x$-scale $= 1$

(b) $[0, 1]$ by $[0, 1]$
$x$-scale $= 0.1$

(c) $[0.7, 0.8]$ by $[0.7, 0.8]$
$x$-scale $= 0.01$

## G  Exercises

**1.** Use a graphing calculator or computer to determine which of the given viewing rectangles produces the most appropriate graph of the function $f(x) = \sqrt{x^3 - 5x^2}$.
   (a) $[-5, 5]$ by $[-5, 5]$       (b) $[0, 10]$ by $[0, 2]$
   (c) $[0, 10]$ by $[0, 10]$

**2.** Use a graphing calculator or computer to determine which of the given viewing rectangles produces the most appropriate graph of the function $f(x) = x^4 - 16x^2 + 20$.
   (a) $[-3, 3]$ by $[-3, 3]$       (b) $[-10, 10]$ by $[-10, 10]$
   (c) $[-50, 50]$ by $[-50, 50]$   (d) $[-5, 5]$ by $[-50, 50]$

**3–14** Determine an appropriate viewing rectangle for the given function and use it to draw the graph.

**3.** $f(x) = x^2 - 36x + 32$

**4.** $f(x) = x^3 + 15x^2 + 65x$

**5.** $f(x) = \sqrt{50 - 0.2x}$

**6.** $f(x) = \sqrt{15x - x^2}$

**7.** $f(x) = x^3 - 225x$

**8.** $f(x) = \dfrac{x}{x^2 + 100}$

**9.** $f(x) = \sin^2(1000x)$

**10.** $f(x) = \cos(0.001x)$

**11.** $f(x) = \sin\sqrt{x}$

**12.** $f(x) = \sec(20\pi x)$

**13.** $y = 10\sin x + \sin 100x$

**14.** $y = x^2 + 0.02\sin 50x$

**15.** (a) Try to find an appropriate viewing rectangle for
   $f(x) = (x - 10)^3 2^{-x}$.
   (b) Do you need more than one window? Why?

**16.** Graph the function $f(x) = x^2\sqrt{30 - x}$ in an appropriate viewing rectangle. Why does part of the graph appear to be missing?

**17.** Graph the ellipse $4x^2 + 2y^2 = 1$ by graphing the functions whose graphs are the upper and lower halves of the ellipse.

**18.** Graph the hyperbola $y^2 - 9x^2 = 1$ by graphing the functions whose graphs are the upper and lower branches of the hyperbola.

**19–20** Do the graphs intersect in the given viewing rectangle? If they do, how many points of intersection are there?

**19.** $y = 3x^2 - 6x + 1$,  $y = 0.23x - 2.25$;
   $[-1, 3]$ by $[-2.5, 1.5]$

**20.** $y = 6 - 4x - x^2$,  $y = 3x + 18$;  $[-6, 2]$ by $[-5, 20]$

**21–23** Find all solutions of the equation correct to two decimal places.

**21.** $x^4 - x = 1$

**22.** $\sqrt{x} = x^3 - 1$

**23.** $\tan x = \sqrt{1 - x^2}$

**24.** We saw in Example 9 that the equation $\cos x = x$ has exactly one solution.
   (a) Use a graph to show that the equation $\cos x = 0.3x$ has three solutions and find their values correct to two decimal places.

---

⊞ Graphing calculator or computer required       **1.** Homework Hints available at stewartcalculus.com

(b) Find an approximate value of $m$ such that the equation $\cos x = mx$ has exactly two solutions.

25. Use graphs to determine which of the functions $f(x) = 10x^2$ and $g(x) = x^3/10$ is eventually larger (that is, larger when $x$ is very large).

26. Use graphs to determine which of the functions $f(x) = x^4 - 100x^3$ and $g(x) = x^3$ is eventually larger.

27. For what values of $x$ is it true that $|\tan x - x| < 0.01$ and $-\pi/2 < x < \pi/2$?

28. Graph the polynomials $P(x) = 3x^5 - 5x^3 + 2x$ and $Q(x) = 3x^5$ on the same screen, first using the viewing rectangle $[-2, 2]$ by $[-2, 2]$ and then changing to $[-10, 10]$ by $[-10,000, 10,000]$. What do you observe from these graphs?

29. In this exercise we consider the family of root functions $f(x) = \sqrt[n]{x}$, where $n$ is a positive integer.
    (a) Graph the functions $y = \sqrt{x}$, $y = \sqrt[4]{x}$, and $y = \sqrt[6]{x}$ on the same screen using the viewing rectangle $[-1, 4]$ by $[-1, 3]$.
    (b) Graph the functions $y = x$, $y = \sqrt[3]{x}$, and $y = \sqrt[5]{x}$ on the same screen using the viewing rectangle $[-3, 3]$ by $[-2, 2]$. (See Example 7.)
    (c) Graph the functions $y = \sqrt{x}$, $y = \sqrt[3]{x}$, $y = \sqrt[4]{x}$, and $y = \sqrt[5]{x}$ on the same screen using the viewing rectangle $[-1, 3]$ by $[-1, 2]$.
    (d) What conclusions can you make from these graphs?

30. In this exercise we consider the family of functions $f(x) = 1/x^n$, where $n$ is a positive integer.
    (a) Graph the functions $y = 1/x$ and $y = 1/x^3$ on the same screen using the viewing rectangle $[-3, 3]$ by $[-3, 3]$.
    (b) Graph the functions $y = 1/x^2$ and $y = 1/x^4$ on the same screen using the same viewing rectangle as in part (a).
    (c) Graph all of the functions in parts (a) and (b) on the same screen using the viewing rectangle $[-1, 3]$ by $[-1, 3]$.
    (d) What conclusions can you make from these graphs?

31. Graph the function $f(x) = x^4 + cx^2 + x$ for several values of $c$. How does the graph change when $c$ changes?

32. Graph the function $f(x) = \sqrt{1 + cx^2}$ for various values of $c$. Describe how changing the value of $c$ affects the graph.

33. Graph the function $y = x^n 2^{-x}$, $x \geq 0$, for $n = 1, 2, 3, 4, 5,$ and 6. How does the graph change as $n$ increases?

34. The curves with equations
$$y = \frac{|x|}{\sqrt{c - x^2}}$$
are called **bullet-nose curves**. Graph some of these curves to see why. What happens as $c$ increases?

35. What happens to the graph of the equation $y^2 = cx^3 + x^2$ as $c$ varies?

36. This exercise explores the effect of the inner function $g$ on a composite function $y = f(g(x))$.
    (a) Graph the function $y = \sin(\sqrt{x})$ using the viewing rectangle $[0, 400]$ by $[-1.5, 1.5]$. How does this graph differ from the graph of the sine function?
    (b) Graph the function $y = \sin(x^2)$ using the viewing rectangle $[-5, 5]$ by $[-1.5, 1.5]$. How does this graph differ from the graph of the sine function?

37. The figure shows the graphs of $y = \sin 96x$ and $y = \sin 2x$ as displayed by a TI-83 graphing calculator. The first graph is inaccurate. Explain why the two graphs appear identical. [*Hint:* The TI-83's graphing window is 95 pixels wide. What specific points does the calculator plot?]

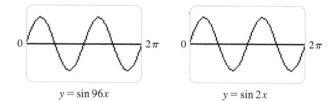

$y = \sin 96x$          $y = \sin 2x$

38. The first graph in the figure is that of $y = \sin 45x$ as displayed by a TI-83 graphing calculator. It is inaccurate and so, to help explain its appearance, we replot the curve in dot mode in the second graph. What two sine curves does the calculator appear to be plotting? Show that each point on the graph of $y = \sin 45x$ that the TI-83 chooses to plot is in fact on one of these two curves. (The TI-83's graphing window is 95 pixels wide.)

## H Complex Numbers

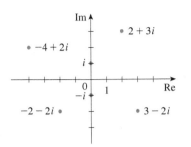

**FIGURE 1**

Complex numbers as points in the Argand plane

A **complex number** can be represented by an expression of the form $a + bi$, where $a$ and $b$ are real numbers and $i$ is a symbol with the property that $i^2 = -1$. The complex number $a + bi$ can also be represented by the ordered pair $(a, b)$ and plotted as a point in a plane (called the Argand plane) as in Figure 1. Thus the complex number $i = 0 + 1 \cdot i$ is identified with the point $(0, 1)$.

The **real part** of the complex number $a + bi$ is the real number $a$ and the **imaginary part** is the real number $b$. Thus the real part of $4 - 3i$ is 4 and the imaginary part is $-3$. Two complex numbers $a + bi$ and $c + di$ are **equal** if $a = c$ and $b = d$, that is, their real parts are equal and their imaginary parts are equal. In the Argand plane the horizontal axis is called the real axis and the vertical axis is called the imaginary axis.

The sum and difference of two complex numbers are defined by adding or subtracting their real parts and their imaginary parts:

$$(a + bi) + (c + di) = (a + c) + (b + d)i$$
$$(a + bi) - (c + di) = (a - c) + (b - d)i$$

For instance,

$$(1 - i) + (4 + 7i) = (1 + 4) + (-1 + 7)i = 5 + 6i$$

The product of complex numbers is defined so that the usual commutative and distributive laws hold:

$$(a + bi)(c + di) = a(c + di) + (bi)(c + di)$$
$$= ac + adi + bci + bdi^2$$

Since $i^2 = -1$, this becomes

$$(a + bi)(c + di) = (ac - bd) + (ad + bc)i$$

**EXAMPLE 1**

$$(-1 + 3i)(2 - 5i) = (-1)(2 - 5i) + 3i(2 - 5i)$$
$$= -2 + 5i + 6i - 15(-1) = 13 + 11i$$

Division of complex numbers is much like rationalizing the denominator of a rational expression. For the complex number $z = a + bi$, we define its **complex conjugate** to be $\bar{z} = a - bi$. To find the quotient of two complex numbers we multiply numerator and denominator by the complex conjugate of the denominator.

**EXAMPLE 2** Express the number $\dfrac{-1 + 3i}{2 + 5i}$ in the form $a + bi$.

SOLUTION We multiply numerator and denominator by the complex conjugate of $2 + 5i$, namely $2 - 5i$, and we take advantage of the result of Example 1:

$$\frac{-1 + 3i}{2 + 5i} = \frac{-1 + 3i}{2 + 5i} \cdot \frac{2 - 5i}{2 - 5i} = \frac{13 + 11i}{2^2 + 5^2} = \frac{13}{29} + \frac{11}{29}i$$

The geometric interpretation of the complex conjugate is shown in Figure 2: $\bar{z}$ is the reflection of $z$ in the real axis. We list some of the properties of the complex conjugate in the following box. The proofs follow from the definition and are requested in Exercise 18.

**FIGURE 2**

---

**Properties of Conjugates**

$$\overline{z + w} = \bar{z} + \bar{w} \qquad \overline{zw} = \bar{z}\,\bar{w} \qquad \overline{z^n} = \bar{z}^n$$

---

The **modulus**, or **absolute value**, $|z|$ of a complex number $z = a + bi$ is its distance from the origin. From Figure 3 we see that if $z = a + bi$, then

$$|z| = \sqrt{a^2 + b^2}$$

Notice that

$$z\bar{z} = (a + bi)(a - bi) = a^2 + abi - abi - b^2i^2 = a^2 + b^2$$

and so

$$z\bar{z} = |z|^2$$

This explains why the division procedure in Example 2 works in general:

$$\frac{z}{w} = \frac{z\bar{w}}{w\bar{w}} = \frac{z\bar{w}}{|w|^2}$$

Since $i^2 = -1$, we can think of $i$ as a square root of $-1$. But notice that we also have $(-i)^2 = i^2 = -1$ and so $-i$ is also a square root of $-1$. We say that $i$ is the **principal square root** of $-1$ and write $\sqrt{-1} = i$. In general, if $c$ is any positive number, we write

$$\sqrt{-c} = \sqrt{c}\; i$$

With this convention, the usual derivation and formula for the roots of the quadratic equation $ax^2 + bx + c = 0$ are valid even when $b^2 - 4ac < 0$:

$$x = \frac{-b \pm \sqrt{b^2 - 4ac}}{2a}$$

**EXAMPLE 3** Find the roots of the equation $x^2 + x + 1 = 0$.

SOLUTION Using the quadratic formula, we have

$$x = \frac{-1 \pm \sqrt{1^2 - 4 \cdot 1}}{2} = \frac{-1 \pm \sqrt{-3}}{2} = \frac{-1 \pm \sqrt{3}\; i}{2}$$

We observe that the solutions of the equation in Example 3 are complex conjugates of each other. In general, the solutions of any quadratic equation $ax^2 + bx + c = 0$ with real coefficients $a$, $b$, and $c$ are always complex conjugates. (If $z$ is real, $\bar{z} = z$, so $z$ is its own conjugate.)

We have seen that if we allow complex numbers as solutions, then every quadratic equation has a solution. More generally, it is true that every polynomial equation

$$a_n x^n + a_{n-1} x^{n-1} + \cdots + a_1 x + a_0 = 0$$

of degree at least one has a solution among the complex numbers. This fact is known as the Fundamental Theorem of Algebra and was proved by Gauss.

**FIGURE 3**

## Polar Form

We know that any complex number $z = a + bi$ can be considered as a point $(a, b)$ and that any such point can be represented by polar coordinates $(r, \theta)$ with $r \geqslant 0$. In fact,

$$a = r \cos \theta \qquad b = r \sin \theta$$

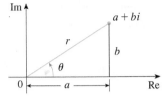

Im $\uparrow$

$a + bi$

$r$

$b$

$\theta$

$0$   $a$   Re

**FIGURE 4**

as in Figure 4. Therefore we have

$$z = a + bi = (r \cos \theta) + (r \sin \theta)i$$

Thus we can write any complex number $z$ in the form

$$z = r(\cos \theta + i \sin \theta)$$

where
$$r = |z| = \sqrt{a^2 + b^2} \qquad \text{and} \qquad \tan \theta = \frac{b}{a}$$

The angle $\theta$ is called the **argument** of $z$ and we write $\theta = \arg(z)$. Note that $\arg(z)$ is not unique; any two arguments of $z$ differ by an integer multiple of $2\pi$.

**EXAMPLE 4**   Write the following numbers in polar form.

(a) $z = 1 + i$  (b) $w = \sqrt{3} - i$

SOLUTION

(a) We have $r = |z| = \sqrt{1^2 + 1^2} = \sqrt{2}$ and $\tan \theta = 1$, so we can take $\theta = \pi/4$. Therefore the polar form is

$$z = \sqrt{2} \left( \cos \frac{\pi}{4} + i \sin \frac{\pi}{4} \right)$$

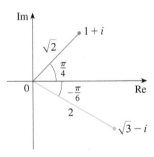

Im $\uparrow$

$1 + i$

$\sqrt{2}$

$\frac{\pi}{4}$

$-\frac{\pi}{6}$

$0$   Re

$2$

$\sqrt{3} - i$

**FIGURE 5**

(b) Here we have $r = |w| = \sqrt{3 + 1} = 2$ and $\tan \theta = -1/\sqrt{3}$. Since $w$ lies in the fourth quadrant, we take $\theta = -\pi/6$ and

$$w = 2 \left[ \cos \left( -\frac{\pi}{6} \right) + i \sin \left( -\frac{\pi}{6} \right) \right]$$

The numbers $z$ and $w$ are shown in Figure 5.

The polar form of complex numbers gives insight into multiplication and division. Let

$$z_1 = r_1(\cos \theta_1 + i \sin \theta_1) \qquad z_2 = r_2(\cos \theta_2 + i \sin \theta_2)$$

be two complex numbers written in polar form. Then

$$z_1 z_2 = r_1 r_2 (\cos \theta_1 + i \sin \theta_1)(\cos \theta_2 + i \sin \theta_2)$$

$$= r_1 r_2 [(\cos \theta_1 \cos \theta_2 - \sin \theta_1 \sin \theta_2) + i(\sin \theta_1 \cos \theta_2 + \cos \theta_1 \sin \theta_2)]$$

Therefore, using the addition formulas for cosine and sine, we have

$$\boxed{1} \qquad z_1 z_2 = r_1 r_2 [\cos(\theta_1 + \theta_2) + i \sin(\theta_1 + \theta_2)]$$

**FIGURE 6**

**FIGURE 7**

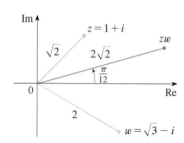

**FIGURE 8**

This formula says that *to multiply two complex numbers we multiply the moduli and add the arguments.* (See Figure 6.)

A similar argument using the subtraction formulas for sine and cosine shows that *to divide two complex numbers we divide the moduli and subtract the arguments.*

$$\frac{z_1}{z_2} = \frac{r_1}{r_2}[\cos(\theta_1 - \theta_2) + i\sin(\theta_1 - \theta_2)] \qquad z_2 \neq 0$$

In particular, taking $z_1 = 1$ and $z_2 = z$ (and therefore $\theta_1 = 0$ and $\theta_2 = \theta$), we have the following, which is illustrated in Figure 7.

$$\text{If} \quad z = r(\cos\theta + i\sin\theta), \quad \text{then} \quad \frac{1}{z} = \frac{1}{r}(\cos\theta - i\sin\theta).$$

**EXAMPLE 5**  Find the product of the complex numbers $1 + i$ and $\sqrt{3} - i$ in polar form.

SOLUTION  From Example 4 we have

$$1 + i = \sqrt{2}\left(\cos\frac{\pi}{4} + i\sin\frac{\pi}{4}\right)$$

and

$$\sqrt{3} - i = 2\left[\cos\left(-\frac{\pi}{6}\right) + i\sin\left(-\frac{\pi}{6}\right)\right]$$

So, by Equation 1,

$$(1 + i)(\sqrt{3} - i) = 2\sqrt{2}\left[\cos\left(\frac{\pi}{4} - \frac{\pi}{6}\right) + i\sin\left(\frac{\pi}{4} - \frac{\pi}{6}\right)\right]$$

$$= 2\sqrt{2}\left(\cos\frac{\pi}{12} + i\sin\frac{\pi}{12}\right)$$

This is illustrated in Figure 8.

Repeated use of Formula 1 shows how to compute powers of a complex number. If

$$z = r(\cos\theta + i\sin\theta)$$

then

$$z^2 = r^2(\cos 2\theta + i\sin 2\theta)$$

and

$$z^3 = zz^2 = r^3(\cos 3\theta + i\sin 3\theta)$$

In general, we obtain the following result, which is named after the French mathematician Abraham De Moivre (1667–1754).

> **2  De Moivre's Theorem**  If $z = r(\cos\theta + i\sin\theta)$ and $n$ is a positive integer, then
> $$z^n = [r(\cos\theta + i\sin\theta)]^n = r^n(\cos n\theta + i\sin n\theta)$$

This says that *to take the nth power of a complex number we take the nth power of the modulus and multiply the argument by n.*

EXAMPLE 6 Find $\left(\frac{1}{2} + \frac{1}{2}i\right)^{10}$.

SOLUTION Since $\frac{1}{2} + \frac{1}{2}i = \frac{1}{2}(1 + i)$, it follows from Example 4(a) that $\frac{1}{2} + \frac{1}{2}i$ has the polar form

$$\frac{1}{2} + \frac{1}{2}i = \frac{\sqrt{2}}{2}\left(\cos\frac{\pi}{4} + i\sin\frac{\pi}{4}\right)$$

So by De Moivre's Theorem,

$$\left(\frac{1}{2} + \frac{1}{2}i\right)^{10} = \left(\frac{\sqrt{2}}{2}\right)^{10}\left(\cos\frac{10\pi}{4} + i\sin\frac{10\pi}{4}\right)$$

$$= \frac{2^5}{2^{10}}\left(\cos\frac{5\pi}{2} + i\sin\frac{5\pi}{2}\right) = \frac{1}{32}i$$

De Moivre's Theorem can also be used to find the $n$th roots of complex numbers. An $n$th root of the complex number $z$ is a complex number $w$ such that

$$w^n = z$$

Writing these two numbers in trigonometric form as

$$w = s(\cos\phi + i\sin\phi) \quad \text{and} \quad z = r(\cos\theta + i\sin\theta)$$

and using De Moivre's Theorem, we get

$$s^n(\cos n\phi + i\sin n\phi) = r(\cos\theta + i\sin\theta)$$

The equality of these two complex numbers shows that

$$s^n = r \quad \text{or} \quad s = r^{1/n}$$

and

$$\cos n\phi = \cos\theta \quad \text{and} \quad \sin n\phi = \sin\theta$$

From the fact that sine and cosine have period $2\pi$ it follows that

$$n\phi = \theta + 2k\pi \quad \text{or} \quad \phi = \frac{\theta + 2k\pi}{n}$$

Thus

$$w = r^{1/n}\left[\cos\left(\frac{\theta + 2k\pi}{n}\right) + i\sin\left(\frac{\theta + 2k\pi}{n}\right)\right]$$

Since this expression gives a different value of $w$ for $k = 0, 1, 2, \ldots, n - 1$, we have the following.

---

**3 Roots of a Complex Number** Let $z = r(\cos\theta + i\sin\theta)$ and let $n$ be a positive integer. Then $z$ has the $n$ distinct $n$th roots

$$w_k = r^{1/n}\left[\cos\left(\frac{\theta + 2k\pi}{n}\right) + i\sin\left(\frac{\theta + 2k\pi}{n}\right)\right]$$

where $k = 0, 1, 2, \ldots, n - 1$.

Notice that each of the *n*th roots of *z* has modulus $|w_k| = r^{1/n}$. Thus all the *n*th roots of *z* lie on the circle of radius $r^{1/n}$ in the complex plane. Also, since the argument of each successive *n*th root exceeds the argument of the previous root by $2\pi/n$, we see that the *n*th roots of *z* are equally spaced on this circle.

**EXAMPLE 7**   Find the six sixth roots of $z = -8$ and graph these roots in the complex plane.

**SOLUTION**   In trigonometric form, $z = 8(\cos \pi + i \sin \pi)$. Applying Equation 3 with $n = 6$, we get

$$w_k = 8^{1/6}\left(\cos \frac{\pi + 2k\pi}{6} + i \sin \frac{\pi + 2k\pi}{6}\right)$$

We get the six sixth roots of $-8$ by taking $k = 0, 1, 2, 3, 4, 5$ in this formula:

$$w_0 = 8^{1/6}\left(\cos \frac{\pi}{6} + i \sin \frac{\pi}{6}\right) = \sqrt{2}\left(\frac{\sqrt{3}}{2} + \frac{1}{2}i\right)$$

$$w_1 = 8^{1/6}\left(\cos \frac{\pi}{2} + i \sin \frac{\pi}{2}\right) = \sqrt{2}\,i$$

$$w_2 = 8^{1/6}\left(\cos \frac{5\pi}{6} + i \sin \frac{5\pi}{6}\right) = \sqrt{2}\left(-\frac{\sqrt{3}}{2} + \frac{1}{2}i\right)$$

$$w_3 = 8^{1/6}\left(\cos \frac{7\pi}{6} + i \sin \frac{7\pi}{6}\right) = \sqrt{2}\left(-\frac{\sqrt{3}}{2} - \frac{1}{2}i\right)$$

$$w_4 = 8^{1/6}\left(\cos \frac{3\pi}{2} + i \sin \frac{3\pi}{2}\right) = -\sqrt{2}\,i$$

$$w_5 = 8^{1/6}\left(\cos \frac{11\pi}{6} + i \sin \frac{11\pi}{6}\right) = \sqrt{2}\left(\frac{\sqrt{3}}{2} - \frac{1}{2}i\right)$$

All these points lie on the circle of radius $\sqrt{2}$ as shown in Figure 9.

**FIGURE 9**

The six sixth roots of $z = -8$

## Complex Exponentials

We also need to give a meaning to the expression $e^z$ when $z = x + iy$ is a complex number. The theory of infinite series as developed in Chapter 11 can be extended to the case where the terms are complex numbers. Using the Taylor series for $e^x$ (11.10.11) as our guide, we define

$$\boxed{4} \qquad e^z = \sum_{n=0}^{\infty} \frac{z^n}{n!} = 1 + z + \frac{z^2}{2!} + \frac{z^3}{3!} + \cdots$$

and it turns out that this complex exponential function has the same properties as the real exponential function. In particular, it is true that

$$\boxed{5} \qquad e^{z_1+z_2} = e^{z_1}e^{z_2}$$

If we put $z = iy$, where *y* is a real number, in Equation 4, and use the facts that

$$i^2 = -1, \quad i^3 = i^2 i = -i, \quad i^4 = 1, \quad i^5 = i, \quad \ldots$$

we get   $e^{iy} = 1 + iy + \dfrac{(iy)^2}{2!} + \dfrac{(iy)^3}{3!} + \dfrac{(iy)^4}{4!} + \dfrac{(iy)^5}{5!} + \cdots$

$$= 1 + iy - \frac{y^2}{2!} - i\frac{y^3}{3!} + \frac{y^4}{4!} + i\frac{y^5}{5!} + \cdots$$

$$= \left(1 - \frac{y^2}{2!} + \frac{y^4}{4!} - \frac{y^6}{6!} + \cdots\right) + i\left(y - \frac{y^3}{3!} + \frac{y^5}{5!} - \cdots\right)$$

$$= \cos y + i \sin y$$

Here we have used the Taylor series for $\cos y$ and $\sin y$ (Equations 11.10.16 and 11.10.15). The result is a famous formula called **Euler's formula**:

**6**

$$e^{iy} = \cos y + i \sin y$$

Combining Euler's formula with Equation 5, we get

**7**

$$e^{x+iy} = e^x e^{iy} = e^x(\cos y + i \sin y)$$

**EXAMPLE 8**  Evaluate:   (a) $e^{i\pi}$   (b) $e^{-1+i\pi/2}$

SOLUTION

We could write the result of Example 8(a) as

$$e^{i\pi} + 1 = 0$$

This equation relates the five most famous numbers in all of mathematics: 0, 1, $e$, $i$, and $\pi$.

(a) From Euler's equation **6** we have

$$e^{i\pi} = \cos \pi + i \sin \pi = -1 + i(0) = -1$$

(b) Using Equation 7 we get

$$e^{-1+i\pi/2} = e^{-1}\left(\cos \frac{\pi}{2} + i \sin \frac{\pi}{2}\right) = \frac{1}{e}[0 + i(1)] = \frac{i}{e}$$

Finally, we note that Euler's equation provides us with an easier method of proving De Moivre's Theorem:

$$[r(\cos \theta + i \sin \theta)]^n = (re^{i\theta})^n = r^n e^{in\theta} = r^n(\cos n\theta + i \sin n\theta)$$

## H   Exercises

1–14  Evaluate the expression and write your answer in the form $a + bi$.

**1.** $(5 - 6i) + (3 + 2i)$

**2.** $\left(4 - \frac{1}{2}i\right) - \left(9 + \frac{5}{2}i\right)$

**3.** $(2 + 5i)(4 - i)$

**4.** $(1 - 2i)(8 - 3i)$

**5.** $\overline{12 + 7i}$

**6.** $\overline{2i\left(\frac{1}{2} - i\right)}$

**7.** $\dfrac{1 + 4i}{3 + 2i}$

**8.** $\dfrac{3 + 2i}{1 - 4i}$

**9.** $\dfrac{1}{1 + i}$

**10.** $\dfrac{3}{4 - 3i}$

**11.** $i^3$

**12.** $i^{100}$

**13.** $\sqrt{-25}$

**14.** $\sqrt{-3}\,\sqrt{-12}$

15–17  Find the complex conjugate and the modulus of the number.

**15.** $12 - 5i$

**16.** $-1 + 2\sqrt{2}\,i$

**17.** $-4i$

**18.** Prove the following properties of complex numbers.
   (a) $\overline{z + w} = \bar{z} + \bar{w}$    (b) $\overline{zw} = \bar{z}\,\bar{w}$
   (c) $\overline{z^n} = \bar{z}^n$,  where $n$ is a positive integer
   [*Hint:* Write $z = a + bi$, $w = c + di$.]

**19–24** Find all solutions of the equation.

**19.** $4x^2 + 9 = 0$

**20.** $x^4 = 1$

**21.** $x^2 + 2x + 5 = 0$

**22.** $2x^2 - 2x + 1 = 0$

**23.** $z^2 + z + 2 = 0$

**24.** $z^2 + \frac{1}{2}z + \frac{1}{4} = 0$

**25–28** Write the number in polar form with argument between 0 and $2\pi$.

**25.** $-3 + 3i$

**26.** $1 - \sqrt{3}\,i$

**27.** $3 + 4i$

**28.** $8i$

**29–32** Find polar forms for $zw$, $z/w$, and $1/z$ by first putting $z$ and $w$ into polar form.

**29.** $z = \sqrt{3} + i, \quad w = 1 + \sqrt{3}\,i$

**30.** $z = 4\sqrt{3} - 4i, \quad w = 8i$

**31.** $z = 2\sqrt{3} - 2i, \quad w = -1 + i$

**32.** $z = 4(\sqrt{3} + i), \quad w = -3 - 3i$

**33–36** Find the indicated power using De Moivre's Theorem.

**33.** $(1 + i)^{20}$

**34.** $\left(1 - \sqrt{3}\,i\right)^5$

**35.** $\left(2\sqrt{3} + 2i\right)^5$

**36.** $(1 - i)^8$

**37–40** Find the indicated roots. Sketch the roots in the complex plane.

**37.** The eighth roots of 1

**38.** The fifth roots of 32

**39.** The cube roots of $i$

**40.** The cube roots of $1 + i$

**41–46** Write the number in the form $a + bi$.

**41.** $e^{i\pi/2}$

**42.** $e^{2\pi i}$

**43.** $e^{i\pi/3}$

**44.** $e^{-i\pi}$

**45.** $e^{2+i\pi}$

**46.** $e^{\pi+i}$

**47.** Use De Moivre's Theorem with $n = 3$ to express $\cos 3\theta$ and $\sin 3\theta$ in terms of $\cos\theta$ and $\sin\theta$.

**48.** Use Euler's formula to prove the following formulas for $\cos x$ and $\sin x$:

$$\cos x = \frac{e^{ix} + e^{-ix}}{2} \qquad \sin x = \frac{e^{ix} - e^{-ix}}{2i}$$

**49.** If $u(x) = f(x) + ig(x)$ is a complex-valued function of a real variable $x$ and the real and imaginary parts $f(x)$ and $g(x)$ are differentiable functions of $x$, then the derivative of $u$ is defined to be $u'(x) = f'(x) + ig'(x)$. Use this together with Equation 7 to prove that if $F(x) = e^{rx}$, then $F'(x) = re^{rx}$ when $r = a + bi$ is a complex number.

**50.** (a) If $u$ is a complex-valued function of a real variable, its indefinite integral $\int u(x)\,dx$ is an antiderivative of $u$. Evaluate

$$\int e^{(1+i)x}\,dx$$

(b) By considering the real and imaginary parts of the integral in part (a), evaluate the real integrals

$$\int e^x \cos x\,dx \qquad \text{and} \qquad \int e^x \sin x\,dx$$

(c) Compare with the method used in Example 4 in Section 7.1.

# Answers to Odd-Numbered Exercises

## CHAPTER 1

**1.** Yes

**3.** (a) 3  (b) $-0.2$  (c) $0, 3$  (d) $-0.8$
(e) $[-2, 4], [-1, 3]$  (f) $[-2, 1]$

**5.** $[-85, 115]$  **7.** No

**9.** Yes, $[-3, 2], [-3, -2) \cup [-1, 3]$

**11.** Diet, exercise, or illness

**13.**

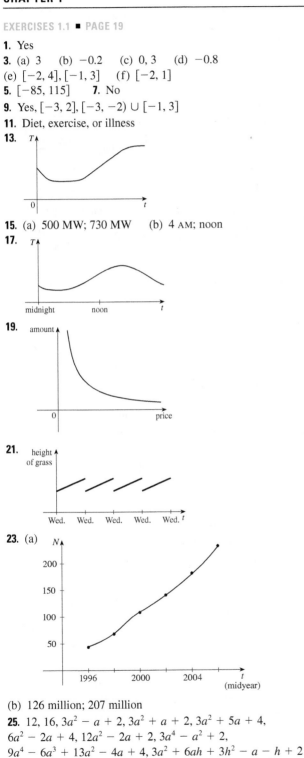

**15.** (a) 500 MW; 730 MW  (b) 4 AM; noon

**17.**

**19.**

**21.**

**23.** (a)

(b) 126 million; 207 million

**25.** $12, 16, 3a^2 - a + 2, 3a^2 + a + 2, 3a^2 + 5a + 4,$
$6a^2 - 2a + 4, 12a^2 - 2a + 2, 3a^4 - a^2 + 2,$
$9a^4 - 6a^3 + 13a^2 - 4a + 4, 3a^2 + 6ah + 3h^2 - a - h + 2$

**27.** $-3 - h$  **29.** $-1/(ax)$

**31.** $(-\infty, -3) \cup (-3, 3) \cup (3, \infty)$  **33.** $(-\infty, \infty)$

**35.** $(-\infty, 0) \cup (5, \infty)$  **37.** $[0, 4]$

**39.** $(-\infty, \infty)$  **41.** $(-\infty, \infty)$

**43.** $[5, \infty)$  **45.** $(-\infty, 0) \cup (0, \infty)$

**47.** $(-\infty, \infty)$  **49.** $(-\infty, \infty)$

**51.** $f(x) = \frac{5}{2}x - \frac{11}{2}, 1 \le x \le 5$  **53.** $f(x) = 1 - \sqrt{-x}$

**55.** $f(x) = \begin{cases} -x + 3 & \text{if } 0 \le x \le 3 \\ 2x - 6 & \text{if } 3 < x \le 5 \end{cases}$

**57.** $A(L) = 10L - L^2, 0 < L < 10$

**59.** $A(x) = \sqrt{3}x^2/4, x > 0$  **61.** $S(x) = x^2 + (8/x), x > 0$

**63.** $V(x) = 4x^3 - 64x^2 + 240x, 0 < x < 6$

**65.** $F(x) = \begin{cases} 15(40 - x) & \text{if } 0 \le x < 40 \\ 0 & \text{if } 40 \le x \le 65 \\ 15(x - 65) & \text{if } x > 65 \end{cases}$

**67.** (a)  (b) $400, $1900

(c)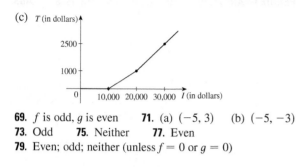

**69.** $f$ is odd, $g$ is even    **71.** (a) $(-5, 3)$    (b) $(-5, -3)$
**73.** Odd    **75.** Neither    **77.** Even
**79.** Even; odd; neither (unless $f = 0$ or $g = 0$)

**EXERCISES 1.2 ■ PAGE 33**

**1.** (a) Logarithmic    (b) Root    (c) Rational
(d) Polynomial, degree 2    (e) Exponential    (f) Trigonometric
**3.** (a) $h$    (b) $f$    (c) $g$
**5.** (a) $y = 2x + b$,
where $b$ is the $y$-intercept.

(b) $y = mx + 1 - 2m$,
where $m$ is the slope.
(c) $y = 2x - 3$

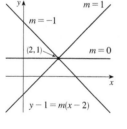

**7.** Their graphs have slope $-1$.

**9.** $f(x) = -3x(x + 1)(x - 2)$
**11.** (a) 8.34, change in mg for every 1 year change
(b) 8.34 mg
**13.** (a)

(b) $\frac{9}{5}$, change in °F for every
1°C change; 32, Fahrenheit
temperature corresponding
to 0°C

**15.** (a) $T = \frac{1}{6}N + \frac{307}{6}$    (b) $\frac{1}{6}$, change in °F for every chirp per
minute change    (c) 76°F
**17.** (a) $P = 0.434d + 15$    (b) 196 ft
**19.** (a) Cosine    (b) Linear
**21.** (a) Linear model is appropriate.

(b) $y = -0.000105x + 14.521$
(c) $y = -0.00009979x + 13.951$

(d) About 11.5 per 100 population    (e) About 6%    (f) No
**23.** (a) Linear model is appropriate.

(b) $y = 0.0265x - 46.8759$    (c) 6.27 m; higher    (d) No
**25.** Four times as bright
**27.** (a) $N = 3.1046A^{0.308}$    (b) 18

**EXERCISES 1.3 ■ PAGE 42**

**1.** (a) $y = f(x) + 3$    (b) $y = f(x) - 3$    (c) $y = f(x - 3)$
(d) $y = f(x + 3)$    (e) $y = -f(x)$    (f) $y = f(-x)$
(g) $y = 3f(x)$    (h) $y = \frac{1}{3}f(x)$
**3.** (a) 3    (b) 1    (c) 4    (d) 5    (e) 2
**5.** (a)    (b)

(c)    (d)

**7.** $y = -\sqrt{-x^2 - 5x - 4} - 1$

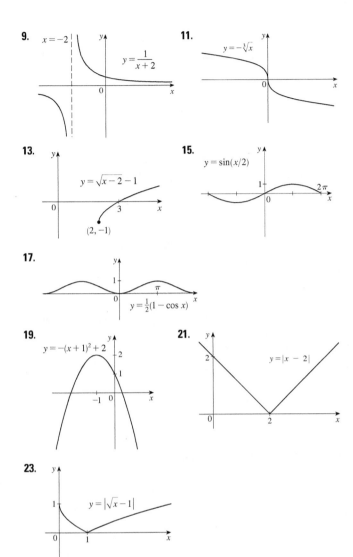

**9.** $x = -2$  $y = \dfrac{1}{x+2}$

**11.** $y = -\sqrt[3]{x}$

**13.** $y = \sqrt{x-2} - 1$  $(2, -1)$

**15.** $y = \sin(x/2)$  $2\pi$

**17.** $y = \frac{1}{2}(1 - \cos x)$  $\pi$

**19.** $y = -(x+1)^2 + 2$

**21.** $y = |x - 2|$

**23.** $y = |\sqrt{x} - 1|$

**25.** $L(t) = 12 + 2 \sin\left[\dfrac{2\pi}{365}(t - 80)\right]$

**27.** (a) The portion of the graph of $y = f(x)$ to the right of the $y$-axis is reflected about the $y$-axis.
(b) $\qquad$ (c)

$y = \sin|x|$ $\qquad$ $y = \sqrt{|x|}$

**29.** (a) $(f + g)(x) = x^3 + 5x^2 - 1,\ (-\infty, \infty)$
(b) $(f - g)(x) = x^3 - x^2 + 1,\ (-\infty, \infty)$
(c) $(fg)(x) = 3x^5 + 6x^4 - x^3 - 2x^2,\ (-\infty, \infty)$
(d) $(f/g)(x) = \dfrac{x^3 + 2x^2}{3x^2 - 1},\ \{x \mid x \neq \pm 1/\sqrt{3}\}$

**31.** (a) $(f \circ g)(x) = 4x^2 + 4x,\ (-\infty, \infty)$
(b) $(g \circ f)(x) = 2x^2 - 1,\ (-\infty, \infty)$
(c) $(f \circ f)(x) = x^4 - 2x^2,\ (-\infty, \infty)$
(d) $(g \circ g)(x) = 4x + 3,\ (-\infty, \infty)$

**33.** (a) $(f \circ g)(x) = 1 - 3\cos x,\ (-\infty, \infty)$
(b) $(g \circ f)(x) = \cos(1 - 3x),\ (-\infty, \infty)$
(c) $(f \circ f)(x) = 9x - 2,\ (-\infty, \infty)$
(d) $(g \circ g)(x) = \cos(\cos x),\ (-\infty, \infty)$

**35.** (a) $(f \circ g)(x) = \dfrac{2x^2 + 6x + 5}{(x + 2)(x + 1)},\ \{x \mid x \neq -2, -1\}$

(b) $(g \circ f)(x) = \dfrac{x^2 + x + 1}{(x + 1)^2},\ \{x \mid x \neq -1, 0\}$

(c) $(f \circ f)(x) = \dfrac{x^4 + 3x^2 + 1}{x(x^2 + 1)},\ \{x \mid x \neq 0\}$

(d) $(g \circ g)(x) = \dfrac{2x + 3}{3x + 5},\ \{x \mid x \neq -2, -\frac{5}{3}\}$

**37.** $(f \circ g \circ h)(x) = 3\sin(x^2) - 2$
**39.** $(f \circ g \circ h)(x) = \sqrt{x^6 + 4x^3 + 1}$
**41.** $g(x) = 2x + x^2,\ f(x) = x^4$
**43.** $g(x) = \sqrt[3]{x},\ f(x) = x/(1 + x)$
**45.** $g(t) = t^2,\ f(t) = \sec t \tan t$
**47.** $h(x) = \sqrt{x},\ g(x) = x - 1,\ f(x) = \sqrt{x}$
**49.** $h(x) = \sqrt{x},\ g(x) = \sec x,\ f(x) = x^4$
**51.** (a) 4 $\quad$ (b) 3 $\quad$ (c) 0 $\quad$ (d) Does not exist; $f(6) = 6$ is not in the domain of $g$. $\quad$ (e) 4 $\quad$ (f) $-2$
**53.** (a) $r(t) = 60t$ $\quad$ (b) $(A \circ r)(t) = 3600\pi t^2$; the area of the circle as a function of time
**55.** (a) $s = \sqrt{d^2 + 36}$ $\quad$ (b) $d = 30t$
(c) $(f \circ g)(t) = \sqrt{900t^2 + 36}$; the distance between the lighthouse and the ship as a function of the time elapsed since noon
**57.** (a) $\qquad$ (b)

$H$ $\qquad$ $V$
$1$ $\qquad$ $120$
$0 \quad t$ $\qquad$ $0 \quad t$

$V(t) = 120H(t)$

(c) $V$
$240$
$0 \quad 5 \quad t$

$V(t) = 240H(t - 5)$

**59.** Yes; $m_1 m_2$
**61.** (a) $f(x) = x^2 + 6$ $\quad$ (b) $g(x) = x^2 + x - 1$
**63.** Yes

**EXERCISES 1.4** ■ PAGE 49

**1.** (a) $-44.4,\ -38.8,\ -27.8,\ -22.2,\ -16.\overline{6}$
(b) $-33.3$ $\quad$ (c) $-33\frac{1}{3}$
**3.** (a) (i) 2 $\quad$ (ii) 1.111111 $\quad$ (iii) 1.010101 $\quad$ (iv) 1.001001
(v) 0.666667 $\quad$ (vi) 0.909091 $\quad$ (vii) 0.990099
(viii) 0.999001 $\quad$ (b) 1 $\quad$ (c) $y = x - 3$
**5.** (a) (i) $-32$ ft/s $\quad$ (ii) $-25.6$ ft/s $\quad$ (iii) $-24.8$ ft/s
(iv) $-24.16$ ft/s $\quad$ (b) $-24$ ft/s
**7.** (a) (i) 4.65 m/s $\quad$ (ii) 5.6 m/s $\quad$ (iii) 7.55 m/s
(iv) 7 m/s $\quad$ (b) 6.3 m/s

**9.** (a) 0, 1.7321, −1.0847, −2.7433, 4.3301, −2.8173, 0, −2.1651, −2.6061, −5, 3.4202; no   (c) −31.4

**1.** Yes

**3.** (a) $\lim_{x \to -3} f(x) = \infty$ means that the values of $f(x)$ can be made arbitrarily large (as large as we please) by taking $x$ sufficiently close to −3 (but not equal to −3).
(b) $\lim_{x \to 4^+} f(x) = -\infty$ means that the values of $f(x)$ can be made arbitrarily large negative by taking $x$ sufficiently close to 4 through values larger than 4.

**5.** (a) 2   (b) 1   (c) 4   (d) Does not exist   (e) 3
**7.** (a) −1   (b) −2   (c) Does not exist   (d) 2   (e) 0
(f) Does not exist   (g) 1   (h) 3
**9.** (a) $-\infty$   (b) $\infty$   (c) $\infty$   (d) $-\infty$   (e) $\infty$
(f) $x = -7, x = -3, x = 0, x = 6$
**11.** $\lim_{x \to a} f(x)$ exists for all $a$ except $a = -1$.
**13.** (a) 1   (b) 0   (c) Does not exist
**15.**           **17.**

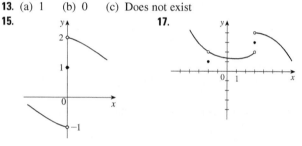

**19.** $\frac{2}{3}$   **21.** $\frac{1}{2}$   **23.** $\frac{1}{4}$   **25.** $\frac{3}{5}$   **27.** (a) −1.5
**29.** $-\infty$   **31.** $\infty$   **33.** $-\infty$   **35.** $-\infty$   **37.** $\infty$
**39.** $-\infty$; $\infty$
**41.** (a) 0.998000, 0.638259, 0.358484, 0.158680, 0.038851, 0.008928, 0.001465; 0
(b) 0.000572, −0.000614, −0.000907, −0.000978, −0.000993, −0.001000; −0.001
**43.** No matter how many times we zoom in toward the origin, the graph appears to consist of almost-vertical lines. This indicates more and more frequent oscillations as $x \to 0$.
**45.** $x \approx \pm 0.90, \pm 2.24$; $x = \pm \sin^{-1}(\pi/4), \pm(\pi - \sin^{-1}(\pi/4))$

**1.** (a) −6   (b) −8   (c) 2   (d) −6
(e) Does not exist   (f) 0
**3.** 105   **5.** $\frac{7}{8}$   **7.** 390   **9.** $\frac{3}{2}$   **11.** 4
**13.** Does not exist   **15.** $\frac{6}{5}$   **17.** −10   **19.** $\frac{1}{12}$
**21.** $\frac{1}{6}$   **23.** $-\frac{1}{16}$   **25.** 1   **27.** $\frac{1}{128}$   **29.** $-\frac{1}{2}$
**31.** $3x^2$   **33.** $\frac{2}{3}$   **37.** 7   **41.** 6   **43.** −4
**45.** Does not exist

**47.** (a)          (b) (i) 1

        (ii) −1
        (iii) Does not exist
        (iv) 1

**49.** (a) (i) 5   (ii) −5   (b) Does not exist
(c)

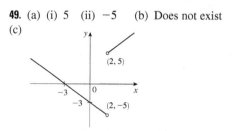

**51.** (a) (i) −2   (ii) Does not exist   (iii) −3
(b) (i) $n - 1$   (ii) $n$   (c) $a$ is not an integer.
**57.** 8   **63.** 15; −1

**1.** 0.1 (or any smaller positive number)
**3.** 1.44 (or any smaller positive number)
**5.** 0.0906 (or any smaller positive number)
**7.** 0.011 (or any smaller positive number)
**9.** (a) 0.031   (b) 0.010
**11.** (a) $\sqrt{1000/\pi}$ cm   (b) Within approximately 0.0445 cm
(c) Radius; area; $\sqrt{1000/\pi}$; 1000; 5; $\approx 0.0445$
**13.** (a) 0.025   (b) 0.0025
**35.** (a) 0.093   (b) $\delta = (B^{2/3} - 12)/(6B^{1/3}) - 1$, where
$B = 216 + 108\varepsilon + 12\sqrt{336 + 324\varepsilon + 81\varepsilon^2}$
**41.** Within 0.1

**1.** $\lim_{x \to 4} f(x) = f(4)$
**3.** (a) $f(-4)$ is not defined and $\lim_{x \to a} f(x)$ [for $a = -2, 2$, and 4] does not exist
(b) −4, neither; −2, left; 2, right; 4, right

**5.**          **7.**

**9.** (a)

**11.** 4   **17.** $f(-2)$ is undefined.

**19.** $\lim_{x \to 1} f(x)$ does not exist.　　**21.** $\lim_{x \to 0} f(x) \neq f(0)$

**23.** Define $f(2) = 3$.　**25.** $(-\infty, \infty)$
**27.** $\left(-\infty, \sqrt[3]{2}\right) \cup \left(\sqrt[3]{2}, \infty\right)$　**29.** $\mathbb{R}$　**31.** $(-\infty, -1] \cup (0, \infty)$
**33.** $x = (-\pi/2) + 2n\pi$, $n$ an integer

**35.** $\frac{7}{3}$　**37.** $\pi/8$
**41.** 0, left　　**43.** 0, right; 1, left

**45.** $\frac{2}{3}$　**47.** (a) $g(x) = x^3 + x^2 + x + 1$　(b) $g(x) = x^2 + x$
**55.** (b) $(0.86, 0.87)$　**57.** (b) 1.434　**63.** None
**65.** Yes

CHAPTER 1 REVIEW ▪ PAGE 94

**True-False Quiz**

**1.** False　**3.** False　**5.** True　**7.** False　**9.** True
**11.** False　**13.** True　**15.** True　**17.** False　**19.** True
**21.** True　**23.** True　**25.** False

**Exercises**

**1.** (a) 2.7　(b) 2.3, 5.6　(c) $[-6, 6]$　(d) $[-4, 4]$
(e) $[-4, 4]$　(f) Odd; its graph is symmetric about the origin.
**3.** $2a + h - 2$　**5.** $\left(-\infty, \frac{1}{3}\right) \cup \left(\frac{1}{3}, \infty\right)$, $(-\infty, 0) \cup (0, \infty)$
**7.** $\mathbb{R}$, $[0, 2]$
**9.** (a) Shift the graph 8 units upward.
(b) Shift the graph 8 units to the left.
(c) Stretch the graph vertically by a factor of 2, then shift it
1 unit upward.
(d) Shift the graph 2 units to the right and 2 units downward.
(e) Reflect the graph about the $x$-axis.
(f) Reflect the graph about the $x$-axis, then shift 3 units upward.

**11.**

**13.**

**15.**

**17.** (a) Neither　(b) Odd　(c) Even　(d) Neither
**19.** (a) $(f \circ g)(x) = \sqrt{\sin x}$, $\{x \mid x \in [2n\pi, \pi + 2n\pi], n \text{ an integer}\}$
(b) $(g \circ f)(x) = \sin \sqrt{x}$, $[0, \infty)$
(c) $(f \circ f)(x) = \sqrt[4]{x}$, $[0, \infty)$
(d) $(g \circ g)(x) = \sin(\sin x)$, $\mathbb{R}$
**21.** $y = 0.2493x - 423.4818$; about 77.6 years
**23.** (a) (i) 3　(ii) 0　(iii) Does not exist　(iv) 2
(v) $\infty$　(vi) $-\infty$　(b) $x = 0$, $x = 2$　(c) $-3, 0, 2, 4$
**25.** 1　**27.** $\frac{3}{2}$　**29.** 3　**31.** $\infty$　**33.** $\frac{4}{7}$
**35.** $-\frac{1}{8}$　**37.** 0　**39.** 1
**45.** (a) (i) 3　(ii) 0　(iii) Does not exist　(iv) 0　(v) 0　(vi) 0
(b) At 0 and 3　(c)

**47.** $[0, \infty)$

PRINCIPLES OF PROBLEM SOLVING ▪ PAGE 102

**1.**

**3.** $f_n(x) = x^{2^{n+1}}$　**5.** $\frac{2}{3}$　**7.** $-4$
**9.** (a) Does not exist　(b) 1
**11.** $a = \frac{1}{2} \pm \frac{1}{2}\sqrt{5}$　**13.** $\frac{3}{4}$　**15.** (b) Yes　(c) Yes; no

# CHAPTER 2

**1.** (a) $\dfrac{f(x) - f(3)}{x - 3}$ (b) $\lim\limits_{x \to 3} \dfrac{f(x) - f(3)}{x - 3}$

**3.** (a) 2 (b) $y = 2x + 1$ (c)

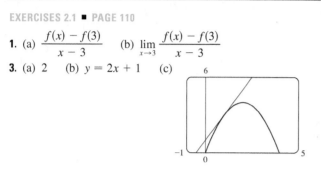

**5.** $y = -8x + 12$ **7.** $y = \frac{1}{2}x + \frac{1}{2}$

**9.** (a) $8a - 6a^2$ (b) $y = 2x + 3$, $y = -8x + 19$

(c)

**11.** (a) Right: $0 < t < 1$ and $4 < t < 6$; left: $2 < t < 3$; standing still: $1 < t < 2$ and $3 < t < 4$

(b)

**13.** $-24$ ft/s

**15.** $-2/a^3$ m/s; $-2$ m/s; $-\frac{1}{4}$ m/s; $-\frac{2}{27}$ m/s

**17.** $g'(0), 0, g'(4), g'(2), g'(-2)$

**19.** $f(2) = 3$; $f'(2) = 4$

**21.**

**23.** $y = 3x - 1$

**25.** (a) $-\frac{3}{5}$; $y = -\frac{3}{5}x + \frac{16}{5}$ (b)

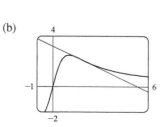

**27.** $6a - 4$ **29.** $\dfrac{5}{(a + 3)^2}$ **31.** $-\dfrac{1}{\sqrt{1 - 2a}}$

**33.** $f(x) = x^{10}$, $a = 1$ or $f(x) = (1 + x)^{10}$, $a = 0$

**35.** $f(x) = 2^x$, $a = 5$

**37.** $f(x) = \cos x$, $a = \pi$ or $f(x) = \cos(\pi + x)$, $a = 0$

**39.** 1 m/s; 1 m/s

**41.**

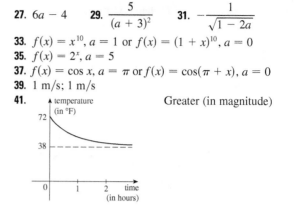

Greater (in magnitude)

**43.** (a) (i) 23 million/year (ii) 20.5 million/year
(iii) 16 million/year
(b) 18.25 million/year (c) 17 million/year

**45.** (a) (i) $20.25/unit (ii) $20.05/unit (b) $20/unit

**47.** (a) The rate at which the cost is changing per ounce of gold produced; dollars per ounce
(b) When the 800th ounce of gold is produced, the cost of production is $17/oz.
(c) Decrease in the short term; increase in the long term

**49.** The rate at which the temperature is changing at 8:00 AM; 3.75°F/h

**51.** (a) The rate at which the oxygen solubility changes with respect to the water temperature; (mg/L)/°C
(b) $S'(16) \approx -0.25$; as the temperature increases past 16°C, the oxygen solubility is decreasing at a rate of 0.25 (mg/L)/°C.

**53.** Does not exist

**1.** (a) $-0.2$ (b) 0 (c) 1 (d) 2
(e) 1 (f) 0 (g) $-0.2$

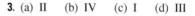

**3.** (a) II (b) IV (c) I (d) III

**5.**

**7.**

**9.**

**11.**

**13.** (a) The instantaneous rate of change of percentage of full capacity with respect to elapsed time in hours

(b)

The rate of change of percentage of full capacity is decreasing and approaching 0.

**15.**

1963 to 1971

**17.** (a) 0, 1, 2, 4  (b) $-1, -2, -4$  (c) $f'(x) = 2x$

**19.** $f'(x) = \frac{1}{2}$, $\mathbb{R}$, $\mathbb{R}$  **21.** $f'(t) = 5 - 18t$, $\mathbb{R}$, $\mathbb{R}$

**23.** $f'(x) = 2x - 6x^2$, $\mathbb{R}$, $\mathbb{R}$

**25.** $g'(x) = -\dfrac{1}{2\sqrt{9-x}}$, $(-\infty, 9]$, $(-\infty, 9)$

**27.** $G'(t) = \dfrac{-7}{(3+t)^2}$, $(-\infty, -3) \cup (-3, \infty)$, $(-\infty, -3) \cup (-3, \infty)$

**29.** $f'(x) = 4x^3$, $\mathbb{R}$, $\mathbb{R}$  **31.** (a) $f'(x) = 4x^3 + 2$

**33.** (a) The rate at which the unemployment rate is changing, in percent unemployed per year

(b)

| $t$ | $U'(t)$ | $t$ | $U'(t)$ |
|------|---------|------|---------|
| 1999 | $-0.2$ | 2004 | $-0.45$ |
| 2000 | $0.25$ | 2005 | $-0.45$ |
| 2001 | $0.9$ | 2006 | $-0.25$ |
| 2002 | $0.65$ | 2007 | $0.6$ |
| 2003 | $-0.15$ | 2008 | $1.2$ |

**35.** $-4$ (corner); 0 (discontinuity)
**37.** $-1$ (vertical tangent); 4 (corner)
**39.**

Differentiable at $-1$; not differentiable at 0

**41.** $a = f, b = f', c = f''$
**43.** $a = $ acceleration, $b = $ velocity, $c = $ position
**45.** $6x + 2$; 6

**47.**

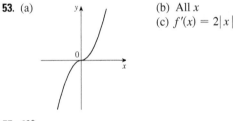

$f'(x) = 4x - 3x^2$,
$f''(x) = 4 - 6x$,
$f'''(x) = -6$,
$f^{(4)}(x) = 0$

**49.** (a) $\frac{1}{3}a^{-2/3}$

**51.** $f'(x) = \begin{cases} -1 & \text{if } x < 6 \\ 1 & \text{if } x > 6 \end{cases}$

or $f'(x) = \dfrac{x-6}{|x-6|}$

**53.** (a)

(b) All $x$
(c) $f'(x) = 2|x|$

**57.** $63°$

EXERCISES 2.3 ■ PAGE 136

**1.** $f'(x) = 0$  **3.** $f'(t) = -\frac{2}{3}$  **5.** $f'(x) = 3x^2 - 4$
**7.** $g'(x) = 2x - 6x^2$  **9.** $g'(t) = -\frac{3}{2}t^{-7/4}$  **11.** $A'(s) = 60/s^6$
**13.** $S'(p) = \frac{1}{2}p^{-1/2} - 1$  **15.** $R'(a) = 18a + 6$

**17.** $y' = \frac{3}{2}\sqrt{x} + \dfrac{2}{\sqrt{x}} - \dfrac{3}{2x\sqrt{x}}$

**19.** $H'(x) = 3x^2 + 3 - 3x^{-2} - 3x^{-4}$

**21.** $u' = \frac{1}{5}t^{-4/5} + 10t^{3/2}$  **23.** $1 - 2x + 6x^2 - 8x^3$

**25.** $V'(x) = 14x^6 - 4x^3 - 6$  **27.** $F'(y) = 5 + \dfrac{14}{y^2} + \dfrac{9}{y^4}$

**29.** $g'(x) = \dfrac{10}{(3-4x)^2}$  **31.** $y' = \dfrac{x^2(3-x^2)}{(1-x^2)^2}$

**33.** $y' = 2v - 1/\sqrt{v}$  **35.** $y' = \dfrac{2t(-t^4 - 4t^2 + 7)}{(t^4 - 3t^2 + 1)^2}$

**37.** $y' = 2ax + b$  **39.** $f'(t) = \dfrac{4 + t^{1/2}}{(2 + \sqrt{t})^2}$

**41.** $y' = (7t^3 + 4t^2 - 2)/(3t^{5/3})$    **43.** $f'(x) = \dfrac{2cx}{(x^2 + c)^2}$

**45.** $P'(x) = na_n x^{n-1} + (n-1)a_{n-1}x^{n-2} + \cdots + 2a_2 x + a_1$

**47.** $45x^{14} - 15x^2$

**49.** (a)    (c) $4x^3 - 9x^2 - 12x + 7$

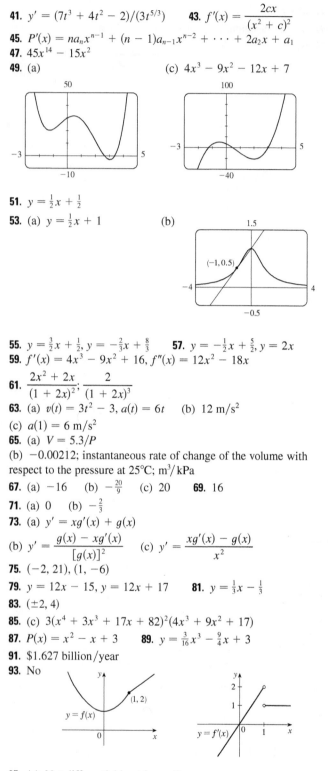

**51.** $y = \frac{1}{2}x + \frac{1}{2}$

**53.** (a) $y = \frac{1}{2}x + 1$    (b)

**55.** $y = \frac{3}{2}x + \frac{1}{2}, y = -\frac{2}{3}x + \frac{8}{3}$    **57.** $y = -\frac{1}{2}x + \frac{5}{2}, y = 2x$

**59.** $f'(x) = 4x^3 - 9x^2 + 16, f''(x) = 12x^2 - 18x$

**61.** $\dfrac{2x^2 + 2x}{(1 + 2x)^2}, \dfrac{2}{(1 + 2x)^3}$

**63.** (a) $v(t) = 3t^2 - 3, a(t) = 6t$    (b) 12 m/s²

(c) $a(1) = 6$ m/s²

**65.** (a) $V = 5.3/P$

(b) $-0.00212$; instantaneous rate of change of the volume with respect to the pressure at 25°C; m³/kPa

**67.** (a) $-16$    (b) $-\frac{20}{9}$    (c) 20    **69.** 16

**71.** (a) 0    (b) $-\frac{2}{3}$

**73.** (a) $y' = xg'(x) + g(x)$

(b) $y' = \dfrac{g(x) - xg'(x)}{[g(x)]^2}$    (c) $y' = \dfrac{xg'(x) - g(x)}{x^2}$

**75.** $(-2, 21), (1, -6)$

**79.** $y = 12x - 15, y = 12x + 17$    **81.** $y = \frac{1}{3}x - \frac{1}{3}$

**83.** $(\pm 2, 4)$

**85.** (c) $3(x^4 + 3x^3 + 17x + 82)^2(4x^3 + 9x^2 + 17)$

**87.** $P(x) = x^2 - x + 3$    **89.** $y = \frac{3}{16}x^3 - \frac{9}{4}x + 3$

**91.** \$1.627 billion/year

**93.** No

**95.** (a) Not differentiable at 3 or $-3$

$f'(x) = \begin{cases} 2x & \text{if } |x| > 3 \\ -2x & \text{if } |x| < 3 \end{cases}$

(b)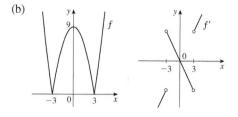

**97.** $a = -\frac{1}{2}, b = 2$    **99.** 6    **103.** 1000    **105.** 3; 1

**EXERCISES 2.4 ■ PAGE 146**

**1.** $f'(x) = 6x + 2 \sin x$    **3.** $f'(x) = \cos x - \frac{1}{2}\csc^2 x$

**5.** $y' = \sec \theta (\sec^2\theta + \tan^2\theta)$

**7.** $y' = -c \sin t + t(t \cos t + 2 \sin t)$

**9.** $y' = \dfrac{2 - \tan x + x \sec^2 x}{(2 - \tan x)^2}$    **11.** $f'(\theta) = \dfrac{\sec \theta \tan \theta}{(1 + \sec \theta)^2}$

**13.** $y' = \dfrac{(t^2 + t)\cos t + \sin t}{(1 + t)^2}$

**15.** $h'(\theta) = \csc \theta - \theta \csc \theta \cot \theta + \csc^2\theta$

**21.** $y = 2\sqrt{3}x - \frac{2}{3}\sqrt{3}\pi + 2$    **23.** $y = x - \pi - 1$

**25.** (a) $y = 2x$    (b)

**27.** (a) $\sec x \tan x - 1$

**29.** $\theta \cos \theta + \sin \theta; 2 \cos \theta - \theta \sin \theta$

**31.** (a) $f'(x) = (1 + \tan x)/\sec x$    (b) $f'(x) = \cos x + \sin x$

**33.** $(2n + 1)\pi \pm \frac{1}{3}\pi$, $n$ an integer

**35.** (a) $v(t) = 8 \cos t, a(t) = -8 \sin t$

(b) $4\sqrt{3}, -4, -4\sqrt{3}$; to the left

**37.** 5 ft/rad    **39.** 3    **41.** 3    **43.** $-\frac{3}{4}$

**45.** $\frac{1}{2}$    **47.** $-\sqrt{2}$    **49.** $-\cos x$    **51.** $A = -\frac{3}{10}, B = -\frac{1}{10}$

**53.** (a) $\sec^2 x = \dfrac{1}{\cos^2 x}$    (b) $\sec x \tan x = \dfrac{\sin x}{\cos^2 x}$

(c) $\cos x - \sin x = \dfrac{\cot x - 1}{\csc x}$

**55.** 1

**EXERCISES 2.5 ■ PAGE 154**

**1.** $\dfrac{4}{3\sqrt[3]{(1 + 4x)^2}}$    **3.** $\pi \sec^2 \pi x$    **5.** $\dfrac{\cos x}{2\sqrt{\sin x}}$

**7.** $F'(x) = 10x(x^4 + 3x^2 - 2)^4(2x^2 + 3)$

**9.** $F'(x) = -\dfrac{1}{\sqrt{1 - 2x}}$    **11.** $f'(z) = -\dfrac{2z}{(z^2 + 1)^2}$

**13.** $y' = -3x^2 \sin(a^3 + x^3)$    **15.** $y' = \sec kx(kx \tan kx + 1)$

**17.** $f'(x) = (2x - 3)^3(x^2 + x + 1)^4(28x^2 - 12x - 7)$

**19.** $h'(t) = \frac{2}{3}(t + 1)^{-1/3}(2t^2 - 1)^2(20t^2 + 18t - 1)$

**21.** $y' = \dfrac{-12x(x^2 + 1)^2}{(x^2 - 1)^4}$

**23.** $y' = (\cos x - x \sin x)\cos(x \cos x)$

**25.** $F'(z) = 1/[(z - 1)^{1/2}(z + 1)^{3/2}]$    **27.** $y' = (r^2 + 1)^{-3/2}$

**29.** $y' = (x \cos\sqrt{1 + x^2})/\sqrt{1 + x^2}$

**31.** $y' = 2\cos(\tan 2x)\sec^2(2x)$    **33.** $y' = 4\sec^2 x \tan x$

**35.** $y' = \dfrac{16 \sin 2x(1 - \cos 2x)^3}{(1 + \cos 2x)^5}$

**37.** $y' = -2\cos\theta \cot(\sin\theta)\csc^2(\sin\theta)$

**39.** $y' = 3[x^2 + (1 - 3x)^5]^2[2x - 15(1 - 3x)^4]$

**41.** $y' = \dfrac{1 + 1/(2\sqrt{x})}{2\sqrt{x + \sqrt{x}}}$

**43.** $g'(x) = p(2r \sin rx + n)^{p-1}(2r^2 \cos rx)$

**45.** $y' = \dfrac{-\pi \cos(\tan \pi x)\sec^2(\pi x)\sin\sqrt{\sin(\tan \pi x)}}{2\sqrt{\sin(\tan \pi x)}}$

**47.** $y' = -2x \sin(x^2);\ y'' = -4x^2\cos(x^2) - 2\sin(x^2)$

**49.** $H'(t) = 3\sec^2 3t,\ H''(t) = 18\sec^2 3t \tan 3t$

**51.** $y = 20x + 1$    **53.** $y = -x + \pi$

**55.** (a) $y = \pi x - \pi + 1$  (b)

**57.** (a) $f'(x) = \dfrac{2 - 2x^2}{\sqrt{2 - x^2}}$

**59.** $((\pi/2) + 2n\pi, 3), ((3\pi/2) + 2n\pi, -1)$, $n$ an integer

**61.** 24    **63.** (a) 30   (b) 36

**65.** (a) $\frac{3}{4}$   (b) Does not exist   (c) $-2$

**67.** $-\frac{1}{6}\sqrt{2}$    **69.** 120    **71.** 96    **73.** $2^{103}\sin 2x$

**75.** $v(t) = \frac{5}{2}\pi \cos(10\pi t)$ cm/s

**77.** (a) $\dfrac{dB}{dt} = \dfrac{7\pi}{54}\cos\dfrac{2\pi t}{5.4}$   (b) 0.16

**79.** $dv/dt$ is the rate of change of velocity with respect to time; $dv/ds$ is the rate of change of velocity with respect to displacement

**81.** (b) The factored form    **85.** (b) $-n\cos^{n-1}x \sin[(n + 1)x]$

EXERCISES 2.6 ■ PAGE 161

**1.** (a) $y' = 9x/y$   (b) $y = \pm\sqrt{9x^2 - 1},\ y' = \pm9x/\sqrt{9x^2 - 1}$

**3.** (a) $y' = -y^2/x^2$   (b) $y = x/(x - 1),\ y' = -1/(x - 1)^2$

**5.** $y' = -\dfrac{x^2}{y^2}$    **7.** $y' = \dfrac{2x + y}{2y - x}$

**9.** $y' = \dfrac{3y^2 - 5x^4 - 4x^3y}{x^4 + 3y^2 - 6xy}$    **11.** $y' = \dfrac{2x + y \sin x}{\cos x - 2y}$

**13.** $y' = \tan x \tan y$    **15.** $y' = \dfrac{y \sec^2(x/y) - y^2}{y^2 + x \sec^2(x/y)}$

**17.** $y' = \dfrac{4xy\sqrt{xy} - y}{x - 2x^2\sqrt{xy}}$    **19.** $y' = \dfrac{y \sin x + y \cos(xy)}{\cos x - x \cos(xy)}$

**21.** $-\frac{16}{13}$    **23.** $x' = \dfrac{-2x^4y + x^3 - 6xy^2}{4x^3y^2 - 3x^2y + 2y^3}$    **25.** $y = \frac{1}{2}x$

**27.** $y = -x + 2$    **29.** $y = x + \frac{1}{2}$    **31.** $y = -\frac{9}{13}x + \frac{40}{13}$

**33.** (a) $y = \frac{9}{2}x - \frac{5}{2}$   (b)

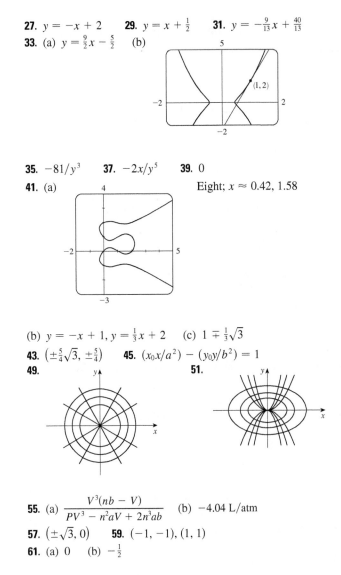

**35.** $-81/y^3$    **37.** $-2x/y^5$    **39.** 0

**41.** (a)                                      Eight; $x \approx 0.42, 1.58$

(b) $y = -x + 1,\ y = \frac{1}{3}x + 2$   (c) $1 \mp \frac{1}{3}\sqrt{3}$

**43.** $\left(\pm\frac{5}{4}\sqrt{3}, \pm\frac{5}{4}\right)$    **45.** $(x_0x/a^2) - (y_0y/b^2) = 1$

**49.**                                      **51.**

**55.** (a) $\dfrac{V^3(nb - V)}{PV^3 - n^2aV + 2n^3ab}$   (b) $-4.04$ L/atm

**57.** $(\pm\sqrt{3}, 0)$    **59.** $(-1, -1), (1, 1)$

**61.** (a) 0   (b) $-\frac{1}{2}$

EXERCISES 2.7 ■ PAGE 173

**1.** (a) $3t^2 - 24t + 36$   (b) $-9$ ft/s   (c) $t = 2, 6$

(d) $0 \le t < 2, t > 6$   (e) 96 ft

(f)

(g) $6t - 24;\ -6$ ft/s$^2$

(h) 

(i) Speeding up when $2 < t < 4$ or $t > 6$; slowing down when $0 \le t < 2$ or $4 < t < 6$

**3.** (a) $-\dfrac{\pi}{4}\sin\left(\dfrac{\pi t}{4}\right)$    (b) $-\frac{1}{8}\pi\sqrt{2}$ ft/s    (c) $t = 0, 4, 8$
(d) $4 < t < 8$    (e) 4 ft
(f)

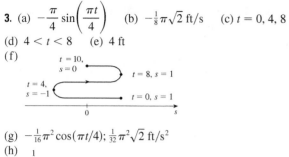

(g) $-\frac{1}{16}\pi^2\cos(\pi t/4);\ \frac{1}{32}\pi^2\sqrt{2}$ ft/s$^2$
(h)

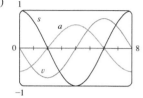

(i) Speeding up when $0 < t < 2, 4 < t < 6, 8 < t < 10$;
slowing down when $2 < t < 4, 6 < t < 8$
**5.** (a) Speeding up when $0 < t < 1$ or $2 < t < 3$;
slowing down when $1 < t < 2$
(b) Speeding up when $1 < t < 2$ or $3 < t < 4$;
slowing down when $0 < t < 1$ or $2 < t < 3$
**7.** (a) 4.9 m/s; $-14.7$ m/s    (b) After 2.5 s    (c) $32\frac{5}{8}$ m
(d) $\approx 5.08$ s    (e) $\approx -25.3$ m/s
**9.** (a) 7.56 m/s    (b) 6.24 m/s; $-6.24$ m/s
**11.** (a) 30 mm$^2$/mm; the rate at which the area is increasing
with respect to side length as $x$ reaches 15 mm
(b) $\Delta A \approx 2x\,\Delta x$
**13.** (a) (i) $5\pi$    (ii) $4.5\pi$    (iii) $4.1\pi$
(b) $4\pi$    (c) $\Delta A \approx 2\pi r\,\Delta r$
**15.** (a) $8\pi$ ft$^2$/ft    (b) $16\pi$ ft$^2$/ft    (c) $24\pi$ ft$^2$/ft
The rate increases as the radius increases.
**17.** (a) 6 kg/m    (b) 12 kg/m    (c) 18 kg/m
At the right end; at the left end
**19.** (a) 4.75 A    (b) 5 A; $t = \frac{2}{3}$ s
**23.** (a) $dV/dP = -C/P^2$    (b) At the beginning
**25.** (a) 16 million/year; 78.5 million/year
(b) $P(t) = at^3 + bt^2 + ct + d$, where $a \approx 0.00129371$,
$b \approx -7.061422$, $c \approx 12,822.979$, $d \approx -7,743,770$
(c) $P'(t) = 3at^2 + 2bt + c$
(d) 14.48 million/year; 75.29 million/year (smaller)
(e) 81.62 million/year
**27.** (a) 0.926 cm/s; 0.694 cm/s; 0
(b) 0; $-92.6$ (cm/s)/cm; $-185.2$ (cm/s)/cm
(c) At the center; at the edge
**29.** (a) $C'(x) = 12 - 0.2x + 0.0015x^2$
(b) \$32/yard; the cost of producing the 201st yard
(c) \$32.20
**31.** (a) $[xp'(x) - p(x)]/x^2$; the average productivity increases as
new workers are added.
**33.** $-0.2436$ K/min
**35.** (a) 0 and 0    (b) $C = 0$
(c) $(0, 0), (500, 50)$; it is possible for the species to coexist.

**1.** $dV/dt = 3x^2\,dx/dt$    **3.** 48 cm$^2$/s
**5.** $3/(25\pi)$ m/min    **7.** (a) 1    (b) 25    **9.** $-18$
**11.** (a) The plane's altitude is 1 mi and its speed is 500 mi/h.
(b) The rate at which the distance from the plane to the station is
increasing when the plane is 2 mi from the station
(c)

(d) $y^2 = x^2 + 1$
(e) $250\sqrt{3}$ mi/h

**13.** (a) The height of the pole (15 ft), the height of the man (6 ft),
and the speed of the man (5 ft/s)
(b) The rate at which the tip of the man's shadow is moving when
he is 40 ft from the pole
(c)

(d) $\dfrac{15}{6} = \dfrac{x + y}{y}$    (e) $\frac{25}{3}$ ft/s

**15.** 65 mi/h    **17.** $837/\sqrt{8674} \approx 8.99$ ft/s
**19.** $-1.6$ cm/min    **21.** $\frac{720}{13} \approx 55.4$ km/h
**23.** $(10{,}000 + 800{,}000\pi/9) \approx 2.89 \times 10^5$ cm$^3$/min
**25.** $\frac{10}{3}$ cm/min    **27.** $6/(5\pi) \approx 0.38$ ft/min    **29.** 0.3 m$^2$/s
**31.** 5 m    **33.** 80 cm$^3$/min    **35.** $\frac{107}{810} \approx 0.132$ $\Omega$/s
**37.** 0.396 m/min    **39.** (a) 360 ft/s    (b) 0.096 rad/s
**41.** $\frac{10}{9}\pi$ km/min    **43.** $1650/\sqrt{31} \approx 296$ km/h
**45.** $\frac{7}{4}\sqrt{15} \approx 6.78$ m/s

**1.** $L(x) = -10x - 6$    **3.** $L(x) = \frac{1}{4}x + 1$
**5.** $\sqrt{1 - x} \approx 1 - \frac{1}{2}x$;
$\sqrt{0.9} \approx 0.95$,
$\sqrt{0.99} \approx 0.995$

**7.** $-0.368 < x < 0.677$
**9.** $-0.045 < x < 0.055$
**11.** (a) $dy = 2x(x\cos 2x + \sin 2x)\,dx$    (b) $dy = \dfrac{t}{\sqrt{1 + t^2}}\,dt$
**13.** (a) $dy = \dfrac{\sec^2\sqrt{t}}{2\sqrt{t}}\,dt$    (b) $dy = \dfrac{-4v}{(1 + v^2)^2}\,dv$
**15.** (a) $dy = \sec^2 x\,dx$    (b) $-0.2$
**17.** (a) $dy = \dfrac{x}{\sqrt{3 + x^2}}\,dx$    (b) $-0.05$
**19.** $\Delta y = 0.64$, $dy = 0.8$

**21.** $\Delta y = -0.1$, $dy = -0.125$

**23.** 15.968  **25.** $10.00\overline{3}$  **27.** $1 - \pi/90 \approx 0.965$

**31.** (a) 270 cm$^3$, 0.01, 1%  (b) 36 cm$^2$, $0.00\overline{6}$, $0.\overline{6}$%

**33.** (a) $84/\pi \approx 27$ cm$^2$; $\frac{1}{84} \approx 0.012 = 1.2\%$

(b) $1764/\pi^2 \approx 179$ cm$^3$; $\frac{1}{56} \approx 0.018 = 1.8\%$

**35.** (a) $2\pi rh\,\Delta r$  (b) $\pi(\Delta r)^2 h$

**41.** (a) 4.8, 5.2  (b) Too large

**CHAPTER 2 REVIEW ■ PAGE 190**

**True-False Quiz**

**1.** False  **3.** False  **5.** True  **7.** False
**9.** True  **11.** False

**Exercises**

**1.** (a) (i) 3 m/s  (ii) 2.75 m/s  (iii) 2.625 m/s
(iv) 2.525 m/s  (b) 2.5 m/s

**3.**

**5.** $a = f$, $b = f''$, $c = f'$

**7.** (a) The rate at which the cost changes with respect to the interest rate; dollars/(percent per year)
(b) As the interest rate increases past 10%, the cost is increasing at a rate of $1200/(percent per year).
(c) Always positive

**9.** The rate at which the total value of US currency in circulation is changing in billions of dollars per year; $22.2 billion/year

**11.** $f'(x) = 3x^2 + 5$  **13.** $4x^7(x+1)^3(3x+2)$

**15.** $\frac{3}{2}\sqrt{x} - \dfrac{1}{2\sqrt{x}} - \dfrac{1}{\sqrt{x^3}}$  **17.** $x(\pi x \cos \pi x + 2 \sin \pi x)$

**19.** $\dfrac{8t^3}{(t^4 + 1)^2}$  **21.** $-\dfrac{\sec^2\sqrt{1-x}}{2\sqrt{1-x}}$

**23.** $\dfrac{1 - y^4 - 2xy}{4xy^3 + x^2 - 3}$  **25.** $\dfrac{2 \sec 2\theta\,(\tan 2\theta - 1)}{(1 + \tan 2\theta)^2}$

**27.** $-(x-1)^{-2}$  **29.** $\dfrac{2x - y \cos(xy)}{x \cos(xy) + 1}$  **31.** $-6x \csc^2(3x^2 + 5)$

**33.** $\dfrac{\cos\sqrt{x} - \sqrt{x}\sin\sqrt{x}}{2\sqrt{x}}$  **35.** $2 \cos \theta \tan(\sin \theta) \sec^2(\sin \theta)$

**37.** $\frac{1}{5}(x \tan x)^{-4/5}(\tan x + x \sec^2 x)$

**39.** $\cos(\tan\sqrt{1 + x^3})(\sec^2\sqrt{1 + x^3})\dfrac{3x^2}{2\sqrt{1 + x^3}}$

**41.** $-\frac{4}{27}$  **43.** $-5x^4/y^{11}$  **45.** 1
**47.** $y = 2\sqrt{3}x + 1 - \pi\sqrt{3}/3$  **49.** $y = 2x + 1$, $y = -\frac{1}{2}x + 1$
**51.** (a) $\dfrac{10 - 3x}{2\sqrt{5 - x}}$  (b) $y = \frac{7}{4}x + \frac{1}{4}$, $y = -x + 8$

(c)

**53.** $(\pi/4, \sqrt{2})$, $(5\pi/4, -\sqrt{2})$  **55.** $y = -\frac{2}{3}x^2 + \frac{14}{3}x$
**59.** (a) 2  (b) 44
**61.** $2xg(x) + x^2g'(x)$  **63.** $2g(x)g'(x)$  **65.** $g'(g(x))g'(x)$
**67.** $f'(x) = g'(\sin x) \cdot \cos x$
**69.** $\dfrac{f'(x)[g(x)]^2 + g'(x)\,[f(x)]^2}{[f(x) + g(x)]^2}$
**71.** $f'(g(\sin 4x))g'(\sin 4x)(\cos 4x)(4)$
**73.** (a) $v(t) = 3t^2 - 12$; $a(t) = 6t$  (b) $t > 2$; $0 \le t < 2$
(c) 23  (d)  (e) $t > 2$; $0 < t < 2$

**75.** 4 kg/m  **77.** $\frac{4}{3}$ cm$^2$/min
**79.** 13 ft/s  **81.** 400 ft/h
**83.** (a) $L(x) = 1 + x$; $\sqrt[3]{1 + 3x} \approx 1 + x$; $\sqrt[3]{1.03} \approx 1.01$
(b) $-0.235 < x < 0.401$
**85.** $12 + \frac{3}{2}\pi \approx 16.7$ cm$^2$  **87.** $\frac{1}{32}$  **89.** $\frac{1}{4}$  **91.** $\frac{1}{8}x^2$

**PROBLEMS PLUS ■ PAGE 194**

**1.** $(\pm\sqrt{3}/2, \frac{1}{4})$  **5.** $3\sqrt{2}$  **9.** $(0, \frac{5}{4})$
**11.** (a) $4\pi\sqrt{3}/\sqrt{11}$ rad/s  (b) $40(\cos \theta + \sqrt{8 + \cos^2\theta})$ cm
(c) $-480\pi \sin \theta\,(1 + \cos \theta/\sqrt{8 + \cos^2\theta})$ cm/s
**13.** $x_T \in (3, \infty)$, $y_T \in (2, \infty)$, $x_N \in (0, \frac{5}{3})$, $y_N \in (-\frac{5}{2}, 0)$
**15.** (b) (i) 53° (or 127°)  (ii) 63° (or 117°)
**17.** $R$ approaches the midpoint of the radius $AO$.
**19.** $-\sin a$  **21.** $(1, -2)$, $(-1, 0)$
**23.** $\sqrt{29}/58$  **25.** $2 + \frac{375}{128}\pi \approx 11.204$ cm$^3$/min

**CHAPTER 3**

**EXERCISES 3.1 ■ PAGE 204**

*Abbreviations: abs, absolute; loc, local; max, maximum; min, minimum*

**1.** Abs min: smallest function value on the entire domain of the function; loc min at $c$: smallest function value when $x$ is near $c$
**3.** Abs max at $s$, abs min at $r$, loc max at $c$, loc min at $b$ and $r$, neither a max nor a min at $a$ and $d$

**5.** Abs max $f(4) = 5$, loc max $f(4) = 5$ and $f(6) = 4$,
loc min $f(2) = 2$ and $f(1) = f(5) = 3$

**7.**

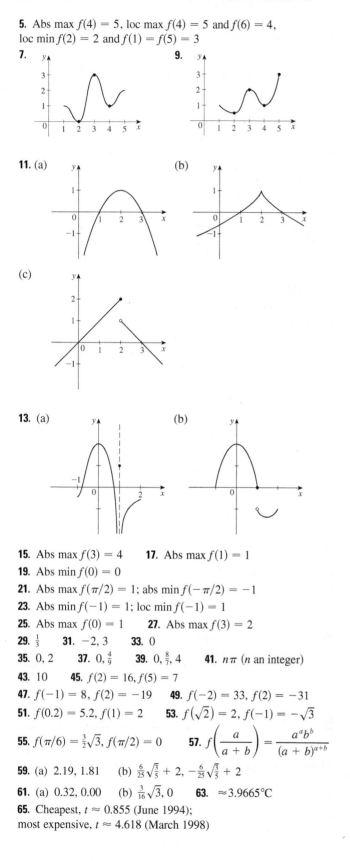

**9.**

**11.** (a)

(b)

(c)

**13.** (a)

(b)

**15.** Abs max $f(3) = 4$     **17.** Abs max $f(1) = 1$

**19.** Abs min $f(0) = 0$

**21.** Abs max $f(\pi/2) = 1$; abs min $f(-\pi/2) = -1$

**23.** Abs min $f(-1) = 1$; loc min $f(-1) = 1$

**25.** Abs max $f(0) = 1$     **27.** Abs max $f(3) = 2$

**29.** $\frac{1}{3}$     **31.** $-2, 3$     **33.** $0$

**35.** $0, 2$     **37.** $0, \frac{4}{9}$     **39.** $0, \frac{8}{7}, 4$     **41.** $n\pi$ ($n$ an integer)

**43.** $10$     **45.** $f(2) = 16, f(5) = 7$

**47.** $f(-1) = 8, f(2) = -19$     **49.** $f(-2) = 33, f(2) = -31$

**51.** $f(0.2) = 5.2, f(1) = 2$     **53.** $f(\sqrt{2}) = 2, f(-1) = -\sqrt{3}$

**55.** $f(\pi/6) = \frac{3}{2}\sqrt{3}, f(\pi/2) = 0$     **57.** $f\left(\dfrac{a}{a+b}\right) = \dfrac{a^a b^b}{(a+b)^{a+b}}$

**59.** (a) $2.19, 1.81$     (b) $\frac{6}{25}\sqrt{\frac{3}{5}} + 2, -\frac{6}{25}\sqrt{\frac{3}{5}} + 2$

**61.** (a) $0.32, 0.00$     (b) $\frac{3}{16}\sqrt{3}, 0$     **63.** $\approx 3.9665°C$

**65.** Cheapest, $t \approx 0.855$ (June 1994);
most expensive, $t \approx 4.618$ (March 1998)

**67.** (a) $r = \frac{2}{3}r_0$     (b) $v = \frac{4}{27}kr_0^3$

(c)

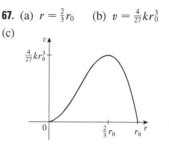

**EXERCISES 3.2** ■ **PAGE 212**

**1.** $2$     **3.** $\frac{9}{4}$     **5.** $f$ is not differentiable on $(-1, 1)$

**7.** $0.3, 3, 6.3$     **9.** $1$     **11.** $\sqrt{3}/9$     **13.** $1$

**15.** $f$ is not continous at $3$     **23.** $16$     **25.** No     **31.** No

**EXERCISES 3.3** ■ **PAGE 220**

*Abbreviations:* inc, increasing; dec, decreasing; CD, concave
downward; CU, concave upward; HA, horizontal asymptote;
VA, vertical asymptote; IP, inflection point(s)

**1.** (a) $(1, 3), (4, 6)$     (b) $(0, 1), (3, 4)$     (c) $(0, 2)$
(d) $(2, 4), (4, 6)$     (e) $(2, 3)$

**3.** (a) I/D Test     (b) Concavity Test
(c) Find points at which the concavity changes.

**5.** (a) Inc on $(1, 5)$; dec on $(0, 1)$ and $(5, 6)$
(b) Loc max at $x = 5$, loc min at $x = 1$

**7.** (a) $3, 5$     (b) $2, 4, 6$     (c) $1, 7$

**9.** (a) Inc on $(-\infty, -3), (2, \infty)$; dec on $(-3, 2)$
(b) Loc max $f(-3) = 81$; loc min $f(2) = -44$
(c) CU on $\left(-\frac{1}{2}, \infty\right)$; CD on $\left(-\infty, -\frac{1}{2}\right)$; IP $\left(-\frac{1}{2}, \frac{37}{2}\right)$

**11.** (a) Inc on $(-1, 0), (1, \infty)$; dec on $(-\infty, -1), (0, 1)$
(b) Loc max $f(0) = 3$; loc min $f(\pm1) = 2$
(c) CU on $\left(-\infty, -\sqrt{3}/3\right), \left(\sqrt{3}/3, \infty\right)$;
CD on $\left(-\sqrt{3}/3, \sqrt{3}/3\right)$; IP $\left(\pm\sqrt{3}/3, \frac{22}{9}\right)$

**13.** (a) Inc on $(0, \pi/4), (5\pi/4, 2\pi)$; dec on $(\pi/4, 5\pi/4)$
(b) Loc max $f(\pi/4) = \sqrt{2}$; loc min $f(5\pi/4) = -\sqrt{2}$
(c) CU on $(3\pi/4, 7\pi/4)$; CD on $(0, 3\pi/4), (7\pi/4, 2\pi)$;
IP $(3\pi/4, 0), (7\pi/4, 0)$

**15.** Loc max $f(1) = 2$; loc min $f(0) = 1$

**17.** Loc min $f\left(\frac{1}{16}\right) = -\frac{1}{4}$

**19.** (a) $f$ has a local maximum at $2$.
(b) $f$ has a horizontal tangent at $6$.

**21.**

**23.**

**25.**

**27.** (a) Inc on $(0, 2)$, $(4, 6)$, $(8, \infty)$;
dec on $(2, 4)$, $(6, 8)$
(b) Loc max at $x = 2, 6$;
loc min at $x = 4, 8$
(c) CU on $(3, 6)$, $(6, \infty)$;
CD on $(0, 3)$
(d) 3
(e) See graph at right.

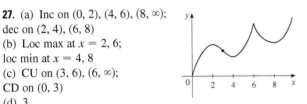

**29.** (a) Inc on $(-\infty, -2)$, $(2, \infty)$; dec on $(-2, 2)$
(b) Loc max $f(-2) = 18$; loc min $f(2) = -14$
(c) CU on $(0, \infty)$, CD on $(-\infty, 0)$; IP $(0, 2)$
(d)

**31.** (a) Inc on $(-\infty, -1)$, $(0, 1)$;
dec on $(-1, 0)$, $(1, \infty)$
(b) Loc max $f(-1) = 3, f(1) = 3$;
loc min $f(0) = 2$
(c) CU on $\left(-1/\sqrt{3}, 1/\sqrt{3}\right)$;
CD on $\left(-\infty, -1/\sqrt{3}\right)$, $\left(1/\sqrt{3}, \infty\right)$;
IP $\left(\pm 1/\sqrt{3}, \frac{23}{9}\right)$
(d) See graph at right.

**33.** (a) Inc on $(-\infty, -2)$, $(0, \infty)$;
dec on $(-2, 0)$
(b) Loc max $h(-2) = 7$;
loc min $h(0) = -1$
(c) CU on $(-1, \infty)$;
CD on $(-\infty, -1)$; IP $(-1, 3)$
(d) See graph at right.

**35.** (a) Inc on $(-\infty, 4)$; dec on $(4, 6)$
(b) Loc max $f(4) = 4\sqrt{2}$
(c) CD on $(-\infty, 6)$; No IP
(d) See graph at right.

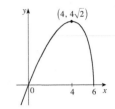

**37.** (a) Inc on $(-1, \infty)$;
dec on $(-\infty, -1)$
(b) Loc min $C(-1) = -3$
(c) CU on $(-\infty, 0)$, $(2, \infty)$;
CD on $(0, 2)$;
IP $(0, 0)$, $\left(2, 6\sqrt[3]{2}\right)$
(d) See graph at right.

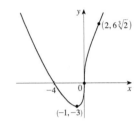

**39.** (a) Inc on $(\pi, 2\pi)$;
dec on $(0, \pi)$
(b) Loc min $f(\pi) = -1$
(c) CU on $(\pi/3, 5\pi/3)$;
CD on $(0, \pi/3)$, $(5\pi/3, 2\pi)$;
IP $\left(\pi/3, \frac{5}{4}\right)$, $\left(5\pi/3, \frac{5}{4}\right)$
(d) See graph at right.

**41.** $(3, \infty)$
**43.** (a) Loc and abs max $f(1) = \sqrt{2}$, no min
(b) $\frac{1}{4}\left(3 - \sqrt{17}\right)$
**45.** (b) CU on $(0.94, 2.57)$, $(3.71, 5.35)$;
CD on $(0, 0.94)$, $(2.57, 3.71)$, $(5.35, 2\pi)$;
IP $(0.94, 0.44)$, $(2.57, -0.63)$, $(3.71, -0.63)$, $(5.35, 0.44)$
**47.** CU on $(-\infty, -0.6)$, $(0.0, \infty)$; CD on $(-0.6, 0.0)$
**49.** (a) The rate of increase is initially very small, increases to a maximum at $t \approx 8$ h, then decreases toward 0.
(b) When $t = 8$   (c) CU on $(0, 8)$; CD on $(8, 18)$   (d) $(8, 350)$
**51.** $K(3) - K(2)$; CD
**53.** $f(x) = \frac{1}{9}(2x^3 + 3x^2 - 12x + 7)$
**55.** (a) $a = 0, b = -1$   (b) $y = -x$ at $(0, 0)$

EXERCISES 3.4 ■ PAGE 234

**1.** (a) As $x$ becomes large, $f(x)$ approaches 5.
(b) As $x$ becomes large negative, $f(x)$ approaches 3.
**3.** (a) $-2$   (b) 2   (c) $\infty$   (d) $-\infty$
(e) $x = 1, x = 3, y = -2, y = 2$
**5.** 0   **7.** $\frac{3}{2}$   **9.** $\frac{3}{2}$   **11.** 0   **13.** $-1$   **15.** 4
**17.** 3   **19.** $\frac{1}{6}$   **21.** $\frac{1}{2}(a - b)$   **23.** $\infty$
**25.** $-\infty$   **27.** $\infty$   **29.** 1
**31.** (a), (b) $-\frac{1}{2}$   **33.** $y = 2; x = 2$
**35.** $y = 2; x = -2, x = 1$   **37.** $x = 5$   **39.** $y = 3$
**41.** $f(x) = \dfrac{2 - x}{x^2(x - 3)}$   **43.** (a) $\frac{5}{4}$   (b) 5

**45.** $y = -1$

**47.** $y = 0$

**49.** $-\infty, -\infty$

**51.** $-\infty, \infty$

**53.**

**55.**

**57.** (a) 0    (b) An infinite number of times

**59.** (a) 0    (b) $\pm\infty$    **61.** 4
**63.** $N \geq 15$    **65.** $N \leq -6, N \leq -22$
**67.** (a) $x > 100$

**EXERCISES 3.5 ■ PAGE 242**

*Abbreviation:* int, intercept; SA, slant asymptote

**1.** A. $\mathbb{R}$    B. $y$-int 0; $x$-int 0, 6
C. None    D. None
E. Inc on $(-\infty, 2)$, $(6, \infty)$;
dec on $(2, 6)$
F. Loc max $f(2) = 32$;
loc min $f(6) = 0$
G. CU on $(4, \infty)$; CD on $(-\infty, 4)$;
IP $(4, 16)$
H. See graph at right.

**3.** A. $\mathbb{R}$    B. $y$-int 0; $x$-int 0, $\sqrt[3]{4}$
C. None    D. None
E. Inc on $(1, \infty)$; dec on $(-\infty, 1)$
F. Loc min $f(1) = -3$
G. CU on $(-\infty, \infty)$
H. See graph at right.

**5.** A. $\mathbb{R}$    B. $y$-int 0; $x$-int 0, 4
C. None    D. None
E. Inc on $(1, \infty)$; dec on $(-\infty, 1)$
F. Loc min $f(1) = -27$
G. CU on $(-\infty, 2)$, $(4, \infty)$;
CD on $(2, 4)$;
IP $(2, -16)$, $(4, 0)$
H. See graph at right.

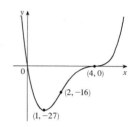

**7.** A. $\mathbb{R}$    B. $y$-int 0; $x$-int 0
C. About $(0, 0)$    D. None
E. Inc on $(-\infty, \infty)$
F. None
G. CU on $(-2, 0)$, $(2, \infty)$;
CD on $(-\infty, -2)$, $(0, 2)$;
IP $\left(-2, -\frac{256}{15}\right)$, $(0, 0)$, $\left(2, \frac{256}{15}\right)$
H. See graph at right.

**9.** A. $\{x \mid x \neq 1\}$    B. $y$-int 0; $x$-int 0
C. None    D. VA $x = 1$, HA $y = 1$
E. Dec on $(-\infty, 1)$, $(1, \infty)$
F. None
G. CU on $(1, \infty)$; CD on $(-\infty, 1)$
H. See graph at right.

**11.** A. $(-\infty, 1) \cup (1, 2) \cup (2, \infty)$
B. $y$-int 0; $x$-int 0    C. None
D. HA $y = -1$; VA $x = 2$
E. Inc on $(-\infty, 1)$, $(1, 2)$, $(2, \infty)$
F. None
G. CU on $(-\infty, 1)$, $(1, 2)$;
CD on $(2, \infty)$
H. See graph at right.

**13.** A. $\{x \mid x \neq \pm 3\}$    B. $y$-int $-\frac{1}{9}$
C. About $y$-axis    D. VA $x = \pm 3$, HA $y = 0$
E. Inc on $(-\infty, -3)$, $(-3, 0)$;
dec on $(0, 3)$, $(3, \infty)$
F. Loc max $f(0) = -\frac{1}{9}$
G. CU on $(-\infty, -3)$, $(3, \infty)$;
CD on $(-3, 3)$
H. See graph at right.

**15.** A. $\mathbb{R}$    B. $y$-int 0; $x$-int 0
C. About $(0, 0)$    D. HA $y = 0$
E. Inc on $(-3, 3)$;
dec on $(-\infty, -3)$, $(3, \infty)$
F. Loc min $f(-3) = -\frac{1}{6}$;
loc max $f(3) = \frac{1}{6}$;
G. CU on $\left(-3\sqrt{3}, 0\right)$, $\left(3\sqrt{3}, \infty\right)$;
CD on $\left(-\infty, -3\sqrt{3}\right)$, $\left(0, 3\sqrt{3}\right)$;
IP $(0, 0)$, $\left(\pm 3\sqrt{3}, \pm\sqrt{3}/12\right)$
H. See graph at right.

**17.** A. $(-\infty, 0) \cup (0, \infty)$    B. $x$-int 1
C. None    D. HA $y = 0$; VA $x = 0$
E. Inc on $(0, 2)$;
dec on $(-\infty, 0)$, $(2, \infty)$
F. Loc max $f(2) = \frac{1}{4}$
G. CU on $(3, \infty)$;
CD on $(-\infty, 0)$, $(0, 3)$; IP $\left(3, \frac{2}{9}\right)$

**19.** A. $\mathbb{R}$    B. $y$-int 0; $x$-int 0
C. About $y$-axis    D. HA $y = 1$
E. Inc on $(0, \infty)$; dec on $(-\infty, 0)$
F. Loc min $f(0) = 0$
G. CU on $(-1, 1)$;
CD on $(-\infty, -1)$, $(1, \infty)$; IP $\left(\pm 1, \frac{1}{4}\right)$
H. See graph at right.

**21.** A. $[0, \infty)$    B. $y$-int 0; $x$-int 0, 3
C. None    D. None
E. Inc on $(1, \infty)$; dec on $(0, 1)$
F. Loc min $f(1) = -2$
G. CU on $(0, \infty)$
H. See graph at right.

**23.** A. $(-\infty, -2] \cup [1, \infty)$
B. $x$-int $-2$, 1
C. None    D. None
E. Inc on $(1, \infty)$; dec on $(-\infty, -2)$
F. None
G. CD on $(-\infty, -2)$, $(1, \infty)$
H. See graph at right.

**25.** A. $\mathbb{R}$    B. $y$-int 0; $x$-int 0
C. About the origin
D. HA $y = \pm 1$
E. Inc on $(-\infty, \infty)$    F. None
G. CU on $(-\infty, 0)$;
CD on $(0, \infty)$; IP $(0, 0)$
H. See graph at right.

**27.** A. $\{x \mid |x| \le 1, x \neq 0\} = [-1, 0) \cup (0, 1]$
B. $x$-int $\pm 1$    C. About $(0, 0)$
D. VA $x = 0$
E. Dec on $(-1, 0)$, $(0, 1)$
F. None
G. CU on $\left(-1, -\sqrt{2/3}\right)$, $\left(0, \sqrt{2/3}\right)$;
CD on $\left(-\sqrt{2/3}, 0\right)$, $\left(\sqrt{2/3}, 1\right)$;
IP $\left(\pm\sqrt{2/3}, \pm 1/\sqrt{2}\right)$
H. See graph at right.

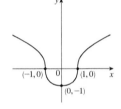

**29.** A. $\mathbb{R}$    B. $y$-int 0; $x$-int 0, $\pm 3\sqrt{3}$    C. About the origin
D. None    E. Inc on $(-\infty, -1)$, $(1, \infty)$; dec on $(-1, 1)$
F. Loc max $f(-1) = 2$;
loc min $f(1) = -2$
G. CU on $(0, \infty)$; CD on $(-\infty, 0)$;
IP $(0, 0)$
H. See graph at right.

**31.** A. $\mathbb{R}$    B. $y$-int $-1$; $x$-int $\pm 1$
C. About $y$-axis    D. None
E. Inc on $(0, \infty)$; dec on $(-\infty, 0)$
F. Loc min $f(0) = -1$
G. CU on $(-1, 1)$;
CD on $(-\infty, -1)$, $(1, \infty)$;
IP $(\pm 1, 0)$
H. See graph at right.

**33.** A. $\mathbb{R}$    B. $y$-int 0; $x$-int $n\pi$ ($n$ an integer)
C. About $(0, 0)$, period $2\pi$    D. None
*E–G answers for $0 \le x \le \pi$:*
E. Inc on $(0, \pi/2)$; dec on $(\pi/2, \pi)$    F. Loc max $f(\pi/2) = 1$
G. Let $\alpha = \sin^{-1}\sqrt{2/3}$; CU on $(0, \alpha)$, $(\pi - \alpha, \pi)$;
CD on $(\alpha, \pi - \alpha)$; IP at $x = 0$, $\pi$, $\alpha$, $\pi - \alpha$
H.

**35.** A. $(-\pi/2, \pi/2)$    B. $y$-int 0; $x$-int 0    C. About $y$-axis
D. VA $x = \pm\pi/2$
E. Inc on $(0, \pi/2)$;
dec on $(-\pi/2, 0)$
F. Loc min $f(0) = 0$
G. CU on $(-\pi/2, \pi/2)$
H. See graph at right.

**37.** A. $(0, 3\pi)$   C. None   D. None
E. Inc on $(\pi/3, 5\pi/3)$, $(7\pi/3, 3\pi)$;
dec on $(0, \pi/3)$, $(5\pi/3, 7\pi/3)$
F. Loc min $f(\pi/3) = (\pi/6) - \frac{1}{2}\sqrt{3}, f(7\pi/3) = (7\pi/6) - \frac{1}{2}\sqrt{3}$;
loc max $f(5\pi/3) = (5\pi/6) + \frac{1}{2}\sqrt{3}$
G. CU on $(0, \pi)$, $(2\pi, 3\pi)$; CD on $(\pi, 2\pi)$;
IP $(\pi, \pi/2)$, $(2\pi, \pi)$
H.

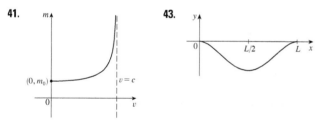

**39.** A. All reals except $(2n + 1)\pi$ ($n$ an integer)
B. $y$-int 0; $x$-int $2n\pi$
C. About the origin, period $2\pi$
D. VA $x = (2n + 1)\pi$
E. Inc on $((2n - 1)\pi, (2n + 1)\pi)$   F. None
G. CU on $(2n\pi, (2n + 1)\pi)$; CD on $((2n - 1)\pi, 2n\pi)$;
IP $(2n\pi, 0)$
H.

**41.**

**43.**

**45.** $y = x - 1$     **47.** $y = 2x - 2$

**49.** A. $(-\infty, 1) \cup (1, \infty)$
B. $y$-int 0; $x$-int 0   C. None
D. VA $x = 1$; SA $y = x + 1$
E. Inc on $(-\infty, 0)$, $(2, \infty)$;
dec on $(0, 1)$, $(1, 2)$
F. Loc max $f(0) = 0$;
loc min $f(2) = 4$
G. CU on $(1, \infty)$; CD on $(-\infty, 1)$
H. See graph at right.

**51.** A. $(-\infty, 0) \cup (0, \infty)$
B. $x$-int $-\sqrt[3]{4}$   C. None
D. VA $x = 0$; SA $y = x$
E. Inc. on $(-\infty, 0)$, $(2, \infty)$;
dec on $(0, 2)$
F. Loc min $f(2) = 3$
G. CU on $(-\infty, 0)$, $(0, \infty)$
H. See graph at right.

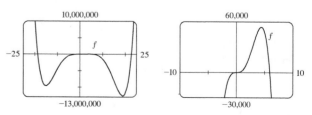

**53.** A. $\mathbb{R}$   B. $y$-int. 1; $x$-int. $-1$
C. None   D. SA $y = 2x + 1$
E. Inc. on $(-\infty, \infty)$   F. None
G. CU on $(-\infty, -\sqrt{3})$,
$(0, \sqrt{3})$;
CD on $(-\sqrt{3}, 0)$, $(\sqrt{3}, \infty)$;
IP $(\pm\sqrt{3}, 1 \pm \frac{3}{2}\sqrt{3})$, $(0, 1)$
H. See graph at right.

**55.**

**59.** VA $x = 0$, asymptotic to $y = x^3$

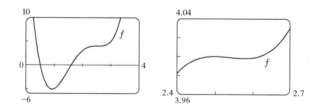

EXERCISES 3.6 ■ PAGE 249

**1.** Inc on $(0.92, 2.5)$, $(2.58, \infty)$; dec on $(-\infty, 0.92)$, $(2.5, 2.58)$;
loc max $f(2.5) = 4$; loc min $f(0.92) \approx -5.12$, $f(2.58) \approx 3.998$;
CU on $(-\infty, 1.46)$, $(2.54, \infty)$; CD on $(1.46, 2.54)$;
IP $(1.46, -1.40)$, $(2.54, 3.999)$

**3.** Inc on $(-15, 4.40)$, $(18.93, \infty)$;
dec on $(-\infty, -15)$, $(4.40, 18.93)$;
loc max $f(4.40) \approx 53,800$; loc min $f(-15) \approx -9,700,000$,
$f(18.93) \approx -12,700,000$; CU on $(-\infty, -11.34)$, $(0, 2.92)$,
$(15.08, \infty)$; CD on $(-11.34, 0)$, $(2.92, 15.08)$;
IP $(0, 0)$, $\approx (-11.34, -6,250,000)$, $(2.92, 31,800)$,
$(15.08, -8,150,000)$

**5.** Inc on $(-\infty, -1.47)$, $(-1.47, 0.66)$; dec on $(0.66, \infty)$; loc max $f(0.66) \approx 0.38$; CU on $(-\infty, -1.47)$, $(-0.49, 0)$, $(1.10, \infty)$; CD on $(-1.47, -0.49)$, $(0, 1.10)$; IP $(-0.49, -0.44)$, $(1.10, 0.31)$

**7.** Inc on $(-1.40, -0.44)$, $(0.44, 1.40)$; dec on $(-\pi, -1.40)$, $(-0.44, 0)$, $(0, 0.44)$, $(1.40, \pi)$; loc max $f(-0.44) \approx -4.68$, $f(1.40) \approx 6.09$; loc min $f(-1.40) \approx -6.09$, $f(0.44) \approx 5.22$; CU on $(-\pi, -0.77)$, $(0, 0.77)$; CD on $(-0.77, 0)$, $(0.77, \pi)$; IP $(-0.77, -5.22)$, $(0.77, 5.22)$

**9.** Inc on $\left(-8 - \sqrt{61}, -8 + \sqrt{61}\right)$; dec on $\left(-\infty, -8 - \sqrt{61}\right)$, $\left(-8 + \sqrt{61}, 0\right)$, $(0, \infty)$; CU on $\left(-12 - \sqrt{138}, -12 + \sqrt{138}\right)$, $(0, \infty)$; CD on $\left(-\infty, -12 - \sqrt{138}\right)$, $\left(-12 + \sqrt{138}, 0\right)$

**11.** Loc max $f(-5.6) \approx 0.018$, $f(0.82) \approx -281.5$, $f(5.2) \approx 0.0145$; loc min $f(3) = 0$

**13.** $f'(x) = -\dfrac{x(x + 1)^2(x^3 + 18x^2 - 44x - 16)}{(x - 2)^3(x - 4)^5}$

$f''(x) = 2\dfrac{(x + 1)(x^6 + 36x^5 + 6x^4 - 628x^3 + 684x^2 + 672x + 64)}{(x - 2)^4(x - 4)^6}$

CU on $(-35.3, -5.0)$, $(-1, -0.5)$, $(-0.1, 2)$, $(2, 4)$, $(4, \infty)$; CD on $(-\infty, -35.3)$, $(-5.0, -1)$, $(-0.5, -0.1)$; IP $(-35.3, -0.015)$, $(-5.0, -0.005)$, $(-1, 0)$, $(-0.5, 0.00001)$, $(-0.1, 0.0000066)$

**15.** Inc on $(-9.41, -1.29)$, $(0, 1.05)$; dec on $(-\infty, -9.41)$, $(-1.29, 0)$, $(1.05, \infty)$; loc max $f(-1.29) \approx 7.49$, $f(1.05) \approx 2.35$; loc min $f(-9.41) \approx -0.056$, $f(0) = 0.5$; CU on $(-13.81, -1.55)$, $(-1.03, 0.60)$, $(1.48, \infty)$; CD on $(-\infty, -13.81)$, $(-1.55, -1.03)$, $(0.60, 1.48)$; IP $(-13.81, -0.05)$, $(-1.55, 5.64)$, $(-1.03, 5.39)$, $(0.60, 1.52)$, $(1.48, 1.93)$

**17.** Inc on $(-4.91, -4.51)$, $(0, 1.77)$, $(4.91, 8.06)$, $(10.79, 14.34)$, $(17.08, 20)$; dec on $(-4.51, -4.10)$, $(1.77, 4.10)$, $(8.06, 10.79)$, $(14.34, 17.08)$; loc max $f(-4.51) \approx 0.62$, $f(1.77) \approx 2.58$, $f(8.06) \approx 3.60$, $f(14.34) \approx 4.39$; loc min $f(10.79) \approx 2.43$, $f(17.08) \approx 3.49$; CU on $(9.60, 12.25)$, $(15.81, 18.65)$; CD on $(-4.91, -4.10)$, $(0, 4.10)$, $(4.91, 9.60)$, $(12.25, 15.81)$, $(18.65, 20)$; IP at $(9.60, 2.95)$, $(12.25, 3.27)$, $(15.81, 3.91)$, $(18.65, 4.20)$

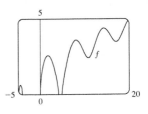

**19.** Max $f(0.59) \approx 1$, $f(0.68) \approx 1$, $f(1.96) \approx 1$; min $f(0.64) \approx 0.99996$, $f(1.46) \approx 0.49$, $f(2.73) \approx -0.51$; IP $(0.61, 0.99998)$, $(0.66, 0.99998)$, $(1.17, 0.72)$, $(1.75, 0.77)$, $(2.28, 0.34)$

**21.** For $c \geq 0$, there is an absolute minimum at the origin. There are no other maxima or minima. The more negative $c$ becomes, the farther the two IPs move from the origin. $c = 0$ is a transitional value.

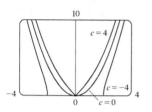

**23.** For $c > 0$, the maximum and minimum values are always $\pm\frac{1}{2}$, but the extreme points and IPs move closer to the $y$-axis as $c$ increases. $c = 0$ is a transitional value: when $c$ is replaced by $-c$, the curve is reflected in the $x$-axis.

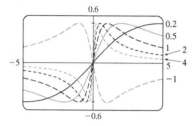

**25.** For $|c| < 1$, the graph has loc max and min values; for $|c| \geq 1$ it does not. The function increases for $c \geq 1$ and decreases for $c \leq -1$. As $c$ changes, the IPs move vertically but not horizontally.

**27.** (a) Positive    (b)

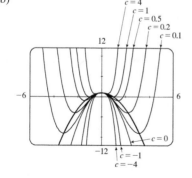

**1.** (a) 11, 12    (b) 11.5, 11.5    **3.** 10, 10    **5.** $\frac{9}{4}$
**7.** 25 m by 25 m    **9.** $N = 1$

**11.** (a)

(c) $A = xy$    (d) $5x + 2y = 750$    (e) $A(x) = 375x - \frac{5}{2}x^2$
(f) $14{,}062.5 \text{ ft}^2$
**13.** 1000 ft by 1500 ft    **15.** $4000 \text{ cm}^3$    **17.** \$191.28
**19.** $\left(-\frac{6}{5}, \frac{3}{5}\right)$    **21.** $\left(-\frac{1}{3}, \pm\frac{4}{3}\sqrt{2}\right)$    **23.** Square, side $\sqrt{2}\,r$
**25.** $L/2, \sqrt{3}\,L/4$    **27.** Base $\sqrt{3}\,r$, height $3r/2$
**29.** $4\pi r^3/(3\sqrt{3})$    **31.** $\pi r^2(1 + \sqrt{5})$    **33.** 24 cm, 36 cm
**35.** (a) Use all of the wire for the square
(b) $40\sqrt{3}/(9 + 4\sqrt{3})$ m for the square
**37.** Height $=$ radius $= \sqrt[3]{V/\pi}$ cm
**39.** $V = 2\pi R^3/(9\sqrt{3})$    **43.** $E^2/(4r)$
**45.** (a) $\frac{3}{2}s^2\csc\theta\,(\csc\theta - \sqrt{3}\cot\theta)$    (b) $\cos^{-1}(1/\sqrt{3}) \approx 55°$
(c) $6s[h + s/(2\sqrt{2})]$
**47.** Row directly to $B$    **49.** $\approx 4.85$ km east of the refinery
**51.** $10\sqrt[3]{3}/(1 + \sqrt[3]{3})$ ft from the stronger source
**53.** $(a^{2/3} + b^{2/3})^{3/2}$    **55.** $2\sqrt{6}$
**57.** (b) (i) \$342,491; \$342/unit; \$390/unit    (ii) 400
(iii) \$320/unit
**59.** (a) $p(x) = 19 - \frac{1}{3000}x$    (b) \$9.50
**61.** (a) $p(x) = 550 - \frac{1}{10}x$    (b) \$175    (c) \$100
**65.** 9.35 m    **69.** $x = 6$ in.    **71.** $\pi/6$
**73.** $\frac{1}{2}(L + W)^2$
**75.** (a) About 5.1 km from $B$    (b) $C$ is close to $B$; $C$ is close to $D$; $W/L = \sqrt{25 + x^2}/x$, where $x = |BC|$
(c) $\approx 1.07$; no such value    (d) $\sqrt{41}/4 \approx 1.6$

**1.** (a) $x_2 \approx 2.3, x_3 \approx 3$    (b) No
**3.** $\frac{9}{2}$    **5.** $a, b, c$    **7.** 1.1785    **9.** $-1.25$    **11.** 1.82056420
**13.** 1.217562    **15.** 0.876726
**17.** $-3.637958, -1.862365, 0.889470$
**19.** 1.412391, 3.057104    **21.** 0.641714
**23.** $-1.93822883, -1.21997997, 1.13929375, 2.98984102$
**25.** 0.76682579    **27.** (b) 31.622777
**33.** (a) $-1.293227, -0.441731, 0.507854$    (b) $-2.0212$
**35.** (1.519855, 2.306964)    **37.** (0.410245, 0.347810)
**39.** 0.76286%

**1.** $F(x) = \frac{1}{2}x^2 - 3x + C$    **3.** $F(x) = \frac{1}{2}x + \frac{1}{4}x^3 - \frac{1}{5}x^4 + C$
**5.** $F(x) = \frac{2}{3}x^3 + \frac{1}{2}x^2 - x + C$    **7.** $F(x) = 5x^{7/5} + 40x^{1/5} + C$

**9.** $F(x) = \sqrt{2}x + C$    **11.** $F(x) = \begin{cases} -5/(4x^8) + C_1 & \text{if } x < 0 \\ -5/(4x^8) + C_2 & \text{if } x > 0 \end{cases}$

**13.** $G(t) = 2t^{1/2} + \frac{2}{3}t^{3/2} + \frac{2}{5}t^{5/2} + C$

**15.** $H(\theta) = -2\cos\theta - \tan\theta + C_n$ on $(n\pi - \pi/2, n\pi + \pi/2)$, $n$ an integer

**17.** $F(t) = 2\sec t + t^{1/2} + C_n$

**19.** $F(x) = x^5 - \frac{1}{3}x^6 + 4$    **21.** $f(x) = x^5 - x^4 + x^3 + Cx + D$

**23.** $f(x) = \frac{3}{20}x^{8/3} + Cx + D$

**25.** $f(t) = -\sin t + Ct^2 + Dt + E$

**27.** $f(x) = x + 2x^{3/2} + 5$    **29.** $f(x) = 4x^{3/2} + 2x^{5/2} + 4$

**31.** $f(t) = 2\sin t + \tan t + 4 - 2\sqrt{3}$

**33.** $f(x) = -x^2 + 2x^3 - x^4 + 12x + 4$

**35.** $f(\theta) = -\sin\theta - \cos\theta + 5\theta + 4$

**37.** $f(x) = 2x^2 + x^3 + 2x^4 + 2x + 3$

**39.** $f(x) = x^2 - \cos x - \frac{1}{2}\pi x$

**41.** 10    **43.** $b$

**45.**

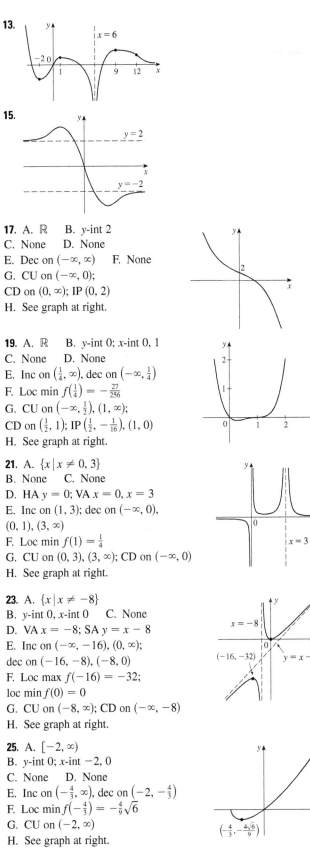

**47.**

**49.**

**51.** $s(t) = 1 - \cos t - \sin t$

**53.** $s(t) = \frac{1}{3}t^3 + \frac{1}{2}t^2 - 2t + 3$

**55.** $s(t) = -10\sin t - 3\cos t + (6/\pi)t + 3$

**57.** (a) $s(t) = 450 - 4.9t^2$    (b) $\sqrt{450/4.9} \approx 9.58$ s
(c) $-9.8\sqrt{450/4.9} \approx -93.9$ m/s    (d) About 9.09 s

**61.** 225 ft    **63.** $742.08    **65.** $\frac{130}{11} \approx 11.8$ s

**67.** $\frac{88}{15} \approx 5.87$ ft/s$^2$    **69.** 62,500 km/h$^2 \approx 4.82$ m/s$^2$

**71.** (a) 22.9125 mi    (b) 21.675 mi    (c) 30 min 33 s
(d) 55.425 mi

**CHAPTER 3 REVIEW ▪ PAGE 276**

**True-False Quiz**

**1.** False    **3.** False    **5.** True    **7.** False    **9.** True
**11.** True    **13.** False    **15.** True    **17.** True    **19.** True

**Exercises**

**1.** Abs max $f(4) = 5$, abs and loc min $f(3) = 1$
**3.** Abs max $f(2) = \frac{2}{5}$, abs and loc min $f(-\frac{1}{3}) = -\frac{9}{2}$
**5.** Abs and loc max $f(\pi/6) = \pi/6 + \sqrt{3}$,
abs min $f(-\pi) = -\pi - 2$, loc min $f(5\pi/6) = 5\pi/6 - \sqrt{3}$
**7.** $\frac{1}{2}$    **9.** $-\frac{2}{3}$    **11.** $\frac{3}{4}$

**13.**

**15.**

**17. A.** $\mathbb{R}$    **B.** $y$-int 2
**C.** None    **D.** None
**E.** Dec on $(-\infty, \infty)$    **F.** None
**G.** CU on $(-\infty, 0)$;
CD on $(0, \infty)$; IP $(0, 2)$
**H.** See graph at right.

**19. A.** $\mathbb{R}$    **B.** $y$-int 0; $x$-int 0, 1
**C.** None    **D.** None
**E.** Inc on $(\frac{1}{4}, \infty)$, dec on $(-\infty, \frac{1}{4})$
**F.** Loc min $f(\frac{1}{4}) = -\frac{27}{256}$
**G.** CU on $(-\infty, \frac{1}{2})$, $(1, \infty)$;
CD on $(\frac{1}{2}, 1)$; IP $(\frac{1}{2}, -\frac{1}{16})$, $(1, 0)$
**H.** See graph at right.

**21. A.** $\{x \mid x \neq 0, 3\}$
**B.** None    **C.** None
**D.** HA $y = 0$; VA $x = 0, x = 3$
**E.** Inc on $(1, 3)$; dec on $(-\infty, 0)$,
$(0, 1)$, $(3, \infty)$
**F.** Loc min $f(1) = \frac{1}{4}$
**G.** CU on $(0, 3)$, $(3, \infty)$; CD on $(-\infty, 0)$
**H.** See graph at right.

**23. A.** $\{x \mid x \neq -8\}$
**B.** $y$-int 0, $x$-int 0    **C.** None
**D.** VA $x = -8$; SA $y = x - 8$
**E.** Inc on $(-\infty, -16)$, $(0, \infty)$;
dec on $(-16, -8)$, $(-8, 0)$
**F.** Loc max $f(-16) = -32$;
loc min $f(0) = 0$
**G.** CU on $(-8, \infty)$; CD on $(-\infty, -8)$
**H.** See graph at right.

**25. A.** $[-2, \infty)$
**B.** $y$-int 0; $x$-int $-2, 0$
**C.** None    **D.** None
**E.** Inc on $(-\frac{4}{3}, \infty)$, dec on $(-2, -\frac{4}{3})$
**F.** Loc min $f(-\frac{4}{3}) = -\frac{4}{9}\sqrt{6}$
**G.** CU on $(-2, \infty)$
**H.** See graph at right.

**27.** A. $\mathbb{R}$   B. $y$-int $-2$
C. About $y$-axis, period $2\pi$   D. None
E. Inc. on $(2n\pi, (2n+1)\pi)$, $n$ an integer; dec. on $((2n-1)\pi, 2n\pi)$
F. Loc. max. $f((2n+1)\pi) = 2$; loc. min. $f(2n\pi) = -2$
G. CU on $(2n\pi - (\pi/3), 2n\pi + (\pi/3))$;
CD on $(2n\pi + (\pi/3), 2n\pi + (5\pi/3))$; IP $\left(2n\pi \pm (\pi/3), -\frac{1}{4}\right)$
H.

**29.** Inc on $\left(-\sqrt{3}, 0\right), \left(0, \sqrt{3}\right)$;
dec on $\left(-\infty, -\sqrt{3}\right), \left(\sqrt{3}, \infty\right)$;
loc max $f\left(\sqrt{3}\right) = \frac{2}{9}\sqrt{3}$,
loc min $f\left(-\sqrt{3}\right) = -\frac{2}{9}\sqrt{3}$;
CU on $\left(-\sqrt{6}, 0\right), \left(\sqrt{6}, \infty\right)$;
CD on $\left(-\infty, -\sqrt{6}\right), \left(0, \sqrt{6}\right)$;
IP $\left(\sqrt{6}, \frac{5}{36}\sqrt{6}\right), \left(-\sqrt{6}, -\frac{5}{36}\sqrt{6}\right)$

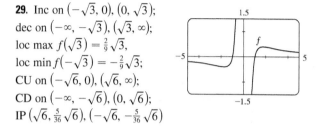

**31.** Inc on $(-0.23, 0), (1.62, \infty)$; dec on $(-\infty, -0.23), (0, 1.62)$;
loc max $f(0) = 2$; loc min $f(-0.23) \approx 1.96$, $f(1.62) \approx -19.2$;
CU on $(-\infty, -0.12), (1.24, \infty)$;
CD on $(-0.12, 1.24)$; IP $(-0.12, 1.98), (1.24, -12.1)$

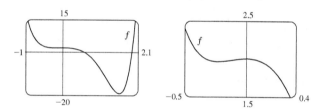

**37.** (a) 0   (b) CU on $\mathbb{R}$   **41.** $3\sqrt{3}\, r^2$
**43.** $4/\sqrt{3}$ cm from $D$   **45.** $L = C$   **47.** \$11.50
**49.** 1.297383   **51.** 1.16718557
**53.** $f(x) = \frac{2}{5}x^{5/2} + \frac{3}{5}x^{5/3} + C$
**55.** $f(t) = t^2 + 3\cos t + 2$
**57.** $f(x) = \frac{1}{2}x^2 - x^3 + 4x^4 + 2x + 1$
**59.** $s(t) = t^2 + \cos t + 2$
**61.**

**63.** No
**65.** (b) About 8.5 in. by 2 in.   (c) $20/\sqrt{3}$ in., $20\sqrt{2/3}$ in.

PROBLEMS PLUS ■ PAGE 280

**5.** $(-2, 4), (2, -4)$   **7.** $\frac{4}{3}$   **9.** $(m/2, m^2/4)$
**11.** $-3.5 < a < -2.5$

**13.** (a) $x/(x^2 + 1)$   (b) $\frac{1}{2}$
**15.** (a) $-\tan\theta \left[\dfrac{1}{c}\dfrac{dc}{dt} + \dfrac{1}{b}\dfrac{db}{dt}\right]$

(b) $\dfrac{b\dfrac{db}{dt} + c\dfrac{dc}{dt} - \left(b\dfrac{dc}{dt} + c\dfrac{db}{dt}\right)\sec\theta}{\sqrt{b^2 + c^2 - 2bc\cos\theta}}$

**17.** (a) $T_1 = D/c_1$, $T_2 = (2h\sec\theta)/c_1 + (D - 2h\tan\theta)/c_2$,
$T_3 = \sqrt{4h^2 + D^2}/c_1$
(c) $c_1 \approx 3.85$ km/s, $c_2 \approx 7.66$ km/s, $h \approx 0.42$ km
**21.** $3/\left(\sqrt[3]{2} - 1\right) \approx 11\frac{1}{2}$ h

## CHAPTER 4

EXERCISES 4.1 ■ PAGE 293

**1.** (a) $L_4 = 33$, $R_4 = 41$

(b) $L_8 \approx 35.2$, $R_8 \approx 39.2$

**3.** (a) 0.7908, underestimate   (b) 1.1835, overestimate

**5.** (a) 8, 6.875   (b) 5, 5.375

(c) 5.75, 5.9375

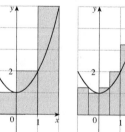

(d) $M_6$

**7.** $n = 2$: upper $= 3\pi \approx 9.42$, lower $= 2\pi \approx 6.28$

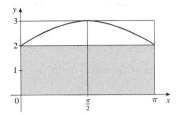

$n = 4$: upper $= \left(10 + \sqrt{2}\right)(\pi/4) \approx 8.96$,
lower $= \left(8 + \sqrt{2}\right)(\pi/4) \approx 7.39$

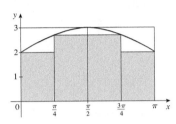

$n = 8$: upper $\approx 8.65$, lower $\approx 7.86$

**9.** 0.2533, 0.2170, 0.2101, 0.2050; 0.2
**11.** (a) Left: 0.8100, 0.7937, 0.7904;
right: 0.7600, 0.7770, 0.7804
**13.** 34.7 ft, 44.8 ft   **15.** 63.2 L, 70 L   **17.** 155 ft

**19.** $\lim\limits_{n\to\infty} \sum\limits_{i=1}^{n} \dfrac{2(1 + 2i/n)}{(1 + 2i/n)^2 + 1} \cdot \dfrac{2}{n}$   **21.** $\lim\limits_{n\to\infty} \sum\limits_{i=1}^{n} \sqrt{\sin(\pi i/n)} \cdot \dfrac{\pi}{n}$

**23.** The region under the graph of $y = \tan x$ from 0 to $\pi/4$
**25.** (a) $L_n < A < R_n$

**27.** (a) $\lim\limits_{n\to\infty} \dfrac{64}{n^6} \sum\limits_{i=1}^{n} i^5$   (b) $\dfrac{n^2(n + 1)^2(2n^2 + 2n - 1)}{12}$   (c) $\dfrac{32}{3}$

**29.** $\sin b$, 1

**EXERCISES 4.2 ■ PAGE 306**

**1.** $-6$
The Riemann sum represents
the sum of the areas of the two
rectangles above the $x$-axis
minus the sum of the areas of
the three rectangles below the
$x$-axis; that is, the net area of the
rectangles with respect to the
$x$-axis.

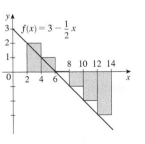

**3.** $-0.856759$
The Riemann sum represents
the sum of the areas of the
two rectangles above the
$x$-axis minus the sum of the
areas of the three rectangles
below the $x$-axis.

**5.** (a) 6   (b) 4   (c) 2
**7.** Lower, $L_5 = -64$; upper, $R_5 = 16$
**9.** 6.1820   **11.** 0.9071   **13.** 0.9029, 0.9018
**15.**

| $n$ | $R_n$ |
|---|---|
| 5 | 1.933766 |
| 10 | 1.983524 |
| 50 | 1.999342 |
| 100 | 1.999836 |

The values of $R_n$ appear to be
approaching 2.

**17.** $\int_2^6 \dfrac{1 - x^2}{4 + x^2}\, dx$   **19.** $\int_2^7 (5x^3 - 4x)\, dx$
**21.** $-9$   **23.** $\frac{2}{3}$   **25.** $-\frac{3}{4}$

**29.** $\lim\limits_{n\to\infty} \sum\limits_{i=1}^{n} \dfrac{2 + 4i/n}{1 + (2 + 4i/n)^5} \cdot \dfrac{4}{n}$

**31.** $\lim\limits_{n\to\infty} \sum\limits_{i=1}^{n} \left(\sin \dfrac{5\pi i}{n}\right) \dfrac{\pi}{n} = \dfrac{2}{5}$

**33.** (a) 4   (b) 10   (c) $-3$   (d) 2
**35.** $\frac{3}{2}$   **37.** $3 + \frac{9}{4}\pi$   **39.** $\frac{5}{2}$   **41.** 0   **43.** 3   **45.** 22.5
**47.** $\int_{-1}^{5} f(x)\, dx$   **49.** 122
**51.** $B < E < A < D < C$   **53.** 15
**59.** $3 \le \int_1^4 \sqrt{x}\, dx \le 6$   **61.** $\dfrac{\pi}{12} \le \int_{\pi/4}^{\pi/3} \tan x\, dx \le \dfrac{\pi}{12}\sqrt{3}$

**63.** $2 \le \int_{-1}^{1} \sqrt{1 + x^4}\, dx \le 2\sqrt{2}$   **71.** $\int_0^1 x^4\, dx$   **73.** $\frac{1}{2}$

**EXERCISES 4.3 ■ PAGE 318**

**1.** One process undoes what the other one does. See the
Fundamental Theorem of Calculus, page 317.

**3.** (a) 0, 2, 5, 7, 3     (d)
(b) (0, 3)
(c) $x = 3$

**5.**

(a), (b) $x^2$

**7.** $g'(x) = 1/(x^3 + 1)$ **9.** $g'(s) = (s - s^2)^8$

**11.** $F'(x) = -\sqrt{1 + \sec x}$ **13.** $h'(x) = -\sin^4(1/x)/x^2$

**15.** $y' = \sqrt{\tan x + \sqrt{\tan x}}\ \sec^2 x$

**17.** $y' = \dfrac{3(1 - 3x)^3}{1 + (1 - 3x)^2}$ **19.** $\frac{3}{4}$ **21.** 63 **23.** $\frac{52}{3}$

**25.** $1 + \sqrt{3}/2$ **27.** $-\frac{37}{6}$ **29.** $\frac{40}{3}$ **31.** 1 **33.** $\frac{49}{3}$

**35.** $\frac{17}{2}$ **37.** 0

**39.** The function $f(x) = x^{-4}$ is not continuous on the interval $[-2, 1]$, so FTC2 cannot be applied.

**41.** The function $f(\theta) = \sec\theta\,\tan\theta$ is not continuous on the interval $[\pi/3, \pi]$, so FTC2 cannot be applied.

**43.** $\frac{243}{4}$ **45.** 2

**47.** 3.75

**49.** $g'(x) = \dfrac{-2(4x^2 - 1)}{4x^2 + 1} + \dfrac{3(9x^2 - 1)}{9x^2 + 1}$

**51.** $h'(x) = -\dfrac{1}{2\sqrt{x}}\cos x + 3x^2\cos(x^6)$

**53.** $(-4, 0)$ **55.** 29

**57.** (a) $-2\sqrt{n}$, $\sqrt{4n - 2}$, $n$ an integer $> 0$

(b) $(0, 1)$, $\left(-\sqrt{4n - 1}, -\sqrt{4n - 3}\right)$, and $\left(\sqrt{4n - 1}, \sqrt{4n + 1}\right)$, $n$ an integer $> 0$ (c) 0.74

**59.** (a) Loc max at 1 and 5; loc min at 3 and 7

(b) $x = 9$

(c) $\left(\frac{1}{2}, 2\right)$, $(4, 6)$, $(8, 9)$

(d) See graph at right.

**61.** $\frac{1}{4}$ **69.** $f(x) = x^{3/2}$, $a = 9$

**71.** (b) Average expenditure over $[0, t]$; minimize average expenditure

**73.** $\ln 3$ **75.** $\pi$ **77.** $e^2 - 1$

**EXERCISES 4.4 ■ PAGE 326**

**5.** $\frac{1}{3}x^3 - (1/x) + C$ **7.** $\frac{1}{5}x^5 - \frac{1}{8}x^4 + \frac{1}{8}x^2 - 2x + C$

**9.** $\frac{2}{3}u^3 + \frac{9}{2}u^2 + 4u + C$ **11.** $\frac{1}{3}x^3 - 4\sqrt{x} + C$

**13.** $\frac{1}{2}\theta^2 + \csc\theta + C$ **15.** $\tan\alpha + C$

**17.** $\sin x + \frac{1}{4}x^2 + C$

**19.** $-\frac{10}{3}$ **21.** $\frac{21}{5}$ **23.** $-2$ **25.** 8 **27.** 36

**29.** $2\sqrt{5}$ **31.** $\frac{256}{15}$ **33.** $1 + \pi/4$ **35.** $\frac{256}{5}$ **37.** 1

**39.** $\frac{5}{2}$ **41.** $-3.5$ **43.** $\approx 1.36$ **45.** $\frac{4}{3}$

**47.** The increase in the child's weight (in pounds) between the ages of 5 and 10

**49.** Number of gallons of oil leaked in the first 2 hours

**51.** Increase in revenue when production is increased from 1000 to 5000 units

**53.** Newton-meters **55.** (a) $-\frac{3}{2}$ m (b) $\frac{41}{6}$ m

**57.** (a) $v(t) = \frac{1}{2}t^2 + 4t + 5$ m/s (b) $416\frac{2}{3}$ m

**59.** $46\frac{2}{3}$ kg **61.** 1.4 mi **63.** 28,320 L

**65.** $4.75 \times 10^5$ megawatt-hours

**67.** $-\cos x + \cosh x + C$ **69.** $\frac{1}{3}x^3 + x + \tan^{-1}x + C$

**71.** $\pi/6$

**EXERCISES 4.5 ■ PAGE 335**

**1.** $-(1/\pi)\cos\pi x + C$ **3.** $\frac{2}{9}(x^3 + 1)^{3/2} + C$

**5.** $-\frac{1}{4}\cos^4\theta + C$ **7.** $-\frac{1}{2}\cos(x^2) + C$

**9.** $-\frac{1}{20}(1 - 2x)^{10} + C$ **11.** $\frac{1}{3}(2x + x^2)^{3/2} + C$

**13.** $\frac{1}{3}\sec 3t + C$ **15.** $\frac{2}{3}\sqrt{3ax + bx^3} + C$

**17.** $\frac{1}{4}\tan^4\theta + C$ **19.** $\frac{1}{15}(x^3 + 3x)^5 + C$

**21.** $-\dfrac{1}{\sin x} + C$ **23.** $\frac{1}{2}(1 + z^3)^{2/3} + C$

**25.** $-\frac{2}{3}(\cot x)^{3/2} + C$ **27.** $\frac{1}{3}\sec^3 x + C$

**29.** $\frac{1}{40}(2x + 5)^{10} - \frac{5}{36}(2x + 5)^9 + C$

**31.** $\frac{1}{8}(x^2 - 1)^4 + C$ **33.** $\frac{1}{4}\sin^4 x + C$

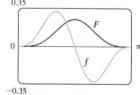

**35.** $2/\pi$ **37.** $\frac{45}{28}$ **39.** 4 **41.** 0

**43.** 3 **45.** $\frac{1}{3}(2\sqrt{2} - 1)a^3$ **47.** $\frac{16}{15}$ **49.** $\frac{1}{2}(\sin 4 - \sin 1)$

**51.** $\frac{1}{6}$ **53.** $\sqrt{3} - \frac{1}{3}$ **55.** $6\pi$

**57.** $\dfrac{5}{4\pi}\left(1 - \cos\dfrac{2\pi t}{5}\right)$ L **59.** 5 **67.** $-\frac{1}{3}\ln|5 - 3x| + C$

**69.** $\frac{1}{3}(\ln x)^3 + C$ **71.** $\frac{2}{3}(1 + e^x)^{3/2} + C$ **73.** $e^{\tan x} + C$

**75.** $\tan^{-1}x + \frac{1}{2}\ln(1 + x^2) + C$ **77.** $-\ln(1 + \cos^2 x) + C$

**79.** $\ln|\sin x| + C$ **81.** 2 **83.** $\ln(e + 1)$ **85.** $\pi^2/4$

**CHAPTER 4 REVIEW ■ PAGE 338**

**True-False Quiz**

**1.** True **3.** True **5.** False **7.** True **9.** True

**11.** False **13.** True **15.** False **17.** False

**Exercises**

**1.** (a) 8      (b) 5.7

**3.** $\frac{1}{2} + \pi/4$    **5.** 3    **7.** $f$ is $c$, $f'$ is $b$, $\int_0^x f(t)\, dt$ is $a$

**9.** 37    **11.** $\frac{9}{10}$    **13.** $-76$    **15.** $\frac{21}{4}$    **17.** Does not exist

**19.** $\frac{1}{3}\sin 1$    **21.** 0    **23.** $\dfrac{1}{2\pi}\sin^2\pi t + C$    **25.** $\frac{1}{2}\sqrt{2} - \frac{1}{2}$

**27.** $\frac{23}{3}$    **29.** $2\sqrt{1 + \sin x} + C$    **31.** $\frac{64}{5}$

**33.** $F'(x) = x^2/(1 + x^3)$    **35.** $g'(x) = 4x^3\cos(x^8)$

**37.** $y' = \dfrac{2\cos x - \cos\sqrt{x}}{2x}$    **39.** $4 \le \int_1^3 \sqrt{x^2 + 3}\, dx \le 4\sqrt{3}$

**43.** 0.280981

**45.** Number of barrels of oil consumed from Jan. 1, 2000, through Jan. 1, 2008

**47.** 72,400    **49.** 3    **51.** $(1 + x^2)(x\cos x + \sin x)/x^2$

**PROBLEMS PLUS ▪ PAGE 342**

**1.** $\pi/2$    **3.** $f(x) = \frac{1}{2}x$    **5.** $-1$    **7.** $[-1, 2]$

**9.** (a) $\frac{1}{2}(n - 1)n$    (b) $\frac{1}{2}[\![b]\!](2b - [\![b]\!] - 1) - \frac{1}{2}[\![a]\!](2a - [\![a]\!] - 1)$

**15.** $2(\sqrt{2} - 1)$

# CHAPTER 5

**EXERCISES 5.1 ▪ PAGE 349**

**1.** $\frac{32}{3}$    **3.** $\frac{4}{3}$    **5.** 19.5    **7.** $\frac{9}{2}$    **9.** $\frac{4}{3}$    **11.** $\frac{8}{3}$    **13.** 72

**15.** $6\sqrt{3}$    **17.** $\frac{32}{3}$    **19.** $2/\pi + \frac{2}{3}$    **21.** $2 - \pi/2$    **23.** $\frac{1}{2}$

**25.** $\frac{59}{12}$    **27.** $\frac{3}{4}$    **29.** $\frac{5}{2}$    **31.** $\frac{3}{2}\sqrt{3} - 1$    **33.** 0, 0.90; 0.04

**35.** $-1.11, 1.25, 2.86; 8.38$    **37.** 2.80123    **39.** 0.25142

**41.** $12\sqrt{6} - 9$    **43.** $117\frac{1}{3}$ ft    **45.** 4232 cm²

**47.** (a) Car A    (b) The distance by which A is ahead of B after 1 minute    (c) Car A    (d) $t \approx 2.2$ min

**49.** $\frac{24}{5}\sqrt{3}$    **51.** $4^{2/3}$    **53.** $\pm 6$    **55.** $\ln 2 - \frac{1}{2}$

**57.** $2 - 2\ln 2$

**EXERCISES 5.2 ▪ PAGE 360**

**1.** $19\pi/12$

**3.** $8\pi$

**5.** $162\pi$

**7.** $4\pi/21$

**9.** $64\pi/15$

**11.** $11\pi/30$

**13.** $2\pi\left(\frac{4}{3}\pi - \sqrt{3}\right)$

**15.** $3\pi/5$

**17.** $10\sqrt{2}\,\pi/3$

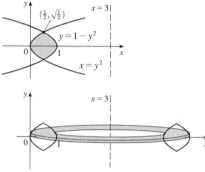

**19.** $\pi/3$     **21.** $\pi/3$     **23.** $\pi/3$
**25.** $13\pi/45$     **27.** $\pi/3$     **29.** $17\pi/45$
**31.** (a) 0.67419     (b) 2.85178
**33.** (a) $2\pi \int_0^2 8\sqrt{1 - x^2/4}\ dx \approx 78.95684$
(b) $2\pi \int_0^1 8\sqrt{4 - 4y^2}\ dy \approx 78.95684$
**35.** $-1.288, 0.884; 23.780$     **37.** $\frac{11}{8}\pi^2$
**39.** Solid obtained by rotating the region $0 \leq x \leq \pi$,
$0 \leq y \leq \sqrt{\sin x}$ about the $x$-axis
**41.** Solid obtained by rotating the region above the $x$-axis bounded
by $x = y^2$ and $x = y^4$ about the $y$-axis
**43.** 1110 cm$^3$     **45.** (a) 196     (b) 838
**47.** $\frac{1}{3}\pi r^2 h$     **49.** $\pi h^2(r - \frac{1}{3}h)$     **51.** $\frac{2}{3}b^2 h$
**53.** 10 cm$^3$     **55.** 24     **57.** $\frac{1}{3}$     **59.** $\frac{8}{15}$
**61.** (a) $8\pi R \int_0^r \sqrt{r^2 - y^2}\ dy$     (b) $2\pi^2 r^2 R$
**63.** (b) $\pi r^2 h$     **65.** $\frac{5}{12}\pi r^3$     **67.** $8 \int_0^r \sqrt{R^2 - y^2}\sqrt{r^2 - y^2}\ dy$

**EXERCISES 5.3 ▪ PAGE 366**
**1.** Circumference $= 2\pi x$, height $= x(x - 1)^2$; $\pi/15$

**3.** $6\pi/7$     **5.** $8\pi$     **7.** $8\pi$     **9.** $4\pi$     **11.** $768\pi/7$
**13.** $16\pi/3$     **15.** $7\pi/15$     **17.** $8\pi/3$     **19.** $5\pi/14$
**21.** (a) $\int_{2\pi}^{3\pi} 2\pi x \sin x\ dx$     (b) 98.69604
**23.** (a) $4\pi \int_{-\pi/2}^{\pi/2} (\pi - x) \cos^4 x\ dx$     (b) 46.50942
**25.** (a) $\int_0^\pi 2\pi(4 - y)\sqrt{\sin y}\ dy$     (b) 36.57476
**27.** 3.68
**29.** Solid obtained by rotating the region $0 \leq y \leq x^4, 0 \leq x \leq 3$
about the $y$-axis
**31.** Solid obtained by rotating the region bounded by
(i) $x = 1 - y^2$, $x = 0$, and $y = 0$, or (ii) $x = y^2$, $x = 1$, and $y = 0$
about the line $y = 3$
**33.** 0, 1.32; 4.05     **35.** $\frac{1}{32}\pi^3$     **37.** $8\pi$     **39.** $4\sqrt{3}\,\pi$
**41.** $4\pi/3$     **43.** $117\pi/5$     **45.** $\frac{4}{3}\pi r^3$     **47.** $\frac{1}{3}\pi r^2 h$

**EXERCISES 5.4 ▪ PAGE 371**
**1.** (a) 7200 ft-lb     (b) 7200 ft-lb
**3.** 4.5 ft-lb     **5.** 180 J     **7.** $\frac{15}{4}$ ft-lb
**9.** (a) $\frac{25}{24} \approx 1.04$ J     (b) 10.8 cm     **11.** $W_2 = 3W_1$
**13.** (a) 625 ft-lb     (b) $\frac{1875}{4}$ ft-lb     **15.** 650,000 ft-lb
**17.** 3857 J     **19.** 2450 J     **21.** $\approx 1.06 \times 10^6$ J
**23.** $\approx 1.04 \times 10^5$ ft-lb     **25.** 2.0 m
**29.** (a) $Gm_1 m_2 \left( \dfrac{1}{a} - \dfrac{1}{b} \right)$     (b) $\approx 8.50 \times 10^9$ J

**EXERCISES 5.5 ▪ PAGE 375**
**1.** $\frac{8}{3}$     **3.** $\frac{45}{28}$     **5.** 29,524/15     **7.** $2/(5\pi)$
**9.** (a) 1     (b) 2, 4     (c)

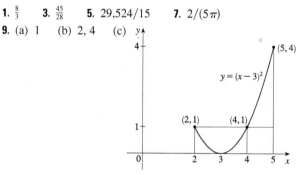

**11.** (a) $4/\pi$     (b) $\approx 1.24, 2.81$
(c) 3

**15.** $\frac{9}{8}$     **17.** $(50 + 28/\pi)°$F $\approx 59°$F     **19.** 6 kg/m
**21.** $5/(4\pi) \approx 0.4$ L

**Exercises**

**1.** $\frac{8}{3}$    **3.** $\frac{7}{12}$    **5.** $\frac{4}{3} + 4/\pi$    **7.** $64\pi/15$    **9.** $1656\pi/5$

**11.** $\frac{4}{3}\pi(2ah + h^2)^{3/2}$    **13.** $\int_{-\pi/3}^{\pi/3} 2\pi(\pi/2 - x)(\cos^2 x - \frac{1}{4})\,dx$

**15.** (a) $2\pi/15$    (b) $\pi/6$    (c) $8\pi/15$

**17.** (a) 0.38    (b) 0.87

**19.** Solid obtained by rotating the region $0 \le y \le \cos x$, $0 \le x \le \pi/2$ about the $y$-axis

**21.** Solid obtained by rotating the region $0 \le x \le \pi$, $0 \le y \le 2 - \sin x$ about the $x$-axis

**23.** 36    **25.** $\frac{125}{3}\sqrt{3}\ \text{m}^3$    **27.** 3.2 J

**29.** (a) $8000\pi/3 \approx 8378$ ft-lb    (b) 2.1 ft

**31.** $f(x)$

**1.** (a) $f(t) = 3t^2$    (b) $f(x) = \sqrt{2x/\pi}$    **3.** $\frac{32}{27}$

**5.** (b) 0.2261    (c) 0.6736 m

(d) (i) $1/(105\pi) \approx 0.003$ in/s    (ii) $370\pi/3$ s $\approx 6.5$ min

**9.** $y = \frac{32}{9}x^2$

**11.** (a) $V = \int_0^h \pi[f(y)]^2\,dy$

(c) $f(y) = \sqrt{kA/(\pi C)}\ y^{1/4}$. Advantage: the markings on the container are equally spaced.

**13.** $b = 2a$

**CHAPTER 9**

**3.** (a) $\frac{1}{2}, -1$    **5.** (d)

**7.** (a) It must be either 0 or decreasing
(c) $y = 0$    (d) $y = 1/(x + 2)$

**9.** (a) $0 < P < 4200$    (b) $P > 4200$
(c) $P = 0, P = 4200$

**13.** (a) III    (b) I    (c) IV    (d) II

**15.** (a) At the beginning; stays positive, but decreases

(c)

**21.** $y = 6x - 15$    **23.** $2x - 3y + 19 = 0$
**25.** $5x + y = 11$    **27.** $y = 3x - 2$    **29.** $y = 3x - 3$
**31.** $y = 5$    **33.** $x + 2y + 11 = 0$    **35.** $5x - 2y + 1 = 0$
**37.** $m = -\frac{1}{3}$,    **39.** $m = 0$,    **41.** $m = \frac{3}{4}$,
        $b = 0$            $b = -2$            $b = -3$

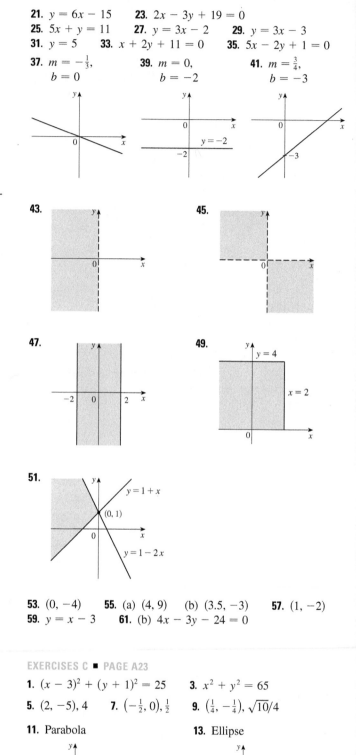

**43.**

**45.**

**47.**

**49.**

**51.**

**53.** $(0, -4)$    **55.** (a) $(4, 9)$    (b) $(3.5, -3)$    **57.** $(1, -2)$
**59.** $y = x - 3$    **61.** (b) $4x - 3y - 24 = 0$

## APPENDIXES

**1.** $18$    **3.** $\pi$    **5.** $5 - \sqrt{5}$    **7.** $2 - x$

**9.** $|x + 1| = \begin{cases} x + 1 & \text{for } x \geq -1 \\ -x - 1 & \text{for } x < -1 \end{cases}$    **11.** $x^2 + 1$

**13.** $(-2, \infty)$

**15.** $[-1, \infty)$

**17.** $(3, \infty)$

**19.** $(2, 6)$

**21.** $(0, 1]$

**23.** $\left[-1, \frac{1}{2}\right)$

**25.** $(-\infty, 1) \cup (2, \infty)$

**27.** $\left[-1, \frac{1}{2}\right]$

**29.** $(-\infty, \infty)$

**31.** $\left(-\sqrt{3}, \sqrt{3}\right)$

**33.** $(-\infty, 1]$

**35.** $(-1, 0) \cup (1, \infty)$

**37.** $(-\infty, 0) \cup \left(\frac{1}{4}, \infty\right)$

**39.** $10 \leq C \leq 35$    **41.** (a) $T = 20 - 10h, 0 \leq h \leq 12$
(b) $-30°C \leq T \leq 20°C$    **43.** $\pm\frac{3}{2}$    **45.** $2, -\frac{4}{3}$
**47.** $(-3, 3)$    **49.** $(3, 5)$    **51.** $(-\infty, -7] \cup [-3, \infty)$
**53.** $[1.3, 1.7]$    **55.** $[-4, -1] \cup [1, 4]$
**57.** $x \geq (a + b)c/(ab)$    **59.** $x > (c - b)/a$

**1.** $5$    **3.** $\sqrt{74}$    **5.** $2\sqrt{37}$    **7.** $2$    **9.** $-\frac{9}{2}$
**17.**    **19.**

**1.** $(x - 3)^2 + (y + 1)^2 = 25$    **3.** $x^2 + y^2 = 65$

**5.** $(2, -5), 4$    **7.** $\left(-\frac{1}{2}, 0\right), \frac{1}{2}$    **9.** $\left(\frac{1}{4}, -\frac{1}{4}\right), \sqrt{10}/4$

**11.** Parabola    **13.** Ellipse

**15.** Hyperbola

**17.** Ellipse

**19.** Parabola

**21.** Hyperbola

**23.** Hyperbola

**25.** Ellipse

**27.** Parabola

**29.** Parabola

**31.** Ellipse

**33.**

**35.** $y = x^2 - 2x$

**37.**

**39.**

**1.** $7\pi/6$   **3.** $\pi/20$   **5.** $5\pi$   **7.** $720°$   **9.** $75°$
**11.** $-67.5°$   **13.** $3\pi$ cm   **15.** $\frac{2}{3}$ rad $= (120/\pi)°$

**17.**                                    **19.**

**21.**

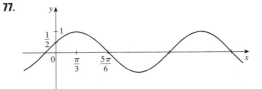

**23.** $\sin(3\pi/4) = 1/\sqrt{2}$, $\cos(3\pi/4) = -1/\sqrt{2}$, $\tan(3\pi/4) = -1$,
$\csc(3\pi/4) = \sqrt{2}$, $\sec(3\pi/4) = -\sqrt{2}$, $\cot(3\pi/4) = -1$
**25.** $\sin(9\pi/2) = 1$, $\cos(9\pi/2) = 0$, $\csc(9\pi/2) = 1$, $\cot(9\pi/2) = 0$,
$\tan(9\pi/2)$ and $\sec(9\pi/2)$ undefined
**27.** $\sin(5\pi/6) = \frac{1}{2}$, $\cos(5\pi/6) = -\sqrt{3}/2$, $\tan(5\pi/6) = -1/\sqrt{3}$,
$\csc(5\pi/6) = 2$, $\sec(5\pi/6) = -2/\sqrt{3}$, $\cot(5\pi/6) = -\sqrt{3}$
**29.** $\cos\theta = \frac{4}{5}$, $\tan\theta = \frac{3}{4}$, $\csc\theta = \frac{5}{3}$, $\sec\theta = \frac{5}{4}$, $\cot\theta = \frac{4}{3}$
**31.** $\sin\phi = \sqrt{5}/3$, $\cos\phi = -\frac{2}{3}$, $\tan\phi = -\sqrt{5}/2$, $\csc\phi = 3/\sqrt{5}$,
$\cot\phi = -2/\sqrt{5}$
**33.** $\sin\beta = -1/\sqrt{10}$, $\cos\beta = -3/\sqrt{10}$, $\tan\beta = \frac{1}{3}$,
$\csc\beta = -\sqrt{10}$, $\sec\beta = -\sqrt{10}/3$
**35.** $5.73576$ cm   **37.** $24.62147$ cm   **59.** $\frac{1}{15}\left(4 + 6\sqrt{2}\right)$
**61.** $\frac{1}{15}\left(3 + 8\sqrt{2}\right)$   **63.** $\frac{24}{25}$   **65.** $\pi/3, 5\pi/3$
**67.** $\pi/4, 3\pi/4, 5\pi/4, 7\pi/4$   **69.** $\pi/6, \pi/2, 5\pi/6, 3\pi/2$
**71.** $0, \pi, 2\pi$   **73.** $0 \le x \le \pi/6$ and $5\pi/6 \le x \le 2\pi$
**75.** $0 \le x < \pi/4, 3\pi/4 < x < 5\pi/4, 7\pi/4 < x \le 2\pi$

**77.**

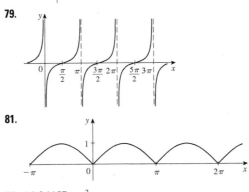

**79.**

**81.**

**89.** $14.34457$ cm$^2$

**EXERCISES E ■ PAGE A38**

**1.** $\sqrt{1} + \sqrt{2} + \sqrt{3} + \sqrt{4} + \sqrt{5}$    **3.** $3^4 + 3^5 + 3^6$

**5.** $-1 + \frac{1}{3} + \frac{3}{5} + \frac{5}{7} + \frac{7}{9}$    **7.** $1^{10} + 2^{10} + 3^{10} + \cdots + n^{10}$

**9.** $1 - 1 + 1 - 1 + \cdots + (-1)^{n-1}$    **11.** $\sum\limits_{i=1}^{10} i$

**13.** $\sum\limits_{i=1}^{19} \dfrac{i}{i+1}$    **15.** $\sum\limits_{i=1}^{n} 2i$    **17.** $\sum\limits_{i=0}^{5} 2^i$    **19.** $\sum\limits_{i=1}^{n} x^i$

**21.** 80    **23.** 3276    **25.** 0    **27.** 61    **29.** $n(n+1)$

**31.** $n(n^2 + 6n + 17)/3$    **33.** $n(n^2 + 6n + 11)/3$

**35.** $n(n^3 + 2n^2 - n - 10)/4$

**41.** (a) $n^4$    (b) $5^{100} - 1$    (c) $\frac{97}{300}$    (d) $a_n - a_0$

**43.** $\frac{1}{3}$    **45.** 14    **49.** $2^{n+1} + n^2 + n - 2$

**EXERCISES G ■ PAGE A53**

**1.** (c)

**3.**

**5.**

**7.**

**9.**

**11.**

**13.**

**15.** (b) Yes; two are needed

**17.**

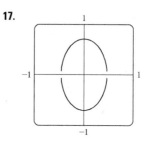

**19.** No    **21.** $-0.72, 1.22$

**23.** 0.65    **25.** $g$

**27.** $-0.31 < x < 0.31$

**29.** (a)

(b)

(c)

(d) Graphs of even roots are similar to $\sqrt{x}$, graphs of odd roots are similar to $\sqrt[3]{x}$. As $n$ increases, the graph of $y = \sqrt[n]{x}$ becomes steeper near 0 and flatter for $x > 1$.

**31.**

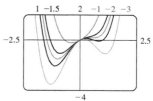

If $c < -1.5$, the graph has three humps: two minimum points and a maximum point. These humps get flatter as $c$ increases until at $c = -1.5$ two of the humps disappear and there is only one minimum point. This single hump then moves to the right and approaches the origin as $c$ increases.

**33.** The hump gets larger and moves to the right.

**35.** If $c < 0$, the loop is to the right of the origin; if $c > 0$, the loop is to the left. The closer $c$ is to 0, the larger the loop.

**EXERCISES H ■ PAGE A61**

**1.** $8 - 4i$    **3.** $13 + 18i$    **5.** $12 - 7i$    **7.** $\frac{11}{13} + \frac{10}{13}i$

**9.** $\frac{1}{2} - \frac{1}{2}i$    **11.** $-i$    **13.** $5i$    **15.** $12 + 5i, 13$

**17.** $4i, 4$  **19.** $\pm\frac{3}{2}i$  **21.** $-1 \pm 2i$

**23.** $-\frac{1}{2} \pm (\sqrt{7}/2)i$  **25.** $3\sqrt{2}\,[\cos(3\pi/4) + i\sin(3\pi/4)]$

**27.** $5\{\cos[\tan^{-1}(\frac{4}{3})] + i\sin[\tan^{-1}(\frac{4}{3})]\}$

**29.** $4[\cos(\pi/2) + i\sin(\pi/2)], \cos(-\pi/6) + i\sin(-\pi/6),$
$\frac{1}{2}[\cos(-\pi/6) + i\sin(-\pi/6)]$

**31.** $4\sqrt{2}\,[\cos(7\pi/12) + i\sin(7\pi/12)],$
$(2\sqrt{2})[\cos(13\pi/12) + i\sin(13\pi/12)], \frac{1}{4}[\cos(\pi/6) + i\sin(\pi/6)]$

**33.** $-1024$  **35.** $-512\sqrt{3} + 512i$

**37.** $\pm1, \pm i, (1/\sqrt{2})(\pm1 \pm i)$  **39.** $\pm(\sqrt{3}/2) + \frac{1}{2}i, -i$

**41.** $i$  **43.** $\frac{1}{2} + (\sqrt{3}/2)i$  **45.** $-e^2$

**47.** $\cos 3\theta = \cos^3\theta - 3\cos\theta\sin^2\theta,$
$\sin 3\theta = 3\cos^2\theta\sin\theta - \sin^3\theta$

# Index

RP denotes Reference Page numbers.
Numbers followed by asterisks denote pages in Sections 6.2*, 6.3*, and 6.4*.

The MTH 132 final exam mainly consists of standard response questions where students must justify their work. In addition to these, the Final Exam may consist of: fill in the blank, true/false, or multiple choice questions.

Most instructors agree that a good way to study for the final is to do lots of problems to help familiarize yourself with all of the concepts covered.

Sections containing similar concepts have been grouped in blue boxes. Most MTH 132 final exam writers agree that the items below contain crucial material for showcasing MTH 132 knowledge and are therefore **very important**. Expect at least one problem from each group on the final exam.

**Important Items from Each Section:**

1.5 - The Limit of a Function

- Be able to determine if a limit of a function exists by looking at a graph or piecewise defined function.

1.6 - Calculating Limits Using the Limit Laws

- Recall the limit laws and how to use algebra to manipulate limits.
- Know the definition and how to calculate one sided limits.
- Recall that $\lim_{x \to a} f(x) = L$ if and only if $\lim_{x \to a^-} f(x) = L = \lim_{x \to a^+} f(x)$.
- Know the Squeeze Theorem.

1.8 - Continuity

- Know the definition of a continuous function.
- Be able to use the Intermediate Value Theorem.

2.4 - Derivatives of Trigonometric Functions

- Know that $\lim_{u \to 0} \dfrac{\sin u}{u} = 1$ and $\lim_{u \to 0} \dfrac{\cos u - 1}{u} = 0$ and be able to use these within limit problems.

**Good Final Exam Review Problems:**

- 1.5.11
- 1.6.2
- 1.6.21
- 1.6.50
- 1.8.36
- 1.8.45
- 1.8.53
- 2.4.39
- 2.4.45

**Important Items from Each Section:**

2.2 - The Derivative as a Function

- Use the limit definition below to formally calculate the derivative of $f$:

$$f'(x) = \lim_{h \to 0} \frac{f(x+h) - f(x)}{h}$$

**Good Final Exam Review Problems:**

- 2.2.21
- 2.2.25
- 2.2.27

**Important Items from Each Section:**

2.3 - Differentiation Formulas

- Recall the power, product, and quotient rules.

2.4 - Derivatives of Trigonometric Functions

- Recall the formulas for the derivatives of all the trig functions. Be able to develop them using the power, product, and quotient rules.

2.5 - The Chain Rule

- Be able to apply the chain rule.

**Good Final Exam Review Problems:**

| | | |
|---|---|---|
| • 2.5.2 | • 2.5.31 | • 2.7.5 |
| • 2.5.17 | • 2.5.42 | • 2.7.7 |
| • 2.5.22 | • 2.5.69 | • 2.7.10 |

---

**Important Items from Each Section:**

2.6 - Implicit Differentiation

- Be able to use a variety of differentiation rules (most importantly the chain rule) to implicitly differentiate.

**Good Final Exam Review Problems:**

| | | |
|---|---|---|
| • 2.6.19 | • 2.6.27 | • 2.6.59 |

---

**Important Items from Each Section:**

2.8 - Related Rates

- Translate real world problems into solvable math equations.

**Good Final Exam Review Problems:**

| | | |
|---|---|---|
| • 2.8.5 | • 2.8.18 | • 2.8.27 |
| • 2.8.15 | • 2.8.20 | • 2.8.31 |

**Important Items from Each Section:**

2.9 - Linear Approximations and Differentials

- Recognize the tangent line as the linearization function.
- Be able to solve for the linearization function.
- Use the linearization function to approximate.

**Good Final Exam Review Problems:**

- 2.9.6
- 2.9.23
- 2.9.27

**Important Items from Each Section:**

3.1 - Maximum and Minimum Values

- Know how to find the critical numbers.
- Classify absolute maximums and minimums on a closed interval using the **Closed Interval Method**.

3.3 - How Derivatives Affect the Shape of a Graph

- Know the **Increasing/Decreasing Test** and the **Concavity Test**.
- Understand how to apply **The First Derivative Test** and **The Second Derivative Test**.
- Know the definition of inflection points and how to find them.

3.4 - Limits at Infinity; Horizontal Asymptotes

- Recall the definition for Horizontal and Vertical Asymptotes.

3.5 - Summary of Curve Sketching

- Be able to complete each of the **Guidelines for Sketching a Curve**:
    - Domain.
    - Intercepts.
    - Asymptotes.
    - Intervals of Increasing/Decreasing.
    - Local Maximums and Minimums.
    - Concavity and Points of Inflection.

**Good Final Exam Review Problems:**

- 3.5.9
- 3.5.16
- 3.5.30

**Important Items from Each Section:**

3.7 - Optimization Problems

- Understand the **Steps in Solving Optimization Problems**:
  - Understand the Problem.
  - Draw a Diagram.
  - Introduce Notation.

**Good Final Exam Review Problems:**

- 3.7.12
- 3.7.32
- 3.7.39
- 3.7.18
- 3.7.35
- 3.7.48

---

**Important Items from Each Section:**

3.9+ - Antiderivatives

- Be able to calculate Antiderivatives.
- Know how to solve for $C$ when given an initial value.

4.2 - The Definite Integral

- Know the properties of the Definite Integral.

4.3 - The Fundamental Theorem of Calculus

- Know and be able to use the two parts of **The Fundamental Theorem of Calculus**.

4.4 - Indefinite Integrals and the Net Change Theorem

- Recognize the table of indefinite integrals as antiderivatives so you don't need to memorize them all.

4.5 - The Substitution Rule

- Know how to apply $u$-substitutions in definite and indefinite integrals.
- Recognize the **Integrals of Symmetric Functions** properties and be able to apply them.

**Good Final Exam Review Problems:**

- 4.5.7
- 4.5.22
- 4.3.9
- 4.5.17
- 4.5.38
- 4.3.18

Most MTH 132 final exam writers agree that the items below contain material that is also **important**. However the appearance of problems below tend to fluctuate more from semester to semester. Students wishing to do well on the final should be familiar with this material as well.

---

**Important Items from Each Section:**

3.2 - The Mean Value Theorem

- Understand the statement and result of The Mean Value Theorem.

**Good Final Exam Review Problems:**

- 3.2.3
- 3.2.9
- 3.2.19

---

**Important Items from Each Section:**

3.8 - Newton's Method

- Recall the formula $x_{n+1} = x_n - \frac{f(x_n)}{f'(x_n)}$ or be able to derive it logically from a picture.

**Good Final Exam Review Problems:**

- 3.8.7
- 3.8.11
- 3.8.34

---

**Important Items from Each Section:**

4.1 - Areas and Distances

- Know how to estimate the area under graphs using:
  - Left Endpoints.
  - Right Endpoints.
  - Midpoints.

**Good Final Exam Review Problems:**

- 4.1.2
- 4.1.3
- 4.1.13

---

**Important Items from Each Section:**

4.2 - The Definite Integral

- Know summation formulas and properties.

- Know the formal definition of the integral: $\int_a^b f(x)dx = \lim_{n\to\infty} \sum_{i=1}^{n} f(x_i^*)\triangle x$ and be able to apply it.

**Good Final Exam Review Problems:**

- 4.2.18
- 4.2.21
- 4.2.22

**Important Items from Each Section:**

5.1 - Area Between Curves

- Know how to find the area between curves by integrating in terms of $x$ or $y$.

**Good Final Exam Review Problems:**

- 5.1.6
- 5.1.11
- 5.1.12

**Important Items from Each Section:**

5.5 - Average Value of a Function

- Know the formula for average value of a function: $f_{ave} = \frac{1}{b-a} \int_a^b f(x)dx$.
- Learn **The Mean Value Theorem for Integrals** and be able to apply it.

**Good Final Exam Review Problems:**

- 5.5.5
- 5.5.7
- 5.5.14

## SPECIAL FUNCTIONS

**Power Functions** $f(x) = x^a$

(i) $f(x) = x^n$, $n$ a positive integer

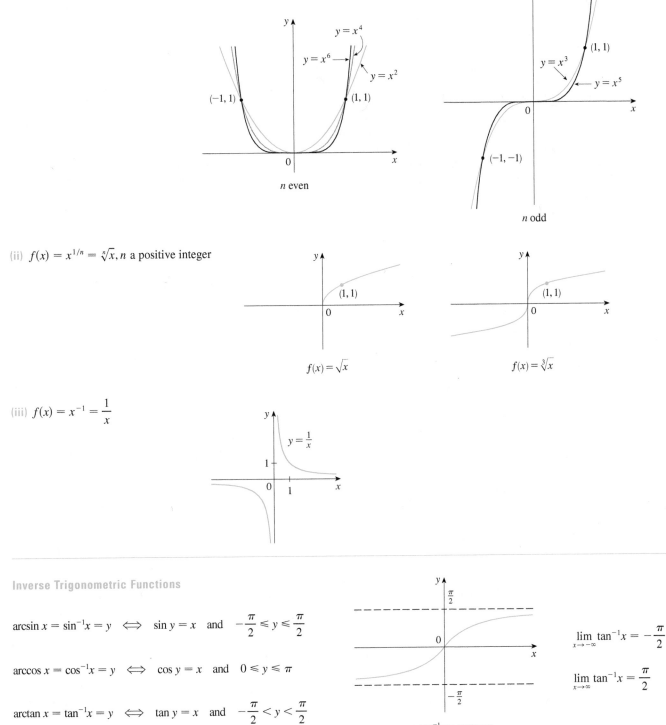

$f(x) = \sqrt{x}$      $f(x) = \sqrt[3]{x}$

(ii) $f(x) = x^{1/n} = \sqrt[n]{x}$, $n$ a positive integer

(iii) $f(x) = x^{-1} = \dfrac{1}{x}$

**Inverse Trigonometric Functions**

$\arcsin x = \sin^{-1}x = y \iff \sin y = x$ and $-\dfrac{\pi}{2} \le y \le \dfrac{\pi}{2}$

$\arccos x = \cos^{-1}x = y \iff \cos y = x$ and $0 \le y \le \pi$

$\arctan x = \tan^{-1}x = y \iff \tan y = x$ and $-\dfrac{\pi}{2} < y < \dfrac{\pi}{2}$

$\lim_{x \to -\infty} \tan^{-1}x = -\dfrac{\pi}{2}$

$\lim_{x \to \infty} \tan^{-1}x = \dfrac{\pi}{2}$

$y = \tan^{-1}x = \arctan x$

## SPECIAL FUNCTIONS

### Exponential and Logarithmic Functions

$$\log_a x = y \iff a^y = x$$

$$\ln x = \log_e x, \quad \text{where} \quad \ln e = 1$$

$$\ln x = y \iff e^y = x$$

#### Cancellation Equations

$$\log_a(a^x) = x \qquad a^{\log_a x} = x$$

$$\ln(e^x) = x \qquad e^{\ln x} = x$$

#### Laws of Logarithms

**1.** $\log_a(xy) = \log_a x + \log_a y$

**2.** $\log_a\left(\dfrac{x}{y}\right) = \log_a x - \log_a y$

**3.** $\log_a(x^r) = r \log_a x$

$$\lim_{x \to -\infty} e^x = 0 \qquad \lim_{x \to \infty} e^x = \infty$$

$$\lim_{x \to 0^+} \ln x = -\infty \qquad \lim_{x \to \infty} \ln x = \infty$$

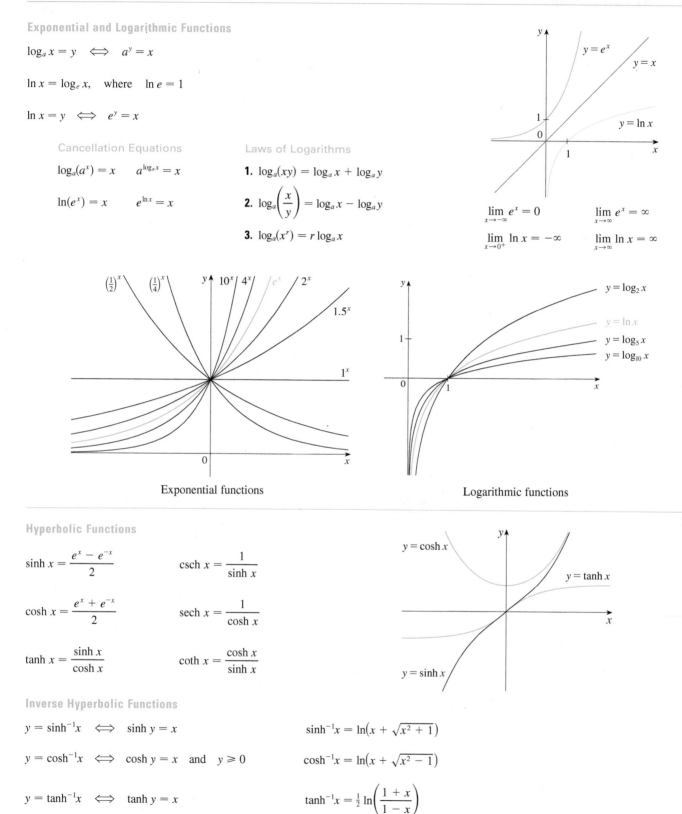

Exponential functions

Logarithmic functions

### Hyperbolic Functions

$$\sinh x = \frac{e^x - e^{-x}}{2} \qquad\qquad \operatorname{csch} x = \frac{1}{\sinh x}$$

$$\cosh x = \frac{e^x + e^{-x}}{2} \qquad\qquad \operatorname{sech} x = \frac{1}{\cosh x}$$

$$\tanh x = \frac{\sinh x}{\cosh x} \qquad\qquad \coth x = \frac{\cosh x}{\sinh x}$$

### Inverse Hyperbolic Functions

$$y = \sinh^{-1}x \iff \sinh y = x \qquad\qquad \sinh^{-1}x = \ln\!\left(x + \sqrt{x^2 + 1}\right)$$

$$y = \cosh^{-1}x \iff \cosh y = x \quad \text{and} \quad y \geq 0 \qquad\qquad \cosh^{-1}x = \ln\!\left(x + \sqrt{x^2 - 1}\right)$$

$$y = \tanh^{-1}x \iff \tanh y = x \qquad\qquad \tanh^{-1}x = \tfrac{1}{2}\ln\!\left(\frac{1 + x}{1 - x}\right)$$

# DIFFERENTIATION RULES

## General Formulas

**1.** $\dfrac{d}{dx}(c) = 0$

**2.** $\dfrac{d}{dx}[cf(x)] = cf'(x)$

**3.** $\dfrac{d}{dx}[f(x) + g(x)] = f'(x) + g'(x)$

**4.** $\dfrac{d}{dx}[f(x) - g(x)] = f'(x) - g'(x)$

**5.** $\dfrac{d}{dx}[f(x)g(x)] = f(x)g'(x) + g(x)f'(x)$   (Product Rule)

**6.** $\dfrac{d}{dx}\left[\dfrac{f(x)}{g(x)}\right] = \dfrac{g(x)f'(x) - f(x)g'(x)}{[g(x)]^2}$   (Quotient Rule)

**7.** $\dfrac{d}{dx}f(g(x)) = f'(g(x))g'(x)$   (Chain Rule)

**8.** $\dfrac{d}{dx}(x^n) = nx^{n-1}$   (Power Rule)

## Exponential and Logarithmic Functions

**9.** $\dfrac{d}{dx}(e^x) = e^x$

**10.** $\dfrac{d}{dx}(a^x) = a^x \ln a$

**11.** $\dfrac{d}{dx}\ln|x| = \dfrac{1}{x}$

**12.** $\dfrac{d}{dx}(\log_a x) = \dfrac{1}{x \ln a}$

## Trigonometric Functions

**13.** $\dfrac{d}{dx}(\sin x) = \cos x$

**14.** $\dfrac{d}{dx}(\cos x) = -\sin x$

**15.** $\dfrac{d}{dx}(\tan x) = \sec^2 x$

**16.** $\dfrac{d}{dx}(\csc x) = -\csc x \cot x$

**17.** $\dfrac{d}{dx}(\sec x) = \sec x \tan x$

**18.** $\dfrac{d}{dx}(\cot x) = -\csc^2 x$

## Inverse Trigonometric Functions

**19.** $\dfrac{d}{dx}(\sin^{-1}x) = \dfrac{1}{\sqrt{1-x^2}}$

**20.** $\dfrac{d}{dx}(\cos^{-1}x) = -\dfrac{1}{\sqrt{1-x^2}}$

**21.** $\dfrac{d}{dx}(\tan^{-1}x) = \dfrac{1}{1+x^2}$

**22.** $\dfrac{d}{dx}(\csc^{-1}x) = -\dfrac{1}{x\sqrt{x^2-1}}$

**23.** $\dfrac{d}{dx}(\sec^{-1}x) = \dfrac{1}{x\sqrt{x^2-1}}$

**24.** $\dfrac{d}{dx}(\cot^{-1}x) = -\dfrac{1}{1+x^2}$

## Hyperbolic Functions

**25.** $\dfrac{d}{dx}(\sinh x) = \cosh x$

**26.** $\dfrac{d}{dx}(\cosh x) = \sinh x$

**27.** $\dfrac{d}{dx}(\tanh x) = \operatorname{sech}^2 x$

**28.** $\dfrac{d}{dx}(\operatorname{csch} x) = -\operatorname{csch} x \coth x$

**29.** $\dfrac{d}{dx}(\operatorname{sech} x) = -\operatorname{sech} x \tanh x$

**30.** $\dfrac{d}{dx}(\coth x) = -\operatorname{csch}^2 x$

## Inverse Hyperbolic Functions

**31.** $\dfrac{d}{dx}(\sinh^{-1}x) = \dfrac{1}{\sqrt{1+x^2}}$

**32.** $\dfrac{d}{dx}(\cosh^{-1}x) = \dfrac{1}{\sqrt{x^2-1}}$

**33.** $\dfrac{d}{dx}(\tanh^{-1}x) = \dfrac{1}{1-x^2}$

**34.** $\dfrac{d}{dx}(\operatorname{csch}^{-1}x) = -\dfrac{1}{|x|\sqrt{x^2+1}}$

**35.** $\dfrac{d}{dx}(\operatorname{sech}^{-1}x) = -\dfrac{1}{x\sqrt{1-x^2}}$

**36.** $\dfrac{d}{dx}(\coth^{-1}x) = \dfrac{1}{1-x^2}$

# TABLE OF INTEGRALS

## Basic Forms

1. $\int u \, dv = uv - \int v \, du$

2. $\int u^n \, du = \frac{u^{n+1}}{n+1} + C, \quad n \neq -1$

3. $\int \frac{du}{u} = \ln|u| + C$

4. $\int e^u \, du = e^u + C$

5. $\int a^u \, du = \frac{a^u}{\ln a} + C$

6. $\int \sin u \, du = -\cos u + C$

7. $\int \cos u \, du = \sin u + C$

8. $\int \sec^2 u \, du = \tan u + C$

9. $\int \csc^2 u \, du = -\cot u + C$

10. $\int \sec u \tan u \, du = \sec u + C$

11. $\int \csc u \cot u \, du = -\csc u + C$

12. $\int \tan u \, du = \ln|\sec u| + C$

13. $\int \cot u \, du = \ln|\sin u| + C$

14. $\int \sec u \, du = \ln|\sec u + \tan u| + C$

15. $\int \csc u \, du = \ln|\csc u - \cot u| + C$

16. $\int \frac{du}{\sqrt{a^2 - u^2}} = \sin^{-1}\frac{u}{a} + C, \quad a > 0$

17. $\int \frac{du}{a^2 + u^2} = \frac{1}{a}\tan^{-1}\frac{u}{a} + C$

18. $\int \frac{du}{u\sqrt{u^2 - a^2}} = \frac{1}{a}\sec^{-1}\frac{u}{a} + C$

19. $\int \frac{du}{a^2 - u^2} = \frac{1}{2a}\ln\left|\frac{u+a}{u-a}\right| + C$

20. $\int \frac{du}{u^2 - a^2} = \frac{1}{2a}\ln\left|\frac{u-a}{u+a}\right| + C$

## Forms Involving $\sqrt{a^2 + u^2}$, $a > 0$

21. $\int \sqrt{a^2 + u^2} \, du = \frac{u}{2}\sqrt{a^2 + u^2} + \frac{a^2}{2}\ln\left(u + \sqrt{a^2 + u^2}\right) + C$

22. $\int u^2 \sqrt{a^2 + u^2} \, du = \frac{u}{8}(a^2 + 2u^2)\sqrt{a^2 + u^2} - \frac{a^4}{8}\ln\left(u + \sqrt{a^2 + u^2}\right) + C$

23. $\int \frac{\sqrt{a^2 + u^2}}{u} \, du = \sqrt{a^2 + u^2} - a\ln\left|\frac{a + \sqrt{a^2 + u^2}}{u}\right| + C$

24. $\int \frac{\sqrt{a^2 + u^2}}{u^2} \, du = -\frac{\sqrt{a^2 + u^2}}{u} + \ln\left(u + \sqrt{a^2 + u^2}\right) + C$

25. $\int \frac{du}{\sqrt{a^2 + u^2}} = \ln\left(u + \sqrt{a^2 + u^2}\right) + C$

26. $\int \frac{u^2 \, du}{\sqrt{a^2 + u^2}} = \frac{u}{2}\sqrt{a^2 + u^2} - \frac{a^2}{2}\ln\left(u + \sqrt{a^2 + u^2}\right) + C$

27. $\int \frac{du}{u\sqrt{a^2 + u^2}} = -\frac{1}{a}\ln\left|\frac{\sqrt{a^2 + u^2} + a}{u}\right| + C$

28. $\int \frac{du}{u^2\sqrt{a^2 + u^2}} = -\frac{\sqrt{a^2 + u^2}}{a^2 u} + C$

29. $\int \frac{du}{(a^2 + u^2)^{3/2}} = \frac{u}{a^2\sqrt{a^2 + u^2}} + C$

## TABLE OF INTEGRALS

Forms Involving $\sqrt{a^2 - u^2}$, $a > 0$

**30.** $\displaystyle\int \sqrt{a^2 - u^2}\, du = \frac{u}{2} \sqrt{a^2 - u^2} + \frac{a^2}{2} \sin^{-1} \frac{u}{a} + C$

**31.** $\displaystyle\int u^2 \sqrt{a^2 - u^2}\, du = \frac{u}{8} (2u^2 - a^2) \sqrt{a^2 - u^2} + \frac{a^4}{8} \sin^{-1} \frac{u}{a} + C$

**32.** $\displaystyle\int \frac{\sqrt{a^2 - u^2}}{u}\, du = \sqrt{a^2 - u^2} - a \ln \left| \frac{a + \sqrt{a^2 - u^2}}{u} \right| + C$

**33.** $\displaystyle\int \frac{\sqrt{a^2 - u^2}}{u^2}\, du = -\frac{1}{u} \sqrt{a^2 - u^2} - \sin^{-1} \frac{u}{a} + C$

**34.** $\displaystyle\int \frac{u^2\, du}{\sqrt{a^2 - u^2}} = -\frac{u}{2} \sqrt{a^2 - u^2} + \frac{a^2}{2} \sin^{-1} \frac{u}{a} + C$

**35.** $\displaystyle\int \frac{du}{u\sqrt{a^2 - u^2}} = -\frac{1}{a} \ln \left| \frac{a + \sqrt{a^2 - u^2}}{u} \right| + C$

**36.** $\displaystyle\int \frac{du}{u^2 \sqrt{a^2 - u^2}} = -\frac{1}{a^2 u} \sqrt{a^2 - u^2} + C$

**37.** $\displaystyle\int (a^2 - u^2)^{3/2}\, du = -\frac{u}{8} (2u^2 - 5a^2) \sqrt{a^2 - u^2} + \frac{3a^4}{8} \sin^{-1} \frac{u}{a} + C$

**38.** $\displaystyle\int \frac{du}{(a^2 - u^2)^{3/2}} = \frac{u}{a^2 \sqrt{a^2 - u^2}} + C$

Forms Involving $\sqrt{u^2 - a^2}$, $a > 0$

**39.** $\displaystyle\int \sqrt{u^2 - a^2}\, du = \frac{u}{2} \sqrt{u^2 - a^2} - \frac{a^2}{2} \ln \left| u + \sqrt{u^2 - a^2} \right| + C$

**40.** $\displaystyle\int u^2 \sqrt{u^2 - a^2}\, du = \frac{u}{8} (2u^2 - a^2) \sqrt{u^2 - a^2} - \frac{a^4}{8} \ln \left| u + \sqrt{u^2 - a^2} \right| + C$

**41.** $\displaystyle\int \frac{\sqrt{u^2 - a^2}}{u}\, du = \sqrt{u^2 - a^2} - a \cos^{-1} \frac{a}{|u|} + C$

**42.** $\displaystyle\int \frac{\sqrt{u^2 - a^2}}{u^2}\, du = -\frac{\sqrt{u^2 - a^2}}{u} + \ln \left| u + \sqrt{u^2 - a^2} \right| + C$

**43.** $\displaystyle\int \frac{du}{\sqrt{u^2 - a^2}} = \ln \left| u + \sqrt{u^2 - a^2} \right| + C$

**44.** $\displaystyle\int \frac{u^2\, du}{\sqrt{u^2 - a^2}} = \frac{u}{2} \sqrt{u^2 - a^2} + \frac{a^2}{2} \ln \left| u + \sqrt{u^2 - a^2} \right| + C$

**45.** $\displaystyle\int \frac{du}{u^2 \sqrt{u^2 - a^2}} = \frac{\sqrt{u^2 - a^2}}{a^2 u} + C$

**46.** $\displaystyle\int \frac{du}{(u^2 - a^2)^{3/2}} = -\frac{u}{a^2 \sqrt{u^2 - a^2}} + C$

## TABLE OF INTEGRALS

Forms Involving $a + bu$

**47.** $\displaystyle\int \frac{u\,du}{a + bu} = \frac{1}{b^2}\left(a + bu - a \ln|a + bu|\right) + C$

**48.** $\displaystyle\int \frac{u^2\,du}{a + bu} = \frac{1}{2b^3}\left[(a + bu)^2 - 4a(a + bu) + 2a^2 \ln|a + bu|\right] + C$

**49.** $\displaystyle\int \frac{du}{u(a + bu)} = \frac{1}{a} \ln\left|\frac{u}{a + bu}\right| + C$

**50.** $\displaystyle\int \frac{du}{u^2(a + bu)} = -\frac{1}{au} + \frac{b}{a^2} \ln\left|\frac{a + bu}{u}\right| + C$

**51.** $\displaystyle\int \frac{u\,du}{(a + bu)^2} = \frac{a}{b^2(a + bu)} + \frac{1}{b^2} \ln|a + bu| + C$

**52.** $\displaystyle\int \frac{du}{u(a + bu)^2} = \frac{1}{a(a + bu)} - \frac{1}{a^2} \ln\left|\frac{a + bu}{u}\right| + C$

**53.** $\displaystyle\int \frac{u^2\,du}{(a + bu)^2} = \frac{1}{b^3}\left(a + bu - \frac{a^2}{a + bu} - 2a \ln|a + bu|\right) + C$

**54.** $\displaystyle\int u\sqrt{a + bu}\,du = \frac{2}{15b^2}(3bu - 2a)(a + bu)^{3/2} + C$

**55.** $\displaystyle\int \frac{u\,du}{\sqrt{a + bu}} = \frac{2}{3b^2}(bu - 2a)\sqrt{a + bu} + C$

**56.** $\displaystyle\int \frac{u^2\,du}{\sqrt{a + bu}} = \frac{2}{15b^3}(8a^2 + 3b^2u^2 - 4abu)\sqrt{a + bu} + C$

**57.** $\displaystyle\int \frac{du}{u\sqrt{a + bu}} = \frac{1}{\sqrt{a}} \ln\left|\frac{\sqrt{a + bu} - \sqrt{a}}{\sqrt{a + bu} + \sqrt{a}}\right| + C,\ \text{ if } a > 0$

$\displaystyle\qquad\qquad\quad = \frac{2}{\sqrt{-a}} \tan^{-1}\sqrt{\frac{a + bu}{-a}} + C,\qquad \text{ if } a < 0$

**58.** $\displaystyle\int \frac{\sqrt{a + bu}}{u}\,du = 2\sqrt{a + bu} + a\int \frac{du}{u\sqrt{a + bu}}$

**59.** $\displaystyle\int \frac{\sqrt{a + bu}}{u^2}\,du = -\frac{\sqrt{a + bu}}{u} + \frac{b}{2}\int \frac{du}{u\sqrt{a + bu}}$

**60.** $\displaystyle\int u^n\sqrt{a + bu}\,du = \frac{2}{b(2n + 3)}\left[u^n(a + bu)^{3/2} - na\int u^{n-1}\sqrt{a + bu}\,du\right]$

**61.** $\displaystyle\int \frac{u^n\,du}{\sqrt{a + bu}} = \frac{2u^n\sqrt{a + bu}}{b(2n + 1)} - \frac{2na}{b(2n + 1)}\int \frac{u^{n-1}\,du}{\sqrt{a + bu}}$

**62.** $\displaystyle\int \frac{du}{u^n\sqrt{a + bu}} = -\frac{\sqrt{a + bu}}{a(n - 1)u^{n-1}} - \frac{b(2n - 3)}{2a(n - 1)}\int \frac{du}{u^{n-1}\sqrt{a + bu}}$

## TABLE OF INTEGRALS

Trigonometric Forms

**63.** $\int \sin^2 u \, du = \frac{1}{2}u - \frac{1}{4}\sin 2u + C$

**64.** $\int \cos^2 u \, du = \frac{1}{2}u + \frac{1}{4}\sin 2u + C$

**65.** $\int \tan^2 u \, du = \tan u - u + C$

**66.** $\int \cot^2 u \, du = -\cot u - u + C$

**67.** $\int \sin^3 u \, du = -\frac{1}{3}(2 + \sin^2 u)\cos u + C$

**68.** $\int \cos^3 u \, du = \frac{1}{3}(2 + \cos^2 u)\sin u + C$

**69.** $\int \tan^3 u \, du = \frac{1}{2}\tan^2 u + \ln|\cos u| + C$

**70.** $\int \cot^3 u \, du = -\frac{1}{2}\cot^2 u - \ln|\sin u| + C$

**71.** $\int \sec^3 u \, du = \frac{1}{2}\sec u \tan u + \frac{1}{2}\ln|\sec u + \tan u| + C$

**72.** $\int \csc^3 u \, du = -\frac{1}{2}\csc u \cot u + \frac{1}{2}\ln|\csc u - \cot u| + C$

**73.** $\int \sin^n u \, du = -\frac{1}{n}\sin^{n-1} u \cos u + \frac{n-1}{n}\int \sin^{n-2} u \, du$

**74.** $\int \cos^n u \, du = \frac{1}{n}\cos^{n-1} u \sin u + \frac{n-1}{n}\int \cos^{n-2} u \, du$

**75.** $\int \tan^n u \, du = \frac{1}{n-1}\tan^{n-1} u - \int \tan^{n-2} u \, du$

**76.** $\int \cot^n u \, du = \frac{-1}{n-1}\cot^{n-1} u - \int \cot^{n-2} u \, du$

**77.** $\int \sec^n u \, du = \frac{1}{n-1}\tan u \sec^{n-2} u + \frac{n-2}{n-1}\int \sec^{n-2} u \, du$

**78.** $\int \csc^n u \, du = \frac{-1}{n-1}\cot u \csc^{n-2} u + \frac{n-2}{n-1}\int \csc^{n-2} u \, du$

**79.** $\int \sin au \sin bu \, du = \frac{\sin(a-b)u}{2(a-b)} - \frac{\sin(a+b)u}{2(a+b)} + C$

**80.** $\int \cos au \cos bu \, du = \frac{\sin(a-b)u}{2(a-b)} + \frac{\sin(a+b)u}{2(a+b)} + C$

**81.** $\int \sin au \cos bu \, du = -\frac{\cos(a-b)u}{2(a-b)} - \frac{\cos(a+b)u}{2(a+b)} + C$

**82.** $\int u \sin u \, du = \sin u - u \cos u + C$

**83.** $\int u \cos u \, du = \cos u + u \sin u + C$

**84.** $\int u^n \sin u \, du = -u^n \cos u + n \int u^{n-1} \cos u \, du$

**85.** $\int u^n \cos u \, du = u^n \sin u - n \int u^{n-1} \sin u \, du$

**86.** $\int \sin^n u \cos^m u \, du = -\frac{\sin^{n-1} u \cos^{m+1} u}{n+m} + \frac{n-1}{n+m}\int \sin^{n-2} u \cos^m u \, du$

$$= \frac{\sin^{n+1} u \cos^{m-1} u}{n+m} + \frac{m-1}{n+m}\int \sin^n u \cos^{m-2} u \, du$$

Inverse Trigonometric Forms

**87.** $\int \sin^{-1} u \, du = u \sin^{-1} u + \sqrt{1 - u^2} + C$

**88.** $\int \cos^{-1} u \, du = u \cos^{-1} u - \sqrt{1 - u^2} + C$

**89.** $\int \tan^{-1} u \, du = u \tan^{-1} u - \frac{1}{2}\ln(1 + u^2) + C$

**90.** $\int u \sin^{-1} u \, du = \frac{2u^2 - 1}{4}\sin^{-1} u + \frac{u\sqrt{1 - u^2}}{4} + C$

**91.** $\int u \cos^{-1} u \, du = \frac{2u^2 - 1}{4}\cos^{-1} u - \frac{u\sqrt{1 - u^2}}{4} + C$

**92.** $\int u \tan^{-1} u \, du = \frac{u^2 + 1}{2}\tan^{-1} u - \frac{u}{2} + C$

**93.** $\int u^n \sin^{-1} u \, du = \frac{1}{n+1}\left[ u^{n+1} \sin^{-1} u - \int \frac{u^{n+1}\, du}{\sqrt{1 - u^2}} \right], \quad n \neq -1$

**94.** $\int u^n \cos^{-1} u \, du = \frac{1}{n+1}\left[ u^{n+1} \cos^{-1} u + \int \frac{u^{n+1}\, du}{\sqrt{1 - u^2}} \right], \quad n \neq -1$

**95.** $\int u^n \tan^{-1} u \, du = \frac{1}{n+1}\left[ u^{n+1} \tan^{-1} u - \int \frac{u^{n+1}\, du}{1 + u^2} \right], \quad n \neq -1$

## TABLE OF INTEGRALS

Exponential and Logarithmic Forms

**96.** $\displaystyle\int ue^{au}\,du = \frac{1}{a^2}(au-1)e^{au} + C$

**97.** $\displaystyle\int u^n e^{au}\,du = \frac{1}{a}u^n e^{au} - \frac{n}{a}\int u^{n-1}e^{au}\,du$

**98.** $\displaystyle\int e^{au}\sin bu\,du = \frac{e^{au}}{a^2+b^2}(a\sin bu - b\cos bu) + C$

**99.** $\displaystyle\int e^{au}\cos bu\,du = \frac{e^{au}}{a^2+b^2}(a\cos bu + b\sin bu) + C$

**100.** $\displaystyle\int \ln u\,du = u\ln u - u + C$

**101.** $\displaystyle\int u^n \ln u\,du = \frac{u^{n+1}}{(n+1)^2}\big[(n+1)\ln u - 1\big] + C$

**102.** $\displaystyle\int \frac{1}{u\ln u}\,du = \ln|\ln u| + C$

Hyperbolic Forms

**103.** $\displaystyle\int \sinh u\,du = \cosh u + C$

**104.** $\displaystyle\int \cosh u\,du = \sinh u + C$

**105.** $\displaystyle\int \tanh u\,du = \ln\cosh u + C$

**106.** $\displaystyle\int \coth u\,du = \ln|\sinh u| + C$

**107.** $\displaystyle\int \operatorname{sech} u\,du = \tan^{-1}|\sinh u| + C$

**108.** $\displaystyle\int \operatorname{csch} u\,du = \ln\left|\tanh\tfrac{1}{2}u\right| + C$

**109.** $\displaystyle\int \operatorname{sech}^2 u\,du = \tanh u + C$

**110.** $\displaystyle\int \operatorname{csch}^2 u\,du = -\coth u + C$

**111.** $\displaystyle\int \operatorname{sech} u\tanh u\,du = -\operatorname{sech} u + C$

**112.** $\displaystyle\int \operatorname{csch} u\coth u\,du = -\operatorname{csch} u + C$

Forms Involving $\sqrt{2au - u^2}$, $a > 0$

**113.** $\displaystyle\int \sqrt{2au-u^2}\,du = \frac{u-a}{2}\sqrt{2au-u^2} + \frac{a^2}{2}\cos^{-1}\!\left(\frac{a-u}{a}\right) + C$

**114.** $\displaystyle\int u\sqrt{2au-u^2}\,du = \frac{2u^2-au-3a^2}{6}\sqrt{2au-u^2} + \frac{a^3}{2}\cos^{-1}\!\left(\frac{a-u}{a}\right) + C$

**115.** $\displaystyle\int \frac{\sqrt{2au-u^2}}{u}\,du = \sqrt{2au-u^2} + a\cos^{-1}\!\left(\frac{a-u}{a}\right) + C$

**116.** $\displaystyle\int \frac{\sqrt{2au-u^2}}{u^2}\,du = -\frac{2\sqrt{2au-u^2}}{u} - \cos^{-1}\!\left(\frac{a-u}{a}\right) + C$

**117.** $\displaystyle\int \frac{du}{\sqrt{2au-u^2}} = \cos^{-1}\!\left(\frac{a-u}{a}\right) + C$

**118.** $\displaystyle\int \frac{u\,du}{\sqrt{2au-u^2}} = -\sqrt{2au-u^2} + a\cos^{-1}\!\left(\frac{a-u}{a}\right) + C$

**119.** $\displaystyle\int \frac{u^2\,du}{\sqrt{2au-u^2}} = -\frac{(u+3a)}{2}\sqrt{2au-u^2} + \frac{3a^2}{2}\cos^{-1}\!\left(\frac{a-u}{a}\right) + C$

**120.** $\displaystyle\int \frac{du}{u\sqrt{2au-u^2}} = -\frac{\sqrt{2au-u^2}}{au} + C$